PHOTOCHEMISTRY
IN
ORGANIZED
AND
CONSTRAINED MEDIA

PHOTOCHEMISTRY

IN

ORGANIZED

AND

CONSTRAINED MEDIA

Edited by

V. Ramamurthy

.4311607

CHEMISTRY

V. Ramamurthy
Central Research and Development
Experimental Station, P.O. Box 80328
The Du Pont Company
Wilmington, DE 19880-0328, U.S.A.

Library of Congress Cataloging-in-Publication Data

Photochemistry in Organized and Constrained Media / edited by V. Ramamurthy.
  P. cm.
  Includes bibliographical references and index.
  ISBN 0-89573-775-2
  1. Photochemistry   2. Solid state chemistry   I. Ramamurthy, V.
  QD715.P458 1991
  541.3'5--dc20   90-26762
                                                                CIP

British Library Cataloguing in Publication Data

Photochemistry in Organized and Constrained Media
  1. Photochemistry
  I. Ramamurthy V.
  541.35
  ISBN 0-89573-775-2

© 1991 VCH Publishers, Inc.

Printed in the United States of America.
ISBN 0-89573-775-2 VCH Publishers
ISBN 3-527-27936-9 VCH Verlgsgesellschaft

Print History:
10 9 8 7 6 5 4 3 2 1

Published jointly by:

VCH Publishers, Inc.   VCH Verlagsgesellschaft mbH   VCH Publishers (UK) Ltd.
220 East 23rd Street   P.O. Box 10 11 61   8 Wellington Court
Suite 909   D-6940 Weinheim   Cambridge CB1 1HW
New York, N Y 10010   Federal Republic of Germany   United Kingdom

Keep away from mere verbal arguments; adopt only those devices which are based on a sound premise.

Adi Sankara (8th century)

Experiments never deceive. It is our judgement that deceives itself because it expects results which experiments refuses.

Leonardo da Vinci (15th century)

Contributors

H. Al-Ekabi
Nulite
511 McCormick Blvd.,
London, Ontario N6W 4C8
Canada.

D. Avnir
Institute of Chemistry
The Hebrew University of Jerusalem
Jerusalem 91904
Israel.

C. Bohne
Steacie Institute of Molecular Sciences
National Research Council
Ottawa, Ontario K1A OR6
Canada.

D. F. Eaton
Central Research and Development
The Du Pont Company
Experimental Station,
Wilmington, DE 19880-0356
U. S. A.

M. Garcia-Garibay
Department of Chemistry
Columbia University
New York, NY 10027
U. S. A.

K. Kalyanasundaram
Swiss Federal Institute of Technology
Institute of Physical Chemistry
EPFL-Ecublens
Lausanne CH-1015
Switzerland.

L. Johnston
Steacie Institute of Molecular Sciences
National Research Council
Ottawa, Ontario K1A OR6
Canada.

W. Jones
Department of Physical chemistry
University of Cambridge
Cambridge, CB2 1EP
United Kingdom.

M. Lahav
Department of Structural Chemistry
The Weizmann Institute of Science
Rehovot, 76100
Israel.

L. Leiserowitz
Department of Structural Chemistry
The Weizmann Institute of Science
Rehovot, 76100
Israel.

R. S. H. Liu
Department of Chemistry
University of Hawaii
Honolulu, HI 96822
U. S. A.

M. Ottolenghi
Institute of Chemistry
The Hebrew University of Jerusalem
Jerusalem 91904
Israel.

T. Osa
Pharmaceutical Institute
Tohuku University
Aobayama, Sendai 980
Japan.

P. R. Pokkuluri
Department of Chemistry
University of British Columbia
Vancouver, BC V6T 1Y6
Canada.

R. Popovitz-Biro
Department of Structural Chemistry
The Weizmann Institute of Science
Rehovot, 76100
Israel.

V. Ramamurthy
Central Research and Development
The Du Pont Company
Experimental Station,
Wilmington, DE 19880-0328
U. S. A.

R. W. Redmond
Steacie Institute of Molecular Sciences
National Research Council
Ottawa, Ontario K1A OR6
Canada.

J. C. Scaiano
Steacie Institute of Molecular Sciences
National Research Council
Ottawa, Ontario K1A OR6
Canada.

J. R. Scheffer
Department of Chemistry
University of British Columbia
Vancouver, BC V6T 1Y6
Canada.

Y. Shichida
Department of Biophysics
Kyoto University
Faculty of Science
Kyoto 606,
Japan.

S. Spooner
Department of Chemistry
University of Rochester
Rochester, NY 14627
U. S. A.

N. J. Turro
Department of Chemistry
Columbia University
New York, NY 10027
U. S. A.

A. Ueno
Department of Bio-engineering and
Bioscience
Tokyo Institute of Technology
Yokahama 227,
Japan.

M. Vaida
Department of Structural Chemistry
The Weizmann Institute of Science
Rehovot, 76100
Israel.

K. Venkatesan
Department of Organic Chemistry
Indian Institute of Science
Bangalore, 560012
India.

C. Vijaya Kumar
Department of Chemistry
University of Connecticut
Storrs, CT 06269-3060
U. S. A.

W. R. Ware
Department of Chemistry
University of Western Ontario
London, Ontario, N6A 5B7
Canada

R. G. Weiss
Department of Chemistry
Georgetown University
Washington, DC 20057
U. S. A.

D. G. Whitten
Department of Chemistry
University of Rochester
Rochester, NY 14627
U. S. A.

Foreword

Photochemistry is an old and distinguished science that has been the subject of resurgent interest during the past thirty years. Photochemical processes have been carried out in media in which molecular motion is restricted since the earliest days of the science and, as with the field in general, this facet of it has quickened and been given new focus in recent decades. In this book Ramamurthy has gathered as authors many of the people who are doing the most interesting and forward-looking research in the field.

It is worth our while to ponder the significance of the title "Photochemistry in Organized and Constrained Media" before launching into the real substance of the volume. Why organized and constrained media? And why photochemistry? And what may it all bring to the magnificent realm of chemical science? These questions assuredly merit an Olympian view; but, since I write from my humble study, I can only offer one man's opinion with a plea of humility.

The phrase "organized and constrained media" is one of those scientific terms having a comfortably diffuse meaning, and I hope that its utility will not be degraded by well-intended, but futile, attempts to give it a precisely defined meaning. Although isotropic liquids certainly have varying degrees of local organization, their behavior will not, by implication, be considered "organized" for purposes of most discussion in this volume. On the other hand, some solutes in isotropic liquids, notably host–guest partners, will be considered organized, because their organizational features are of interest for purposes under consideration. In other words, "organization" lies in the eye of the beholder as much or more than in the structure of the reference system.

To a dedicated photochemist or spectroscopist there are many reasons for carrying out photoreactions in organized media. Molecules may be held in more or less rigid orientations that promote, or discourage, specific reactions of excited states. The prohibition of diffusion in rigid media (which for many purpose might be regarded as isotropic!) may allow the preservation for study of certain species— excited states, highly strained molecules, free radicals, etc. At one time low-temperature matrix isolation was known as "the poor man's flash photolysis" but the method has survived in a time when even ultrafast flash studies have become commonplace because some studies, eg., of long-lived radical pairs, can be done in rigid media at low temperature but not at all by flash kinetic techniques. However, the cryoscopic method of prolonging the lifetime of interesting species is now put in competition with methods based upon generation and preservation of the long-lived materials of interest at room temperature and above in secure molecular nests such as the cavities in zeolites or the interior of molecular clathrating agents.

A number of bimolecular photoreactions, especially certain [2+2] cycloadditions, can be uniquely carried out in crystalline solids, a fact that has been responsible for the evolution of the elegant field of topologically controlled photochemistry by Schmidt and his co-workers. There are important logistic problems which have prevented development of topo-photo-chemistry as a powerful synthetic method but the methodology is there to be exploited, and occasionally it is. An example with much promise is the asymmetric reactions of molecules, which by themselves are achiral, when they are resident in chiral crystals. A much more common "use" of photochemistry in organized media is as a tool in the study of reaction mechanisms. The restraint of internal molecular motions can impose conformational restrictions on molecules in ordered systems which, by favoring or disfavoring certain photoprocesses, can reveal intimate details of reaction mechanisms. Intriguing examples include the imaginative study of the delicate control of chemical and physical behavior of the excited states of molecules of photoisomerization of visual pigments gained through analysis of the role played by steric guidance by associated proteins.

The interest of chemists in general has turned to organized media for many reasons. Perhaps the most global of these is the rising interest in materials science, a filed in which structure–property relations depend on supramolecular properties of matter. The properties of crystalline lattices, molecular crystals, liquid crystals, monolayers and other surface films are determined as much by the organization of assemblies of molecules as by the inherent properties of the molecules themselves. Serious efforts are underway to use photochemical, especially photoelectrochemical, properties of surface films on solids for information storage and readout. Some progress has been made in this branch of the science and technology of materials for the "information age," but much of the promise remains to be realized.

Photochemistry and spectroscopy have been used to reveal details of the structures of organized materials such as membranes, liquid crystals and Langmuir-Blodgett films which are not readily susceptible to absolute structure determination by x-ray crystallography. (I should note in passing that advanced techniques, such as grazing angle x-ray diffraction, are now allowing definitive study of some of these systems.) Such research is not simple. A material such as a liquid crystal is not as robust as most crystalline solids. The enthalpies of transition among mesophases or from mesophase to isotropic phase are smaller that enthalpies of fusion of most crystalline solids. The inclusion of guests in such structures may impose steric ordering on the solute molecules with consequent modulation of their photochemical and spectroscopic properties. However, the solute in the "softly" organized host may cause extensive disordering *or reordering* of the host material. The photochemical behavior of the guest in such a system may be frustratingly complex, but laden with (difficult to interpret) *information about the structure of the system.* To extract that information will require deconvolution of a maze of fact and conjecture concerning reaction mechanisms, quantum yields, product distribution, luminescent behavior, and phase relationships in the solvent–solute systems.

The basic attraction of photochemistry in organized media is obvious. It allows one to organize a system and then carry out reactions within the assembly in a timely fashion and with minimum disturbance of the structure that has been assembled. This option is not usually possible through the use of thermal reactions having significant activation energies, because heating to promote the reaction will often destroy the organized structure. On the other hand, thermal reactions having little or no activation energy will normally have been finished before the requisite structure can be assembled and put in place. However, not all is easy even with well-behaved photoreactions, since the change in composition of the material as a reaction progresses may alter the structure drastically. Photodimerization by [2+2] cycloaddition in crystalline solids commonly results in reduction of the samples to noncrystalline powders well before the reactions have gone to completion. Related problems are encountered with reactions in most other organized media, with the notable exception of host–guest molecular complexes and reactions in the cavities of very stable crystalline solids.

Study and exploitation of photochemistry in organized media is a field, if not in its infancy, barely past adolescence. This fine group of chapters should set the stage for what is likely to by a dynamic showcase of new science and useful technology in the near future.

George S. Hammond

Georgetown University
December 24, 1990

Preface

Most books published during the last ten years on photochemistry point out that the field has reached a stage of maturity. Yet photochemistry continues to interest many scientists, perhaps, because, as pointed out in the Vedas, "all that exists was born from the Sun." In someways, understanding light-initiated processes is understanding life itself. Nobody can be convinced that we have understood life itself.

During the last ten to fifteen years a group of talented and dedicated photochemists migrated to other areas of Chemistry and have explored those fields with a photochemical perspective. Their discoveries have benefitted photochemists as well as chemists with interest in other areas. In this monograph, these "emigrant photochemists" narrate what they have discovered in these other lands of opportunity. The discoveries described herein are truly interdisciplinary in nature although their descriptions bear a heavy photochemical tinge due to their root.

Over the years, the material described in this book has appeared in the literature under various titles—photochemistry in microheterogeneous systems, photochemistry in nanoscopic reactors, photochemistry in organized media, photochemistry in confined media, photochemistry in oriented systems, photochemistry in nonhomogeneous media, photochemistry in anistropic media and supramolecular photochemistry. The title for this book *Photochemistry in Organized and Constrained media* has not been chosen with any intention of replacing the other titles the individual authors have favored. As pointed out by George Hammond in the Foreword, the meaning of the words "organized" and "constrained" have been purposely kept broad and diffuse. The main theme, in my understanding, has been to discover a magical "reaction cavity" that can be utilized by chemists to make molecules behave (chemically and physically) the way the chemists would like them to. Traditionally chemists have varied the characteristics of the reaction cavity (in solution) by varying the polarity, viscosity etc. of the solvent. In recent times the solvent medium has been replaced with other media, such as micelles, crystals, outer and inner surfaces, liquid crystals, organic and inorganic hosts, monolayer and bilayer films and naturally occurring polymers such as polypeptides and polynucleotides. The achievement of above goal requires an increased understanding of these new and unique media and photochemical techniques have proven most useful in this endeavor.

We have attempted to provide a comprehensive survey of photochemical and companion photophysical studies in various media. This book can serve as a source of reference and as a supplementary textbook in photochemistry courses. A major topic that has not been covered here is photochemistry and photophysics in micelles. Because of the extensive coverage photochemistry in micelles has received in several recent reviews and in a monograph and because of the constraints on the size of this book, an editorial decision was made not to include this topic. It is hopeed that this will not offend micelle lovers and the photochemists who have expended considerable effort in exploring micelles.

Every chapter was reviewed by at least three independent reviewers in addition to Paul de Mayo and George Hammond who read all the chapters and provided comments. I appreciate all of them for the help. Among the many who assisted me in this process, I would like to mention the names of Dave Eaton, Linda Johnston and John Scheffer on whom I heavily depended. Special thanks to George Hammond who readily agreed to my request of providing a foreword. As you may note he has been busy on a Christmas eve.

The original plan of having the book typeset had to be abandoned once the number of pages exceeded our initial expectation. Getting the book in camera ready format from the diskettes—prepared in various word processing programs and formatted to different computers—supplied by the authors was undertaken by me without realizing the time demanded by this task. Help provided by Bill Bushy, Jayne Allen, Melinda Delawski, Rajee Ramamurthy, and Dave Sanderson eased my frustration. Copyeditor Lindsay Ardwin and Mark Sacher of VCH Publishers set the rules for maintaining a uniformity among chapters and for resisting the temptation to use every font, every size and every style available on the computer. Their help is appreciated.

Several people have been responsible for enabling me to edit a volume of this size and quality. My parents Vaidhyanathan and Jayalaksmi instilled in me a desire to learn, to study and to be disciplined – these are the foundation upon which my scientific career is based. Bob Liu, Paul de Mayo, and Nick Turro, who continue to point out my strengths and weaknesses and provide inspiration, taught me the fundamentals of photochemistry and the value of research. My good friends Venkatesan and Dave Eaton remain as a source of strength and encouragement at times of need. I owe my deepest appreciation to my wife Rajee and son Pradeep who have shown interest, patience, and concern on my projects and have tolerated me over the years.

This book would not be in your hands if not for the support I have received from the authors. I consider myself fortunate for having coordinated the efforts of the many distinguished chemists who readily agreed to contribute to this volume. The authors are to be admired for their patience and cooperation in revising the chapters following the suggestions made by the reviewers. Their cooperation and understanding at every stage have been remarkable. This editorial task could not have been undertaken and completed without the assistance and support provided by the Du Pont Company. It is indeed a pleasure to thank all the authors and the Du Pont Company.

I wish I were gifted with greater vision and greater ability so that I could have done this work better than I have done. I am thankful, however, for what I have been able to do. It is hoped that this volume will be of considerable value to students, teachers, and researchers.

V. Ramamurthy
April 15, 1991

Contents

PHOTOCHEMISTRY
IN
ORGANIZED
AND
CONSTRAINED MEDIA

Chapter 1

Thinking Topologically about Photochemistry in Restricted Spaces

Nicholas J. Turro and Miguel Garcia-Garibay

Department of Chemistry, Columbia University, New York, NY, USA.

Contents

1. Introduction

Topology is a branch of mathematics that is concerned with identifying the sameness of mathematical forms that appear to be different upon simple geometrical inspection.[1] Topology provides mathematical tools for determining whether two or more mathematical forms are the same or not by way of rules that correspond to a procedure which attempts to map the topologically relevant properties of one form on to another. If the mapping is one to one (i.e., each topological feature of form A may be mapped onto form B and vice versa), the forms A and B are declared topologically identical. In the scientific and real worlds we are often concerned with identifying the sameness of objects, techniques, methods, etc., and topology provides the means to do so.

The mathematical definition of a topological space attempts to extract the most common structural essence of familiar and intuitive spaces while being as universal as possible. The definition of a topological space emphasizes the notion of nearness and neighborhood relationships and avoids the concepts of metric relationships such as angle and distance between points in the space; hence, the description of topology as rubber-sheet geometry.[2] Topological figures may be flexibly stretched or distorted as long as they are not cut and glued back together. Flexible distortions do not disturb the fundamental character or the topological space since nearby points remain neighbors and sequenced points remain in the same sequence.

The term *structure* describes a geometric or mathematical object whose arrangement of components and connections generates a general character for the whole object. The structure possesses both local and global characteristics. Structures are the *sine qua non* of scientific thinking because they provide the paradigms on which scientists construct techniques, make underpinning of the measurements, decide on variables, and interpret observations.[3] It has been said that we think in generalities but we live in details. This thought might be paraphrased for scientific activities as *we think topologically, but we experiment geometrically*.[3]

Topological geometry has been described as a qualitative but precise geometry because it is precise in defining the topological features of a geometric form. Geometry is at the heart of chemical thinking, so that it is natural to ask whether topological geometry can be of use to chemists. In the authors' view, organic chemistry has flourished because organic chemists have traditionally thought topologically, i.e, qualitatively but *precisely*.[3,4] In this account, we present a description of topological methods in terms that should appeal to chemists and that can be employed to analyze problems involving microheterogeneous systems and restricted reaction spaces. First some examples will be given of how topology works for geometric forms. This will be followed by examples that apply topological thinking to the supramolecular level of chemical analysis.

1.1. Geometry in Chemical Thinking. Euclidean versus Topological Geometry

Euclidean geometry involves the study of certain properties of mathematical figures in space.[5] To a geometrician not all properties of a figure are of interest. He is only interested in the pertinent geometrical properties of the figure. When viewing a triangle, for example, the color of the triangle, the material on which the triangle is represented, the physical width of the lines representing the edges of the triangle, etc., are not of concern. The viewer is capable of intellectually idealizing the lines and vertices of the figure. What then are the pertinent geometrical properties of a geometric figure? In Euclidean geometry the answer lies in the search for *congruence*. To a geometrician, two geometrical figures are exactly the same if they are congruent. Establishing congruence is a purely intellectual enterprise that detaches itself from the constraints of physically representing figures on a piece of paper or on a blackboard. Two figures are termed congruent or the same if a defined intellectual transformation or mapping allows one figure to be placed on the other so that the two figures exactly coincide in a one-to-one relationship in all geometric properties. As a result of the rules of mapping, each member of a family of congruent figures shares all geometric properties that are commonly possessed by the family. In topology sameness is established by applying a similar idea termed *homeomorphism*.[1,2] Two figures are homeomorphic if an intellectual transformation or mapping allows one figure to be placed on the other so that the two figures exactly coincide in all topological properties.

The essential difference between Euclidean and topological geometry is made clear by the rules for establishing congruence and homeomorphism. In Euclidean geometry the mapping or placing procedure involves *isometries* (rotation, translation, and reflection) *that conserve the size and shape of the geometric figure*. Two figures are recognized to be the same by performing isometric transformations which show that the figures are congruent. In Euclidean geometry, we are allowed to move a figure only by applying motions that do not change the metric properties of distance and angle relations between any two points of the figure. Translational motion of a Euclidean figure in three-dimensional space does not change its metric properties.

In topological geometry the establishment of homeomorphism involves a mapping that is continuous and *preserves sequence and connectivity. No regard is given to the metric relationships between the points of the figure*. As a result, two topological figures are the same or homeomorphic whenever the points of one figure may be continuously mapped onto those of the other by performing elastic motions that preserve sequence and connectivity. Just as color is irrelevant to geometric congruence, *size and shape are of no topological significance*. Thus, to a topologist the two dimensional figures A, B, C, and D are the same (Figure 1). The reason they are the same is that each may be mapped onto the other by elastic transformations which preserve sequence and connectivity. In spite of their obvious difference as geometric figures in the Euclidean sense, in the *topological sense* the figures are identical. What then is topologically common in the figures A, B, C, and D? The answer is that they are all simple connected curves possessing an inside, an outside and a boundary.

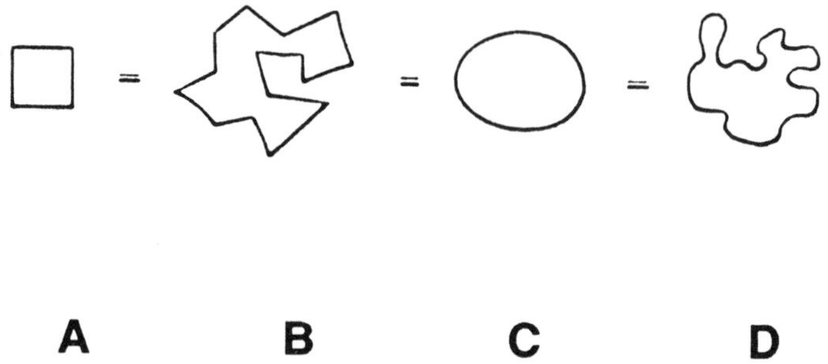

A   **B**   **C**   **D**

Figure 1. All simply connected closed curves, such as line A, B, C, and D, are homeomorphic in spite of their obvious geometrically incongruent structures.

For simple chemical systems, Euclidean geometry may work well and a complete quantitative analysis may be possible. *For more complex systems, it is usually best to initiate examination by resorting to topological geometries, identifying possible homeomorphisms to other well-investigated systems, and then testing for local and then global congruence with the unknown system.* To the extent that homeomorphisms can be established, the topological aspects of the unknown system are established. Armed with this information, the kinds of experiments, the kinds of mechanisms, the kinds of parameters, the kinds of variables, etc., that can be employed to investigate the unknown system legitimately and confidently are apparent to the topologically understanding experimentalist. *The power of such an attitude is that the techniques, parameters, mechanisms, and methods of analysis relate to universal geometries and not to specific experiments.* After identifying the topology of an unknown system under investigation and finding a homeomorphic system that has been well characterized, the experimentalist can move a set of ideas and techniques in a whole and confident manner to examine the unknown system.

1.2. Relationship of Euclidean and Topological Geometry to Chemical Structures

We are accustomed to the analysis of concrete, readily visualizable geometric figures embedded in Euclidean space of one, two, or three dimensions. How can figures embedded in an abstract topological space be transformed into figures in one-, two-, or three-dimensional space that are as intuitive as Euclidean figures?

First we note that although topology does not concern itself with the concept of quantitative distance relationships, composition (the number and kinds of distinct points in a figure) and constitution (the way the points are connected in a figure) are at the heart of topological spaces. In fact, one of the easiest ways to visualize a figure in a topological space is to transform it into a *graph*.[3,4,6]

Chemists come into contact with graph theory at an early stage when they are confronted with hydrocarbon isomers. They quickly become accustomed to writing down the graph of a hydrocarbon structure and identifying when two graphs (and their corresponding structures) are different or identical. In a molecular graph, the vertices represent atoms (or groups of atoms considered as a unit) and the lines represent bonds. *To a topologically thinking chemist the points represents the molecular composition of the graph while the network of connecting vertices represents the molecular constitution of the graph.* The graph of course has nothing to say about the molecular configuration or the molecular conformation of a structure. In spite of this limitation, the power of the graphical representation of molecules is its generality. The points in the graph could represent any atom or even groups of atoms. Following simple rules and without any fancy mathematics, the chemist can represent and systematize an enormous amount of information about molecular structure by employing graphs. The graph of a molecule may be viewed as an expression of the topological concept of special neighborhood relationships which are described by the valence of an atom. This topological concept of valency or bond order was the foundation of the first formulation of structural theory. At this level, the actual arrangement of the atoms in space, their bond lengths and bond angles, need not be specified, as is true of any topological figure. *A major question we wish to explore in this essay is whether chemistry in constrained spaces can be viewed in terms of some topological structures, such as the graphs which so usefully represent molecules.* The hope is that the fundamental usefulness of topological concepts which have proved to be useful at the molecular level will be equally powerful at the supramolecular level.

2. Interfaces, Surfaces, Organized Assemblies, Restricted Spaces, and Other Supramolecular Structures

What do the structures of micelles, cyclodextrins, liquid crystals, polymers, DNA, zeolites, porous silica, and other microheterogeneous systems have in common? From the standpoint of Euclidean geometry the answer would appear to be, "very little." However, from the standpoint of topological geometry these apparently structurally distinct systems have much in common. To the extent that the topological features of these systems are correctly identified, their sizes, shapes and chemical aspects are irrelevant and each system can be considered to be *locally or globally homeomorphic*. A feature that is common to all these systems is that they all have the ability of restricting the rotational and translational molecular freedom of the substrate by virtue of relatively long-lasting nonbonded associations.[7] This feature may include highly specific binding in preorganized systems such as enzymes, relatively static media such as molecular crystals, as well as nonspecific and often less organized environments such as glassy matrices and amorphous surfaces. Micelles, liquid crystals, inclusion complexes, porous solids, and other systems of current interest may lie somewhere in between.

From the standpoint of a photochemist concerned with photoreactions in restricted spaces, the numerous alternatives of binding can be broadly classified in terms that depend on the freedom enjoyed by the substrate during the time scales of interest in photochemical phenomena. Probes that are tightly bound at the molecular level or that experience high local translational or rotational barriers, for instance, can be considered to be in a *closed environment*. Closed environments do not allow for mass transport through their boundaries during (at least) the time scale in which the events of interest take place. Closed environments (Figure 2) may include systems such as molecular crystals,[8] cyclodextrin complexes,[9] zeolite supercages and channels,[10] micelles,[11,12] and hydrophobic pockets of folded polymers.[12a]

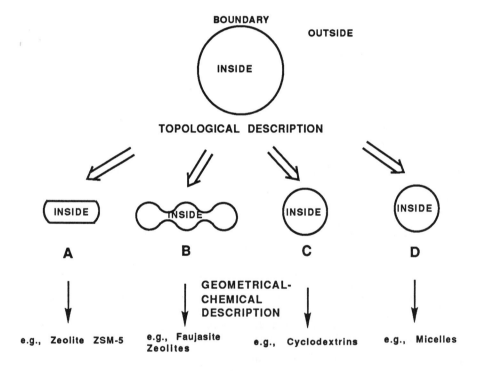

Figure 2. Zeolites, cyclodextrins, and micelles can be commonly associated with the topology of closed structures.

Probes that are weakly bound and have low local translational or rotational barriers may have much molecular mobility left and can be considered to be in *open environments*. Open environments (Figure 3) may allow for mass transport to occur during the time scale of the events of interest and can be commonly found in the external surfaces of amorphous solids such as porous glasses,[13] clays,[14] and silica gel[12,15], on the surfaces of a number of interesting macromolecules including proteins,[16] DNA,[17] and other simpler polyelectrolytes[12a,18]; on the external surface of micelles,[11,12] membranes,[19] and in liquid crystals.[20]

Closed and open environments can be represented by some of the simplest topological spaces, a sphere, an infinite plane, and an infinitely extending line segment. As shown in Figures 2 and 3, the properties of these spaces can also be represented conveniently in two dimensions with a closed curve and an infinitely extending line segment. These structures have the property of cutting space into two pieces. The "inside" and the "outside" are the spaces found in the case of the sphere and the closed curve. The "top" side and "bottom" side, are the relevant spaces in the case of the plane and the line segment. These spaces possess, in addition to "sideness," the topological property of a boundary.

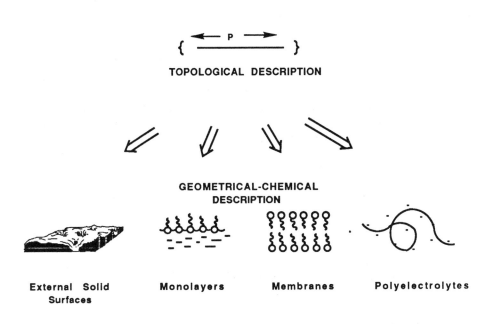

TOPOLOGICAL DESCRIPTION

GEOMETRICAL-CHEMICAL DESCRIPTION

External Solid Surfaces   Monolayers   Membranes   Polyelectrolytes

Figure 3. Examples of topologically open environments. This classification is based on association and mass-transport criteria. External surfaces, monolayers, membranes, and polyelectrolytes allow for relatively efficient diffusion along a two-dimensional surface.

2.1. Control of the Reaction Molecularity by Different Topological Spaces

Depending on the topological structure of the system and on the affinity of the probes for the different spaces available, different supramolecular systems may exert a controlling effect on the molecularity of the reaction by sorting the prospective reactants within the same or different topological spaces. Partition of the probes between these spaces may result in catalytic or inhibitory effects. Keeping the reactants together in the same space of a closed environment, for instance, may have the effect of speeding slow bimolecular reactions. Literature examples[11,12] involving enhanced photodimerization, excimer formation, etc., which fall into this category are ubiquitous. Inhibition of competing bimolecular processes by keeping the reactants apart could encourage and enhance the efficiency of slower unimolecular reactions. Room temperature phosphorescence in organized media under conditions where oxygen would act as an efficient quencher is a phenomenon in which this type of effect is manifested.[12a] Partition of the probe between different topological spaces is a general phenomenon whose relevance has been demonstrated often in connection with energy-transfer studies in micellar media (Figure 4).[11b,12a] Partition of probes in this case depends primarily on the hydrophilic and lipophilic properties and electrostatic charges of the donor and the acceptor.

Important differences may arise in the use of photophysical and photochemical probes in the study of organized media since the topological spaces of a given system need not be identical under photochemical and photophysical conditions. Bimolecular photophysical processes such as energy transfer through a dipole mechanism and long-range electron transfer may be included in this category. Processes resulting from electronic energy transfer in zeolites, inclusion complexes, and mixed crystals may occur despite physical barriers that would make a bimolecular chemical reaction impossible.

2.2. Organized Media and Topological Isomerism

The location of a reactant (inside, boundary or outside) in a supramolecular system may be considered a manifestation of topological isomerism. A reactant found in the inside space of a closed system, for instance, is part of a topological structure that is different from those in which it is found in the outside or at the boundary (Figure 5a). The difference between the three structures is in fact highly reminiscent of the structural isomerism found in organic compounds as a result of atoms being joined with different connectivity. Consider for example the molecular

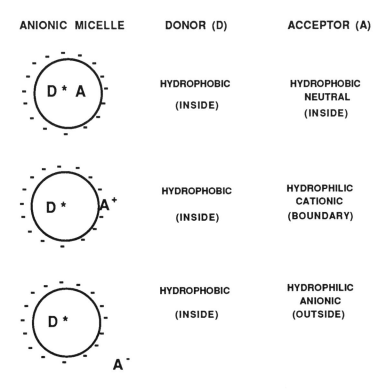

Figure 4. Sorting of donor–acceptor pairs in the different spaces of an anionic micelle based on hydrophilicity and electrostatic considerations.

graphs of the three C_5 isomers, *n*-pentane, isopentane, and neopentane (Figure 5b). An immediate analogy between the systems in Figures 5a and 5b comes from the fact that one can obtain three isomeric structures by changing the connectivity of the elements in their respective graphs. The analogy between the two systems can be better appreciated if we perform yet another transformation that brings the two systems into a closer dimension. For this purpose we label the outside, the boundary and the inside with the letters O, B, and I, and we identify the probe with the letter P as shown in Figure 5c. This representation clearly suggests that in order to interconvert the inside to the outside isomers one has to break the current connectivity in a manner similar to the way one needs to break a C—C bond in the case of the alkane structural isomers. It should be pointed out that the number of reactants and the number of different topological spaces will increase the number of topological isomers much in the same way that the number of carbon atoms increases the number of possible structural isomers represented by different molecular graphs (Figure 5d).

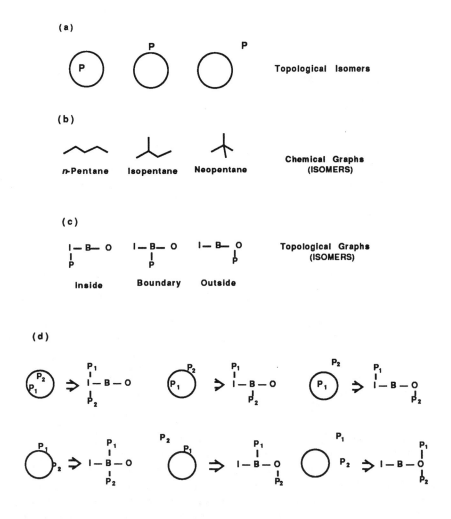

Figure 5. Analogies between supramolecular and molecular isomerism. The connectivity between the probe and its environment defines the supramolecular graph.

2.3. Isomerization and Reactivity of Topological Structures. Applications of the Curtin-Hammet Principle to Supramolecular Chemistry

The concept of topological isomerism becomes particularly attractive in situations where the reactant molecules can be found in different topological spaces. The thermodynamics and kinetics of the binding equilibria, which pertain to most of these systems, can be understood within the rather broad and well-understood framework of isomerization processes (Figure 6). One of the perspectives of this interpretation is that supramolecular chemistry of highly dynamical systems can benefit from the paradigms developed from such areas as structure–reactivity correlations and conformational analysis.

Within the realm of supramolecular chemistry and restricted environments, the equilibria and dynamics of different topological isomers range over a wide spectrum of energy barriers. Decomplexation of a strongly bound inclusion complex and melting of a solid-state reactant, for instance, can be regarded as isomerization processes with very high energy barriers. Weakly bound complexes and molecules adsorbed at specific sites on solid surfaces, on the other hand, may have relatively short association lifetimes and may be able to isomerize with relative ease during the time scale of the measurable photochemical events. The reactivity determined by topological isomerizations in a supramolecular system can be compared to the conformational control commonly observed in reactions where unimolecular rearrangements compete with bimolecular reaction pathways. This is illustrated by considering the alternative reaction routes of the 5-hexenyl radical (Figure 6a) which can undergo intermolecular free radical reactions in a direct manner and in competition with free radical reactions following intramolecular cyclization.[21] From this point of view, chemical processes occurring in supramolecular systems can be analyzed within the context of the Curtin-Hammett kinetics[22] and extensions derived therefrom which have been so successfully applied to molecular systems.[23] The basic Curtin-Hammett kinetic scheme relates the product ratio (or reaction rate ratio) arising from two equilibrating isomers (Figure 6b), each displaying a different reactivity, to their energy difference and to the relative free energy of the competing transition states.[23] Application of the Curtin-Hammett principle to the chemical reactivity of supramolecular systems needs to be addressed from the standpoint of the dynamics and reactivity of both the substrate and the supramolecular assembly.

The time scale for common photochemical and photophysical processes spans over twelve orders of magnitude, i.e., from ca. 10^{-12} s to ca. 10 s.[24] The dynamics of the supramolecular assembly can also span over a similarly wide temporal range. Restricted environments range from relatively static media, such as crystalline and noncrystalline solids, to highly dynamic entities, such as micelles and liquid crystals. When considering the dynamics of the reacting system, however, the dynamics of both the substrate and the medium must be included in the picture. An organic molecule adsorbed on the external surface of an amorphous solid, for instance, may not necessarily be as restricted as when it is found in a micelle or in a liquid crystal.

When the rate of change of the medium (as observed by the probe) is much slower or much faster than the rates of reaction, a single topological representation may suffice to describe the reactivity observed in that particular system. In molecular crystals, for instance, the rate of change of the medium is much slower than most chemical reactions.[8] In the solid state, all the photochemically relevant correlation times of the substrates, excited states and reactive intermediates evolve within a unique topological structure given by the locally closed environment (Figure 7a) of the reaction cavity.[25]

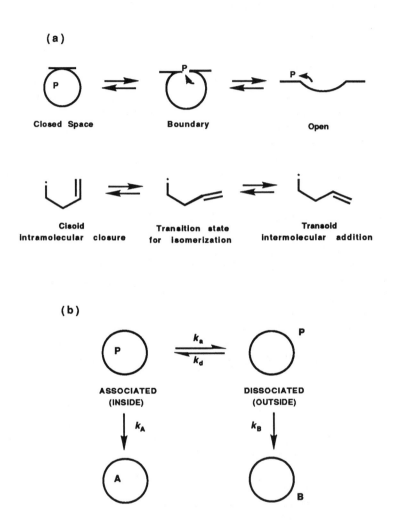

Figure 6. (a) Dynamics of supramolecular systems from the point of view of topological isomerizations. (b) Applications of the Curtin-Hammet principle to the reactivity of interrelated topological isomers.

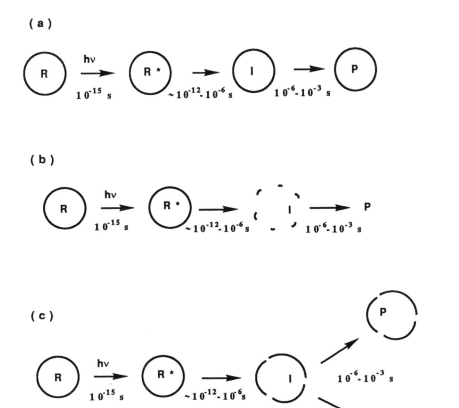

Figure 7. The interplay of size and time relationships on the topological structure of supramolecular systems and restricted media.

When the rate of change of the medium is relatively close to the rates of chemical reactions, the substrate may experience a variable effect at every point along the reaction coordinate. Consideration of the dynamics of the probe and of the dynamics of the environment are absolutely essential to understanding the reactivity of the system. This situation is topologically identical to that depicted in Figure 7b and 7c and is found in common photochemical reactions (10^{-9} to 10^{-3} s) occurring in micelles, liquid crystals, zeolites, and in other restricted environments. A common source of dynamic heterogeneity found in supramolecular photochemistry is therefore due to complexation–decomplexation equilibria. Whether it is because of structural changes occurring in the environment (such as cleavage of a micelle) or simply because of the low dissociation barriers of an equilibrium system (separation of a host–guest complex), or by the superseding of local adsorption interactions (in a porous solid, a clay, or a zeolite), the dynamics of the medium may be perceived in terms of isomerizations into locally open interfaces.

3. A Probe for Closed and Open Topological Spaces: Photochemistry of Dibenzyl Ketones in Restricted Spaces

The use of dibenzyl ketone (DBK) and its derivatives as probes for studying the effect of organized media and restricted environments follows from their ability to report on events that cover relatively large spatial and dynamical ranges.[26]   The correlation times in the photochemical processes of dibenzyl ketones are closely related to the rates commonly found in the topological isomerization processes of a number of organized media and restricted environments. In this section we illustrate how the photobehavior of DBK reflects the topological and geometrical features of the environment in which the photochemistry occurs. Several key features of the mechanism that have facilitated this are also described. Readers are advised to consult other chapters in this book for detailed presentation of results as we only utilize examples from various media without going into the details.

Irradiation of DBK in unrestricted environments such as solution media is characterized by the formation of benzyl radicals following α cleavage and decarbonylation processes (Scheme 1).[27] The termination process of these radicals is generally the uncorrelated (random) coupling to yield 1,2-diphenylethane. Photolysis of unsymmetrically substituted derivatives (represented as ACOB) results in the formation of three radical coupling products, AA, AB, and BB, in a ratio of 1:2:1 (Scheme 2). In contrast, when escape of the radicals is impeded by confining environments, the radical recombination probability is modified by favoring the formation of geminate products of the AB type. It has also been observed that recombination of the primary radical pair can sometimes occur before decarbonylation (or escape), giving rise to rearranged ketone photoproducts and recovered starting material (Schemes 1 and 2).[28] Recombination of the primary radical pair generates 1-phenyl-*ortho*-methyl acetophenone (*o*-MAP) and 1-phenyl-*para*-methtyl acetophenone (*p*-MAP) by ortho and para coupling, respectively, in the case of DBK. Thus, in using DBKs as probes one measures the yields of diphenyl ethanes and rearranged products and in certain cases the isotope enrichment factor (see below). The cage effect, defined as

$$\text{Cage Effect} = \frac{AB - (AA + BB)}{(AA + AB + BB)}$$

is a measure of restriction (a factor dependent on topology) and is calculated from the yields of diphenyl ethanes. The isotope enrichment factor, α, defined as {rate of disappearence of ^{12}C ketone/rate of disappearence of ^{13}C ketone}, is measured from quantum yields of disappearence of the ^{12}C and ^{13}C ketones.

The accepted mechanism for the formation of products upon uv excitation of DBK is shown in Scheme 1.[27,28] The singlet excited state produced by light absorption undergoes intersystem crossing quantitatively and rapidly to the triplet state, which then cleaves quantitatively to a spin- and composition-correlated triplet geminate radical pair, $^3[C_6H_5CH_2CO \cdot \ \cdot CH_2C_6H_5]GP$. In this case the elemental atomic composition is correlated to that of the starting ketone, i.e., each atom in the geminate radical pair is correlated with each atom of the parent ketone. In addition, *the electron spin state of the geminate radical pair is correlated to that of the starting parent ketone*, i.e., the orbitally uncoupled electron spins are parallel and correspond to a triplet state. The geminate radical pair also possesses a *constitutional correlation identical to that of the parent ketone in all respects except for the disconnection of the bond, which has undergone α cleavage.*

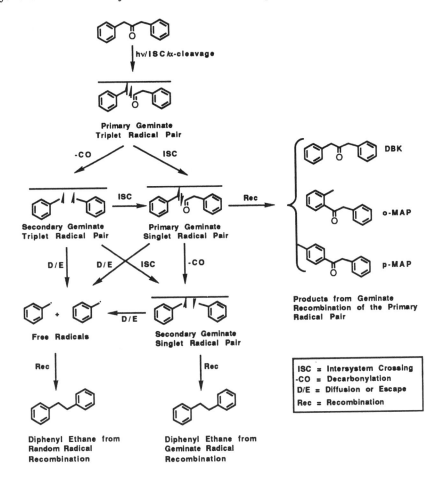

Scheme 1

As time goes on the three correlations of spin, composition, and constitution will be lost: (1) as the triplet relaxes to the singlet or to uncorrelated doublets, (2) as the geminate composition becomes lost as the result of diffusional separation and randomization of the radicals in the global space, and (3) as the constitution correlation becomes lost when the $C_6H_5CH_2CO\cdot$ radical loses CO via bond cleavage. The time scales for each correlation loss are different and depend on several factors, including the reaction conditions and the reaction space. Thus, the type of "space" explored during each correlation time will depend on the time scales of the loss of correlation and the correlation time for exploring the space.

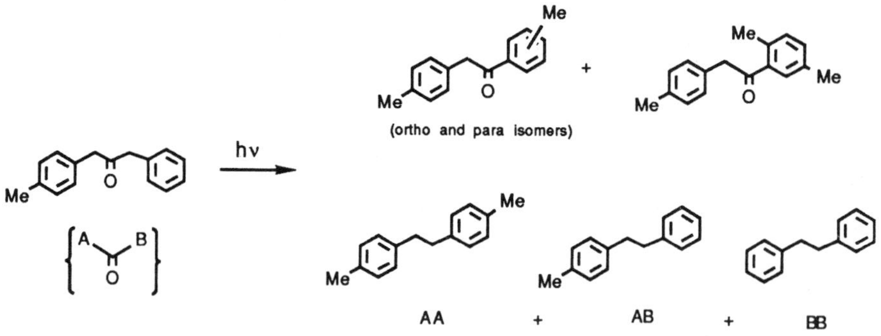

(ortho and para isomers)

AA   +   AB   +   BB

Scheme 2

The loss of spin correlation is a strong function of both (1) the intramolecular magnetic fields due to nuclear hyperfine coupling and to spin–orbit coupling of the odd electrons in the radical pair and (2) to the separation of the radical centers in space.[29] While the strength of hyperfine coupling in the radical pair is distance independent, the effectiveness of hyperfine coupling to induce intersystem crossing is strongly distance dependent. The strength of spin–orbit coupling is strongly distance dependent, being related to the magnitude of orbital overlap between the radical centers, and falling off roughly exponentially as a function of the radical pair separation. The magnitude of the singlet–triplet energy gap determines the efficiency of intersystem crossing by any mechanism and the larger the energy gap the less efficient the intersystem crossing. The singlet–triplet energy gap is inversely related to the distance between the radical centers. Although the situation is obviously complex, the essential points are readily comprehended and qualitative effects are easy to identify. For example, if the size of the restricted space is very small relative to the size of the radical pair, the singlet triplet gap will be large (enforced orbital overlap) and the intersystem crossing will be slow. On the other hand, if the size of the restricted space is large relative to the size of the radical pair the triplet–singlet separation will tend toward zero and either hyperfine or spin–

orbit-induced intersystem crossing can be effective. However, if the size of the space is too large, allowing for mass transport to occur, separated radical pairs will tend to lose their compositional correlation as they spread throughout the space and encounter radicals generated from photolysis of other parent ketones. Finally, the constitutional and compositional correlation will be lost if the primary geminate pair is separated long enough that loss of carbon monoxide occurs.

Various confined media in which a large number of photochemical and photophysical investigations have been carried out come under four general categories based on the intersystem crossing mechanism that operates in the case of DBK when included in these media. Such an analysis of the results of DBK also helps to classify the media into open and closed topologies. We find that closed topologies can be divided into several categories based on the geometrical features of the media. Intersystem crossing of the triplet geminate radical pair generated in a confined medium can occur while it is still confined in a restricted environment or after diffusion and separation of the two fragments (the radical pair converts into two uncorrelated doublets). Adsorption of dibenzyl ketones on solid surfaces such as that of the zeolite NaA represents an example of the latter situation.[30] The relatively large inside spaces of the zeolite NaA are completely inaccessible to DBK molecules because of the small size (~4 Å) of the accessing windows. Photolysis of *para*-MeDBK adsorbed onto zeolite NaA was found to give a statistical distribution of the three diphenyl ethanes as observed in isotropic solution media (Table 1).[27,28] Under these conditions, with a topologically open environment, diffusion of the radical pair results in decarbonylation followed by re-encounter of the randomized radicals.

The rate of intersystem crossing in more restricting media may depend on three factors:[29,31] (1) the extent of spin–orbit coupling (SOC) in the pair (which is a function of separation), (2) the strength of the hyperfine coupling (HFC) in the pair (which is independent of separation), and, (3) the efficiency of mixing of the triplet and singlet states by one of these mechanisms. If the space available to the radical pair is too small, the radical pair may separate only slightly (e.g., ~ 2–3 Å), and the singlet–triplet gap will be too large for hyperfine interactions to induce intersystem crossing. When the singlet–triplet energy gap is forced to remain large by the close proximity of the two radical termini, the rate of intersystem crossing will be effectively too slow by both the spin-orbit and the hyperfine coupling mechanisms. However, the key point is that if enough time (ca. 100 ns, based on solution data)[31] elapses due to the delay in intersystem crossing, the phenacetyl radical ($C_6H_5CH_2CO$) will decarbonylate (the second cleavage reaction is a spin-independent process), and the radical pair will lose its constitutional correlation. Intersystem crossing may now occur in the secondary radical pair and recombination can give compositionally correlated diphenyl ethane (DPE$_c$) as the product (AB will be the only product in the case of ACOB). Molecular crystals[13] and solid-state cyclodextrin complexes,[32] where separation of the primary radical pair is opposed by strong structural barriers, are examples of such topologically closed, geometrically rigid environments (Table 1). The photochemical reactivity of DBK in these closed spaces results in the formation of compositionally correlated diphenyl ethane as the only product.

Table 1. Representative Examples of the Product Distribution from Photolysis of DBK in Various Environments

Media	Reactant	Percentage Yield DPE	RP	C.E[a]	Ref.
Benzene	pMeDBK	100%	—	0%	15[a]
	α–MeDBK	100%	—	0%	15[a]
2-propanol	p-Mep'-MeO	100%	—	0%	15[c]
Porous Silica					
22 Å	pMeDBK	100%	—	~22[b]	15[a]
40 Å		—		~15[b]	15[a]
95 Å		—		~10[b]	15[a]
139 Å		—		~10[b]	15[a]
Merck (35–70)	pMe,p'MeO-				
(10%[c]/20°C)		100%	—	23	15[c]
(10%[c]/-55°C)		100%	—	52	15[c]
(10%[c]/-165°C)		100%	—	>90	15[c]
(20°C)	DBK	~97%	~3	—	15[c]
β-Cyclodextrin					
Aq. complex	DBK	19	80	—	33
	α–MeDBK	40	60	80	33
Solid complex	DBK	100	—	100	33
	α–MeDBK	98	2	96	33
HDTCl–Micelles	DBK	~98	~2	33[d]	11
	p-MeDBK	100	—	59[d]	11
	4,4'-di-t-Bu	100	—	95[d]	11
Porous glass	α–MeDBK	100	—	38	13
Liquid crystals					
CCl/CN[e], Chol[f]	p-MeDBK	100	—	18	20[a,b]
CCl/CN, Iso[g]		100	—	14	20[a,b]

[a] *Cage effect (%).*
[b] *Cage effect at large coverage values; c) Based on monolayer coverage.*
[d] *Cage effect measured by quenching of the escape radicals with CuCl$_2$.*
[e] *Cholesteryl chloride/cholesteryl nonanoate, 35/65, w/w.*
[f] *Chloesteric phase.*
[g] *Isotropic phase.*

If the size of the confining space is large enough to allow for separation and efficient intersystem crossing, but not too large to let the radical pair diffuse away and decarbonylate, products of primary radical recombination can be formed. Recombination can give the starting ketone (DBK) as well as the ortho and para coupling products 1-phenyl-*ortho*-methyl-acetophenone and 1-phenyl-*para*-methylacetophenone (Scheme 1; no diphenylethanes will be formed). Such a situation has been observed in aqueous β-cyclodextrin complexes where photolysis of DBK has been found to result in the nearly exclusive formation of *p*-MAP (Table 1).[33] We classify these systems as having a closed topology with a tight geometry.

Another possibility comes from closed but geometrically large environments. Here, the primary radical pair may separate far enough in space so that intersystem crossing can compete with decarbonylation. This situation is encountered when the photolysis of DBK is carried out in micelles. Under these conditions, intersystem crossing occurs by means of the hyperfine coupling mechanism. As per this mechanism, magnetic isotope effects (^{13}C vs. ^{12}C) can give rise to a discriminatory effect on the intersystem crossing which is manifested in the products of primary radical recombination.[11a] Notably, the products from primary radical recombination containing ^{13}C on the carbonyl carbon can be favored, thus giving rise to a methodology for isotope separation (Scheme 3).[13,26]

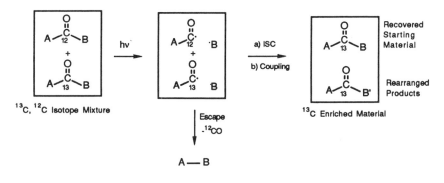

Scheme 3

While the details of this process are not discussed here (for details see ref. 26), it is important to note that intersystem crossing in the ^{13}C-containing radical pair is naturally favored over the spinless ^{12}C-containing carbonyl group of the other radical pair. Isotope enrichment has been found in the recovered DBK and in the rearranged product (*p*-MAP).[11] Reaction media in which the above topological phenomenon occurs include micelles,[11] porous glass,[13] porous silica[15], liquid crystals,[20] and emulsions.[34] These reactions are characterized by the formation of variable amounts of rearranged products and diphenylethanes. In the case of ACOB, diphenylethanes AA, AB, and BB will be formed in ratios different from 1:2:1, and the starting material recovered will display significant isotope enrichment factor α (Table 2). Systems such as these possess a closed topology with a loose geometry.

Table 2. ^{13}C Isotope Enrichment Factor (α) from Photolysis of DBK in Various Media

Ketone	Medium	α	ref.
DBK	Benzene	1.04	11[a]
DBK	HDTCl[a] micelles	1.68	11[a]
DBK	Silica (22 Å)[b]	1.21	15[a]
DBK	Silica (40 Å)[b]	1.22	15[a]
DBK	Silica (95 Å)[b]	1.21	15[a]
DBK	Zeolite mordenite[c]	~1.15—1.05	30
DBK	Zeolite NaA[c]	~1.01	30
DBK	Zeolite NaY[c]	~1.06	30
DBK	Zeolite NaX[c]	~1.10	30
DBK	BS[d] liquid crystals (s)[e]	1.09	20[a]
DBK	BS liquid crystals (i)[f]	1.03	20[a]

[a] *Hexadecyltrimethylammonium chloride.*
[b] *The number in parentheses refers to the average pore diameter.*
[c] *The value of α was shown to vary with ketone loading; larger variations were only found in mordenite.*
[d] *n-Butyl stearate.*
[e] *Smectic phase.*
[f] *Isotropic phase.*

Based on the above analysis, the following conclusions can be drawn regarding the connectivity between the photobehavior of DBK and the topological and geometrical features of the media in which the reaction occurs:

Open topology

-Intersystem crossing through spin uncorrelated-pair (doublet)
-No rearrangement; only diphenylethane product (AA, AB, and BB in the ratio of 1:2:1 from ACOB)
e.g., zeolite 4A

Closed topology–rigid geometry

-Very slow intersystem crossing
-No rearrangement; only diphenylethane product (only AB from ACOB)
e.g., cyclodextrin solid complex; crystals

Closed topology–tight geometry
      -Efficient intersystem crossing through spin-orbit coupling
      mechanism; no isotope effect
      -Rearrangement product only
      e.g. cyclodextrin complex in aqueous medium
Closed topology–loose geometry
      -Intersystem crossing through hyperfine coupling mechanism;
      presence of isotope effect
      -Rearrangement product and diphenylethanes reflecting cage effect
      e.g. (at least partially in) silica surface, micelles, porous glass,
      liquid crystals, and emulsions

The distinction between closed and open environments is based on strictly topological arguments and is relatively straightforward. The distinction between the various closed environments distinguished by DBK is clearly based on geometrical arguments (size). This mixed classification immediately identifies possible homeomorphisms, while it also helps to distinguish certain geometrical aspects that may determine a degree of congruence between different supramolecular structures. Concepts such as "tight" and "loose" remain fuzzy and clearly depend on both the sizes of the probe and its environment.

3.1. The Influence of Geometrical and Chemical Variables on Topological Structures

The topological structures of most supramolecular systems are determined by several factors resulting from the chemical and geometrical properties of the reactants and the medium. Supramolecular systems, however, vary from those that have relatively limited degrees of structural freedom such as molecular crystals, to those such as micelles where a large number of experimental (geometrical and chemical) variables can be manipulated. In the latter case the topological structure will depend on the features added on to the basic system. In the case of micelles, for instance, structural variations can be carried out both on the micelle and on the probe. Changes in the charge and nature of the polar heads and on the chain lengths of the surfactant molecules may have pronounced effects on the binding properties of a given substrate. Probe molecules can be charged, neutral, hydrophilic, large, or small. Addition of electrolytes to the outside aqueous phase may also result in pronounced topological structural effects. Another example would be the surfaces of silica gel. Here, variables such as the pore size, the number and nature of the silanol groups, and the occurrence of a dry gaseous interface or a solvent slurry may influence the topologies of the structure in which the probe is associated. To illustrate the dependence and the interrelations between the topological structure and the chemical and geometrical properties of a given system, we can select a single restricted system and then vary its geometric properties in a systematic fashion. This approach is illustrated in the following section with the faujasite-type zeolites as an example. In this process we have analyzed our results on the basis of our thinking that for more complex systems, it is usually best to initiate examination by resorting to topological geometries, identifying possible homeomorphisms to other well-investigated systems, and then testing for local and then global congruence with the unknown system.

4. Topological Spaces in the Faujasite Zeolites

Structure: We consider first the structure of the faujasite zeolites to explore the number and properties of the various topological levels available in the system. At the constitutional level, the structure of the zeolites is reminiscent of a highly cross-linked and stereoregular polyanion (Figure 8). The framework structure is constituted by an infinitely extending three-dimensional network of oxygen-sharing AlO_4^- and SiO_4 tetrahedra (Figure 8b).[35] It is in fact interesting to note that, at least ideally, every microscopic zeolite particle not only is a single crystal but can also be considered a single macromolecular entity. The sodalite unit (Figure 8c), represented as a cubooctahedron with Al or Si at the vertices and oxygen links at the edges, is the building block of the faujasite structure. When ten sodalite units link as shown in Figure 8c, the faujasite supercage structure having four open "windows" and a large internal volume is formed. At the next structural level, supercages link sharing their window structures and forming a three–dimensional void–space network. A two-dimensional representation including the manner in which the supercages link to each other in a given plane is shown in Figure 8d. The size of the supercages is approximately 13 Å, while the size of the windows is approximately 8 Å (for further details on the structure of zeolites see section 3.1 of Chapter 10).

Topological Spaces: The internal void space of the faujasites can be varied systematically[10] by varying the number density of the cations in the supercage (M^+ or M^{2+}), by varying the composition of the zeolite, and by the addition of physisorbed guest molecules.[10,36] The composition, or Si/Al ratio, determines the nature of the MX (Si/Al < 1.4) and MY (Si/Al > 1.6) zeolites.[35] The number of cations depends on the composition because each tetravalent aluminum atom requires one M^+ compensating cation. Systematic variations in the size of the cation in a zeolite of fixed composition modify the size of the internal space. In spite of the Si/Al ratio difference, the MX and MY zeolites possess the same internal and external topology.

In the broadest sense, the space occupied by the probes interacting with the faujasite zeolites can be classified into external or internal spaces. These spaces, corresponding to either closed or open topologies, on further examination can be subdivided as illustrated in Figure 9. The *first space* (O) is given by the absence of binding or adsorption interaction and would be determined by its immediate external interface, the gas phase or a solution in the case of a solvent slurry. The *second space* (B; open) is constituted by the external zeolite surface which is the location expected for probes whose sizes and shapes are mismatched with respect to the dimensions of the zeolite internal structure. The probe is free to explore the external surface with no geometric constraints within the time scale of a chemical

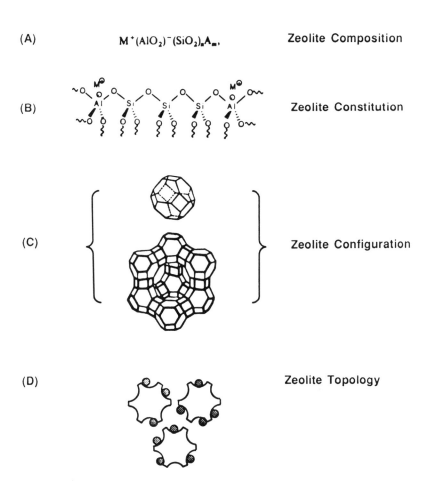

$$M^+(AlO_2)^-(SiO_2)_xA_m,$$

(A)   Zeolite Composition

(B)   Zeolite Constitution

(C)   Zeolite Configuration

(D)   Zeolite Topology

Figure 8. The structure of the faujasites from chemical composition (A), to constitution (B), to configuration (C), and, (D) to topolgical representation. The empty space of a supercage connecting three of its four neighbors and indicating the exchangeable cations (shaded) is represented in D.

process. The *third space* (I_g; globally closed) possesses a toroidal topology and is constituted by the entire pore lattice. This space is occupied by probes that are not tightly held at specific sites mainly because of their relatively small sizes and weak adsorption forces. These probes can explore the inside surface with no geometric constraints during the time scale of a chemical process. In the absence of interparticle diffusion, every zeolite crystallite may be considered a single isolated closed system. The *fourth topological space* (I_l; locally closed) of the zeolite system is constituted by the local environment of the supercages and can be occupied by organic compounds that interact relatively strongly with the framework, with the cations in the framework, or with other guest molecules. However, if the time scale for reaction is long, the probe originally confined in I_l will diffuse to the globally closed space, I_g. This process represents an example of topological isomerism. Partition of a probe within the above spaces is expected to be determined by factors that depend both on the properties of the probe (size, structure, loading) and on the properties of the zeolite (structure, preparation, and contents in the cavities). Organic compounds with kinetic diameters smaller than ~8 Å, including DBK and a number of other common photochemical probes, tend to occupy the internal spaces of the zeolite structure. While the outside (O) and the boundary (B) should always be considered and occasionally have been found to have a determining role, our discussion here will only pertain to the inside spaces (I_g and I_l) with which DBK has been found to be preferentially associated.

Homeomorphisms Between the Faujasites and Other Restricted Media: Even though the detailed topological features of the faujasites are quite unique, broad similarities between the topological and geometrical spaces identified in the case of faujasites and the ones we discussed for other organized media in Section 3 are obvious. Therefore, once we identify a closely related system (homeomorphic and congruent) we should be able to predict the photobehavior of DBK included in faujasite zeolites.

Among the systems discussed so far, aqueous cyclodextrin complexes and micelles possess topological and geometrical properties that are fairly similar to those identified for the internal spaces (I_l) of the faujasites. Cyclodextrins, for instance, possess a small cavity with a free volume of ~300 Å^3, while micelles have fairly large closed spaces with sizes on the order of 15,000 Å^3. Such size differences, reflected in the photobehavior of DBK (Tables 1 and 2), allowed us to associate a tight geometry with the cyclodextrin complex and a loose geometry with the micelles. In comparison, a single supercage of X and Y zeolites possesses a free volume in the range of ~800 Å^3. This number is closer to that of the cyclodextrin complex and when DBK is confined to a single cage (locally closed) one would expect a behavior qualitatively similar to that observed in β-cyclodextrin. Quantitative differences, if observed, would not be surprising as there are not only differences in the cavity sizes, but also in specific interactions that may exist between the host and the guest. One would also predict that when DBK is allowed to migrate within the internal structure of the faujasites (globally closed space) it may exhibit a behavior qualitatively similar to that in micelles. Here, quantitative differences may be also expected as the size and shape of the space explored by the excited state of DBK and by the primary radical pair are distinctively different. With an average displacement of about 30 supercages before decarbonylation occurs,[37] the volume available to a benzyl radical may extend up to ~24,000 Å^3 (compare this with a micellar volume of ~15,000 Å^3).

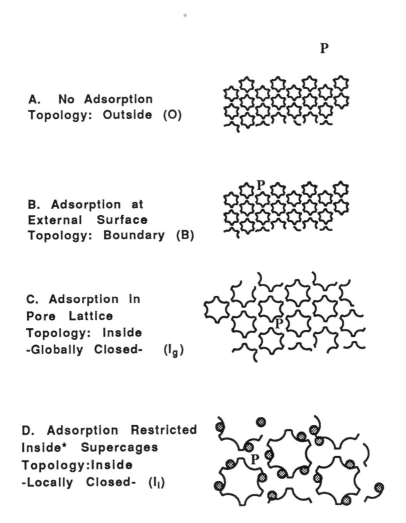

A. No Adsorption
Topology: Outside (O)

B. Adsorption at
External Surface
Topology: Boundary (B)

C. Adsorption In
Pore Lattice
Topology: Inside
-Globally Closed- (I_g)

D. Adsorption Restricted
Inside* Supercages
Topology:Inside
-Locally Closed- (I_l)

Figure 9. The four distinct topological spaces of the faujasites. The 6-connectivity between adjacent supercages on a given plane is emphasized. The position of the probe refer to the spaces occupied or explored during the time required for the processes of interest.

The dynamics of the excited state of DBK and of its subsequent radical fragments can evolve within the local supercage environment or within the global zeolite pore lattice with quite impressive chemical differences.[30,36] Radical escape from the spherical supercage environment (local) into the toroidal faujasite pore lattice (global), constitutes, as discussed in a previous section, an isomerization process. The rate of this isomerization process and the relative contribution of the two structural levels (I_l and I_g) toward the overall reaction is determined by the lifetime of the reactive intermediates and by factors that modify the intracrystalline self-diffusion of the probes.[38,39] Some of these factors include the size and nature of the cations, the presence of unreactive additives (spectators), the temperature of the system, and the loading level of the reactant.[40] Therefore, in each case, the relative contribution from reactions occurring at the local and global levels will depend upon the above factors. At one extreme limit, reactions occurring entirely within the local supercage environment are expected to give, as pointed out earlier, a product distribution similar to that observed in cyclodextrins. At the other extreme, where the reactions occur entirely in the pore lattice, free diffusion of the probe will result in fast randomization of the reactive intermediates. We now analyze the results of the photobehavior of DBK obtained under various conditions in faujasites on the basis of the above discussion involving globally and locally closed topological spaces (see Chapter 10 for a detailed discussion of other photoreactions within zeolites).

4.1. The Photochemistry of DBK in the Faujasite Zeolites
4.1.1. The Effect of the Cations[10b,36]

The supercages of the X and Y zeolites contain cations located at three distinct sites.[35] Of these, the type I are present within the sodalite cages and the other two, type II and type III, are within the supercages. Since sodalite cages are too small, the preferred position for the guests included in zeolites X and Y are the supercages. Therefore, only cations present within supercages are expected to influence the reactivity of the guests. While type II cations are located within the framework of the supercages, type III occupy the void space of the supercage and are expected to influence the reactivity of the guests by both steric and electronic factors. In this section we concern ourselves with the steric influence of the cations on the photoreactivity of DBK included in X and Y zeolites. The ionic diameters of the alkali ions, $Li^+ = 1.4$ Å, $Na^+ = 1.9$ Å, and $K^+ = 2.7$ Å, increase by a factor of two in going from Li^+ to K^+. Such an increase would be expected to reduce the free volume available within the supercage where MX zeolites should display a greater effect due to the large number of exchangeable cations. Greater reactivity differences should be observed in the MX zeolites if volume alone is the determining factor (see Chapter 10).

The results obtained upon photolysis of evacuated samples of DBK under conditions of a relatively low loading (2% w/w) (which represents a nominal occupancy of ~15%) in the cation exchanged X and Y zeolites are shown in Table 3. The first important observation comes from the fact that the product distribution depends significantly on the cation in the MX zeolites, while it remains relatively constant in the MY zeolite series. The expectations regarding the size of the environment and the partition of the probe between the two distinct topologies available in the medium can be appreciated upon closer examination of the results. When steric interactions dominate, as it is postulated in the present case, large cations would localize the probe at a single site more efficiently than the smaller ones. Therefore, large cations should increase the relative contribution of reaction from locally closed spaces. In fact, an increase in the yields of isomeric products in going from LiX to KX reveals the increasing importance of the locally closed environment (I$_1$). As per our expectations, in the case of KX, the yield of the isomers clearly approaches that observed in the cyclodextrin complexes (Table 1).

The formation of substantial amounts of products of primary radical recombination in a microscopically closed system would suggest that a substantial cage effect should be obtained. Furthermore, a parallel correlation may be expected between the yields of primary geminate products and the yield of secondary geminate recombination (cage effect). In this context, studies of the cage effect in faujasite zeolites have been carried out by using the labeled product analysis methodology.[41] The results of the cage effect obtained with isotopically labeled, DBK-d$_5$ at 2% w/w loading (~15% occupancy) are also shown in Table 3.[42] The yields of DPE-d$_5$, DPE-d$_{10}$, and DPE-d$_0$ determined by gc-ms were used to calculate the cage effect. The most important observations relating to the cage effect (Table 3) is that it increases from Li to K in the MX zeolites, and that it remains constant in the MY zeolite series. The trend followed by the cage effect in MX, once again, indicates an increasing contribution of a locally closed topology with an increase in the size of the cation.

Table 3. Product Distribution^a and Cage Effect from Photolysis of DBK (2% w/w) in Ion- exchanged Faujasites

Zeolite	Percentage Yield			Cage Effect^b
	DPE	o-MAP	p-MAP	
LiX	80	3	16	5
NaX	55	17	26	22
KX	40	40	16	73
LiY	100	0	0	19
NaY	95	0	5	17
KY	94	2	4	19

^{a} Calculated error limit, 10%.
^{b} Cage effect measured under the same conditions in a separate experiment by using DBK-d$_5$.

The above dependence of the reaction site (local vs. global) on the cation would also be expected to be reflected on the type of intersystem crossing mechanism that operates on the triplet primary radical pair. Variations in intersystem crossing mechanism can be experimentally monitored through magnetic field effects (MFE) and magnetic isotope effects (MIE). Recent results from experiments in this area in the case of DBK further support the above interpretation of the photochemical results in terms of local vs. global environments. A global topology would provide a large distance of separation, whereas a local topology would provide a short distance of separation between the two radical centers. The nature of the intersystem crossing mechanism that operates would depend on this distance in the following manner:

1. The tightest environments (3–5 Å separation between radical centers) should enforce a large singlet–triplet energy gap enforcing the spin-orbit-coupling (SOC) intersystem crossing mechanism. Large MFE are expected to occur.

2. In relatively loose environments (5–7 Å separation between radical centers), the hyperfine coupling mechanism (HFC) should be dominant and large MIE should be observed.

3. The largest available spaces (>10 Å separation between radical centers) may facilitate intersystem crossing through spin- and magnetic-field independent processes such as spin–lattice relaxation. Under such situations no MFE or MIE are expected.

It should be noted that besides allowing for a large interradical separation, a large space within the local supercage environment would facilitate escape of the radicals into the global space. Radical coupling would be expected to occur both from radicals escaping into the global space and from radicals enclosed within the local environments. However, only the radical coupling reactions that occur within the closed spaces, where the triplet radical pair can maintain its spin correlation, will be subject to the rules for intersystem crossing. Separation of the triplet primary radical pair would result in the formation of uncorrelated doublets, which are insensitive to the magnetic field and magnetic isotope effects.

The magnetic field and magnetic isotope effects in the MX and MY zeolites have been analyzed by examining the product distribution from isotopically labeled (^{13}C and ^{2}H) DBK in faujasites both at the earth's magnetic field and at 2000 G.[43] The results in Table 4 can be summarized as follows: (1) there is no difference in the product ratio (within the experimental error) for any of the variables for the MY and LiX zeolites; (2) there are significant ^{13}C and ^{2}H isotope and magnetic field effects upon photolysis in NaX; (3) there is a significant magnetic field effect on the product ratio for photolysis of DBK in KX (even though there is no magnetic isotope effect); and (4) the results observed with DBK-d_{10} closely parallel the results observed with ^{13}C-DBK.

Table 4. Magnetic Isotope and Field Effects on the Product Distribution from Photolysis of DBK in MX Zeolites[a]

System	DPE	Percentage Yield o-MAP	p-MAP
LiX			
DBK-^{12}CO	81 [85][c]	3 [3]	16 [12]
DBK-^{13}CO (90%)[b]	81 [85]	3 [2]	16 [13]
DBK-d_{10} (95%)	79	2	19
NaX			
DBK-^{12}CO	56 [65]	17 [13]	26 [22]
DBK-^{13}CO (90%)	27 [33]	37 [25]	36 [42]
DBK-d_{10} (95%)	67 [59]	10 [13]	23
KX			
DBK-^{12}CO	40 [68]	40 [14]	16 [18]
DBK-^{13}CO (90%)	45 [62]	32 [13]	20 [25]
DBK-d_{10} (95%)	45	35	20

[a] Results for MY zeolites are similar under all conditions (900–100% yield of DPE and very little isomers).
[b] Isotope content of the sample.
[c] Product yield in the presence of 2000 G external magnetic field.

In the case of LiX and MY zeolites, substantial yields of decarbonylation products, relatively modest cage effects (Table 3), and small amounts of rearranged products (0–20%) emphasize the importance of the globally closed zeolite spaces. The lack of magnetic field and magnetic isotope effects in the products of primary radical recombination are also consistent with a relatively large space where the radical centers can separate and explore the global environment. In contrast, in KX zeolites large yields of rearranged products are obtained. Large magnetic field effects and no magnetic isotope effects in this zeolite are consistent with most coupling reactions occurring in a tight local environment where escape is difficult and enforced orbital overlap is important. These results suggest that spin–orbit coupling, which is isotope independent, may be the intersystem crossing mechanism in KX zeolites. Finally, in agreement with our expectations, photolyses in NaX give evidence of local and global effects that are intermediate between those observed in LiX and MY, on the one hand, and KX, on the other. The local interradical distance and the ease of escape are expected to be intermediate between them. As the local space is also expected to be larger, intersystem crossing can be controlled by nuclear–electron hyperfine couplings (which are isotope dependent) as well as by strong external magnetic fields.[43]

4.1.2.   Loading Effects

Radicals escaping the restrictions imposed by the local environments diffuse along the pore structure from supercage to supercage until they find another radical with which to couple. Since steady-state irradiation produces a very low concentration of radicals at all times during photolysis, the most probable encounters should involve ground-state reactant molecules (such encounters, however, result in no alternative reactions). At very low loadings there will be no obstruction of the pore network and diffusion within the globally closed space will be relatively free. Increasing loading values, on the other hand, will tend to congest the global space, enforcing a locally closed environment. Therefore, the yields of rearrangement products and the cage effect are expected to increase with substrate loading. The influence of the additives (unreactive spectators and increased reactant loading) on the topology of the system, however, is expected to depend on the relative diffusion coefficients of the spectator and the reactive intermediates.

4.1.3. Percolation in the Faujasite Pore Lattice

If the diffusion coefficient of the spectator (including unreactive additives or the reactant) is much lower than that of the radicals, the spectator molecules will appear to be static and will effectively be part of the zeolite lattice. The global space of a partially occupied zeolite interior will appear to possess "occupation structure". Under these conditions, if deposition of the probe occurs with random probability, the randomization of the probe (benzyl radicals) becomes a *bond percolation* problem.[44]   Clusters of spectator molecules will grow larger with increasing loading levels of adsorbate while pockets of empty space will grow smaller (Figure 10). The probability of entrapping two geminate radicals within such an empty pocket increases as the average pocket size decreases very rapidly with increasing loading values.[45]

In contrast, when the spectator molecules have diffusion coefficients that are similar or larger than those of the radical intermediates, the global lattice space will not possess a rigid structure. Randomization of the geminate radicals will become a special case of a *correlated random walk in dynamically disordered system*.[45]   The transition (crossover) between a static and a dynamic disposition of the spectators changes the topological structure with the emergence of a new topological isomerization phenomenon. High loading levels of a rapidly diffusing spectator are not expected to induce large yields of geminate recombination products as the loss of structural correlation of the lattice (given by the dynamics of the spectator) would approach the loss of structural correlation of the probe (given by the rate of escape of the radical pair). New escape pathways become available to the radicals as the structural correlation of the lattice is lost and pockets in the lattice could now be methaphorically described as rapidly diffusing "bubbles" of empty space.

LOADING

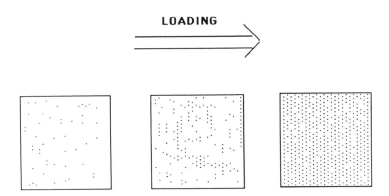

Figure 10. Decrease in the size of empty pockets in the pore lattice with increasing adsorbate concentration. The occupancy level causes changes to the globally closed topological space of the zeolites.

Consideration of the *combined* effects of the local and the global spaces in inducing the recombination of geminate radicals (e.g., cage effect) can lead to situations such as those in Figure 11. Variations in the cage effect with respect to loading can vary from those where no cage effect is observed at all loading values (case I), to those where a constant effect can be obtained (case II), and to situations where a variable cage effect is obtained (cases III and IV). The behavior represented by case I in Figure 11 is clearly solution-like and represents the limit of the "well-stirred reactor." [46] The diffusion coefficient of both the probe and the spectator need to be sufficiently large so that randomization of the system occurs efficiently within the lifetime of the reacting intermediates. Case II represents a situation where the cage effect can be interpreted in terms of a strictly local cage effect where the contents of the neighboring cages are of no consequence to the final result (as in micelles, cyclodextrins, etc.). Case III represents a situation where there is no cage effect intrinsically attributable to the properties of the local environment but an increase in substrate loading results in increasing cage effect values. The diffusion coefficient of the probe should be much larger than that of the spectators such that recombination can occur after diffusion of the radicals within an empty pocket. Finally, case IV represents a hypothetical system where both local and global cage effects contribute to the observed cage effect value at all loading levels.

Before analyzing the results obtained in the zeolites NaX and NaY it is important to point out that the faujasites have a diamond-like pore structure where every supercage is equivalent to a carbon atom and the channels between every connecting window are equivalent to the C—C bonds. An interesting aspect about the diamond structure is that it has bond percolation threshold $P_c = 0.39$, which is relatively large for a three-dimensional lattice.[45] This value can be interpreted as the percent of lattice sites that need to be empty in order to have infinite channels across the ideally infinite lattice (the bond percolation cluster). While the value of

P_C may appear of relatively little interest, it represents a property of the lattice that relates to quantities such as the mean cluster perimeter[44] (the average size of an empty pocket) which are of more relevance to the present problem. An important aspect of percolation theory is that different lattices (square, cubic, diamond, honeycomb, etc.) have different percolation properties from which different topological behaviors can be expected.

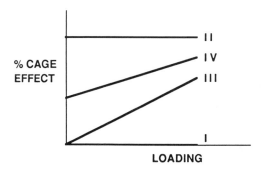

Figure 11. Changes in the percent cage effect as a function of loading for ideal situations (see text).

4.1.4. The Effect of Loading on NaX and NaY Zeolites

The variation of the cage effect with respect to loading studied in the case of the NaX and NaY zeolites using DBK-d5 is shown in Figure 12.[42] Loading levels were changed from 4 to 40% occupancy levels, which correspond to close neighborhood relationships involving about three empty intermediate supercages, at the lowest loading, and up to two neighboring molecules in every four closest supercages.

The photochemical results plotted in Figure 12a compare the variations in the percent cage effect versus DBK loading levels for the zeolites NaX and NaY. In Figures 12b and 12c, the yields of diphenylethane and rearranged products are also plotted against the loading levels of DBK for NaX and NaY, respectively. In the latter figures, a distinction has been made between the DPE formed by recombination of geminate partners (DPE$_{local}$) and that formed after randomization in the global space (DPE$_{random}$). The yields of DPE local and DPE random were calculated as follows:

$$DPE_{local} = DPE_{total} \, [\%cage\ effect]$$

and

$$DPE_{random} = DPE_{total} - DPE_{local}$$

The most important observations are summarized as follows: (1) The cage effect increases with loading, rapidly and to a large value (70%) in NaX, while it increases only marginally and up to about 20% in NaY (Figure 12a); (2) the cage effect starts from a very low value of ~4% in NaX, while in NaY the starting cage effect is ~12% (Figure 12a); (3) an increase in the cage effect in NaX (Figure 12b) and NaY (Figure 12c) correlates with a decrease in the yields of DPE formed in random encounters (DPE_{random}); (4) the yield of rearrangement products correlates directly with the cage effect values in both zeolites (Figures 12b and 12c); and (5) the effect of loading in the yield of geminate products, either primary or secondary, is greater in NaX than in NaY.

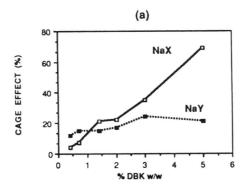

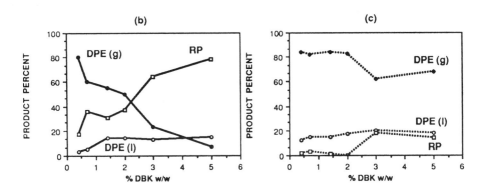

Figure 12. Product analysis as a function of loading in faujasites (a) percent cage effect vs. loading for NaX and NaY; product distribution [DPE_{random} = circles, DPE_{local} = triangles, and Rearranged Products = squares] vs. loading in NaX (b) and in NaY (c).

The cage effect versus loading behavior observed in NaX and NaY resembles that of models III and IV discussed above (Figure 11). The diffusion of the probe (benzyl radical) and the diffusion of the starting material DBK-d5 are expected to be different in the two zeolites. Restrictions in the diffusional freedom of DBK-d5 are expected to be larger in NaX because of the smaller size of the available space (higher cation density). The DBK-d5–NaX system approaches the percolation limit[44,45] (case III, Figure 11) and the recombination probability of both the primary and the secondary radicals increases with loading at the expense of escape and randomization pathways. In the case of zeolite NaY, a very small variation in the product ratio upon changes in the loading values indicates a relatively fast randomization of the lattice structure. This situation is clearly analogous to that in the model described in line IV in Figure 11. The diffusional freedom of DBK-d5 is large and occupied neighboring cages become open at a fast rate so that benzyl radicals find an opportunity to escape even at relatively large loading values. The cage effect observed in the system results almost entirely (very small loading effect) from recombination encouraged at the local supercage environment. Radical recombination in the DBK-d5–NaY system occurs from reactions in the locally closed (cage effect) and globally closed spaces (fast randomization). At the latter level, radicals behave as correlated random walkers in a dynamically disordered system.[45]

4.1.5. The Effect of Additives

The addition of additives other than the reactant can be understood on the basis of random walks in a percolating lattice developed in the previous section. An important aspect, however, is that by modifying the size and diffusional properties of the adsorbates one may engineer a desired set of properties for the corresponding topological spaces. This postulate has been illustrated by studying the effects of benzene in the zeolite NaX. The loading level employed in this case,[43] is 35% w/w and corresponds to about seven molecules per supercage, insuring multiple spectator occupancy at the supercage level. It can be expected that under these conditions the space available to DBK should approach that which is observed in tight systems such as cyclodextrins. However, the high mobility of the adsorbates, which determines the compositional correlation time for the environment, should also be considered.

The experimental results obtained with samples having a constant loading of 2% w/w (~15% occupancy) DBK-d5 are summarized in Table 5. Results obtained under reduced pressure (10^{-4} torr) are also included in the table for comparison. As expected, the cage effect increases in both zeolites NaX and NaY when the spectator is added into the system. In agreement with the loading experiments, the effect of the spectator on the cage effect and on the relative product yields (DPE_{random}, DPE_{local}, and RP) is greater in NaX than in NaY. An increase in cage effect in both zeolite systems correlates with a decrease in the relative yields of DPE_{random} and with an increase in the relative yields of rearranged ketone products (RP). These observations are consistent with a change in the reaction environment from globally closed to locally closed spaces as induced by the unreactive spectators.

Table 5. Effect of Unreactive Additives on the Product Yields^a from Photolysis of DBK-d$_5$

System	Percentage Yield			Cage Effect^e
	DPE$_1^b$	DPE$_r^c$	RP^d	
NaX	14	50	37	22
NaX–Benzene^f	7	1	92	82
NaY	17	83	0	17
NaY–Benzene^f	31	53	16	37

^{a} Estimated error ±5%.
^{b} Geminate (local) DPE.
^{c} Random DPE.
^{d} Rearranged products.
^{e} Cage effect (%).
^{f} Average of six molecules per supercage.

5. Concluding Remarks

It is intuitively recognized that most of the concepts developed and successfully applied to molecular systems should apply to supramolecular entities. Topology provides a mathematical framework based on which the complicated structures found in supramolecular systems may be described and classified. The importance of taking into account the time scale for reaction and the time scale of the binding interaction in such classification has been stressed. *Supramolecular structures and restricted environments owe their complexity, at least in part, to the fact that the dynamics of the reactant and the dynamics of the medium often evolve simultaneously. Excited states and reactive intermediates may experience a continuously changing influence as the geometric and topological properties of the medium evolve through their various structural correlations.* Heterogeneity in supramolecular systems, which is often associated with structural diversity, can also be due to the kinetic dispersion resulting from concomitant and interrelated changes in the dynamics of the medium and the dynamics of the probe. This complex situation can be dealt with the use of the Curtin and Hammet principle. Topology provides representations that place the complexity of supramolecular systems and restricted media within the same intellectual framework that now we use confidently for the covalently bound molecular systems.

Acknowledgments

The authors thank the NSF, the AFSOR and the DOE for their generous support of this research. MGG gratefully acknowledges Dr. V. Ramamurthy for many suggestions and insightful comments.

References

1. P. Alexandroff, *Elementary Concepts of Topology*, Dover, New York, 1961.
2. B. H. Arnold, *Intuitive Concepts in Elementary Topology*, Prentice-Hall, Englewood Cliffs, 1961.
3. N. J. Turro, *Angew. Chem., Int. Ed. Engl.*, **1986**, *25*, 882.
4. Topology in Chemistry: (a) V. I. Sokolov, *Russ. Chem. Rev.*, **1972**, *42*, 452.
    (b) D. M. Walba in *Chemical Applications of Topology and Graph Theory*, Studies in Physical and Theoretical Chemistry, Vol. 28, Elsevier, Amsterdam, 1983.
    (c) D. M. Walba, *Tetrahedron*, **1985**, *41*, 3161.
5. D. Hilbert and S. Cohn-Vossen, *Geometry and the Imagination*, Chelsea, England, 1952.
6. D. H. Rouvray, *Chem. Brit.*, **1974**, *11*, 10.
7. (a) J. M. Lehn, *Science*, **1985**, *227*, 849.
    (b) J. M. Lehn, *Pure Appl. Chem.*, **1978**, *50*, 871.
    (c) J. M. Lehn, *Angew. Chem., Int. Ed. Engl.*, **1989**, *27*, 89.
8. (a) J. R. Scheffer, M. Garcia-Garibay, and N. Omkaram, *Org. Photochem.*, **1987**, *8*, 249.
    (b) V. Ramamurthy and K. Venkatesan, *Chem. Rev.*, **1987**, *87*, 433.
9. V. Ramamurthy and D. F. Eaton, *Acc. Chem. Res.*, **1988**, *21*, 300.
10. (a) N. J. Turro, *Pure Appl. Chem.*, **1986**, *58*, 1219.
    (b) N. J. Turro and Z. Zhang, in *Photochemistry on Solid Surfaces*, M. Anpo and T. Matsura, eds., Elsevier, Amsterdam, 1989, p. 197.
11. (a) N. J. Turro, G. S. Cox, and M. A. Paczkowsky, *Top. Curr. Chem.*, **1985**, *129*, 57.
    (b) J. K. Thomas, *Chem. Rev.*, **1980**, *80*, 283.
    (c) G. von Bunau and T. Wolff, *Adv. Photochem.*, **1988**, *14*, 273.
    (d) N. Ramnath, V. Ramesh and V. Ramamurthy, *J. Photochem.*, **1985**, *31*, 75.
12. (a) K. Kalyanasundaram, *Photochemistry in Microheterogeneous Systems*, Academic Press, Orlando, Fl, 1987.
    (b) V. Ramamurthy, *Tetrahedron*, **1987**, *42*, 1598.
13. B. H. Baretz and N. J. Turro, *J. Am. Chem. Soc.*, **1983**, *105*, 1310.
14. (a) J. K. Thomas, *Acc. Chem. Res.*, **1988**, *21*, 275.
    (b) A. G. Cairns-Smith and H. Hartman, *Clay Minerals and the Origin of Life*, Cambridge University Press, Cambridge, 1986.
15. (a) N. J. Turro, *Tetrahedron*, **1987**, *43*, 1589.
    (b) P. de Mayo and L. J. Johnston, in *Preparative Chemistry Using Supported Reagents*, P. Laszlo, ed., Academic Press, San Diego, 1987, Ch. 4.
    (c) B. Fredrerick, L. J. Johnston, P. de Mayo, and S. K. Wong, *Can. J. Chem.*, **1984**, *62*, 403.
16. (a) J. Terner and M. A. El-Sayed, *Acc. Chem. Res.*, **1985**, *18*, 331.

(b) G. McLendon, *Acc. Chem. Res.*, **1988**, *21*, 160.

17. C. V. Kumar, J. K. Barton, and N. J. Turro, *J. Am. Chem. Soc.*, **1985**, *107*, 5518.
18. (a) G. L. Duveneck, C. V. Kumar, N. J. Turro, and J. K. Barton, *J. Phys. Chem.*, **1988**, *92*, 2028.
    (b) P. Chandar, P. Somasundaran and N. J. Turro, *Macromolecules*, **1988**, *21*, 950.
19. (a) H. Ti Tien, *Bilayer Lipid Membranes: Theory and practices*, Marcel Dekker, New York, 1974.
    (b) H. Ti Tien, in *Photosynthesis in Relation to Model Systems*, J. Barber, ed., Elsevier, Amsterdam, 1979.
20. (a) D. A. Hrovat, J. H. Lieu, N. J. Turro, and J. D. G. Weiss, *J. Am. Chem. Soc.*, **1984**, *106*, 7033.
    (b) D. A. Hrovat, J. H. Lieu, N. J. Turro, and R. G. Weiss, *J. Am. Chem. Soc.*, **1984**, *106*, 529.
    (c) R. G. Weiss, *Tetrahedron*, **1988**, *44*, 3413.
21. A. L. J. Beckwith and K. U. Ingold in *Rearrangements in Ground and Excited States*, Vol. 1, P. de Mayo, ed., Academic Press: New York, 1980, Ch. 4.
22. D. Y. Curtin, *Rec. Chem. Prog.*, **1954**, *15*, 111.
23. J. I. Seeman, *Chem. Rev.*, **1983**, *83*, 83.
24. N. J. Turro, *Modern Molecular Photochemistry*, Benjamin/Cummins, Menlo Park, 1978.
25. M. D. Cohen, *Angew. Chem., Int. Ed. Engl.*, **1971**, *14*, 386.
26. (a) N. J. Turro, *Proc. Natl. Acad. Sci. U. S. A.*, **1983**, *80*, 609.
    (b) G. F. Lehr and N. J. Turro, *Tetrahedron*, **1981**, *37*, 3411.
    (c) N. J. Turro and B. Kraeutler, *Acc. Chem. Res.*, **1980**, *13*, 369.
27. (a) P. S. Engel, *J. Am. Chem. Soc.*, **1970**, *92*, 6074.
    (b) W. K. Robins and R. H. Eastman, *J. Am. Chem. Soc.*, **1970**, *92*, 6076.
28. N. J. Turro and G. C. Weed, *J. Am. Chem. Soc.*, **1983**, *105*, 1861.
29. C. Doubleday, Jr., N. J. Turro, and J. F. Wang, *Acc. Chem. Res.*, **1989**, *22*, 199.
30. N. J. Turro and P. Wan, *J. Am. Chem. Soc.*, **1985**, *107*, 678.
31. (a) I. R. Gould, B. H. Baretz, and N. J. Turro, *J. Phys. Chem.*, **1987**, *91*, 925.
    (b) N. J. Turro, I. R. Gould, and B. H. Baretz, *J. Phys. Chem.*, **1983**, *87*, 531.
    (c) L. Lunazzi, K. U. Ingold, and J. C. Scaiano, *J. Phys. Chem.*, **1983**, *87*, 529.
32. B. N. Rao, N. J. Turro, and V. Ramamurthy, *J. Org. Chem.*, **1986**, *51*, 460.
33. B. N. Rao, M. S. Syamala, N. J. Turro, and V. Ramamurthy, *J. Org. Chem.*, **1987**, *52*, 5517.
34. (a) N. J. Turro, M. F. Chow, C. J. Chung, and C. H. Tung, *J. Am. Chem. Soc.*, **1983**, *105*, 1572.
    (b) N. J. Turro and K. Arora, *Macromolecules*, **1986**, *16*, 42.
    (c) W. Saenger, *Angew. Chem., Int. Ed. Engl.*, **1980**, *19*, 344.
35. (a) D. W. Breck, *Zeolite Molecular Sieves*, John Wiley, New York, 1974.
    (b) J. V. Smith, *Chem. Rev.*, **1988**, *88*, 149.
36. N. J. Turro and Z. Zhang, *Tetrahedron Lett.*, **1987**, *28*, 5517.
37. The value of ~30 supercages is obtained assuming that the diffusion coefficient of the benzyl radical is the same as that of toluene ($\sim 10^{-10}$ m^2 s^{-1})[40] and that the lifetime of the phenacetyl radical is 100 ns.[31]
38. J. Karger and H. Pfeifer, *Zeolites*, **1987**, *7*, 90.
39. D. M. Ruthven and I. H. Doetsht, *A.I.Ch.E.*, **1976**, *22*, 882.
40. D. M. Ruthven, *Principles of Adsorption and Adsorption Processes*, John Wiley, New York, 1984.

41. (a) T. Koenig and H. Fisher in *Free Radicals*, J. Kochi, ed., Wiley, New York, 1973, p. 157.
    (b) R. M. Noyes, *J. Am. Chem. Soc.*, **1955**, *77*, 2042.
    (c) R.M. Noyes, *J. Am. Chem. Soc.*, **1956**, *78*, 5386.
    (d) R. Keptain, *Adv. Free Radical Chem.*, **1975**, *5*, 381.
42. Z. Zhang, Ph.D Thesis, Columbia University, New York, 1989.
43. N. J. Turro and Z. Zhang, *Tetrahedron Lett.*, **1989**, *30*, 3761.
44. D. Stauffer, *Introduction to Percolation Theory*, Taylor and Francis, London, 1985.
45. R. Heifer and R. Orbach in *Dynamical Processes in Condensed Molecular Systems*, J. Klafter, J. Jortner, and A. Blumen, eds., World Scientific, Teaneck, NJ, 1989.
46. G. Zumofen, A. Blumen, and J. Klafter, *J. Chem. Phys.*, **1985**, *82*, 3198.

Chapter 2

Photophysical Probes for Microenvironments

K. Kalyanasundaram

**Institute of Physical Chemistry,
Swiss Federal Institute of Technology at Lausanne,
Lausanne, Switzerland.**

Contents

1. Introduction

A broad overview of various photophysical probes that have been found to be useful in studies of microheterogeneous systems will be presented in this chapter. This chapter, with that by Bohne, Redmond and Scaiano (Chapter 3), provides background information related to the use of photophysical probes in obtaining information about confined media. A slight overlap between the two chapters has been unavoidable. This chapter concerns itself with the steady-state techniques whereas Chapter 3. emphasizes the use of time-resolved techniques. This chapter also considers the photophysical properties of the probes, their solvent dependence, and how they are indispensable in the studies of structure and dynamics of large aggregated systems. Detailed discussions on the results obtained in various microheterogeneous systems are not made, as they form the subject of discussion of other chapters in this volume. The field is currently very active, as evidenced by the growing number of publications that appear and the international conference sections devoted to advances. Because of limitations of space and due to the availability of several review articles,[1-16] the discussion of photophysics here is minimal and select publications are cited that illustrate their applications. The choice is certainly arbitrary and I apologize for omissions.

2. Organized Media and Their Microenvironments
2.1. Various Types of Microheterogeneous Systems

During the last decade, in addition to studies on highly ordered molecular crystals, a growing number of organized assemblies and host systems of increasing complexity have been investigated by a number of physical methods. These microorganized systems can be divided into three broad types: 1) *organized assemblies,* composed of surfactants and lipids such as micelles, vesicles, liposomes, mono-, and multilayers, microemulsions, and liquid crystals; 2) *supramolecular host* systems, with large cavities capable of accommodating small guest molecules and ions such as zeolites, clays, cyclodextrins, and crown ethers, and 3) *adsorbed molecules on reactive and nonreactive surface,s e.g.,* silica, porous glass, and semiconductors. Some of these systems are generally called "microheterogeneous" to indicate their heterogeneous character at the microscopic level with the presence of charged interfaces separating distinct hydrophobic and hydrophilic domains. In spite of the large number of molecules they contain, these aggregates are rather small, usually of colloidal dimensions. Quantitative studies on macroscopically heterogeneous systems have been difficult due to intense light scattering. Microheterogeneous systems are composed of particles usually much smaller than the wavelength of light and hence form clear/translucent solutions having no such scattering problems for photochemical and spectroscopic studies. In aggregated systems and host-guest assemblies, there are several static and dynamic properties that are of interest. Table 1 presents a listing of various structural parameters and dynamic properties pertinent to studies of microheterogeneous systems.

Table 1. Static and Dynamic Properties of Microheterogeneous Systems of Interest

Property	Example
Composition and Structural aspects of the host	
Aggregation	Critical concentrations for onset of aggregation (e.g., cmc), aggregation number
Surface	Polarity, surface charge, potential, pK_a, extent of water penetration, counterion binding
Bulk/interior	Fluidity/microviscosity, rotational diffusion and order, accessibility to solutes, distance between various groups
Dynamical processes associated with the host	
Aggregation, breakup	Rates of micelle dissolution, vesicle fusion, lipid–surfactant exchange
Dynamical processes associated with the guest molecules	
Solubilization	partition of solutes between phases, solute entry and exit rates, exchange with those in the bulk
Access, mobility	lateral/translational diffusion, bimolecular quenching

2.2. Types of Microenvironments and Their Features

Hydrophobic Pockets, cages or domains: For most of the surfactant–lipid-based aggregates and organized assemblies, one is interested in the detailed architecture of the inner hydrophobic part of the host. Similar hydrophobic pockets with restricted or no access to water molecules are also found in the interiors of such hosts as zeolites and cyclodextrins. Two of the interesting properties of these pockets are: (1) *fluidity/rigidity of the inner hydrophobic region,* and (2) *pocket/cavity size.* The term "microviscosity" or "microfluidity" is often used to distinguish the viscosity of the inner part from that of the bulk solvent in which the host aggregates are present. There are several luminescence probe methods of the steady-state and time-resolved types that measure the ease of rotational or translational diffusion of solubilized probes (e.g., fluorescence polarization, excimer, formation and bimolecular luminescence quenching) to deduce the microviscosity. For lipid bilayers, the so-called "*order parameter* " (reciprocal of viscosity) is of much interest. We will return later to some of the problems encountered in the evaluation of microviscosity and order parameters. The size of the hydrophobic cavity will largely determine the dimensions of the probe molecules they can incorporate. In host systems such as cyclodextrins and zeolites, phosphorescence probes of different sizes have been used to demonstrate selective solubilization of one solute over another (discriminate binding).

Inner Water Pools/Pockets of Reversed Micelles, Microemulsions: Reversed micelles can solubilize large amounts of water, forming aqueous cores in which various reactions can be carried out [e.g., sodium bis(2-ethylhexyl) sulfosuccinate – aerosol OT in heptane]. Studies have shown that at low concentrations of water ([water] / [surfactant] $\leq$ 12), water molecules of the aqueous core have properties such as polarity, viscosity, and structure different from those of bulk water.[17,18] This is because the water molecules are tightly bound around the sodium and sulfonate ions to form the hydration shell and differ significantly from bulk water. Microemulsions are related microheterogeneous phases resulting from an admixture of surfactants, water, hydrocarbons, and alcohols. Depending on the proportion of water to hydrocarbon in the bulk mixture, these are classified either as oil-in-water or water-in-oil microemulsions. Activity and acidity of the water pools are the properties of interest.

Hydrophilic Surfaces (Lipid-Water Interface): There is also considerable interest in studies on the hydrophilic domains of organized assemblies and host systems. From a fundamental as well as a practical point of view, the interface has received a great deal of attention. Properties of interface that are of interest are the polarity, electrical potential, pH, etc. There are several methods of determining the "polarity" of the interface region. The interpretation of the term "polarity" in these types of systems can pose problems.[19] Probes whose luminescence is dependent on the polarity of the solvent and/or pH are often used to obtain estimates of "effective dielectric constant" for the interface region. Caution needs to be exercised in the use of different probes. Ionic or highly polar probes can preferentially seek polar regions of the host aggregate. Data on the polarity of the inner regions provide information on the extent of water penetration in the host and in the folding up, if any, of charged head groups.

3. Luminescence Probe Analysis
3.1. General Aspects

The basic idea behind the use of luminescence probes is that certain types of molecules display a selective affinity for a unique site on a "supramolecular assembly" and the structural and dynamic properties of the host system are reflected in the luminescence properties of the probe (guest). The various *luminescence parameters* that are used in this context are the location (λ_{max}), intensity (Φ), lifetime (τ), and polarization (P) of the emission. The success and the ambiguities associated with the luminescence probe analysis very much depend on two factors: understanding of the excited-state interactions (mechanisms) that affect the above-listed excited-state parameters and knowledge of the probe location and its possible perturbation, if any, in the macromolecule or aggregate.

Two principal attractions of the luminescence probe methods are the *extremely high sensitivity* of the technique and a *large dynamic range of time scales* that can be explored. With the availability of modern luminescence spectrometers capable of detecting weak luminescence from compounds present at ppb level (parts per billion or nanomolar concentrations), most often there is no need to use probes at concentrations more than a few micromoles. The concentrations required are certainly orders of magnitude less than those required in other physical methods such as those based on magnetic resonance (spin labels for example). Low levels of probe loading are essential if one is concerned about perturbation of the host by the guest molecules.

The majority of photophysical probes that have been examined are organic molecules and, for these, there are two types of luminescence with different origins and spectral and decay characteristics: fluorescence that accompanies the decay of the singlet excited state and phosphorescence associated with the triplet excited state. Fluorescence is comparatively short-lived with lifetimes typically in the range of $10^{-6} - 10^{-12}$ s. Phosphorescence occurs generally at longer time scales, with typical $\tau \approx 10^{-6} - 10^{-1}$ s. Together, the two types of emission span a dynamic range of 10^{13} and allow monitoring of both the static and dynamic properties of the aggregates. On fluorescence time scales, most of the aggregates can be considered as rigid host systems carrying the solubilized probe molecules. Hence fluorescence techniques allow monitoring of static features of the host assemblies and/or related fast kinetic processes. On slower phosphorescence time scales, there can be exit and reentry of solutes as well as total collapse and reformation of the host aggregate (see Chapters 1 and 3 for details).

Most luminescence studies (including polarization) involve measurement of emission maxima and quantum yield of emission under steady state illumination. Time-resolved measurements of luminescence if they can be carried out are highly desirable (see Chapters 3 and 13). As will be indicated later, photophysical processes associated with the luminescence probes are complex and only time-resolved measurements can indicate whether some of the inherent assumptions made in the interpretation of probe emission spectral shifts and intensity variations are valid.

3.2. Introduction and Location of Probe

Topological features of various host systems suggest the presence of very different microenvironments separated by short distances. Therefore, emission properties can change even with small variations in the solubilization sites. Hence, before one can interpret luminescence properties of incorporated probes, one needs to have precise information on the location of the probe in the host aggregate (in the bulk interiors or exterior or at or near the surface) and on possible perturbations of the host if any caused by the probe. The probe can be introduced either as a freely moving (noncovalently bound) probe or as a covalently bound integral component of the host by appropriate synthetic procedure. Both have their own merits and demerits. Figure 1 illustrates these different methods of introducing the probe.

1. As freely moving/non-bound molecules incorporated
   into the host assembly e.g., pyrene in lipid bilayers

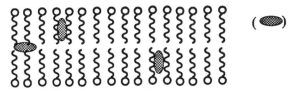

2. As covalently bound/derivatized component of the host assembly.
   a) e.g., dansyl chloride reacted with surface or pendant amino groups

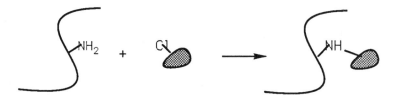

   b) chromophoric units introduced as a derivatized version of the host
      e.g., long chain derivatives of pyrene co-solubilized in lipid vesicles

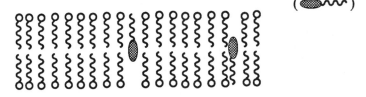

   c) chromophoric unit built into the host as an integral component
      by elaborate synthetic procedure. e.g., polymers with pendant
      pyrenyl groups, crown ethers compounds containing aromatic groups

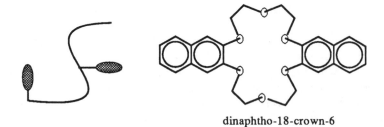

dinaphtho-18-crown-6

Figure 1. Different ways of introducing probe molecules into host assemblies.

Freely Moving Non-covalently Bound Probe: If one uses freely moving, noncovalently linked probes, information on their location needs to be based on some *a priori* knowledge on the structural details of the host, possible modes of solute solubilization in the host, and additional information obtained by other physical methods. As regards possible solubilization sites, neutral/apolar molecules may be located, on the average, in the hydrophobic domains. Polar or ionic probe molecules would seek hydrophilic domains and/or adsorb to the surface with their nonpolar portion in the hydrophobic core. Physical methods such as high-resolution nmr, esr, and X-ray have been employed to obtain information on the location of luminophores in host systems. It should be borne in mind that assignments of probe location to be in the "inner core" or at the "interface" are only approximate guides and precise discussions are meaningful only for solutes that are quite small compared to the overall dimension of the host aggregate.

Luminescence properties of the probe itself can be an indicator of the local environment, if information on possible solubilization sites is available. For example, fluorescence properties are sensitive to isotopic constitution of the solvent. Hence, solvent isotope effects (replacement of D_2O for H_2O) can be used to locate solubilization of probes near the surface where the probe can have access to solvent molecules or the extent of water penetration in the host system. We will return to a more elaborate discussion of solvent effects on luminescence and their applications. An important point to note is the complementary nature of the two approaches. Knowledge on the probe properties and location allows information on the microenvironments found in the host. Studies of probe behavior in well-characterized hosts allow information on the excited state behavior under selective microenvironments and/or constraints. Herein lies the advantages and demerits of luminescence probes.

Covalently bound probe: Here one uses a chromophoric molecule that carries a reactive unit which reacts with known functional groups of the host system to form a stable covalent linkage. Thus the mode of derivatization already gives some indication of the probe location in the host. This procedure is quite common in studies of macromolecules such as proteins and membranology (whole cells), although they have not been fully exploited in studies of simple microheterogeneous systems.

Various types of reagents carrying chromophoric groups are commercially available for derivatization with specific functional groups of the host systems. These include: (a) *sulfhydryl reagents*, chromophores that carry acetamide, succinimide, maleimide, or aziridine units to react with thiol groups of amino acids such as cysteine to form stable thioethers; (b) *activated carboxylic or sulfonic acids* such as dansyl chloride and fluorescein isothiocyanate that react with amines at basic pH (>9); (c) *alkyl halides* that react with amines or phenol groups of the hosts; (d) *hydrazine derivatives* that react with aldehyde or ketone functions of host to form conjugates of Schiff's base structure.

Another method of introducing the probe covalently is to functionalize the host systems. Here the probe is introduced by an appropriate synthetic route to be an integral component of the host. As in the other methods, the synthetic design allows some prior knowledge of the chromophore location in the host. There have been a growing number of studies on micellar, vesicular, and membrane assemblies that have profitably employed specifically synthesized surfactant, lipid molecules that carry a chromophoric unit. Examples of this kind are the surfactant derivatives of indole, coumarin, and pyrene lecithin. Similarly, by elegant synthetic routes, derivatized versions of hosts such as crown ethers and cyclodextrins with pendant chromophores have been made. Such functionalized systems with chromophoric groups which are localized either in the hydrophobic portion or at the surface have provided novel information on the dynamic aspects of host interactions with the environment and also lead to unusual chemical processes.

A complementary approach in the organization of reactants at the microscopic level is to take advantage of the inherent ability of the hosts to organize the location and distribution of solutes. For example, host systems can be prepared by appropriate synthetic procedures to carry quencher, heavy atom or chelating ions at locally high concentrations to promote one particular photophysical process over the other. Typical examples are anionic micelles with halide counterions (with enhanced concentration of quenchers); cationic micelles or zeolites exchanged with rare earth ions such as Eu^{3+}, Tb^{3+}; or even transition metal ions such as Ru^{2+} capable of further chelating ligands. Quantitative analysis as a function of loading or ion exchange can provide insights on the host architecture.

3.3. Probe Distribution and Movement

Numerous studies have established that, irrespective of the finer details of the host architecture, the observed photophysics and photochemical behavior of the probes are quite different from those observed in homogeneous solvent systems. In studies of microheterogeneous systems, there are a number of factors that need to be considered in the interpretation of the experimental results. Some of these are listed below.

Probe Distribution: Quantitative interpretation of efficiency and kinetics of photoprocesses requires knowledge of the local concentration of the reactants. While the concentration term is well defined in homogeneous solvents, it is less so in heterogeneous systems. Due to the microheterogeneity of the host system, the probe distribution often is not uniform. Multiple site distribution would result clearly in kinetics and effects far different from those in simple systems (see Chapter 13 for details). In systems with a finite number of hosts (molecules or aggregates), description of the solute distribution using stochastic models is a useful approach.

Probe Mobility: Unlike in simple homogeneous systems, the reactant molecules can find themselves in pockets or cages where their movement is restricted to a limited volume or even to restricted dimensions. Most of the phenomenological theories of reaction kinetics and diffusion in homogeneous media are based on uniform distribution of solutes and isotropic diffusion (rotational and translational) over an infinite volume. Anisotropic (hindered) rotational diffusion in a cone or cylinder probably is the only type of movements the probe can execute in some media. Neutral molecules located in the inner lipid bilayers or ionic reactants adsorbed onto the surfaces of ionic micelles may be forced to diffuse in restricted dimensions. Lateral diffusion in the former case is a unique situation to encounter.

Conformational Restraints: In microenvironments where there are restrictions on the probe mobility and diffusion, the molecules may be forced to assume a different orientation or conformation atypical of homogeneous media, and this may influence strongly the efficiency of those photoprocesses that require reactants to assume a preferred geometric orientation or conformation, e.g., excimer formation and photocycloadditions. The local structures and the mode of solubilization can promote or inhibit these processes.

Presence of Local Electric Fields: A common feature in most of the organized assemblies composed of simple surfactant or lipid molecules is the presence of charged interfaces. Properties of this interface such as electrical potential, polarity, or effective dielectric constant and their influence on chemical and photochemical processes have been of interest. Photoreactions of ionic solutes solubilized at/near the interfaces are subject to strong local electric fields.

Analysis of the nature of the solute distribution, their mobility, the nature of the diffusion, and their kinetic aspects (role of dimensionality, fractal aspects of heterogeneous kinetics, etc.) in microheterogeneous systems are some of the areas being investigated intensely in several laboratories. Recent monographs[20-22] give an excellent introduction to some of the problems faced and ways of analyzing them (also see Chapters 12 and 13).

4. Fluorescence Probes of Various Types, Their Principles and Applications

Table 2 lists some of the commonly used luminescence probes along with indications of related photoprocesses and areas wherein they may be applied. Figures 2–4 present the chemical structures of some of these probes. Table 3 presents some spectral data on the absorption and luminescence of some of the probes. Mention may be made here of two chemical firms that specialize in the synthesis and marketing of luminescence probes for various applications: Molecular Probes Inc.,[23] and Lambda Probes, Inc.[24]

Table 2. Various types of photophysical probes for studies of microenvironments (only the chromophoric groups are indicated and their derivatives)

Process	Typical probes	Applications
Fluorescence probes		
Ham effect	Pyrene, anthracene	Polarity, critical concentration for aggregation
Solvent relaxation around excited-state	ANS, NPN, dansyl derivatives, Pyranine, alkoxycoumarins, naphthols, carboxyfluorescein	Polarity
Acid–base reactions		Surface pH/potential
Fluorescence polarization	DPH, NPN, parinaric acid, 2-MeAn, perylene, anthroyl fatty acids	Rotational diffusion, microviscosity
Inter/intramolecular excimer	pyrene	Lateral diffusion, phase changes, microviscosity
Fluorescence quenching	Any fluorophore with $t_\eta \geq 10$ ns	Accessibility of solutes, counterion binding, exchange.
H-bonding effects	Xanthenes, methyl salicylate	Surface acidity, acid-base reactions
Donor-Acceptor molecules with TICT	DMABN, cyanobiphenyl	Polarity
Inversion of $(n,\pi^*),(\pi,\pi^*)$ states	Pyrene-3-CHO	Polarity
Solvent isotope effects	Ru(bpy)$_3^{2+}$	Water accesibility
Phosphorescence probes		
Internal heavy atom effect	Br-pyrene, Br-naphthalene	Dynamics of host-guest interactions (counterion binding,exchange)
External heavy atom effect	Most arenes and heterocyclics	

Table 3. Absorption, emission spectral properties on some of the common photophysical probes in solution at room temperature. Data taken from Molecular Probes, R.R. Haughland, Molecular Probes, Inc., Oregon, USA, 1989.

Probe	Mol.Wt	Solvent	Abs.Max(nm) & $e(mM^{-1}cm^{-1})$	Em.Max(nm)	F_{em}	$\tau(ns)$
1. Pyrene						
Pyrene (py)		Cyclohx	340	378,398	0.58	450
Py-(CH2)3-COOH (PBA)	288	MeOH	339(40)	377		
Py-(CH2)11-COOH	400	MeOH	341(40)	377		
1,3 bis-(1-Py)propane		MeOH	344(80)	378		
Pyrenyl-1-(CH2)4NMe3Br	396	MeOH	341(45)	377,397		
Pyrenyl-1-SO3- Na+ (PSA)	304	MeOH	342(34)	376	>0.3	62
Pyrenyl-1-CHO	230	MeOH	361(20)	452		
Pyranine		acid	403(9.4)	511	>0.3	
"		base	454(21)	513		
Pyrenyl-(SO3)4-(4Na+)	610	H2O	375(54)	404		
2. Aminonaphthalene						
1,8-ANS	299	MeOH	370(6.8)	482		<5
2,6-ANS	299	MeOH	319(25)	422		<5
2,6-TNS	335	MeOH	351(27)	437		
2,6-MANS	335	MeOH	315(21)	439		<5
NPN	219	MeOH	336(7.6)	424	0.22	4.8
Dansyl-X		hexane	335(4.5)	463	>0.1	12.2
3. Polyene, polyene acid						
DPH	232	MeOH	348(80)	429		<10
cis-Parinaric acid	276	MeOH	318(74)	410	0.015	1.3
trans-Parinaric acid	276	MeOH	313(84)	410	0.009	≤1

Table 3. Continued

Probe	Mol.Wt	Solvent	Abs. Max (nm) & ε (mM⁻¹cm⁻¹)	Em. Max (nm)	F_{em}	τ(ns)
4. Coumarin						
7-(OH)-4-(Me)-coumarin	234	acid	323(16)	386		
		base	358(19)	447	>0.3	<5
7-(OH)-4-(C17)-coumarin	234	acid/MeOH	325(17)	385		
		base/MeOH	367(22)	450	>0.3	<5
5. Xanthene						
Dodecyl acridine orange	514	MeOH	493(57)	520		
Rhodamine 6G-C18ester	703	MeOH	528(145)	552		
5-(N-hexadecanoyl)eosin	901	MeOH	520(100)	550		
6. Carbocyanine						
DiOC16(3)	928	MeOH	551(127)	575		
DiIC18(3)	936	MeOH	547	571		
DiSC18(3)	928	MeOH	546	575		
7. Arenes, aroyl acids						
2-Me Anthracene		MeOH	365,385	395,420	≈0.1	4.41
2-(9-anthroyl)stearic acid	504	MeOH	360(7.5)	463		≈11
9-(9-anthroyl)stearic acid	504	MeOH	359(7.6)	475		≈11
8. Fluorescein						
5-COO-fluorescein	376	H₂O(pH8)	489(74)	518	>0.3	<5
6-COO-fluorescein	376	H₂O(pH8)	489(73)	516	>0.3	<5
5-(N-hexadecanoyl)-fluorescein	585	MeOH	495(79)	521		
9. Others						
Nile red	290	MeOH	551(40)	636		
Prodan	227	MeOH	360(18)	497		
4-alkylaminostilbene		MeOH	440	540		
DABMN	197	MeOH	429(53)	492		

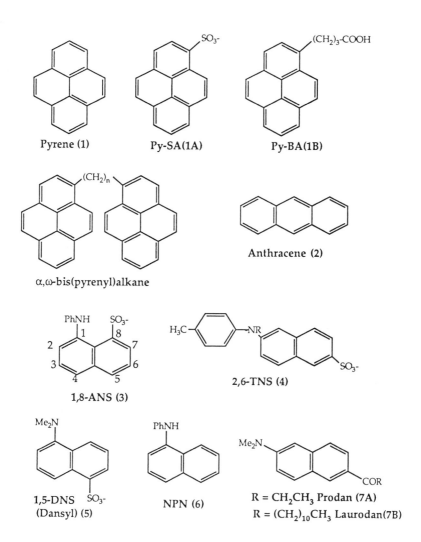

Figure 2. Chemical structures of luminescence probes: I.

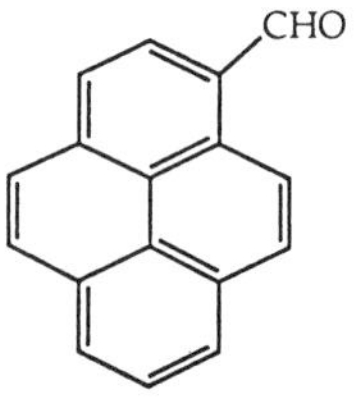

Py-CHO (8)

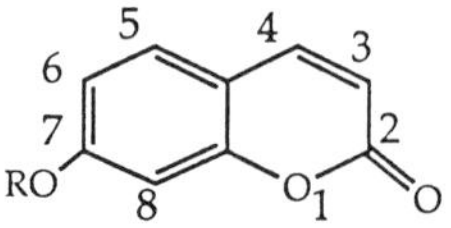

7-alkyloxycoumarin (9)

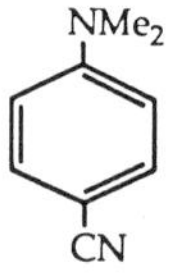

DMABN (11)

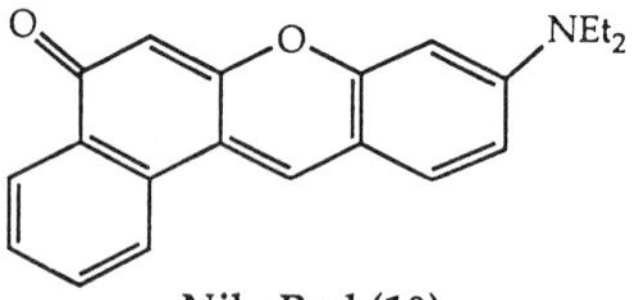

Nile Red (10)

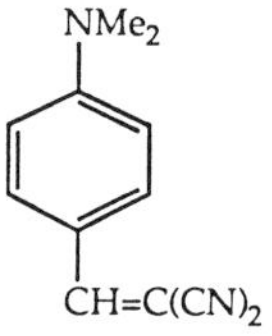

DMABMN (11A)

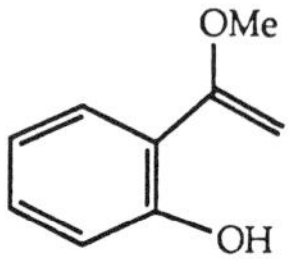

Methyl salicylate (13)

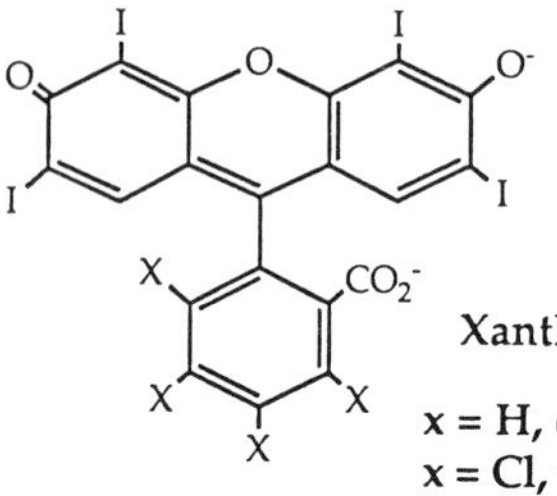

Xanthene (12)

x = H, erythrosin
x = Cl, rose bengal

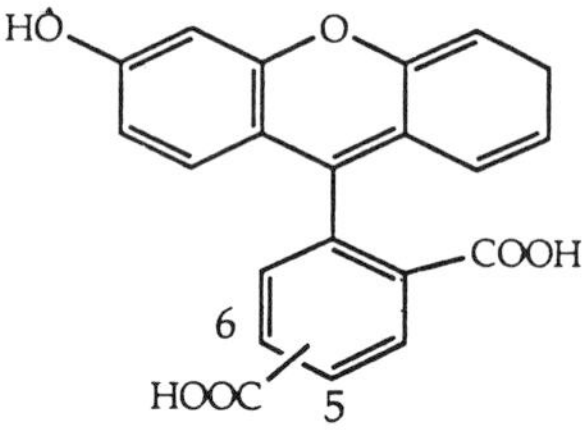

5(6)-carboxyfluorescein (14)

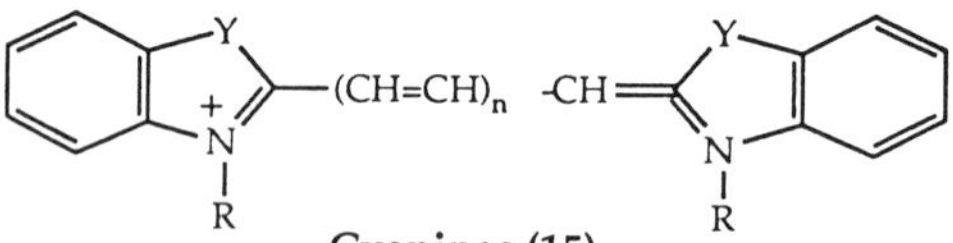

Cyanines (15)

Y = O   oxacarbocyanine;   n =1 cyanine
Y = S   thiocarbocyanine;   n= 3 carbocyanine
Y = iPr indocarbocyanine;   n = 5 dicarbocyanine

Figure 3. Chemical structures of luminescence probes: II.

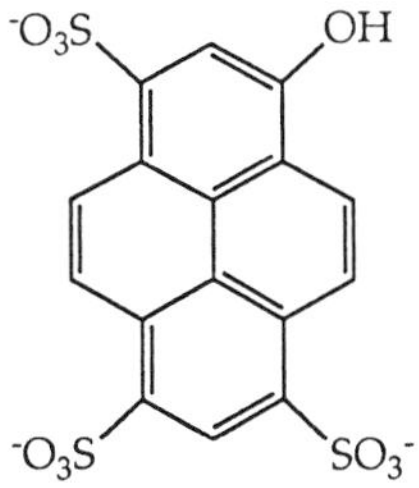

Pyranine (16)

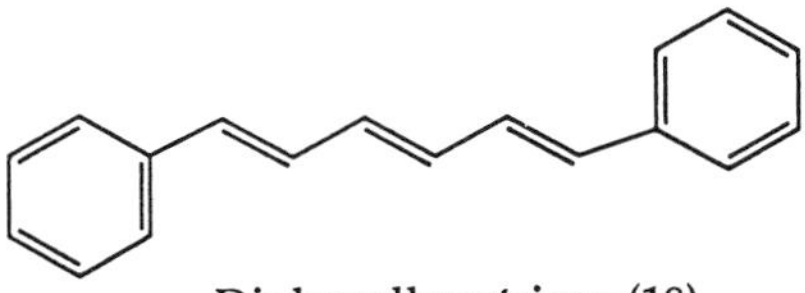

Diphenylhexatriene (18)

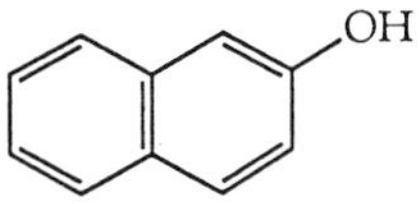

2-naphthol (17)

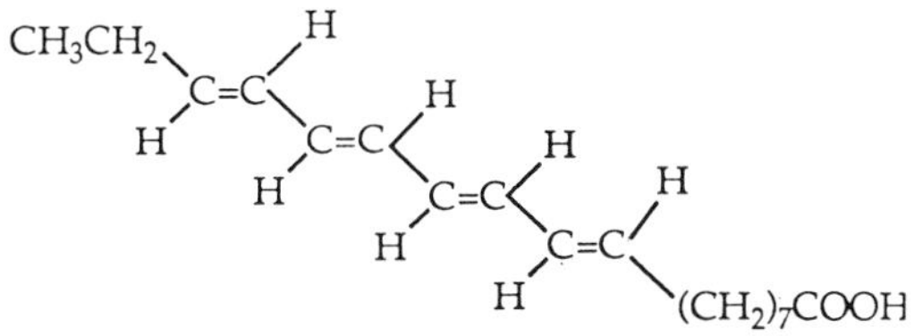

trans-parinaric acid (19)

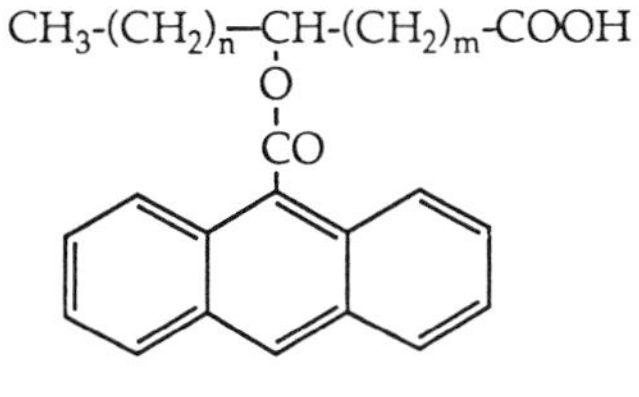

Anthroyloxy fatty acid (20)

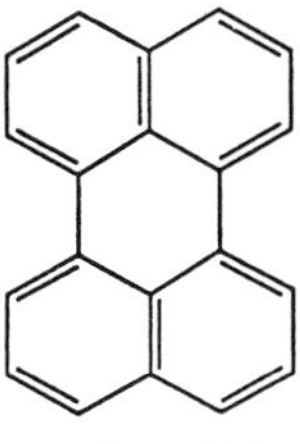

perylene (21)

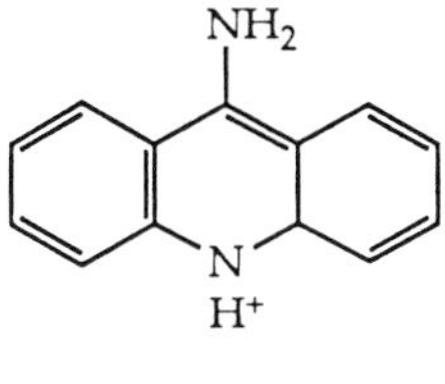

9-amino acridine

Figure 4. Chemical structures of luminescence probes: III.

4.1. Fluorescence Probes Based on Environmental Effects

Luminescence characteristics of an organic molecule may be altered by the environment for a variety of reasons. Environmental effects on the luminescence of probes can be divided under the following categories:

   a. Effects on vibronic band intensities of fluorescence (Ham effect)
   b. Effects of medium relaxation around the excited state
   c. Effects on (π,π^*) and (n,π^*) states and their interconversion
   d. Effects on "twisted internal charge-transfer" states of donor–acceptor molecules
   e. Effects via inter- and intramolecular hydrogen bonding
   f. Effects through environmental rigidity

4.1.1. Probes Based on the Ham Effect on the Vibronic Fine Structure of Fluorescence

Condensed aromatic molecules such as pyrene (1) and anthracene (2) show fairly well-resolved fluorescence spectra in solution at ambient temperature. The vibronic bands are clearly distinguishable to allow quantitative measurements of their exact location and intensity variations with solvent. Variation in the nature of the solvent leads to large variations in the intensities of the vibronic bands without much spectral shift and the phenomenon is known as the *Ham Effect*. As illustrated in Figure 5 for pyrene, due to the Ham effect, the forbidden vibronic bands in weak electronic transitions show marked enhancements under the influence of solvent polarity.[25-29]

Although the Ham effect has been noted in a large number of arenes, the most popular probe for applications in microheterogeneous media has been pyrene. If we number the principal vibronic bands observed in the room temperature fluorescence as I to V, then band III (0–737 cm^{-1} transition at ~382.9 nm) is strong and shows minimal variations in intensity. Band I (0 – 0 transition) located at ~ 372.4 nm shows significant intensity enhancement in polar solvents. Thus the peak ratio (I/III) can serve as a sensitive guide of the environmental micropolarity. Various quantitative correlations of the intensity variation with single or multiple solvent polarity scales have only been of limited success. Detailed interpretation of the magnitude of the (I/III) ratio can be complex due to the intricate solute–solvent interactions but the appeal is in the use of relative intensity as a probe of local environmental changes. Fluorescence lifetime also shows pronounced solvent effects but, in the absence of expensive lifetime measurement equipment, the Ham effect involving simple measurements of the relative band intensity serves as a poor man's tool for studies of organized assemblies and host–guest interactions. There have been numerous studies of this kind and some of these are covered in Chapter 3.

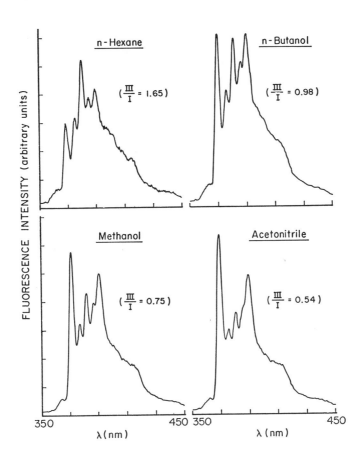

Figure 5. The Ham effect on pyrene.

4.1.2. Solvent Relaxation Around the Excited State ("Polarity" Probes)

If the dipole moment of the excited state is significantly different from the ground state's, then the solvent molecules will reorganize to provide a configuration of lowest free energy (equilibrium state). A consequence is that the emission maxima and yield depend on the "polarity" of the solvent and the relative change in the dipole moment of the chromophore associated with the excitation process. In a fluid solution, the relaxation of solvent molecules around the excited state will be completed in times less than the lifetime of the emitting state. Thus absorption will lead to a metastable Franck-Condon state, but emission will occur from the equilibrium state. This gives rise to a separation of the 0–0 bands in the absorption and emission. Figure 6 presents a schematic representation of various energy levels associated with such processes.

Solvent effects on electronic transitions can arise from specific polarization or by bulk dielectric effects. The theory of solvent dielectric effects on the transition energies (dielectric continuum theory) has been developed principally by McRae, Bayliss, Marcus, and Ooshika.[30-34] In the absence of polarization effects, the dipole moment of the excited state can be determined from the dependence of the of the electronic transitions energy (absorption and emission) on the solvent.

Several theoretical relationships can describe the solvent dependence of the separation of the 0–0 bands in absorption and emission (ΔE_T). The one proposed by McRae has the form:

$$\Delta E_T = (2/hc) \, [(\mu_{ex} - \mu_{gs})^2/a^3] \, [\, (\varepsilon-1)/(\varepsilon+2) - (n^2-1)/n^2+1] + \text{const} \qquad (1)$$

where μ_{gs} and μ_{ex} represent the dipole moments of the solute molecule in the ground and excited states, respectively; a is the effective cavity radius, ε is the static dielectric constant; and n is the solvent refractive index at zero frequency respectively. In many cases the second term, which is the result of induced dipole–dipole interaction between solute and solvent can be neglected. From the measured spectral shifts in a series of solvents of known n and ε and a plot of ΔE vs. the first term in the above equation, the $\Delta \mu$ value of the solute is determined. If the two dipole moments of the solute (ground and excited states) or $\Delta \mu$ values are known, an estimate of the "solvent polarity" can be obtained.

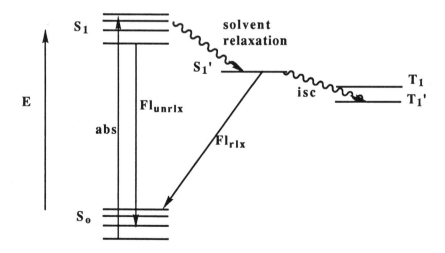

Figure 6. Scheme for solvent relaxation around the excited state. S_o refer to the ground state molecule. S_1, S_1' and T_1, T_1' refer to the unrelaxed(Franck-Condon) and relaxed form of the singlet and triplet excited state respectively.

Most often due to the difficulties involved in implementing a rigorous approach, an empirical procedure of comparison of the emission spectra of the probe in different solvents with those obtained in the organized media is used. One difficulty with this procedure is that the microenvironment of the probe in these systems may not have the necessary fluidity to relax around the excited state during the lifetime of this state. Experiments by Brand, Ware, and others[35,36] have demonstrated that the solvent shift is time dependent and can be reduced by lowering the temperature in viscous media. That is, "orientation constraint" will in general lead to a blue shift in emission as has been demonstrated for *N*-phenyl-naphthylamine.[37] In the absence of "time relaxed" experiments, therefore, the spectral shifts alone may not be a reliable indicator of the polarity of the microenvironment or the location of the probe.

Examples: 1-Anilinonaphthalene-8-sulfonate (ANS, **3**) and related molecules [2-(p-toluidinyl)-naphthalene-6-sulfonate (TNS, **4**), dansyl chloride (**5**), *N*-phenyl-naphthylamine (NPN, **6**] have been landmarks among various fluorescence probes.[38-41] Slavik[38] has presented a comprehensive review of the photophysics of ANS and related molecules and their applications in biomembrane research. Upon going from water to hexane, the fluorescence quantum yield of ANS increases from 0.004 to 0.98 and the emission maximum shifts from 515 to 454 nm.[42] Seliskar and Brand[43] have made a detailed analysis of the solvent-induced spectral shifts and deduced large differences in the dipole moments of the excited and ground states. Some estimates of $\Delta\mu$ values are as follows (in debye units): 19 (2,6-DNS), 40 (2,6-ANS), 46 (2,6-MANS), and 44 (2,6-TNS). Although in early days solvent interactions with a polar twisted internal charge-transfer state was considered as a possible mechanism, the solvent relaxation model is presently accepted as the origin of variation in the emission quantum yields and spectral shifts.[44-46] A completely hydrophobic version of ANS is *N*-phenylnaphthylamine (**6**) (NPN). NPN also shows high fluorescence enhancements in hydrophobic environments and is used widely in fluorescence polarization studies.

Weber and co-workers have introduced a series of alkanoyl derivatives of dimethylaminonaphthalene unit as potential fluorescent probes for studies of lipophilic environments.[47-50] These probes are characterized by less quantum yield variation on environment but large spectral dependence on polarity. Prodan (6-propionyl-2-dimethylaminonaphthalene, **7A**) and Laurodan (6-dodecanoyl-2-dimethylaminonaphthalene, **7B**), for example, show large environment-dependent emission spectral shifts: emission maximum blue shifts with solvents of decreasing polarity in the order: 531 nm (H_2O) > 496 nm (EtOH) > 452 nm (acetone) > 421nm (benzene) > 401 (cyclohexane). The absorption, however, shows only a small solvent dependence. 2,6-Dipyrenoylpyridine is another example of solvent-dependent fluorescent marker. Fluorescence maximum shifts from 470 nm in hexane to 570 nm in acetonitrile.[52b]

4.1.3. Probes Based on Solvent Effects on (n,π*) and (π,π*) States

Occurrence of low-lying closely spaced (n,π*) and (π,π*) levels is a common feature in the photophysics of organic heterocycles. Due to the nature of the electronic transitions, (π,π*) excited states are considerably more sensitive to solvent polarity than (n,π*) states. As a consequence, in molecules that have the (n,π*) state as the lowest energy in nonpolar solvents, there can be an inversion in the nature of the lowest excited state in polar solvents (cf. Figure 7). Such changes can result in changes in emission yield and emission maxima. The extent of red shift of the (π,π*) emission depends on the polarity of the solvents. Fluorescence probes such as pyrene-3-carboxaldehyde and 7-alkoxycoumarin show this type of behavior.

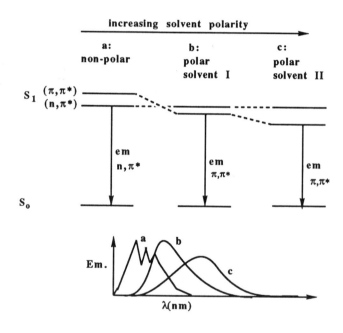

Figure 7. Scheme for solvent-induced inversion of (n,π*), (π,π*) states. Emission spectra labelled a, b and c refer to the spectrum obtained in a non-polar solvent (case a), polar solvent I (case b) and polar solvent II(case c) respectively (cf. discussion in the text).

Examples: The solvent-dependent fluorescence behavior of pyrene-3-carboxaldehyde (**8**) was first reported by Förster et al. in 1960.[51] In nonpolar solvents such as *n*-heptane, the fluorescence spectrum is structured with peaks around the 400–420 nm region and the quantum yield of emission is quite low (<0.001). This particular fluorescence is due to an n–π^* transition. However, on increasing the polarity of the solvent media, the π,π^* level that lies slightly above the nπ^* level is brought below the n,π^* level by solvent relaxation during the lifetime of the excited state. Thus in polar solvents, the π,π^* state becomes the emitting state. For solvents with dielectric constant values of $\varepsilon \geq 10$, the emission maximum shows a linear red–shift with ε of the solvent.[52] Dederens *et al.*[53] have noted that the emission maxima plotted as a function of solvent polarity correlate differently with the dielectric constant of the solvent depending on whether it is protic or aprotic. As a neutral probe that tends to be solubilized at the interfaces, **8** can be used to obtain polarity estimates. Das et al. have carried out a detailed photophysical study of the probe.[54]

7-Alkoxycoumarins (**9**) are hydrophobic probes with fluorescence at unusually long wavelengths and having large Stokes shift. They are nonfluorescent in benzene but are highly fluorescent in water. A detailed analysis of the solvent effects has shown a poor correlation between the relative fluorescence intensities in various solvents, solvent–water mixtures, and the macroscopic properties (the polarity parameters) of the medium such as dielectric constant Z and $E_T(30)$ values.[55] The fluorescence originate from the lowest excited singlet π,π^* state and competes with a radiationless process consisting mainly of intersystem crossing to the triplet state. The variations in fluorescence intensity attributed to the change in the energy levels of (π,π^*) and (n,π^*) states by the solvent.[56, 57]

Nile Red (**10**) is another probe that is becoming increasingly popular in studies of organized assemblies such as membranes and proteins.[58,59] The dye is almost nonfluorescent in water and polar solvents but undergoes a large absorption and emission shift to shorter wavelengths upon transfer to a hydrophobic environment. The emission is considerably more intense in hydrocarbon solvents. For example the emission maximum shifts from ~670 nm in water to 640 nm (MeOH), 605 nm (acetone), 560 nm (xylene) and 525 nm (heptane).

4.1.4. Donor–Acceptor Molecules with Twisted Internal Charge-Transfer State

Numerous organic molecules with electron donor and acceptor groups linked by a single bond react in the excited state with the formation of a highly polar state with mutually perpendicular conformation of the D$^+$ and A$^-$ subunits (Figure 8).[60-64] A classical example of this kind is 4,4'-dimethylaminobenzonitrile (DMABN, **11**). The anomalous fluorescence of DMABN was first noted by Lippert *et al.*[60] Since then these systems have been the subject of extensive photophysical and spectroscopic investigations and quantum mechanical and thermodynamic analyses. The existence of a twisted intramolecular charge-transfer state (TICT), of the type shown below is now well established. Rettig, in a recent review[64] lists over 55 compounds that generate this kind of TICT upon optical excitation.

A common feature of these molecules is the existence of a single broad fluorescence in nonpolar solvents centered around 350 nm and dual fluorescence in polar organic solvents. A linear correlation is observed in plots of emission energy of the longer wavelength band vs. solvent polarity parameter $E_T(30)$. Plots of emission energy vs. solvent dielectric constant (ε) also exhibits linearity at ε values above 10. The two emissions are generally believed to occur from two conformationally distinct singlet excited states—the short wavelength emission from a planar state and the long-wavelength emission from a twisted internal charge transfer state (TICT). The TICT excited state is believed to be formed from the planar excited state by twisting around the phenyl–nitrogen bond and appears to be induced by solvent–solute interactions in polar solvents.

Recently N,N-dialkylaminobenzylidene malononitrile (DMABMN, **11A**) has been used for similar reasons.[65,66] The S_1 state of this class of compounds has a CT π,π^* and undergoes predominantly radiationless decay via molecular relaxation. The molecular relaxation can be slowed by increasing the molecular rigidity of the chromophore, increasing the solvent viscosity or decreasing the ambient temperature. Since the fluorescence maximum ($\lambda_{max, fl}$) is only sensitive to the solvent polarity and the fluorescence quantum yield (ϕ_{fl}) only sensitive to the viscosity of the solvent, the fluorescence of this class of compound simultaneously offers information on both the polarity and the viscosity for the microenvironment.

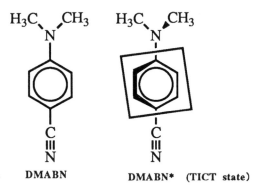

Figure 8. Structures of ground and TICT excited state of DMABN

The fluorescence properties of cyanobiphenyl derivatives are also strongly dependent on the solvent polarity in fluid solution.[67] The fluorescence is assigned to occur from the planar ^1L_a state to the nonplanar ground state with the transition moment parallel to the long axis of the molecule. The red shift of the emission in polar solvents has been assigned to an orientation relaxation process of the solvent cage in the electric field of the excited solute that has higher dipole moment in the excited state than the ground state.

4.1.5. Hydrogen-Bonding Effects

There are a number of fluorescent probe molecules such as xanthene dyes (**12**) (e.g., rose bengal and erythrosin) whose absorption and emission properties are particularly sensitive to the hydrogen-bonding character of the solvent.[68-70] The clearest manifestation of this effect is the drastic increase in the fluorescence lifetime in going from protic to polar aprotic media. The effect is useful in probing the hydrogen-bonding capabilities of various microenvironments. This effect is attributed to the solvent effect on the S–T energy gap. Intersystem crossing rates are quite sensitive to the S–T energy gap and hence decreasing energy gap with increasing solvent polarity favors intersystem crossing (Figure 9).

Methyl salicylate (MS, **13**) exhibits dual fluorescence in most organic solvents: one at 450 nm and the other at $\approx$ 335, 360 nm).[71] Following the initial suggestion by Weller,[71] it has been established that the different emitting species arise from different ground-state conformers derived via inter- and intramolecular hydrogen bonding (Figure 10).[72-75] Conformers II and III fluoresce at 330 and 360 nm, respectively ,with the excitation spectra different from that of the emission at 450 nm corresponding to I. These results indicate that the fluorescence of MS is dependent on ground-state conformation and that there is no interconversion between the conformers within the excited-state lifetime.

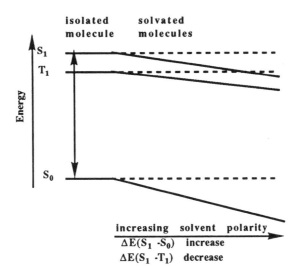

Figure 9. Scheme for excited-state interaction in xanthene dyes showing the relative variation of the energies of the singlet (S_1) and triplet (T_1) excited state with increasing polarity of the solvent (Solvent effect).

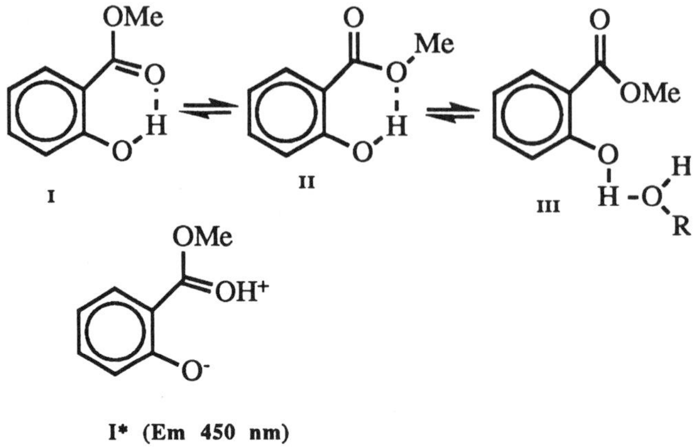

Figure 10. Hydrogen-bonded ground-state conformers of methyl salicylate.

The fluorescence spectrum of MS changes quite dramatically with the hydrogen-bonding ability of the solvent. With increasing hydrogen-bonding ability of the solvent, conformer III is favored over conformers I and II. This change in ground-state conformational equilibrium causes an increase in the emission intensity at 330 nm at the expense of the emissions at 360 and 450 nm bands. Solvent hydrogen bonding to the solute also changes the uv absorption maximum. Therefore, there are three spectroscopic parameters of MS that are indicative of ground-state conformation. Turro et al. have demonstrated a novel application of MS to infer the geometry of inclusion complexes with cyclodextrin.[76]   Data collected on the conformational preference indirectly provide information on the hydrogen-bonding ability of the solvent.

4.1.6. Probes for Environmental Rigidity

Stilbenes and higher polyenes in general represent a group of molecules whose fluorescence properties are very sensitive to the environmental rigidity. *trans*-Stilbene, for example, is only weakly fluorescent ($\phi = 0.05$) and *cis*-stilbene is nonfluorescent in fluid solutions at room temperature. But in rigid environments, however, both compounds are strongly fluorescent ($\phi \approx 0.75$). Both temperature and environmental rigidity contribute to this enhancement. *trans*-Stilbene fluorescence quantum yield increases from 0.05 observed in fluid organic solvents to 0.15 in viscous glycerol. Presumably twisting about the C=C bonds is inhibited in more viscous solvents. Whitten and co-workers have used successfully several surfactant derivatives of stilbenes as probes.[77] For a C_{16}-derivative of stilbene, the fluorescence quantum yield has been reported to increase from 0.04 measured in methylene chloride to 0.48 upon solubilization in mixed cationic–anionic micelles. Similar environmental effects on the fluorescence of diphenylhexatriene (DPH, 18) and parinaric acid (PA, 19) have also been observed. In the case of DPH, the all-trans isomer alone is fluorescent and in a highly viscous medium where the cis--trans isomerization is inhibited, the fluorescence quantum yields increase nearly to unity.

4.2. Applications of Fluorescent pH Indicators
4.2.1. Electrical Potential of the Interfaces

An important structural component of organized assemblies is the interface. Numerous studies have shown that the interfacial solvent properties can be very different from bulk aqueous solution properties.[78] Attempts to measure the electrical potential at the surface of a charged membrane or a similar interface dates back to 1940 when Hartley and Roe used pH indicators adsorbed onto charged micelles.[79] The "apparent" shift of pK_a measured in micellar solutions as compared to pure aqueous solution was attributed to a change in the "local interfacial" proton activity $a_{H^+}^i$ at the surface of charged micelles as compared to that in bulk water $a_{H^+}^w$. This pH shift was related to an interfacial potential, Ψ, by equation (2):

$$a_{H^+}^i = a_{H^+}^w \ \exp \ (- F\Psi/RT) \tag{2}$$

where F is the Faraday constant, R is the general gas constant, and T is the temperature. However, the equilibrium of an indicator bound to a surface may be affected not only by an electrostatic potential, but also by dielectric constant. Mukerjee and Banerjee pointed out in 1964[80] that the apparent shift of pK_a includes a shift of the intrinsic pK_a such that an electrical potential or "local pH shift" is related only to a difference of the "apparent" pK_a and intrinsic "interfacial" pK_a according to equation (3):

$$pK_a^{obs} - pK_a^i \ = \ - F\Psi/RT \tag{3}$$

In fact significant pK shifts of indicators are observed even in nonionic micelles.
In order to calculate an apparent electrostatic potential for the charged systems, Fromherz[81,82] proposed that the apparent pK_a^0 in a neutral monolayer (or nonionic micelle) be taken as a reference according to equation (4):

$$pK_a^{obs} - pK_a^0 \ = \ - F\Psi/RT \tag{4}$$

This procedure identifies the interfacial pK_a^i of equation (3) with the apparent pK_a^0 and implies that the environmental effect on the intrinsic pK_i is similar in the charged and uncharged interfaces. Because of the simplicity of equation (4) and the ease with which pK_a^{obs} values for interfacially located acid–base indicators can be measured, there have been numerous studies of acid–base equilibria to determine the electrostatic potential at charged interfaces. Grieser, Drummond, and co-workers[83,84] recently have questioned the above procedures of using nonionic micelles as reference systems of zero electrical potential. Drummond recently has presented[85] a very comprehensive and critical review of this subject.

Examples: One of the major applications of fluorescence probes is as indicators of solution and intracellular pH. Most of these probes are phenolic derivatives, which undergo absorption shifts to longer wavelengths with considerable fluorescence enhancement in basic solution. Coumarin, fluorescein, and their derivatives, are examples of this type of indicator. 7-Hydroxycoumarin (9) (also known as umbelliferone) is a well-known fluorescent indicator. Excitation wavelength shifts of these indicators permit pH to be measured from the ratio of the excitation intensities at a particular wavelength. Due to their better retention in host systems such as membrane cells, Carboxyfluoresceins (14) are often employed instead of fluorescein itself as fluorescent pH indicators.

Carbocyanines (15) are a series of organic probes for interfacial potentials. These dyes are strongly colored with molar absorbances (ϵ) > 10^5 $M^{-1}cm^{-1}$ and also are highly fluorescent. Fluorescence of carbocyanines generally increase on going to ethanol from water. They are also useful in the study of lateral diffusion and lipid exchange.

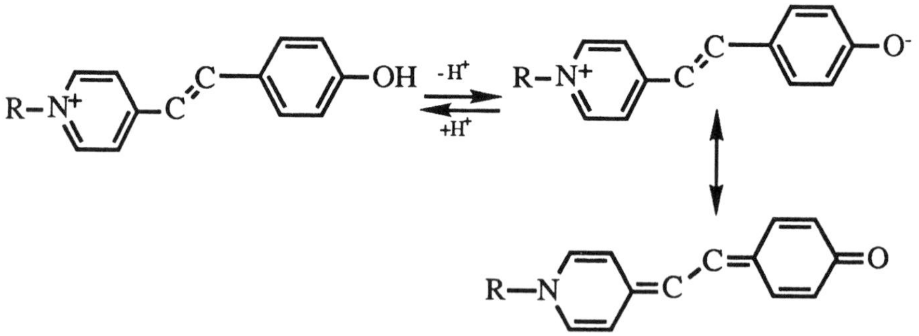

Figure 11. Structures involved in the solvatochromism of HOED

Solvatochromic dyes such as 1-alkyl-[4-(oxocyclohexadienylidene)-ethylidene] 1,4-dihydro-pyridine (HOED), and 5- and 6-hydroxyquinoline(R-HQ) have been used to study interfacial properties of several surfactant aggregate systems.[84] The

solvatochromism of molecules such as HOED is due to ⸱
state of these molecules are much less dipolar than the g
there is a solvent-associated variation in the relative co
extreme resonance forms to the ground-state structure (Figur
polarity of the solvating medium therefore shifts the balan
resonance hybrid structure in the direction of the nonpolar q
resulting in a decrease in the energy difference between the ground and excited
states.

4.2.2. Surface pH and Acid–Base Equilibria of the Excited State at Interfaces

Examination of the acid–base behavior of dye molecules[86] constitute an interesting way to examine the pH of the interface region and it's influence on the kinetics of acid–base equilibria. Aryl alcohols such as naphthols and pyranine have been used to investigate the proton-accepting character of the microenvironment at probe binding sites. It is well known that the pK_a of aromatic alcohols is lower in the excited state than in the ground state. For most of these probes at neutral pH absorption will be due to the protonated species. However, in the excited state equilibrium will favor formation of the ionized species. Depending on the rate of proton transfer, emission can be observed from the protonated species, the ionized species, or both (Figure 12). If the rate of proton transfer is of the same order of magnitude as the fluorescence decay, then the proton transfer can be monitored directly by means of time-resolved emission measurements.

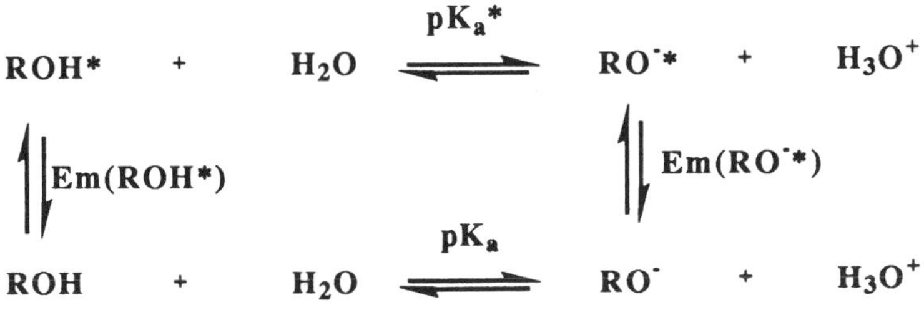

$$\text{ROH*} \quad + \quad \text{H}_2\text{O} \quad \underset{}{\overset{pK_a*}{\rightleftharpoons}} \quad \text{RO}^{\cdot}\text{*} \quad + \quad \text{H}_3\text{O}^+$$

$$\updownarrow \text{Em(ROH*)} \qquad\qquad \updownarrow \text{Em(RO}^{\cdot}\text{*)}$$

$$\text{ROH} \quad + \quad \text{H}_2\text{O} \quad \underset{}{\overset{pK_a}{\rightleftharpoons}} \quad \text{RO}^{-} \quad + \quad \text{H}_3\text{O}^+$$

Figure 12. Scheme for acid–base equilibria in the ground and excited state.

Examples: Molecules such as 8-hydroxy-1,3,6-pyrenetrisulfonate (known as pyranine) (16),[87-94] 2-naphthol (17)[95] and acridines (18)[96] have been used extensively to study the surface pH and kinetics of acid–base reactions in various organized assemblies such as normal and reversed micelles. In aqueous solution at pH values ≤ 6, the absorption spectrum of pyranine (16) is characteristic of the acidic form ROH alone (max. 405 nm). At higher pH values (pH ≥ 8), an

ption band at 445 nm appears, revealing the presence of the basic form RO⁻
the ground state. In highly acidic solutions, blue fluorescence with a single
maximum at 445 nm is observed. Increase of pH leads to partial conversion of
ROH* to RO⁻* before returning to the ground state. Thus an additional band at 510
nm appears, characteristic of green fluorescence from the basic form RO⁻*. An
isoemissive point is observed at 490 nm. In aqueous solution, the pK_a and pK_a^*
values have been estimated to be 7.5 and 0.5, respectively.

At neutral pH in water, 2-naphthol (pK_a = 9.4 and pK_a^* = 2.8) shows
fluorescence from both ROH* and RO⁻*. Studies of fluorescence of solubilized
molecules when they are in organized media such as ionic micelles have shown
that the charged interfaces have a pronounced effect on the acid–base behavior of the
dye. Due to their negative surface charge, anionic micelles accumulate protons at
the surface and hence lower the ground-state pK_a but raise the excited-state pK_a^*.
Cationic micelles have an inverse effect. Acid–base reactions of amines can also be
employed in similar studies. Acridine (with pK_a = 5.5 and pK_a^* = 10.6) has
received some scrutiny in this context.

4.3. Depolarization of Fluorescence as a Measure of Rotational Diffusion and Microviscosity

Depolarization of fluorescence is an important process widely used in studies
of microheterogeneous systems.[97] When a fluorescent molecule is excited by
polarized light, its emission is maximally polarized if, during its excited state-
lifetime, the probe does not change its position or orientation. However if the
molecule is not rigidly held, Brownian motions will tend to remove the orientation
imposed by the polarized radiation. The existence of polarization is due to the fact
that electronic transitions are determined by transition moments involving electric
moment vectors and wave functions that have unique symmetries and orientation.
The electric dipole moment determines the direction along which charge is
displaced in a molecule undergoing electronic transition. Fluorescence
depolarization in general is determined by one of the following ways: 1) steady
state measurements on degree of polarization (p_0); 2) direct time-resolved decay
measurements of the polarized components, and 3) phase-shift difference
measurements of the polarized components. The steady-state method is based on
photoselection. Here, one excites the sample with polarized light and measures the
emission intensity along two perpendicular directions.   Only the polarization
measurements based on steady-state experiments are described here (for time-
resolved measurements see Chapter 3).

For plane-polarized light excitation, the polarization of fluorescence (p) and the
fluorescence anisotropy (r) are given by:

$$p = [I_v - I_h] / [I_v + I_h] \tag{5}$$
$$r = [I_v - I_h] / [I_v + 2I_h] \tag{6}$$

where I_v and I_h are the fluorescence intensities observed through a polarizer oriented vertically or horizontally to the plane of polarization of the exciting beam. In time-resolved anisotropy experiments the anisotropy measured in the above manner needs to be corrected for instrumental artefacts (for details see Chapter 3). For randomly oriented but rigidly held molecules, p takes the characteristic value p_0:

$$p_0 = [3\,Cos^2\theta - 1]/[Cos^2\theta + 3] \qquad (7)$$

where θ is the angle between the emission and absorption oscillators. If the absorption and emission vectors are parallel ($\theta = 0$), $p_0 = 0.5$. If these vectors are perpendicular ($\theta = 90°$), then $p_0 = -0.33$. In practice, p_0 has a value between these two limiting cases.

In the absence of any motion, some fluorescence anisotropy exists which is called the fundamental anisotropy, r_0:

$$r_0 = [3\,Cos^2\theta - 1]/5 \qquad (8)$$

In simple hydrodynamic treatments of the Perrin type, the ratio (r_{obs}/r_0) is related to the microviscosity η of the microenvironment in which the excited probe undergoes tumbling motion:

$$(r_{obs}/r_0) = 1 + (kT\tau/\eta V_0) \qquad (9)$$

Here r_{obs} and r_0 are the measured and limiting fluorescence anisotropies, respectively, k is the Boltzmann constant, τ is the excited-state lifetime, and V_0 is the effective volume of the tumbling sphere. The limiting anisotropy (r_0) values are often determined at the same excitation wavelength in a highly viscous solvent such as mineral oil.

In order to have a proper appreciation for the quantity "microviscosity η" it is useful to digress a little on the hydrodynamic treatments of Brownian motion of solutes that lead to viscosity. Focusing on the solute, hydrodynamic treatments consider the solvent medium to be continuous (or at least the size of the solute to be quite large as compared to the solvent). This approximation, known as the "stick-boundary condition," is quite appropriate for the Brownian motion of large colloidal particles (even the aggregated system carrying the probe) and supramolecular systems such as proteins. On the other hand, for unsolvated, flat aromatic hydrocarbons such as perylene or methylanthracene, the rotational diffusion may largely be determined by the slipping (or partial slip) hydrodynamic boundary conditions. It is questionable whether the viscosity concept is appropriate at all for the slip (or partial slip) conditions. In structurally anisotropic media such as lipid vesicles or liposomes, the above considerations become important. Nevertheless, following the early pioneering work of Weber et al. probing the hydrophobic regions of micelles and lipid vesicles, there have been numerous

68 Kalyanasundaram

investigations of steady-state anisotropy in model systems. Often they extrapolate
the measured anisotropies into microviscosities.

In contrast to free rotation in aqueous media, motion in organized structures
such as vesicles or membranes is limited in angular range. The surrounding
architecture usually imposes certain restrictions on the orientations of the probe.
Moreover, friction within the structure often reduces the rate of reorientational
motion from the value that would be expected in aqueous media. Evidence for such
a situation can be found in the fluorescence anisotropy decay not relaxing to zero
value for most of the probes. Analysis of optical anisotropy decay with appropriate
models can provide information on the structural (range) and dynamical (rate)
aspects. A wobbling-in-cone model has been proposed by Kinoshita et al [99,100]
for the motion of rod-shaped probe such as diphenylhexatriene in lipid membranes.
The model assumes that the major axis of the probe wobbles uniformly within the
cone of semiangle θ_c with a wobbling diffusion constant D_w. The value of these
parameters can be estimated from an experimental anisotropy decay $r(t)$. For
macroscopically isotropic systems such as multiple domain liquid crystals and
vesicle suspensions, the order parameter $\langle S \rangle$ is defined as the square root of the
ratio of the asymptotic anisotropy, r_∞ to the zero-time (or limiting) anisotropy, r_0:

$$\langle S \rangle = (r_\infty / r_0)^{1/2} \tag{10}$$

With respect to the theoretical value of $r_0 = 0.4$, values in the range 0.362–
0.395 have been measured experimentally. The parameter γ, defined as the ratio of
relaxation time ϕ for rotational diffusion and τ, the fluorescence lifetime of the
probe of the probe, $\gamma = (\phi / \tau)$, satisfies the relation:

$$\gamma = (r_s - r_\infty) / (r_0 - r_s) \tag{11}$$

where r_s, is the steady-state fluorescence anisotropy. For an isotropic liquid such as
mineral oil, with $r_\infty = 0$ as an (good) approximation, then, ϕ can be translated into
a microviscosity η using classical hydrodynamic expressions of the Perrin type.
There have seen several extensions and modifications of this "wobbling cone "
model by various authors.[101-107]

Examples: All-*trans*-diphenylhexatriene (DPH) (18) is probably the most
popular probe for fluorescence polarization studies.[108] The cis–trans isomerization
of DPH is accompanied by loss of fluorescence intensity and the process strongly
depends on the viscosity of the medium. The quantum yield of emission
approaches unity in a highly viscous medium that inhibits the isomerization. DPH
has been popular as a probe due to two key features: strong absorption at 355 nm
($\varepsilon \approx 8000$ $M^{-1}cm^{-1}$) and a high, nearly constant r_0 value of 0.362. The transition
dipole of fluorescence and the $S_0 \leftarrow S_1$ absorption band lay close to parallel to the
long axis of the molecule. Hence the fluorescence depolarization reflects almost
exclusively the angular displacement of this axis. Being a pure hydrocarbon, DPH
readily distributes itself in the hydrophobic regions of the host. There have been

reports recently raising concerns on the effective use of DPH. These relate to the photobleaching upon uv excitation and of DPH fluorescence excitation with a transition dipole moment not parallel to the dipole moment of the strongly absorbing near uv–visible electronic excitation.

Hudson and co-workers have introduced[109-111] a variation of DPH in the form of *cis*- and *trans* -parinaric acid (**19**). As an unsaturated fatty acid, parinaric acid mixes freely with the host constituents in lipid aggregates composed of natural or synthetic lipids. The linear trans isomer is believed to be preferentially associated with the solid phase of the lipid bilayers, while the cis isomer partitions between the solid and liquid crystalline phases. Abrupt changes in the fluorescence intensity and polarization degree have been observed for the solubilized probe during lipid phase transitions, vesicle fusion, and lateral phase separation of lipids.

Waggoner, Stryer, and co-workers have introduced[112-116] a series of anthroyl fatty acids (**20**) as potential probes. By appropriate synthetic procedure, the flurophore unit can be attached to various sites of a long-chain fatty acid, thus allowing it to be positioned at various depths in a bilayer structure. Anisotropy and fluorescence quantum yield increase as the fluorophore moves deeper into the bilayer. Simple aromatic molecules that have found use as probes are perylene (**21**) and 2-methylanthracene (**2A**).[117,118] These molecules are also sufficiently hydrophobic to be almost exclusively associated within the lipid bilayers. The acid–base indicator pyranine has also been used extensively as the polarization probe for the interface region of organized assemblies.

5. Phosphorescence Probe Analysis

In contrast to fluorescence, phosphorescence emission from the triplet state of molecules is less commonly observed in neat solvents at room temperature. Studies of triplet states were restricted to monitoring of transient absorption changes via flash photolysis or emission in low-temperature glasses. In the late 1970's, it was discovered that solubilization of arenes in micellar aggregates or spotting on solid substrates such as filter paper led to ready observation of phosphorescence at room temperature (RTP).[119-122] Since 1980, a renaissance in RTP research has produced an abundance of techniques for observation of phosphorescence at room temperature that the technique has been adopted as a powerful analytical tool.[123-125] Systems of interest to readers of this volume where RTP has been reported include in addition to micelles, host systems such as cyclodextrins, zeolites, and solid surfaces. Comprehensive monographs describing the features of RTP in various host systems and their analytical applications have become recently available.[125]

In fluid solution, triplet-state formation, if not spontaneous, can be facilitated by external heavy atoms residing in close proximity of the lumiphor. Once populated, radiative deactivation from the triplet state requires a high degree of rigidity of the microenvironment and minimal interactions with any quencher species present in the solution. A variety of organized assemblies and hosts apparently satisfy such requirements. For example, intense RTP has been observed from arenes solubilized in sodium lauryl sulfate (SDS) micelles in the presence of Tl^+ or Ag^+ ions. Heavy atom effects are known to have a strong distance dependence and clearly the micelles increase the proximity of the heavy atom and probe molecules and present locally high concentration of the heavy atom species. 1,2-Dibromoethane cosolubilized or coincluded heavy atom molecule in micellar and cyclodextrin (CD) solutions helps to observe RTP from numerous arenes and N,O-heterocyclics.[126-128] Hamai has recently used cosolubilized brominated alcohols (2-bromoethanol, 2,3-dibromo-1-propanol) to observe RTP from acenaphthene in aqueous solutions of β-Cyclodextrin.[129] RTP can also be observed from internal heavy-atom-containing molecules such as 1-bromo-naphthalene and related arene derivatives in deoxygenated solutions. Femia and Cline Love synthesized a bromo-substituted β-cyclodextrin, heptakis(6-bromo-6-deoxy-β-cyclodextrin) and have observed RTP of several aromatic hydrocarbons when they are included in this CD cavity.[130] In host systems such as cyclodextrins or zeolites, the host cavity has specific dimensions and hence, in solute mixtures, the RTP is highly selective.

Room temperature phosphorescence from a number of arenes and olefins has been observed recently when they are incorporated into a series of heavy-atom-exchanged zeolites.[131] For example, naphthalene when included in the X-type Faujasites M^+X^- exchanged with Rb^+ or Cs^+ exhibits intense RTP with short singlet and triplet lifetimes. In faujasites exchanged with Li^+, Na^+, or K^+ ions, intense fluorescence and weak phosphorescence with long singlet and triplet lifetimes have been observed. The microscopic organization in zeolites is so efficient that RTP is observable even from olefins (e.g., *trans*-stilbene) and polyenes (e.g.,1,6-all-*trans*-diphenyl-hexatriene) for which phosphorescence is normally weak even in the presence of conventional heavy atom solvents (for details see Chapter 10). Filter papers have been used extensively to promote solid surface RTP. A relationship was found between ϕ_p and the modulus (softness or stiffness) of cellulose. The modulus of the filter paper, which is related to the hydrogen-bonding network in the paper, is an important factor in obtaining high ϕ_p for an adsorbed phosphor on filter paper.

Formulations and methods described earlier for fluorescence probes, such as polarization of emission (for monitoring rotational diffusion) and quenching by small and medium sized molecules (for solute accessibility) are also applicable to phosphorescence probes.[132-134] For probes tumbling in fluid solutions at ambient temperature, rotational correlation time is several tens of nanoseconds and can be studied easily by time-dependent fluorescence anisotropy. In viscous solutions and for probes embedded in rigid media, the rotational correlation time is in the range of µs to ms and these are most easily studied by transient triplet/phosphorescence anisotropy. Luminescence, although not a prerequisite for photophysical investigations, certainly allows ready, low-cost investigations. In the absence of luminescence, transient absorption/flash photolysis techniques permit studies of triplet state processes.

The long-lived nature of the phosphorescence emission along with ready observation of RTP allow quantitative studies of the slow static and dynamic processes (>μs) of the organized assemblies and host–guest systems. An important application of the RTP is in the determination of solute entry/exit rates associated with the solubilization process. Figure 13 presents schematically the principles of these experiments.[135] Solubilization is a dynamical process and there exists a finite possibility of the excited-state triplet probe's being quenched (by quenchers restricted to the aqueous phase) during its residence in the aqueous phase. Quenching of the excited state occurs in competition with reentry of the probe. With increasing concentration of the quencher, the observed phosphorescence lifetime decreases until the exit rate becomes the controlling step in the quenching process. At this point, further increase in the quencher concentration will not change the phosphorescence lifetime. Quantitative analysis using suitable kinetic models allows extraction of the exit, and entry rate constants. Entry and exit rate constants of haloaromatics have been determined in micelles and cyclodextrins using this procedure.[135-138]

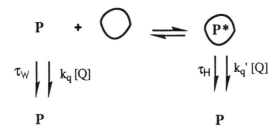

Figure 13. Scheme for the excited-state quenching in micellar media. t_W and t_H refer to the probe lifetime in water and host respectively. k_q and k_q' refer to the rate constant for quenching in the aqueous and host environment respectively.

There have been a number of studies using phosphorescence probes in proteins.[132-134] Intrinsic phosphorescence probes in these systems are amino acids (indole or tryptophan) and porphyrins (in heme proteins). Numerous porphyrins including chlorophylls a and, b and Zn-porphyrins show both fluorescence and phosphorescence. The heme component of proteins such as cytochrome c, myoglobin, and hemoglobin are iron porphyrins that are nonluminescent. The porphyrin can be made luminescent by removal of iron. Replacement of iron by Zn, Pd, Pt or Os ions in the porphyrin core has led to synthesis of phosphorescent proteins and protein hybrids. There have been a number of studies of long-distance electron transfer recently using such systems.[139] Alternate popular phosphorescence probes have been xanthene dyes (eosin and erythrosin in particular) covalently bound to the protein. For example, rotational motion of protein components of serum lipoproteins have been examined using erythrosin-labeled lipoproteins.

6. Inorganic Complexes and Ions as Probes

Even though the luminescence probes described earlier are all organic, there are a few inorganic complexes and ions that can serve as photophysical probes. Among simple metal ions, mention can be made of lanthanide ions (rare earths), which are known to luminesce in solution at ambient temperature.[140] Mentioned below are some representative applications. Efficient energy transfer from solubilized arenes to surface adsorbed Tb^{3+} ions has been shown to occur in reversed micellar system.[141] Luminescence from Eu^{3+} and Tb^{3+} that have been exchanged on montmorillonite and hectorite have been used to monitor extent of hydration in these clay minerals.[142] Tb^{3+} is known to form a complex with dipicolinic acid that is four orders of magnitude more strongly fluorescent than the parent terbium ion. Fusion of lipid vesicles has been studied by following the complex fluorescence growth that occurs upon mixing of two vesicle populations, one loaded with dipicolinic acid and the other with terbium ions.[143]

Uranyl ion (UO_2^+) luminesces in solution and the excited state is an excellent oxidant. Luminescence of uranyl ion-exchanged pillared clays and zeolites have been used to probe the nature of the active site distribution and source of selectivity in these catalysts.[144] Demas *et all*. have used the solvent sensitivity of metal-to-ligand charge-transfer excited-state emission (LMCT) in $Re(CO)_3(Cl)(bpy)$ and $Ru(bpy)_3^{2+}$ to probe micellar aggregates.[145] Excited-state quenching of adsorbed polypyridyl complexes by small molecules has been used to study interfacial charge effects and solute accessibility in micelles, polyelectrolytes, and ion-exchange membranes.[4]

The most important application of inorganic complexes is as photosensitizers of electron-transfer and energy-transfer processes. Most popular are the transition metal polypyridyl complexes [e.g., tris(bipyridyl)ruthenium(II), $Ru(bpy)_3^{2+}$] and metalloporphyrins [e.g., tetrakis(*N*-methyl-4-pyridyl)porphyrinato zinc(II), $ZnTMPyP^{4+}$].$Cr(bpy)_3^{3+}$ shows moderate phosphorescence in solution at room temperature and can serve as a cationic phosphorescence probe. Following the initial reports of visible light-induced cleavage of water to H_2 and O_2 using surfactant derivatives of $Ru(bpy)_3^{2+}$, there have been large number of photochemical studies of polypyridyl complexes and porphyrins in various kinds of organized assemblies and confined media. Another reason to study photoredox processes in organized media is to achieve inhibition or retardation of undesirable thermal back electron transfers that follow light-induced electron transfer. By suitable spatial separation of the reactants/products and introducing charged interfaces, it has been shown possible to increase significantly the cage escape yield of photoredox products. Recent reviews of some of these studies are available.[146]

7. Conclusions

Application of luminescence probes to studying the fascinating microworld of organized media as reviewed here will continue to grow. The list of chromophores capable of serving as luminescence probes is constantly increasing; so also are their applications. It is obvious from the discussion on various luminescence probes that the excited-state interactions with the environment are quite varied for different probes. In a good majority of the cases, the solvent–probe interactions responsible for the observed effects have been fully elucidated by time-resolved spectroscopy. The domain of application of different probes is vast. As mentioned in several places, the interpretation of luminescence data in different organized media requires a clear understanding of the probe photophysics. Fluorescence probes are useful in making structural comparisons and in determining local structure variation of the organized media.

References

1. N. J. Turro, *Modern Molecular Photochemistry*, Benjamin, Menlo Park, CA, 1978.
2. J. R. Lakowicz, *Principles of Fluorescence Spectroscopy*, Plenum Press, New York, 1983.
3. J. K. Thomas, *Chemistry of Excitation at the Interfaces*, American Chemical Society, Washington, D.C., 1984.
4. K. Kalyanasundaram, *Photochemistry in Microheterogeneous Systems*, Academic Press, New York, 1987.
5. *Organic phototransformations in Microheterogeneous Systems*, M. A. Fox, ed., American Chemical Society, Washington, D. C., 1982.
6. *Organic Chemistry in Anisotropic Media*, V. Ramamurthy, J. R. Schaffer and N. J. Turro, eds., *Tetrahedron*, **1987**, *43*, issue 7.
7. L. M. Loeb, *Spectroscopic Membrane Probes*, Vols. I-III, CRC Press, New York, 1988.
8. (a) R. Zana in *Surfactant Solutions: New Methods of Investigations*, R. Zana, ed., Marcel Dekker Inc., New York, 1987.
   (b) J. K. Thomas, *J. Phys. Chem.*, **1987**, *91*, 267.
9. K. Kalyanasundaram, *Chem. Soc. Rev.*, **1978**, *7*, 453.
10. N. J. Turro, M. Grätzel and A. M. Braun, *Angew. Chem. Int. Edn. Engl.*, **1980**, *19*, 675.
11. D. G. Whitten, J. C. Russell and R. H. Schmehl, *Tetrahedron*, **1982**, *38*, 2455.
12. A. Azzi, *Quart. Rev. Biophys.*, **1975**, *8*, 237.
13. G. K. Radda and J.Vanderkooi, *Biochem. Biophys. Acta*, **1972**, *265*, 509.
14. M. Shinitzky and Y. Barenholtz, *Biochem. Biophys. Acta*, **1978**, *515*, 367
15. *Membrane Spectroscopy*, Mol. Biol. Biochem. Biophys. Ser., Vol. 31, E. Grell, ed., Springer-Verlag, Berlin, 1983.
16. a) V. Ramamurthy, *Tetrahedron*, **1986**, *42*, 5753.
   b) N. Ramnath, V. Ramesh and V. Ramamurthy, *J. Photochem.*, **1985**, *31*, 75.
17. M. Wong, T. Nowak and J. K. Thomas, *J. Am. Chem. Soc.*, **1977**, *99*, 4730.
18. H. F. Eicke, *Top. Curr. Chem.*, **1980**, *87*, 86.
19. (a) Ch. Reichardt, *Solvents and Solvent Effects in Organic Chemistry*, 2nd ed., VCH Publishers, Weinheim, 1988 and references cited therein.

(b) for an extensive discussion of various solvent polarity scales see, T. R. Griffith and D. C. Pugh, *Coord. Chem. Rev.*, **1979**, *29*, 129.

(c) for a recent, critical review of various types of interactions of the solvent with the ground and excited state leading to solvatochromic shifts see, P. Suppan, *J. Photochem. Photobiol.*, **1990**, *50*, 293.

20.   *Kinetics of Non-homogeneous Processes*, G.R. Freeman, ed., John Wiley, New York, 1987.

21.   (a) *Molecular Dynamics in Restricted Geometries*, J. Klafter and J. M. Drake, eds., John Wiley, New York, 1989.

(b) *The Fractal Approach to Heterogeneous Chemistry: Surfaces, Colloids and Polymers*, D. Avnir, ed., John Wiley, Chicester, 1989.

22.   *Kinetics and Catalysis in Microheterogeneous Systems*, M. Grätzel and K. Kalyanasundaram, eds., Marcel Dekker, New York, 1991.

23.   Molecular Probes Inc., P.O. Box 22010, Eugene, Oregon, 97402, USA.

24.   Lambda Probes, Grottenhofstrasse 3, A-8053 Graz, Austria.

25.   A. Nakajima, *Bull. Chem. Soc. Japan*, **1971**, 44, 3272.
       A. Nakajima, *Spectrochim. Acta*, **1974**, *30A*, 360.
       A. Nakajima, *J. Lumin.*, **1971**, *11*, 429.

26.   K. Kalyanasundaram and J. K. Thomas, *J. Am. Chem. Soc.*, **1977**, *99*, 2039.

27.   J. R. Cardinal and P. Mukerjee, *J. Phys. Chem.*, **1978**, *82*, 1614.

28.   D. C. Dong and M. A. Winnik, *Photochem. Photobiol.*, **1982**, *35*, 17.

29.   P. Lianos and S. Georghiou, *Photochem. Photobiol.*, **1979**, *29*, 843; **1979**, *30*, 355.

30.   G. W. Robinson, *J. Chem. Phys.*, **1967**, *46*, 572.

31.   E. G. McRae, *J. Phys. Chem.*, **1957**, *61*, 562.

32.   N. S. Bayliss, *J. Chem. Phys.*, **1950**, *18*, 292.

33.   Y. Ooshika, *J. Phys. Soc. Jpn.*, **1954**, *9*, 594.

34.   R. A. Marcus, *J. Chem. Phys.*, **1963**, *39*, 1734; **1964**, *43*, 1261.

35.   C. J. Seliskar and L. Brand, *Science*, **1971**, *171*, 799.

36.   W. R. Ware, S. K. Lee, G. J. Brant and P. P. Chou, *J. Chem. Phys.*, **1971**, *54*, 4729.

37.   M. L. Bhaumik and R. Hardwick, *J. Chem. Phys.*, **1963**, *39*, 1595.

38.   J. Slavik, *Biochem. Biophys. Acta*, **1982**, *694*, 1.

39.   G. M. Edelman and W. O. Mc Clure, *Acc. Chem. Res.*, **1968**, *1*, 65.

40.   W. O. Mc Clure and G. M. Edelman, *Biochemistry*, **1966**, *5*, 1908.

41.   L. Brand and J. R. Guhlke, *Ann. Rev. Biochem.*, **1972**, *41*, 843.

42.   L. Stryer, *J. Mol. Biol.*, **1965**, *13*, 482.

43.   C. J. Seliskar and L. Brand, *J. Am. Chem. Soc.*, **1971**, *95*, 5414.

44.   R. A. Auerbach, J. A. Synowiec and G. W. Robinson, *Chem. Phys.*, **1980**, *14*, 215.

45.   P. J. Sadkowski and G. R. Fleming, *Chem. Phys.*, **1980**, *54*, 79.

46.   K. H. Grellman and U. Schmidt, *J. Am. Chem. Soc.*, **1982**, *104*, 6267.

47.   G. Weber and F. J. Farris, *Biochem*istry, **1979**, *18*, 3075.

48.   R. B. McGregor and G. Weber, *Ann. N.Y. Acad. Sci.*, **1981**, *366*, 140.

49.   J. R. Lakowicz, D. R. Bevan, M. P. M. Maliwal, H. Cherek and A. Balter, *Biochemistry*, **1983**, *22*, 5714.

50.   F. R. Prendergast, M. Meyer, G. L. Cabron and S. Ida, *J. Biol. Chem.*, **1983**, *258*, 7541.

51.   K. Bredereck, Th. Forster, and H. G. Oenstein in *Luminesence of Organic and Inorganic Materials*, H. P. Kallman and G. M. Spruch, eds., John Wiley, New York, 1960.

52.   (a) K. Kalyanasundaram and J. K. Thomas, *J. Phys. Chem.*, **1977**, *81*, 2176.

(b) C. V. Kumar and L. M. Tolosa, unpublished results.

53. J. C. Dederens, L. C. Oosemans, F. C. De Schryver and A. van Dormael, *Photochem. Photobiol.*, **1979**, *30*, 443.
54. C. V. Kumar, S. K. Chattopadhayoy and P. K. Das, *Photochem. Photobiol.*, **1983**, *38*, 141.
55. K. Muthuramu and V. Ramamurthy, *J. Photochem.*, **1984**, *26*, 57.
56. W. W. Mantulin and P. S. Song, *J. Am. Chem. Soc.*, **1973**, *95*, 5123.
57. P. S. Song, M. L. Harter, T. A. Moore and W. C. Henderson, *Photochem. Photobiol.*, **1971**, *14*, 521.
58. D. L. Sackett and J. Wolff, *Anal. Biochem.*, **1987**, *167*, 228.
59. P. Greenspan and S. O. Fowler, *J. Lipid Res.*, **1985**, *26*, 781.
60. E. Lippert in *Luminescence of Organic and Inorganic Materials*, H. P. Kallman and G. M. Spruch, eds., John Wiley, New York, 1962, p. 271.
61. E. Lippert, W. Luder and H. Boos, in *Advances in Molecular Spectroscopy*, A. Mangini, eds., Pergamon, Oxford, 1962, p. 443.
62. Z. R. Grabowski and J. Dobkowski, *Pure Appl. Chem.*, **1983**, *55*, 245.
63. E. Lippert, W. Rettig, V. Bonacic-Koutecky, F. Heisel and J. A. Miehe, *Adv. Chem. Phys.*, **1987**, *68*, 1.
64. W. Rettig, *Angew. Chem. Int. Ed. Engl.*, **1983**, *15*, 971.
65. R. Loutfy and K. Y. Law, *J. Phys. Chem.*, 84, 2803 **1980**; *Chem. Phys. Lett.*, **1980**, *75*, 545.
66. K. Y. Law, *Photochem. Photobiol.*, **1981**, *33*, 799.
67. C. David, E. Szalai and D. Baeyens-Volant, *Ber. Bunsenges. Phys. Chem.*, **1982**, *86*, 710.
68. G. R. Fleming, A. E. W. Knight, J. M. Morris, R. J. S. Morrison and G. W. Robinson, *J. Am. Chem. Soc.*, **1977**, *99*, 4306.
69. L. E. Cramer and K. G. Spears, *J. Am. Chem. Soc.*, **1978**, *100*, 221.
70. W. Yu, F. Pellegrino, M. Grant and R. R. Alfano, *J. Chem. Phys.*, **1977**, *67*, 1766.
71. A. Weller, *Z. Electrochem.*, **1956**, *60*, 1144.
72. W. Klopfer and G. Naundor, *J. Lumin.*, **1974**, *8*, 457.
73. K. Sandros, *Acta Chem. Scand.*, **1976**, *30A*, 761.
74. D. Ford, P. J. Thistlewhite and G. J. Woolfe, *Chem. Phys. Lett.*, **1980**, *69*, 246.
75. K. K. Smith and K. J. Kaufman, *J. Phys. Chem.*, **1978**, *82*, 2286; **1981**, *85*, 2895.
76. G. S. Cox and N. J. Turro, *Photochem. Photobiol.*, **1984**, *40*, 185.
77. B. R. Suddaby, P. E. Brown, J. C. Russel and D. G. Whitten, *J. Am. Chem. Soc.*, **1985**, *107*, 5609.
78. F. Tokiwa, *Adv. Coll. Interface Sci.*, **1972**, *3*, 389.
79. G. S. Hartley and J. W.Roe, *Trans. Faraday. Soc.*, **1940**, *36*, 101.
80. P. Mukerjee and K. Banerjee, *J. Phys. Chem.*, **1964**, *68*, 3567.
81. P. Fromherz and B. Masters, *Biochem. Biophys. Acta.*, **1974**, *356*, 270.
82. M. S. Fernandez and P. Fromherz, *J. Phys. Chem.*, **1977**, *81*, 1755.
83. C. J. Drummond and F. Grieser, *Photochem. Photobiol.*, **1987**, *45*, 2604.
84. C. J. Drummond, F. Grieser and T. W. Healy, *J. Phys. Chem.*, **1988**, *92*, 2604 and references cited therein.
85. C. J. Drummond, Ph.D. Thesis. Univ. of Melbourne, Australia, 1989.
86. (a) J. F. Ireland and P. A. H. Wyatt, *Adv. Phys. Org. Chem.*, **1976**, *12*, 131.
    (b) A. Weller, *Z. Phys. Chem.*, *(Wiesbaden)*, **1978**, *17*, 224; *Progr. Reac. Kinet.*, **1961**, *1*, 187.
87. M. Gutman, D. Huppert and E. Pines, *J. Am. Chem. Soc.*, **1981**, *103*, 3709.
88. D. Huppert and E. Kolodney, *Chem. Phys.*, **1981**, *663*, 401.
89. K. K. Smith, K. J. Kaufman, D. Huppert and M. Gutman, *Chem. Phys. Lett.*, **1979**, *64*, 522.
90. E. Pines and D. Huppert, *J. Phys. Chem.*, **1983**, *87*, 4471.

91. G. D. Correll, R. N. Cheser, F. Nome and J. H. Fendler, *J. Am. Chem. Soc.*, **1978**, *100*, 1254.
92. N. R. Clement and J. M. Gould, *Biochemistry*, **1981**, 20, 1534.
93. E. Bardez B. T. Goguillon, E. Keh and B. Vakur, *J. Phys. Chem.*, **1984**, *88*, 1909.
94. M. J. Politi and J. H. Fendler, *J. Am. Chem. Soc.*, **1984**, *106*, 265.
95   U. Klein and M. Hauser, *Z. Phys. Chem. Neue Folge*, **1974**, *90*, 215.
96. M. P. Pileni and M. Grätzel, *J. Phys. Chem.*, **1980**, *84*, 2403.
97. G. Weber, *J. Chem. Phys.*, **1971**,*55*, 2399; *Ann. Rev. Biophys. Bioengg.*, **1972**, *1*, 553.
98. F. Hare and C. Lusson, *Biochem. Biophys. Acta*, **1977**, *467*, 262.
99. K. Kinoshita, S. Kawato and A. Ikegami, *Biophys. J.*, **1977**, *20*, 289; **1982**, *37*, 461.
100. A. Ikegami, K. Kinoshita, T. Kouyama and S. Kawato, in *Structure, Dynamics & and Biogenis of Biomembranes*, R. Sato and S. Ohnishi, eds., Plenum Press, New York, 1982, p. 1.
101. C. Zannoni, *Mol. Phys.*, **1981**, *42*, 1303.
102. L. W. Engel and F. G. Prendergast, *Biochemistry*, **1981**, *20*, 7338.
103. G. Lipari and A. Szabo, *J. Chem. Phys.*, **1981**, *75*, 2971.
104. F. Jähnig, *Proc. Natl. Acad. Sci. U. S. A.*, **1979**, *76*, 6361.
105. H. Pottel, W. van der Meer and W. Herreman, *Biochem. Biophys. Acta*, **1983**, *730*, 181.
106. W. van der Meer, H. Pottel, W. Herreman, M. Ameloot, H. Hendrickx and H. Schroder, *Biophys. J.*, **1984**, *46*, 515 and 525.
107. F. Hare, *Biophys. J.*, **1983**, *42*, 205.
108. B. S. Hudson, B. E. Kohler and K. Schulten, in *Excited States,* Vol. 6, E. C. Lim, ed., Academic Press, New York, l982, p. 241.
109. L. A. Skalar, B. S. Hudson and R. D. Simoni, *Biochemistry*, **1977**, *16*, 813 and 819.
110. R. Welti and D. F. Silbert, *Biochemistry*, **1982**, *21*, 5685.
111. P. K. Wolber and B. S. Hudson, *Biochem*istry, **1981**, *20*, 2800.
112. A. S. Waggoner and L. Stryer, *Proc. Natl. Acad. Sci. U. S. A.*, **1970**, *67*, 579.
113. A. S. Waggoner, *Ann. Rev. Biophys. Bioengg.*, **1979**, *8*, 47.
114. E. Blatt and W. H. Sawyer, *Biochem. Biophys. Acta*, **1985**, *822*, 43.
115. E. Blatt, W. H. Sawyer and K. P. Ghiggino, *Aust. J. Chem.*, **1983**, *36*, 1079.
116. J. Eisenger and J. Flores, *Biophys. J.*, **1983**, *41*, 367.
117. M. Shinitzky, A. C. Dianoux, C. Gitler and G. Weber, *Biochem*istry, **1971**, *10*, 2106.
118. U. Cogan, M. Shinitzky, G. Weber and T. Nishidia, *Biochemistry*, **1973**, *12*, 521.
119. K. Kalyanasundaram, F. Grieser and J. K. Thomas, *Chem. Phys. Lett.*, **1977**, *51*, 501.
120. N. J. Turro, K. C. Liu, M. F. Chow and P. Lee, *Photochem. Photobiol.*, **1977**, *27*, 523.
121. N. J. Turro and M. Aikawa, *J. Am. Chem. Soc.*, **1980**, *102*, 4866.
122. R. Humphry-Baker, Y. Moroi and M. Grätzel, *Chem. Phys. Lett.*, **1978**, *58*, 207.
123. S. Scypinski and L. J. Cline Love, *Internat. Lab.*, **1984**, *14*, 61; *Amer. Lab.*, **1984**, 55 and references cited therein.
124. L. J. Cline Love, J. G. Habarta and J. G. Dorsey, *Anal. Chem.*, **1984**, *56*, 1133A and references cited therein.
125. T. Vo-Dinh, *Room Temperature Phosphorimetry in Chemical Analysis,* John Wiley, New York, 1984.

126. (a) L. J. Cline Love, M. Skrilec and J. G. Habarta, *Anal. Chem.*, **1980**, *52*, 754; **1981**, *53*, 437.
    (b) M. Skrilec and L. J. Cline Love, *Anal. Chem.*, **1981**, *52*, 1559; *J. Phys. Chem.*, **1981**, *85*, 2047.
    (c) L. J. Cline Love and R. Weinberger, *Spectrochim. Acta*, **1983**, *38B*, 1421.
    (d) R.A. Femia and L.J. Cline Love, *Anal. Chem.*, **1984**, *56*, 327; *Spectrochim. Acta*, **1986**, *42A*, 1239.
127. M. R. Richmond and R. J. Hurtubise, *Anal. Chem.*, **1989**, *61*, 2643.
128. J. M. Bello and R. J. Hurtubise, *Anal. Chem.*, **1989**, *59*, 2395.
129. S. Hawai, *J. Am. Chem.Soc.*, **1989**, *111*, 3954.
130. R. A. Femia and L.J. Cline Love, *J. Phys. Chem.*, **1985**, *89*, 1897.
131. V. Ramamurthy, J. V. Caspar, D. R. Corbin and D. F. Eaton, *J. Photochem. Photobiol.*, **1989**, *50*, 157;
    V. Ramamurthy, J. V. Caspar and D. R. Corbin, *Tetrahedron Lett.*, **1990**, *31*, 1097.
    V. Ramamurthy, J. V. Caspar, D. R. Corbin, B. D. Schlyer and A. H. Maki, *J. Phys. Chem.*, **1990**, *94*, 3391;
    J. V. Caspar, V. Ramamurthy and D. R. Corbin, *Coord. Chem. Rev.*, **1990**, *97*, 225.
132. N. E. Geacintov and H. C. Brenner, *Photochem. Photobiol.*, **1989**, *50*, 841.
133. R. D. Ludeschar, *Spectroscopy*, **1989**, *5*, 20.
134. J. M. Vanderkooi and J. W. Berger, *Biochem. Biophys. Acta*, **1989**, *976*, 1.
135. M. Almgren, F. Grieser and J. K. Thomas, *J. Am. Chem. Soc.*, **1979**, *101*, 279.
136. N. J. Turro, J. D. Bolt, Y. Kuroda and I.Tabushi, *Photochem. Photobiol.*, **1982**, *35*, 69.
137 N. J. Turro, G. S. Cox and X. Li, *Photochem. Photobiol.*, **1983**, *37*, 149.
138. N. J. Turro, T. Okubo and G. J. Chung, *J. Am. Chem. Soc.*, **1982**, *104*, 1789.
139. a) H. B. Gray, *Chem. Soc. Rev.*, **1986**, *15*, 17.
    b) S. L. Mayo, W. R. Ellis, R. J. Crutchley and H. B. Gray, *Science*, **1984**, *233*, 948.
140. *Lanthanide Probes in Life, Chemical and Earth Sciences*, J. C. Bunzli and G. R. Choppin, eds., Elsevier, Amsterdam, 1989.
141. H. F. Eicke and P. Zinsli, *J. Coll. Int. Sci.*, **1978**, *65*, 131.
142. F. Bergaya and H. van Damme, *J. Chem. Soc., Faraday Trans.*, II, **1983**, *79*, 505.
143. J. Wilschutt, N. Duzgunes and D. Papahadjopoulos, *Biochemistry*, **1980**, *19*, 6011.
144. S. L. Suib, J. F. Tanguay and M. L. Occelli, *J. Am. Chem. Soc.*, **1986**, *108*, 6972.
145. B. L. Hauenstein, W. J. Dressick, T. B. Gilbert, J. N. Demas and B. A. DeGraff, *J. Phys. Chem.*, **1984**, *88*, 1902 and 3337.
146. (a) K. Kalyanasundaram, *Coord. Chem. Rev.*, **1982**, *46*, 159.
    (b) K. Kalyanasundaram, E. Pelizetti and M. Grätzel, *Coord. Chem. Rev.*, **1986**, *69*, 57.
    (c) *Energy Resources Through Photochemistry and Catalysis*, M. Grätzel, ed., Academic Press, New York, 1983.

Chapter 3

Use of Photophysical Techniques in the Study of Organized Assemblies

C. Bohne, R. W. Redmond, and J. C. Scaiano

Steacie Institute for Molecular Sciences,[#]
National Research Council,
Ottawa, Ontario, Canada.

Contents

[#] Issued as NRCC-31529.

1. Introduction

This chapter deals with the experimental photochemical and photophysical techniques and methodologies that are commonly employed in the study of organized systems. The first part presents an overview of the experimental methods used; more often than not these methods are not specific to organized systems but are widely employed in their study. It is impossible to be exhaustive and/or present enough detail to be useful to the practitioner of these techniques. It is however hoped that the information provided will help those who use data acquired with these techniques understand better how these data are obtained and what their limitations are. There are abundant detailed reviews dealing with the techniques and appropriate references are provided.

The second part of this chapter discusses the ways in which the data acquired with the techniques mentioned above can be employed to learn about the characteristics of organized systems. From these techniques we can learn about polarity of the environment, aggregation, mobility within a given microphase, and mobility between microphases. A case in point is that of micelles, where photochemical and especially photophysical techniques may tell us about polarity, aggregation number, and critical micelle concentration, microviscosity, and exit/entry rates from the micelle. These methods have been studied and employed in great detail in the case of micelles, but in other organized systems these ideas are less developed; notably, in solid systems (e.g., zeolites, silica gel, or inclusion compounds) our knowledge is very limited, although the technique of time-resolved diffuse reflectance that Wilkinson and his group have pioneered over the last 10 years is a major and highly promising development.[1]

We have not tried to tabulate the values (e.g., polarities or exit rate constants from micelles) available in the literature. We have simply selected a few examples that allow us to illustrate the type and quality of data that can be obtained. Many of the examples involving liquid phase work concentrate on micelles as a result of the availability of quantitative data pointed out above. There is no doubt that the application of these methods to other organized systems is not only possible, but indeed desirable.

2. Experimental Techniques
2.1. Laser Flash Photolysis

Several time–resolved techniques based on the direct monitoring of the absorption or emission from a photogenerated short-lived reaction intermediate are of common use in the study of organized systems; among these, conventional flash photolysis, laser (or nanosecond) flash photolysis, and picosecond techniques cover about 12 orders of magnitude in time scales, from seconds to picoseconds. In this section we will only deal with laser flash photolysis and, in a latter section, with a variation of the technique (time-resolved diffuse reflectance) which is very valuable with opaque samples. Many of the ideas included herein are common to other techniques where the intermediates are generated by nonphotochemical methods (e.g., pulse radiolysis, stopped flow, or T-jump).

The first application of laser flash photolysis techniques can be found in a 1966 report by Lindqvist.[2,3] While there have been numerous improvements to the technique, the basis of the method remains much the same. One can readily identify four classes of components: (1) the excitation source used to generate the intermediates, (2) the monitoring (or analyzing) beam, (3) the detection system and (4) the data acquisition/processing system. We outline below the more common choices for each one of these parts of a laser photolysis system.

Excitation source: This is always a pulsed laser. Among the parameters that one must take into account in the selection of a laser are the energy per pulse, the repetition rate, the pulse duration, and the wavelengths available. Typically one needs between 4 and 40 mJ per pulse to carry out experiments conveniently. Lower energies can be used, but make system alignment very critical and may limit what type of chemical systems can be examined. On the other hand, very high energies (over 100 mJ/pulse can be readily achieved with some lasers) make the system very prone to multiphoton processes. Pulse durations between 2 and 20 ns are common and usually adequate. Most lasers in the market today can be pulsed much faster than required for laser flash photolysis; e.g., in our experiments we employ a frequency of 1 Hz, and actually use every second or third pulse for data acquisition. In fact, a more common problem is that some lasers do not operate reliably (e.g., pulse energy being irreproducible) at low frequencies. Table 1 illustrates the types of lasers that are commonly employed, including the more common harmonics. There are, of course, lasers with characteristics well outside those indicated here.

Monitoring Beam: The most common source is a xenon arc lamp which provides an adequate (though clearly not ideal) spectral distribution and can be readily operated in pulsed mode. Lamps between 75 and 1000 W are quite adequate; lamps of different ratings differ substantially in the size of the arc. Thus, large changes in power are not matched by comparable changes in brilliance. Given the small size of the 'pinhole' where the monitoring light enters the sample

(0.5 – 2 mm) there is little gain with lamp ratings over 150 W. Typically the lamp can be in a vertical position in a conventional housing, or horizontally in a housing containing an elliptical mirror. The latter are far more efficient (i.e., higher output intensity) but are somewhat less stable and the beam has a small dark spot in the center. Both types are commercially available and are frequently used in a pulsed mode. That is, the lamp intensity is increased by a factor of 20–100 for a few milliseconds. This improves the absorption vs. emission discrimination and facilitates monitoring in the uv region; naturally, it also shortens substantially the lifetime of the lamp.

Pulsing of the analyzing beam is essential in the case of diffuse reflectance (*vide infra*).

Table 1. Typical Characteristics of Common Pulsed Laser Sources

Laser	Type	λ (nm)	Duration (ns)	Energy mJ/pulse	Notes
Nitrogen	Gas	337.1	1–10	1–10	
Excimer	Gas	193	5–20	<500	a
		248		<500	a
		308		<150	a
		351		<150	a
Ruby	Solid State	694	10–20	1–50	
		347			
Nd/YAG	Solid State	1060	2–25	1000–2000	
		530		100–1000	
		355		10–200	
		266		5–100	
Dye (laser pumped)	Liquid	Variable	<20	0.1–100	b
Dye (Flash pumped)	Liquid	Variable	>100	<2000	c

[a] Uses highly corrosive gases.
[b] Duration is a function of pump.
[c] High volumes of dye solution required

Detection system: The detection system can be either time-resolved or spectrally resolved. Time-resolved detection systems usually consist of a (single pass) monochromator and a fast photomultiplier (PMT); typically only part of the dynode chain is used, the rest of the PMT dynods being connected to ground; 5 -7 dynode chains are quite common. The signals are normally terminated into 50 or 93 ohms and compatible cabling employed throughout. Higher termination resistors can be employed but they affect the rise time of the system. The PMT signals are normally fed into a digitizer or scope (see below). Transient spectra can be obtained using a point-by-point type of approach by recording traces at various wavelengths.

Spectrally resolved data acquisition can be carried out with a gated (and normally intensified) optical multichannel analyzer (OMA). In these systems a diode array (usually 512 or 1024 diodes) is placed at the output from a spectrograph. By appropriate timing of the gate opening with respect to the laser pulse it is possible to obtain spectra at different delays following laser excitation.

Data acquisition/processing: Time-resolved transient signals are fed into a transient digitizer and then transferred to a computer; typical sampling rates should exceed 100 MHz and the bandwidth should be in the 20-MHz to 1-GHz range. Among the desirable features in a digitizer are GPIB control interface, minimum 8-bit resolution, record size of 512 points or more and pretrigger; the last one may not be available in the faster commercial digitizers. Some local processing of the data (averaging, accumulation, smoothing) is advantageous but far from essential.

The data is transferred from the digitizer to a computer that can also be employed for experiment control and data processing. The alternatives here are common to many other types of computerized work and will not be discussed in any detail.

The transient absorption signals are derived from small variations (tens of millivolts) in the PMT output which usually consists of a few hundred millivolts. The best resolution is obtained by offsetting the PMT signals and by recording independently its total output with an appropriate A/D converter.

2.2. Laser Photolysis—Time-Resolved Diffuse Reflectance

Much of what has been stated in the previous section applies also to diffuse reflectance measurements as well, except that in this case the opaque samples need to be excited from the front. Figure 1 illustrates a typical arrangement. The technique has been developed largely by Wilkinson's group and has been discussed on a number of occasions.[1,4] In this case the data are not treated in terms of absorbances as in transmission experiments; rather, the Kubelka-Munk treatment [5] for strongly scattering samples is employed, i.e.

$$F(R) = \frac{K}{S} = \frac{(1-R)^2}{2R} \qquad (1)$$

where $K = 2\varepsilon C$, ε is the extinction coefficient and C the concentration, S is the scattering coefficient, and R the diffuse reflectance. For small changes in reflectance during transient experiments the relative change in reflected light, Δr, is frequently a suitable parameter that can be taken as proportional to transient concentration and facilitates kinetic analysis of the data.

$$\Delta r = \frac{\Delta J}{J_o} \qquad (2)$$

where J_0 is the intensity of the light reflected before the transient is produced (i.e., prelaser pulse) and ΔJ is the change in reflected intensity.

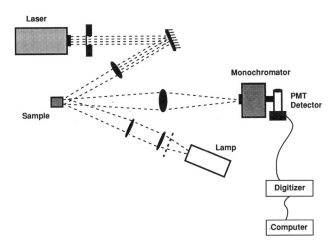

Figure 1. Schematic representation of a laser system for time-resolved diffuse reflectance studies.

As mentioned above, time-resolved diffuse reflectance studies require a pulsed monitoring beam and have a few limitations in comparison with transmission studies. In particular, the high sensitivity to emission from the sample (particularly fluorescence) makes measurements at very short times difficult, and the scattering of the laser pulse frequently requires at least 25-nm separation between the excitation and monitoring wavelengths (7–10 nm being typical in transmission studies).

Finally, the simple kinetics that are frequently encountered in solution studies are a rarity in solid samples. Kinetic analysis is usually complex and a subject of ongoing research.

2.3. Time-Correlated Single-Photon Counting

Accurate and reliable measurement and analysis of fluorescence lifetimes are of extreme importance in photophysical and photobiological systems. There are several methods by which this information can be obtained. Phase modulation techniques, based on the lifetime-dependent phase angle of the emission, utilize detection in the frequency domain; pulse sampling techniques, employing short pulse excitation with rapid response electronics such as fast response photomultiplier and streak camera, exhibit real-time detection; pump-probe methods are based on a sampling of the fluorescence emission at various delays relative to the excitation pulse, which is then built up to give a profile of the fluorescence decay. All of these techniques are outlined in detail by O'Connor and Phillips[6] and Holden[7] and have found particular applications, but by far the most general and widely used method for time-resolved fluorescence studies is time-correlated single-photon counting (SPC), which is described in more detail below.

Time-correlated single-photon counting (SPC) has its basis in the fact that the probability distribution (in time) for emission of a single photon must follow the statistical distribution of all the emitted photons. The use of accurate statistical counting of single-photon emission events occurring on a very large number of excitation pulses allows a histogram to be built up that reflects exactly the time profile of fluorescence emission from the sample.

Apparatus : The typical SPC apparatus generally consists of a high repetition rate, low-intensity excitation source in combination with rapid response, high-sensitivity detectors and accurate timing electronics. Such a setup is illustrated in Figure 2.

Suitable flash lamps or lasers may be used as excitation sources for SPC experiments. Lamps can be operated at a rate approaching 100 kHz with their intensity and spectral output being dependent on the gas (H_2, D_2, or N_2) used to fill the space between the electrodes. For such lamps pulse durations are typically around 2 ns. The desired high repetition rate of the source necessitated by the statistical nature of the experiment limits the type of laser that may be used. The mode-locked argon ion laser on its own or when used to pump a dye-laser, is suitable when used in conjunction with pulse selection using a Pockel cell or cavity dumper to limit the repetition rate to a useable frequency of <1 MHz.

Lasers have the advantage of monochromaticity, higher intensity and shorter duration (in the picosecond domain) pulses with respect to discharge lamp sources.

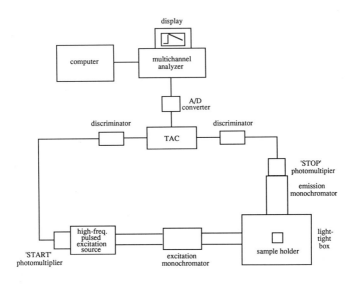

Figure 2. Schematic representation of a simplified single-photon counting apparatus.

The sample is contained within a light-tight box to limit detection of background radiation, with detection normally being carried out perpendicular to the excitation of the sample. For broad-band lamp excitation, wavelength selection for both excitation and emission can be achieved by interspersing monochromators in the respective optical paths or less precisely by using cutoff or interference filters. For monochromatic laser sources obviously the excitation monochromator is unnecessary.

The detector used should combine rapid response with very high sensitivity for use in single-photon detection. Photomultipliers have been extensively used in this context although more recently avalanche photodiodes[8] and microchannel plates[9] have been adopted particularly for picosecond studies where their faster response is an obvious advantage.

A description of the modulated electronics used in SPC is given here, outlining the precise timing of the delay between the trigger of the excitation source and the detection of the emitted photon. The triggering of the excitation source is detected optically using a photomultiplier pickup that generates an electrical signal correlated with the excitation pulse. This signal travels to a discriminator (see below) and then to the vital time-to-amplitude converter (TAC), thus initiating the charging of a capacitor resulting in a voltage ramp which is linear with time. On detection of the emitted photon the ramp is halted and the voltage output of the TAC is given a value by the analog to digital converter and a count is stored in the appropriate channel of the multichannel analyzer (MCA) corresponding to the delay of the emitted photon with respect to the excitation pulse. With the high frequency of detected events a statistical distribution of the fluorescence emission is quickly accumulated that corresponds to the time dependence of the intensity of fluorescence from which the lifetime information can be obtained.

For accurate analysis of the data it must be ensured that the apparatus is operating under strictly single-photon conditions. If the emission intensity is too high then there may actually be two or more photons impinging on the detector per excitation pulse. As the detector only times the arrival of the first photon a distortion of the distribution, termed "pulse pileup" will occur. The occurrence of multiphoton events can be reduced by the use of discriminators in the excitation and/or detection electronics which results in rejection of multiphoton detected events. A practical rule of thumb that may be applied to avoid multiphoton effects is to arrange the conditions such that the rate of counting of fluorescence photons is kept to a level <2% of the rate at which excitation pulses are generated.

Convolution of Signal: As the excitation pulse is not infinitely short but has a finite duration the measured decay curve does not represent the true decay but rather the true decay convoluted with the time profile of the excitation pulse; i.e., molecules are still being excited by the "tail" of the excitation pulse while those excited at the beginning of the pulse have decayed to a significant extent. In order to obtain the true fluorescence decay profile the measured signal must be deconvoluted against the instrument response function (IRF). The IRF is generated by detection of scattered excitation photons using a colloidal scattering medium such as silica, milk, or the commercial product Ludox in water. IRFs of photomultipliers generally show a slight wavelength dependence and thus the IRF should be measured at the sample emission wavelength. A detailed description of the deconvolution procedure and methods is inappropriate here and the reader is directed to available literature on this topic.[6] Special precautions need to be taken in the case of weak, long-lived signals.[10]

Fitting of Fluorescence Decay Profiles: SPC traces are normally recorded using semilogarithmic graphics whereby deviations from exponential behavior are quickly apparent. Fluorescence decay functions can be generally described by an exponential or sum of exponentials described by the equation

$$G(t) = \sum_{i=1}^{n} a_i \exp(-t/\tau_i) \tag{3}$$

Numerous methods may be used for the calculation of the lifetimes (τ) and pre-exponential factors (a_i). The most versatile and commonly used analysis is that of least squares fitting using an iterative reconvolution process. By this method the measured instrument response function, described by $R(t)$ is fixed and convoluted along with a variable $G(t)$ function to give a simulated decay function $S(t)$.

$$S(t) = \int_{0}^{t} R(t) G(t - t') \, dt' \tag{4}$$

The simulated decay, $S(t)$, is then compared to the measured decay, $I(t)$, and the chi-squared (χ^2) value is calculated using

$$\chi^2 = \sum_{i=n_1}^{n_2} \frac{[I(t_i) - S(t_i)]^2}{I(t_i)} \tag{5}$$

where n_1 and n_2 are the limits of the analysis range. In the iterative reconvolution process the parameters in $G(t)$ are varied and optimized to achieve a minimum value in χ^2, the point by point deviations between experimental and simulated decay curves. Normally n_1 will be the channel with maximum count and n_2 a channel where the count reaches a background level. In order for the analysis to be statistically sound the fluorescence data should span an intensity range of at least four orders of magnitude, especially where multiexponential decays are being analyzed. By far the most effective algorithm used to reach the χ^2 minimum is the Marquardt theorem[11] by which all the fitting parameters are simultaneously varied. The algorithm adjusts the increments in the fitting parameters to reflect the steepness of the χ^2 surface around the initial guesses entered by the operator. As the minimum is approached the incremental change is reduced and the χ^2 minimum is rapidly approached.

A more recent and complex method for fitting the decay parameters is the global analysis technique (see also Chapter 13).[12, 13] By this method several decays are analyzed simultaneously under conditions where the decay parameters (τ_i) are common to all curves. An example of this technique is in the detection of decays at different wavelengths from a sample containing multicomponent fluorescence where spectral overlap of the fluorophores exists.

Statistical Evaluation of Curve Fitting: A variety of statistical tests exist for the evaluation of the compatibility between experimental and simulated decay profiles. The statistical deviation in the number of counts $I(t_i)$ in channel i is given by

$$\sigma_i = [I(t_i)]^{1/2} \tag{6}$$

The reduced chi-squared parameter χ_R^2 is related to χ^2 by

$$\chi_R^2 = \frac{\chi^2}{(n_2 - n_1 + 1) - p} \tag{7}$$

where $(n_2 - n_1 + 1)$ is the number of data points analyzed and p is the number of variable parameters in the fitting function. For Poisson statistics χ_R^2 is ≥ 1 and values of χ_R^2 between 0.9 and 1.3 are acceptable. The χ_R^2 parameter, however, contains no information on the error distribution within the analyzed range of data points. It is of importance to establish that the errors are distributed randomly around the fitted curve and that no systematic deviations exist. A plot of the weighted residuals $r(t_i)$ may be used to visualize the error distribution where

$$r(t_i) = \frac{I(t_i) - S(t_i)}{[I(t_i)]^{1/2}} \tag{8}$$

In this context the autocorrelation function, $C(j)$, is useful. It describes the correlation between the residuals $r(t_i)$ and $r(t_{i+j})$ and is defined by

$$C(j) = \frac{\frac{1}{m} \sum_{i=n_1}^{n_1+m-1} r(t_i)\, r(t_{i+j})}{\frac{1}{n_3} \sum_{i=n_1}^{n_2} [r(t_i)]^2} \tag{9}$$

where $n_3 = n_2 - n_1 + 1$, $j = n_3 / 2$, and $m = n_3 - j$. Autocorrelation plots should show an initial value of 1.0 and random dispersion of the residuals around the $C(j) = 0$ line. A poor fit is demonstrated by sigmoidal distribution around $C(j) = 0$ or a plot that fails to cross this line. The Durbin-Watson parameter (DW) also extracts information from the weighted residuals and is used to test for serial correlation in the data.[14, 15]

$$DW = \frac{\sum_{i=n_1+1}^{n_2} [r(t_i) - r(t_{i-1})]^2}{\sum_{i=n_1}^{n_2} [r(t_i)]^2} \qquad (10)$$

DW values should be greater than 1.7, 1.75, and 1.8 for mono-, bi-, and triexponential decays, respectively. Another test commonly used to assess the randomness of the residual distribution around the fit is the runs test that compares the signs of the residuals of adjacent points to that expected from a true random distribution.

It should be stressed that reliance on statistics alone in determining fluorescence lifetime parameters should be avoided. For a given experimental decay it can be demonstrated that many different parameter values may give satisfactory statistical analyses. More certainty the correctness of the analysis requires a larger accumulation of counts in the SPC experiment. Up to $10^5 - 10^6$ counts in the maximum channel should result in a more accurate determination of the decay parameters. In addition, errors in evaluation of fluorescence lifetimes may result from an inaccurately calibrated SPC system and it is wise to check the performance of the SPC apparatus and analytical programs by using some of the many fluorescence lifetime standards reported in the literature.[16]

2.4. Time-Resolved Fluorescence Anisotropy

The single-photon counting technique may also be adapted for use in measurements of time-dependent fluorescence depolarization or anisotropy. From these experiments information is obtained on the rotational relaxation rates in excited molecules.[6]

Excitation of an assembly of fluorescent molecules results in the induction of anisotropy within the system because molecules that have their absorption transition moment oriented in the same direction as the polarization vector of the exciting light will have a higher probability of excitation. If the exciting light is plane-polarized the fluorescence emission transition moment should have the same directional properties as the exciting light as both transitions generally involve the same states.

In fluid media, where the molecules are not fixed, the anisotropy of the system is time dependent, more specifically, dependent on the respective lifetimes of rotational relaxation (τ_r) and fluorescence (τ_f). If τ_f is longer than τ_r then all the initially induced anisotropy (τ^0_r) will quickly be lost due to molecular rotation. If τ_r is much longer than τ_f then τ^0_r will be retained for the duration of the fluorescence decay. For time-dependent anisotropy studies the best situation is when τ_r and τ_f are comparable in lifetime such that significant changes in anisotropy will occur during the fluorescence decay.

The fluorescence anisotropy as a function of time is defined by

$$r(t) = \frac{I_{\parallel}(t) - I_{\perp}(t)}{I_{\parallel}(t) + 2I_{\perp}(t)} \tag{11}$$

where $I_{\parallel}$ and $I_{\perp}$ are the fluorescence intensities detected parallel and perpendicular to the plane of polarization of the exciting light.

For a single exponential fluorescence decay it has been demonstrated that

$$I_{\parallel}(t) = \exp(-\frac{t}{\tau_f}) \; (1-2r_0 \exp(-\frac{t}{\tau_r})) \tag{12}$$

and

$$I_{\perp}(t) = \exp(-\frac{t}{\tau_f}) \; (1 - r_0 \exp(-\frac{t}{\tau_r})) \tag{13}$$

where r_0 is the initial anisotropy.

The time dependence of the total fluorescence emission, $F(t)$, is given by

$$F(t) = I_{\parallel}(t) + 2I_{\perp}(t) \tag{14}$$

and by

$$F(t) = F_0 \exp(-\frac{t}{\tau_f}) \tag{15}$$

where F_0 is the initial fluorescence intensity.

Similarly, the time dependence of the fluorescence anisotropy, $r(t)$, is given by

$$r(t) = r_0 \exp(-\frac{t}{\tau_r})$$ (16)

As $F(t)$ and $r(t)$ depend only on τ_f and τ_r, respectively, the lifetimes may be separated.

It should be stressed that apparent polarization of the emitted fluorescence may not be due solely to molecular effects but to a certain extent may arise from the apparatus itself, particularly the optical elements. This instrumental anisotropy, usually termed the G factor, must be corrected for in the evaluation of the time-dependent anisotropy of the sample under study. The G factor is defined as the ratio of transmission of vertically polarized light to horizontally polarized light through the apparatus and the corrected anisotropy is then given by

$$r = \frac{(I_{\parallel}/I_{\perp}) - G}{(I_{\parallel}/I_{\perp}) + 2G}$$ (17)

Experimental: The intensity of emitted fluorescence parallel ($I_{\parallel}$) and perpendicular ($I_{\perp}$) to the plane of polarization of the incident light can be measured by inserting polarizers in the excitation and emission optical paths. The time dependence of the anisotropy, r(t), may be studied by measuring the fluorescence decays of the parallel and perpendicular components of the emission. This apparently simple measurement requires much more care than would be expected. As the measurement of $r(t)$ involves the measurement of a difference between $I_{\parallel}$ and $I_{\perp}$, the high accuracy of measurement required necessitates the accumulation of very large numbers of counts for the individual component decays. As the time needed for such an accumulation will be relatively long, instrumental factors such as excitation source instability, detection drift, and even probe instability must be taken into consideration. The instrumental deviations may be surmounted by continuously switching between detection of $I_{\parallel}$ and $I_{\perp}$ and the recording of the IRF during the accumulation of the data.

Correction for G Factor: the measurement of the G factor is of paramount importance in correcting for instrumentally induced anisotropy which does not arise from the sample under study. Several alternatives have been suggested for the elimination of the G factor problem[6] and are briefly outlined below.

   1. Tail Matching. When $\tau_f > \tau_r$ it is obvious that at longer time scales, with respect to τ_r, the time dependence of $I_{\parallel}$ and $I_{\perp}$ should be identical. If this is not observed experimentally then the mismatch is considered to be due to the nonunity value of G. The curves can thus be normalized in the tail region to correct for the G factor.

2. Leading edge matching. Where the transition dipoles of absorption and emission have identical orientation then the angle θ between the dipoles is zero with the result that from theory $r_0 = 0.4$. In experimental systems where $r_0 \neq 0.4$ then the discrepancy is assumed to be due to the G factor, and again normalization may be carried out. This is potentially risky as it is not usually certain that for any given system r_0 will equal 0.4.

3. For a rapidly rotating probe molecule where $\tau_r \ll \tau_f$ both $I_{\parallel}(t)$ and $I_{\perp}(t)$ should be equal. This approach allows direct measurement of the G factor.

4. As the total fluorescence decay of a fluorophore can be described by

$$F(t) = I_{\parallel}(t) + 2I_{\perp}(t) \qquad (18)$$

it is possible to take the independently measured $I_{\parallel}(t)$ and $I_{\perp}(t)$ functions and combine them with a suitable weighting factor ($=G$ factor) to exactly reproduce the fluorescence decay profile measured under nonpolarized conditions.

Deconvolution in Fluorescence Anisotropy: The deconvolution process, as outlined previously, is especially critical in anisotropy experiments where a large proportion of the information may be contained in the data points in the first small segment of the measured profiles. Various methods have been adopted for deconvolution combined with suitable fitting analyses and have been outlined by O'Connor and Phillips.[6] The difficulties encountered in obtaining satisfactory analyses of the experimental data underline the problematical nature of the experimental measurement of time-dependent anisotropy.

2.5. Detection of ir Luminescence from Excited Singlet Molecular Oxygen

Photosensitization of the lowest excited singlet state of molecular oxygen, O_2 $(^1\Delta_g)$, occurs, with a varying degree of efficiency, through triplet energy transfer from a variety of molecules provided the triplet energy of the sensitizer lies above the energy of O_2 $(^1\Delta_g)$ of 22.4 kcal mol^{-1}. The spin-forbidden relaxation of O_2 $(^1\Delta_g)$ to its triplet ground state is relatively slow and results in a lifetime of the excited state from a few μs to several hundred ms, dependent on solvent. A recent review of physical and chemical parameters is given by Gorman and Rodgers.[17] Deactivation of the excited singlet occurs almost exclusively by nonradiative means; however, a very small fraction ($<.01$) can decay by emission of a photon. This luminescence is centered at 1.27 μm and it is only with the relatively recent advent of detectors sensitive in this spectral region that detection of the weak emission from O_2 $(^1\Delta_g)$ has been possible.[18-25]

Prior to the development of luminescence detection, a host of chemical trapping methods were (and still are) used to indicate the involvement of the highly reactive O_2 ($^1\Delta_g$) in chemical reactions. Such indirect techniques are essentially dependent on the specificity of the trapping species for O_2 ($^1\Delta_g$) (in not all cases has this been conclusively demonstrated) and quantitative measurements are additionally complicated by fairly complex kinetic considerations. Detection of the ir luminescence, however, is a direct fingerprint for involvement of O_2 ($^1\Delta_g$) and can be used to determine both the efficiency of photosensitized generation of O_2 ($^1\Delta_g$) and the rate constants for reaction of O_2 ($^1\Delta_g$) with any given substrate molecule. Both steady-state and time-resolved techniques have been developed but for kinetic measurements the latter is required.

The apparatus for the detection of O_2 ($^1\Delta_g$) luminescence is relatively straightforward and many such systems exist throughout the world. Figure 3 shows the particular setup that we employ at present in our laboratory, this being fairly representative. A pulsed laser is used to excite an aerated sample of the sensitizer and detection is carried out perpendicular to the direction of laser propagation using an ir-sensitive germanium photodiode. A silicon filter is placed between the sample and the detector in order to remove interfering light of wavelengths < 800 nm. The current generated in the diode is passed through a load resistor and the resultant voltage is then suitably amplified and recorded using a digitizer. In order to improve the S/N ratio of these inherently noisy signals averaging of a suitable number of laser shots is routinely carried out. Such set ups have also been used to record phosphorescence in the infra red region from polyenes.[26]

Quantum yields of O_2 ($^1\Delta_g$) formation (Φ_Δ) are evaluated with reference to a standard system for which Φ_Δ has been accurately determined, such as phenazine.[27] Typically, the slope of the dependence of the amplitude of the initial signal on the emission decay (I_0) on laser pulse energy is taken as a measure of Φ_Δ and the actual Φ_Δ of the sample is obtained from the ratio of slopes obtained for the standard and sample under conditions of matched absorbance at the laser wavelength and in the same solvent. In the absence of quenching species the decay of O_2 ($^1\Delta_g$) is monoexponential with a lifetime that is dependent on the solvent. For kinetic studies two types of experimental approaches are possible;

1. O_2 ($^1\Delta_g$) is generated by the triplet state of a molecule, the ground state of which then quenches O_2 ($^1\Delta_g$) by a physical or chemical process.
2. O_2 ($^1\Delta_g$) is generated from a sensitizer whose ground state is inert toward O_2 ($^1\Delta_g$) which is then quenched by the ground state of an added substrate.

In both cases the slope of the plot of observed rate constant of O_2 ($^1\Delta_g$) decay against the quencher concentration yields the rate constant for quenching of O_2 ($^1\Delta_g$) by the substrate (k_q). It was demonstrated early in O_2 ($^1\Delta_g$) studies that deuteration of solvent results in a lengthening of the O_2 ($^1\Delta_g$) lifetime, by approximately an order of magnitude.[28] Thus, use of a deuterated solvent extends the range over which the quenching behavior can be observed and is useful for the study of substrates that have relatively low k_q values.

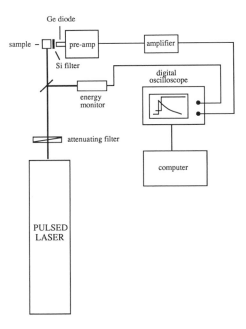

Figure 3. Schematic representation of apparatus used in time-resolved singlet oxygen luminescence detection experiments.

2.6. Photoacoustic and Photothermal Techniques

Irradiation of a sample by a pulse of light results in a rapid local deposition of some of the absorbed energy in the form of heat due to relaxation of excited states by nonradiative pathways, e.g., internal conversion or intersystem crossing. This release of heat within a small volume of solution gives rise to a number of different phenomena, which fall into two general categories with associated detection techniques.[29-31]

1. *Photoacoustic Effects.* Rapid local release of heat in a sample results in the generation of a pressure or acoustic wave which travels outward through the solution and can be detected by a suitable pressure-sensitive detector coupled to the sample.

2. Photothermal Effects. The local deposition of heat on excitation with a Gaussian source results in a temperature and therefore, density and refractive index gradient centered along the axis of the laser propagation. The solution in this region behaves as a divergent lens and any light that passes coaxially along this volume of the sample becomes defocused leading to a decrease in its irradiance at the far field. This is the so-called thermal blooming or thermal lensing effect. Alternatively a beam of light that overlaps the excitation beam at an angle can be refracted resulting in a spatial translation of the "probe" beam profile. This effect is the basis for the photothermal beam refraction (deflection) and "mirage effect" methods.[30]

Both types of detection essentially probe the amount of energy given up in the form of heat following excitation of an absorbing sample and as such are useful tools for the investigation of efficiencies of nonradiative relaxation processes. Representative apparatus for both photoacoustic and photothermal detection systems are outlined below.

Photoacoustic: Photoacoustic studies can be carried out using either a modulated continuous wave (cw) or pulsed laser source for irradiation of samples. The alternative excitation sources require their own types of apparatus which in turn yield different types of information.

1. The modulated cw photoacoustic apparatus utilizes a modulator (e.g., mechanical chopper) to obtain the desired frequency in combination with a monochromator to achieve narrow bandwidth excitation of a sample (liquid or solid) to which an acoustic detector is coupled. Of common use in this experimental approach is a microphone that senses the pressure changes occurring in the gas phase above the sample caused by the periodic heating and cooling due to the modulated excitation. The signal from the detector passes to a lock-in amplifier along with a reference signal from the modulator and the resulting signal is passed to the computer for data analysis.

The "action" spectrum of the sample toward generation of the photoacoustic response is analogous to a fluorescence excitation spectrum and is obtained by plotting the signal as a function of the excitation wavelength (varied using the monochromator) and is an important tool for studying heterogeneous samples containing more than one chromophore. For such a spectrum to be obtained a correction for the spectral output of the lamp should be carried out; usually this is done by generating the photoacoustic (PA) spectrum of a totally absorbing "black" sample, such as carbon black ink, which is proportional to the output intensity at each wavelength. An advantage of the PA technique is that the sample to be studied need not necessarily be transparent. Indeed the probing of a heterogeneous sample in terms of chromophore distribution may be carried out by the process of *depth profiling*;[30] as the fraction of the sample under the surface which will actually be probed is dependent on the chopping frequency, the variation in signal (and action spectrum) with chopping frequency will reflect the chromophoric distribution under the surface. This has great importance, for example, in layered biological samples.

2. Laser-induced photoacoustic signals give information on the energy given up by non-radiative processes within a detection time window, determined by the detector response frequency and a parameter known as the acoustic transit time (τ_a), which is the time taken for the acoustic wave generated to traverse the laser beam radius.[31] Whichever of these has the longest time constant will determine the overall detection window. Typical detectors used are of piezoelectric ceramic material or a foil such as PVF2. The detector is usually coupled to the sample directly or through a solid interface (microscope slide, quartz cuvette). The incident acoustic wave will be transmitted to the transducer surface and a voltage proportional to the amplitude of the acoustic wave and thus, the amount of heat liberated within the detection time window, will be developed. The technique affords the possibility of sensitive calorimetric study of the nonradiative processes occurring on relaxation of an excited sample. For such experiments a calorimetric reference sample is required, typically one that will releases all its absorbed energy nonradiatively within the detection time window.[31] For liquid samples the signal amplitudes of sample and reference under optically matched conditions *and in the same solvent* reveals the fraction of the energy absorbed by the sample that is released as heat within the detection window. From this measurement, and providing the overall relaxation scheme is known a variety of thermodynamic and calorimetric data can be elucidated. Figure 4 depicts representative apparatus for both cw and pulsed photoacoustic experiments.

Photothermal: The technique of dual-beam time-resolved thermal lensing (TRTL) is described here as this is one of the most commonly applied examples of a photothermal method.[31, 32] The power of the lens, i.e., the extent to which the probe beam irradiance at the detector is reduced, is proportional to the amount of energy given up as heat. The temporal behavior of the thermal lensing signal is determined by the rate constants for the possible nonradiative transitions contributing to the overall heat evolution. Figure 5 shows a typical TRTL apparatus consisting of a pulsed laser excitation source and a cw probe beam, aligned to overlap coaxially within the liquid sample. The probe beam should be of a wavelength that is transparent to the sample in both its ground and the excited states. The decrease in the irradiance at the center of the probe beam, following pulsed irradiation, is measured using a photodiode in the far field, and the time-dependent behavior of the thermal lens is reflected in the temporal signal from the photodiode. The signal may be analyzed in a simple fashion to give information on quantum yields of relaxation processes energy contents of excited states and lifetimes of transient species.[27a, 33-36]

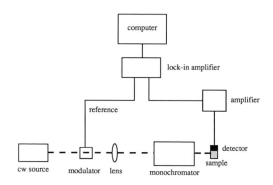

Modulated CW Excitation with Photoacoustic Detection

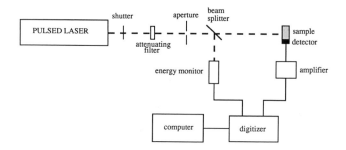

Pulsed Excitation with Photoacoustic Detection

Figure 4. Schematic representation of apparatus used in photoacoustic detection experiments using modulated CW (upper) and pulsed laser (lower) excitation sources.

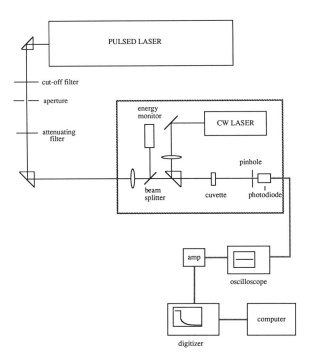

Figure 5. Schematic representation of apparatus used in typical dual-beam, time-resolved thermal lensing experiments.

The calorimetric information which can be obtained using PAS can, in principle, also be measured using TRTL. The latter has the advantage of kinetic information on microsecond transient lifetimes, whereas kinetic studies using PAS can only be achieved by complex deconvolution of signals (some novel approaches yielding enhanced time resolution are now being employed),[37-39] a process that is practically limited to transient species of lifetime < 1 μs. TRTL, as described here, is obviously not suitable for the study of solid samples, which may, however, be studied using alternative photothermal techniques such as the mirage effect.[30] Both techniques can be used to differentiate between heat release which occurs on a time scale that is "fast" or "stored" with respect to the response of the detection Analysis of the data is based on the application of a simple energy balance diagram:

$$E_{abs} = \alpha E_{abs} + E_{st} + E_f \tag{19}$$

where E_{abs} is the absorbed energy, αE_{abs} is that fraction released as heat within the response time of the technique, E_{st} is the stored heat, and E_f is the fraction lost through radiative relaxation. TRTL has an advantage, in this respect, as PAS does not "see" E_{st}. For samples where a slowly decaying transient exists ($t > 1$ μs) the TRTL signal can be directly analyzed to obtain its lifetime and from the ratio of the amplitudes due to the "slow" and overall heat release the product of the quantum yield and the energy content of the long-lived transient may be evaluated.[27a, 35, 36] The latter measurement is absolute in nature, not requiring the use of a reference, such as in PAS, and has been used to produce primary standards for use in comparative photophysical techniques.[27b]

3. Studies of the Microenvironment

The probing of various properties of microenvironments will be outlined with a brief description of measurements concerning the probing of polarity, microviscosity and induction of optical activity. These properties are most commonly studied by employing fluorescence measurements (steady-state, single-photon counting and depolarization). A brief discussion may also be found in Chapter 2.

The following criteria are desirable when choosing a molecule to probe for a specific property:

1. The sensitivity of the probe to the property being measured (e.g., polarity, microviscosity) should be well characterized in homogeneous solution in order to facilitate the interpretation of results in constrained media. Ideally, the change measured should only depend on the property under study and not be influenced by additional effects, e.g., in the case of probes of polarity no specific probe–solvent interactions should occur.

2. The feature of the probe that changes should be easily measurable. In the case of a fluorescent probe its lifetime, emission quantum yield or the position of the emission maxima should change in media with different properties.

3. When photophysical measurements are performed the probe should be chemically stable.

3.1. Polarity of the Environment

Pyrene has been successfully employed as a probe of polarity and microviscosity in a variety of constrained media. The vibronic fine structure of pyrene monomer emission shows a strong solvent dependence.[40,41] In polar media the 0–0 band is enhanced by a mechanism involving vibronic coupling similar to the Ham effect in the absorption spectra of benzene.[42] This solvent dependence of the pyrene emission is expressed as a marked change of the ratio between the first

(0–0) and third emission bands (I/III ratio). This phenomenon is related to symmetry, and asymmetric substitution of the rings decreases the utility of this method.

Figure 6 shows pyrene emission spectra in different solvents.[41] The I/III ratio varies from 0.5 in nonpolar solvents up to 1.87 in water. Good agreement is obtained for values determined by different research groups and differences in reported values are normally less than 10%.[40] When measuring the I/III ratio highly resolved emission spectra are required. It is advisable to use narrow emission slits, if possible with corresponding bandwidths smaller than 0.5 nm.[43] The I/III ratio can be significantly altered in the case of emission spectra obtained with poor resolution. A comparison of this ratio to known values from the literature is recommended. Alternatively, a self-calibrated I/III scale can be used for pyrene in different solvents.

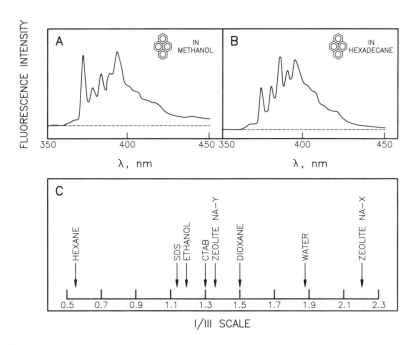

Figure 6. Fluorescence spectra of pyrene in (A) methanol and (B) hexadecane (adapted from ref. 40) and (C) I/III scale for pyrene in homogeneous solvents,[40] micelles,[41] and zeolites.[45]

The sensitivity of the vibronic emission spectrum of pyrene was initially employed to sense the polarity of micelles.[41] The I/III ratio for pyrene in sodium dodecyl sulfate (SDS) is 1.11, a value close to that in alcohols, showing that pyrene experiences an environment that is less polar than water, the bulk solvent, but more polar than typical hydrocarbons. The I/III pyrene scale has been applied to test for polarity in a variety of constrained media such as vesicles (liposomes), cyclodextrins,[44] and zeolites.[45] In egg phosphatidylcholine and dipalmitoylphosphatidylcholine liposomes below the transition temperature, pyrene senses an environment similar to butanol and propanol, respectively. Above the transition temperature a less polar environment is sensed.[46] In this same work the enhancement of the weakly allowed electronic transition in the emissions of 1,12-benzoperylene and naphthalene were used to sense the environment in the liposomes, which are slightly different from that observed with pyrene. This example illustrates the need to keep in mind that the polarity measured can be very dependent on the probe location and, thus, the validity of extrapolations to the whole microenvironment should be carefully considered.

The application of the vibronic spectrum of pyrene to test for polarity is not restricted to fluid media. For example, the photophysical properties of pyrene in X and Y type zeolites were tested suggesting a polar media, zeolite X being more polar than Y.[45] (See also Chapter 10).

The dependence of emission quantum yields and/or emission maxima of fluorescent molecules can also be used in probing for polarity. Pyrene-3-carboxyaldehyde and 1-anilino-8-naphthalenesulphonate (ANS) are examples of this class of compounds. The pyrene derivative has been used to test polarity in micelles by studying, for example, the effect of additives[47] or high pressure[48] on the environment sensed by the probe. Although the change in emission quantum yields were correlated to polarity, a study[49] which shows that these changes could also be related to structural parameters of the solvent shows how careful one should be in using probes that have not been fully characterized in homogeneous solvents.

The quantum yield and fluorescence maxima of ANS are very dependent on the solvent, the yield being very low in polar (e.g., 0.004 in water) but increasing in apolar solvents (e.g., 0.5 in hexane). This probe was used in testing the environment of AOT reversed micelles when the volume of the water core was varied.[50] Increasing the water content results in a decrease in the fluorescence quantum yield of ANS and a red shift of the emission maximum indicative of an increase in polarity, although even for a core size of 73 Å the polarity is still lower than in homogeneous water. The fluorescence lifetime is also dependent on polarity showing that the same property can be probed using a different technique. In small clusters of AOT the dipolar rotation of water determines the fluorescence properties of ANS and not the polarity, indicating again that care has to be taken in the interpretation of the results.[50]

Several other probes can be used to test the polarity or accessibility of specific molecules such as water to microenvironments. For example phenolbetaine ET(30) is a well-known probe of solvent polarity due to the large differences in dipole moment between the ground and excited states.[51] This molecule has been used in micelles, microemulsions and phospholipid bilayers. ET(30) is a charged molecule that is always solubilized at the aqueous interface and has been used successfully to test for the effect of salt addition, surfactant chain length and concentration, counterion and temperature.[52] Various inorganic materials can also be used as probes. Lanthanides, like Eu^{2+} are useful probes to sense the amount of water present in the microenvironment. Eu^{2+} luminescence in cryptate complexes increases drastically when compared to the emission detected in water, indicating that the ion is protected from water upon complexation.[53] An interesting combined fluorescence emission and thermal lensing study of Tb^{3+} in AOT reverse micelles[54] demonstrated that the signals obtained from both techniques were dramatically enhanced when Tb^{3+} was in the aqueous core of the reverse micelle compared to neat water. In thermal lensing, signal amplitudes are solvent dependent and water is the worst solvent with regard to its thermal–physical properties and hence the signal is weak. In reverse micelles, however, the solvent is predominantly hydrocarbon, with more favorable properties, and the thermal lensing signal is thus enhanced. Fluorescence is enhanced in the reverse micelle due to its protection from free water in the relatively small AOT reverse micelles employed in the study.

3.2. Local Viscosity or "Microviscosity"

The microviscosity of constrained media can be determined by different methods. As an example, we will describe the fluorescence depolarization technique (see section on time-resolved spectroscopy above) and methods where the intra-molecular formation of excimers yields information about microviscosity.[55-57]

Under certain conditions fluorescence depolarization can be related to the viscosity sensed by the probe. A detailed description of the technique was given by Weber et al.[58] The molecular anisotropy (r) and degree of polarization (p) are given by equations 17 (see above) and 20.

$$p = \frac{(I_{||}/I_{\perp}) - G}{(I_{||}/I_{\perp}) + G} \tag{20}$$

where $I_{||}$ and $I_{\perp}$ are the fluorescence intensities observed with the polarizers oriented parallel and perpendicular to the plane of polarization and G is a correction factor that takes into account different detection sensitivities for light with different polarizations (see also section 2.4). The limiting values for r and p (r_0 and p_0) are required and can be obtained from measurements in very viscous media. For spherical fluorescent probes the depolarization degree (r_0/r) can be directly related to the viscosity by

$$\frac{r_o}{r} = 1 + \frac{kT\tau}{\mu\upsilon} \tag{21}$$

where k is the Boltzman constant, T is the absolute temperature, μ is the viscosity, υ is the effective volume of the fluorescent sphere, and τ is the average lifetime of the excited state. The probes usually employed for depolarization measurements are not spherical but planar and in- and out-of-plane rotations contribute to a varying degree to the depolarization. The above equation can be used when the molecule is excited at a wavelength for which p_0 is close to 0.5.[58] The depolarization in a constrained medium has to be compared to a calibration curve of the depolarization in homogeneous solution at different viscosities. This comparison assumes that the effective volume is the same, which means that no specific interaction between the probe and the constrained medium should occur as this would tend to increase the rotation time of the probe.

Probes for fluorescence depolarization should have the following characteristics:[58]

1. Rigid structure to avoid depolarization due to rotation of its side structures.

2. High and, if possible, constant p_0 values over the range of excitation wavelength employed in order to avoid small shifts in the absorption spectra which would affect the measurement.

3. The lifetime of the excited state should be short enough to avoid the possible contribution of the rotation of the constrained medium to the depolarization. For studies in which viscosities between 1 and 100 cP are being tested, e.g., micelles, the fluorescence lifetime should be between 1 and 8 ns.

4. High extinction coefficients and emission quantum yields to increase signal/noise ratios.

5. A minimum overlap of emission and absorption to eliminate depolarization due to reabsorption.

2-Methylanthracene and perylene are suitable probes for these experiments. Using this technique a microviscosity of 17 to 50 cP was obtained for several micelles.[58]

The rate constant for excimer formation is viscosity dependent and has been exploited to probe the microviscosity of microheterogeneous media. An early study[59] investigated formation of pyrene excimers distributed in micelles and estimated an internal viscosity of 150 cP, a value much higher than that determined using fluorescence depolarization. Later, several authors[55-57] studying the intramolecular formation of excimer in bridged compounds measured microviscosity values of tens of cP, values that confirmed those obtained by fluorescence depolarization. The higher value obtained from the formation of pyrene excimers is probably an artifact due to the distribution of these molecules in the micellar population. This example shows again that when extrapolating from homogeneous solution to constrained media one has to keep in mind that different factors, in this case probe distribution, can play a key role in the phenomenon being studied.

The microviscosities measured by any of the two methods above, depolarization and excimer formation, are a function of the solubilization site of the probe. These differences can be significant; for example, diphenylpropane and dipyrenylpropane, both of which form intramolecular excimers, sensed microenvironments in SDS with microviscosities of 4 and 19 cP, respectively.[55]

trans-Stilbene has also been used to determine microviscosities. The fluorescence quantum yield of *trans*-stilbene increases as the viscosity decreases. The competitive deactivation pathway, rotation leading to the perpendicular excited state that deactivates nonradiatively, has a small activation barrier that is viscosity dependent. Several stilbene probes with the stilbene located at different positions in the hydrophobic chain have been used in micelles or vesicles. In the former different probes sense similar, moderately viscous, media, whereas in vesicles a more complex behavior is observed, indicating a more organized structure in the latter case.[60]

3.3. Induced Optical Activity

Different constrained media require probes for different properties. For example, for cyclodextrin inclusion complexes chirality is a suitable property as these complexes have very defined interactions. The induced chirality of pyrene or 1,3-dinaphthylpropane dimers in γ-cyclodextrins was detected in their circular dichroism and circularly polarized fluorescence spectra. In the case of the pyrene dimers the chirality is left-handed, whereas a right-handed chirality is observed for the intramolecular 1,3-dinaphthylpropane dimer. The "handedness" of the chirality of the pyrene dimer changes when included in 6-*O*-α-maltosyl-γ-cyclodextrin, this change being explained on the basis of a different location of the dimer.[61,62] These experiments show that techniques measuring optical activities can yield information about very defined complexes.

These examples show that polarity, microviscosity, and induced optical activity are some properties of constrained media that can be investigated using photophysical probes. Any of the experimental techniques described in the first part of this chapter can in principle be used. For example, the absorption maximum of triplet xanthone is dependent on the solvent polarity[63]; laser flash photolysis was used to probe the acidity in Nafion membranes[64] and diffuse reflectance was employed to yield information about the hydrophobic zeolite, silicalite.[65] Clearly different techniques and suitable probes will have to be developed to deal with different constrained media.

4. Determination of Aggregation Properties
4.1. Equilibrium Constants in Systems with Defined Stoichiometry

For a complex with known stoichiometry, where the complex and the free molecule have different spectral properties, the equilibrium constant can be obtained from a plot of the spectral changes (ΔI) vs. the concentration of the complexing molecule ([B]) (i.e., cyclodextrin). For a 1:1 complex ΔI is expressed by the linear Benesi-Hildebrand plot:[66]

$$A \; + \; B \; \underset{\longleftarrow}{\overset{K_{eq}}{\rightleftharpoons}} \; (AB) \tag{22}$$

$$\frac{1}{\Delta I} = \frac{1}{K_{eq} \Delta C \, [A]_T} \frac{1}{[B]_T} + \frac{1}{\Delta C \, [A]_T} \tag{23}$$

where ΔC is the difference in the factor relating ΔI to concentrations (i.e., absorption coefficients, emission quantum yields, or molar ellipticities) for A and B and for (AB). This procedure has been used extensively to obtain binding constants for cyclodextrin complexes. Non-linear fits can be employed for stoichiometries which differ from 1:1 complexation. A drawback in using the procedure above is that a complexation stoichiometry has to be assumed in order to obtain the mathematical expression employed in the fit to the experimental data. In some cases, such as studies of pyrene inclusion in γ-cyclodextrin (γ-CD), several complexes with different stoichiometries were observed (1:1, 1:2, and 2:2, pyrene:CD) (ref. 67 and references therein) requiring the use of additional methods to describe this system.

When different ligand molecules L are bound to equivalent and noninteracting binding sites of a macromolecule P the equilibrium constant K_e and number of binding sites n can be obtained from a Scatchard plot, which was initially developed for the binding of small ligands to proteins.[68] The following equilibrium is assumed:

$$P \; + \; nL \; \underset{\longleftarrow}{\overset{K_{eq}}{\rightleftharpoons}} \; PL_n \tag{24}$$

Ligand L must have a measurable characteristic I that changes upon complexation. From the measured values in solution I_0, when completely bound to P (I_{inf}) and at intermediate P concentrations (I_p), the bound fraction of L (X) is given by

$$X = \frac{I_p - I_o}{I_{inf} - I_o} \qquad (25)$$

The value of I_{inf} is obtained when an excess of P is added. Substituting $R = X[L]/[P]$ in the equation for the equilibrium constant shown in equation (24),

$$\frac{R}{(1 - X)[L]} = nK_{eq} - RK_{eq} \qquad (26)$$

The K_{eq} is obtained from the slope, and the number of binding sites is determined from the intercept-to-slope ratio. A similar expression can be written when the macromolecule has two or more classes of binding sites with significantly different equilibrium constants.[69]

4.2. Aggregation Parameters of Micelles

Critical Micelle Concentration: Micelles are formed spontaneously above a certain surfactant concentration called the critical micelle concentration (cmc). It is usually assumed that below the cmc only monomer surfactants are present in solution, whereas above the cmc any increase of surfactant concentration leads to more micelles with little variation of the free monomer concentration. This relationship is expressed by

$$[M] = \frac{[\text{surfactant}] - \text{cmc}}{N} \qquad (27)$$

where [M] is the micelle concentration and N is the aggregation number, i.e., the number of surfactant molecules that make up one micelle.

Techniques sensitive to physicochemical changes in the solution due to micelle formation can be used to determine the cmc; these include nonphotophysical techniques such as the measurement of surface tension.

A variety of probe molecules have been used to determine the cmc of micelles, most of which have fluorescence parameters sensitive to solvent polarity. For example, a marked decrease of the I/III emission ratio of pyrene is observed at the cmc of SDS[41]; the emission quantum yield of 2-p-toluidinylnaphthalene-6-

sulfonate increases when SDS micelles are formed[70] or a shift of the fluorescence maximum is observed for 11-[3-[hexyl-1-indiyl]]undecyltrimethylammonium bromide upon formation of cetyltrimethylammonium bromide micelles (CTAB).[71]

To determine correct values for the cmc the probes employed have to be mostly solubilized in the micellar phase (i.e., high partitioning coefficient) and special care has to be taken to ensure that no aggregation between the probe and surfactant monomers occurs below the cmc. In this case the change of environment sensed by the probe could be mistaken for micelle formation. This phenomenon was observed in the case of ANS, which gives correct cmc values for nonionic micelles but forms aggregates with positive surfactants at concentrations below the cmc.[72]

Determination of Partition Coefficients: The concentration of a given probe in water and in the micellar phase must be known for the quantitative analysis of data obtained from photophysical techniques to be possible. Incorporation of probes into micelles can be treated as the partitioning between two phases, the micelles being visualized as a pseudophase. In this model the distribution of molecules follows the Poisson distribution (for a discussion of different models see ref. 44). Partitioning is a different phenomenon from binding. In the latter process only a finite number of binding sites is available and saturation at high concentrations of ligands is expected. Although the terms *binding* and *partition constants* were both used to describe the distribution of molecules in micelles the latter should be preferred (see discussion for vesicles where both phenomena occur).

Several of the mathematical treatments that will be described throughout this text rely on some or all of the processes shown in Scheme 1, where P is a probe molecule and Q a quencher, both able to partition between the aqueous (subscript w) and micellar (subscript m) phases. The expressions for the partition constant of the quencher (K), the mean occupation number (n), and the total quencher concentrations ([Q]) are also shown.

Different methods are available to obtain the values for partition constants. Encinas and Lissi[73] described a method that is independent of the quenching mechanism, the probe lifetime, the quencher residence time, and the distribution statistics. Assuming that the fraction of the solution that is occupied by the micelles is small compared to the aqueous phase and substituting equations (35) and (36) in (37) the total quencher concentration is expressed by

$$[Q] = \frac{n}{K} + n[M] \tag{38}$$

$$P_m \ (P_w) \xrightarrow{\ h\nu\ } P_m^* \ (P_w^*) \tag{28}$$

$$P_m^* \xrightarrow{\ 1/\tau_o \ (k_o)\ } P_m \tag{29}$$

$$P_w^* \xrightarrow{\ (1/\tau_o)_w\ } P_w \tag{30}$$

$$P_m^* \ \underset{k_{p+}}{\overset{k_{p-}}{\rightleftharpoons}} \ P_w^* \ + \ M \tag{31}$$

$$P_m^* \ + \ Q_m \xrightarrow{\ k_q\ } P_m \ + \ Q_m \tag{32}$$

$$P_w^* \ + \ Q_w \xrightarrow{\ k_{qw}\ } P_w \ + \ Q_w \tag{33}$$

$$Q_w \ + \ M \ \underset{k_-}{\overset{k_+}{\rightleftharpoons}} \ Q_m \tag{34}$$

$$K \ = \ \frac{[Q_m]}{[Q_w]\,[M]} \ = \ \frac{k_+}{k_-} \tag{35}$$

$$n \ = \ \frac{[Q_m]}{[M]} \tag{36}$$

$$[Q] \ = \ [Q_m] + [Q_w] \tag{37}$$

Scheme 1

The ratio of fluorescence intensities in the absence and presence of quencher (I_0/I) of a probe, which resides only in the micelles, are plotted according to the usual Stern-Volmer equation (I_0/I vs. [Q]). Quenching plots are obtained for different micelle concentrations. I_0/I is directly related to the occupation number (n) of the quencher and the plot of [Q] vs. [M] at a constant I_0/I value (i.e. constant n) should be linear as expected from equation (38) with K being obtained from the slope-to-intercept ratio. Figure 7 shows an example for the quenching of 2,3-dimethylnaphthalene by 2,5-dimethyl-2,4-hexadiene in SDS micelles.[73] A variety of partitioning coefficients were determined using this method; the values ranging from 50 M^{-1} for acetone[74] in SDS to 1.1 x 10^4 M^{-1} for dibenzoylperoxide in SDS[73] and in CTAC micelles.[75] This method was also applied to determine the effect of the water pool size on the solubilization of neutral molecules in AOT/H2O/n-heptane reversed micelles. For some of the molecules the method had to be adapted to account for three phases, i.e., organic phase, interface, and water pool.[76]

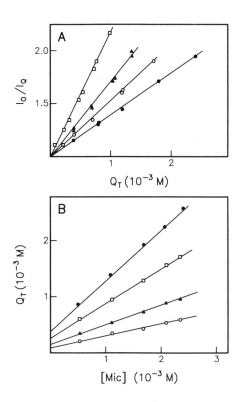

Figure 7. Stern-Volmer plots (A) and Q_T vs. [micelle] plots (B) for the quenching of 2,3-dimethylnaphthalene by 2,5-dimethyl-2,4-hexadiene in SDS. SDS concentrations on the Stern-Volmer plots (A): (□) 0.04 M, (▲) 0.075 M, (O) 0.11 M ,and (●) 0.15,M. I_0/I_Q values for the plots in B are: (●) 2.0, (□) 1.65, (▲) 1.35 and (O) 1.2. (Adapted from ref. 73.)

In an alternative method the partitioning of a probe molecule between water and the micellar phase can be determined through quenching experiments when the probe is quenched exclusively in water.[77,78] The fluorescence emission intensities of the probe in the absence (I_0) and presence (I) of quencher can be related to the fraction of the probe in the micelles (x), the fraction in water (y), the emission quantum yields of the probe in water (Φ_w) and in the micelle (Φ_m), the quenching rate constant (k_{qw}), and the lifetime in water [$(\tau_0)_w$]

$$\frac{I_0}{I_0 - I} = \left(\frac{x}{y} \frac{\Phi_m}{\Phi_w} + 1 \right) \left(1 + \frac{1}{k_{qw} (\tau_0)_w [Q]} \right) \tag{39}$$

When the emission quantum yields in water and in the micelle are known or are assumed equal the value of x/y can be determined from the slope of [$I_0/(I_0 - I)$] vs. $1/[Q]$ and the partition constant is given by $K = x/y[M]$. In the original work[77] [M] corresponds to the concentration of the empty micelles but due to the Poisson distribution the total concentration should be used.[79] At low occupancy numbers the different definitions for [M] do not alter the value for the partitioning constant. This method was used in the determination of the partition coefficient of 4-(1-pyrene)butyrate [$(2.40 \pm 0.15) \times 10^4$ M^{-1}][77] and naphthalene [$(4 \pm 1) \times 10^4$ M^{-1}][79] in SDS micelles.

Another method to determine the partition constant of a probe is based on the observation that certain probes have very different lifetimes in water and in micelles.[80] Measurements of the fluorescence intensities in water (I_0), at 100% incorporation in the micelles (I_{inf}), and at intermediate micelle concentrations (I_t) allows the following expression to be applied:[79]

$$\frac{I_{inf} - I_0}{I_t - I_0} = 1 + \frac{1}{K[M]} \tag{40}$$

This method assumes that the emission quantum yield is independent of surfactant concentration. In some cases, such as CTAB, the probe emission is quenched by the counter-ion and the equation above is not applicable.

A partition coefficient of $(3.0 \pm 0.5) \times 10^4$ M^{-1} was measured for naphthalene in SDS,[79] a value that compares well with that obtained by the same authors employing the method described by equation 39 (see above).

A special case of binding to micelles is counterion binding to ionic surfactants. A pseudophase ion-exchange model was used to obtain the expressions related to the dynamic quenching of a probe completely incorporated in the micelle.[81] This model assumes a fast counterion redistribution, no probe-to-probe interaction, and only quenching by the local concentration of counterions. Counterion exchange coefficients between different counterions were obtained without knowledge of the absolute value of the quenching constant.[81]

Determination of Aggregation Numbers: Quenching studies have been extensively used to determine not only the aggregation parameters of micelles, but also kinetic parameters such as entry and exit rates (see also sections 4.3 and 5.1). Probes and quenchers with different partitioning properties were used. The different cases outlined by Yekta *et al.*[82] will be presented to summarize the various experimental situations encountered. A Poisson distribution is assumed for the partitioning of probes and quenchers, which assumes that the entrance probability of a quencher molecule into a micelle is independent on how many quencher molecules are already in this particular micelle. The probes are hydrophobic and are exclusively located in the micelle and their excited state has a lifetime which is short compared to the exit rate [i.e., equations (30) and (31) in Scheme 1 do not contribute to the mechanism]. The probe concentration should be low enough to avoid multiple occupancy. The entry and exit rates of the quenchers are assumed independent of how many molecules the micelle contains and the total quenching rate in the micelle is given by nk_q. All the necessary equations are shown in Scheme 1. Four different cases arise:

Case 1. The quencher is exclusively incorporated in the micelle and only the static quenching mechanism is operative.[83, 84] All probes in micelles that also contain quenchers are readily quenched upon excitation, so that the overall intensity of the probe is reduced, but not its lifetime [equations (28), (29), (32) and $k_{qm} >> 1/\tau_0$] (case 1, Figure 8). Under these conditions

$$\ln\left(\frac{I_0}{I}\right) = \frac{[Q]}{[M]} = \frac{N[Q]}{[\text{surfactant}] - \text{cmc}} \tag{41}$$

This method is a very convenient way to obtain aggregation numbers, but one has to be sure that the quenching occurs exclusively through a static mechanism. It is recommended to check by time-resolved experiments whether this is indeed the case. Otherwise the quenching is described by the equations shown in case 3.

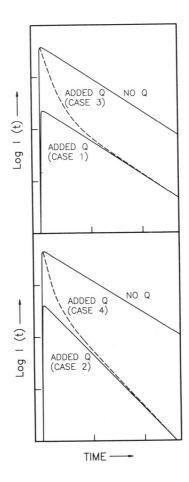

Figure 8. Schematic representation of the different cases for the fluorescence decays of excited states in the presence of quenchers that are exclusively (cases 1 and 3) or partially (cases 2 and 4) incorporated in the micelles. Static quenching operates in cases 1 and 2 and a dynamic process is responsible for the quenching in cases 3 and 4. (Adapted from ref. 82.)

Case 2. The quencher is partially solubilized and quenching inside the micelle is static. Two components contribute to the quenching: a static one for the quenchers inside the micelle and a dynamic one due to the quencher entering the micelle after excitation of the probe. The latter is responsible for the reduction of the probe's lifetime [equations (28), (29), (32), (34) and $k_{qm} >> 1/\tau_o$] (case 2, Figure 8).The observed lifetime for a given micelle concentration is given by

$$\frac{1}{\tau} = \frac{1}{\tau_o} + \frac{k_+ [Q]}{1 + K[M]} \tag{42}$$

where the second term on the right-hand side relates to dynamic quenching; an increase in lifetime, at constant [Q], is observed with increasing micelle concentration. The equation for steady-state measurements is given by

$$\frac{I_o}{I} = (1 + k_+ \tau_o [Q]) e^{[K[Q]/(1 + K[M])]} \tag{43}$$

The linear term corresponds to the dynamic and the exponential term to the static quenching, respectively.

Case 3. The quencher is exclusively micellized and dynamic quenching occurs. The lifetime of the excited probe molecule depends on the quencher population, i.e., how many quencher molecules are in a particular micelle [equations (28), (29) and (32)] (case 3, Figure 8). For this case time-resolved rather than steady-state experiments are more suitable and the fluorescence decay [I(t)] is expressed by

$$\ln\left(\frac{I(t)}{I_o}\right) = -\left[\frac{t}{\tau_o} + \frac{[Q]}{[M]}\left(1 - e^{-k_q t}\right)\right] \tag{44}$$

at infinite time the above expression is reduced to

$$I(t) = I_0 e^{-[n + t/\tau_o]} \tag{45}$$

the aggregation number can thus be calculated from the value of n obtained from the intercept of ln $[I(t)/I_0]$ vs. t.

A frequently cited example of this mechanism is the study of pyrene excimer formation in order to obtain values for the aggregation numbers of micelles.[85, 86] Excimer formation is a special case of quenching in which the singlet pyrene is quenched by a ground-state molecule, both in the same micelle. The monomer emission decay is fitted by equations (44) or (45), where the quenching constant (k_q) is the excimer formation constant (k_a). This method assumes that excimer emission is much faster than excimer dissociation. Due to the high pyrene concentrations that have to be employed, care should be taken to avoid formation of microcrystalline pyrene, which also shows excimer emission. This is apparently a problem when studying perfluorocarbon micelles, where the solubility of pyrene is low.[87]

Case 4. Partially micellized quencher and the dynamic quenching mechanism is operative.[88, 89] This case is the more general one [Equations (28), (29), (32) and (34)] (case 4, Figure 8) and depending on the conditions the equations below are reduced to those described in cases 1 to 3. The general form of the fluorescence decay is given by

$$\ln\left(\frac{I(t)}{I_o}\right) = -At + B\left(e^{-Ct} - 1\right) \tag{46}$$

where

$$A = \frac{1}{\tau_o} + \left(\frac{k_q k_+}{(k_q + k_-)}\frac{1}{1 + K[M]}\right)[Q] \tag{47}$$

$$B = \left(\frac{k_q^2 k_+}{(k_q + k_-)^2 k_-}\frac{1}{1 + K[M]}\right)[Q] \tag{48}$$

$$C = k_q + k_- \tag{49}$$

If the quenching rate constant is larger than the exit rate constant of the quencher from the micelle ($k_q \gg k_-$), then the above expressions for the term B is reduced to[90]

$$B = \frac{k_+}{k_-}\left(\frac{1}{1 + K[M]}\right)[Q] \tag{50}$$

Inverting the expression and using equation 35.

$$\frac{1}{B} = \left(\frac{1}{K} + [M]\right) [Q] \tag{51}$$

By fitting the decay data to equation (46) at different quencher concentrations the values of B obtained are plotted according to equation (51) to determine N. The aggregation numbers for SDS micelles were determined from the quenching of 1-methylpyrene by Cu^{2+} ions.[90]

When intermicellar exchange of the quencher occurs during the lifetime of the excited probe the rate for exchange (k_e) has to be incorporated and the terms A, B, and C are given by [90, 91]

$$A = \frac{1}{\tau_o} + \left(\frac{k_q}{k_q + k_e[M] + k_-}\right)\left(\frac{k_+ + k_e K[M]}{1 + K[M]}\right)[Q] \tag{52}$$

$$B = \left(\frac{k_q}{k_q + k_e[M] + k_-}\right)^2 \left(\frac{1}{\frac{1}{K} + [M]}\right)[Q] \tag{53}$$

$$C = k_q + k_e[M] + k_- \tag{54}$$

When $k_q \gg k_- + k_e[M]$, equation (51) above is valid and the aggregation number can be calculated from the decay parameter B determined at different quencher concentrations. This model assumes that k_- is independent of the micelle concentration. These equations were applied to the fluorescence decay of 1-methylpyrene in the presence of several metal ions in order to determine the aggregation number of SDS micelles.[91]

Aggregation numbers of micelles can also be determined by absorption measurements.[92] All the probe molecules (P; in this particular study, selenopyronine dyes) are excited by a microsecond flash pulse and the decay of the triplets state population is monitored. Triplets in micelles containing more than one probe molecule in the triplet state decay very fast by a bimolecular process (triplet–triplet annihilation) and after the flash only micelles containing one triplet will be detected. The concentration of triplets [T], obtained from the extinction coefficient at a certain wavelength, is measured and the aggregation number is obtained from the following expression:

$$[M] = \frac{[P]}{\ln\left(\frac{[T]}{[P]}\right)} \tag{55}$$

The advantages of this method are that the bimolecular and unimolecular processes differ by three orders of magnitude and are therefore easily separated. For the equation above to be applicable one has to be sure that all the probe molecules are being excited. This was the case in the study by Turro et al.[92] for selenopyronines where the values of N obtained by the absorption method are the same as those obtained with fluorescence techniques.

4.3. Aggregation Parameters in Systems Different from Micelles

Other microheterogeneous systems are not as easy to describe quantitatively as micelles and mathematical descriptions are still being developed. These systems are more heterogeneous and cannot be treated with a pseudophase model. In vesicles, both partitioning and binding processes can occur in contrast to micelles where only partitioning is observed.[93] To differentiate between partitioning and binding the method of Encinas and Lissi, described above, was extended taking into account the possibility of binding K_b, to a number of different binding sites, p.[93] An apparent association constant K_{app} is measured by the method of Encinas and Lissi and this value is related to the partition K and binding constant K_b by

$$K_{app} = VK + \frac{pK_b}{1 + K_b[Q_w]} \tag{56}$$

where V is the molar volume of the micellar or vesicular phase. If binding is an important factor K_{app} should depend on the added quencher concentration. A constant value of K_{app} was observed for Triton-X micelles, but binding occurs to egg phosphatidylcholine vesicles. Using the expression above the partition and binding constants as well as the number of binding sites were determined.[93]

Association of molecules to constrained media is frequently heterogeneous and non-random. For example, it was shown from fluorescence decay curves that the association of aromatic hydrocarbons, such as pyrene, naphthalene, and anthracene, to silica gel occurs at "preferred sites", this adsorption being dependent on the pretreatment and on other coadsorbed species.[94-97]

New approaches are currently being sought to analyze fluorescence decay curves in microheterogeneous systems. For example, the energy-transfer process measured by the fluorescence decay of dyes in vesicles can be fitted to a superposition of two- and three- dimension equations of the Förster formalism.[98] These authors also suggest that the distribution of dyes is nonrandom, but that adsorption sites are spatially correlated leading to formation of colonies of dyes; i.e., the distribution follows a fractal structure. It is clear that the mathematical formalism developed for micelles cannot be applied directly to other systems and further investigation will be necessary to determine suitable quantitative treatments.

5. Mobility within the Microphase

Mobility can be sampled by systems that are viscosity dependent. Thus, bimolecular reactions, such as energy- or electron-transfer processes, triplet–triplet annihilation (TTA) or excimer formation are suitable to test for mobility in microheterogeneous systems.

Fluorescence quenching is the technique most widely used to study energy- or electron- transfer processes, but any experimental approach that detects donor or acceptor steady-state concentrations, excited-state decays, or the appearance of a product can be employed. Due to confinement in microheterogeneous systems a reduction of the rate order is observed and first-order rate constants that correspond to the average diffusion time for the encounters between donors and acceptors can be measured. For example, the energy transfer from triplet N-methylphenothiazine to trans-stilbene in CTAB micelles was monitored at the triplet–triplet absorption of the thiazine and a quenching rate of 1.5×10^7 s^{-1} was obtained.[99] The average diffusion time for the encounter (70 ns) is calculated from the reciprocal of this quenching rate.

The dimensionality of the quenching process (second, third, or fractal dimensions) has to be known in order to obtain absolute diffusion coefficients from the quenching rates. An alternative approach is to compare relative values. In a recent study[100] the relative mobilities of oxygen in reversed micellar solutions, expressed as pseudo[unimolecular rate constants, and accessibility of O_2 to excited states were obtained from quenching experiments. The quenching efficiencies of donors located in different portions of the microphase were compared to the efficiencies in homogeneous solutions. The donors employed were several pyrene probes, that are located either at the micellar interface or in the water pool and tris(4,4-dicarboxylate-2,2-bipyridine)ruthenium II ($Ru(bpy)_3^{2+}$) which is located at the interface. The pseudo-unimolecular rate constants were defined as

$$k_q = \frac{1}{\tau_{air}} - \frac{1}{\tau_{N_2}} \qquad \text{or} \qquad k_q = \frac{1}{5}\left(\frac{1}{\tau_{O_2}} - \frac{1}{\tau_{N_2}}\right) \quad (57)$$

where τ_{air}, τ_{N_2} and τ_{O_2} are the lifetimes under air, N_2 or O_2, respectively.

As the solubility of O_2 in water is smaller than in the organic phase oxygen needs to diffuse toward the excited molecule. For the pyrene probes located at the interface of the AOT–H_2O–isooctane system the quenching rate constant by oxygen has an intermediate value between the one observed in neat isooctane or water. When isooctane is substituted by dodecane a smaller quenching rate constant is obtained. This decrease can be accounted for by the higher viscosity of the isooctane. The magnitude of the quenching rate constant is related to the accessibility of O_2 to the excited molecule. For example the k_q value for pyrene sulfonate (PS) in AOT–H_2O–isooctane ($R = 20$, where $R = $ [H_2O]/[AOT]) is 2.3 x 10^7 s^{-1}, whereas for pyrene tetrasulfonate (PTS) the value is smaller than 1.5 x 10^6 s^{-1}; PS is located at the micellar interface and PTS in the water pool. The smaller quenching rate constant for the latter reflects that oxygen has to traverse the interface which acts as a barrier to reach the excited PTS molecule. The quenching of Ru(bpy)$_3$$^{2+}$ by O_2 is less efficient than that for pyrenes. In this case k_q is affected less by changes in the viscosity of the organic solvent because the quenching is not solely determined by diffusion.

A different kind of mobility, exchange of probe molecules between micelles, can be measured when the exchange rate is fast compared to the quenching process. This is the case for reversed micelles where the exchange happens after the collision and fusion of two micelles.[101] Using stopped-flow absorption spectroscopy to study electron-transfer, proton-transfer and metal–ligand complexation processes, exchange rate constants (k_e) for AOT–H_2O–hydrocarbon systems at several temperatures and different R ratios were determined. For example, the k_e value for AOT–H_2O–heptane at 20°C and $R = 10$ is (1.0 ± 0.2) x 10^7 M^{-1} s^{-1}. This value is considerably smaller than the diffusion-controlled limit calculated by the Smoluchovski equation (1.7 x 10^{10} M^{-1} s^{-1}), indicating that fusion does not occur at every encounter between two micelles.[101]

A critical evaluation considering the limitations of fluorescence quenching experiments to determine aggregation numbers and k_e values in AOT–H_2O–n-hexane reversed micelles has been published.[102] The fluorescence decay curves were fitted to equation (46) where parameters A, B and C are given by

$$A = \frac{1}{\tau_0} + \left(\frac{k_q \, k_e}{k_q + k_e[M]} \right) [Q] = \frac{1}{\tau_0} + S_2 [Q] \qquad (58)$$

$$B = \left(\frac{k_q}{k_q + k_e[M]}\right)^2 \frac{[Q]}{[M]} = S_3[Q] \qquad (59)$$

$$C = k_q + k_e[M] \qquad (60)$$

The rate constants are related to Scheme 1 above. The aggregation number can be obtained accurately only when $k_e[M] < k_q$. On the other hand, the value of k_e is obtained when $k_e[M] > k_q$. As the quenching efficiency is dependent on the size of the water pool a switch between the two boundary conditions can occur when the R values and/or surfactant concentrations are varied. Other restrictions in the application of this equation are that it does not account for the exchange of more than one quencher molecule during the lifetime of the donor and the fact that at high exchange rates the distribution of molecules does not follow the Poisson distribution. The inadequacy in fitting the fluorescence decay curves in the presence of quencher equation (46) is shown by the fact that the parameters S_2, S_3 and C do not show the expected dependence on the surfactant concentration. Further, the application of the above boundary conditions is very dependent on the excited state–quencher system employed. The quenching rate has to be high enough to avoid breakdown of the Poisson distribution by micellar exchange.[103] The k_e values obtained in this study are $>10^9$ M^{-1} s^{-1}, i.e., two orders of magnitude higher than those determined in stopped flow experiments[101] (see above) or those determined in the quenching of excited PTS by Fremy's salt (1.3 x 10^7 M^{-1}s^{-1}).[104] This difference has not been interpreted adequately and further experiments are necessary. Careful analysis should be made for different systems and, when the right boundary conditions apply, equation (46) can be used to measure k_e values, such as in the case of the exchange between didodecylammonium chloride–H$_2$O–apolar solvents micelles.[105] We note the need to be careful when applying a methodology successfully used for one kind of microheterogeneous system (e.g., normal micelles) to another (e.g., reversed micelles).

Triplet–triplet annihilation (TTA) processes, which are generally diffusion controlled in homogeneous solution, were used to obtain information about the mobility in microheterogeneous phases. The TTA of 1-bromonaphthalene in CTAB was followed by the triplet decay at 425 nm.[106] The exchange of probe molecules between micelles is negligible in the time range measured and TTA is due exclusively to triplets residing in the same micelle. The TTA rate measured in CTAB micelles is (1.4 ± 0.2) x 10^7 M^{-1} s^{-1}. This value compares well to that obtained in energy-transfer studies in the same micelle (see above).[99]

TTA was also observed after excitation of acridine on pretreated silicas.[107] The delayed fluorescence due to the TTA process and the triplet decay were measured with diffuse reflectance techniques (vide supra). The delayed fluorescence and the second-order decay rate constant for the triplet are dependent on the laser dose, surface coverage of absorbate and temperature of the silica pretreatment. For silica pretreated at 300°C the two dimensional bimolecular rate constant was determined to be 8 x 10^{13} dm^2 mol^{-1} s^{-1}. When the silica was pretreated at temperatures higher than 500°C no TTA process was observed, suggesting that the acridine molecules have reduced mobility on this solid support. After low-temperature pretreatment Si–OH (silanol) groups predominate, whereas after high-temperature pretreatment Si—O—Si (siloxane) prevail, suggesting that silanol groups promote the mobility of acridine.

The rate constant for pyrene excimer formation in homogeneous solution is close to the diffusion limit. The formation of pyrene excimers in cetyltrioxyethylene sulfate micelles was studied by time-resolved fluorescence and an excimer formation rate constant (k_a) of (9 ± 1) x 10^6 M^{-1} s^{-1} determined.[86] This value compares well with that obtained for energy transfer in CTAB micelles, especially considering that cetyltrioxyethylene sulfate micelles are somewhat bigger than CTAB micelles.[99]

The lateral diffusion in several microorganized systems other than micelles was determined by studying pyrene excimer formation, some examples being membranes, vesicles,[108-111] and spread monolayers.[112] Pyrenedodecanoic acid (PDA) excimer formation was studied in dipalmitoyllecithin (DPL) and dipalmitotylphosphatidic acid (DPA) pure and mixed vesicles.[111] The ratio of excimer (I') to monomer (I) emission is proportional to the encounter rate and the lateral diffusion (D_L).[110, 111] Figure 9 shows the variation with temperature of I'/I of PDA in DPA vesicles.[111] The hysteresis observed at the pH studied is due to the simultaneous presence of one and two negative charges on the lipid molecules. At high temperatures (>60°C) the proportion of excimers increases due to the higher fluidity of the vesicle. The higher excimer content observed below the transition temperature is due to clustering of the pyrene molecules. The D_L values for DPA (pH 9.0) and DPL, both at 60°C, are 1.7 and 0.8 x 10^{-7} cm^2 s^{-1}, respectively. In mixed DPA and DPL vesicles the formation of rigid lipid clusters is observed upon addition of Ca^{2+} ions. Cluster formation leads to an increase in excimer emission as pyrene molecules are squeezed out of the rigid regions and the local PDA concentration increases in the liquid phase.

Excimer formation has also been used to test for the lateral diffusion in mitochondrial membranes[108] where the D_L value for pyrene is 3 x 10^{-8} cm^2 s^{-1}. This value compares well with values determined using epr and nmr techniques.

Lateral diffusion of PDA molecules in oleic acid monolayers was treated by the random walk model and Monte Carlo simulations. The D_L value obtained in this study was 1×10^{-6} cm^2 s^{-1}, substantially higher than values obtained for DPA and DPL vesicles. These examples show that a qualitative description of lateral diffusion and the observation of increases or decreases of fluidity are easily monitored by pyrene excimer formation. On the other hand, their quantitative treatment is not straightforward, as attested to by the variations in lateral diffusion values reported.

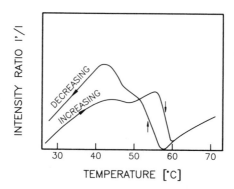

Figure 9. Transition curve of the excimer (I') - to - monomer (I) intensity ratio for DPA vesicles at pH 7.8 observed at increasing and decreasing temperatures. (Adapted from ref. 111.)

Pyrene excimer formation has been extensively used to characterize silica surfaces. A critical review on this subject has recently been published (see also Chapters 8, 12 and 13).[113] The dynamic formation of excimers is a measure of the mobility on silica surfaces and only this aspect of these photophysical studies will be addressed. Assuming that only one kind of monomer and one kind of excimer are present, the steady-state intensity ratio is given by

$$\frac{I'}{I} = \frac{k_f'}{k_f} \frac{k_a[M]}{\left(k_d + \dfrac{1}{\tau_o'}\right)} \tag{61}$$

where I', I, k_f', and k_f are the emission intensities and radiative rate constants of excimer and monomer, k_a and k_d, are the excimer formation and dissociation rates, τ_o' is the excimer lifetime and [M] the monomer concentration. Thus, for a dynamic process a higher excimer intensity indicates a higher pyrene mobility. The excimer emission can be due to two different mechanisms, one in which they are formed by a dynamic process that depends on diffusion and the other which is due to the presence of dimers in the ground state. To differentiate between these, time-resolved experiments have to be performed. The decay of the monomer and excimer emissions is given by

$$I_M(t) = A_{11} e^{-\lambda_1 t} + A_{12} e^{-\lambda_2 t} \tag{62}$$

$$I_D(t) = A_{21} e^{-\lambda_1 t} + A_{22} e^{-\lambda_2 t} \tag{63}$$

The solution for the reciprocal decay times is[114]

$$\lambda_{1,2} = \frac{1}{\tau_{1,2}} = \frac{1}{2} \left\{ (X + Y) \pm \left[(Y - X)^2 + 4k_a[M] k_d \right]^{0.5} \right\} \tag{64}$$

where X and Y are given by

$$X = k_a[M] + \frac{1}{\tau_o} \tag{65}$$

$$Y = k_d + \frac{1}{\tau_o} \tag{66}$$

In the absence of ground-state dimers the initial excimer concentration is zero and $A_{22}/A_{21} = -1$. When ground-state derived excimers are present $A_{22}/A_{21} > -1$. Identical values for τ_1 and τ_2 have to be obtained from the analysis of the monomer and excimer decays.

The information on excimer formation (k_a) is contained in the τ_2 value (excimer growth) when $X > Y$. Under this condition τ_2 is shortened at higher pyrene concentrations. When $X < Y$ the lifetime for excimer growth (τ_2) is independent of the pyrene concentration and information about k_a can be obtained from τ_1. Thus, care has to be taken when analyzing the kinetics for excimer growth and conclusions about the excimer formation rate constant can only be drawn when the condition $X > Y$ is met.

In the characterization of the pyrene mobility on silica the contributions of the dynamic process and ground-state dimers were evaluated from the excimer decay.[95,97] Dimers are responsible for all the excimer emission of pyrene on dry silica. One possibility for excimer emission in solid samples is the formation of microcrystalline pyrene. This is not the case in this study as the excimer emission maximum for pyrene on silica is very different from that for the pyrene crystal powder. The presence of ground-state dimers is also indicated by a red shift of the excitation spectra monitored at the excimer emission wavelength when compared to that monitored at the monomer emission.

In the presence of long-chain alcohols pyrene mobility was observed. The effect of methanol and decanol on the A_{21} and A_{22} values, the ratio of these preexponential parameters, the lifetimes, and the excimer-to-monomer emission intensity ratio are shown in Table 2.

With the addition of methanol an increase in excimer emission is observed but this is due to the formation of dimers as no kinetic growth was detected. Only at high decanol concentrations is a growing-in of the excimer emission observed, indicating the occurrence of a dynamic process. Pyrene in decanol moves through a layer of a solvent that has long enough hydrocarbon chains to create a fluid medium. In the case of methanol the solvent does not create a fluid medium to allow the movement of pyrene molecules. The excimer emission of pyrene in decanol-covered silica is temperature dependent and the relative contribution of the dynamic process decreases at lower temperatures,[115] showing that the mobility of pyrene decreases.

Table 2. Effect of Methanol and 1-Decanol on the Excimer Emission of Pyrene Adsorbed on Silica Gel

Alcohol, mol g^{-1} of SiO_2	A_{21}	A_{22}	A_{22}/A_{21}	τ_1 ns	τ_2 ns	I'/I
Methanol						
0	0.06	0.08	1.33	77	15	0.21
3.75×10^{-5}	0.35	0.15	0.43	68	4.54	0.33
5.0×10^{-4}	0.35	0.22	0.63	64	3.33	0.46
1.2×10^{-4}	0.17	0.18	1.06	68	7.14	0.57
1-Decanol						
0	0.06	0.09	1.5	77	16	0.21
1.3×10^{-4}	0.12	0.38	3.17	103	3.2	0.32
1.3×10^{-3}	0.48	-0.37	-0.77	155	45	0.37

Intramolecular formation of pyrene excimers has also been explored to study the mobility on silica and reversed-phase silica surfaces.[116] No dynamic intramolecular excimer formation is observed in the absence of a monolayer of coadsorbed 1-octanol; a result that parallels those described above. In the case of derivatized reversed-phase silica dynamic excimer generation takes place even in the absence of coadsorbed solvents.

Information about mobilities in microheterogeneous systems can also be obtained with experimental techniques that are not based on bimolecular interactions. One of these techniques is the fluorescence photobleaching recovery, which was used to determine the lateral diffusion in membranes and vesicles.[117-121] A probe molecule evenly distributed in the microphase is irradiated with an intense laser which causes the photochemical bleaching of the probe. The same laser, attenuated by several orders of magnitude, irradiates the bleached area to detect the fluorescence recovery due to the increase of probe concentration. This recovery can be due to diffusion or flow mechanisms. Only the former yields information about the lateral diffusion coefficients. Using this technique the effect of temperature and addition of cholesterol on the lateral diffusion of probe molecules in membranes and vesicles was determined.

5.1. Entry/Exit Dynamics

The partition of a given substrate between a microphase or host and the main, continuous medium is determined by the rates with which the molecule moves between these environments. As with many of the other systems in this article, the vast majority of the data in the literature concerns micellar systems; the material presented here is no exception.

The techniques employed for the study of entry/exit kinetics in micellar systems depend largely upon the phase in which the probes are predominantly located. Thus, in quenching studies (from which most of these data are derived), the type of analysis depends on the location of excited state and quencher. We deal briefly with a few specific cases below.

Mostly Micellized Probe and Aqueous-Bound Quencher: Among the many papers on this subject we have selected two articles[79, 122] that provide examples of the exit of aromatic hydrocarbons and carbonyl compounds from various micelles, particularly SDS. Thus, if we assume that in a first approximation our probe is almost exclusively located in the micelles and that only the small fraction in the aqueous phase is available for quenching, equations (28), (29), (31) and (33) of Scheme 1 apply. The observed decay rate constant of the triplet is given by

$$k_{exp} = k_o + \frac{k_{p-} k_{qw} [Q]}{k_{qw} [Q] + k_{p+} [M]} \tag{67}$$

As the concentration of aqueous quencher increases the experimental rate constant (k_{exptl}) for excited state decay (normally monitored by luminescence or transient absorption) tends to $k_0 + k_{p-}$. Thus, the value of k_{p-} can be determined by obtaining the limiting value of k_{expt} and subtracting the rate constant for decay in the absence of quencher. A much better approach is to plot the data in a reciprocal fashion, as shown in equation (68):

$$\frac{1}{k_{exp} - k_o} = \frac{1}{k_{p-}} + \frac{k_{p+} [M]}{k_{p-} k_{qw} [Q]} \tag{68}$$

Nitrite ions frequently have been employed as quenchers in the case of anionic micelles.[79,122] The rate constants for exit obtained are a function of the hydrophobicity of the probe, and in the case of arenes correlate well with their boiling points.[79] Representative values of k_{p-} are 2.5×10^5 s^{-1} for naphthalene[79] and 3×10^6 s^{-1} for propiophenone.[122]

Mostly Micellized Probe and Micellized (but Mobile) Quencher: In this case the excited probe and quencher can meet as a result of either static or dynamic processes. The latter can only occur as a result of quencher mobility. Such a mechanism takes place in the quenching of some triplet arenes (e.g., phenanthrene) by small dienes.[123] Thus, quenching events occur in the micelles as a result of a "visit" by the quencher. This mechanism corresponds to equations (28), (29), (32) and (34) in Scheme 1 and is analogous to case 2 described in section 4.2. If we assume that quenching occurs with 100% probability if the probe and quencher meet in the same micelle, then, the experimentally determined rate constant for excited probe decay is related to the quenching concentration according to equation (69):

$$k_{exp} = k_o + \frac{k_- [Q]}{[M]} \tag{69}$$

where it was assumed that the fraction of quencher in the water phase can be neglected and therefore $k_+ [M] > k_-$.

When the amount of quencher molecules in the aqueous phase cannot be neglected the experimental rate constant is expressed by equation (42) and the values of k_+ and k_- can be obtained from studies of the kinetics of triplet decay as a function of quencher and surfactant concentrations.[123]

A number of other approaches also lead to kinetic data on the exit from micelles. For example, studies of magnetic field effects on photogenerated radical pairs led to values of k_- for free radicals.[124,125] Exit/entry rate constants can also be determined from quenching studies in which the fluorescence decay curve is fitted to equation (46).[90,91] Using this latter approach rate constants for fragment detachment from micelles were determined (ref. 126 and references therein)

The dynamics of singlet oxygen in heterogeneous media have been studied in detail largely by Rodgers and co-workers, initially using chemical trapping experiments [25, 127-131] but with the advent of direct luminescence detection the latter technique was adopted.[132-134] Studies in a variety of biphasic (aqueous/nonaqueous) heterogeneous systems such as normal micelles,[127-130,132] reverse micelles,[131-133] lipid vesicles,[25] and Nafion powders[134] demonstrated the general applicability of kinetic analysis based on Scheme 2.

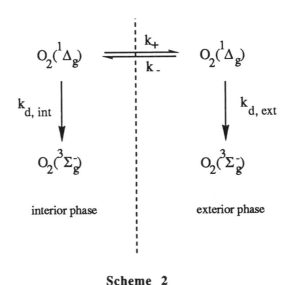

Scheme 2

In all systems it was shown that the decay of $O_2(^1\Delta_g)$ was monoexponential due to the fact that the $O_2(^1\Delta_g)$ on generation rapidly equilibrated itself between the internal and external phases (with partition coefficient $K = k_- / k_+$) on a time scale much faster than the solvent dependent decay of $O_2(^1\Delta_g)$ in either internal or external phases; i.e. $k_+, k_- \gg k_{int}, k_{ext}$. This resulted in the exclusive measurement of monoexponential decay of the luminescence as rapid equilibration dictates that both phases decay with the same observed rate constant. The measured value of the $O_2(^1\Delta_g)$ lifetime in these systems is determined by the fractional composition of the system (aqueous to nonaqueous phases) and the partition coefficient for $O_2(^1\Delta_g)$ between the two phases. Kinetic analysis allows evaluation of the partition coefficient and, for example, in AOT–n-heptane reverse micelles a value of 0.11 was determined[132] which is almost identical to that exhibited by ground-state oxygen in these phases. Additionally, by selecting the site of sensitizer and quencher they showed that $O_2(^1\Delta_g)$ could easily travel from one phase of the system to the other and even from the core of one unit (micelle, vesicle) to another to facilitate reaction. $O_2(^1\Delta_g)$ has also been sensitized and detected in thin films[135] and polymer matrices;[136,137] $O_2(^1\Delta_g)$ dynamics in the latter proved complex, being nonexponential and showing dependence on sensitizer and its concentration. In glassy polymers $O_2(^1\Delta_g)$ kinetics were shown to reflect that of the triplet sensitizer (long lived in glassy polymer) but as the polymer was made less glassy the triplet decay increased and monoexponential decay of $O_2(^1\Delta_g)$ was observed.[137]

Similar approaches to those outlined above can also be employed for other organized systems and inclusion complexes. Not surprisingly, the difficulties in the analysis of the data and their interpretation grow as the system becomes more complex.

6. Final Remarks

In preparing this chapter we have tried to provide some basic information on the photophysical and photochemical techniques employed in the study on organized assemblies. Clearly our coverage is not (and could not be in the limited space available) comprehensive. We have also tried to outline some of the more common approaches to the determination of equilibrium and kinetic parameters employing these techniques; quite frequently the examples are heavily weighted toward micellar systems. This bias reflects the availability of reported data, as well as the relative simplicity of micellar systems, which makes them ideally suited for brief examples.

Finally, some of the techniques described in the first part of this article have been barely mentioned in Section 3; this reflects our feeling that these techniques (e.g., photothermal methods), which have received little attention in organized systems in solution, do in fact have an enormous potential. We hope that by including them here they may attract the attention of those interested in these problems. Similarly, diffuse reflectance techniques have the potential to become useful not only in solid systems, but also in opaque solution samples.

References

1. F. Wilkinson and G. Kelly in *Handbook of Organic Photochemistry*, Vol. 1, J. C. Scaiano, ed., 1989, CRC Press, Boca Raton, FL, p. 293.
2. L. Lindqvist, *C. R. Hebd. Seances Acad. Sci., Ser. C.*, **1966**, *263*, 852.
3. J. C. Scaiano, *Acc. Chem. Res.*, **1983**, *16*, 234.

4.   D. Oelkrug, W. Honnen, F. Wilkinson and C. J. Willsher, *J. Chem. Soc., Faraday Trans. II*, **1987**, *83*, 2081.
5.   P. Kubelka, *J. Opt. Soc. Am.,* **1948**, *38*, 448.
6.   D. V. O'Connor and D. Phillips, *Time-Correlated Single Photon Counting*, Academic Press, Orlando, FL, 1984.
7.   D. A. Holden in *Handbook of Organic Photochemistry*, Vol. 1, J. C. Scaiano, ed., 1989, CRC Press, Boca Raton, FL, p. 261.
8.   S. Cova, A. Longoni and A. Andreoni, *Rev. Sci. Instrum.,* **1981**, *52*, 408.
9.   T. Murao, I. Yamazaki and K. Yoshihara, *Appl. Optics* ., **1982**, *21*, 2297.
10. S. W. Snyder, J. N. Demas and B. A. DeGraff, *Anal. Chem.*, **1989**, *61*, 2704.
11. D. W. Marquardt, *J. Soc. Ind. Appl. Math.*, **1963**, *11*, 431.
12. J. R Knutson, J. M. Beechem and L. Brand, *Chem. Phys. Lett.*, **1983**, *102*, 501.
13. J. M. Beechem, M. Ameloot and L. Brand, *Anal. Instrum.*, **1985**, *14*, 379.
14. J. Durbin and G. S. Watson, *Biometrika*, **1950**, *37*, 409.
15. J. Durbin and G. S. Watson, *Biometrika*, **1951**, *38*, 159.
16. D. F. Eaton in *Handbook of Organic Photochemistry*, Vol. I, J. C. Scaiano, ed., CRC Press, Boca Raton, Florida, 1989, p. 231.
17. A. A. Gorman and M. A. J. Rodgers in *Handbook of Organic Photochemistry*, Vol. II, J. C. Scaiano, ed., CRC Press, Boca Raton, FL, 1989, p. 229.
18. A. A. Krasnovskii, *Photochem. Photobiol.*, **1979**, *29*, 29.
19. I. M. Byteva and G. P. Gurinovich, *J. Lumin.*, **1979**, *21*, 17.
20. K. I. Salokhiddinov, I. M. Byteva and B. M. Dzhagarov, *Opt. Spectrosc.*, **1979**, *47*, 881.
21. A. U. Khan and M. Kasha, *Proc. Natl. Acad. Sci. USA*, **1979**, *76*, 6047.
22. J. R. Hurst, J. D. McDonald and G. B. Schuster, *J. Am. Chem. Soc.*, **1982**, *104*, 2065.
23. J. G. Parker and W. D. Stanbro, *J. Am. Chem. Soc.*, **1982**, *104*, 2067.
24. P. R. Ogilby and C. S. Foote, *J. Am. Chem. Soc.*, **1982**, *104*, 2069.
25. M. A. J. Rodgers and P. T. Snowden, *J. Am. Chem. Soc.*, **1982**, *104*, 5541.
26. V. Ramamurthy, J. V. Caspar, D. R. Corbin, B. Schyler and A. H. Maki, J. Phys. Chem., **1990**, *94*, 3391.
27. a. R. W. Redmond and S. E. Braslavsky in *Photosensitization: Molecular, Cellular and Medical Aspects*, G. Moreno, R. H. Pottier and T. G. Truscott, eds., Springer-Verlag, Berlin, 1988, p. 93.
     b. R. W. Redmond and S. E. Braslavsky, *Chem. Phys. Lett.*, **1988**, *148*, 523.
28. P. B. Merkel and D. R. Kearns, *J. Am. Chem. Soc.*, **1972**, *94*, 7244.
29. C. K. N. Patel and A. C. Tam, *Rev. Mod. Phys.*, **1981**, *53*, 517.
30. A. C. Tam, *Rev. Mod. Phys.*, **1986**, *58*, 381.
31. S. E. Braslavsky and K. Heihoff in *Handbook of Organic Photochemistry*, Vol. I, J. C. Scaiano, ed., CRC Press, Boca Raton, FL, 1989, p. 237.
32. H. L. Fang and R. L. Swofford in *Ultrasensitive Laser Spectroscopy*, D. S. Kliger, ed., 1983, Academic Press, New York.
33. K. Fuke, A. Hasagawa, M. Ueda and M. Itoh, *Chem. Phys. Lett.*, **1981**, *84*, 176.
34. K. Fuke, M. Ueda and M. Itoh, *J. Am. Chem. Soc.*, **1983**, *105*, 1091.
35. G. Rossbroich, N. A. Garcia and S. E. Braslavsky, *J. Photochem.*, **1985**, *37*, 37.
36. R. W. Redmond, K. Heihoff, S. E. Braslavsky and T. G. Truscott, *Photochem. Photobiol.*, **1987**, *45*, 209.
37. J. E. Rudzki, J. L. Goodman and K. S. Peters, *J. Am. Chem. Soc.*, **1985**, *107*, 7849.
38. K. Heihoff, S. E. Braslavsky and K. Schaffner, *Chem. Phys. Lett.*, **1986**, *131*, 183.
39. J. L. Goodman, K. S. Peters, H. Misawa and R. A. Caldwell, *J. Am. Chem. Soc.*, **1986**, *108*, 6803.

40. D. C. Dong and M. A. Winnik, *Photochem. Photobiol.*, **1982**, *35*, 17.
41. K. Kalyanasundaram and J. K. Thomas, *J. Am. Chem. Soc.*, **1977**, *99*, 2039.
42. M. Koyanagi, *J. Molec. Spec.*, **1968**, *25*, 273.
43. K. W. Street Jr. and W. E. Acree Jr., *Analyst*, **1986**, *111*, 1197.
44. K. Kalyanasundaram, *Photochemistry in Microheterogeneous Systems*, Academic Press, Orlando, FL, 1987.
45. X. Liu, K. K. Iu and J. K. Thomas, *J. Phys. Chem.*, **1989**, *93*, 4120.
46. P. Lianos, A. K. Mukhopadhyay and S. Georghiou, *Photochem. Photobiol.*, **1980**, *32*, 415.
47. K. Kalyanasundaram and J. K. Thomas, *J. Phys. Chem.*, **1977**, *81*, 2176.
48. N. J. Turro and T. Okubo, *J. Phys. Chem.*, **1982**, *86*, 159.
49. J. C. Dederen, L. Coosemans, F. C. De Schryver and A. Van Dormael, *Photochem. Photobiol.*, **1979**, *30*, 443.
50. M. Wong, J. K. Thomas and M. Grätzel, *J. Am. Chem. Soc.*, **1976**, *98*, 2391.
51. K. Dimroth and C. Reichardt, *Z. Anal. Chem.*, **1966**, *215*, 344.
52. K. A. Zachariasse, N. V. Phuc and B. Kozankiewicz, *J. Phys. Chem.*, **1981**, *85*, 2676.
53. N. Sabbatini, M. Ciano, S. Dellonte, A. Bonazzi and V. Balzani, *Chem. Phys. Lett.*, **1982**, *90*, 265.
54. C. D. Tran, *SPIE Fluorescence Detection I*, **1988**, *910*, 66.
55. K. A. Zachariasse, *Chem. Phys. Lett.*, **1978**, *57*, 429.
56. J. Emert, C. Behrens and M. Golden, *J. Am. Chem. Soc.*, **1979**, *101*, 771.
57. N. J. Turro, M. Aikawa and A. Yekta, *J. Am. Chem. Soc.*, **1979**, *101*, 772.
58. M. Shinitzky, A. C. Dianoux, C. Gitler, and G. Weber, *Biochemistry*, **1971**, *10*, 2106.
59. H. J. Pownall and L. C. Smith, *J. Am. Chem. Soc.*, **1973**, *95*, 3136.
60. B. R. Suddaby, P. E. Brown, J. C. Russell and D. G. Whitten, *J. Am. Chem. Soc.*, **1985**, *107*, 5609.
61. K. Kano, H. Matsumoto, S. Hashimoto, M. Sisido and Y. Imanishi, *J. Am. Chem. Soc.*, **1985**, *107*, 6117.
62. K. Kano, H. Matsumoto, Y. Yoshimura and S. Hashimoto, *J. Am. Chem. Soc.*, **1988**, *110*, 204.
63. J. C. Scaiano, *J. Am. Chem. Soc.*, **1980**, *102*, 7747.
64. D. Weir and J. C. Scaiano, *Tetrahedron*, **1987**, *43*, 1617.
65. F. Wilkinson, C. J. Willsher, H. L. Casal, L. J. Johnston and J. C. Scaiano, *Can. J. Chem.*, **1986**, *64*, 539.
66. H. A. Benesi and J. H. Hildebrand, *J. Am. Chem. Soc.*, **1949**, *71*, 2703.
67. S. Hamay, *J. Phys. Chem.*, **1989**, *93*, 6527.
68. G. Scatchard, *Ann. N.Y. Acad. Sci.*, **1949**, *51*, 660.
69. G. Sudlow, D. J. Birkett and D. N. Wade, *Mol. Pharmacol.*, **1975**, *11*, 824.
70. H. C. Chiang and A. Lukton, *J. Phys. Chem.*, **1975**, *79*, 1935.
71. N. E. Schore and N. J. Turro, *J. Am. Chem. Soc.*, **1974**, *96*, 306.
72. R. C. Mast and L. V. Haynes, *J. Colloid Interface Sci.*, **1975**, *53*, 35.
73. M. V. Encinas and E. A. Lissi, *Chem. Phys. Lett.*, **1982**, *91*, 55.
74. W. J. Leigh and J. C. Scaiano, *J. Am. Chem. Soc.*, **1983**, *105*, 5652.
75. M. V. Encinas and E. A. Lissi, *Photochem. Photobiol.*, **1983**, *37*, 251.
76. M. V. Encinas and E. A. Lissi, *Chem. Phys. Lett.*, **1986**, *132*, 545.
77. F. H. Quina and V. G. Toscano, *J. Phys. Chem.*, **1977**, *81*, 1750.
78. E. Abuin and E. A. Lissi, *J. Phys. Chem.*, **1980**, *84*, 2605.
79. M. Almgren, F. Grieser and J. K. Thomas, *J. Am. Chem. Soc.*, **1979**, *101*, 279.
80. R. R. Hautala, N. E. Schore and N. J. Turro, *J. Am. Chem. Soc.*, **1973**, *95*, 5508.
81. E. Abuin, E. Lissi, N. Blanchi, L. Miola and F. H. Quina, *J. Phys. Chem.*, **1983**, *87*, 5166.

82. A. Yekta, M. Aikawa and N. J. Turro, *Chem. Phys. Lett.*, **1979**, *63*, 543.
83. N. J. Turro and A. Yekta, *J. Am. Chem. Soc.*, **1978**, *100*, 5951.
84. P. P. Infelta, *Chem. Phys. Lett.*, **1979**, *61*, 88.
85. S. S. Atik, M. Nam and L. A. Singer, *Chem. Phys. Lett.*, **1979**, *67*, 75.
86. P. P. Infelta and M. Grätzel, *J. Chem. Phys.*, **1979**, *70*, 179.
87. K. Kalyanasundaram, *Langmuir*, **1988**, *4*, 942.
88. M. Tachiya, *Chem. Phys. Lett.*, **1975**, *33*, 289.
89. P. P. Infelta, M. Grätzel and J. K. Thomas, *J. Phys. Chem.*, **1974**, *78*, 190.
90. J. C. Dederen, M. V. D. Auweraer and F. C. De Schryver, *Chem. Phys. Lett.*, **1979**, *68*, 451.
91. J. C. Dederen, M. V. D. Auweraer and F. C. De Schryver, *J. Phys. Chem.*, **1981**, *85*, 1198.
92. E. Fanghanel, W. Ortman, K. Behrmann, S. Willscher, N. J. Turro and I. R. Gould, *J. Phys. Chem.*, **1987**, *91*, 3700.
93. E. Blatt and W. H. Sawyer, *Biochem. Biophys. Acta*, **1985**, *822*, 43.
94. R. K. Bauer, R. Borenstein, P. de Mayo, K. Okada, M. Rafalska, W. R. Ware and K. C. Wu, *J. Am. Chem. Soc.*, **1982**, *104*, 4635.
95. R. K. Bauer, P. de Mayo, W. R. Ware and K. C. Wu, *J. Phys. Chem.*, **1982**, *86*, 3781.
96. R. K. Bauer, P. de Mayo, K. Okada, W. R. Ware and K. C. Wu, *J. Phys. Chem.*, **1983**, *87*, 460.
97. R. K. Bauer, P. de Mayo, L.V. Natarajan and W. R. Ware, *Can. J. Chem.*, **1984**, *62*, 1279.
98. I. Yamazaki, N. Tamai and T. Yamazaki, *J. Phys. Chem.*, **1990**, *94*, 516.
99. G. Rothenberger, P. P. Infelta and M. Grätzel, *J. Phys. Chem.*, **1979**, *83*, 1871.
100. M. Saez, E. B. Abuin and E. A. Lissi, *Langmuir*, **1989**, *5*, 942.
101. P. D. I. Fletcher, A. M. Howe and B. H. Robinson, *J. Chem. Soc., Faraday Trans. I*, **1987**, *83*, 985.
102. A. Verbeeck and F. C. De Schryver, *Langmuir*, **1987**, *3*, 494.
103. E. Gelade and F. C. De Schryver, *J. Am. Chem. Soc.*, **1984**, *106*, 5871.
104. S. S. Atik and J. K. Thomas, *J. Am. Chem. Soc.*, **1981**, *103*, 3543.
105. A. Verbeeck, G. Voortmans, C. Jackers and F. C. De Schryver, *Langmuir*, **1989**, *5*, 766.
106. G. Rothenberger, P. P. Infelta and M. Grätzel, *J. Phys. Chem.*, **1981**, *85*, 1850.
107. D. Oelkrug, S. Uhl, F. Wilkinson and C. J. Willsher, *J. Phys. Chem.*, **1989**, *93*, 4551.
108. J. M. Vanderkooi and J. B. Callis, *Biochemistry*, **1974**, *13*, 4000.
109. H. J. Galla and J. Luisetti, *Biochim. Biophys. Acta*, **1980**, *596*, 108.
110. H. J. Galla and E. Sackmann, *Biochim. Biophys. Acta*, **1974**, *339*, 103.
111. H. J. Galla and E. Sackmann, *J. Am. Chem. Soc.*, **1975**, *97*, 4114.
112. T. Loughran, M. D. Hatlee, L. K. Patterson and J. J. Kozak, *J. Chem. Phys.*, **1980**, *72*, 5791.
113. K. A. Zachariasse in *Photochemistry on Solid Surfaces*, M. Anpo and T. Matsuura, eds., 1989, Elsevier, Tokyo, p. 48.
114. J. B. Birks, *Photophysics of Aromatic Molecules*, Wiley-Interscience, London, 1970.
115. P. de Mayo, L.V. Natarajan and W. R. Ware, *J. Phys. Chem.*, **1985**, *89*, 3526.
116. D. Avnir, R. Busse, M. Ottolenghi, E. Wellner and K. A. Zachariasse, *J. Phys. Chem.*, **1985**, *89*, 3521.
117. D. Axelrod, D. E. Koppel, J. S. Schlessinger, E. Elson and W. W. Webb, *Biophys. J.*, **1976**, *16*, 1055.
118. D. E. Koppel, D. Axelrod, J. S. Schlessinger, E. Elson and W. W. Webb, *Biophys. J.*, **1976**, *16*, 1315.

119. E. S. Wu, K. Jacobson and D. Papahadjopoulos, *Biochemistry*, **1977**, *16*, 3936.
120. J. Teissie, J. F. Tocanne and A. Baudras, *Eur. J. Biochem.*, **1978**, *83*, 77.
121. D. E. Koppel, *Biophys. J.*, **1979**, *28*, 281.
122. J. C. Scaiano and J. C. Selwyn, *Can. J. Chem.*, **1981**, *59*, 2368.
123. J. C. Selwyn and J. C. Scaiano, *Can. J. Chem.*, **1981**, *59*, 663.
124. J. C. Scaiano, E. B. Abuin and L. C. Stewart, *J. Am. Chem. Soc.*, **1982**, *104*, 5673.
125. C. H. Evans and J. C. Scaiano, *J. Am. Chem. Soc.*, **1990**, *112*, 2694.
126. A. Malliaris, J. Lang, J. Sturm and R. Zana, *J. Phys. Chem.*, **1987**, *91*, 1475.
127. A. A. Gorman, G. Lovering and M. A. J. Rodgers, *Photochem. Photobiol.*, **1976**, *23*, 399.
128. A. A. Gorman and M. A. J. Rodgers, *Chem. Phys. Lett.*, **1978**, *55*, 52.
129. I. B. C. Matheson and R. Massoudi, *J. Am. Chem. Soc.*, **1980**, *102*, 1942.
130. B. A. Lindig and M. A. J. Rodgers, *Photochem. Photobiol.*, **1981**, *33*, 627.
131. I. B. C. Matheson and M. A. J. Rodgers, *J. Phys. Chem.*, **1982**, *86*, 884.
132. P. C. Lee and M. A. J. Rodgers, *J. Phys. Chem.*, **1983**, *87*, 4894.
133. M. A. J. Rodgers and P. C. Lee, *J. Phys. Chem.*, **1984**, *88*, 3480.
134. P. C. Lee and M. A. J. Rodgers, *J. Phys. Chem.*, **1984**, *88*, 4385.
135. I. M. Byteva, G. P. Gurinovich, O. L. Golomb and V. V. Karpov, *Chem. Phys. Lett.*, **1983**, *97*, 167.
136. P. R. Ogilby, K. K. Iu and R. L. Clough, *J. Am. Chem. Soc.*, **1987**, *109*, 4746.
137. R. L. Clough, M. P. Dillon, K. K. Iu and P. R. Ogilby, *Macromolecules*, **1989**, *22*, 3620.

Chapter 4

Bimolecular Photoreactions in Crystals

K. Venkatesan

**Department of Organic Chemistry,
Indian Institute of Science, Bangalore, India.**

V. Ramamurthy

**Central Research and Development Department,
The Du Pont Company, Wilmington, DE, USA.**

Contents

1. Introduction

The occurrence of photoreactions in the crystalline state was recognized and widely reported by the end of the last century.[1] However, enthusiasm to pursue studies in this area did not exist because little was known at that time about the nature and structure of crystals. Spectacular developments have been made in the study of the structure of molecules, especially in methods of structure determination via x-ray crystallography, so it has now become possible to investigate in depth "structure–reactivity correlations" in the solid state. In this chapter we provide a brief summary of bimolecular photoreactions in crystals. Several recent reviews[2, 3] provide a number of examples of photodimerization in the solid state. Our emphasis in this chapter is toward conceptual developments and crystal engineering. We also touch upon the questions concerning the mechanism of photodimerization in crystals. *Basic understanding that has been reached with respect to bimolecular reactions in crystals is expected to be of value in other solid-state reactions such as unimolecular reactions in crystals and reactions of host–guest complexes in the solid state.*

2. Conceptual Developments
2.1. Topochemical Principles

As early as 1889, Liebermann observed the dimerization of olefins in crystals.[4] In 1918 Kohlschutter proposed that the nature and properties of the products of solid-state reactions are governed by the fact that they take place within or on the surface of the solid.[5] Bernstein and Quimby in 1943 interpreted the formation of α-truxillic and β-truxinic acids from two types of cinnamic acid crystals as a crystal lattice-controlled reaction.[6] From crystallographic investigations, pioneered by Schmidt and his co-workers during the early 1960s, of a large number of cinnamic acids (which exhibit a rich variety of polymorphic forms and photochemical reactivity patterns) emerged the important set of "topochemical rules" connecting the configuration of the product and the crystal structure of the reactant.[7] In Table 1 a list of most of the cinnamic acids whose behavior has been investigated in the solid state is provided. The important results obtained by analyzing the solid state behavior of these are the following:

1. The product formed is governed by the environment rather than by the intrinsic reactivity of the reactive bonds in the crystalline state.

2. The proximity and degree of parallelism of the reacting centers are crucial for the dimerization.

3. There is a one-to-one relationship between the configuration and symmetry of the product with the symmetry between the reactants in the crystal.

Table 1. Photodimerization of Cinnamic Acids in the Solid State

Compound	Nature of Packing	Nature of Dimer	Yield (%)	Notes
Cinnamic Acid (CA)	α	Anti H-T	74	a
	β	Syn H-H	80	a
ortho-hydroxy CA	α	Anti H-T	90	a
meta-hydroxy CA	α	Anti H-T	76	a
para-hydroxy CA	α	Anti H-T	78	a
ortho-methoxy	α	Anti H-T	-	a
ortho-ethoxy CA	α	Anti H-T	93	a
	β	Syn H-H	-	a
ortho-propyloxy CA	α	Anti H-T	94	a
ortho-isopropyloxy CA	α	Anti H-T	97	a
ortho-allyloxy CA	α	Anti H-T	93	a
ortho-methyl CA	α	Anti H-T	83	a
para-methyl CA	α	Anti H-T	95	a
ortho-nitro CA	β	Syn H-H	27	a
meta-nitro CA	β	Syn H-H	60	a
para-nitro CA	β	Syn H-H	70	a
ortho-choloro CA	β	Syn H-H	85	a
meta-choloro CA	β	Syn H-H	70	a
para-choloro CA	β	Syn H-H	71	a
ortho-bromo CA	β	Syn H- H	82	a
meta-bromo CA	β	Syn H- H	91	a
para-bromo CA	β	Syn H-H	90	a
5-bromo-2-hydroxy CA	β	Syn H-H	30	a
5-choloro-2-methoxy	β	Syn H-H	85	a
5-bromo-2-methoxy	β	Syn H- H	50	a
2,4-dicholoro CA	β	Syn H-H	78	a
2,6-dicholoro CA	β	Syn H-H	70	a
3,4-dicholoro CA	β	Syn H-H	60	a
3,4-methylene dioxy CA	β	Syn H-H	74	b
3,4-dimethoxy CA	α	Anti H-T	-	b
α-acetylamino CA	α	Anti H-T	-	c
para-formyl CA	β	Syn H-H	-	d

a Ref. 7.
b Ref. 52.
c Ref. T. Iwamoto, S. Kashio and M. Haisa, *Acta. Cryst.*, **1989**, *C45*, 1753.
d Ref. 39c.

Although there had been sporadic reports relating to solid-state photodimerization earlier, it must be said that the systematic and thorough studies by Schmidt and co-workers laid the foundation for the flowering of this field (Scheme 1).

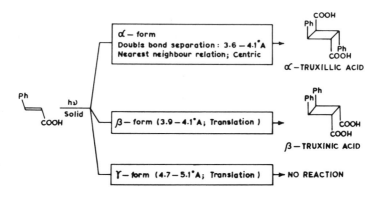

Scheme 1

According to topochemical principles, reactions in the solid state take place with minimum atomic movements. Further, as understood originally the principle implies that dimerization can be expected to take place only if the reactive double bonds are within the correct distance (the bonds being separated by not more than 4.2 Å in the case of dimerization) and the bonds are parallel. These conditions ensure near perfect orientation of the p_z atomic orbitals on each of the four reacting centers.

The topochemical rules have enabled investigators working in organic solid state photochemistry to rationalize their observations of a large number of [2+2] photodimerization reactions. Striking examples of some of the observations on different organic systems are dimerizations of fumaryl derivatives,[8] acenapthalic acid,[9] heterocyclic analogues of *trans*-cinnamic acid,[9] butadiene derivatives,[10] benzylidene cyclopentanones,[11] and coumarins.[12] For a few examples see Schemes 2 and 3 and for an extensive list of reactions see refs. 2 and 3. In addition to cinnamic acids several substituted coumarins and benzylidene cyclopentanones have been investigated in the solid state. A list of reactive compounds investigated are provided in Tables 2 and 3. A collection of selected articles by Schmidt has also appeared in the form of a monograph.[13]

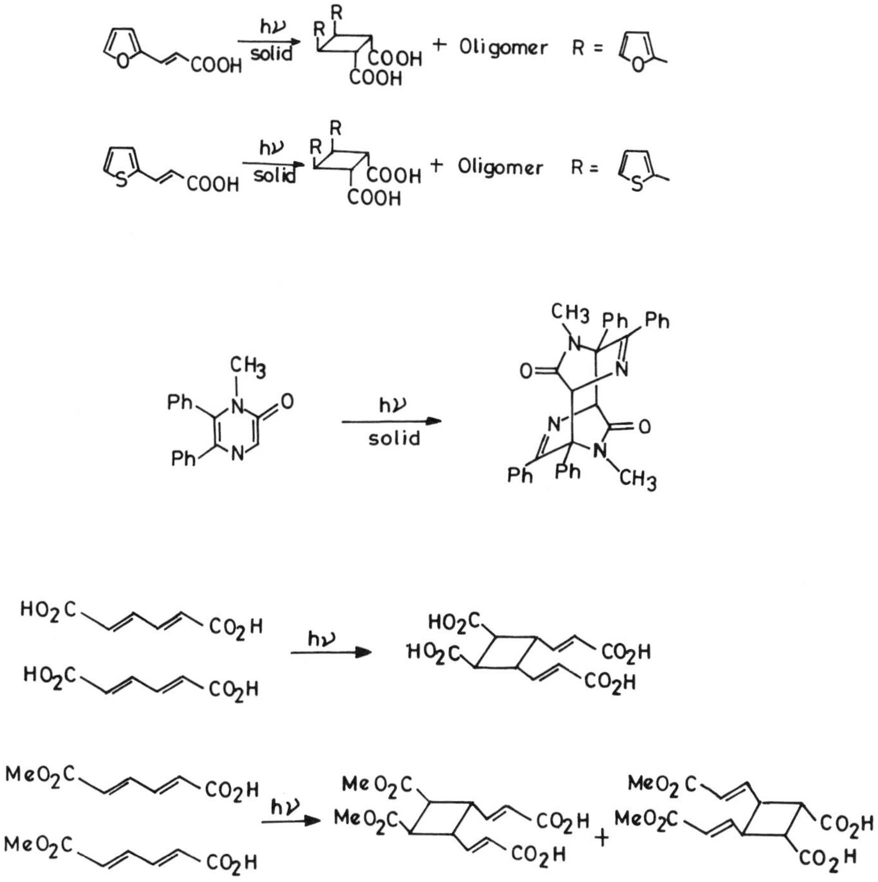
hν
solid
COOH
R
+ Oligomer
R =
hν
solid
COOH
R
+ Oligomer
R =
CH3
Ph
N
O
Ph
N
hν
solid
CH3
Ph
Ph
N
O
N
N
O
N
Ph
Ph
CH3
HO2C
CO2H
HO2C
CO2H
hν
HO2C
HO2C
CO2H
CO2H
MeO2C
CO2H
MeO2C
CO2H
hν
MeO2C
MeO2C
CO2H
CO2H
+
MeO2C
MeO2C
CO2H
CO2H

Scheme 2

Table 2. Photodimerization of benzyl benzylidenecyclopentanone[a]

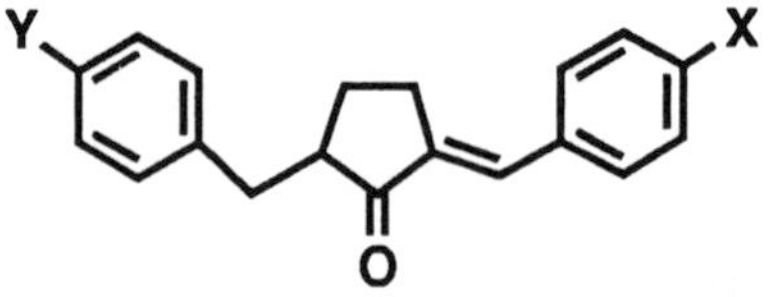

Compound	Nature of Dimer	Yield (%)	Characteristic of the Reaction
X = Y = H	Anti H-T	100	Single-crystal–Single-crystal
X = p-Br, Y = H	Anti H-T	100	Single-crystal–Single-crystal
X = H, Y = p-Cl	Anti H-T	100	Single-crystal–Single-crystal
X = H, Y = p-Me	Anti H-T	100	Single-crystal–Single-crystal
X = H, Y = P-Br	Anti H-T	100	Single-crystal–Single-crystal

[a] Ref. C. R. Theocharis in *The Chemistry of Enones*, S. Patai and Z. Rappoport, eds., John Wiley, NewYork, 1989, p. 1133.

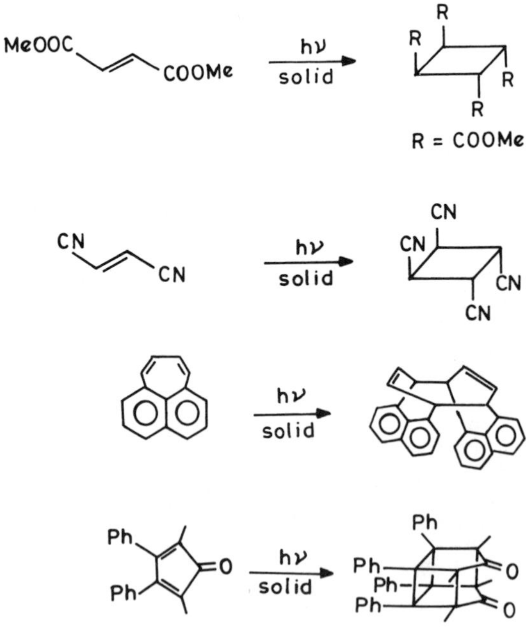

Scheme 3

Table 3. Photodimerization of Coumarins in the Solid State[a]

(syn head-head)   (anti head-head)

(syn head-tail)   (anti head-tail)

Coumarins	Nature of Packing	% Yield	Nature of Dimer	Notes
Coumarin	γ		Three dimers	
6-Chloro Coumarin	β	100	Syn H-H	a
7-Chloro Coumarin	β	70	Syn H-H	a
4-Methyl-6-Chloro Coumarin	β	50	Syn H-H	a
4-Methyl-7-Chloro Coumarin	β	80	Syn H-H	a
4-Chloro Coumarin	NT[d]	25	Anti H-H and Syn H-T	a
7-Methyl Coumarin	NT[d]	65	Syn H-H	a
6-Methoxy Coumarin	β	60	Syn H-H	a
7-Methoxy Coumarin	–	90	Syn H-T	a
8-Methoxy Coumarin	α	50	Anti H-T	a
6-Acetoxy Coumarin	β	70	Syn H-H	a
7-Acetoxy Coumarin	β	90	Syn H-H	a
4-Methyl-7-Acetoxy Coumarin	β	80	Syn H-H	a
6-Flurocoumarin	β	100	Syn H-H	b
7-Flurocoumarin	β	100	Syn H-H	b
7-Iodocoumarin	β	40	Syn H-H	b
6-Bromocoumarin	β	90	Syn H-H	c
7-Bromocoumarin	β	100	Syn H-H	c

[a] Ref. N. Ramasubbu, K. Gnanaguru, K. Venkatesan and V. Ramamurthy, *J. Org. Chem.*, **1985**, *50*, 2337.
[b] R ef. N. S. Begum, V. Amerendra Kumar and K. Venkatesan, Unpublished results.
[c] Ref. P. Venugopalan and K. Venkatesan, Unpublished results.
[d] NT: Non-topochemical or defect initiated.

2.2. Nontopochemical Photodimerization: Role of Defects

At a stage when it was considered that one could understand product formation via the topochemical rules, the photochemical behavior of 9-cyanoanthracene and 9-anthraldehyde was at variance.[14] Whereas head–head photodimers were expected based on their crystal structures, what was obtained was the head–tail isomer (Scheme 4). While the topochemical rules were being formulated, in a communication in 1966, which is of great significance in solid-state organic photochemistry, Craig and Sarti-Fantoni[14] showed that reactions of 9-cyanoanthracene and 9-anthraldehyde took place at defects or surfaces or in zones that were disordered. From interference contrast and fluorescence microscopy investigations, Thomas et al.[15] characterized the nature of the defects. Luminescence studies by Ludmer[16] and Ebeid and Bridge[17] further provided support for the importance of defects in the reactivity. Since the appearance of these papers demonstrating the importance of defects in solid-state dimerization, considerable information on the role of defects, particularly dislocations, has accumulated on the reactions of organic crystals. For example, a systematic chemical and crystallographic study by Schmidt and co-workers[18] brought to light many cases of substituted anthracenes behaving in a nontopochemical fashion (Scheme 5). Examples shown in Scheme 5 clearly illustrate that unlike cinnamic acid derivatives, the stereochemistry of the product dimer from anthracenes cannot be predicted on the basis of crystal packing.

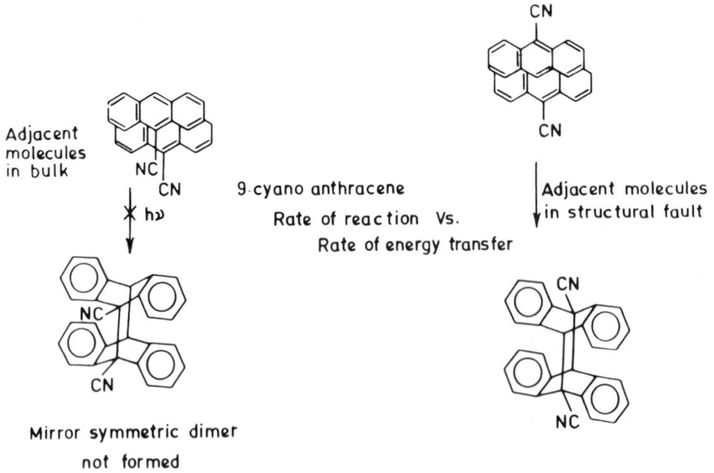

Adjacent molecules in bulk

9-cyano anthracene
Rate of reaction Vs.
Rate of energy transfer

Adjacent molecules in structural fault

Mirror symmetric dimer
not formed

Scheme 4

The crystal structure of anthracene shows that no molecules are separated by <4 Å , yet upon irradiation a dimer is readily formed.[19] Chandross and Ferguson[20] originally suggested that the free surface, rather than the bulk of the crystalline anthracene, was the locus of the reaction. Thomas and Williams,[21] using novel etch-pit studies, showed that crystal defects may function as the preferred centers for reaction, since anthracene molecules have their excitation energies slightly reduced when they are displaced from regular lattice sites. However, later studies[22] via electron microscopy revealed the coexistence inside "normal" anthracene crystals of regions of a metastable phase and showed that moderate stress produced crystallites of a new phase in coherent contact with the parent crystal matrix. In the new phase (space group P1), the C9 $\cdots$ C9' distance is 4.2 Å, whereas in the original crystal it is 4.5 Å. Other examples in which defects have been proposed as the loci for photodimerization have been reported in the literature.[23] Excellent reviews have appeared dealing with various kinds of structural imperfections, their characterizations by physical techniques, and their relevance to photo-transformations.[24]

Obviously, rationalization of the observed products via nontopochemical processes is by no means straightforward. Diverse techniques such as electron microscopy, optical microscopy, chemical etching, atom–atom potential calculations, and interference contrast and fluorescence microscopy have been brought into the arsenal to identify the type of defect. The potential of x-ray diffraction topography with the recent availability of intense x-ray beams from synchrotron sources[25] could play an important future role in characterizing the nature of defects.

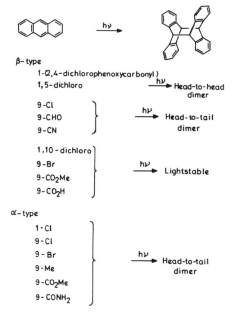

Scheme 5

It has been observed in the photoreactions taking place nontopochemically that there is, in general, only one product obtained. This suggests that even in the so-called defect sites there is some order within a microscopic region. Two exceptions to this general observation have come to light recently. In 4-chlorocoumarin crystals two products (anti head–head and syn head–tail) have been obtained, the total yield being 25%.[26] Crystal packing shows that the reaction cannot be topochemical in origin. In contradistinction to the earlier reports that coumarin is photostable in the solid state[12] while dimerizing in solution,[27] it has been observed that upon irradiation in the solid state coumarin produces three dimers (syn head–tail, syn head–head and anti head–head) in a total yield of about 20% after 48 h of irradiation.[28] Careful experimentation has shown that the reaction occurs only in the solid phase. The crystal packing as revealed by x-ray structural analysis shows that the reaction is clearly nontopochemical. The occurrence of more than one isomer in 4-chlorocoumarin as well as coumarin dimerization is intriguing and remains unexplained. The presence of more than one type of defect, each having its own packing arrangement, cannot be ruled out. The possibility of long-range disorder of the molecular orientation in the crystals,[22] as observed in anthracene and clearly discerned from single-crystal structure analysis, is eliminated in these two structures. It is also significant that most compounds that exhibit nontopochemical photobehavior are conformationally rigid and essentially planar in the ground state.

Studies on the detection and characterization of defects and their role in the photochemistry of crystals should not be interpreted in terms of the breakdown of the topochemical principles discovered by Schmidt and Cohen from x-ray results. x-Ray crystallographic results pertain to the macroscopic ordered region of the crystal and say nothing about the packing mode in microscopic defect regions.

2.3. The Concept of the Reaction Cavity

The above discussion on topochemical rules and defects leaves the impression that reactions in the crystalline state can be rationalized on the basis of defects when molecules are not suitably (topochemically) oriented for reaction in the crystal. However, this is not generally the case. While the origin of reactions in crystals where the molecules are not topochemically oriented may be understood on the basis of defects, one is at a loss to rationalize situations where molecules do not react even when they are topochemically oriented. Indeed there are such examples in the literature (Table 4 and Scheme 6).[29] A qualitative concept developed by Cohen[30] has been of value in this context and is presented in this section. This concept also facilitates understanding cases where reaction occurs in spite of a less than ideal arrangement of reacting partners, and these will be discussed in a later section.

Table 4. Examples of Exceptions to Original Topochemical Principles Regarding Distance[a,b]

Compound	Packing Type	Double Bond Distance	Reactivity	Nature of Dimer
Methyl-*p*-iodo cinnamate (1)	β-type	4.3 Å	Yes	mirror symmetric
7-Chlorocoumarin (2)	β-type	4.45 Å	Yes	syn head-head
Eteretinate (3)		4.4 Å	Yes	-
p-Formyl cinnamic acid (4)	β-type	4.83 Å	Yes	mirror symmetric
Distyryl pyrazine (5)		4.19 Å	No	-
Enone (6)		3.79 Å	No	-
4-Hydroxy-3-nitrocinnamate (7)		3.78 Å	No	-
Benzylidene-dl-pipertone (8)		4.0 Å	No	-
(+) 2,5-Dibenzylidene-3-methyl cyclopentanone (9)		3.87 Å	No	-
2-Benzylidene cyclopentanone (10)		4.14 Å	No	-
o-Cl Benzylidene-*dl*-Piperitone		3.94 Å	No	-

[a] For structures of compounds see Scheme 6.
[b] For details see G. S. Murthy, P. Arjunan, K. Venkatesan and V. Ramamurthy, *Tetrahedron*, **1987**, *43*, 1225.

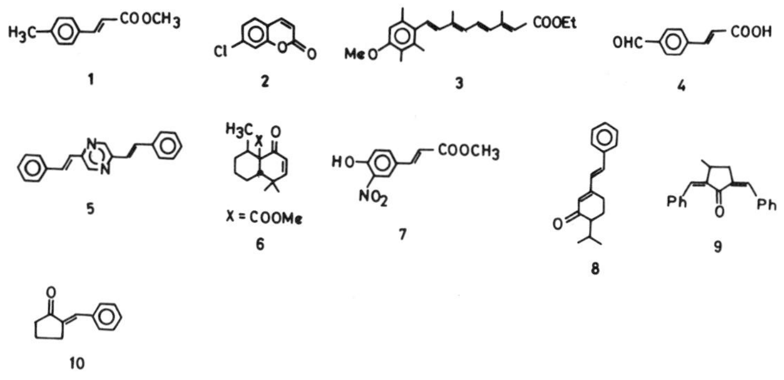

Scheme 6

Once a compound has been crystallized, the template, either for good or otherwise, has been cast for the reaction. The topochemical postulate derives from this point. However, the postulate lacks precision in the following details: (1) Do the immediate neighbors of the reacting partners have any role to play? (2) Does the postulate consider the changes in the molecular geometry upon excitation? In order to take these into account at the phenomenological level, Cohen proposed the idea of the reaction cavity.[30] The cavity or cage is the space in the crystal occupied by the reacting partners. The atomic movements following the reaction exert pressures on the cavity wall, which becomes distorted. However, the close packing works against large-scale changes in shape, so that only minimal change can occur (Figure 1). This concept has been of help in qualitatively understanding the course of a variety of solid-state reactions. Cohen et al. have used this concept to understand the geometries of the excimers from polyaromatics in the crystal.[31] Scheffer and co-workers have made use of this argument to gain insight into the mechanism of intramolecular photorearrangements of enones.[32]

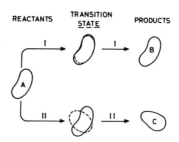

Figure 1. The reaction cavity of a favorable and an unfavorable reaction in the crystalline state (ref. 30).

The usefulness of this concept is readily apparent when applied to photostable crystals which would be expected to be otherwise on the basis of topochemical principles. The separation distances between reactive double bonds are less than 4.2 Å in compounds **5–10** (Scheme 6) listed in Table 4 but they do not undergo dimerization upon photolysis. The exceptional situations in all these cases can be understood qualitatively by invoking the "reaction cavity" concept.

In this context results of the lattice energy calculations performed on these systems have been revealing. These calculations were performed using the computer program WMIN developed by Busing[33] on a large number of photodimerizable olefins.[29] It may be stressed that in these calculations only the relative values within a series are meaningful in view of the many approximations made. It is important to point out that the program WMIN does not allow the environment of the reactants to be kept fixed while allowing the reactants to undergo movements. Although the calculations have been carried out using the ground-state geometry with the dispersion constants appropriate to the ground state, the results provide some insight. For example, in the case of 7-chloro-coumarin (for packing arrangement see Figure 2) it is calculated that the rise in the lattice energy to achieve the ideal geometry (see Table 4) for the translated pair (separated by 4.45 Å) is 177 kcal mol^{-1}, whereas for the centrosymmetric pair (separated by 4.12 Å) the energy increase is as large as 18,083 kcal mol^{-1}. This shows that the reaction pathway leading to the experimentally observed syn head–head dimer is energetically more favorable than the anti head–head isomer. Calculations performed on enone 6 also reveal that there is a significant increase in energy (1504 kcal mol^{-1}) as the pair moves along the dimerization reaction coordinate. These numbers are to be compared with the values obtained for ideally oriented systems which undergo reaction (e.g., 6-acetoxycoumarin, ~8.3; 7-acetoxycoumarin, ~167.5; 8-methoxycoumarin, ~45 kcal mol^{-1}). The examples provided in Table 4 suggest that in spite of favorable arrangement, neighboring molecules may not permit any movement by the reacting partners. These calculations clearly show the role of immediate neighbors on the reactivity of crystals. Although the approach via potential energy considerations allow us to rationalize the photobehavior of the compounds mentioned above, it does not bring out in quantitative terms all aspects implied in the reaction cavity concept. In particular it must be noted that the calculations represent the ground-state situation.

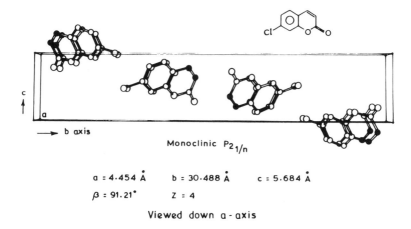

Figure 2. Packing arrangement of 7-chloro coumarin in the crystalline state. Note the presence of two pairs of reactive 7-chloro coumarin.

While in the examples discussed above the cavity *restricts* the motion of reacting molecules, there are cases in which some amount of motion is tolerated within this cavity. In other words, the presence of void space around the reacting partners, the size of which may vary from system to system, can favor reaction between less than ideally oriented pairs. Gavezzotti[34] has developed a computer program that allows us to calculate packing density maps. Such maps depict the void zones or holes or channels in the crystal structure. Starting from the premise that crystal reactivity requires the availability of free space around the reaction site, Gavezzotti has employed packing density maps to interpret a variety of solid-state reactions. This methodology developed by Gavezzotti can be an invaluable aid for understanding the photobehavior of organic crystals.

2.4. The Concept of Dynamic Preformation: Role of Photoinduced Lattice Relaxation

The static concept of preorganization does not correspond to reality in as much as it does not take into account the changes caused by molecular excitation. Excitation of molecules to higher electronic levels brings about changes, among other things, to the geometry and polarizability of molecules. For example, it is well known that formaldehyde undergoes pyramidalization upon excitation with a corresponding change in dipole moment.[35a] For olefins the preferred minimum energy configuration in the excited state is the perpendicular (orthognal p orbitals) rather than planar form.[35a] It is also established that for some aromatics, dimeric complexes, i.e., excimers, are stabilized with respect to monomers in the excited state.[35b] Such differences in geometry and polarizability between the ground and the reactive state (excited) is expected to have subtle consequences for the topochemical postulates based on ground-state properties. Craig was the first to recognize this aspect of solid-state dimerization.[36]

In the ground state the crystal is expected to be homogeneous and the forces operating between molecules in the crystals are expected to be uniform. However, upon excitation the crystal will contain two types of molecules, most in the ground state and a few in the excited state. The forces operating between an excited molecule and its neighbors differ from those operating between a ground-state molecule and its surroundings. The change in polarizability upon excitation increases the attractive part of the intermolecular force, while the repulsive part remains, initially, unchanged. *The localized excitation produces a particular type of local instability of the lattice configuration which may lead to large molecular displacements.* The displacements may favor the formation of excimers and photodimers in crystals. Starting from these arguments, Craig and coworkers[36] have carried out an incisive theoretical investigation of this problem and have shown that a short-term lattice instability created upon excitation has the effect of driving one molecule close to a neighbor, thus promoting excimer or exciplex formation. With information regarding the change in polarizability on excitation from the ground state available for anthracene, lattice energy calculations for anthracene, 9-cyanoanthracene, and 9-methylanthracene were carried out by Craig and Mallett.[36b] In the calculation of the potential energy between one excited and one ground-state molecule Craig and Mallett used the following expression:

$$V_E = -a^*/r^6 + b/r^{12} \tag{1}$$

Here the dispersive constant a^* corresponds to the molecule in its excited state and is greater than a for the molecule in the ground state. The calculation for 9-cyanoanthracene showed that, for a short period after excitation, an excited molecule can be displaced away from its equilibrium crystal lattice position into an unsymmetrical local structure, with the excited molecule closer to one neighbor in the stack of molecules than to the other. In such a model there is a transient preformation of an excimer not evident in the equilibrium local structure. For 9-methylanthracene there is also significant movement on excitation. Similar observations have been noted from the calculations of the mixed crystal of 9-methoxyanthracene and 9-cyanoanthracene as well as 9-cyano and 9,10-dimethyl-anthracene mixed crystals.[36] It may be added that the calculations were performed with a fixed environment. To summarize their important results, short-term lattice instability caused by photoexcitation can have the effect of driving one molecule close to a neighbor so as to cause a photochemical reaction. If a reaction proceeds in the short time available before general lattice relaxation to the equilibrium ground-state structure it can be said to be dynamically preformed.

The important message of the investigations by Craig and co-workers is that it is of the utmost importance to consider the dynamic properties of lattices (caused by photoexcitation) to understand the processes involved in photochemical reactions in crystals. This also implies that the dimerization may occur within a reaction cavity under conditions where the molecules are less than ideally oriented. The driving force to bring the pair into proper orientation will be provided by electronic excitation energy and the increased attractive interaction energy in the excited state.

2.5. Topochemical Rules as Refined by the "Reaction Cavity" and "Photoinduced Lattice Instability" Concepts

Recent studies carried out on the photodimerization of several crystalline olefins have warranted some modifications with respect to the conditions on the question of parallelism of double bonds and the distance criterion as they relate to the principle of least motion. Before discussing these points it is useful to define a few relevant geometrical parameters. These additional geometrical parameters as defined below are introduced to identify the relative orientation of π-orbitals involved in the dimerization. Relative arrangement of the double bonds and the orientation of the π orbitals can be identified through the following geometrical parameters: the center–center distance; angles θ_1, θ_2 and θ_3; and the displacement of the double bonds with respect to each other (Figure 3).[12a] Angle θ_1 corresponds to the rotation of one double bond with respect to the other when projected down the line perpendicular to the plane containing one of the double bonds and atoms connected to this double bond. θ_2 corresponds to the obtuse angle of the parallelogram formed by the double bonds $C_{3'}=C_{4'}$ and $C_3=C_4$, and θ_3 is the angle between the least-square plane through the atoms C_3, C_4, $C_{3'}$, and $C_{4'}$ and that passing through atoms $C_{2'}$, $C_{3'}$, $C_{4'}$, and $C_{10'}$. For the best overlap of the π orbitals of the reacting partners, the values for θ_1, θ_2, and θ_3 must be 0, 90, and 90°, respectively. Kearsley has used a simple numerical description for the overlap of the reactive orbitals to rationalize the reactivity of several compounds

(Figure 4).[37] The orbital overlap is defined as the sum of the distances between reacting orbital lobes 1 and 2 (SUM = lobe 1 + lobe 2). Lobe 1 in the above equation corresponds to the distance between T and T' of the reacting p orbitals; lobe 2 corresponds to that between the other pair of the p orbitals (Figure 4). According to Kearsley the distance between T and T' of the reacting orbitals is a description of the amount of orbital overlap and to some extent their orientation with respect to each other. The above two approaches attempt to describe geometrically the relative orientation of the reacting π orbitals. We show below that while both of these descriptions widen the scope of the topochemical rules further, the real criterion for dimerization depends on the relative flexibility of the immediate surroundings (reaction cavity).

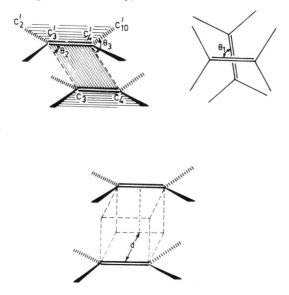

Figure 3. Geometrical parameters used in the relative representation of reactant double bonds (ref. 12a).

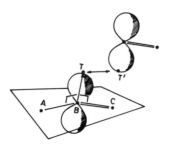

Figure 4. Geometrical parameters used to measure the overlap between the reactant double bond π-orbitals (ref. 37).

Table 5 (see Scheme 7 for compound structures) records the calculated values of the above angles and Kearsley's overlap parameter (SUM) for a few compounds whose photochemical behavior has been investigated in the crystalline state.[38] It is seen from Table 5 that although the deviations from the ideal values for θ_1, θ_2, and θ_3 are significant, most of the crystals are photoreactive. Particularly noteworthy are 7-methoxycoumarin (14)[12a] and 2,5-dibenzylidene cyclopentanone (15)[11, 39a] with the values for θ_1 being 67.5° and 56°, respectively. It is remarkable that although the relevant olefinic π orbitals are not overlapping in their ground state geometry both are photoreactive. These cases in which the nonparallel alignment of the π orbitals does not inhibit photoreactivity indicate that there must be enough freedom for the reactive molecules to undergo the necessary movements to reorganize in their respective crystal lattices to allow dimerization to occur.

Table 5. Examples of Exceptions to Original Topochemical Principles Regarding Parallelism of Double Bonds[a, b]

Compound	Angle θ_1	Dimerization	SUM
Methyl m-bromocinnamate (11)	38.2°	No	3.19
1,1'-Trimethylene-bis-thymine (12)	6°	Yes	1.68
2,2 (2,5)-Benzoquinophane (13)	3°	Yes	-
7-Methoxycoumarin (14)	67.5°	Yes	3.56
2,5-Dibenzylidene cyclopentanone (15)	56°	Yes	1.86
1,4-Dicinnamoyl benzene (16)	28.5°	Yes	2.41
1-Methyl-5,6-diphenyl pyrazine-2-one (17)	24°	Yes	2.86

[a] For structures of compounds see Scheme 7.
[b] For details see G. S. Murthy, P. Arjunan, K. Venkatesan and V. Ramamurthy, *Tetrahedron*, **1987**, *43*, 1225 and S. K. Kearsley in *Organic Solid State Chemistry*, G. R. Desiraju, ed., Elsevier, Amsterdam, 1987, p. 69.

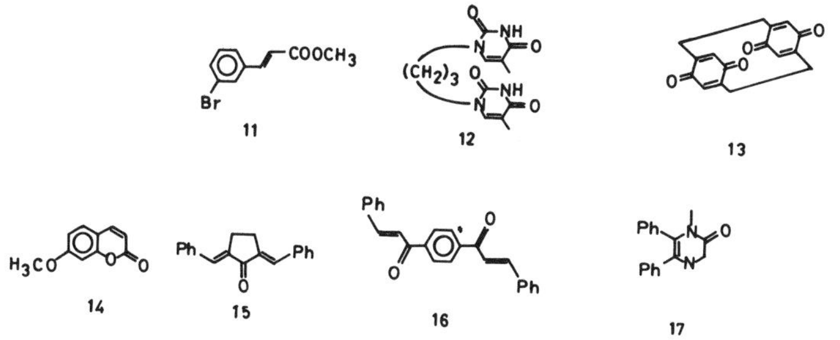

Scheme 7

There are also examples in which violation of the original distance criterion does not prevent reaction (Table 4). A perusal of Table 4 and Figure 2 shows that in 7-chlorocoumarin[12e] the incipient dimer pair molecules are related by translation with the reacting groups being separated by 4.45 Å, a distance that would be expected to be unfavorable for the reaction. This is all the more striking as the centrosymmetrically related double bonds in the crystal are separated by only 4.12Å. It has been observed that crystals of etretinate (3)[39b] dimerize yielding two dimers (Scheme 8). The center–center distance for the two sets of dimerizable bonds in etretinate are 3.8 Å and 4.4 Å, the latter being outside the presently accepted limit. The case of *para*-formylcinnamic acid (4)[39c,d] is most unusual as the double bond separation is as large as 4.83 Å. However the plane-to-plane perpendicular distance between the reacting molecules is only 3.88 Å. It is clear that the dictum requiring the reacting double bonds to be within 4.2 Å is no longer operational. Further experimental observations may very well stretch the upper limit.

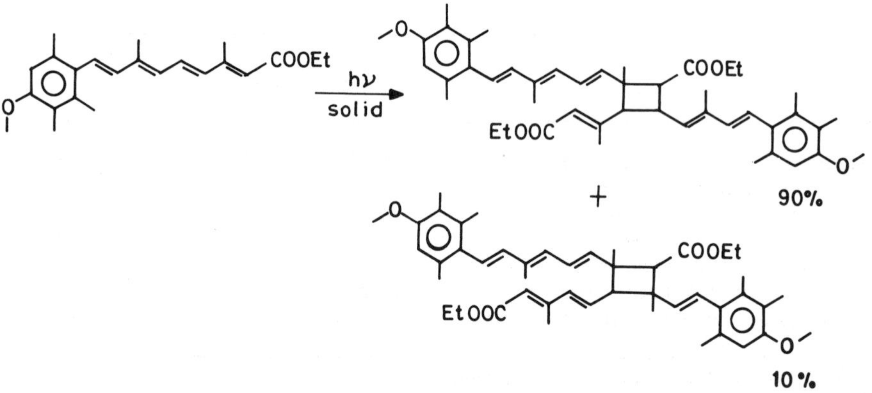

Scheme 8

Ad hoc rationalization has been offered to explain every one of anomalous cases. It is our opinion that all these examples which appear anomalous in the light of topochemical rules can be understood on a unified conceptual basis if one combines the reaction cavity concept of Cohen[30] with the photorelaxability concept introduced by Craig.[36] Dimerization may be considered as taking place in a "microcavity" in the bulk crystal, it being the host and the reactive pair the guest. The size and shape of the cavity and the interactions between the "guest reactants" and the host lattice will determine whether the nontopochemically arranged molecules will be permitted to undergo the motion necessary to reach a topochemical arrangement. Light absorption provides energy for such a motion. Therefore, most examples in which dimerization occurs in spite of poor topochemical arrangement (in Schmidt's sense) can be understood on the basis of lattice flexibility.

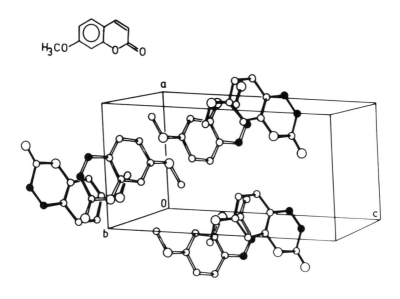

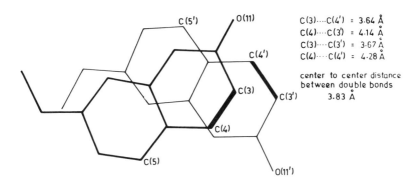

C(3)····C(4') = 3·64 Å
C(4)····C(3') = 4·14 Å
C(3)····C(3') = 3·67 Å
C(4)····C(4') = 4·28 Å

center to center distance
between double bonds
3·83 Å

Perspective view of the asymmetric unit (crystal coordinates)
(z axis through center of mass perpendicular to the plane of the molecule)

Figure 5. Packing arrangement of 7-methoxycoumarin in the crystals: unit cell representation (top) and one molecule projected over the other (below).

We illustrate this point with two revealing examples, 7-methoxycoumarin[12f] and *meta*-bromomethylcinnamate (11).[38a] According to bond overlap criterion of Kearsley, both of these would be expected to react since their SUM is less than 3.5. In spite of unfavorable arrangement (the reactive double bonds are rotated by 65° with respect to each other and the center-to-center double-bond distance is 3.83 Å, see Figure 5), photodimerization occurs in crystals of 7-methoxycoumarin to give the syn head–tail isomer. Indeed it has been observed that the reaction is very fast, with a yield of about 95% of the dimer after about 10 h of irradiation. Defects are not the loci for the reaction. The orientational flexibility of the molecules in the crystal lattice has been shown to be feasible by lattice energy calculations. Indeed, the energy increase needed to bring the two reactant molecules together to obtain the right isomer is about 200 kcal mol^{-1}, roughly the same order of magnitude as for many photoreactive crystals with favorably oriented pairs. It may be emphasized that the calculations have been performed using the dispersion constants applicable to the ground state rather than to an excited molecule. In spite of the very severe approximations in the approach employed, the results obtained throw light on the influence of environment in the course of the reactions in crystals and shows that the product obtained is one for which presumably the transition state involves *the least change in the shape of the reaction cavity*. This example when compared to the photoinertness of *meta*-bromomethylcinnamate is interesting. In this case, similar to 7-methoxycoumarin, the double bonds are not ideally oriented for topochemical dimerization. The distance between the centers of adjacent double bonds is 3.93 Å, but the double bonds are not parallel. They make an angle of 28° when projected down the line joining the centers of the bonds. In light of the observations on the behavior of 7-methoxycoumarin one could expect this molecule also to undergo dimerization in the solid state. However, it is photostable. The reason becomes obvious when one examines the results of lattice energy calculations. The energy increase to align the molecules parallel to each other in a geometry suitable for dimerization is enormous (6726 kcal mol^{-1}) when compared to 7-methoxycoumarin. Such a large increase in the lattice energy will not favor reorganization of the molecule to result in photodimerization. Thus "dynamic preformation" would favor dimerization but may well be resisted by the nearest neighbors. In other words, the crystal as a whole has a role to play - it is both an observer and an actor.

It is clear that in addition to relative atomic positions, relative orientation of the reactive π orbitals must be monitored to assess the feasibility of dimerization in the solid state. Less than ideal atomic and orbital orientations can still give rise to dimerization if the surrounding lattice can tolerate motions that would steer the molecules to proper mutual orientation. In order for the dimerization to occur within the cavity under conditions where the molecules are not so ideally oriented, there must be a driving force that brings the pair into proper orientation. The driving force is the attractive force between an excited and a ground-state molecule. Even for such less than ideal topochemical reactions it is reasonable to expect that there will be upper limits beyond which reaction does not take place. It is our opinion that there will be distance and geometry limitations beyond which an excited molecule will not be able to reach out, attract, and alter the geometry of a nearby molecule. This will depend, among other things, on the nature ($n\pi^*$, $\pi\pi^*$, etc.) and the spin (singlet, triplet) of the reactive excited state and on the type of molecules involved in dimerization. It is important to note that predictions concerning excited-state reactivity are made based on accurate ground-state geometries and packing arrangements obtained crystallographically. Accurate predictions are possible only if the difference in geometry between the ground and the reactive excited states is taken into account. In this context, it is of interest to note that the possibility of obtaining x-ray crystal structures of molecules in excited states is being considered.[40]

3. Crystal Engineering

From the above discussion it is clear that in order for a reaction to occur in the crystalline state one has to have the molecules preorganized in the desired pattern in the crystals. This is easier said than done. The central problem in organic solid-state photochemistry is concerned with steering molecules so as to obtain an organic crystal structure of a predetermined form. Schmidt[41] has termed this operation "crystal engineering." One of the major problems encountered here is lack of complete understanding of the intra- and intermolecular interactions leading to the observed crystal packing. Indeed, the factors called into play are so subtle that even a minor change in the molecule can result in major changes in crystal structure. However, this very fact could be exploited in producing the desired packing mode. If one had a complete understanding of the ways in which inter- and intramolecular interactions control packing of molecules in crystals it would be feasible to design template groups, perhaps of temporary attachment, to the functional molecules to guide photochemically reactive groups into appropriate juxtaposition in crystals. In order to bring the reactive molecules into proper orientations three distinct strategies have been employed. These are intramolecular substitution, mixed-crystal formation, and more recently inclusion within host structures. We briefly discuss each one of these below.

3.1. Intramolecular Substitution

In this approach the parent molecule that fails to pack in a desired arrangement is intramolecularly substituted with groups that are known to steer the packing arrangement through weak "intergroup" interactions. Hopefully, under such conditions the "altered parent" molecules will pack in the desired arrangement and undergo the expected photoreaction in the crystal. It is necessary to realize that one is no longer dealing with the same parent molecule and the molecule undergoing reaction is altered by an intramolecular substitution. This strategy will be profitable only if one can utilize a group that can be easily attached before crystallization and detached after the reaction. At this stage, no such group has been identified.

Crystal packing is controlled by a large number of less than obvious weak forces that are not often considered by chemists utilizing isotropic solvents as reaction media. Some of these include C—H···O, X···X (where X is a halogen), S···S, S···X, and C=O···X interactions. Crystallographers in recent times have become interested in identifying a common packing pattern in similarly substituted molecules and in understanding the relative importance of the weak forces in controlling the packing pattern.[42] This is facilitated by the large amount of information available in the Cambridge Crystallographic Data Base.[43] This approach has resulted in identifying a few groups that may be suitable for steering dimerization reactions. It is important to realize that the identified group may not be universal and may be useful or useless depending on the presence of other functionalities in the molecule. One may be well advised to examine the literature carefully before embarking on a tedious synthesis of a substituted "parent" molecule. At this stage there is no easy solution to this problem.

3.1.1. Halogen Atom Substitution

Schmidt and co-workers[44] recognized quite early that monochloro substitution and especially dichloro substitution in aromatic molecules tend to steer molecules in crystal lattices with a short axis of ~4 Å, the so called β structure. Sparked by their initial important observations several groups working in the area of solid-state photochemistry investigated the effectiveness of the chloro group to engineer different molecular systems to attain the β packing mode.[2] A few examples are provided in Scheme 9. Thomas and co-workers have employed chloro substitution in the 2-benzyl-5-benzylidene cyclopentanone framework (Table 2).[11] Recent results on the photodimerization of coumarins in the solid state are noteworthy (Table 3).[12] It has been observed that coumarin undergoes photodimerization nontopochemically, yielding three dimers.[28] However, all of the five chlorocoumarins investigated underwent clean dimerization in the solid state.[12] Syn head–head dimers were obtained in 6-chloro-, 7-chloro-, 4-methyl-6-chloro- and 4-methyl-7-chlorocoumarins as a direct consequence of their β-packing structure, although 4-chlorocoumarin reacted nontopochemically. The role of intra- and interstack Cl···Cl interactions in the stabilization of crystal structures has been selectively reviewed by Sarma and Desiraju.[45a]

There have been several theoretical studies reported on the nature of Cl···Cl interactions.[46] From the crystal structure data for Cl_2,[47] Br_2,[48] and I_2,[49] it has been observed that the intermolecular contacts between Cl···Cl, Br···Br, and I···I are much shorter than the sum of the van der Waals radii, indicating the presence of specific attractive interactions. This has been confirmed from the analyses of the packing arrangement of a large number of chloro-substituted organic molecules. Making use of the Cambridge Data Base statistical analyses of Cl···Cl distances in organic structures have been carried out independently by us[12d] and by Desiraju.[45] The geometrical parameters used in our analyses are shown in Figure 6 and the results of these analyses are given in Figure 7. It is noteworthy from the plot of N (No. of interactions) vs. d (Cl···Cl) shown in Figure 7 that when $\chi \approx 0°$ most of the Cl⋯Cl distances lie within a narrow range of 3.8–4.0 Å, whereas the range is broad (3.5–4.2 Å) when $\chi \approx 180°$ (for the definition of χ see Figure 6). The observed smaller width for $\chi \approx 0°$, which corresponds to the β-type packing (in topochemical terms), may be attributed to the additional attractive interactions between the aromatic rings in this packing mode. It may be mentioned that the intermolecular Cl⋯Cl contact in the structure of Cl_2 is as short as 3.27 Å, whereas in general Cl⋯Cl contact distance in the chloro substituted aromatics is much longer than this value (Figure 7). It appears from the experimental data available so far that chlorine is a good steering group, although there are some failures. For example attempts to use it for achieving a β structure in the case of benzylidene-*dl*-piperitone have not been successful.[50] Whereas both *para*-chloro and *para*-bromo derivatives give rise to the centrosymmetric α structure, the *ortho*-chloro derivative is photostable (Scheme 10). This seems to indicate that molecular shape, in the present case nonplanar, may also dictate the type of packing modes so as to attain overall close packing of molecules in the unit cell overriding the specific halogen⋯halogen interaction. From the crystallographic and theoretical studies on Br_2 and I_2 it appears that both bromine and iodine could also be useful steering groups but not much work has been done in this direction (see Tables 1 and 2). From the work of Schmidt and Cohen[7] and the more recent studies on bromo coumarins[51] there are indications that bromine may also be a good steering group to produce β-type structures.

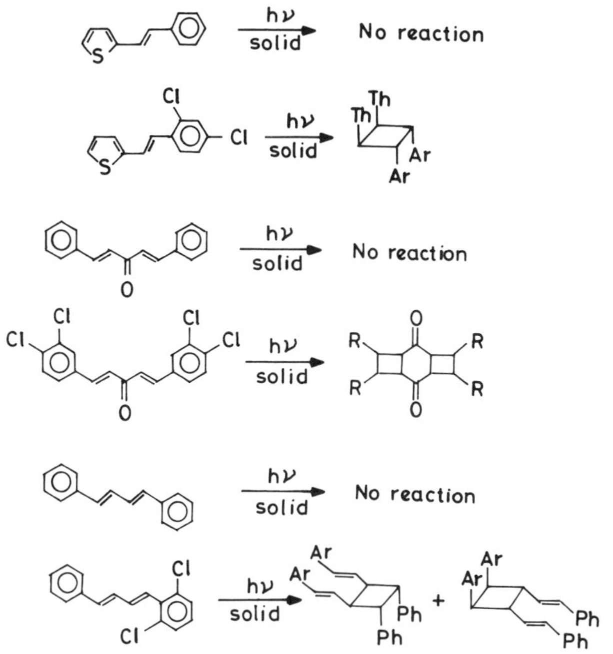

Scheme 9

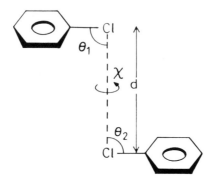

Figure 6. Geometrical parameters analysed for Cl- - - Cl interactions of aromatic compounds

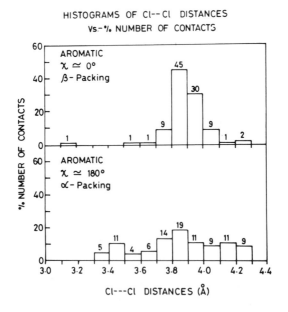

Figure 7. Histograms of Cl- - -Cl distances vs. percentage number of contacts.

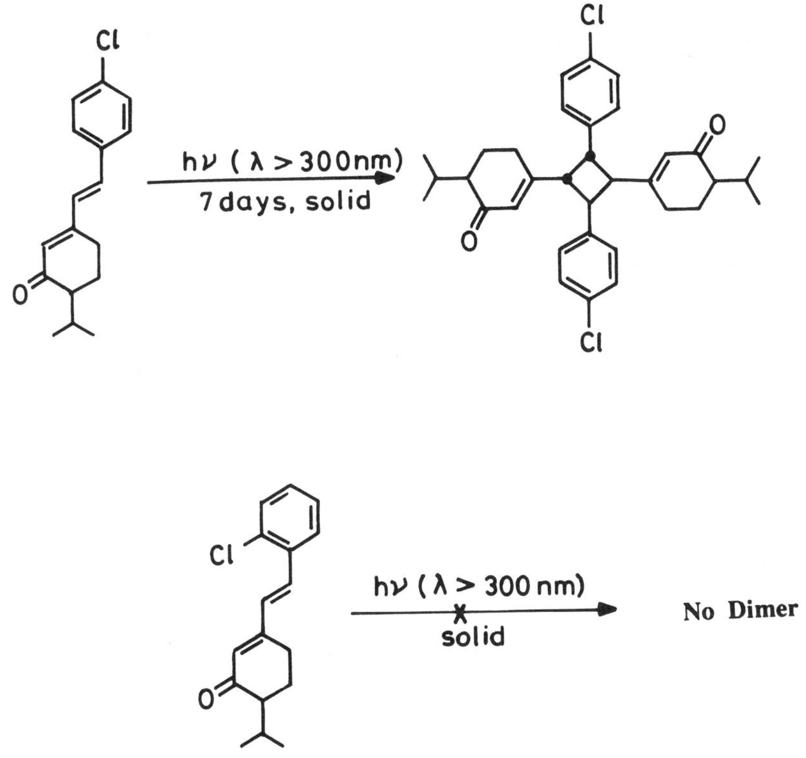

Scheme 10

3.1.2. Methylenedioxy and Methoxy Groups

Desiraju et al. observed that the presence of a methylenedioxy substituent in a planar aromatic molecule tends to favor the β structure.[52] For example, it has been shown that in a series of methylenedioxycinnamic acid derivatives, there is a preference for attaining the β-packing mode. In a more recent study Desiraju and Kishan[53] have examined the crystal packing of a few more oxygenated aromatic compounds (methylenedioxy and dimethoxy compounds) and have emphasized the importance of the presence of a critical number of oxygen atoms and C—H···O interactions to produce layered structures with a short axis of approximately ~4.0 Å. It is worth noting the earlier report by Gnanaguru et al. on the steering capability of methoxy group based on structure–photoreactivity studies on 6-methoxy-, 7-methoxy-, and 8-methoxycoumarin crystals (Table 3).[12a] Whereas the 6-methoxy derivative yields syn head–head dimer, the 7-methoxy case gives syn head–tail and the 8-methoxy anti head–tail dimers, each one giving different isomers and none of them having β structure. While the crystal structure of the 6-methoxy derivative is not available yet, the crystal structures of the 7-methoxy and 8-methoxy derivatives do not reveal the presence of C—H···O interactions. It is not obvious at this stage how general the methoxy or methylenedioxy groups can be in providing β structure necessary for dimerization reaction.

3.1.3. Acetoxy Substitution

The use of the acetoxy group in crystal engineering was investigated by substituting it at different positions in coumarins.[12b,54] Of the five acetoxy-coumarins investigated, 6-acetoxycoumarin, 7-acetoxycoumarin, and 4-methyl-7-acetoxycoumarin crystallized in the β structure yielding syn head–head dimer upon irradiation. However 4-acetoxycoumarin and 4-methyl-6-acetoxycoumarin were photostable. Analysis of acetoxy···acetoxy interactions in the acetoxy compounds retrieved from the Cambridge Data Base (1983) shows that the majority of them are anti dipolar.[54] The fact that in all the three photolabile acetoxycoumarins one obtains β packing may mean that not only the interactions between the substituents (in this case the acetoxy groups) but also the total interactions between the parent molecules are of crucial importance in reaching the final observed packing.

Interactions involving the overlap of an ester group of one molecule with the benzene ring of another have been utilized to steer acrylic acid esters into packing arrangements suitable for solid-state polymerizations. Green, Addadi, and Lahav have utilized the ester functionality as a steering group in their elegant asymmetric synthesis of chiral dimers and polymers from benzene-1,4-diacrylates.[55,56] In these cases an attractive interaction between the carboxy and phenyl groups of adjacent molecules has been proposed to be responsible for juxtaposing the double bonds at a separation distance of ~4 Å. From all these observations it appears that it is worth exploring the use of acetoxy group in other planar aromatics to achieve β-type structures.

3.1.4. Sulfur as a Steering Group

In many of the crystal structures containing divalent sulfur, the sulfur atoms show electrophilic–nucleophilic interaction with short intermolecular contacts[57] less than the sum of the van der Waals radius of sulfur atoms (1.85 Å). Motivation for using the directionality in S···S interactions for the purpose of crystal engineering was examined by Desiraju and Nalini.[58] However the crystal structure of the thione (18) with R = H does not posses β packing, while the chloro derivative (R = Cl in 18) acquires the β structure (Scheme 11). The intriguing aspect is that there are neither Cl···Cl nor S···S interactions in the crystal structure, the presumed guiding principles to attain the observed β-packing mode. It has been observed that the cis isomer of bis (butoxycarbonyl)-substituted tetrathiafulvalene (19) (Scheme 12) upon irradiation of the crystals yields dimers, whereas the trans isomer is photostable (Scheme 12).[59] x-Ray crystallographic analysis of the photolabile cis isomer does not show any S···S interaction, as in 18.[60] It is not unreasonable to conclude from these observations that, although based on very limited number of observations, the sulfur atom is not likely to be of general application in crystal engineering.

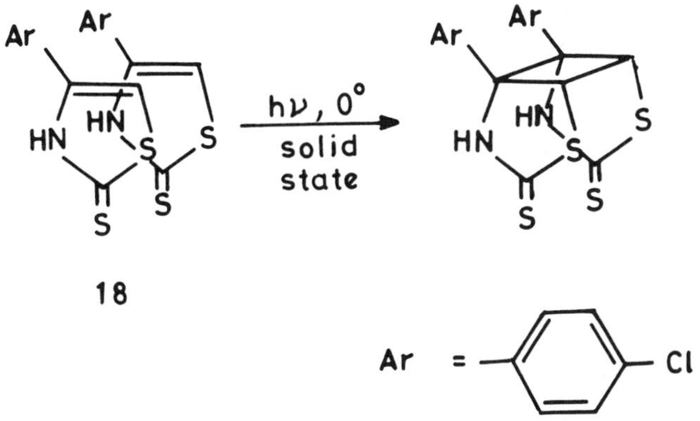

Scheme 11

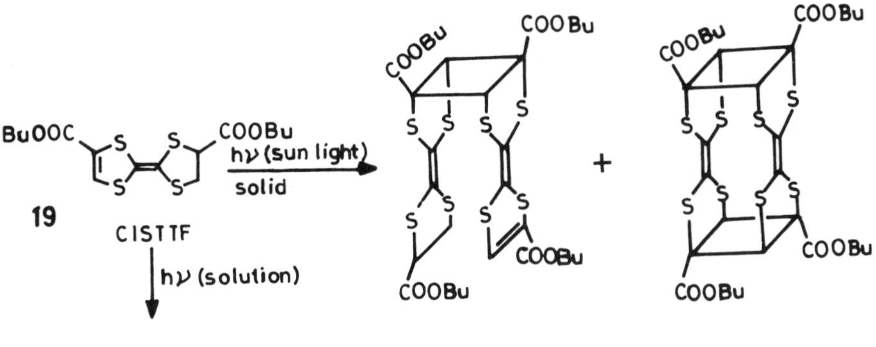

Scheme 12

The above examples suggest that there is still a large area of darkness in our knowledge of parameters of importance for achieving crystal engineering with certainty. Except for the chloro as a steering group for which there is a great body of experimental evidence all other claims are based on very limited data. The large amount of information available in the Cambridge Data Base has not been fully exploited to identify the chemical groups that may be of value in crystal engineering. Intuition combined with the Cambridge Data Base will provide new leads.

3.2. Mixed-Crystal Formation

Yet another approach to achieving photoreaction in crystals is to adopt the strategy of forming a mixed crystal between two closely similar molecules whereby the unreactive parent molecule is steered into a geometry favorable for reaction. With crystal reactivity in mind, mixed-crystal formation, a strategy slightly different from host–guest inclusion, was explored initially by the group led by Schmidt.[61] Cocrystallization is often a difficult task to achieve and depends on several factors, including solvent of crystallization, temperature, and presence of minor impurities. Although donor–acceptor-type mixed crystals are easily obtained, these are not useful to engineer solid-state reactions as they tend to form crystals of definite stoichiometry and often result in quenching of the excited state or in undesired photoreactions between the donor and the acceptor. The basic strategy in the preparation of a mixed crystal involves finding a closely related molecule that will form a solid solution (mixed crystal) over a wide range of compositions (varying ratios of the two materials). Molecules possessing nearly the same size and shape are the ideal choice as partners for mixed-crystal or solid solution formation. In this context the similar sizes of chloro and methyl and of fluoro and hydrogen are to be noted.

Mixed crystals have been prepared between several monomers that are themselves individually reactive in the crystalline state.[61] These include mixed crystals between monomers that crystallize individually in the α form ($\alpha+\alpha$) and between those of the β form ($\beta+\beta$). Also, mixed crystals have been obtained between two monomers in which one crystallizes in the α (e.g., *p*-methylcinnamic acid) and the other in the β form (*p*-chlorocinnamic acid) individually ($\alpha+\beta$). In all these cases, photolysis gave heterodimers in addition to homodimers (Scheme 13). Although these examples are not directly relevant from the point of view of crystal engineering, they provide basic information on the requirements for formation of mixed crystals. Also, they offer a unique method of obtaining reaction between two different monomers in the crystalline state. Some of these examples are shown in Scheme 14.

From the point of view of crystal engineering the important mixed crystal would be between a reactive and an unreactive pair in which the inherently unreactive molecule is steered to undergo reaction. Indeed a mixed crystal of a photostable compound with a photolabile one has been found to be photoreactive. For example, an equimolar solid solution of an β-type monomer, *para*-cholorocinnamic acid, and a γ monomer, *para*-methoxycinnamic acid, yielded three α dimers, i.e., dimers from both *para*-choloro- and *para*-methoxycinnamic acids in addition to a heterodimer (Scheme 15). The use of mixed crystals for engineering solid-state reaction has also been applied in the case of benzyl benzylidene cyclopentanones (Scheme 16)[62] and other systems.[63] Whereas **20** is photoreactive, compound **21** is photostable. However, in the mixed crystal of **20** and **21** both are photoreactive. This is attributed to the combined influence of both electronic and volume factors.[62b] This can be considered as an example of crystal engineering of **21**.

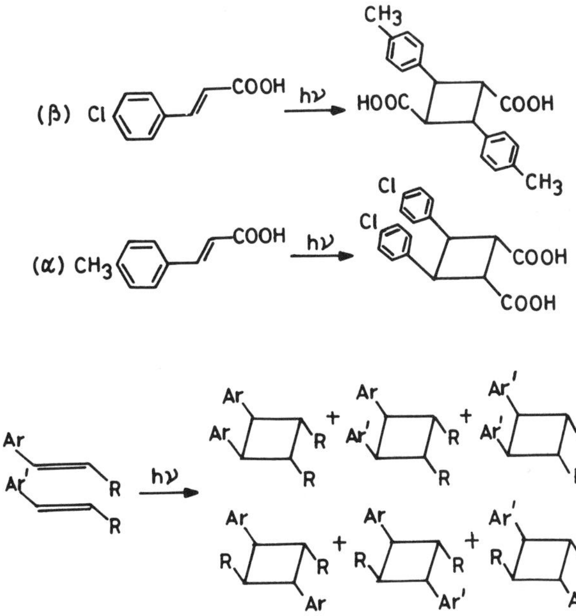

Scheme 13

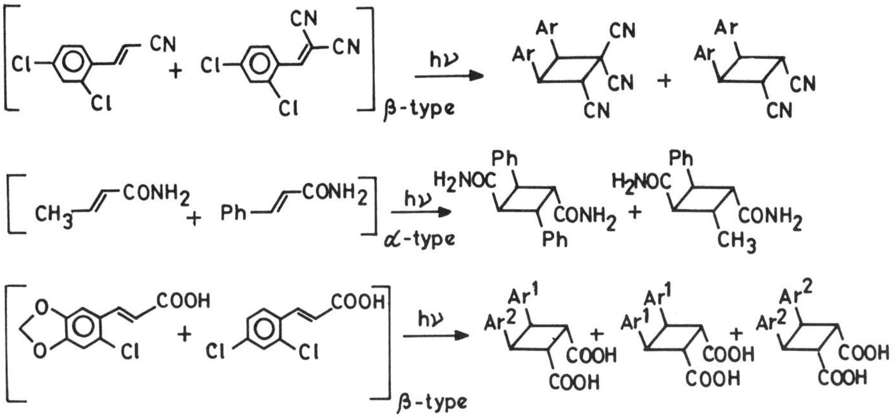

Scheme 14

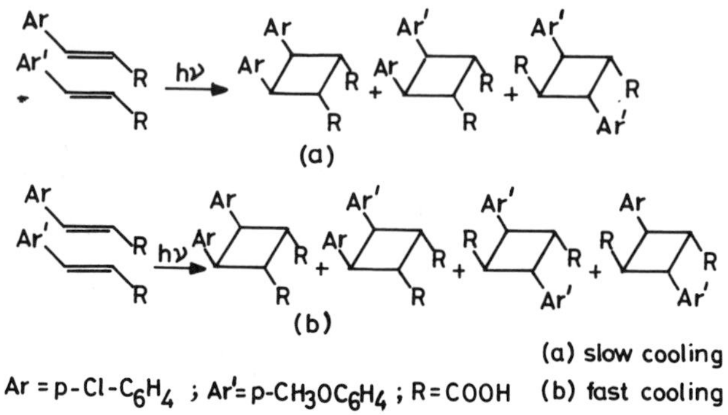
Ar
Ar'
R
hν
(a)
(b)
(a) slow cooling
(b) fast cooling
Ar = p-Cl-C6H4 ; Ar'= p-CH3OC6H4 ; R = COOH

Scheme 15

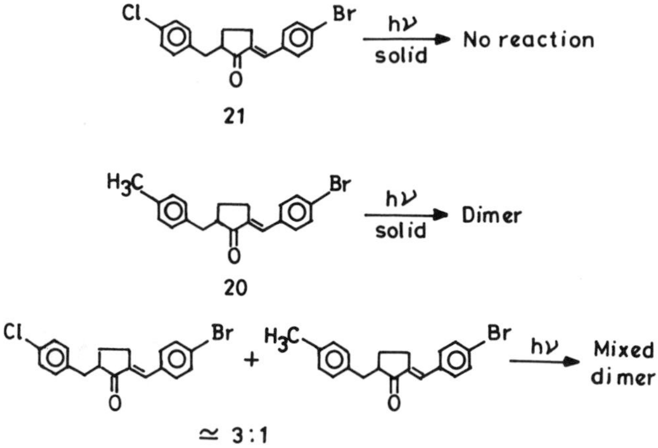
Cl
Br
hν
solid
No reaction
21
H3C
Dimer
20
Mixed
dimer
≃ 3:1

Scheme 16

On the basis of the similar atomic sizes of fluorine and hydrogen atoms, an organic compound and its fluorinated analogue have a high probability of forming a mixed crystal.[64] In this context formation of solid solution between **22** and **23**,[65a] and between 1,3-dimethyl-5-fluorouracil (**25**) and 1,3-dimethyluracil (**24**)[65b] is noteworthy.

22: X = H

23: X = F

24

25

3.3. Host–Guest Complexation[66]

From what has been discussed above it is clear that the problem in organic solid-state photochemistry, namely crystal engineering, remains still to be resolved completely. Yet another approach to this problem is to design crystalline molecular host–guest compounds. Unlike the mixed-crystal strategy discussed above, in this approach one selects a compound as the host based on its known ability to include other molecules. In this strategy a host crystal is chosen such that the packing of the host molecules in the crystalline state will enable the potentially reactive guest molecules to pack within the host lattice in a manner that will facilitate reaction. This approach is different from the mixed-crystal method discussed above in that the host lattice is generally inert and generally packs loosely so that a guest molecule can be accommodated easily. The partners need not have similar size and shape. The literature contains three distinctly different approaches that fall into the class of host–guest inclusion strategy. In the first, inorganic materials are used that are known to form weak complexes with an organic molecule and that upon crystallization yield cocrystals in which the molecules are pre-arranged for reaction. In the second approach, molecules are used that form channels or cages upon crystallization. In some cases the presence of these void spaces in the crystal structures becomes evident only in the presence of guests. These molecules do not form inclusion complexes in solution. Examples of this class include urea, Dianin's compound, deoxycholic acid, etc. In the third set, host molecules are utilized that by the nature of their three dimensional architecture contain cavities in which they can include guest molecules. These can form complexes both in solution and in the solid state. Some of these examples include cyclodextrins, cryptates, spherands, and calixiranes.[67] Although these have been extensively used as hosts in solution studies, they are not yet popular in the solid state. Photoreactions of host–guest complexes form the subject matter of two other chapters (Chapters 7 and 16) in this monograph and only for the sake of completeness a brief discussion follows.

An extremely instructive example belonging to the first type mentioned above which was discovered quite some time ago is the cocrystallization of α,β-unsaturated ketones with mercuric chloride.[68] $HgCl_2$ itself crystallizes in a cell of dimensions $a = 5.96$, $b = 12.74$, and $c = 4.33$ Å. Coumarin which is nontopo-chemically reactive in the solid state,[28] forms a 1:1 complex with mercuric chloride, and photolysis of the complex gives the topochemically expected syn head–head dimer. Coumarin crystallizes with a shortest axis of 5.68 Å. Crystals of the mercuric chloride complex have a repeat distance of 4.03 Å along the *c* axis, resulting in a double bond separation of 4.03 Å. Similarly, cinnamaldehyde and benzalacetophenone give crystalline 1:1 complexes with mercuric chloride having a 4 Å double-bond separation. Recently, Lewis and co-workers reported the photodimerization of Lewis acid complexes of alkyl cinnamate esters with $SnCl_4$ and BF_3.[69] The exclusive formation of the syn head–tail dimer upon irradiation of the crystalline $SnCl_4$ complex is in accord with the postulates of topochemical control of solid-state dimerization. The molecular structure of the 2:1 complex (ethyl cinnamate:$SnCl_4$) displays the expected octahedral geometry with Sn at the center of inversion. The distance between the reactive double bonds in the infinite stacks of esters is 4.023 Å for one symmetry-related pair and 4.125 Å for the second symmetry-related pair, well within the range of values observed for photodimerizable cinnamic acids. Efficient solid-state photodimerization to yield α-truxillate dimers is observed also for the 2:1 $SnCl_4$ complexes of the trans isomers of methyl and *n*-propyl cinnamates and methyl α-methyl cinnamate. Another illustrative example is provided by $(UO_2)Cl_2$ complexes with *trans, trans*-dibenzylidene acetone.[70] As early as 1910, Praetorius and Kohn reported[71] that the uranyl chloride complex of dibenzylidene acetone yields the truxillic acid-type dimer upon irradiation as a solid material. This is to be contrasted with the photostability of dibenzylidene acetone in the absence of uranyl chloride. Light-sensitive 2:1 complexes possess two pairs of ethylenic bonds adjacent to each other within a distance of 4.1 Å. The examples so far reported provide optimism and certainly encourage further exploration of this technique.

Examples belonging to the second type of inclusion strategy are provided by Toda.[72, 73] Toda and Akagi reported in 1968 that diacetylene diol (**26**, see Scheme 17) forms crystalline stoichiometric inclusion complexes with a variety of small molecules.[74] The features that contribute to complex formation are hydrogen bonding with the two OH groups, the linear nature of the acetylenic bond, and π interactions with the aryl ring. The large groups at the end of the linear chain act as spacers preventing the hosts from packing closely. Several structural analogues of the parent diacetylene diol have recently been synthesized and reported to form channel inclusion complexes. A point of interest to this chapter is the remarkable use of this host in dimerization reactions.

Irradiation of powdered complexes of benzylidene acetophenone **29** with the nonchiral host **26** gave a single photoproduct (>80% yield) which has been characterized as a syn head–tail dimer (Scheme 17).[75] It is important to note that irradiation of **29** in solution gives a mixture of cis and trans isomers of **29** and polymer and in the solid state (pure crystals) a mixture of stereoisomeric photodimers in low yields. The x-ray crystal structure of the complex of **26** with benzylidene acetophenone **29** has recently been reported. Benzylidene acetophenone in the absence of the host matrix crystallizes in two polymorphic modifications and the center-to-center distances between the double bonds are 5.2 and 4.8 Å in polymorphs **A** and **B**. A remarkable effect of **26** is to bring the two reactive molecules closer in the inclusion complex. The molecules of the guest are packed in parallel pairs related by an inversion center. As a result, the planes of the double bonds are parallel and the center-to-center distance is 3.862 Å (Figure 8). The arrangement enables the photodimerization to give the syn head–tail dimer. Several derivatives of **29** were found to photodimerize when included in the crystal matrix of **26**. Similarly inclusion complexes of 2-pyridone and 9-anthraldehyde as guest molecules with **26** give photodimers in the solid state.[75] In this context, yet another host molecule that has been successfully utilized by Toda and Kaftory is 2,5-diphenyl hydroquinone **27**.[74] Dibenzylidene acetone **30** when included in the latter, yields the syn head-tail dimer in good yield (Scheme 18). Use of the host in this dimerization reaction becomes evident when one realizes that the guest dibenzylidene acetone yields the all trans dimer in solution and is light stable in the solid state. Photodimerization of a few coumarins with host system **26** and its chiral derivative **28** has also been examined recently.[77] It has been found that inclusion complexes of coumarin, 7-methoxycoumarin, and 7-methylcoumarin with the chiral host **28** give the mirror symmetric dimer syn head–head. The packing arrangement of coumarin complex with **28** is shown in Figure 9. The host **28**-coumarin complex (ratio 1:2) crystallizes in the space group P2₁2₁2₁. The two crystallographically independent molecules are held in space via O-H---O hydrogen bonds between the carbonyl oxygens of the coumarin and the two hydroxyl groups of the host molecule. When one of the coumarin molecules is transalated along the a-axis, the reactive double bonds of the crystallographically independent coumarins come within a distance of about 3.8 Å. Inclusion complex of **26** with 7-methoxycoumarin, however, gives the syn head–tail isomer as observed in the crystals of the guest molecule alone. Whereas 7-methylcoumarin dimerizes when complexed with **28**, the complexation fails when **26** is used as the host.[76] The results are summarized in Table 6.

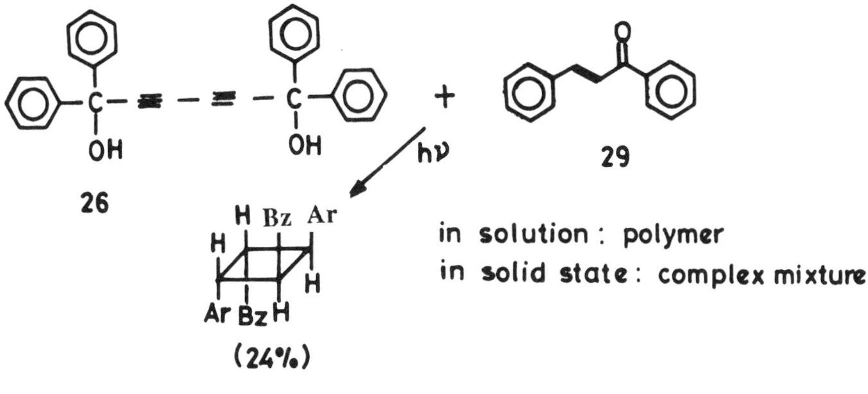

26

29

$h\nu$

in solution : polymer
in solid state : complex mixture

(24%)

Scheme 17.

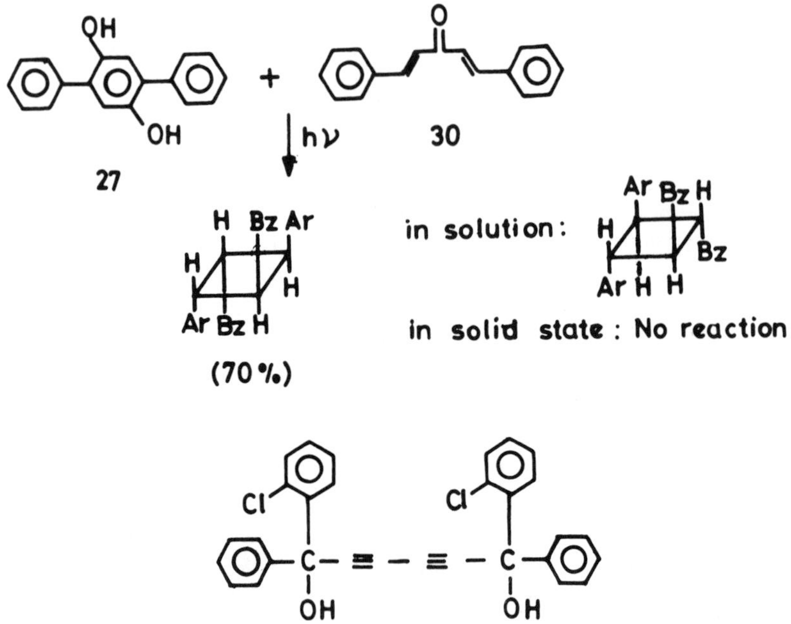

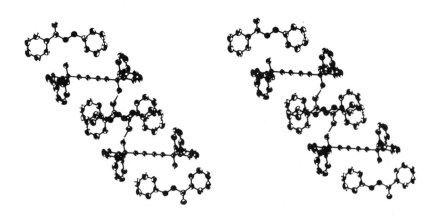

Scheme 18

Figure 8. Packing arrangement of the complex between benzylidene acetophenone and **26**.

Table 6. Photolysis of Coumarin Complexes in the Solid State[a]

Coumarin	Guest	H : G Ratio	Nature of Dimer	Yield of Dimer
Coumarin	β CD	1:2	Syn H-H	65%
	2 6	1:2	No dimer	
	2 8	1:2	Syn H-H	100
7 Methoxy	β CD	1:1	Nil	
	γ CD	1:2	Syn H-H & Syn H-T	65% 35%
	2 6	1:1	Syn H-T	90%
	2 8	1:1	Syn H-H	66%
4-Methyl-6-Chloro	γ	?	Syn H-H	70%
7 Methyl	β	1:1	No dimer	-
	γ	1:2	Syn H-H	80%
	2 6	1:2	No dimer	-
	2 8	1:2	Syn H-H	100
4,7 Dimethyl	β	2:2	Anti H-H	95%
	γ	1:2	Syn H-H & Syn H-T	57% 28%
	2 6	1:2	No dimer	-
	2 8	1:2	No dimer	-
4,6 Dimethyl	β	2:2	Syn H-H	48%
	γ	1:2	Syn H-H	75%
4-Methoxy	β	1:1	No dimer	-
4-Chloro	β	1:2	No dimer	-
	γ	1:2	No dimer	-
	2 6	1:2	No dimer	-
	2 8	1:2	No dimer	-

[a] Ref. J. Narasimha Moorthy, K. Venkatesan and R. G. Weiss, unpublished results.

As an example of the last class we briefly mention how cyclodextrin (α, β, and γ), a molecule possessing large cavity (diameter 6–9 Å), can be utilized to engineer dimerization reactions in the solid state. This has very recently been attempted by forming complexes of coumarins with both β- and γ-cyclo-dextrins.[28,78] Coumarin complexes with β-cyclodextrin and yields syn head–head dimer upon irradiation. The results of photolysis are summarized in Table 6.

It is clear that there is much promise in this area, where design and synthesis of new host systems of different cavity sizes, conformational flexibility, and ability to hold guest molecules with favorable intermolecular interactions such as hydrogen bonding, hydrophobic interactions, etc., can be highly rewarding (see Chapter 7).

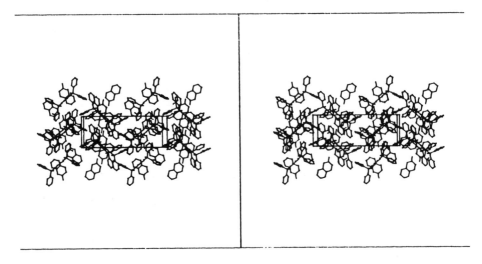

Figure 9. Packing arrangement of the coumarin complex with **28** in the crystalline state.

3.4. General Comments on Crystal Engineering

From what has been discussed above it is clear that there is no simple trick available to make a molecule react if it fails to crystallize with the required packing. "Intramolecular substitution" results in a new compound with entirely different electronic and steric properties. "Mixed crystals" upon photolysis give rise to additional products (heterodimers and homodimers from the host crystal). "Host–guest inclusion" is no longer a simple photoreaction of a crystalline monomer. An elegant and an ideal way to crystal engineer the unreactive crystal form would be to steer it to a polymorph favorable for reaction (different from the one unfavorable for the reaction). Although classical studies by Schmidt and Cohen[7] on cinnamic acid and further development of topochemical principles were based on the very existence of polymorphism, efforts in the direction of steering the molecules towards a particular crystal modification has only just been initiated.[79]

4. Mechanistic Questions

In this section we briefly summarize certain aspects of the mechanism of dimerization originating at nondefect regions of crystals. Studies on the dimerization reactions of aromatic molecules such as anthracenes, in which the dimerization is often proved to be initiated at defect sites, are not included. Although photofragmentation of crystalline peroxides has been studied in depth with mechanistic questions in mind[80a] no such detailed mechanistic investigations have been carried out on other solid-state reactions.[80b]

4.1. Yield, Quantum Yield, and Reactive State

Most often the yields of photodimers from crystals are reported for low initial conversions. Some of these reports result from the apprehension that formation of dimers within the monomer crystal would destroy the crystal packing and result in nontopochemical products at high conversions. α-Cinnamic acid[7] and some of the coumarins[12] crystallizing in β-packing are known to give topochemically controlled dimers in high yields (>80%, Tables 1 and 2). Maximum yield of the dimer that can be obtained is important to measure since it informs us about how well the product dimer is accommodated within the monomer crystal (see below).

Rarely does one measure the quantum yield of solid-state reactions owing to difficulties associated in carrying out such experiments. In this context, the reported quantum yield of dimerization of α-cinnamic acid (0.59) by Ito and Matsuura is important.[81] Recently Zimmerman,[82] Ito and Matsuura,[81] and de Mayo[83] have provided methods for measuring the quantum yields of photoreactions in the solid state. The reactive state is almost never reported for dimerization reactions in crystals. Sensitization and quenching studies are rarely carried out. Although singlet energy migration (exciton coupling etc.) in crystals is well established, there is not much literature on the triplet energy transfer within crystals. In this context, recent reports by Scheffer and co-workers.[84] on triplet initiated reactions in crystals and by Ito et al.[85] on triplet quenching of Norrish type II reactions in crystals are noteworthy.

4.2. Progress of Dimerization: Dimer within Monomer Crystals

The question of how dimerization proceeds in the monomer crystals was originally addressed by Schmidt, who envisioned two possibilities.[7] These are single-crystal to polycrystalline and single-crystal to single-crystal dimerizations.

Single-Crystal to Solid Solution to Polycrystalline: In this case, the product phase goes into solid solution in the lattice of the monomer and then as the dimer concentration rises, the solubility limit is exceeded, and the new phase precipitates. Cinnamic acids, coumarins and most of the dimerization examples presented in this chapter belong to this class. x-Ray powder diagrams in the case of cinnamic acid show a gradual and complete loss of long-range order and an eventual appearance of an ordered product phase.[7,86] There is no evidence yet as to whether the product phase separates out of the parent phase at specific or at random sites. Once the original monomer crystal breaks down due to contamination by the product dimer, the photodimerization may no longer be controlled by the initial packing. Since the solubility limit of the dimer in the monomer phase will vary with the reactant molecule, one might expect the maximum yield of topochemical dimer also to vary with the reactant molecule.

Single-Crystal to Single-Crystal Dimerizations:[87] There are a few examples in which the dimerization proceeds through a series of solid solutions of varying composition and is under topochemical control throughout. In these examples there is topotactic relationship between these solid solution phases. In simple terms these reactions are single-crystal to single-crystal transformations. These transformations are easily visualized in terms of an elegant drawing by Escher shown in Figure 10. Structure–photodimerization correlation studies on 2-benzyl-5-benzylidene cyclopentanone **31**, its bromo derivative and other molecules of this class (Scheme 19, see also Table 2) undergo single-crystal to single-crystal dimerization.[88] Other examples include the photopolymerization of distyrylpyrazine and diactylenes.[89]

Figure 10. A drawing of Escher to illustrate the single crystal to single crystal transformation.

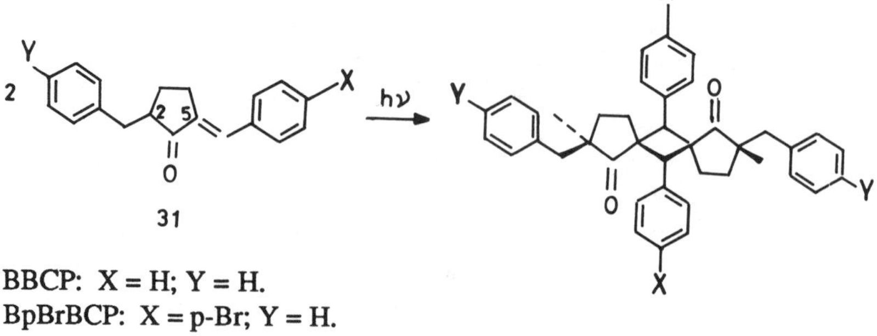

a) BBCP: X = H; Y = H.
b) BpBrBCP: X = p-Br; Y = H.

Scheme 19

In the crystals of **31a**, the neighboring molecules are related by a center of symmetry with the reactive double bonds separated by 4.1 Å. Photolysis of crystals of **31a** yields single-crystals of its dimer (Scheme 19). The fact that the product is crystalline indicates that there is a definite crystallographic relationship between the parent and the daughter phases. Indeed the maximum change in unit cell parameters between the monomer and the dimer is only about 0.7%. By careful control of the rate at which dimerization takes place in **31b** it was possible to retain a homogeneous single-crystal–single-crystal dimerization reaction. Why the single-crystal to single-crystal photodimerization has been observed only in the benzylidene cyclopentanones (**31**) and not in other molecules studied so far is an important question to be addressed. Examination of the literature examples of dimerization in crystals reveal that one of the basic conditions for single-crystal to single-crystal transformation is that the formation of the dimer should not introduce too much strain in the monomer crystals. This could be expected if the volume of the dimer is very nearly the same as that of the reacting monomer pairs and the reacting double bonds are well poised for dimerization with $\theta_1 = 0°$, $\theta_2 = \theta_3 = 90°$. Further, there should not be strong intermolecular forces (such as hydrogen bonding) in the crystal. All these conditions are met in **31**. In the case of **31** the reactive double bond is essentially at the central part of the molecular framework. During the course of the dimerization it is this part of the molecule which undergoes a large movement with the peripheral part of the molecule remaining essentially at the same position as illustrated in Figure 11. Obviously in rigid molecular systems such as coumarins one cannot hope to achieve this condition. In the case of cinnamic acids and similar molecules the presence of strong hydrogen bonding in the crystal does not allow sufficient relaxation of the dimer within the monomer crystals. This results in disruption of the crystal packing and formation of amorphous product.

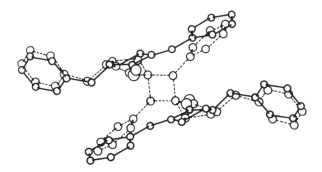

Figure 11. Composite digram comparing the packing of the molecular units within the monomer and dimer crystal structures of BBCP (ref. 87).

Reversible photodimerizations in crystals are expected to find wide application in molecular-scale devices.[90] Successful application can be found only for systems that are truly single-crystal–single-crystal transformations. In this context it is worth examining systems that fulfill the above criteria, namely the absence of strong intermolecular forces in the crystal packing, matching molecular sizes for the monomer and the dimer and the presence of reactive double bonds in the middle of the molecule with anchoring groups at the terminal positions.

4.3. Effect of Temperature

Based on qualitative considerations that at any given temperature the atoms and molecules in a crystal are not at rest, one would expect the percentage yield and the rate of reaction to depend upon the temperature of irradiation. For example Hasegawa et al.[91] studied the photodimerization of methylchalcone-4-carboxylate (32, m.p. 124°C) at different temperatures and for different irradiation times (Scheme 20). The maximum yield occurs at about -10°C, the yield increasing with increasing irradiation time.

The photochemical conversion of 1,4-dicinnamoyl benzene, (33), at different temperatures shows that the tricyclic dimer yield reaches a maximum at 25°C, reaction taking place even at -17°C (Scheme 21).[92] It has also been observed that the percent yield of dimer in 7-methoxycoumarin shows systematic variation with temperature.[29] Dimerization of 2-benzyl-5-benzylidene cyclopentanone also shows a temperature dependence.[93] While it dimerizes readily at room temperature it fails to dimerize at 77 K. This has been interpreted to mean that at 77 K the overlap between the potentially reactive orbitals is insufficient to drive the reaction. Based on empirical calculations, Kearsley has suggested[37] that a difference of ~1 Å in maximal amplitude of vibration of double bonds between low and room temperature (at room temperature ~2 Å, whereas at 77 K ~1 Å) is responsible for the difference in reactivity. On the other hand, 2,4-dichloro-*trans*-stilbene is reported to dimerize even at 77 K.[94] In the limited number of examples where the temperature effect has been investigated there is a certain temperature range in each case at which the yield is maximum. This could mean that the required molecular motions for dimerization differ in each case and are favored at different temperatures in different monomers. Intuitively one would expect the effect of temperature to be large in systems that require larger movement to acquire a favorable topochemical geometry. Therefore, the systems shown in Table 5 may be expected to show a greater temperature dependence than the ones in which the molecules are ideally oriented. No data exist to verify this proposition.

The origin of the temperature effects discussed above for topochemical reactions is different from the well-established temperature effects in defect-initiated reactions.[95] In defect-initiated dimerization (e.g., anthracenes, parent coumarin, and acenaphthylenes) the dimer yield is sensitive to the temperature of irradiation, and this is attributed to the increase of defects in the crystal.

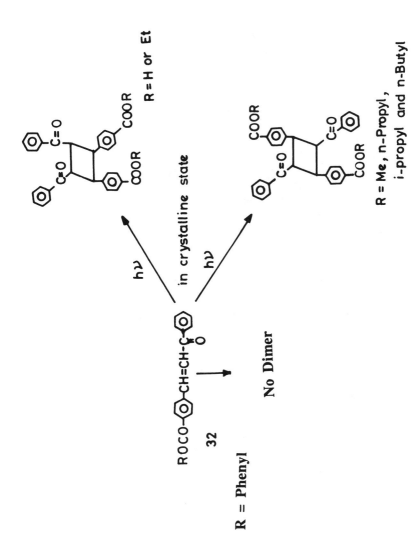

Scheme 20

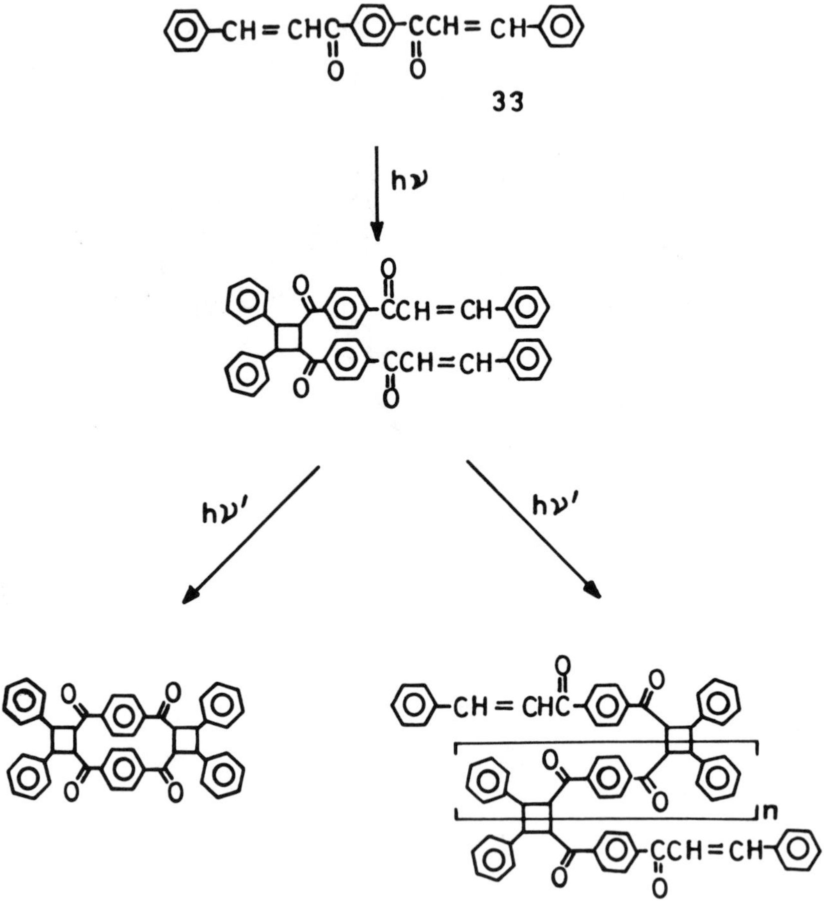

Scheme 21.

4.4. Reaction Pathway: Excimer and Diradical

Dimerizations in isotropic solvents are often established to proceed through excimers and 1,4-diradicals.[35] Generally excimer geometry is considered to be present along the path from a monomer to a dimer. Whether the excimer geometry is a minimum on this reaction coordinate and whether it is detectable by its emission depends on the system. In defect-initiated solid-state dimerizations (anthracenes), excimer emission has been recorded and the excimer is suggested to be present along the reaction coordinate connecting the monomer and the dimer.[31] In addition, in several aromatic crystals (naphthalene, pyrene, perylene, etc.) excimer emissions have been recorded.

Involvement of an excimer as an intermediate during the dimerization of 2,4-dichloro *trans*-stilbene and 1-(2,6-dichlorophenyl)-4-phenyl-*trans, trans*-1,3-butadiene has been proposed based on the observed excimer emission.[96] Parent systems, namely *trans*-stilbene and *trans,trans*-diphenyl-1,3-butadiene neither showed excimer emission nor dimerized in the crystalline state. Further, 2,4-dichloro-3'-methyl-*trans*-stilbene, which is dimorphic (pair and stack crystals), gave centrosymmetric and mirror symmetric dimers from the two forms respectively. Also the excimer emission spectra differed slightly between the two dimorphs (Figure 12). Excepting for these systems excimer involvement in topochemical dimerization of olefins has not been explored. It may be worthwhile to explore the presence of excimers, especially in systems that undergo dimerization in spite of poor topochemical alignment (Table 5).

Diradicals have not been identified as intermediates in any of the dimerization processes that occur in crystals. Diradical involvement in dimerization processes that occur in isotropic solvents is often inferred from the stereochemical scrambling in the resulting product dimer. Since almost all dimerization in crystals proceeds with stereospecificity there is no necessity to invoke 1,4-diradicals as intermediate in dimerization reactions. However, although the authors themselves did not consider this as a possibility, there are a few examples in which the products obtained can be rationalized on the basis of diradical intermediates. Schmidt et al. found that irradiation of a number of crystalline *cis*-cinnamic acids led to products known to arise from lattice controlled [2+2] photodimerization of the corresponding trans isomers.[97] In a few cases trans isomers were isolated. Based on the crystal and molecular structures of several *cis*-cinnamic acid derivatives, Schmidt et al. suggested that the mechanism of cis-trans isomerization was through formation of a metastable complex between an excited molecule and its nearest neighbor (Scheme 22). It was suggested that the intermediate complex failed to dimerize because it would have resulted in a very sterically crowded all-cis-cyclobutane derivative. Can this metastable intermediate be a 1,4-diradical? The occurrence of geometric isomerization can be rationalized readily on the basis of a 1,4-diradical intermediate. It should be cautioned, however, that the geometric isomerization can occur without the involvement of a second molecule if the reaction cavity is sufficiently large to allow molecular rotation.

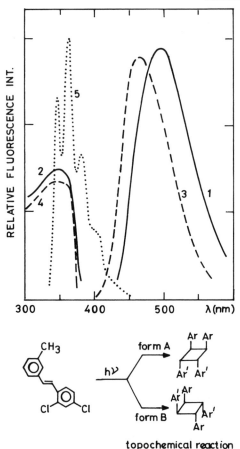

topochemical reaction

Figure 12. Emission and excitation spectra of 2,4-dichloro-3'methyl-*trans*-stilbene at 77 K. Curves 1 and 2: emission and excitation spectra of crystal form A; Curves 3 and 4: emission and excitation spectra of crystal form B; Curve 5: emission spectrum of glassy ethanol solution.

A reinvestigation of Shechter and co-workers original work[98] on the photodimerization of β-nitrostyrene led Desiraju and Pedireddi[99] to conclude on the basis of x-ray structural data that crystals of β-nitrostyrene are disordered. Irradiation of β-nitrostyrene crystal give two dimers. One of these is the result of reaction in the ordered regions of the crystal (α packing). The minor isomer appears to result from the reaction between a trans and a cis isomer of β-nitrostyrene. Desiraju and Pedireddi have rationalized the results on the basis of geometric isomerization occurring in the disordered regions (25%) of the crystal where there is enough empty space around the monomer. No cis isomer was isolated. It is quite possible that the minor isomer results from the 1,4-diradical in which rotation of the C—C bond could be facile. From the above brief presentation it is clear that it is not unreasonable to expect a 1,4-diradical intermediate during dimerization in crystals. Two of the reasons for their not having been postulated more often may be that (1) most often one is dealing with a singlet reaction (consequently a singlet 1,4-diradical with a very short lifetime) and (2) that there is not much void space around the reaction site to allow rotation of the C—C bond (so that the 1,4-diradical's presence can result in sterochemically scrambled products). There is a higher probability of observing 1,4-diradical intermediates in crystals where the reaction originates from triplet instead of singlet excited states and where the reaction cavity is large enough to allow rotation. Spectroscopic identification of 1,4-diradical intermediates in crystals should be attempted.

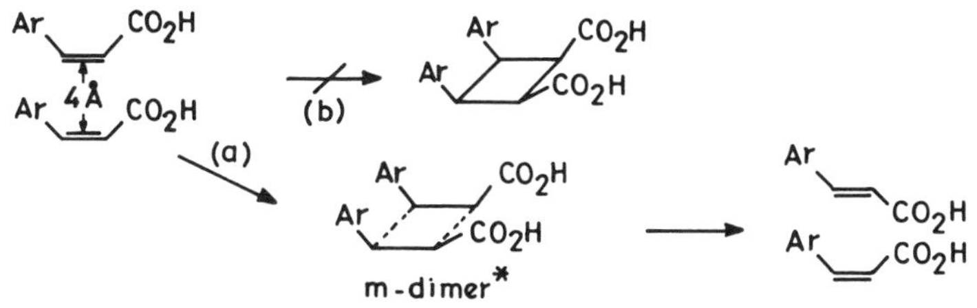

Scheme 22

5. Concluding Remarks

In spite of the active efforts put forth and some significant developments in this area, it must be conceded that it has not yet attracted the attention of mainstream organic chemists. This is mainly owing to the fact that the principal problem, namely to be able to preorganize the molecules in the lattice the way one would like to have, has not been fully surmounted. Our understanding of the various forces which control the crystal packing is highly fragmentary. However, the situation is not as hopeless as it may sound. Now the solid-state chemists working in the area have at their disposal a gold mine in the Cambridge Database which contains a wealth of information stored in it. Concerted efforts must go into analyzing the Database from two points of view: (1) to discover such of those molecular groups or fragments which have strong propensity to come close together in the lattice and (2) to discover the systematics, if any, in the packing modes of planar as well as nonplanar molecules. Most recently a beginning has been made in this direction.[42,100,101] Results from such studies would be of immense value for identifying potential steering groups and also should lead us to a better understanding of the forces operating between molecules. It seems likely that there may not be unique steering group of universal applicability. The investigator will have to possess in his arsenal a few different steering groups which he may have to exploit in his efforts to achieve the desired packing mode and hence the dimer or polymer of desired stereochemistry. Yet another pathway open to the researcher where some beginning has been made recently is the strategy of employing suitable host systems for encapsulating guest molecules and performing photoreactions (see also Chapter 7).

Although this chapter is devoted entirely to dimerization reactions in the solid-state, one should not get the impression that is all there is to it; Chapters 5 and 6 in this monograph reveal that exciting and novel reactions have been discovered in the crystalline-state recently. Dimerization reactions in the solid-state occupy a unique position as they have been subjected to extensive and systematic study by several groups. Still mechanistic studies are not that many and this area needs careful scrutiny. With more and more sophisticated techniques becoming routinely available, one would predict that mechanistic problem will attract future attention. There is an exciting possibility that the use of x-rays from a synchrotron source in conjunction with the Laue technique may permit us to gain a deeper understanding of the geometry of the molecule in its excited state and the reaction pathway.[102]

The following powerful words of Mendel Cohen written in 1976, which are valid even today, illustrate that dimerization in the crystalline media is only a pointer to something to come.[103] "As yet topochemistry has dealt with crystals, the epitome of organized systems. But the same principles must be involved, at least in part, in less organized systems. One can think of liquid crystals, membranes, micelles, molecular aggregates, monomolecular films, surfaces, intercalate compounds, macromolecules, and so on, all either synthetic or natural. The range of such systems is vast, and they are of significance not only to chemist, but also to the biologist, soil and material scientists. Applications of topochemical ideas in this area are just beginning. It is probable that it is in this direction that topochemistry is destined to make its most significant contribution."

Acknowledgement

Our grateful thanks go to our co-workers who have contributed immensely to the progress of this chapter. Collaboration with them has been and still is a source of great pleasure. KV thanks the Council of Scientific and Industrial Research, India for financial support of this project.

References

1.   H. Stobbe, F. K. Steinberger, *Chem. Ber.*, **1922**, *55*, 2225.
      Rubber, *Chem. Ber.*, **1902**, *35*, 2411.
      G. Ciamician, P. Silber, *Chem. Ber.*, **1902**, *35*, 4128.
2.   V. Ramamurthy, K. Venkatesan, *Chem. Rev.*, **1987**, *87*, 433.
3.   (a) J. M. Thomas, S. E. Morsi and J. P. Desvergne, *Adv. Phy. Org. Chem.*, **1977**, *15*, 63.
      V. E Shklover, T. V. Timofeeva and Y. T. Struchkov, *Russian Chem. Rev.*, **1986**, *8*, 55.

B. S. Green, R. Arad-Yellin and M. D. Cohen, *Top. Stereochem.*, **1986**, *16*, 131.

J. R. Scheffer, M. Garcia-Garibay and O. Nalamatsu, *Org. Photochem.*, **1987**, *8*, 249.

C. R. Theocharis in *The Chemistry of Enones*, S. Patai and Z. Rappoport, eds., John Wiley, NewYork, 1989, p. 1133.

(b) For an extensive list of reviews see ref. 4 - 13 in ref. 2 above and ref. 3 - 27 in Chapter 5 of this volume.

(c) Since our last review (ref. 2) the following articles provide further examples of photodimerization in the solid state: A. Baracchi, S.Chimichi, F. De Sio, C. Polo, P. Sarti- Fantoni and T. Torroba, *Heterocycles*, **1989**, *29*, 2023.

J. Reisch, N. Ekiz-Gucer, M. Takacs and G. Henkel, *Liebigs Ann. Chem.*, **1989**, 595.

G. Kaupp and G. Behmann, *Chem. Ber.*, **1988**, *121*, 2135.

S. Chimichi, P. Sarti-Fantoni, G. Coppini, F. Perghem and G. Renzi, *J. Org. Chem.*, **1987**, *52*, 5124.

G. R. Desiraju, J. Bernstein, K. V. Rada Kishan and J. A. R. P. Sarma, *Tetrahedron Lett.*, **1989**, *30*, 3029.

4. C. Libermann, *Chem. Ber.*, **1889**, *22*, 124 and 782.

5. H. W. Z. Kohlshutter, *Anorg. Allg. Chem.*, **1918**, *105*, 121.

6. H. I. Bernstein and W. C. Quimby, *J. Am. Chem. Soc.*, **1943**, *65*, 1845.

7. M. D. Cohen and G. M. J. Schmidt, *J. Chem. Soc.*, **1964**, 1996.

M. D. Cohen, G. M. J. Schmidt and F. I. Sonntag, *J. Chem. Soc.*, **1964**, 2000.

G. M. J. Schmidt, *J. Chem. Soc.*, **1964**, 2014.

8. T. Sadeh and G. M. J. Schmidt, *J. Am. Chem. Soc.*, **1982**, *84*, 3970.

9. M. Lahav and G. M. J. Schmidt, *J. Chem. Soc. B.*, **1967**, 239.

10. M. Lahav and G. M. J . Schmidt, *J. Chem. Soc. B.*, **1967**, 312.

B. S. Green, M. Lahav and G. M. J. Schmidt, *J. Chem. Soc. B.*, **1971**, 1552.

11. H. Nakanishi, W. Jones and J. M. Thomas, *Chem. Phys. Lett.*, **1980**, *71*, 44.

H. Nakanishi, W. Jones, J. M. Thomas, M. B. Hursthouse and M. Motevalli, *J. Chem. Soc., Chem. Commun.*, **1980**, 611.

W. Jones, H. Nakanishi, C. R. Theocaris and J. M. Thomas, *J. Chem. Soc., Chem. Commun.*, **1980**, 610.

J. M. Thomas, *Nature (London)*, **1981**, *289*, 633.

H. Nakanishi, W. Jones, J. M. Thomas, M. B. Hursthouse and M. J. Motevalli, *J. Phys. Chem.*, **1981**, *85*, 3636.

W. Jones, S. Ramdas, C. R. Theocaris, J. M. Thomas and N. W. Thomas, *J. Phys. Chem.*, **1981**, *85*, 2594.

12. (a) K. Gnanaguru, N. Ramasubbu, K. Venkatesan and V. Ramamurthy, *J. Org. Chem.*, **1985**, *50*, 2337.

(b) N. Ramasubbu, K. Gnanaguru, K. Venkatesan and V. Ramamurthy, *Can. J. Chem.*, **1982**, *60*, 2159.

(c) M. M. Bhadbhade, G. S. Murthy, K. Venkatesan and V. Ramamurthy, *Chem. Phys. Lett.*, **1984**, *109*, 259.

(d) K. Gnanaguru, G. S. Murthy, K. Venkatesan and V. Ramamurthy, *Chem. Phys. Lett.*, **1984**, *109*, 255.

(e) K. Gnanaguru, N. Ramasubbu, K. Venkatesan and V. Ramamurthy, *J. Photochem.*, **1984**, *27*, 355.

(f) N. Ramasubbu, T. N. Guru Row, K. Venkatesan, V. Ramamurthy and C. N. R. Rao, *J. Chem. Soc., Chem. Commun.*, **1982**, 178.

13. For a collection of selected papers by Schmidt see *G. M. J. Schmidt et al. Solid State Photochemistry*, D. Guinsburg, ed., Verlag Chemie, Weinheim, 1976.

14. D. P. Craig and P. Sarti-Fantoni, *J. Chem. Soc., Chem. Commun.*, **1966**, 742.

15.   M. D. Cohen, Z. Ludmer, J. M. Thomas and J. O. Williams, *J. Chem. Soc.,*
      Chem. Commun., **1969,** 1172.
      M. D. Cohen, Z. Ludmer, J. M. Thomas and J. O. Williams, *Proc. R. Soc.,*
      London, A, **1971,** *324,* 459.
16.   Z. Ludmer, *Chem. Phys.,* **1977,** *26,* 113.
      Z. Ludmer, *J. Lumin.,* **1978,** *17,* 1.
17.   E. Z. M. Ebeid and N. J. Bridge, *J. Chem. Soc., Faraday Trans. 1.,* **1984,** *80,*
      1131.
18.   J. C. J. Bart and G. M. Schmidt, *Isr. J. Chem.,* **1971,** *9,* 429.
      E. Heller and G. M. J. Schmidt, *Isr. J. Chem.,* **1971,** *9,* 449.
19.   R. Luther and F. Wright, *Z. Phys. Chem.,* **1905,** *51,* 297.
      B. Stevens, R. Dickinson and R. R. Sharpe, *Nature (London),* **1964,** *204,* 876.
      E. A. Chandross and J. Ferguson, *J. Chem. Phys.,* **1966,** *405,* 3564.
      M. O'Donnel, *Nature (London),* **1968,** *218,* 460.
      H. Bouas-Laurent, R. Lapouyade and J. G. Faugere, *C. R. Hebd. Seances Acad.*
      Sci., Ser. C., **1967,** *265,* 506.
20.   E. A. Chandross and J. Ferguson, *J. Chem. Phys.,* **1966,** *45,* 3564.
21.   J. M. Thomas and J. O. Williams, *J. Chem. Soc. Chem. Commun.,* **1967,** 432.
22.   G. M. Parkinson, M. J. Goringe, S. Ramdas, J. O. Williams and J. M. Thomas, *J.*
      Chem. Soc., Chem. Commun., **1978,** 134.
23.   M. D. Cohen, I. Ron, G. M. J. Schmidt and J. M. Thomas, *Nature (London),*
      1969, *224,* 167.
24.   W. Jones and W. Thomas, *Prog. Solid State Chem.,* **1979,** *12,* 101.
      J. O. Williams, *Sci. Prog. (Oxford),* **1977,** *64,* 247.
      J. O. Williams, J. M. Thomas in *Surface and Defect Properties of Solids,*
      Specialist Periodical Reports, Vol. 2, Chemical Society, London, 1973, p. 229.
      J. M. Thomas, J. O. Williams in *Surface and defects properties of solids:*
      Specialist Periodical Reports, Vol. 1, Chemical Society, London, 1972, p. 129.
      J. M. Thomas and J. O. Williams, *Prog. Solid State Chem.,* **1971,** *6,* 119.
25.   B. K. Tanner, *X-ray Diffraction Topography,* Pergamon Press, Oxford, 1976.
      D. B. Sheen and J. N. Sherwood, *Chem. in Britain,* **1986,** 535.
26.   P. Arjunan, K. Gnanaguru, V. Ramamurthy and K. Venkatesan in *Natural Products*
      Chemistry, R. I. Zaderski and J. J. Skolik, eds., Elsevier, Amsterdam, 1985, p.
      347.
27.   H. Morrison, H. Curtin and T. Medowell, *J. Am. Chem. Soc.,* **1966,** *88,* 5415.
28.   J. Narasimha Moorthy, K. Venkatesan and R. G. Weiss, unpublished results.
29.   G. S. Murthy, P. Arjunan, K. Venkatesan and V. Ramamurthy, *Tetrahedron,*
      1987, *43,* 1225.
30.   M. D. Cohen, *Angew. Chem. Int. Ed. Engl.,* **1975,** *14,* 386.
31.   V. Yakhot, M. D. Cohen and Z. Ludmer, *Adv. Photochem.,* **1979,** *11,* 489.
32.   S. Ariel, S. Askari, J. R. Scheffer, J. Trotter and L. Walsh, *J. Am. Chem. Soc.,*
      1984, *106,* 5726.
      S. Ariel, S. Askari, J. R. Scheffer, J. Trotter and L. Walsh in *Organic*
      Phototransformations in Non-homogeneous Media, M. A. Fox, ed., ACS
      Symposium series 278, American Chemical Society, Washington, D. C., 1985,
      p. 243.
33.   W. R. Busing,"WMIN", a computing programme to model molecules and crystals
      in terms of potential energy function, Oak Ridge National Laboratory, Oak
      Ridge, TN, 1981.
34.   A. Gavezzotti and M. Simonetta in *Organic Solid State Chemistry,* G. R.
      Desiraju, ed., Elsevier, Amsterdam, 1987, p. 391.
      R. Destro and A. Gavezzotti in *Structure and Properties of Molecular Crystals,*
      M. Pierrot, ed., Elsevier, Amsterdam, 1990, p. 161.
      A. Gavezzotti and M. Simonetta, *Chem. Rev.,* **1982,** *82,* 1.

A. Gavezzotti, *J. Am. Chem. Soc.*, **1983**, *105*, 5220.
A. Gavezzotti, *J. Am. Chem. Soc.*, **1985**, *107*, 962.
A. Gavezzotti, *Nouv. J. Chim.*, **1982**, *6*, 443.
A. Gavezzotti, *Tetrahedron*, **1987**, *43*, 1241.
A. Gavezzotti, *Acta Cryst.*, **1987**, *B43*, 559.
35. (a) N. J. Turro, *Modern Molecular Photochemistry*, The Benjamin Cummings, New York, 1978.
    (b) D. F. Eaton and D. A. Pensak, *J. Phys. Chem.*, **1981**, *85*, 2760.
36. (a) M. A. Collins and D. P. Craig, *Chem. Phys.*, **1981**, *54*, 305.
    (b) D. P. Craig and C. P. Mallet, *Chem. Phys.*, **1982**, *65*, 129.
    (c) D. P. Craig, R. N. Lindsay and C. P. Mallet, *Chem. Phys.*, **1984**, *89*, 187.
37. S. K. Kearsley in *Organic Solid State Chemistry*, G. R. Desiraju, ed., Elsevier, Amsterdam, 1987, p. 69.
38. (a) L. Leisorowitz and G. M. J. Schmidt, *Acta Crystallogr.*, **1965**, *18*, 1058.
    (b) N. J. Leonard, R. S. Mc Credie, M. W. Loyue and R. L. Cundall, *J. Am. Chem. Soc.*, **1973**, *95*, 2320.
    (c) J. K. Frank and I. C. Paul, *J. Am. Chem. Soc.*, **1973**, *95*, 2324.
    (d) H. Imgartinger, R. D. Aekes, W. Rebafka and H. A. Stabb, *Angew. Chem. Int. Ed. Engl.*, **1974**, *13*, 674.
39. (a) C. R. Theocaris, W. Jones, J . M. Thomas, M. Motevalli and M. B. Hursthouse, *J. Chem. Soc. Perkin Trans.* 2, **1984**, 71.
    (b) K. H. Pfoertner, G. Englert and P. Schoenholzer, *Terahedron*, **1987**, *43*, 1321.
    (c) F. Nakanishi, H. Nakanishi, M. Tsuchiya and M. Hasegawa, *Bull. Chem. Soc. Jpn.*, **1976**, *49*, 3096.
    (d) H. Nakanishi, M. Hasegawa and T. Mori, *Acta Crystallogr. Sect.* **1985**, *C41*, 70.
40. L. Leiserowitz and G. M. J. Schmidt, *Acta Crystallogr.*, **1965**, *18*, 1058.
    Z. Ludmer, G. E. Berkovic and L. Zeiri, *Tetrahedron*, **1987**, *43*, 1579.
41. (a) G. M. J. Schmidt, *Pure Appl. Chem.*, **1971**, *27*, 647.
    (b) For a popular presentation of crystal engieneering in nature see: S. Mann, New Scientist, **1990**, March, 42.
42. G. R. Desiraju, *Crystal Engineering: The Design of Organic Solids*, Elsevier, Amsterdam, 1989; M. C. Etter, *Acc. Chem. Res.*, **1989**, *23*, 120.
43. Cambridge Crystallographic Data Center. University Chemical Laboratories, Lensfield Road, Cambridge, England.
44. M. D. Cohen, B. S. Green, Z. Ludmer and G. M. J. Schmidt, *Chem. Phys. Lett.*, **1970**, *7*, 486.
    B. S. Green and G. M. J. Schmidt, *Tetrahedron Lett.*, **1970**, 4249.
    M. D. Cohen, A. Elgavi, B. S. Green, Z. Ludmer and G. M. J. Schmidt, *J. Am. Chem. Soc.*, **1972**, *94*, 6776.
    M. D. Cohen, G. M. J. Schmidt and F. I. Sonntag, *J. Chem. Soc.*, **1964**, 2000.
    G. M. J. Schmidt, *J. Chem. Soc.*, **1964**, 2014.
45. (a) J. A. R. P. Sarma and G.R. Desiraju, *Acc. Chem. Res.*, **1986**, *19*, 222.
    (b) J. A. R. P. Sarma and G. R. Desiraju, *Chem Phys. Lett.*, **1985**, *117*, 160.
    G. R. Desiraju, J. A. R. P. Sarma and T. S. R. Krishna, *Chem Phys. Lett.*, **1986**, *131*, 124.
    G. R. Desiraju and R. Parthasarathy, *J. Am. Chem. Soc.*, **1989**, *111*, 8725.
46. K. Yamasaki, *J. Chem. Soc. Jpn.*, **1962**, *17*, 1262.
    S. C. Nyburg and N. G. Wong, *Proc. Roy. Soc. A.*, **1979**, *29*, 367.
    D. E. Williams and H. S. W. Leh-Yeh, *Acta. Crystallogr.*, **1985**, *A41*, 296.
47. E. D. Stevens, *Mol. Phys.*, **1979**, *37*, 27.
48. B. Vonnegut and B. E. Warren, *J. Am. Chem. Soc.*, **1936**, *58*, 2459.
49. F. B. Van Bolhuis, P. B. Koster and T. Migchlesan, *Acta. Cryst.*, **1967**, *23*, 90.

50.   P. Venugopalan and K. Venkatesan, *Acta Crystallogr.*, in press.
51.   P. Venugopalan, T. Bharathi and K. Venkatesan, *J. Chem. Soc., Perkin II.*, in press.
52.   G. R. Desiraju, R. Kamala, B. Hanuma Kumari and J. A. R. P. Sarma, *J. Chem. Soc., Perkin Trans.2*, **1984**, 181.
53.   G. R. Desiraju and K. V. Radhakishan, *J. Am. Chem. Soc.*, **1989**, *111*, 4838.
54.   G. S. Murthy, V. Ramamurthy and K. Venkatesan, *Acta Crystallogr.*, **1988**, *C44*, 307.
55.   B. S. Green, M. Lahav and D. Rabinovich, *Acc. Chem. Res.*, **1979**, *12*, 191.
56.   L. Addadi and M. Lahav, *Pure Appl. Chem.*, **1979**, *51*, 1269.
57.   R. E. Rosenfield, R. Parthasarathy and J. D. Dunitz, *J. Am. Chem. Soc.*, **1977**, *99*, 4860.
       T. N. Guru Row and R. Parthasarathy, *J. Am. Chem. Soc.*, **1981**, *103*, 477.
58.   V. Nalini and G. R. Desiraju, *J. Chem. Soc. Chem. Commun.*, **1986**, 1030.
       V. Nalini and G. R. Desiraju, *Tetrahedron*, **1987**, *43*, 1313.
59.   N. Kreitsberga Ya, E. E. Liepin'sh, I. B. Mazheika, Heiland Ya and O. Neiland, *J. Org. Chem., U.S.S.R., Eng. Transl.*, **1986**, *22*, 367.
60.   P. Venugopalan and K. Venkatesan, *Bull. Chem. Soc., Jpn.*, **1990**, *63*, 2368.
61.   J. D. Hung, M. Lahav, M. Luwich and G. M. J. Schmidt, *Isr. J. Chem.*, **1972**, *10*, 585.
       M. D. Cohen, R. Cohen, M. Lahav and P. L. Nie, *J. Chem. Soc., Perkin Trans.*, **1973**, *2*, 1095.
       B. S. Green and L. Heller, *J. Org. Chem.*, **1974**, *39*, 1960.
62.   (a) C. R. Theocaris, G. R. Desiraju and W. Jones, *J. Am. Chem. Soc.*, **1984**, *106*, 3606.
       W. Jones, C. R. Theocaris, J. M. Thomas and G.R. Desiraju, *J. Chem. Soc., Chem. Commun.*, **1983**, 1443.
       (b) C. R. Theocaris, W. Jones, M. Motevalle and M. B. Hursthouse, *J. Cryst. Spect. Res.*, **1982**, *12*, 377.
63.   J. A. R. P. Sarma and G. R. Desiraju, *J. Am. Chem. Soc.*, **1986**, *108*, 2791.
       J. A. R. P. Sarma and G. R. Desiraju, *J. Chem. Soc.*, Perkin Trans.2, **1985**, 1905.
64.   P. Goldman, *Science*, **1969**, *164*, 1123.
65.   (a) C. R. Theocaris, S. E. Hopkin, A. M. Clark and M. J. Godolen, *Solid State Ionics*, in press.
       (b) T. Tago, N. Yamamoto and K. Machida, *Bull. Chem. Soc., Jpn.*, **1989**, *62*, 354.
66.   For an extensive list of reviews on host-guest chemistry in solid state and in solution see Chapters 7 and 16 respectively of this volume.
67.   J. M. Lehn, *Angew. Chem. Int. Ed. Engl.*, **1988**, *27*, 90.
       D. J. Cram, *Angew. Chem. Int. Ed. Engl.*, **1986**, *25*, 1039.
       C. D. Gutsche, *Calixarenes*, Royal Society of Chemistry, London, 1989.
68.   J. Bregman, K. Osaki, G. M. J. Schmidt and F. I. Sonntag, *J. Chem. Soc.*, **1964**, 2021.
69.   F. D. Lewis, J. D. Oxman and J. C. Hoffman, *J. Am. Chem. Soc.*, **1984**, *106*, 466.
       E. Gavuzzo, F. Mazza and E. Giglio, *Acta Crystallogr*, **1974**, *B30*, 1351.
       F. D. Lewis, S. L. Quillen, P. D.Hale and J. D. Oxman, *J. Am. Chem. Soc.*, **1988**, *110*, 1261.
70.   N. W. Alcock, P. de Meester and T. J. Kemp, *J. Chem. Soc., Perkin Trans. 2*, **1979**, 921.
71.   P. Praetorius and F. Kohn, *Ber. Dksch. Chem. Ges*, **1910**, *43*, 2744.
72.   F. Toda, *Top. Curr. Chem.*, **1988**, *149*, 212.
73.   F. Toda and K. Akagi, *Tetrahedron Lett.*, **1968**, 3605.

74. M. Kaftory, K.Tanaka and F. Toda, *J. Org. Chem.*, **1985**, *50*, 2154.
75. M. Kaftory, *Tetrahedron*, **1987**, *43*, 1503.
76. J. Narasimha Moorthy and K. Venkatesan, unpublished results.
77. F. Toda, *J. Inclu. Pheno. Mol. Recog. Chem.*, **1989**, *7*, 247.
78. Y. Tanaka, S. Sasaki and A. Kobayashi, *J. Inclu. Phenomenon.*, **1984**, *2*, 851.
79. E. Staab, L. Addadi, L. Leiserowitz and M. Lahav, *Adv. Mater.*, **1990**, *2*, 40.
80. (a) M. Hollingsworth and J. M. McBride, *Adv. Photochem.*, **1990**, *15*, 279.
    (b) E. V. Boldyreva, *Reactivity of Solids*, in press.
81. Y. Ito and T. Matsuura, *J. Photochem. Photobiol. A.*, **1989**, *50*, 141. Y. Ito and T. Matsuura, *Tetrahedron Letters*, **1988**, *29*, 3087.
82. H. E. Zimmerman and M. J. Zuraw, *J. Am. Chem. Soc.*, **1989**, *111*, 2358 and 7974.
83. S. Lazare, P. de Mayo and W. R. Ware, *Photochem. Photobiol.*, **1981**, *34*, 187.
84. M. Garcia-Garibay, J. R. Scheffer, J. Trotter and F. Wierko, *Tetrahedron Letters*, **1987**, *28*, 1741.
85. Y. Ito, H. Ito, M. Ino and T. Matsuura, *Tetrahedron Letters*, **1988**, *29*, 3091. Y. Ito, *Photochemistry on Solid Surfaces*, M. Anpo and T. Matsuura, eds., Elsevier, Amsterdam, 1989, p. 469.
86. L. Addadi, M. Cohen, M. Lahav and L. Leiserowitz, *J. Chim. Physi.*, **1986**, *83*, 831.
87. C. R. Theocaris and W. Jones in *Organic Solid State Chemistry*, G. R. Desiraju, ed., Elsevier, Amsterdam, 1987, p. 47.
88. D. A. Whitting, *Chem. and Ind.*, **1920**, 1411.
    D. A. Whitting, *J. Chem. Soc. C*, **1971**, 3396.
    H. Nakanishi, W. Jones, J. M. Thomas, M. B. Hurthouse and M. Motevalli, *J. Chem. Soc., Chem. Commun.*, **1980**, 611.
    W. Jones, H. Nakanishi, C. R. Theocaris and J. M. Thomas, *J. Chem. Soc., Chem. Commun.*, **1980**, 610.
    J. M. Thomas, *Nature (London)*, **1981**, *289*, 633.
    H. Nakanishi, W. Jones, J. M. Thomas, M. B. Hursthouse and M. Motevalli, *J. Phys. Chem.*, **1981**, *85*, 3636.
    W. Jones and C. R. Theocaris, *J. Cryst. Spectrosc. Res.*, **1987**, *14*, 447.
    M. Hasegawa, S. Kato, K. Saigo, S. R. Wilson, C. L. Stern and I. C. Paul, *J. Photochem. Photobiol.*, **1988**, *41*, 385.
89. Y. Sasada, H. Nakanishi, M. Hasegawa, *Bull. Chem. Soc. Jpn.*, **1971**, *44*, 1262.
    H. Nakanishi, M. Hasegawa, Y. Sasada, *J. Polym. Sci., Part A-2*, **1972**, *10*, 1537.
90. R. D. Rieke, G. O. Page, P. M. Hudnall, R. W. Arhart, T. W. Bouldin, *J. Chem. Soc. Chem. Commun.*, **1990**, 38.
91. M. Hasegawa, H. Arioka, H. Harashina, M. Nohara, M. Kabo, T. Nishikubo, *Isr. J. Chem.*, **1985**, *25*, 302.
92. M. Hasegawa, K. Saigo, T. Mori, H. Uno, M. Nohara and H. Nakanishi, *J. Am. Chem. Soc.*, **1985**, *107*, 2788.
93. J. Swlatklewicz, G. Elsenhardt, P. N. Prasad, J. M. Thomas, W. Jones and C. R. Theocharis, *J. Phys. Chem.*, **1982**, *86*, 1764.
94. M. D. Cohen, B. S. Green, Z. Ludmer and G. M. J. Schimdt, *Chem. Phys. Lett.*, **1970**, *7*, 486.
95. D. P. Craig, *personal communication*, 1989.
96. M. D. Cohen A. Elgavi, B. S. Green, Z. Ludmer and G. M. J. Schmidt, *J. Am. Chem. Soc.*, **1972**, *94*, 6776.
    R. Cohen, Z. Ludmer and V. Yakhot, *Chem. Phys. Lett.*, **1975**, *34*, 271.
    A. Warshell and Z. Shakked, *J. Am. Chem. Soc.*, **1975**, *97*, 5679.

97.   J. Bergman, K. Osaki, G. M. J. Schmidt and F. I. Sonnatag, *J. Chem. Soc.*, **1964**, 2021.
       G. M. J. Schmidt in *Reactivity of the Photoexcited Organic Molecule*, Interscience, New York, 1967, p. 227.
98.   D. B. Miller, P. W. Flanagan and H. Shechter, *J. Am. Chem. Soc.*, **1972**, *94*, 3912.
99.   G. R. Desiraju and V. R. Peddireddi, *J. Chem. Soc., Chem. Commun.*, **1989**, 1112.
100.   A. Kitaigordsky, *Acta Crystallogr.*, **1965**, *18*, 589.
       A. Kitaigordsky, *Molecular Crystals and Molecules*, Academic press, New York, 1973.
       D. E. William, *J. Chem. Phys.*, **1967**, *45*, 3770; *Acta Cryst.*, *A*, **1974**, *30*, 71.
       A. Gavezotti and M. Simonetta, *Acta Crystallogr.*, **1977**, *b33*, 447.
101.   A. Gavezzotti, *Acta. Crystalogr.*, **1990**, *B 46*, 275.
       A. Gavezzotti, *J. Phys. Chem.*, **1990**, *94*, 4319.
102.   J. M. Preses, J. R. Grover, A. Kvick and M. G. White, *American Scientist*, **1990**, *78*, 424.
       M. M. Harding, Chem. Brit., **1990**, 956.
103.   M. D. Cohen in *G. M. J. Schmidt et al., Solid State Photochemistry*, D. Guinsburg, ed., Verlag Chemie, Weinheim, 1976, p. 233.

Chapter 5

Unimolecular Photoreactions of Organic Crystals: The Medium Is the Message

John R. Scheffer and Phani Raj Pokkuluri

Department of Chemistry,
University of British Columbia,
Vancouver, Canada.

Contents

1. Introduction

In his book entitled *Understanding Media*,[1a] the late Marshall McLuhan coined the phrase "the medium is the message" to refer to the effects that modern electronic methods of communication, most notably television, have had on our culture. It was McLuhan's thesis that it was not so much the content of what was presented on television—soap opera versus grand opera, for example—that was most important, but the very existence of television that has profoundly changed our perceptions and way of life. To quote McLuhan, "The 'medium is the message' because it is the medium that shapes and controls the scale and form of human action."[1b]

The word "medium" in the present review has, of course, an entirely different meaning from McLuhan's. We use it to refer to the matter in which a chemical reaction is carried out, and we are specifically concerned here with the *crystalline* medium. There are, however, parallels between McLuhan's use of the word medium and ours. In common with McLuhan's ideas concerning electronic media, we contend that the simple fact that the crystalline environment exists as a medium for carrying out chemical reactions is beginning to have, and will increasingly have, a profound effect on the field of chemistry in general, and on the specific subject of this review, organic photochemistry. Like McLuhan's perception of television, we suggest that the details of what occurs in the chemical reactions of crystals is, for the purposes of this review, less important than the realization that the crystalline medium provides *an entirely new method of investigating chemical reactivity*. The fantastic diversity of molecular organization present in organic crystals provides an almost limitless variety of reaction media in which to study chemical processes, and the message of this review is that such media will very likely continue to control the form of an increasing number of chemists' actions for some time to come.

Despite an interesting history that dates back to the early 1800s,[1c] the investigation of the chemistry of organic solids is a subject that is still in its infancy relative to the study of organic chemistry in solution. One possible reason why more organic chemists have not ventured into this field lies in the difficulty associated with obtaining the necessary x-ray crystal structures that are an integral part of such studies. With the recent advent of relatively inexpensive automated data collection instruments and associated structure-solving computer software, this impediment is rapidly being removed. The organic solid-state chemist is now faced with the pleasant prospect of being able to browse through a wide-open field where many of the normal rules of solution-phase organic chemistry do not apply and where many of the chemical reactions he encounters are new and unusual. Furthermore, through crystallography and other solid-state spectroscopic techniques such as magic angle spinning solid-state ^{13}C NMR, one has a penetrating insight into the structural details of the reactants and their surroundings immediately prior to reaction. One of the more intriguing aspects of the chemistry of organic crystals is that it is sometimes possible to study the same reaction in more than one crystalline modification or polymorphic form. Polymorphs may differ from one another not only in their packing arrangements but also in subtle variations in the conformations of the constituent molecules. Any reactivity differences that are noted between polymorphs can frequently be ascribed to these structural differences, thereby allowing a deeper insight into the relationship between structure and reactivity in organic chemistry.

Being a new subject, most of the descriptions of solid-state chemical reactions have been, and continue to be, of the "show and tell" variety. That is, an unexpected result is obtained, and the main task of the experimentalist is to explain the results in an after the fact fashion. This is the natural first step in establishing the ground rules for a new field. Part of the reason for the present unpredictability of solid-state chemical reactivity is in our inability to predict the packing arrangement of organic molecules in crystals. This is a particularly severe impediment to designing *bimolecular* solid-state reactions, where the geometric and distance requirements normally associated with such processes may not be met in the crystal system under study (see Chapter 4). For this reason, we decided some years ago to investigate *unimolecular* chemical reactions in the solid-state, the advantage being that packing effects, while still important, do not usually dictate the success or failure of the experiment. A second positive feature associated with the study of unimolecular reactions in the solid-state is that, although crystal packing is difficult to predict, one can usually rely on the fact that the molecules that make up the crystal lattice will be found in or near their minimum energy *conformations*.[2] Often these conformations can be predicted using the principles of conformational analysis, so that if one has an idea of how a given conformer should behave chemically, a certain amount of predictability becomes available in the solid-state. Because most conformational equilibria are severely restricted in the crystalline phase, unimolecular processes in this medium tend to be restricted to a single, low-energy conformer, a situation that is quite different from that present in isotropic liquid phases, and one that can give rise to very different chemical results in the two media. We shall encounter examples of this effect in the sections that follow.

Consistent with the rapidly growing nature of the field, many review articles, monographs, book chapters, and symposia in print dealing with solid-state organic chemistry have appeared in the last 10 years.[3-27] The present article, however, will be limited to a discussion of unimolecular *photorearrangement* reactions in crystals, the author's particular research interest. Because a thorough review on the same topic by Scheffer, Garcia-Garibay, and Nalamasu appeared in 1987,[22] the present article will cover papers published since late 1986 as well as material that was inadvertently omitted in the original review. In the interests of completeness, however, and by way of introduction to new results on projects for which only preliminary data were available at that time, the next section of this article will be devoted to a brief, selective summary of material covered in our 1987 review,[22] material that is no doubt biased in favor of the author's own work (see also Chapters 4 and 6 for other aspects of solid-state photochemistry).

2. Brief, Selective Review of Material Published prior to 1987

The Diels-Alder reaction between *p*-quinones and 1,3-dienes (Scheme 1) gives adducts whose solid-state photochemistry has been studied in considerable detail.[23,28-33] By varying the nature and location of the substituents on both the diene and the dienophile, including 2,3 as well as 6,7-benzo substitution, over 20 so-called tetrahydronaphthoquinones have been prepared and investigated. Crystallography reveals that these compounds invariably crystallize in conformations that are basically *cis*-decalin-like, but with half-chair rather than full-chair six-membered rings. An idealized drawing of this conformation for the parent butadiene–benzoquinone adduct (4) is shown in Scheme 2. In this conformation there are two allylic hydrogen atoms on the cyclohexene ring that are favorably oriented for hydrogen atom transfer to the photoexcited ene-dione chromophore. One is located on carbon atom number 8 and can be abstracted by the adjacent oxygen to give biradical 5. The other hydrogen atom is situated on carbon atom 5 and can be abstracted by carbon atom 2 to afford biradical 6. As indicated in Scheme 2, different excited states are thought to be responsible for these sometimes competing processes.

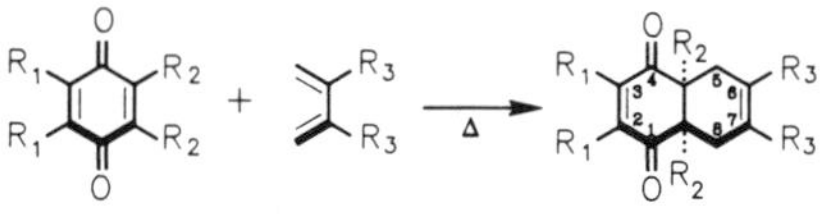

Scheme 1

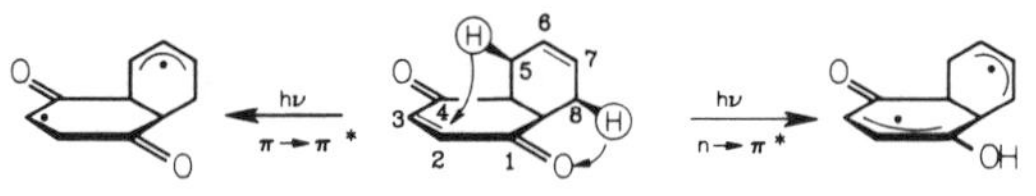

Scheme 2

In principle, biradical 5 is capable of four modes of intramolecular coupling [C1···C6, C1···C8, C3···C6 and C3···C8] and biradical 6 of two [C3···C5 and C3···C7]. Of these, however, only two are geometrically feasible: C1···C6 bonding in 5 and C3···C7 bonding in 6, and the compounds resulting from these modes of collapse are the major photoproducts in the solid-state. In solution, however, unlike the solid-state, conformational isomerization of biradicals 5 and 6 can compete with closure, and this can lead to different products. For example, a half-chair to half-chair ring inversion of 5 leads to a species in which the radical centers at C3 and C8 are within bonding distance of one another, and photoproducts derived from this pathway are commonly observed in solution. A specific example of the type of behavior described above is found in the photochemistry of ene-dione 7 (Scheme 3).[30] Photolysis of this material in the solid-state affords 8 exclusively (the product of C1···C6 bonding), and while **8** is still the major photoproduct in solution, significant amounts (ca. 25%) of diketone **9**, the product of C3···C8 bonding, are formed in this medium.

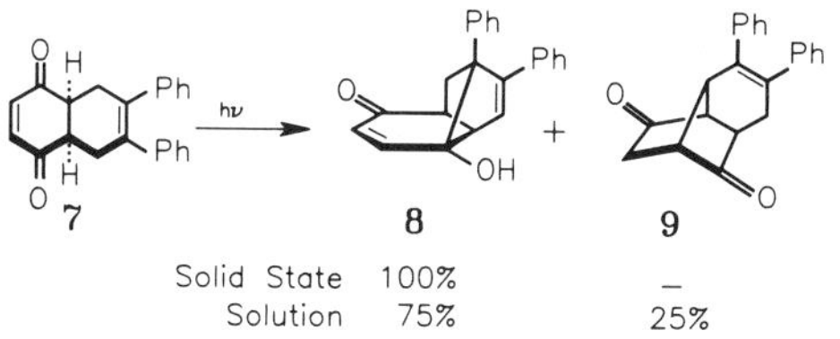

Solid State	100%	—
Solution	75%	25%

Scheme 3

An interesting variation on the structure–reactivity relationships described above has been reported by Weisz, Kaftory, Vidavsky and Mandelbaum.[33]   They showed that irradiation of ene-dione **10** in ethyl acetate solution affords a mixture of the intramolecular [2+2] cycloaddition product **11** plus the unusual rearrangement product **12** (Scheme 4); in contrast, photolysis of crystals of **10** gave only **12**. x-Ray crystallography again provided the key to understanding these differences. The solid-state conformation of **10**, which is similar to that of all the other tetrahydronaphthoquinones whose crystal structures have been determined, places a δ-hydrogen atom in the vicinity of O1. Abstraction of Hδ by O1 followed by C1···C6 bonding leads to photoproduct **12** by a process that is very similar to that observed previously (e.g., **7** → **8**, Scheme 3). The solid-state photoreaction fails for analogues of **10** in which the seven-membered rings are replaced by five- and six-membered rings. This was ascribed to unfavorable O1···Hδ abstraction distances in the crystal.[33]   Only ene-dione **10**, with an abstraction distance of 2.7 Å, was photoactive; the others, which had O···H distances ranging from 3.2 to 4.8 Å, were inert.

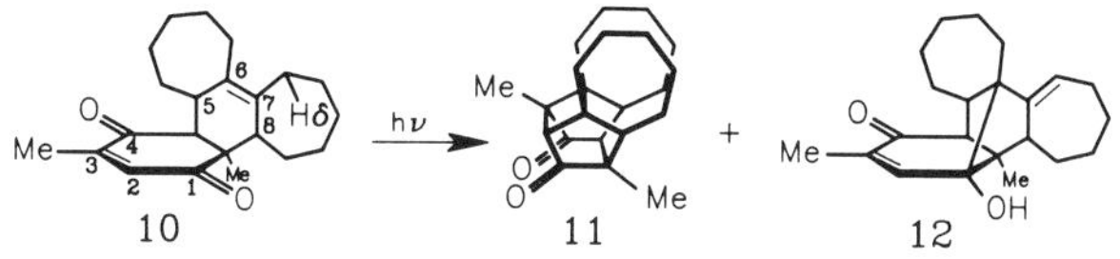

Scheme 4

A simple chemical modification of the Diels-Alder adducts discussed above, namely sodium borohydride reduction of one of the two equivalent carbonyl groups, transforms the ene-dione chromophore into a 4-α-hydroxy- and/or 4-β-hydroxycyclohex-2-en-1-one system.   As in the ene-dione series, we have investigated the solid-state as well as the solution-phase photochemistry of many of these so-called tetrahydronaphthoquinols and have correlated the results with the conformation and packing of the molecules in the crystal as determined by x-ray crystallography.[34-37] Scheme 5 summarizes the basic findings of these studies. Regardless of the stereochemistry at C4 (L = larger substituent, usually OH; S = smaller substituent, usually H), irradiation in isotropic liquid media invariably leads to cage compounds resulting from intramolecular [2+2] cycloaddition. In contrast, photolysis of crystals of epimers 13 and 16 affords the tricyclic ketones 14 and 17, respectively.

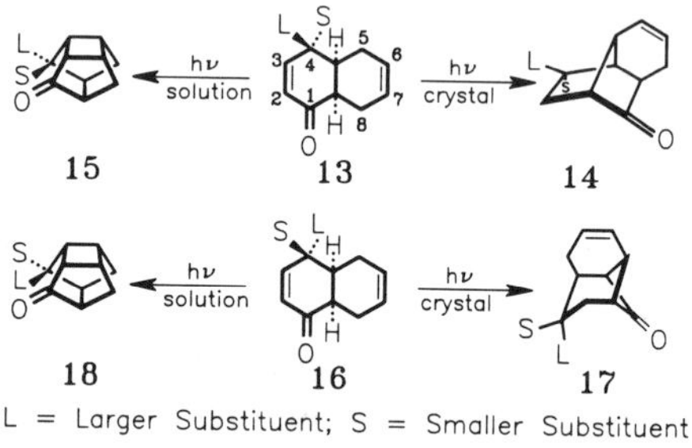

L = Larger Substituent; S = Smaller Substituent

Scheme 5

x-Ray crystallographic studies revealed the reason for the different photochemical behavior of enones 13 and 16 in the solid-state. The explanation is that, because of their epimeric relationship at C4, these compounds crystallize in, and are restricted to, *different conformations*, and these conformations exhibit different photoreactivity. We have termed this "conformation-specific" photochemistry. In solution, on the other hand, compounds 13 and 16 are free to explore several conformations during their excited-state lifetimes. One of these, not present in the solid-state, brings the C2-C3 and C6-C7 double bonds into proximity, and the ensuing rapid [2+2] cycloaddition forms cage compounds 15 and 18.

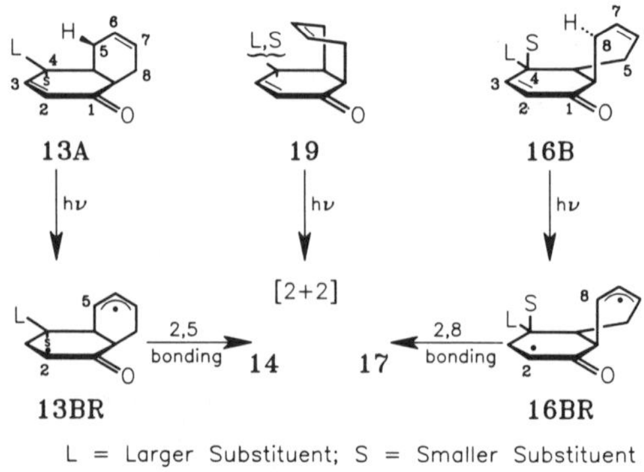

L = Larger Substituent; S = Smaller Substituent

Scheme 6

Scheme 6 summarizes the structure–reactivity relationships in more detail. Enones of type **13** crystallize in conformations (**13A**) that place the bulkier group at C4 in the pseudoequatorial position. The solid-state photochemistry involves initial transfer of the allylic hydrogen at C5 to C3 to form biradical **13BR**, which then closes to the observed product **14**. With the opposite configuration at C4, enones of type **16** invariably crystallize in the alternative half-chair conformation **16B** in order to maintain the pseudoequatorial nature of the larger group.

The photoreactivity of this conformer also involves hydrogen atom abstraction by the β-carbon atom of the α,β-unsaturated ketone, but in this case it is the C8 allylic hydrogen atom that is transferred. This produces biradical **16BR**, which undergoes closure to form the observed photoproduct **17**.

A final point concerns the nature of the conformer that leads to intramolecular [2+2] photocycloaddition in solution. Since neither **13A** nor **16B** is capable of this process, the logical candidate is **19** (Scheme 6). Although undoubtedly a minor constituent of the equilibrium mixture due to eclipsing, conformer **19** can account for the observed solution-phase photochemistry through rapid [2+2] cycloaddition of its parallel and close-lying double bonds. Recent work from our laboratory which corroborates this mechanistic picture will be presented in Section 3.2.

An exception to the general rule that enones of type **13** and **16** undergo internal hydrogen atom transfer to the β carbon of the α,β-unsaturated ketone upon photolysis in the solid-state is found in compound **20** (Scheme 7). Irradiation of crystals of this material was shown to give tricyclic ketone **22**, most likely through the mechanism depicted, which involves initial hydrogen atom transfer to the α carbon of the photoexcited enone.[36,37] Why is this pathway followed, particularly when it requires formation of what is presumably the less stable biradical **21**? We suggest that a unique crystal lattice steric effect is operative in this case. The crystal packing diagram for enone **20** reveals that the methyl group on the β-carbon atom is sterically encumbered by a second methyl group from a neighboring molecule lying directly below it. This prevents hydrogen atom transfer to Cβ, as the downward motion of the attached methyl group would drive it into its neighbor. In contrast, the crystal provides void space below the methyl group at Cα, resulting in no steric impediment to hydrogen transfer and pyramidalization at this center. Experiments bearing on this hypothesis will be presented in Section 3.2.

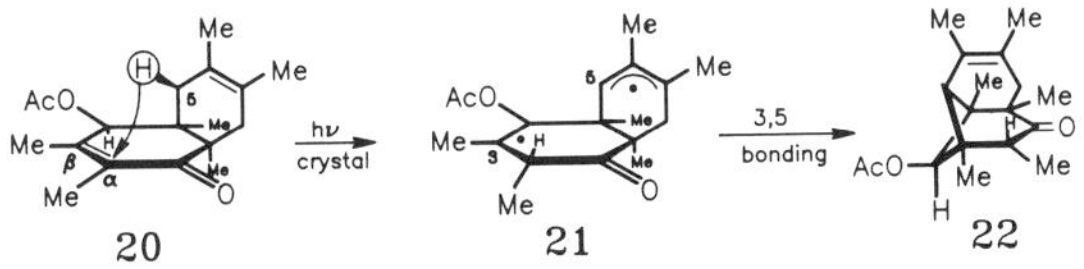

Scheme 7

Several additional hydrogen atom abstraction-initiated photoreactions have been investigated in the solid-state, both in our own[38-41] as well as other laboratories.[42-46] Our work has centered primarily on *p*-substituted α-cycloalkylacetophenone derivatives of general structure **23** (Scheme 8). These compounds undergo the well-known Norrish type II photoreaction[47] in the crystalline state, and our goals in this research were (a) to provide, through crystallography, experimental evidence on the geometric requirements for hydrogen abstraction, and (b) by assuming that the 1,4-biradicals generated in the solid-state have the same basic shapes as their ground-state precursors, to correlate biradical structure and reactivity.

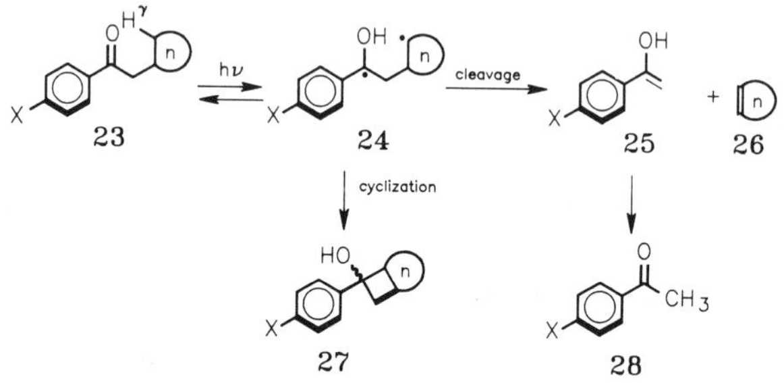

Scheme 8

A striking feature that emerged from this work was the finding that, contrary to expectation, boatlike rather than chairlike six-membered transition-state hydrogen abstraction geometries are prevalent in the solid-state.[39] A drawing of the solid-state conformation of α-cyclohexyl-*p*-chloroacetophenone (**29**) is shown in Scheme 9. In this conformation, the cyclohexane ring is in a nearly perfect chair form to which the ketone-containing side chain is equatorial. This leads to an obvious geometric advantage for abstraction of the *equatorial* γ-hydrogen atom $(O \cdots H_e = 2.6 \text{ Å})$ through a boatlike geometry rather than the axial hydrogen $(O \cdots H_a = 3.8 \text{ Å})$.

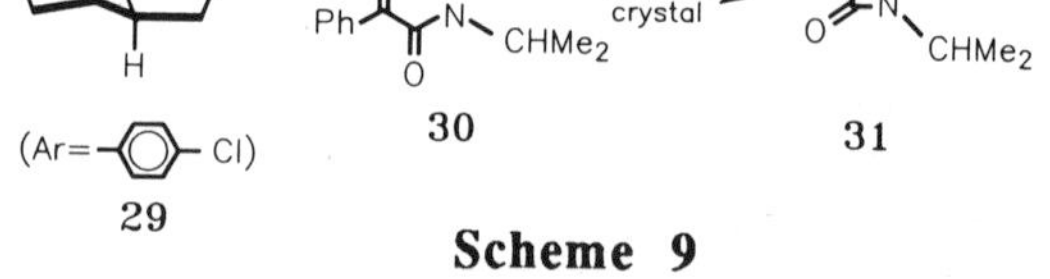

Scheme 9

Such abstraction distances refer, of course, to the situation in the ground state, and one must ask what significance they have for reactions that occur in excited states. This question has been discussed in some detail,[48] and this discussion will not be repeated here except to say that such measurements are, at the very least, valid as rough guidelines for predicting intramolecular hydrogen abstraction photoreactivity. Based on 17 separate crystallographic C=O⋯H distance measurements,[20,22] the upper limit for intramolecular hydrogen atom abstraction appears to be approximately 3.1 Å, somewhat greater than the sum of the van der Waals radii for oxygen and hydrogen (2.7 Å). Even though exceptions to this rule will undoubtedly be found, just as exceptions to Schmidt's rules for solid-state [2+2] photocycloaddition reactions are coming to light,[19,49,50] the 3.1 Å rule should lead to a greater degree of predictability for internal hydrogen atom abstraction reactions provided that the O···H distances for the molecules in question can be estimated by using molecular models or molecular force field calculations. Additional examples of the validity of the 3.1 Å rule will be presented in later sections of this review.

One of the earliest investigations of a Norrish type II-like photoreaction in the solid-state was reported by Aoyama et al.[42,43] These authors showed that the so-called α-oxoamide **30** (Scheme 9) underwent photolysis in the crystalline state to give high yields of the β-lactam derivative **31**; hydrogen transfer also occurred upon irradiation in solution, but in this case the intermediate had sufficient freedom of motion to undergo a rather complex rearrangement, and relatively little of the simple, direct closure product **31** was produced. Similar medium effects were noted for the tetrahydronaphthoquinones and quinols discussed earlier, and as such, form the true central theme of this review.

No review of solid-state unimolecular photoreactivity would be complete without mentioning the pioneering work of McBride and co-workers on crystalline diacyl peroxides.[51-55] As an example, consider the case of bis(3,3,3-triphenylpropanoyl)peroxide (**32**, Scheme 10). By detailed analysis of the esr spectra of irradiated samples, McBride et al.[52-54] were able to determine which of the three phenyl groups (nonequivalent by virtue of their different environments in the anisotropic medium) of the neophyl radical (**33**) underwent migration in the solid-state. The factor controlling migration in this case was postulated to be the anisotropic "local stress" of a recently liberated molecule of carbon dioxide trapped in an unfavorable lattice site. It was suggested that the stress was transmitted mainly to one side of the migration terminus, thus inclining the radical carbon in the opposite direction, more toward one phenyl group (the migrating group) than the other two. Such effects appear to be the rule rather than the exception in the solid-state photochemistry of diacyl peroxides and are often more important than topochemical effects arising from crystal packing and molecular conformation.[60]

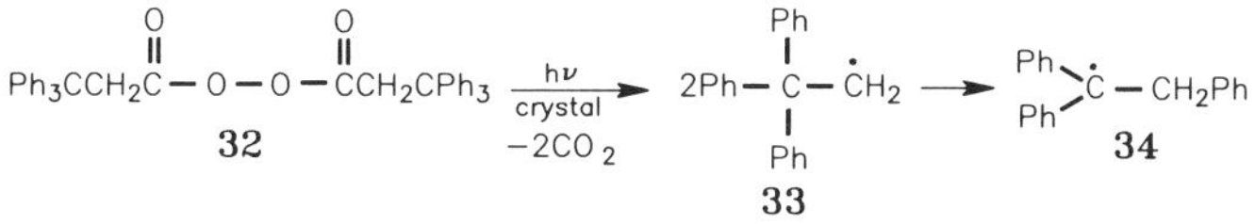

Scheme 10

A final example in this brief overview comes from the elegant work of Ohashi and co-workers on the x-ray-induced racemization of several optically active cobaloxime complexes (structure **35**, Scheme 11).[56-58] The racemization mechanism presumably involves rupture of the carbon–cobalt bond, inversion of the cyanoethyl group, and rebonding. Because of their single-crystal–single-crystal (topotactic) nature, these reactions can be followed by x-ray crystallography and their relative rates determined. These rates are then correlated with hypothetical reaction cavity volumes calculated from the crystal structure data. As one would expect, the comparison shows that the faster rate is associated with the larger reaction cavity. This represents the first truly quantitative correlation of a solid-state reaction rate with a crystal lattice structural parameter.

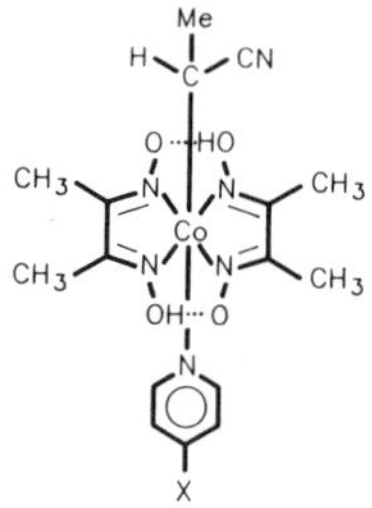

Scheme 11

3. Review of the Current Literature

As mentioned previously, the primary emphasis of this section will be on papers published during the 3 years 1987–1989; earlier papers that were inadvertently omitted from the original review[22] will also be included. The organization of this section conforms to that of the original review.

3.1. Studies on Tetrahydronaphthoquinones[59-67]

Ariel and Trotter have investigated the conformational properties of tetrahydronaphtho-quinones.[59,60] Owing to the cis ring junction, such compounds are capable of half-chair–half-chair ring inversion in solution (conformers **A** and **B**, Scheme 12). These conformers can be interconverted *via* a higher energy conformation, such as **C**, with eclipsed 4a/8a bridgehead substituents. In the solid-state, however, such motions are topochemically forbidden, and this difference in conformational mobility is thought to be responsible for the different photochemical results observed in the two media. For symmetrically substituted compounds ($R_1 = R_1'$; $R_2 = R_2'$; $R_3 = R_3'$), conformers **A** and **B** are enantiomeric and hence isoenergetic. Ariel and Trotter[59,60] studied a non-symmetrically substituted trimethyl ene-dione in which $R_1 = R_2 = R_2' = H$ and $R_1' = R_3 = R_3' = CH_3$. In this case conformers **A** and **B** bear a diastereomeric, non-isoenergetic relationship. Remarkably, the crystal structure of this latter compound showed that both conformers **A** and **B** are present in equal amounts in the solid-state. In none of the 20 odd tetrahydronaphthoquinones studied previously[23,28-33] was this found to be the case.

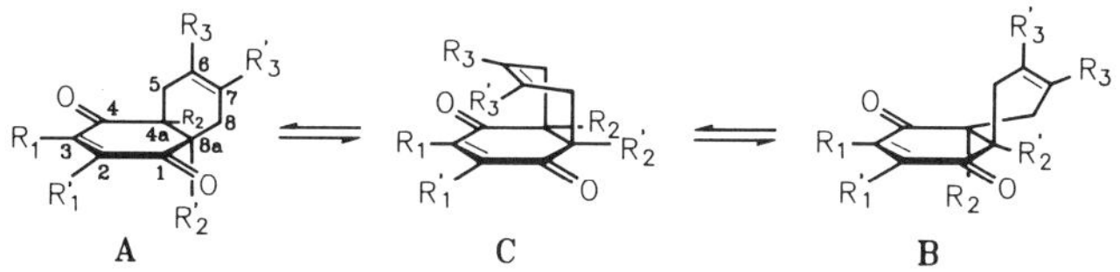

Scheme 12

The conformational energies of unsubstituted **A** and **B** were estimated by molecular mechanics methods based on the atomic coordinates from the x-ray crystal structure of the trimethyl compound. Similarly, the energy of conformer **C** was calculated from the crystal structure of a derivative in which an ethano bridge across C5 and C8 locks the molecule in an eclipsed conformation. Energy minimization preserves **A** and **B**, which are confirmed as the most stable conformers, but results in major changes in **C**, with the final energy-minimized conformations being either **A** or **B**. The intermediate points in these energy minimizations allow mapping of the complete conformational energy surface, which can be constructed as a function of two parameters: the ring junction torsion angle (which changes from -60 to +60° in converting **A** to **B**), and total displacement of C6 and C7 from a mean plane through C5—C4a—C8a—C8. The calculated energy barrier of 33 kJ mol^{-1} compares well with an experimental value of $\Delta G^* \approx 37$ kJ mol^{-1} for a derivative bearing methyl groups at positions 4a and 8a.

The conformational aspects of tetrahydronaphthoquinone photochemistry are nicely illustrated by the work of Ariel et al. on the aromatic ene-dione **36** (Scheme 13).[61,62] As shown, direct irradiation of **36** in isotropic liquid phases gives a temperature-dependent mixture of photoproducts **39** and **41**, whereas photolysis of crystals of ene-dione **36** gives only product **41**, regardless of the temperature employed. The solid-state conformation of **36**, as well as the structure and stereochemistry of photoproduct **41**, were established by x-ray crystallography.[62] Sensitization/quenching studies showed that **39** is singlet-derived and **41** is triplet derived.

The explanation advanced for the photochemical results was as follows: the excited singlet state of **36** reacts through intramolecular abstraction of H$_b$ by the carbonyl group at C1 to afford biradical **37**. This biradical is geometrically incapable of biradical closure and therefore undergoes conformational isomerization to biradical **38**, which *can* close by C3···C8 bonding to give **39**. Alternatively the singlet excited state of ene-dione **36** can intersystem cross to the lowest triplet state (likely π,π^* in nature) followed by reaction of the latter via transfer of H$_a$ from C5 to C2. Closure of the biradical so produced (**40**) leads to photoproduct **41**. It has been suggested that photoproduct **39** is not formed in the solid-state owing to topochemical restriction of the required conformational isomerization of

196   *Scheffer and Pokkuluri*

37 to **38** (cf., topochemical restriction of **A** → **B**, Scheme 12). No such restrictions exist in solution, however, and **39** is formed in this medium. Also included in this work are kinetic studies that establish the source of the temperature dependence of the solution phase photochemistry, as well as speculations on the origins of the reactivity differences between the singlet and triplet excited states of **36**.

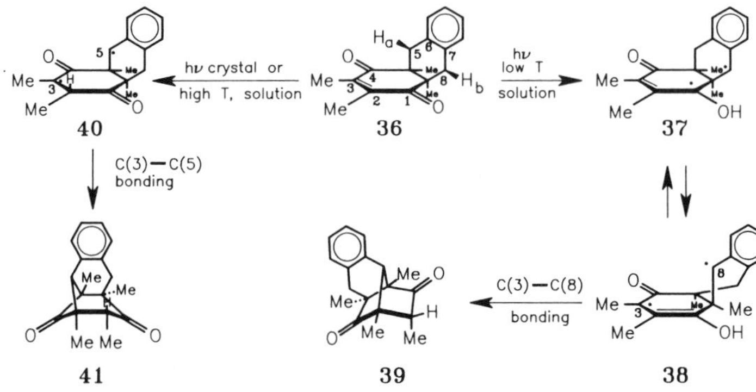

Scheme 13

With the crystal structures of ene-dione **36** and its solid-state photoproduct **41** in hand, the postulated reaction pathway in the crystalline phase was simulated by computer to see if it is in fact least motion in character and compatible with the surrounding lattice.[62]   The following motions served to describe the process: (1) transfer of H5 to C2 and (2) reduction of the C3...C5 distance of 3.13 Å to a final bonded distance of 1.57 Å by (a) folding about C4a—C8a by 20°, (b) folding about C1···C4 by 50°, and (c) partial pyramidalization of C3. Scheme 14 shows stereoviews of the reacting molecule at various stages along the reaction coordinate. Examination of the steric interactions between the reactant molecule and its lattice of unreacted neighbors indicates the development of unfavorable H···H intermolecular contacts during reaction. However, reaction also results in some free space becoming available in the lattice, and the short H···H contacts can be relieved by a movement of the whole reacting molecule by about 1.0 Å along the -z axis. The overall conclusion from these simulations, which are the first of their type to be reported, is that formation of photoproduct **41** in the solid-state is topochemically favorable, both inter- and intramolecularly.

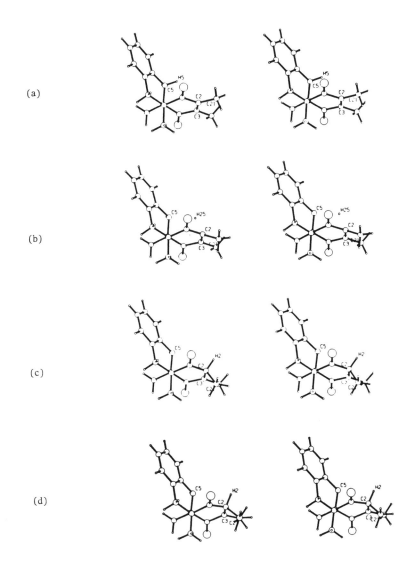

Scheme 14. Simulated Reaction Pathyway for the Conversion of Ene-dione **36** to Solid State Photoproduct **41**. Step (i): Transfer of H(5) to C(2) to form biradical **40** with accompanying pyramidalization at C(2): (a) starting material, ene-dione **36**; (b) half-way through H transfer; (c) biradical **40** (change of hybridization at C(5) is ignored). Step (ii): Stepwise reduction of the C(3)...C(5) distance of 3.133(6) Å to a final bonded distance of 1.570(3) Å via: folding about C(4a)-C(8a) by 20°, folding about C(1)...C(4) by 50°, partial pyramidalization at C(3), and maintaining approxmiate planarity of carbonyl functions: (d) 20% of motions applied [C(3)...C(5) = 2.896 Å];

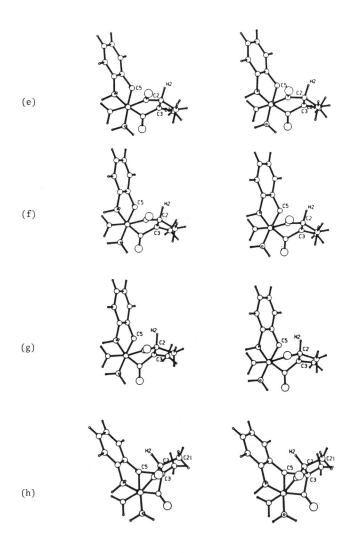

(e)

(f)

(g)

(h)

Scheme 14 (continued). (e) 40% [2.633 Å]; (f) 60% [2.347 Å]; (g) 80% [2.044 Å]; (h) 100% [1.729 A]. Step (iii): MMP2 minimization of the energy of the simulated photoproduct [(h), 100%] results in minor adjustments of the conformation, with C(3)-C(5) adjusting to 1.563 Å, compared with measured values of 1.569, 1.570 Å.

In our original review of unimolecular solid-state photoreactivity,[22] we discussed in some detail our work on the photorearrangement reactions of tetrahydronaphthoquinones **42–45** (Scheme 15). This material needs no repetition here except to point out that crystallographic details have since been published for reactants **42**,[63] **43**,[64] and **44**.[65] A more detailed account of the solid-state and solution-phase photochemistry of ene-dione **43** has also appeared,[66] and a very recent paper[67] describes the photochemical behavior of the parent tetrahydronaphtho-quinone molecule **45** when dissolved in a solid polymer film matrix.

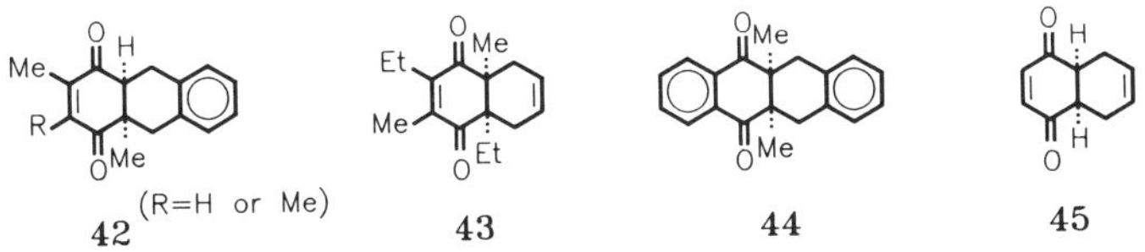

Scheme 15

3.2. Studies on Tetrahydronaphthoquinols[68-73]

The majority of this section of the review will deal with recent work from the University of British Columbia solid-state group on the crystallography[68-70] and photochemistry[71] of the tetrahydronaphthoquinol derivatives **46, 49, 52,** and **54**. The structures of these compounds along with the structures of their solution-phase and solid-state photoproducts are shown in Scheme 16. The results conform to the structure–reactivity relationship pattern outlined for other tetrahydronaphthoquinols in Section 2, namely that the solid-state photoreactions are conformation specific, with the conformations being determined by the configuration at C4 and the relative steric bulk of the substituents at this position. As an example, application of these principles to tetrahydronaphthoquinol **46** is presented in Scheme 16.

As shown by x-ray crystallography, compound **46** crystallizes in its lowest energy, *cis*-decalin-like conformation, which places the bulkier of the two substituents at C4 (the methyl group) in the pseudoequatorial position. This conformation, designated **46A** in Scheme 17, is analogous to that termed **A** in the tetrahydronaphthoquinone series discussed in section 3Å. Photoexcitation results in the transfer of the H5 benzylic hydrogen atom to C3 (C3···H5 distance = 2.80 Å), and closure of the resulting biradical affords the observed solid-state photoproduct **47**. In solution, on the other hand, conformational isomerization between **46A** and **46B** is facile, and it is the latter that is suggested to give rise to the liquid phase photoproduct **48**. This occurs through transfer of H8 to C3 followed by biradical closure and internal hemiketal formation. Even though **46B** is undoubtedly the minor conformer present, it apparently reacts much faster than major conformer **46A** owing to a more favorable six-membered, as opposed to five-membered, hydrogen atom abstraction geometry. It will be recalled that, for the tetrahydro-naphthoquinols discussed earlier, intramolecular [2+2] photocycloaddition between the 2,3 and 6,7 double bonds is the predominant solution phase reaction;[34-37] such a process is obviously disfavored in the case of the 6,7-benzannulated compounds discussed here.

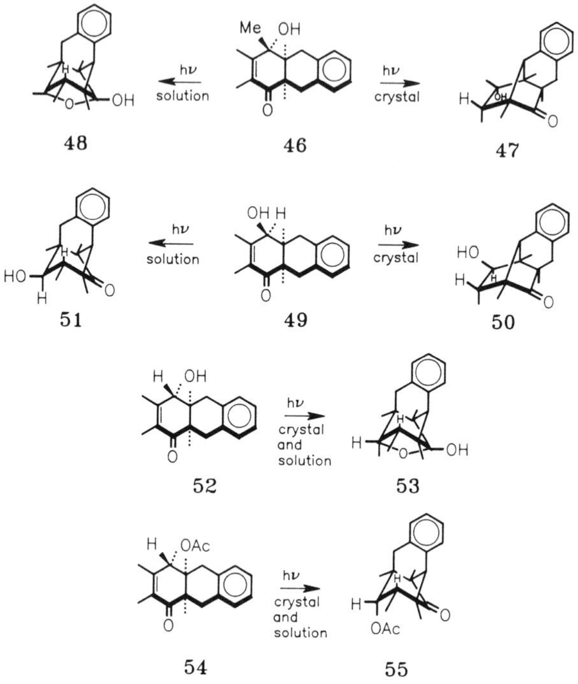

Scheme 16

An analysis similar to that presented above rationalizes the photochemistry of enones **49**, **52**, and **54**.[71] Two of these (compounds **52** and **54**) crystallize in B-type conformations and therefore give the same photoproducts in the solid-state as observed in solution. The fourth member of the series, enone **49**, crystallizes in a type A conformation and behaves similarly to **46**.

Kinetic analysis of the solution phase photochemical results according to the principles set forth by Lewis[74] indicates that equilibration of A-type and B-type conformers is rapid in the *excited state* as well as the ground state. As applied to tetrahydronaphthoquinol **46** (Scheme 17), this predicts that it may be possible to obtain some of the solid-state photoproduct **47** in solution provided the rate of conformational isomerization of **46A*** to **46B*** is reduced to the point that formation of biradical **46BRA** becomes competitive. Such experiments have been carried out, not with compound **46**, but with a related tetrahydronaphthoquinol.[72] These results will not be discussed further, as they relate more to the solution phase than the solid-state. They do, however, support the general mechanistic picture of tetrahydronaphthoquinol photochemistry outlined in Schemes 6 and 17.

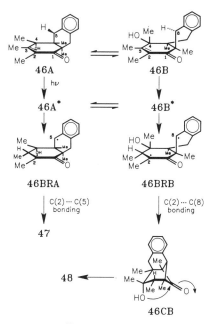

Scheme 17

Finally, we note a paper by Gudmundsdottir and Scheffer[73] that is relevant to previous work on the solid-state photochemistry of tetrahydronaphthoquinol acetate **20** (Scheme 7). It will be recalled that the abnormal solid-state photoreactivity of this material was postulated to arise from a novel *inter*molecular crystal lattice steric effect. Gudmundsdottir and Scheffer showed that enone **20** behaves normally (i.e., undergoes intramolecular H5 to C3 hydrogen transfer) when irradiated in solid polymer film matrices. This was taken as evidence in support of the postulated mechanism, the reasoning being that the polymer matrix provides a medium that is solid-state-like but that lacks the specific intermolecular contact responsible for the unique photoreactivity of pure crystalline **20**.

3.3. Photorearrangement of Di-π-methane, Cyclohexenone, and Cyclohexadienone Systems[25,75-91]

Scheffer and Trotter and co-workers have carried out an extensive investigation of the solid-state and solution-phase photochemistry of the 9,10-ethenoanthracene (or "dibenzobarrelene") system,[75-84] and it is this work that is reviewed in the first part of this section. For ease of synthesis, most of the compounds studied had ester substituents attached at C11 and C12, the carbon atoms comprising the so-called vinyl double bond. Ciganek[92] was the first to study the photochemistry of this class of compounds in solution; he showed that irradiation of the dimethyl diester **56a** in isotropic liquid phases leads to a product having the interesting "dibenzosemibullvalene" structure **57a** (Scheme 18), a reaction that is now recognized as an example of a very general type of process termed the di-π-methane photorearrangement.[93] Evans et al. showed that compound **56a** also photorearranges smoothly to the same product in the solid-state, and these authors then went on to study the corresponding isopropyl ester **56b**.[75]

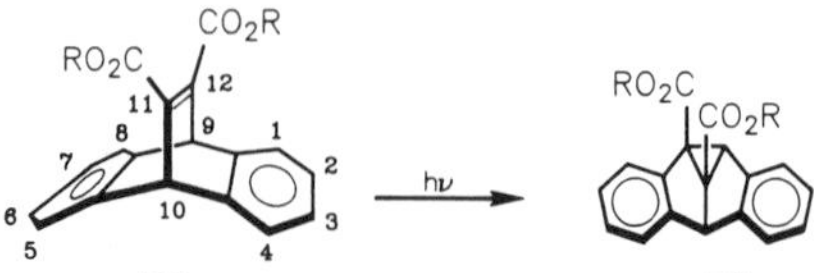

(a) R=Me
(b) R=iPr

Scheme 18

   Crystals of **56b** are dimorphic; recrystallization from ethanol provides prisms in the racemic space group Pbca, and recrystallization from the melt affords material in the chiral space group $P2_12_12_1$. In the latter case, the molecules in the crystal adopt identical homochiral conformations in a process that has been termed second-order spontaneous resolution.[94] The stage was thus set for an interesting experiment, namely to see whether the $P2_12_12_1$ crystal chirality could induce, through a stereospecific solid-state photorearrangement, optical activity in the chiral product **57b**. Such proved to be the case. Within the limits of the chiral shift reagent method, photoproduct **57b** was obtained *optically pure* in the $P2_12_12_1$ crystal irradiations; as expected, photolysis of **56b** in solution or in its Pbca crystal modification gave racemic material. The results with compound **56b** parallel the pioneering work of Lahav and co-workers on the induction of optical activity in *bimolecular* solid-state photoreactions of achiral molecules that spontaneously crystallize in chiral space groups (see Chapter 6 for details).[25,95,96]

   The commonly accepted mechanism of the di-π-methane photorearrangement of compounds such as **56b** involves initial, product-determining interaction between one of the two vinyl carbon atoms and one of the four nearby aromatic carbon atoms.[97,98] One of these pathways (path I) is depicted in Scheme 19, and it is readily apparent that there are three additional and equivalent pathways (II–IV) for a total mechanistic degeneracy of four. The realization that two of these pathways (I and II) lead to one photoproduct enantiomer, whereas pathways III and IV give the other, tells us that for the $P2_12_12_1$ dimorph of compound **56b**, there is complete discrimination between (I + II) and (III + IV) in the solid-state. Pathways (I + II) and (III + IV) were differentiated experimentally by determining (through anomalous dispersion x-ray crystallography) the absolute configuration of the molecules in a reactant crystal and correlating this with the absolute configuration of the photoproduct generated by irradiation of that same crystal.[25,76] This established that the $P2_12_12_1$ crystalline form of compound **56b** rearranges exclusively through the (I + II) mode. The crystal structures of the reactant and its photoproduct give us a clue as to why this should be so. Based on the reactant conformation in the solid-state, paths III and IV appear to involve severe intramolecular steric interactions between the bulky ester groups; such is not the case for paths I and II. Along the same lines, path II was concluded to be preferred over path I for two reasons: (1) path II involves initial bond formation at the vinyl carbon atom bearing the ester group that is in a freer lattice environment, and (2) assuming a topochemical process, path II (but not path I) produces the photoproduct in a conformation similar to that in which it is found in its own crystal lattice. These solid-state absolute configuration correlations are the first of their kind to be reported.

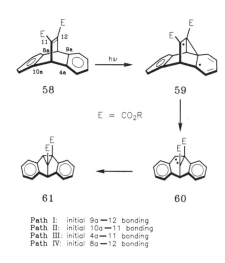

Path I: initial 9a—12 bonding
Path II: initial 10a—11 bonding
Path III: initial 4a—11 bonding
Path IV: initial 8a—12 bonding

Scheme 19

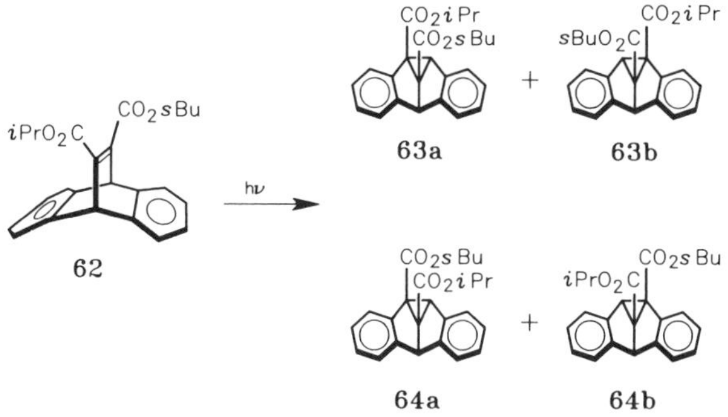

Scheme 20

In a related study, Garcia-Garibay et al.[77] investigated the solid-state and solution-phase photochemistry of the mixed isopropyl/(±)-*sec*-butyl diester **62** (Scheme 20). x-Ray crystallographic studies established that this material is isostructural with diisopropyl diester **56b** and crystallizes in the chiral space group P2$_1$2$_1$2$_1$ with disorder in the sec-butyl group. It is precisely this disorder, which is well known for *sec*-butyl groups, that allows the racemic material to adopt a chiral space group. Di-π-methane photorearrangement of dibenzobarrelene **62** can give rise to four possible diastereomeric products that can be designated as either **63** or **64** on the basis of the location of the nonequivalent ester groups or as **a** or **b** on the basis of the configuration of the substituted dibenzosemibullvalene ring system (Scheme 20). Without going into detail, it was found possible to determine accurately both the **63:64** and **a:b** ratios for the solid-state and solution photolyses.[77] Both ratios were near unity (as expected) in solution, but in the solid-state, high regio- and enantioselectivity (up to 90%) was observed. The results demonstrate once again the power of the crystal lattice in directing unimolecular chemical processes along certain selected pathways.

It has been recognized for some time that results such as those described above for compounds **56b** and **62**, in which there is an enantioselective solid-state chemical reaction of an achiral or racemic compound that crystallizes in a chiral space group, provide not only a plausible explanation for the prebiotic origin of optical activity, but also an attractive method of asymmetric synthesis.[96] Toward this latter end, Chen et al.[78] reasoned that a more general approach would be to incorporate a resolved chiral substituent in the prochiral reactant. This would ensure a chiral space group, and following solid-state photoreaction, removal of the chiral "handle" would leave a partially or fully resolved photoproduct depending on the degree of asymmetric induction. Experiments along these lines were conducted on several dibenzobarrelene derivatives,[78] two of which are described below.

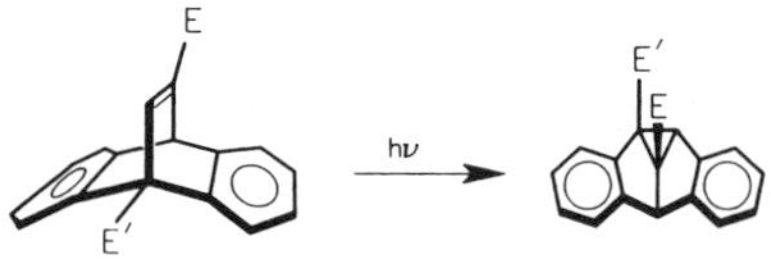

$$\begin{pmatrix} a \end{pmatrix} E = CO_2Me; \ E' = CO_2(-)-menthyl$$
$$\begin{pmatrix} b \end{pmatrix} E = CO_2(-)-menthyl; \ E' = CO_2Me$$

Scheme 21

The dibenzobarrelene derivatives **65a** and **65b** (Scheme 21), in which the chiral handle is a (–)-menthyl ester substituent located at either C9 or C11, were prepared and photolyzed in the solid-state and solution.[78] For both compounds, reaction gives only the regioisomer shown owing to preferential initial bonding at the unsubstituted vinyl carbon atom (more stable biradical intermediate). Two diastereomers are possible in each case, and it was found that the diastereoselectivity went from 60:40 in solution to 20:80 in the solid-state for compound **65a** but remained at 50:50 in both media for **65b**. These results were interpreted in terms of a *molecular* asymmetric inductive effect of the chiral handle, which is exhibited in solution, and an *environmental* effect in the solid-state that is due to the presence of the chiral crystal lattice. It was suggested that these effects may either reinforce or oppose one another, the latter situation apparently being the case for compound **65a**, where the diastereoselectivity is reversed in the solid-state compared to solution. The lack of diastereoselectivity observed for dibenzo-barrelene derivative **65b** in solution was attributed to the greater distance in this compound between the site of reaction and the chiral handle; the nondiastereoselective photoreaction of **65b** in the solid-state illustrates clearly that the asymmetric inductive power of a chiral organic crystal lattice need not be high.

Differences in the *regioselectivity* of the di-π-methane photorearrangement in going from the liquid to the solid phase also tell us much about the forces that control chemical reactivity in crystals. In addition to compound **62** discussed above, several other dibenzobarrelene derivatives bearing nonequivalent ester substituents at C11 and C12 were investigated by the UBC solid-state group.[79,80] In one study, the methyl ester substituent was held constant while the second ester group was varied from ethyl to *tert*-butyl (compounds **67a–d**, Scheme 22). In solution, there was a small preference in all four cases for formation of what was suggested to be the less hindered photoproduct **69**, but in the solid-state, the results were unpredictable. The **69:71** ratio in this medium varied with R in the following way: 45:55 (R = ethyl), 93:7 (R = isopropyl), 99:1 (R = (±)-*sec*-butyl) and 15:85 (R = *t*-butyl).

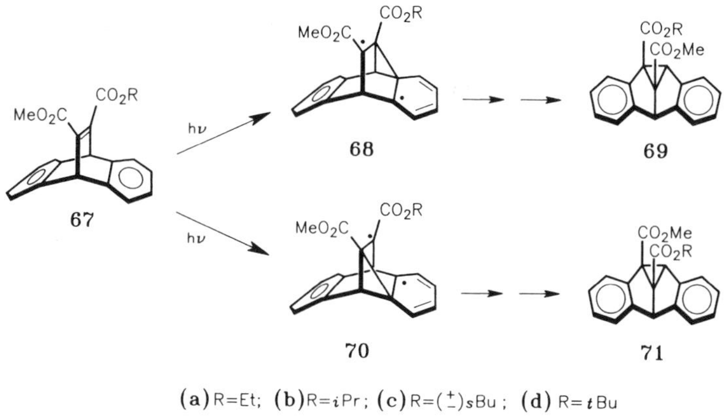

(a) R=Et; (b) R=*i*Pr; (c) R=(±)*s*Bu; (d) R=*t*Bu

Scheme 22

Two explanations were considered for the solid-state results. The first, subsequently found to be incompatible with the experimental data, was that the preference for formation of either biradical **68** or **70** depends on which ester group is better oriented by the crystal lattice for radical stabilization by resonance. The second explanation involved the possibility that steric effects between the reacting molecule and its lattice neighbors may be the controlling factor. Reasoning that it is the ester substituent attached to the bridging vinyl carbon atom that moves most during the initial stages of reaction, Scheffer et al.[80] calculated the intermolecular non-bonded repulsion energies resulting from a computer simulation of these motions. In the case of compound **67b**, for example, the calculations showed that it is more expensive in terms of repulsion energy to move the CO_2Me substituent than the CO_2iPr substituent, a finding that is in accord with the experimental results.

An important point that emerges from the work described above is that substituent effects in the solid-state are not as regular as they tend to be in solution. Exchanging one substituent for another often leads to a completely different packing arrangement and steric environment, and this can result, as we have seen for diesters **67a–d**, in very different chemical behavior for closely related compounds.

Other substituents may be tested against one another for their effect on di-π-methane regioselectivity exactly as described above, and in one such study Garcia-Garibay et al. investigated a carboxylic acid group versus an ester group (compound **72**, Scheme 23).[81] Interestingly, the solid-state photochemical results were found to be very different from those observed in solution. In dilute benzene solutions, photoproduct **73** predominated (**73:74** = 83:17), whereas in the solid-state there was a nearly complete preference for regioisomer **74** (**73:74** = 5:95). This was attributed to a difference in hydrogen bonding that exists in solution and the crystal. In the latter medium, there is strong intermolecular hydrogen bonding between carboxylic acid groups. This, it was suggested, "anchors" the carboxylic acid group and hinders the motions necessary for initial bonding at the vinyl carbon atom to which the carboxylic acid substituent is attached. As a result, bonding at the other vinyl position and formation of photoproduct **74** is favored in the crystal. Infrared studies showed that in solution, there is a competition between inter- and intramolecular hydrogen bonding, and the suggestion was made that the latter situation presents conditions favorable for formation of photoproduct **73**. In accordance with this mechanism, it was found that as the proportion of cyclic dimer was increased in benzene by increasing the reactant concentration, the proportion of the solid-state photoproduct **74** increased.

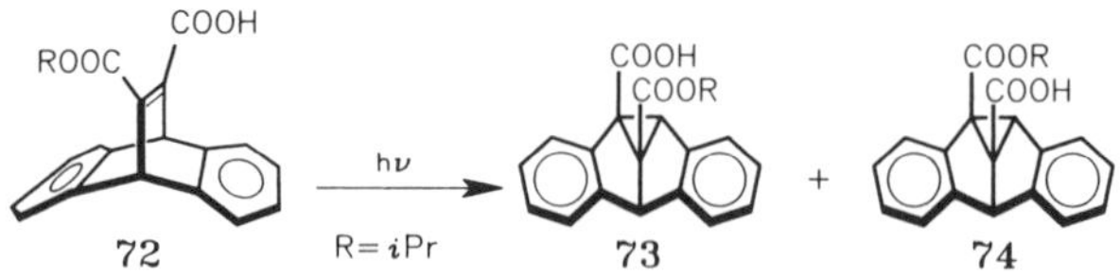

Scheme 23

If, as suggested above, crystal packing effects control di-π-methane regioselectivity in the solid-state, then it follows that regioselectivity may be different at the surface of a crystal compared to the bulk. Such an effect has been documented by Pokkuluri et al.[82] By increasing the surface area of a crystalline, nonsymmetrically substituted dibenzobarrelene derivative by grinding, it was found that the regioselectivity of its di-π-methane photorearrangement was significantly altered compared to that observed in large, carefully grown single crystals. These authors also pointed out that surface reactivity is particularly likely for photoinduced reactions of crystals that absorb strongly at the photolysis wavelength, where a simple Beer-Lambert calculation reveals that most of the incident radiation will be absorbed near the surface. For this reason, investigations of crystal photoreactivity should be conducted on samples of widely differing surface area, and where bulk reactivity is desired, photolysis wavelengths near the absorption tail should be used.

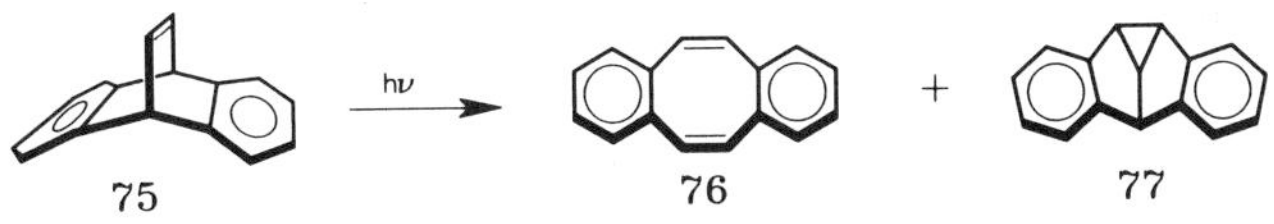

Scheme 24

In a related study, Garcia-Garibay et al.[83] studied the solid-state photochemistry of dibenzobarrelene itself (**75**, Scheme 24). This is an interesting case, as earlier studies had established that the solution-phase photochemistry of dibenzobarrelene is multiplicity-dependent, the singlet leading to dibenzo-cyclooctatetraene (**76**) and the triplet giving rise to the normal di-π-methane product, dibenzosemibullvalene (**77**).[92,99,100] The goals of the solid-state studies were to see if whether multiplicity dependence is maintained in the crystalline phase and to explore the use of triplet energy sensitizers in this medium. Experimentally, it was found that photolysis of solid solutions consisting of small amounts of the triplet energy sensitizer xanthone (1% mol/mol) in dibenzobarrelene *does* lead to exclusive formation of the triplet product **77**, whereas direct irradiation of crystals of dibenzobarrelene gives mainly the singlet photoproduct **76** (**76:77** = 4:1). This represents the first example of the use of sensitizers to direct the course of a multiplicity-dependent photorearrangement in the solid-state. In order to determine whether energy transfer could be occurring in liquid regions of the sensitizer–reactant phase diagram, experiments were conducted with γ-phenylbutyrophenone as the sensitizer. Separate studies had shown that this compound undergoes the Norrish type II reaction in liquid phases to afford acetophenone, but is light stable in the solid-state; it thus acts as a solid-state "indicator." Gas chromatographic analysis of the sensitizer–reactant–photoproduct mixture showed no trace of acetophenone.

A particularly novel result was obtained by Pokkuluri et al.[84] during their study of the solid-state photochemistry of the 9,10-dimethyldibenzobarrelene derivative **78** (Scheme 25). Irradiation of crystals of compound **78** gave mainly the unexpected diester **79** plus traces of the cyclooctatetraene derivative **80** and the di-π-methane photoproduct **81**; in acetone (solvent and triplet energy sensitizer) only the latter was formed, and in benzene an approximately 1:1 mixture of **80** and

81 was obtained. The mechanism by which photoproducts **79** and **80** are formed was suggested to involve formation of biradical **82** followed by either double ester migration or Grob fragmentation (Scheme 25). Particularly noteworthy is the fact that COT derivative **80** has a structure different from that expected on the basis of well-established mechanistic studies in the monobenzobarrelene series.[101-104] These have shown that monobenzocyclooctatetraenes are formed with substituent patterns that are consistent with a mechanism involving intramolecular $2\pi+2\pi$ photocycloaddition followed by thermal reorganization of the resulting cage compound. Such a mechanism applied to compound **78** predicts the formation of a dibenzo-COT with *mirror* symmetry, rather than with the C_2 symmetry actually observed. Pokkuluri et al. suggested that the bridgehead methyl groups of **78** disfavor the $2\pi+2\pi$ pathway by a steric effect and favor formation of biradical **82** through their inductive effect; similar behavior was noted for dibenzobarrelene derivatives with a single methyl or phenyl group at C9.

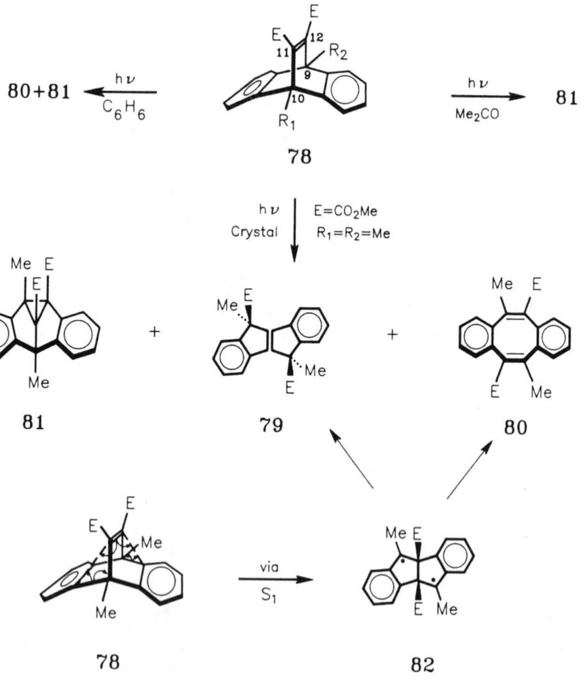

Scheme 25

Turning now to a different topic, we discuss three recent papers by Matsuura and co-workers on the solid-state photochemistry of some substituted 2,5-cyclohexadienones.[85-87] The majority of this work was concerned with the 4-methoxy derivatives **83a** and **83b** (Scheme 26). These compounds were found react very efficiently in the solid-state to give the same products observed in solution, namely the lumiketones **84** and **85** and the secondary photoproducts **86** and **87**. The latter two compounds are not formed at photolysis wavelengths > 400 nm because **84** and **85** do not absorb in that region. At wavelengths of 290 nm and greater (Pyrex filter), **84** and **85** do absorb, and their independent photolysis in both the solid-state and solution was studied.

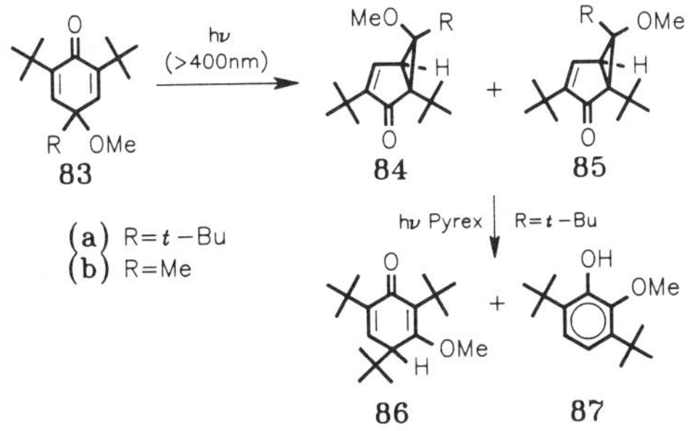

Scheme 26

Matsuura's work on these compounds is noteworthy in several respects. For one thing, it represents one of the first reports of the determination of solid-state quantum yields. These were carried out by depositing a thin crystalline film of the material to be photolyzed on the inside of a Pyrex test tube by evaporation of an ether solution and then irradiating the samples using a merry-go-round apparatus. A more detailed description of this technique has appeared recently.[105] Interestingly, the quantum yield studies showed that photoproducts **84a** and **85a** have approximately equal quantum yields of formation in the solid-state (0.24 and 0.22, respectively), but very different values in solution (0.56 and 0.012). Qualitatively similar results were observed in the case of the methyl derivative **83b**. A second point of interest concerns a comparison of the solid-state photoreactivity of compounds **83a** and **83b** with that of α-santonin. Over 20 years ago, Matsuura et al. had shown that α-santonin, a naturally occurring cross-conjugated cyclohexadienone derivative, undergoes an unusual solid-state photorearrangement to afford a novel cyclopentadienone derivative, which then dimerizes under the reaction conditions.[106] This was in contrast to the formation of lumisantonin in solution, a classic early example of the so-called lumiketone photorearrangement. The crystalline phase rearrangement to cyclopentadienone was suggested to be topochemically controlled, that is to involve fewer and less drastic atomic and molecular motions compared to lumisantonin formation. Although crystallographic evidence on this point is lacking, the failure of cyclohexadienones **83a** and **83b** to give cyclopentadienones in the solid-state was attributed to their "loose crystal lattice structure," thereby allowing the favored but sterically more demanding process of lumiketone formation.

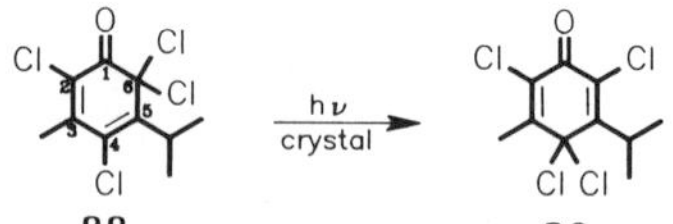

Scheme 27

An interesting solid-state reaction in which a linearly conjugated 2,4-cyclo-hexadienone derivative is converted photochemically to a substituted cross-conjugated cyclohexadienone was reported by Decoret et al.[88]   As outlined in Scheme 27, crystals of the 6,6-dichloro-substituted dienone **88** react photochemically (and thermally) to afford the 4,4-dichloro isomer **89**. Based on the crystal structure of compound **88**, it was suggested that the chlorine is transferred *inter*molecularly from the 6-position of one reactant molecule to the 4-position of a lattice neighbor, the C6—Cl···C4 distance being 4.6 Å. In accord with this postulate, packing potential energy calculations at various stages along the reaction coordinate in which the C6—Cl bond is lengthened show that the chlorine atom moves preferentially toward the C4 position of an adjacent molecule.

Baro et al. have reported the case of a β, γ-unsaturated ketone that photo-rearranges in the solid-state.[89]   As outlined in Scheme 28, irradiation of crystals of bridged diketone **90** affords the fused diketone **91**. The reaction occurs in solution as well, and continued irradiation in either medium at 509 nm leads to decarbonylation. It will be noted that there are two possible photorearrangement products, **91a** and **91b**. Owing to high moisture sensitivity, however, neither could be isolated in pure form, and it was assumed that a mixture was present. Because of this, the interesting question of whether **91a** and **91b** are formed in different ratios in the solid-state and solution could not be answered.

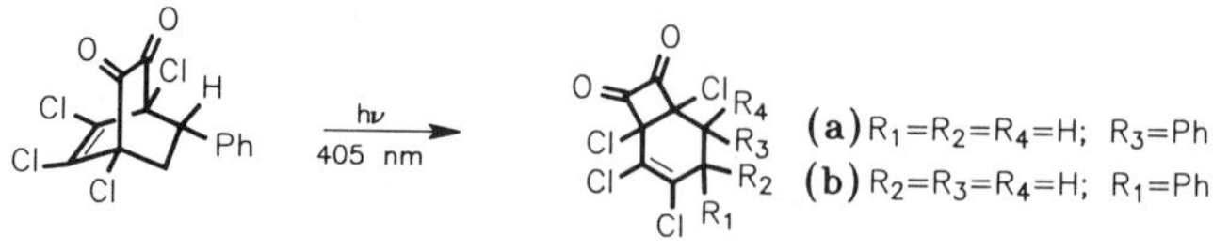

Scheme 28

Finally we turn to a discussion of the recent and important work of Zimmerman and Zuraw on the solid-state photochemistry of several *acyclic* di-π-methane systems;[90,91] these papers also include interesting results on the crystalline-phase photoreactivity of some cyclohexenone derivatives. Altogether 11 compounds were prepared and photolyzed and the results correlated with the reactant x-ray crystal structures. As it is beyond the scope of this review to discuss all 11 compounds in detail, the results obtained with four representative examples (compounds **92, 94, 96** and **98**, Scheme 29) will be summarized.

Irradiation of diene **92** in the solid-state affords the novel product **93** in which the cyano-substituted double bond has undergone intramolecular [2+2] cycloaddition to one of the vinyl phenyl groups. In benzene, however, none of this material is formed, its place being taken by conventional di-π-methane type photoproducts. The photochemistry of triene **94** is also quite different in the solid-state compared to solution. In the former medium, vinylcyclopentene **95** is the sole photoproduct, whereas di-π-methane reactivity is observed exclusively in the latter. The formation of **95** was unexpected and involves a novel "tri-π-methane" mechanism which will be discussed in Scheme 30.

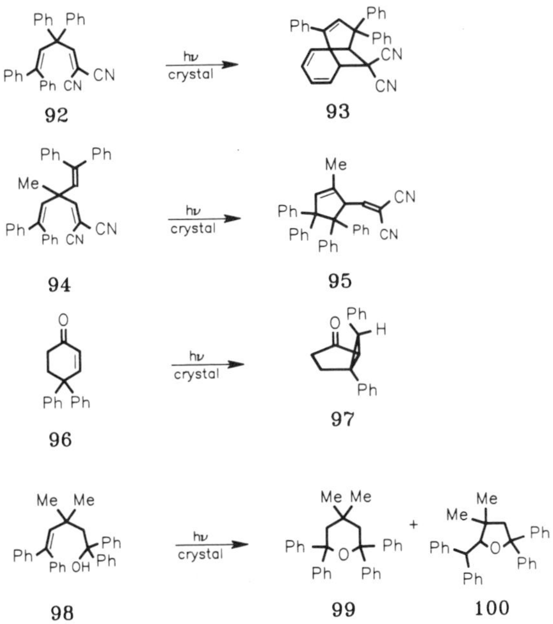

Scheme 29

4,4-Diphenylcyclohex-2-en-1-one (**96**), whose solution-phase photochemistry was established by Zimmerman and Wilson over 25 years ago,[107] also photorearranges in the solid-state. In this case, the major product (compound **97**) is the same in both media, the difference being that in solution small amounts of the epimer of **97** as well as a third photoproduct are also formed. Finally, these articles report the unusual solid-state photocyclization reaction of unsaturated alcohol **98**; in solution, compound **98** undergoes a novel long-range phenyl migration reaction.

In order to rationalize the differences between the results in the solid-state and solution, Zimmerman and Zuraw devised three reactivity parameters. The first of these was ΔM, a least motion parameter defined as the sum of all non-hydrogen atom displacements in proceeding from the x-ray structure of the reactant to the superimposed product or branch point species. The branch point was defined as the point along the reaction coordinate (usually corresponding to a biradical intermediate) that determines the partitioning between the possible photoproducts. The structures of the branch point species were calculated using modified molecular mechanics methods. In order to get an estimate of the volume changes accompanying photoreaction, a second parameter was defined. This was ΔV, the volume of the photoproduct or branch point species not in common with the reactant when the two are superimposed. Finally, a parameter termed ΔS was devised. This is a measure of the interference (volume overlap) between the photoproduct or branch point species and its nearest neighbors when optimally situated in a crystal lattice position originally occupied by the reactant. Values for these parameters were calculated for the solution phase photoreactions as well as the solid-state processes, the idea being that if least motion, volume changes or lattice interference are important in controlling reactivity in the crystal, then the values of ΔM, ΔV, and ΔS in this medium should be less than those in solution.

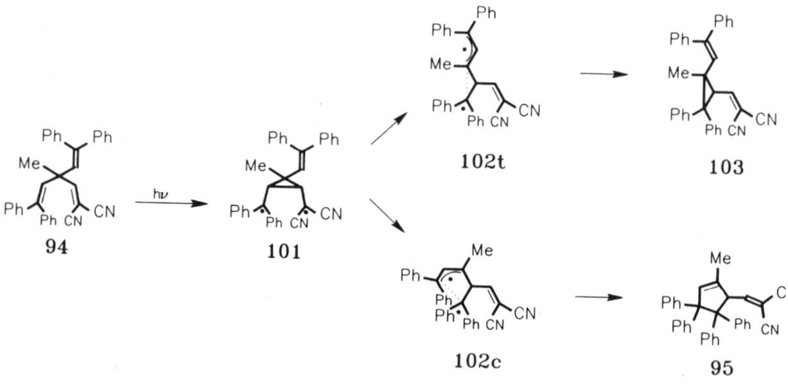

Scheme 30

Let us illustrate the application of these reactivity parameters to one example, that of compound **94**. As outlined in Scheme 30, triene **94** photorearranges to vinylcyclopentene **95** in the solid-state and to four divinylcyclopropane derivatives in solution, one of which has structure **103**. The probable mechanism by which photoproducts **95** and **103** are formed is shown in Scheme 30. For geometric reasons, Zimmerman and Zuraw suggested that formation of **95** (the solid-state photoproduct) proceeds through the *cisoid* biradical **102c**. The *transoid* form of this biradical, **102t**, is incapable of five-ring closure and was therefore suggested to be the precursor of cyclopropane derivative **103** which is formed in solution. The calculated values of all three reactivity parameters support this picture in that they are substantially lower for formation of biradical **102c** ($\Delta M = 0.24$ Å atom^{-1}, $\Delta V = 0.06$, and $\Delta S = 0.01$) than for production of biradical **102t** (1.87 Å atom^{-1}, 0.26, and 0.10, respectively). Thus the basic reason for the preferential formation of compound **95** in the solid-state is that the reactant **94** in its own crystal lattice more closely resembles the *cisoid* biradical conformation **102c** than the *transoid* form **102t**.

A final note on the work by Zimmerman and Zuraw concerns their measurement of solid-state quantum yields. In order to correct for scattered light not absorbed by the sample, a special cell was designed which consists of a sample compartment surrounded by actinometer solution except for a narrow opening for the light beam. In general, it was found that the solid-state quantum yields were considerably lower than their solution phase counterparts. For example, photoproduct **95** is formed from triene **94** in the solid-state with a quantum yield of 0.0008, whereas the total di-π-methane reaction quantum yield in solution is 0.18.

3.4. The Norrish Type II and Related Reactions[75,87,108-129]

As has been pointed out earlier in connection with the di-π-methane photorearrangement of dibenzobarrelene derivative **56b** (Scheme 18), the discovery of solid-state reactions that convert crystal chirality into permanent molecular chirality in high optical yield without the use of resolved chiral reagents is rare and intriguing, and we begin this section by discussing an example of such a process that occurs in a Norrish type II photoreaction. α-(3-Methyladamantyl)-*p*-chloroacetophenone (**104**, Scheme 31) spontaneously crystallizes in the chiral space group P2$_1$2$_1$2$_1$, and irradiation of single crystals of this material leads, via the Norrish type II reaction,[47] to good yields (albeit in low conversion) of the cyclobutanol derivative **105**.[75,108] The optical purity of **105** was determined by nmr chiral shift reagent methods, which indicated an enantiomeric excess of approximately 80%; as expected, compound **105** isolated from solution-phase photolysis of **104** was racemic. Although there is no a priori reason to believe that chemical reactions of perfect chiral crystals need occur with 100% stereospecificity, in the case of ketone **104** the crystal was sticky following photolysis, and we attribute the less than quantitative optical yield to partial sample melting.

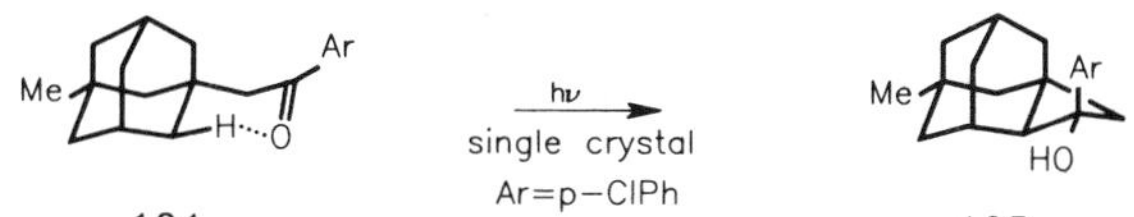

Scheme 31

Other α-adamantylacetophenone derivatives that crystallize in more common centrosymmetric space groups have also been studied, both in the solid-state and solution.[108,109] These studies will not be detailed here other than to point out that, in general, such compounds lead to 100% cyclobutanol formation owing to the strain involved in forming the photoelimination product, adamantene. In one case, that of the *p*-chloro derivative, dimorphic crystals were obtained, and each dimorph gave rise to a significantly different mixture of stereoisomeric cyclobutanols when irradiated in the solid-state.[109] In a closely related study, the x-ray crystal structure and Norrish type II photochemical behavior of α-exo-2-norbornyl-*p*-chloroacetophenone has been reported.[110]

The UBC solid-state chemistry group has carried out a particularly wide ranging study of the crystallography and solid-state photochemistry of acetophenone derivatives substituted in the α-position by *mono*cyclic rings ranging in size from cyclobutyl to cyclooctyl.[111-120] The most thoroughly studied of these are the *p*-chloro derivatives, and we devote the next few paragraphs of our review to this topic. Scheme 32 shows the compounds studied and their photoproducts.

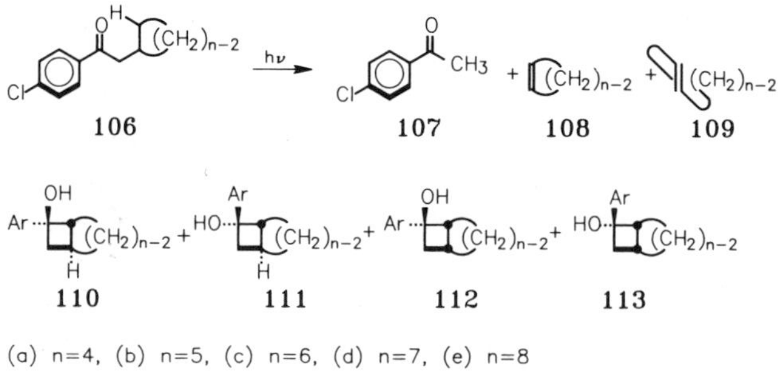

(a) n=4, (b) n=5, (c) n=6, (d) n=7, (e) n=8

Scheme 32

The key findings of this study were that (1) as the cycloalkyl group ring size is increased from four to eight, the extent of cyclization (photoproducts **110 – 113**) increases (from 8% to 82% of the total photoproduct mixture) at the expense of type II cleavage (photoproducts **107 –109**), (2) the cyclobutanol products are predominantly trans-fused (**110 and 111**), and (3) the small amount of cyclooctene formed from photolysis of compound **106e** has largely the cis geometry. These trends were found to hold for both the solid-state and solution phases.[111]

The photochemical results were interpreted with the help of the crystal and molecular structures for each ketone, a key assumption being that the initially formed 1,4-biradical intermediates have the same basic geometry in the solid-state as their ketonic precursors. In every case, the biradicals thus arrived at have poor geometries for either cleavage or closure, and must undergo substantial bond rotations in order to align the orbitals properly for reaction. It was suggested that bond rotation at C4 in the (–) direction leads mainly to cleavage to *cis*-cycloalkene and enol (normal 1,4-biradical behavior), and that rotation in the opposite direction leads predominantly to trans-fused cyclobutanol formation (cleavage to *trans*-cycloalkene being disfavored). This is illustrated in Scheme 33. Why then is there such a difference in reactivity with ring size? The answer was suggested to lie in the different conformations of the biradicals involved and to arise from the different extent of rotation required at C4 in each case. For the four- and five-membered ring cases, only -45° and -65° rotations are required for optimum orbital alignment, whereas 135° and 115° rotations are required in the (+) direction. The former motions are favored topochemically as well as thermodynamically, the result being that cleavage is favored, as is observed experimentally. In the case of the six- and seven-membered ring homologues, approximately equal rotations in either direction are required, a circumstance that is in accord with the fact that approximately equal amounts of cyclization and cleavage are observed. The eight-membered ring case is just the reverse of the situation with the four- membered ring compound; rotation at C4 of +45° is favored topochemically over -135° rotation, which accounts for the observed preference for cyclization to the trans-fused cyclobutanol derivatives **110e** and **111e**. A computer simulation of the reaction pathway for the conversion of cyclooctyl ketone **106e** to cyclobutanol **110e** supports this interpretation.[114] The theory presented above accounts nicely for the seeming paradox that photolysis of ketones **106a–e** affords predominantly trans-fused cyclization products on the one hand and mainly *cis*-cycloalkene cleavage products on the other. These ideas also provide a nice explanation for the formation of *trans*-fused products in the solution phase [2+2] photocycloaddition of olefins to cyclohexenones.[130] Biradicals that close to *trans*-fused products or cleave to starting enone plus olefin explain the results without requiring the intermediacy of *trans*-cyclohexenones.

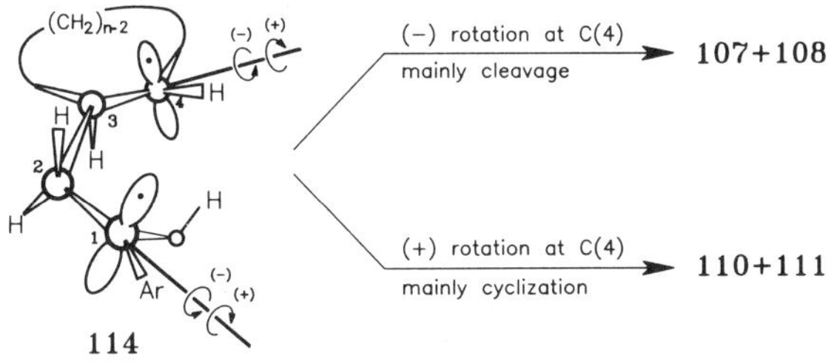

Scheme 33

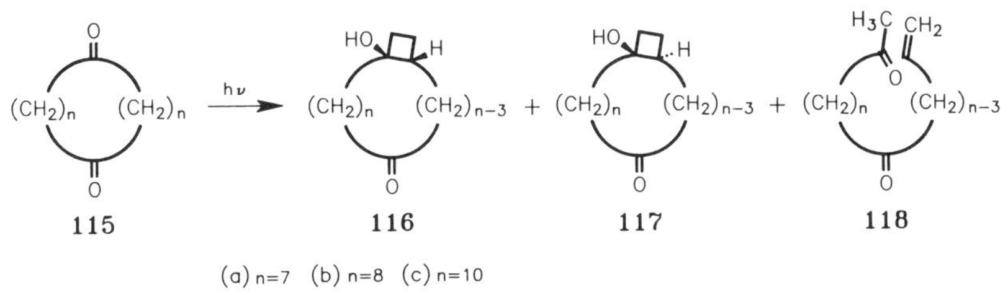

(a) n=7 (b) n=8 (c) n=10

Scheme 34

Very recently, Lewis et al.[121] have reported the solid-state (and solution phase) photochemistry of the three macrocyclic diketones **115a–c** (Scheme 34). The motive behind this study was to provide further evidence on the preferred geometry for intramolecular hydrogen atom abstraction and to extend the correlation of biradical behavior with molecular structure. As shown in Table 1, striking differences between the solid-state and solution-phase photoproduct ratios were observed. These differences can be summarized as follows: (1) cyclization is favored over cleavage in the solid-state, (2) the stereochemistry of cyclization in the solid-state depends on ring size, and (3) the photochemistry of diketone **115a** is strongly temperature dependent in the solid-state but not in solution.

Table 1. Photoproduct Percentages as a Function of Medium and Temperature

Diketone	Medium	Temp (°C)	%116	%117	%118
115a	Crystal	20	89	10	1
	Crystal	40	17	20	63
	Solution	20	22	35	43
	Solution	40	19	30	51
115b	Crystal	20	3	84	13
	Solution	20	17	42	41
115c	Crystal	20	4	91	5
	Solution	20	10	34	56

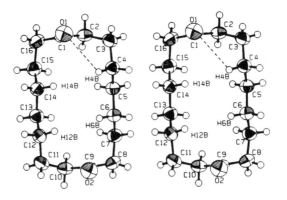

Scheme 35

As with the acetophenone derivatives discussed above, the solid-state photochemical results are explicable in terms of the molecular conformations of the diketones as determined by x-ray crystallography. This indicates that the diketones react in a topochemical, conformation-specific manner in the crystal, and that the different product ratios obtained in solution are the result of reaction from other conformational isomers present in this medium. As a specific example, consider the 16-membered ring diketone **115a**, whose conformation in the solid-state is depicted stereoscopically in Scheme 35. The dotted line indicates the closest C=O···H$_\gamma$ contact (2.7 Å), and abstraction of this hydrogen followed by biradical closure with retention of configuration at both the carbonyl carbon and the γ-carbon leads, in a topochemical fashion, to the experimentally observed cis-fused cyclobutanol **116a**; involvement of any of the other three accessible γ-hydrogens H14B, H12B or H6B leads to the same conclusion. The striking change in the solid-state photoproduct percentages for diketone **115a** between 20° and 40° was attributed to a solid–solid phase change in this region. Differential scanning calorimetry indicated a transition temperature of 34°C, well below the melting temperature of 86°. The high-temperature phase evidently mimics the situation in solution to a considerable degree, and unpublished solid-state nmr studies support this conclusion.[131]

As mentioned in the brief review of the pre-1987 literature, α-oxoamides undergo Norrish type II type photorearrangement in the solid-state. The *N,N*-diisopropyl derivative **30** (Scheme 9), originally studied by Aoyama et al.,[42] has been reinvestigated recently by Toda and co-workers[122-124] and found to crystallize spontaneously in a chiral space group and to photorearrange in this medium to the chiral β-lactam derivative **31** in 93% ee. Crystals of each antipode of **30** were prepared on a relatively large scale by seeding during recrystallization from benzene, and this allowed the controlled synthesis of each enantiomer of photoproduct **31**. We note in passing a brief mention of the solid-state photoreactivity of *N,N*-dibenzyl-2-benzoylpropionamide, a so-called β-oxoamide.[125] In benzene or on silica gel, this material photorearranges efficiently to the corresponding five-membered lactam via δ-hydrogen abstraction; however, only traces of this product are formed in the crystal, more than 95% of the starting material being recovered.

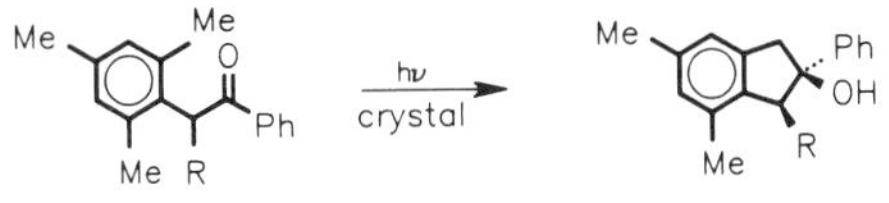

119   120

Scheme 36

Wagner and Zhou have recently reported the solid-state photochemistry of a series of α-mesityl and α-(*o*-tolyl)acetophenone derivatives.[126,127] Upon irradiation in the solid-state, these compounds (e.g., **119**, Scheme 36) undergo δ-hydrogen atom abstraction from an *ortho* methyl group followed by 1,5-biradical cyclization to afford 2-indanol derivatives. In solution, alternative reactions such as enol ether formation and radical cleavage predominate, and the authors suggest that these processes are topochemically forbidden in the crystal.

Finally, we discuss the interesting work of Ito and Matsuura on the type II photochemistry of a series of *p*-substituted 2,4,6-triisopropylbenzophenone derivatives.[87,128] Altogether, seven compounds of general structure **121** were studied, and all except one (R = COOMe) underwent essentially quantitative photocyclization to the corresponding cyclobutanols **122** in the crystalline state (Scheme 37). By using the same procedures employed in their studies of crystalline phase enone photorearrangements,[86,105] Ito and Matsuura determined the type II quantum yields in the solid-state and compared them with those measured in solution. In the solid-state, a steady *decrease* in Φ_{CB} (from 0.56 to ~ 0.001) was noted as the *p*-substituents were changed from electron donating (OMe) to electron withdrawing (COOMe). In solution, however, Φ_{CB} changed relatively little over the same range (0.48 to 0.13). Based on known substituent effects on the lowest triplet states of aromatic ketones in solution, these data were interpreted by the authors as indicating reaction through an n,π* state in the liquid phase and a π,π* in the crystal; topochemical arguments based on x-ray crystal structure data for two of the ketones were also advanced to support this conclusion.

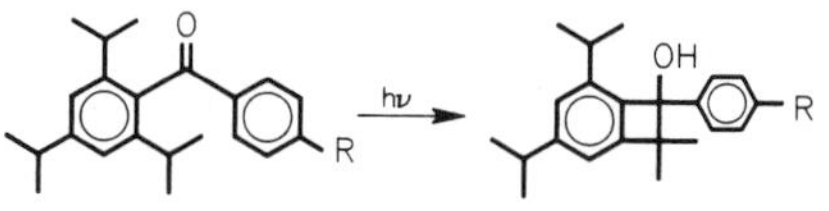

121   122

R = OMe, *t*−Bu, Me, H, Cl, CF_3, CO_2Me

Scheme 37

We close this section of the review with a brief discussion of what these new studies tell us about the preferred geometry of intramolecular hydrogen atom abstraction. It will be recalled that previous studies had established a rough upper limit of 3.1 Å for this process.[48] To paraphrase Woodward's famous statement concerning the Woodward-Hoffman rules, there are no exceptions! All of the type II photoactive compounds discussed in this section whose crystal structures were determined have C=O···H abstraction distances of less than 3.1 Å. It is interesting to compare this number with what little is known concerning the distances over which *inter*molecular hydrogen atom abstraction can occur. In a paper on the solid-state photochemistry of 4,4-dimethylbenzophenone, Ito et al.[129] reported that transfer of a methyl hydrogen from one molecule to the carbonyl oxygen of another took place over a distance of 3.3 Å. By contrast, 4-methylbenzophenone was photostable under the same conditions despite a much shorter intermolecular

C=O···H contact of 2.7–2.8 Å. This was ascribed to the inability of the radical pair produced by abstraction to couple (4.4–4.5 Å separation); the corresponding radical pair separation in 4,4-dimethylbenzophenone was 3.9 Å. Similar conclusions were reached by a group at the Weizmann Institute studying the photochemical hydrogen abstraction reactions of a series of ketones as guests in deoxycholic acid inclusion complexes (see Chapter 6).[132,133] In these cases, the approximate upper limits for hydrogen atom transfer and radical coupling were 3.5 and 4.2 Å, respectively. What is responsible for this difference in the upper limit for intra- versus intermolecular abstraction? It is interesting to speculate that this may reflect the slight but significant translational freedom that molecules in a crystal possess, so that intermolecular abstraction may actually occur at distances less than the crystallographically determined values. Clearly, further experimental evidence on this point is required before a firm conclusion can be reached.

The α-mesityl and α-(*o*-tolyl)acetophenone derivatives studied by Wagner and Zhou[126,127] deserve special mention. The X-ray crystal structure of one of the mesityl derivatives showed a poor angular arrangement for hydrogen abstraction in that the angle τ, defined as the degree by which the target hydrogen atom lies outside the mean plane of the carbonyl group, was 61°. The ideal value for τ is 0° because of the involvement of the oxygen n orbital in the abstraction process. Despite this poor geometry, efficient photocyclization was observed in the crystal. Ito and Matsuura reported similar behavior in their compounds for values of τ in the 50-60° range,[128] and several examples of solid-state type II photoreactivity at high τ angles have also been reported by Scheffer.[48] Wagner and Zhou[126,127] have suggested that the rate constant for hydrogen atom abstraction, k_H, may vary with the angle τ according to $\cos^2 \tau$. If this is the case (and Wagner and Zhou provide preliminary evidence that it may be), then k_H is diminished by only a factor of 4 in going from $\tau = 0°$ to $\tau = 60°$.

3.5. Photofragmentation Reactions[134-142]

The initial part of this section of the review will be devoted to a discussion of the solid-state photochemistry of a series of diacyl peroxides.[134-139] Much of the work described was either based on, or performed by, McBride and co-workers. The first paper to be discussed concerns the solid-state photochemistry of bis(3,3,3-triphenylpropanoyl)peroxide (**32**, Scheme 10), a brief account of which was given in Section 1. It will be recalled that photolysis of this material leads to formation of a pair of neophyl radicals (TT*) that are stressed by being pushed apart by the two new carbon dioxide molecules between them. Subsequent warming or further photolysis gave five additional radical pairs, one of which (TT) is the relaxed form of the initial pair, the remaining pairs being the result of 1,2-phenyl migration (neophyl rearrangement). Three of the four solid-state neophyl rearrangements involve a 1,2-shift of the phenyl group that has the most favorable torsional angle for migration, but in the case of TT*, nontopochemical behavior resulting from the stress of a newly liberated carbon dioxide molecule causes migration of a different phenyl group, the one toward which the methylene radical is pushed by the CO_2.

This mechanism was worked out by Walter and McBride primarily on the basis of crystallographic studies and the detailed analysis of the esr spectra of irradiated single crystals of triphenylpropanoyl peroxide.[52-54] In an effort to provide further insight into the reaction, Gavezzotti[134] carried out a crystal potential energy calculation and packing analysis of its initial stages. The main goal of this work was to locate the carbon dioxide molecules, which are of course undetectable by esr spectroscopy. The calculations indicated that the newly liberated CO_2's can indeed occupy vacant space between the two neophyl radicals, and based on the geometry of the system, which seemed to indicate little steric hindrance to topochemical phenyl migration, the radical pair was assigned the relaxed structure TT. Additional conclusions from the calculations were that the packing coefficient for triphenylpropanoyl peroxide is rather low, that photofragmentation may therefore be easier than thought and that crystals with sizeable domains of reacted molecules could be obtained.

A second theoretical approach to the understanding of solid-state diacyl peroxide photochemistry was described by Kearsley and McBride.[135] In this work a special molecular mechanics program was devised which incorporates a "surface walking" algorithm to help determine the reaction path among the possible intermediates that can be formed upon warming a partially photolyzed single crystal. The system chosen for investigation was crystalline acetyl benzoyl peroxide (**ABP**), which earlier work[143,144] had shown photolyzes to carbon dioxide and a benzoyloxy–methyl (**BM**) radical pair. This pair either reacts to form methyl benzoate or liberates a second molecule of CO_2 to leave a phenyl–methyl (**PM**) pair which collapses to toluene. These processes are shown in Scheme 38.

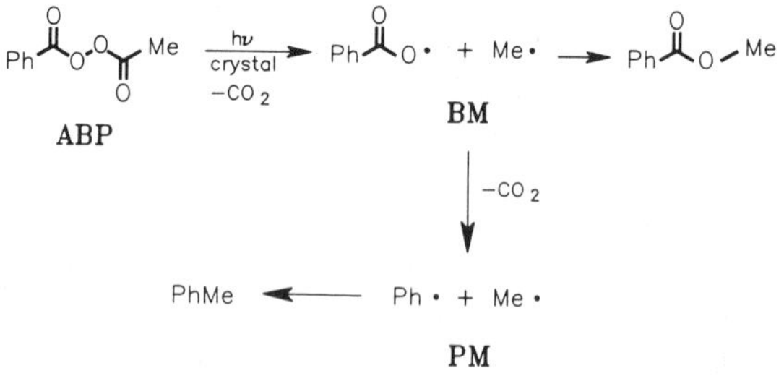

Scheme 38

One of the remarkable features of this work was the finding, based on careful analysis of the **g** tensor of **BM** and of the ^{17}O hyperfine splittings from specifically labeled **ABP** molecules, that the benzoyloxy radical undergoes a nontopochemical 30° in-plane rotation that preferentially exposes its peroxy oxygen atom to the methyl radical, thus accounting for the predominant formation of PhCO—O^{18}Me from PhCO—O^{18}—O^{18}COMe. The driving force for this rotation was suggested to be the non-bonded repulsion between the benzoyloxy radical and the newly liberated CO_2 molecule (local stress), a theory that was subsequently supported by detailed studies with benzoyl peroxide.[52] Strikingly, the molecular mechanics calculations of Kearsley and McBride on **ABP** were able to reproduce this feature of the postulated reaction mechanism. The calculations were performed by holding the surrounding, intact **ABP** molecules at their crystallographic positions and applying the surface walking algorithm to the fragments within the reaction cavity. The results of the calculations could be visualized qualitatively in stereopair drawings of space-filling models showing the intersections of the van der Waals envelopes of the CO_2 molecules and the radical fragments with each other as well as with the atoms of the surrounding cage molecules. Gavezzotti has also analyzed the **ABP** crystal system by using his packing potential energy calculations and free volume analysis approach[145] and found results that are qualitatively consistent with the mechanism postulated by McBride and co-workers.

The next paper to be discussed deals with the role of *long-range* stress in solid-state peroxide photoreactions.[136] The compound studied was diundecanoyl peroxide, and the mode of analysis was by FTIR spectroscopy of the liberated CO_2. Previous work on this system using single crystal esr spectroscopy[146] had established a sequence of events similar to that observed for other diacyl peroxides, namely formation of an initial pair (pair A) of ordered decyl radicals plus 2 mols of CO_2; upon warming, radical pair A was successively transformed into two other well-defined radical pair intermediates (B and C) before radical pair collapse to give structure D. The reaction sequence was unusual in that in going from A to C, the radical pairs are pushed further and further apart by the CO_2 molecules before, finally, the CO_2 is squeezed out of the molecular layers into the loosely packed interface and the radicals combine to form eicosane and other products.

Each radical pair identified by esr was associated with a characteristic FTIR spectrum in the 2300 cm^{-1} region for the liberated CO_2. From the frequencies of the CO_2 antisymmetric stretching bands, aspects of the local stress that the trapped molecules were experiencing could be assessed. A key feature of this work was the finding that, as the extent of photolysis was varied from 0 to 0.1%, there was a sudden appearance at 0.06% of ir peaks in pair C and especially in D indicating particularly high local stress. Strikingly, the effect was enhanced by warming the irradiated crystal to about 140 K, but annealing above 280 K caused its disappearance. These results were interpreted as being due to a spreading of the stress field around each photolysis site until, in C and D, adjacent stress fields overlap and reinforce one another. Since A and B show no abnormal spectral changes at conversions of 0.06%, it was inferred that the influence is produced when the crystal is warmed and the stress field expands during plastic deformation. In a graphic analogy, the situation was likened to a nest of concentric balloons with the strongest, highest pressure balloon in the middle: popping the central balloon causes the pressure at the center to fall, but the pressure further out increases. A rough analysis based on random decomposition indicates that the perturbing influence can extend over distances of at least 6 to 10 molecules in the crystal. While such effects are likely to be greater for solid-state reactions that generate several fragments, Hollingsworth and McBride point out that analogous influences might be expected from any process that changes the size or shape of the reaction cavity. Furthermore, since even very low-decomposition levels can have an effect, they warned that caution must be exercised when interpreting product distributions from solid-state reactions in terms of ideal crystal packing.

In a recent review article on the use of FTIR spectroscopy in solid-state chemistry,[137] Hollingsworth and McBride reported that the mechanism of the decomposition of diundecanoyl peroxide is a good deal more complicated than originally thought. For example, although esr gave no evidence for any intermediates between A and B, very careful FTIR measurements employing the technique of subtracting the spectra of the major components from the spectra of mixtures revealed the presence of as many as seven separate pairs of CO_2 molecules during this transition! The basic conclusions from the original work were, however, unchanged. This paper also reports the effect that altering the structure of the parent diundecanoyl peroxide molecule has on the photodecomposition pathway. Structural changes that were made include the introduction of halogen at the terminal methyl group, homologation, the generation of lattice vacancies through the formation of solid solutions of diundecanoyl peroxide with peroxides of shorter chain length, and specific deuteration. A detailed analysis of the CO_2 pairs formed in the solid-state photolysis of acetyl benzoyl peroxide is also included.

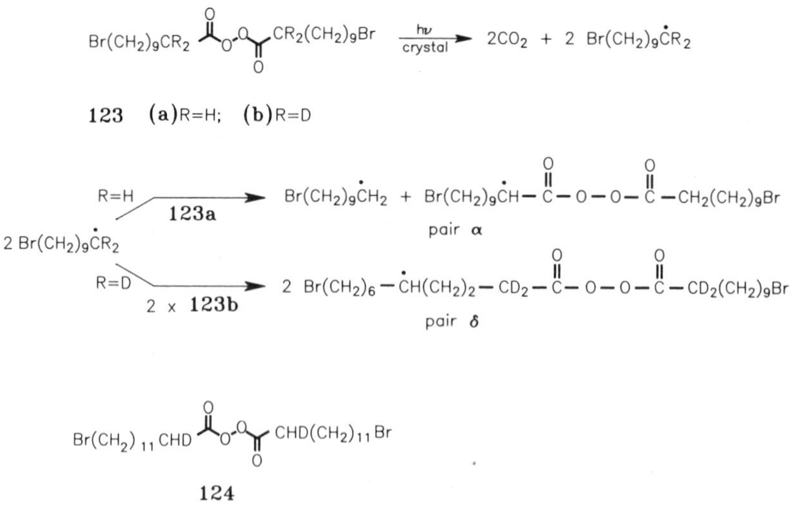

123   (a) R=H;   (b) R=D

Scheme 39

The final paper from the Yale group that we shall consider in the present review is concerned with the photodecomposition of crystals of 11-bromoundecanoyl peroxide and some of its deuterated analogues (Scheme 39).[138] Photolysis of the parent peroxide **123a** at 80 K leads to an initial pair of primary 10-bromodecyl radicals plus two molecules of CO_2. Upon warming to 100 K, one of these radicals abstracts a hydrogen atom from a neighboring intact peroxide molecule to form a secondary/primary radical pair (pair α). Identification of pair α was based on zerofield and hyperfine splittings in esr spectra of undeuterated and specifically deuterated single crystals. In contrast, irradiation of crystals of α-deuterated 11-bromoundecanoyl peroxide (**123b**) failed to give pair α, affording instead at 140 K pair δ, a secondary–secondary radical pair formed by transfer of a δ-hydrogen atom from a neighboring molecule to each of the initial radicals. This result was interpreted as being due to an isotope effect that changes the preferred site of abstraction from α (C· to $H_α$ distance 3.10 Å) to δ (C· to $H_δ$ distance 3.33 Å).

Support for this interpretation was provided by a very clever experiment. *Monodeuterated* 13-bromotridecanoyl peroxide **124**, whose solid-state photochemistry is very similar to that of the other compounds studied, was synthesized in optically active (S,S) form. Based on the knowledge that this material crystallizes in one or the other of the enantiomorphous space groups $P4_32_12$ or $P4_12_12$, Feng and McBride reasoned that in the former space group the α-hydrogen atom should be in a position favorable for abstraction, and these crystals should lead to pair α. In the latter space group, however, because geminal hydrogen atoms exchange roles between space groups, *deuterium* occupies the abstractable position, and isotopic retardation should lead to formation of pair δ. This is exactly what was found experimentally. One way of thinking about the situation is that, because these systems combine molecular chirality and crystal chirality, the two crystals are diastereomeric, and like other diastereomers, can react differently (see also Chapter 6).

Next we discuss an interesting paper by Lomolder and Schafer concerning the solid-state photochemistry of several unsymmetrical diacyl peroxides in which one half of the molecule contains two or more chiral centers of known absolute configuration.[139] Following previous work on *achiral* mixed peroxides, which showed that it is possible to obtain unsymmetrical coupling products with high selectivity in the solid-state,[147] Lomolder and Schafer prepared four peroxides based on various derivatives of tartaric or gluconic acid. The goal of the work was to determine whether the configurational integrity of the α-carbon atom is maintained during the solid-state coupling reaction. The results for compound **125**, which are typical, are shown in Scheme 40. At -60°C crystals of peroxide **125** give moderate (44–64%) yields of the coupling product **126** in which the configuration at the α-carbon is largely unchanged (de = 89–95.6%).

Raising the temperature to -20°C causes a serious diminution in both chemical yield (25%) and diastereomeric excess (35.7%). An interesting aspect of this work concerns the way in which the solid-state irradiations were carried out. This was done by adding an ether solution of the peroxide to a large excess of cold (-60°C) petroleum ether (in which the peroxide is insoluble) and photolyzing the resulting suspension. Overall, the results are significant in that they offer a potentially general procedure whereby carboxylic acids from the chiral pool can be decarboxylated and alkylated with retention of optical activity. Attempts to extend the process to mixed peroxides of various protected amino acids were, however, unsuccessful owing to peroxide decomposition during synthesis.

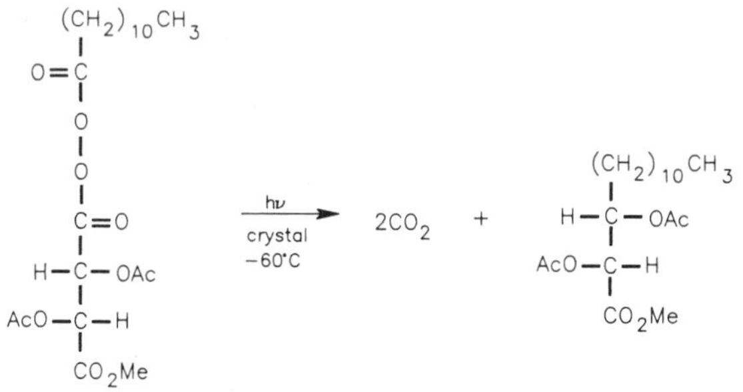

125   126

Scheme 40

Some novel solid-state photofragmentation reactions have been reported by Baro et al.[89,140] The first of these involves the bridged bicyclic α-diketone **90** which, as mentioned earlier, photorearranges in the crystalline phase to a mixture of the fused α-diketones **91a** and **91b** (Scheme 28).[89] Further irradiation of this mixture at 509 nm leads to the loss of 2 mols of carbon monoxide with the concomitant formation of the corresponding diene, 1-phenyl-2,3,4,5-tetrachloro-2,4-cyclohexadiene. The reaction occurs with high efficiency both in the solid-state and solution. The same paper[89] reports the solid-state photochemistry of 9,10-dihydroanthracene-9,10-dicarboxylic anhydride (**127**, Scheme 41). Irradiation of this material at 254 nm in various media including CH_2Cl_2, KBr matrices and thin solid films leads to the formation of CO, CO_2, and anthracene. The course of the solid-state reaction could be monitored in two ways: by measuring the increase in pressure in the photolysis cell as a function of time and (in KBr matrices) by recording the ir spectrum of the sample at various stages of photolysis. In the latter experiments, CO_2 could be detected by a strong peak at 2325 cm^{-1} which disappeared upon crushing and recompressing the pellet. For some reason, however, absorptions due to CO were not detected, even though in other systems this compound has been trapped in KBr matrices and identified by ir spectroscopy.

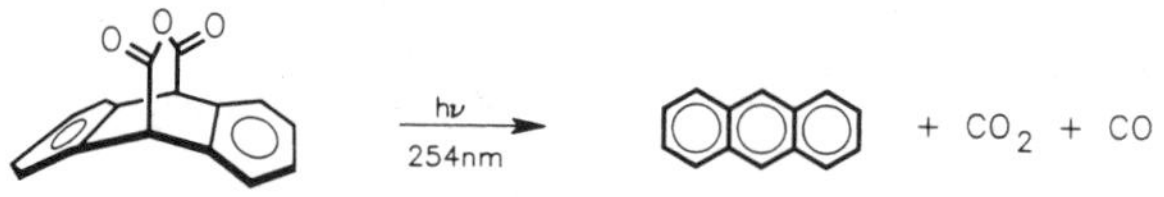

127

Scheme 41

The second paper by Baro et al.[140] deals with the solid-state photochemistry of 1,3-diphenyltriazene (**128**, Scheme 42); the solution phase photochemistry of this compound is described in some detail in a separate publication.[141] As outlined in Scheme 42, in both media nitrogen is eliminated, but the fate of the phenyl/anilino radical pair so produced is quite different. In the solid-state, radical combination leads to high yields (ca. 85%) of diphenyl amine (**129**), whereas in cyclohexane solution hydrogen atom abstraction from the solvent leads to benzene and aniline as the major photoproducts (30–35%), with a host of byproducts being formed as well. Such cage effects are of course well documented in the work of McBride and others already referred to in this and the previous review.[22] As in their work on anhydride **127**, Baro and co-workers were able to follow the crystalline phase photoreaction of 1,3-diphenyltriazene by infrared spectroscopy of irradiated KBr pellets and by measuring the pressure due to the liberated nitrogen as a function of irradiation time.

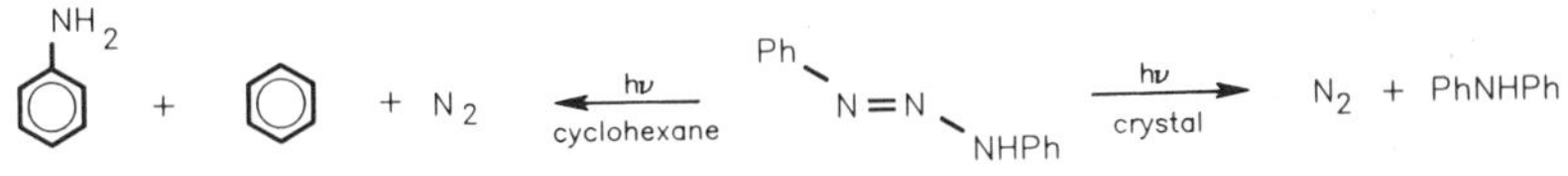

Scheme 42

The final paper to be discussed in the photofragmentation section of the review is concerned with the mechanism of the photodecomposition of the explosive compound HMX (**130**, Scheme 43).[142] Single crystals of this material were irradiated at 77 K by using the output from an unfiltered 500-W mercury flashlamp and then analyzed by esr spectroscopy at different alignments of the crystal with respect to the direction of the applied magnetic field. This indicated that paramagnetic ·NO_2 molecules are formed that are highly ordered within the HMX crystal lattice. Puzzlingly, however, signals due to the other component of the radical pair were missing. Based on a comparison of the direction of the maximum value of the ^{14}N hyperfine coupling tensor for the liberated ·NO_2 with the two independent, crystallographically determined N—N bond directions in HMX, it was concluded that the N3—N4 bonds cleave preferentially. This is in accord with expectation in that the crystal structure shows that the N3—N4 bond is longer and more closely packed than the N1—N2 bond and hence should cleave more easily. The results with HMX differ from those found previously for the closely related explosive RDX (**131**); RDX undergoes photolytically induced fission of all three of its N—N bonds in the solid-state, a finding that was attributed to its lower molecular symmetry in the crystal and its higher ring strain.[148]

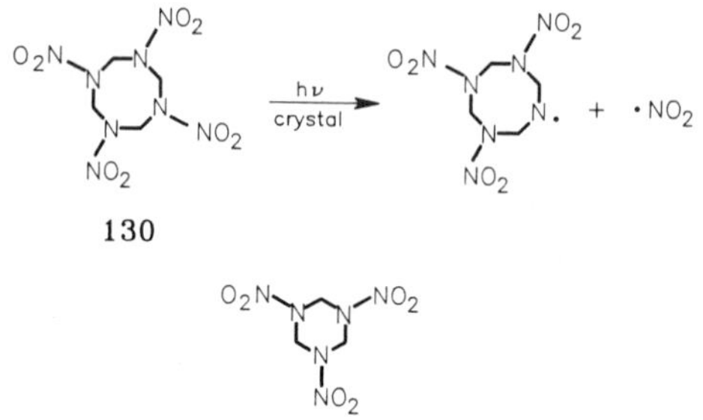

130

131

Scheme 43

3.6. Cis, Trans Photoisomerization [149-151]

For obvious reasons, the topochemically demanding process of cis,trans photoisomerization of crystalline olefins is rare, and as pointed out in our earlier review,[22] is most often found in strained alkenes that are significantly nonplanar to begin with. Such is the case with the first example to be discussed, cis-1,2-di-(1-naphthyl)ethylene. This compound was shown by x-ray crystallography to be twisted about the aliphatic double bond by 44.1°, and in accord with this relatively large distortion, Aldoshin et al. demonstrated that crystals of this material undergo facile cis to trans photoisomerization.[149] In contrast, the corresponding trans isomer, which is twisted by only 18.3°, is photochemically inert in the solid-state in the absence of oxygen. Crystals of the trans compound are denser (1.229 g cm^{-1}) than the cis isomer (1.177 g cm^{-1}), and packing potential energy calculations indicate that the trans lattice is more stable than the cis by ca. 5.6 kcal/mol. This was cited by the authors as the reason for the one-way photoisomerization in the solid-state. Small (2–5 μm) crystals of the cis compound appear to retain their integrity during photoisomerization, but larger crystals react with cracking.

The second paper we shall discuss in this section describes the E/Z photoisomerization of the aromatic retinoid **132** (Scheme 44) in the solid-state. Pfoertner, Englert, and Schoenholzer[150] studied this compound along with the corresponding ethyl ester, monoethyl amide and free carboxylic acid. In this group, only the diethyl amide **132**·0.5 H_2O underwent geometric isomerization; crystals of the ester photodimerized without isomerization, and the other two compounds were photochemically inert. Interestingly, only the terminal double bond of amide **132** isomerized in the solid-state, although upon prolonged photolysis, the crystal melted and isomerization at the other double bonds was observed. It is not clear why compound **132** photoisomerizes and the others do not, particularly since **132** is perfectly arranged in the crystal for [2+2] photocycloaddition (3.66 Å separation between terminal double bonds of adjacent molecules). The authors suggest that the increased excitation energy of the amide relative to the ester may be responsible, and that this effect is enhanced by hydrogen bonding between the incorporated water molecules and the amide carbonyl groups.

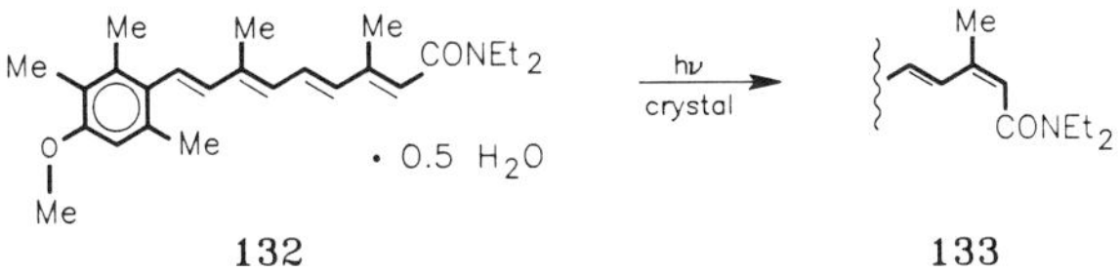

Scheme 44

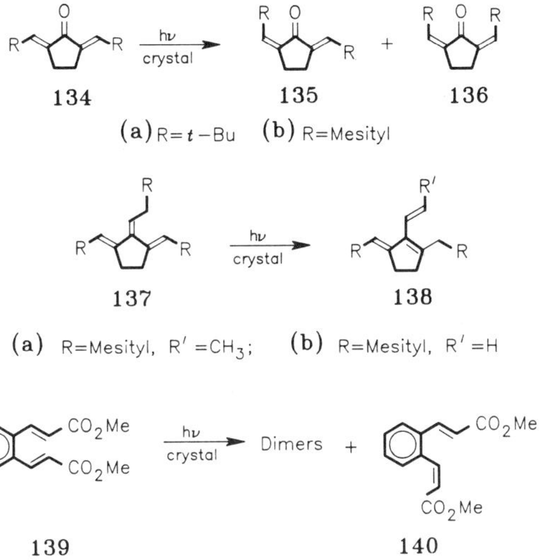

Scheme 45

Finally we mention a paper by Kaupp, Frey, and Behmann that describes some interesting solid-state photoisomerizations.[151] Among a number of other compounds that undergo [2+2] photodimerization in the solid-state, Kaupp et al. studied the 2,5-dimethylene-1-cyclopentanone derivatives **134a** and **134b** (Scheme 45). Thin crystalline films of these compounds were irradiated to low conversions at 0° and found to afford the geometric isomers **135** and **136**. The product percentages were as follows: **135a:136a** = 13:2 and **135b:136b** = 8:9. The related trienes **137a** and **137b** also underwent E/Z photoisomerization in the solid-state, but the ultimate photoproducts in this case were **138a** (68%) and **138b** (23%), the result of 1,5-sigmatropic rearrangement. Although these product studies were not accompanied by crystallographic investigations, the authors speculated that the bulky *t*-butyl and mesityl groups provided enough "free room" in the crystals for reaction to occur. It seems likely that double bond distortion in the ground state plays a role as well. One additional example of solid-state *cis,trans* photoisomerization was reported by Kaupp, this being the case of *trans,trans*-diester **139**. In addition to undergoing intermolecular photodimerization, crystals of this material were found to afford low yields of the corresponding cis,trans isomer **140**.[151]

3.7. Solid-state Photochromism[152-172]

Photochromism is the light-induced transformation of a chemical species A to a higher energy species B whose absorption spectrum (color) differs from that of A. By definition, the process must be reversible, and this occurs most often thermally, although in some cases it can be reversed photochemically as well.[173] The phenomenon has been observed in a variety of reaction media, but in this review we restrict ourselves to photochromism in crystals. For a historical background and summary of work on the subject prior to 1987, the reader is referred to our earlier review on solid-state photochemistry.[22]

Among the most thoroughly studied solid-state photochromic compounds are the Schiff bases formed by condensation of various salicylaldehydes with anilines, aminopyridines, thenylamines, and benzylamines, and the next few paragraphs will deal with three recent papers by Hadjoudis and co-workers on this topic.[152-154] No fewer than 93 compounds were tabulated and their photochromic and thermochromic properties recorded in the solid-state, in solution, and in polymer films. Before discussing these results, a brief reminder of the basic features of salicylaldehyde Schiff base photochromism and thermochromism is in order. As illustrated using the example of the Schiff base of salicylaldehyde with aniline (compound **141**, Scheme 46), thermochromism is associated with a conversion of the colorless, so-called enol form **141a** to the colored, *cis*-keto tautomer **141b**. In the photochromic compounds, the same hydrogen transfer occurs, but it is accompanied by isomerization to the *trans*-keto form **141c**. In the crystalline state, thermochromism and photochromism are found to be mutually exclusive; that is, a given compound is either photochromic or thermochromic but not both. Traditionally, this has been ascribed to the conformation and packing of the Schiff bases in the solid-state. The thermochromic crystals are planar as a result of strong intramolecular hydrogen bonding and pack in stacks with short intermolecular contacts (ca. 3.5 Å) normal to the molecular planes. In contrast, photochromic crystals are made up of molecules in which the aniline ring is significantly twisted (40–50°) out of planarity with the rest of the molecule. As a result, these molecules pack in looser, so-called open structures that lack aromatic ring stacking and provide sufficient lattice freedom for stabilization of the keto form by isomerization to a trans configuration (**141c**). This situation is reminiscent of the solid-state E/Z photoisomerization of olefins that have twisted ground state conformations.

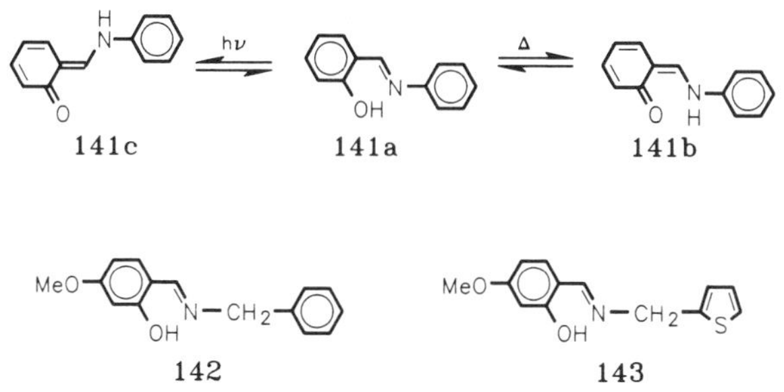

141c   141a   141b

142   143

Scheme 46

The main feature of the recent work by Hadjoudis et al.[152-154] is the finding that certain Schiff bases, for example the *p*-methoxy-substituted benzylamine and thenylamine derivatives **142** and **143** (Scheme 46), exhibit *both* thermochromic and photochromic behavior, a property that does not fit with the explanation given above. The x-ray crystal structure of compound **143** shows that the thiophene ring is substantially twisted relative to the rest of the molecule, and that the molecules pack in an open structure that should lead to photochromism but not thermochromism. The authors thus concluded that molecular planarity is *not* the factor that governs the photochromism and thermochromism of the Schiff bases of salicylaldehyde. They suggest instead, without elaboration, that it is the substituent-dependent electron density on the imino nitrogen atom that determines photochromic vs. thermochromic behavior—compounds with high electron density at this position being thermochromic and those with low electron density photochromic. They conclude, however, by saying that more examples are needed to prove (or disprove) this hypothesis.

Aziridines form an important class of solid-state photochromic compounds that was overlooked in our original review. Early reports that crystals of aryl- and aroyl-substituted aziridines are photochromic were due to Cromwell and co-workers,[155,156] and these papers were followed some time later by similar reports from Padwa and Hamilton[157] and Heine et al.[158] The most recent work in this area is due to Trozzolo and co-workers[159] and DoMinh,[160] who showed that a wide range of monocyclic and bicyclic aziridines become colored when their crystals are irradiated. Depending on the substituents present and their stereochemistry, the colors span the visible spectrum, and upon warming slowly, the colors fade and the aziridine is reformed. As depicted in Scheme 47, the molecular basis for these observations was suggested by Trozollo to be due to ring opening (disrotatory if concerted) to the highly resonance-stabilized azomethine ylide species **145**. In accord with this hypothesis, substituting positions X or Y with *p*-nitro groups leads to especially facile formation of colored intermediates.

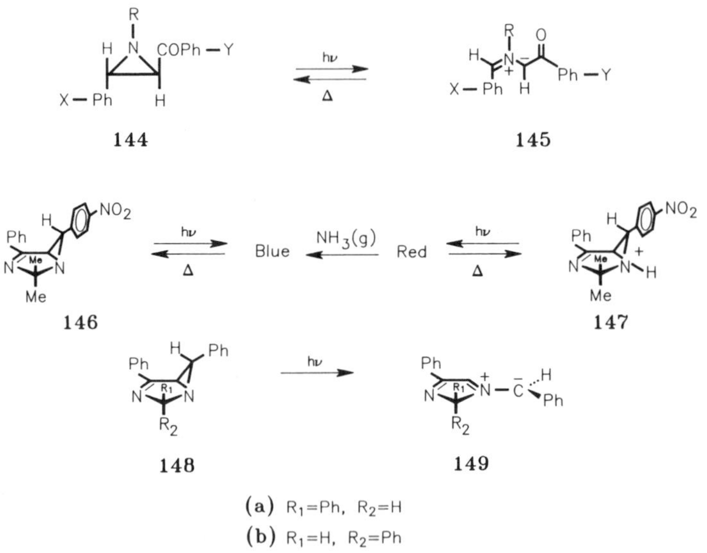

144   145

146   147

148   149

(a) R_1=Ph, R_2=H
(b) R_1=H, R_2=Ph

Scheme 47

Bicyclic aziridines behave similarly to their monocyclic analogues. For example, photolysis of crystals of the colorless compound **146** at 77 K leads to an intense blue color that can be erased by irradiation in the visible ($\lambda > 550$ nm) or by heating.[159] Interestingly, crystals of the oxalic acid *salt* of aziridine **146** (i.e., **147**) are also photochromic and give a red rather than a blue coloration. In a very neat experiment, it was demonstrated that the red color is due to the protonated form of the blue azomethine ylide. This was done by passing a stream of anhydrous ammonia gas over the red crystals, which thereupon turned blue!

In an interesting piece of work, DoMinh[160] showed that the epimeric aziridines **148a** and **148b** photolyzed in solution to give a common colored azomethine ylide, presumably the planar species **149** (Scheme 47), and that the less stable *cis*-aziridine **148a** isomerized (in part) to the trans isomer **148b**. In contrast, there was no isomerization in the solid-state and each epimer exhibited photochromism of a different color. The solid-state results were explained by suggesting that each epimer reacts to give a different, nonplanar azomethine ylide and that, owing to the restraints of the crystalline medium, these species cannot interconvert or revert to the aziridine of the opposite configuration.

Based on a report by Lottermoser in 1896 that crystals of 2,2,4,6-tetra-phenyldihydro-1,3,5-triazene (**150**, R_1 = Ph, R_2 = H, Scheme 48) are photochromic,[161] Maeda and co-workers reinvestigated this and several related compounds in which the groups R_1 and R_2 were varied.[162-165] Based on spectroscopic studies, it was concluded that those compounds in which R_2 = H exist in an equilibrium mixture of the tautomeric forms **150a** (major) and **150b** (minor), both in solution and the solid-state. Subsequently it was found that the tetraphenyl derivative (R_1 = Ph, R_2 = H) forms crystalline 1:1 inclusion complexes with a variety of organic solvents and that certain of these solvates (acetone, chloroform, and THF) are photochromic (colorless to red) and others (ethanol, 2-propanol and propylamine) are not; the unsolvated crystalline material was also found to exhibit colorless to red photochromism.

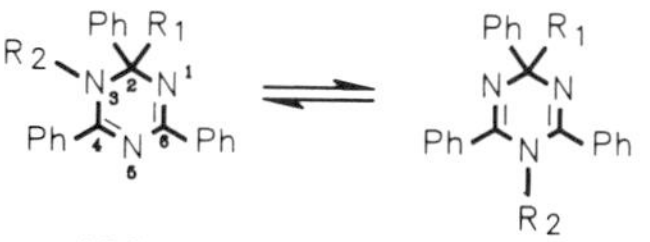

Scheme 48

In an attempt to understand the photochromism, the crystal structures of the solvated and unsolvated tetraphenyl triazene were determined. In all cases, the molecular structure corresponded to the 2,3-dihydro form (tautomer **a**), the difference being that in the case of the hydrogen-donating solvates, the triazene and alcohol (or amine) molecules are connected alternately by N3—H···O and N5···H—O hydrogen bonds to form a ribbon along the *b* axis. This is not possible in the case of the unsolvated triazene or its hydrogen-accepting solvates, and these systems are all photochromic. Thus, while it seems clear that hydrogen-donating solvents destroy photochromism, the mechanism by which this occurs is not clear.

In the course of their photochromic studies, Maeda et al. noted that the unsolvated tetraphenyl triazene discussed above (compound **151**, Scheme 49) underwent an irreversible photochemical change upon prolonged irradiation in the solid-state.[165] The photoproducts, formed in low yield, were identified as the triphenyl-substituted heteroaromatic compounds **155** (major) and **154** (minor); similar results were obtained in solution. A speculative mechanism suggested by Maeda is shown in Scheme 49. It is initiated by a di-π-methane photorearrangement to form compound **152**, which then undergoes an aziridine ring opening of the type discussed earlier to give the zwitterion **153**, a species that is tentatively identified as the photochromic species. The mechanisms by which intermediates **152** and **153** are converted into photoproducts **154** and **155** (loss of C_6H_5N and C_6H_6, respectively) are, however, far from clear.

Maeda et al. have also studied the photochemistry of several closely related tetra- and tri-aryl 1,4-dihydropyridines.[166] Crystals of these compounds, e.g., the tetraphenyl derivative **156**, Scheme 49, were originally reported to be photochromic (colorless to violet) by Peres de Carvalho[174] and, as in the case of the triazenes discussed above, were found by Maeda to undergo irreversible solid-

state color changes. Based on studies in solution,[166] the ultimate photoproduct
was shown to be 2,3,4,6-tetraphenylpyridine (**157**), and a di-π-methane-initiated
mechanism similar to that suggested for the triazenes was postulated. The species
responsible for the violet color was not identified, but it is interesting to note that
it was also formed during an X-ray crystal structure determination.

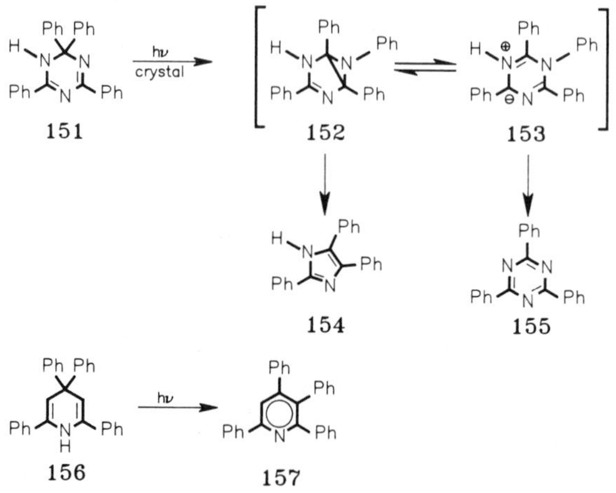

Scheme 49

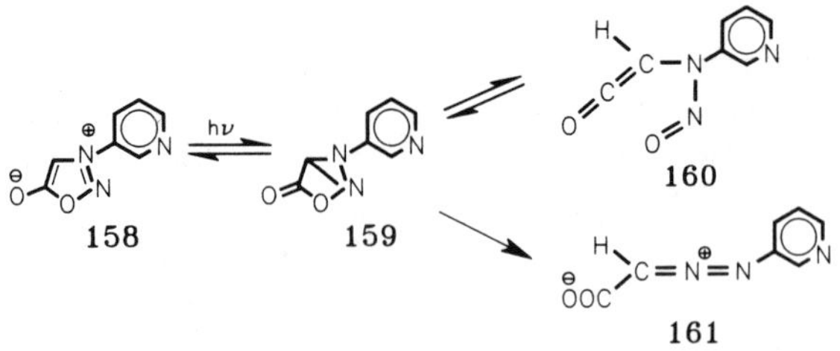

Scheme 50

The so-called sydnones, e.g., 3-(3-pyridyl)sydnone (**158**, Scheme 50),
constitute another class of organic compounds that exhibit photochromism in the
solid-state. Originally studied by Tien and Hunsberger,[175,176] the colorless to
blue photochromism of sydnone **158** was reinvestigated by Trozzolo et al.[159] and
more recently by Nespurek, Bohm, and Kuthan.[167] Trozzolo and co-workers
showed that irradiation (λ > 300 nm) of compound **158** in a KBr disc at 77 K for
2 min caused a sharp decrease in the uv absorption band at 315 nm, but no blue
coloration. Subsequent warming of the pellet in the dark led to the development of
the blue color, whose intensity was dependent on the irradiation time. These
results were interpreted in terms of a photochemical ring closure of **158** to the

colorless diaziridine **159** followed by thermal ring opening of the latter to form the betain **160**, the presumed colored species. In support of this assignment, no blue color was developed in the presence of the excellent dipolariphile, dimethyl acetylenedicarboxylate. The colored species was assigned a different structure by Nespurek et al.[167] Based on comparisons of the calculated and experimentally observed uv-vis spectra, these authors concluded that the blue species has the nitroso-ketene structure **161**. Both groups agree on the probable intermediacy of diaziridine **159**, and neither group was able to provide direct infrared evidence in support of their structure.

Indenone oxide derivatives, for example 2,3-diphenylindenone oxide (**162**, Scheme 51), have been intensively studied in solution from the point of view of their reversible, colorless to red photochromic behavior as well as their interesting and irreversible photorearrangement processes.[177] A recent paper by Hadjoudis and Pulima[168] reports that compound **162** is also photochromic in the solid-state, and the authors go on to describe the photochromism of several related compounds. Additional and essentially isolated cases in which reversible solid-state photochromism has been noted are also shown in Scheme 51.

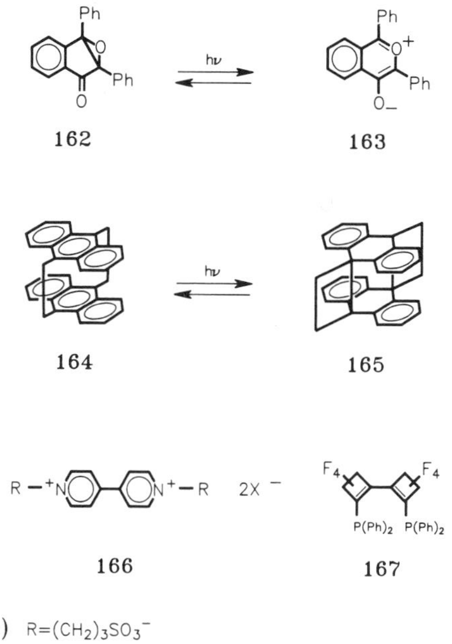

$$162 \quad \rightleftharpoons^{h\nu} \quad 163$$

$$164 \quad \rightleftharpoons^{h\nu} \quad 165$$

$$R-^{+}N\bigcirc-\bigcirc N^{+}-R \quad 2X^{-} \qquad 167$$

166

(a) $R=(CH_2)_3SO_3^-$
(b) $R=CH_2Ph, X=p-MeC_6H_4SO_3^-$

Scheme 51

These include the orange to colorless transformation of crystals of the novel paracyclophane **164**,[169] the colorless to blue reaction of the viologen derivatives

166a–b (attributed to viologen radical cation formation[170]) and the yellow to red coloration of crystals of the fluorinated ditertiary phosphine **167** (suggested to arise from excited-state planarization and increased conjugation).[171] The final paper we mention in this section is by Gusten et al.[172] and concerns the interesting use of photoacoustic spectroscopy in the measurement of the absorption spectra as well as the photochemical and thermal kinetics of some photochromic compounds in the solid-state. Standard photochromic systems were investigated, and it was shown that the method could be adapted to follow the photochemical reactions of organic molecules adsorbed on silica gel.

3.8. Miscellaneous Reactions[178-192]

We begin this section of the review by drawing attention to a recent article by Ohashi[178] that provides an excellent overview of his work on the x-ray-induced racemization of a series of optically active cobaloxime complexes, a topic that was briefly discussed in Section 1 and more thoroughly covered in our previous review.[22] As mentioned there, the mechanism of racemization is thought to involve homolytic Co—C bond cleavage, followed by radical recombination to give the inverted configuration. The Co—C bond dissociation energy has been determined to be 117–122 kJ mol^{-1},[193] and esr spectra of irradiated crystals are consistent with a homolytic dissociation mechanism.[194] An interesting question arises as to the nature of the excitation responsible for the dissociation process. Based on the fact that the racemization rate of the crystalline complex by x-rays is far lower than that by visible light, Ohashi speculates that secondary radiation produced by the interaction of the x-rays with the Co atom may be responsible for the bond cleavage.

Two conference abstracts dealing with the solid-state chemistry of cobaloxime complexes have appeared recently. The first, due to Uchida and Dunitz,[179] describes packing potential energy calculations designed to elucidate the reaction pathways and activation energy differences for the x-ray-induced racemizations. While no details are given, the results are said to provide some insight into the racemization mechanism and to be generally compatible with the experimental results. The second abstract is by Ohashi and co-workers[180] and describes the solid-state photochemistry of some cobaloxime complexes whereby a β-cyanoethyl group attached to cobalt is isomerized to its α-cyanoethyl isomer. The rate of the process in KBr matrices was followed by infrared spectroscopy and found to depend not only on the size of the reaction cavity, but also on the conformation of the β-cyanoethyl group in the crystal and on the presence of an intermolecular hydrogen bond between the nitrile nitrogen and a hydrogen atom of a neighboring ligand.

Next we discuss two papers by Porte et al. that deal with the solid-state photochemistry of nitroso compounds, specifically (-)-2-chloro-2-nitrosocamphane (**168**, Scheme 52)[181] and its close relative, (+)-10-bromo-2-chloro-2-nitroso-camphane (**171**).[182] This represents an extension of previous work by this group on the crystalline phase photoreactivity of humulene nitrosite[195] and caryophyllene nitrosite.[196] Considerable work had been done in the 1950's on the ORD and CD spectra of optically active nitroso compounds as well as on their solution phase photochemistry. The goal of the solid-state work was to provide further information on the complex sequence of reactions that occurs when these compounds are photolyzed. Irradiation of polycrystalline samples of compounds 168 and 171 (n,π* λ_{max} = ca. 670 nm) with red light causes the formation of nitroxide radicals that can be detected by esr spectroscopy. Based on the agreement between the observed and calculated spectra, the initially formed nitroxide radicals in each case were assigned the structures 170 and 173. At this point, the crystals are still intact, but continued irradiation leads to crystal melting and to formation of a host of diamagnetic species as well as other nitroxide radicals. Nitroxide radical 170 is thought to be formed through a Beckmann-like rearrangement of species 169; in contrast, the *endo*-nitroso compound 171 was suggested to react via homolytic loss of nitric oxide followed by combination of the resulting free radical 172 with a molecule of starting material. Although a crystal structure of nitroxide 171 was known,[197] no attempt was made by Porte et al. to draw any structure-reactivity correlations.

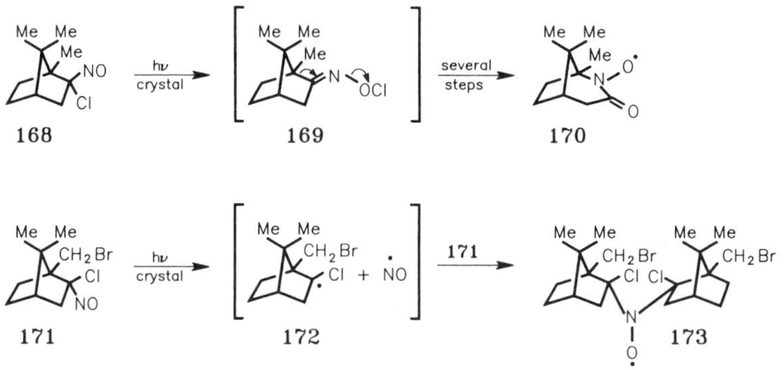

Scheme 52

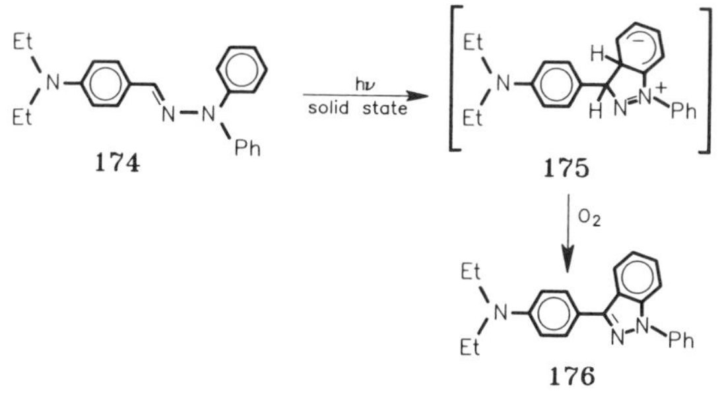

Scheme 53

Pacansky et al.[183] have studied the solid-state photorearrangement of *p*-diethylaminobenzaldehyde diphenylhydrazone (**174**, Scheme 53), a commercially important photoconductor. Pyrex-filtered irradiation of thin solid films of this material in air led to a quantitative conversion to the indazole derivative **176**. A different product was detected (but not identified) when the reaction was run in vacuo; upon exposure of these films to air, indazole **176** was formed. The authors suggest the mechanism outlined in Scheme 53, which involves cyclization to zwitterion **175** (presumed to be the intermediate detected above) followed by air oxidation. The x-ray crystal structure of starting material **174** shows that one of the *N*-phenyl rings is in a position favorable for photocyclization. Interestingly, neither solution phase photolysis of hydrazone **174** nor exposure of solid films of this material to an electron beam[184] gave any trace of indazole **176**.

Reisch et al.[185] have recently reported the photochemistry of the heterocyclic compound **177** (Scheme 54) in the solid-state and in aqueous solution. Photolysis of crystals of this material in air leads to the products (**178–180**) shown in Scheme 54. After 20 days of irradiation of 6 g of starting material, 140 mg of **178**, 40 mg of **179** and 25 mg of **180** were isolated; none of these products is formed in the aqueous phase photolysis. No reasons were advanced for the observed medium-dependent reactivity differences, nor were any crystallographic data presented.

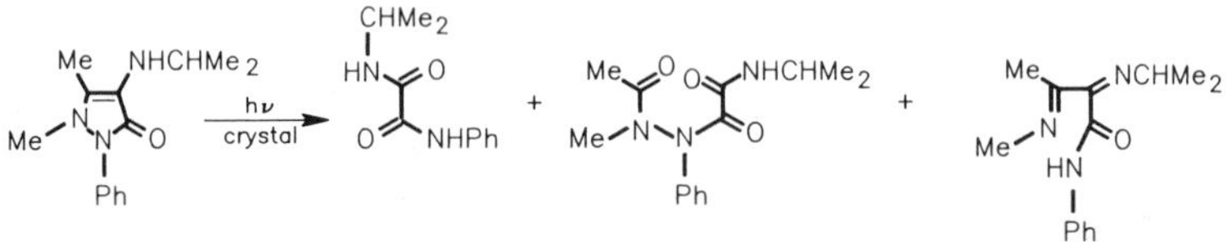

Scheme 54

Investigations of the solid-state stability of pharmaceutically relevant compounds toward light, heat, moisture, etc. (of which the study discussed above is an example), is an active field (as a recent book by Byrn[8] attests) because such information is necessary in designing solid dosage forms of drug substances. In the next few paragraphs we briefly discuss five recent papers that describe the solid-state photochemistry of pharmaceuticals.[186-190]

The first two papers deal with the analgesic phenazone (**181**, Scheme 55,) and several of its derivatives[186] and ubidecarenone[187] (**182**, used in the treatment of angina). These papers discuss in some detail the rates with which the title compounds degrade photochemically in the solid-state and go on to describe the changes in the color and crystallinity of the compounds as a function of photolysis wavelength and light intensity, reaction temperature and sample grain size. The exact chemical reactions responsible for these changes are, however, unknown, as no product studies were reported.

Methadone (**183**), a narcotic substitute, has been shown by Reisch and Reisch[188] to photoreact in the solid-state to give a mixture of propanal and the tertiary amine **185** (Scheme 55). Although intramolecular abstraction of a hydrogen atom from the β-CH$_2$ group by the carbonyl oxygen was suggested as the primary photochemical step, it seems more likely that a mechanism involving type I cleavage followed by transfer of a methylene hydrogen atom to the acyl radical is operative. Reisch et al. have also investigated the solid-state photochemistry of the steroids testosterone (**186a**) and methyltestosterone (**186b**).[189] Irradiation of the former as a powder in the presence of silica gel gave low yields of the disproportionation products **187** and **188** (Scheme 55). Based on an x-ray crystal structure by Roberts et al.,[198] an intermolecular hydrogen abstraction mechanism was proposed. Under the same conditions, methyltestosterone was reported to give the diketone **189**. Finally we mention a paper by Jochym et al.[190] that describes the photochemistry of suspensions of the sodium salts of several barbituric acid derivatives (**190**, Scheme 55) in paraffin oil. Based on the infrared spectra of the suspensions, a ring opening to the isocyanates of general structure **191** was proposed. Addition of trapping agents, for example alcohols or amines, caused the infrared bands associated with these species to decay.

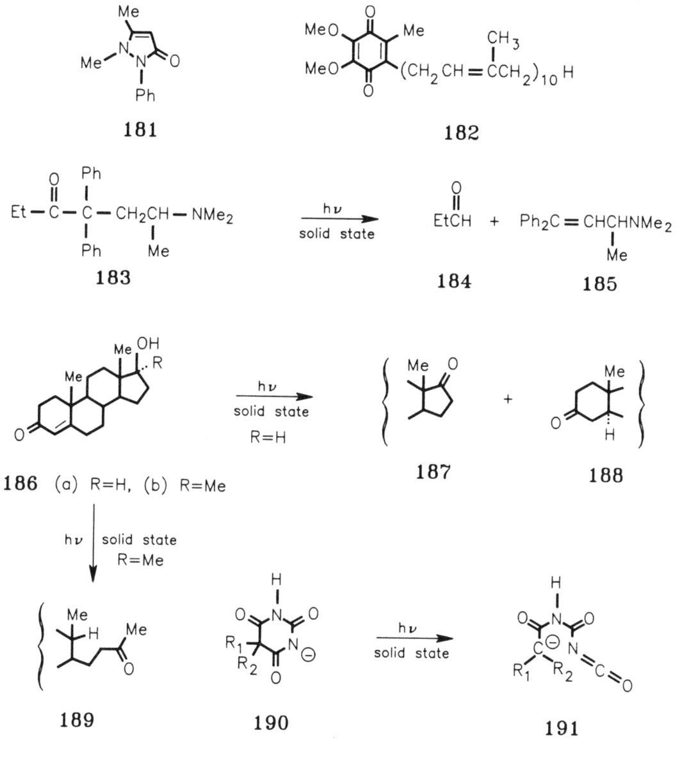

Scheme 55

Theocharis et al. have studied the solid-state photoreactivity of a series of benzylidenecyclopentanone derivatives and their complexes with transition metal cations.[191] Many of these compounds undergo intermolecular [2+2] photodimerization in the solid-state, but some are unreactive and others, for example compounds **192** and **193** (Scheme 56), unexpectedly react via decarboxylation (**192**) or dehydration (**193**) of the malonate group. The evolved CO_2 could be detected by FTIR spectroscopy, and from the overall sharpness of the spectrum of **192** following photolysis, it was concluded that crystallinity was maintained during decarboxylation.

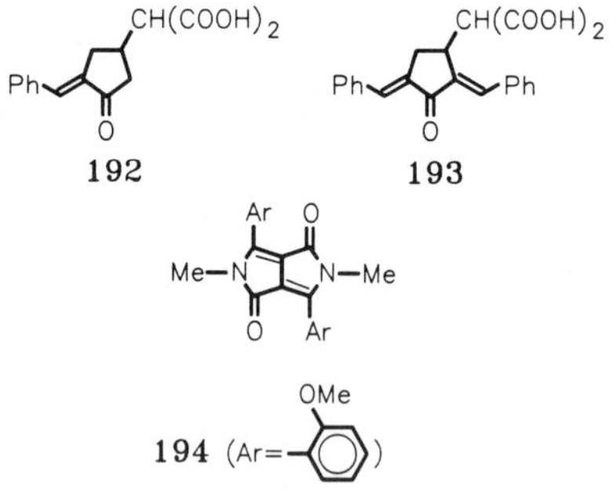

Scheme 56

The emission of light through fluorescence, while not involving the rearrangement of one molecule to another, is nevertheless a unimolecular process of an excited state, and the final paper that we shall consider in this review reports an interesting study of the influence of crystal packing on the solid-state fluorescence of the diketopyrrolopyrrole **194** (Scheme 56).[192] This material crystallizes in two packing arrangements, a more stable, yellow form that is strongly fluorescent in the solid-state and a slightly less stable (by 1.5 kcal mol^{-1}) orange modification, which exhibits only weak fluorescence. The crystal structure of each dimorph was determined, and this showed that in the weakly fluorescing crystals, neighboring molecules are stacked almost directly on top of one another with an interplanar separation of 3.81 Å. In contrast, the fluorescent crystals are packed with a much larger separation (6.18 Å) between the planes of the diketopyrrolopyrrole rings. The authors suggest that in the closely packed, weakly fluorescent crystals, there is strong lattice vibrational–electronic coupling leading to internal conversion without emission. They go on to argue that because of the greater distance between chromophores in the other dimorph, this mechanism is inoperative, and strong fluorescence similar to that present in dilute solution is observed.

4. Concluding Remarks

The message to be gained from the results presented above is clear: *the crystalline medium is capable of exerting a dramatic control over organic photoreactivity; in some cases, the liquid- and solid-phase results are completely different. The solid-state thus provides a medium for the discovery of new chemical reactions.* Furthermore, since molecules in crystals are constrained in their motions by the high viscosity of the medium, it can be assumed in many cases that the intermediate(s) and transition state(s) in the reaction under study resemble the starting materials, and since the shape as well as the environment of the starting materials can be determined accurately through x-ray crystallography (in stereo, no less!), detailed structure–reactivity correlations of a precision not possible in solution become available.

We close this review with a look to the future. Imagine an experimental technique by which stereodiagrams of the starting materials, intermediates and products in a chemical reaction could be recorded in real time — a technique that would provide a step-by-step visualization of organic reaction mechanisms. Sound fanciful? Such experiments may be feasible in the not too distant future through the application of synchrotron source x-ray crystallography[199] to chemical reactions that occur in the organic crystalline phase. For this reason, among many other more immediate reasons such as its direct relevance to the field of materials science, the study of chemical processes that occur in solids is an important and worthwhile undertaking and will continue to be so for many years to come.

5. Note Added in Proof

After the completion of this manuscript, we became aware of two additional classes of organic molecules that exhibit photochromism in the solid-state. The first of these is a 2,4,5-triphenylimidazolyl dimer, first reported by Hayashi and Maeda[200] and shortly thereafter by Zimmerman, Baumgartel, and Bakke[201] and assigned the N—N structure **195a** (Scheme 57). In a subsequent paper by White and Sonnenberg,[202] this material was assigned the C—N structure **195b**. Crystals of this substance, which are pale yellow, become reddish purple when irradiated at room temperature (or heated to 170°C), and all three research groups agree that the color is due to dissociation of the dimer to a pair of imidazolyl radicals.[203] Interestingly, the C—C dimer (not shown) has also been isolated and found to be nonphotochromic in the solid-state; its crystals however, are thermochromic as well as piezochromic, that is, they become colored upon grinding.[203] Several *ortho-*, *meta-*, and *para*-substituted triarylimidazolyl dimers were also prepared and their solid-state photochromic, thermochromic, and piezochromic properties recorded.[202,203]

The second case concerns the solid-state photochromism of the 1-aryloxyanthraquinone derivatives **196 a–d** (Scheme 57).[204] When crystals of these four compounds were irradiated at 365 nm, new absorptions in the 458–496 nm region were produced, and upon standing at room temperature in the dark, a second colored species that absorbs around 600 nm was detected. After prolonged standing in the dark, the original spectrum of the starting material was restored. Tajima et al.[204] interpreted these results in terms of an initial conformational change of the molecules to produce species **197**, which was presumed to absorb in the shorter wavelength visible region, followed by 1,5-aryl migration to form species **198**, which was suggested to have the longer wavelength visible absorption. Infrared measurements in support of these assignments were reported.

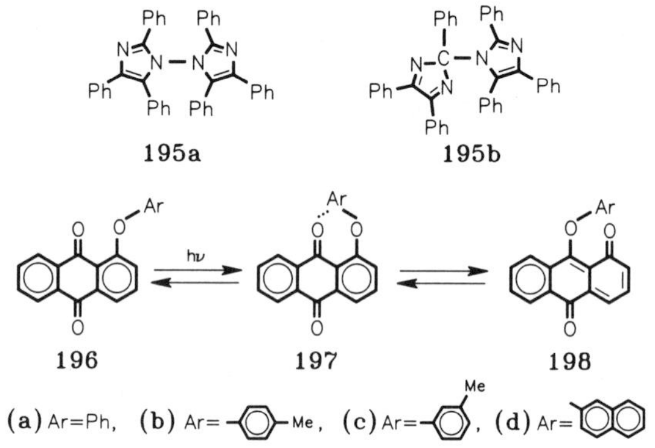

Scheme 57

Acknowledgments

We thank the Natural Sciences and Engineering Research Council of Canada and the United States Petroleum Research Fund for continuing support of our solid-state chemistry program at UBC. We also thank Professor James Trotter and his research group for their invaluable crystallographic contributions to the work described from UBC. It is a special pleasure to acknowledge the warm hospitality extended to JRS during a 2-month visit to the Department of Structural Chemistry at the Weizmann Institute of Science, Rehovot, Israel, where this review was written. Financial assistance during this period in the form of a Killam Senior Fellowship and a UBC-Weizmann Institute Exchange Program Travel Grant to JRS is also gratefully acknowledged. Finally, we thank Elizabeth Varty and Susan Rollinson for doing the drawings.

References

1. (a) Marshall McLuhan, *Understanding Media*, McGraw Hill, New York, 1964.
   (b) ibid., p. 9.
   (c) According to a fascinating article by H. D. Roth, *Angew. Chem. Int. Ed. Engl.*, **1989**, *28*, 1193, the first investigation of an organic photoreaction in the solid state was due to H. Trommsdorff, *Ann. Chem. Phar.*, **1834**, *11*, 190, who reported that crystals of santonin turn yellow and cleave when exposed to sunlight. This problem was reinvestigated in 1968 by Matsuura et al. (ref.106).

2. J. D. Dunitz, *X-Ray Analysis and the Structure of Organic Molecules*, Cornell University Press, Ithaca, NY, 1979, pp. 312-318.

3. L. Addadi, S. Ariel, M. Lahav, L. Leiserowitz, R. Popovitz-Biro and C. P. Tang, *Chemical Physics of Solids and their Surfaces*, Specialist Periodical Reports, Vol. 8, M. W. Roberts and J. M. Thomas, eds., The Royal Society of Chemistry, London, 1980, Ch. 7.

4. J. R. Scheffer, *Acc. Chem. Res.*, **1980**, *13*, 283.

5. J. M. Thomas and W. Jones in *Reactivity of Solids*, Vol. 2, K. Dyrek, J. Haber and J. Nowotny, eds., Elsevier, Amsterdam, 1980, p. 551.

6. D. Y. Curtin and I. C. Paul, *Chem. Rev.*, **1981**, *81*, 525.

7. A. Gavezzotti and M. Simonetta, *Chem. Rev.*, **1982**, *82*, 1.

8. S. R. Byrn, *The Solid State Chemistry of Drugs*, Academic Press, New York, 1982.

9. M. Hasegawa, *Adv. Polym. Sci.*, **1982**, *42*, 1.

10. M. Hasegawa, *Chem. Rev.*, **1983**, *83*, 507.

11. C. R. Theocharis, *The Chemistry of Enones*, S. Patai and Z. Rappoport, eds., John Wiley, New York, 1989, p. 1133.

12. J. Trotter, *Acta Crystallogr.*, **1983**, *B39*, 373.

13. G. R. Desiraju, *Endeavour*, **1984**, *8*, 201.

14. J. A. R. P. Sarma and G. R. Desiraju, *Acc. Chem. Res.*, **1986**, *19*, 222.

15. G. A. Vinogradov, *Russ. Chem. Rev.*, **1984**, *53*, 77.

16. V. M. Misin and M. I. Cherkashin, *Russ. Chem. Rev.*, **1985**, *54*, 562.

17. V. E. Shklover and T. V. Timofeeva, *Russ. Chem. Rev.*, **1985**, *54*, 619.

18. B. S. Green, R. Arad-Yellin and M. D. Cohen., *Top. Stereochem.*, **1986**, *16*, 131.

19. V. Ramamurthy and K. Venkatesan, *Chem. Rev.*, **1987**, *87*, 433.

20. *Organic Solid State Chemistry*, G. R. Desiraju, ed., Elsevier, Amsterdam, 1987.

21. *Organic Chemistry in Anisotropic Media*, Tetrahedron Symposia-in-Print Number 20, V. Ramamurthy, J. R. Scheffer and N. J. Turro, eds., Vol. 43, 1987, issue 7.

22. J. R. Scheffer, M. Garcia-Garibay and O. Nalamasu in *Organic Photochemistry*, Vol. 8, A. Padwa, ed., Marcel Dekker, New York, 1987, p. 249.

23. J. R. Scheffer and J. Trotter in *The Chemistry of the Quinonoid Compounds*, Vol. 2, Part 2, S. Patai and Z. Rappoport, eds., Wiley, New York, 1988, p. 1199.

24. R. Lamartine, *Bull. Soc. Chim. France*, **1989**, 237.

25. *Photochemistry on Solid Surfaces*, M. Anpo and T. Matsuura, eds., Elsevier, Amsterdam, 1989.

26. L. Addadi, M. Cohen, M. Lahav and L. Leiserowitz, *J. Chim. Phys.*, **1986**, *83*, 831.

27. G. R. Desiraju, *Crystallogral Engineering: The Design of Organic Solids*, Elsevier, New York, 1989.

28. O. Diels and K. Alder, *Chem. Ber.*, **1929**, *62*, 2362.

29.   R. C. Cookson, E. Crundwell, R. R. Hill and J. Hudec, *J. Chem. Soc.*, **1964**, 3062.
30.   J. R. Scheffer and A. A. Dzakpasu, *J. Am. Chem. Soc.*, **1978**, *100*, 2163.
31.   S. Ariel, S. H. Askari, J. R. Scheffer and J. Trotter, *Tetrahedron Lett.*, **1986**, *27*, 783.
32.   S. Ariel, S. Evans, C. Hwang, J. Jay, J. R. Scheffer, J. Trotter and Y. F. Wong, *Tetrahedron Lett.*, **1985**, *26*, 965.
33.   A. Weisz, M. Kaftory, I. Vidavsky and A. Mendelbaum, *J. Chem. Soc., Chem. Commun.*, **1984**, 18.
34.   W. K. Appel, Z. Q. Jiang, J. R. Scheffer and L. Walsh, *J. Am. Chem. Soc.*, **1983**, *105*, 5354.
35.   T. J. Greenhough, J. R. Scheffer, A. S. Secco, J. Trotter and L. Walsh, *Isr. J. Chem.*, **1985**, *25*, 297.
36.   S. Ariel, S. Askari, J. R. Scheffer, J. Trotter and L. Walsh, *J. Am. Chem. Soc.*, **1984**, *106*, 5726.
37.   S. Ariel, S. Askari, J. R.Scheffer, J. Trotter and L. Walsh in *Organic Phototransformations in Nonhomogeneous Media*, M. A. Fox, ed., American Chemical Society, Washington, D.C., 1985, Ch. 15.
38.   J. R. Scheffer, J. Trotter, N. Omkaram, S. V. Evans and S. Ariel, *Mol. Crystal. Liq. Crystal.*, **1986**, *134*, 169.
39.   S. Ariel, V. Ramamurthy, J. R. Scheffer and J. Trotter, *J. Am. Chem. Soc.*, **1983**, *105*, 6959.
40.   S. V. Evans, N. Omkaram, J. R. Scheffer and J. Trotter, *Tetrahedron Lett.*, **1985**, *26*, 5903.
41.   S. V. Evans, N. Omkaram, J. R. Scheffer and J. Trotter, *Tetrahedron Lett.*, **1986**, *27*, 1419.
42.   H. Aoyama, T. Hasegawa and Y. Omote, *J. Am. Chem. Soc.*, **1979**, *101*, 5343.
43.   H. Aoyama, M. Sakamoto, K. Kuwabara, K. Yoshida and Y. Omote, *J. Am. Chem. Soc.*, **1983**, *105*, 1958.
44.   S. Mohr, *Tetrahedron Lett.*, **1980**, *21*, 593.
45.   S. Mohr, *Tetrahedron Lett.*, **1979**, *20*, 3139.
46.   P. J. Wagner, B. P. Giri, J. C. Scaiano, D. L. Ward, E. Gabe and F. L. Lee, *J. Am. Chem. Soc.*, **1985**, *107*, 5483.
47.   P. J. Wagner in *Rearrangements in Ground and Excited States*, Vol. 3, P. de Mayo, ed., Academic, New York, 1980, Ch. 20.
48.   J. R. Scheffer in *Organic Solid State Chemistry*, G. R. Desiraju, ed., Elsevier, Amsterdam, 1987, p. 1.
49.   G. S. Murthy, P. Arjuanan, K.Venkatesan and V. Ramamurthy, *Tetrahedron*, **1987**, *43*, 1225.
50.   S. K. Kearsley in *Organic Solid State Chemistry*, G. R. Desiraju, ed., Elsevier, Amsterdam, 1987, Ch. 3.
51.   N. J. Karch, E. T. Koh, B. L. Whitsel and J. M. McBride, *J. Am. Chem. Soc.*, **1975**, *97*, 6729.
52.   J. M. McBride, *Acc. Chem. Res.*, **1983**, *16*, 304.
53.   D. W. Walter and J. M. McBride, *J. Am. Chem. Soc.*, **1981**, *103*, 7069.
54.   D. W. Walter and J. M. McBride, *J. Am. Chem. Soc.*, **1981**, *103*, 7074.
55.   M. D. Hollingsworth and J. M. McBride, *J. Am. Chem. Soc.*, **1985**, *107*, 1792.
56.   T. Kurihara, A.Uchida, Y.Ohashi, Y. Sasada and Y. Ohgo, *J. Am. Chem. Soc.*, **1984**, *106*, 5718.
57.   Y. Ohashi, K. Yanagi, T. Kurihara, Y. Sasada and Y. Ohgo, *J. Am. Chem. Soc.*, **1982**, *104*, 6353.
58.   Y. Ohashi, A. Uchida, Y. Sasada and Y. Ohgo, *Acta Crystallogr.*, **1983**, *B39*, 54.
59.   S. Ariel and J. Trotter, *Acta Crystallogr.*, **1988**, *B44*, 538.

60. S. Ariel and J. Trotter, *Zeitschr. Krist.*, 1988, *185*, 253.
61. S. Ariel, S.H. Askari, J.R. Scheffer, J. Trotter and F. Wireko, *J. Am. Chem. Soc.*, 1987, *109*, 4623.
62. S. Ariel, S. H. Askari, J. R. Scheffer, J. Trotter and F. Wireko, *Acta Crystallogr.*, 1987, *B43*, 532.
63. S. Ariel and J. Trotter, *Acta Crystallogr.*, 1987, *B43*, 563.
64. S.V. Evans, C. Hwang and J. Trotter, *Acta Crystallogr.*, 1989, *C45*, 148.
65. S. Ariel and J. Trotter, *Acta Crystallogr.*, 1986, *C42*, 1804.
66. S. Ariel, S. Askari, S. V. Evans, C. Hwang, J. Jay, J. R. Scheffer, J. Trotter, L. Walsh and Y. F. Wong, *Tetrahedron*, 1987,*43*, 1253.
67. J. R. Scheffer and A.D. Gudmundsdottir, *Mol. Crystal. Liq. Crystal. Inc. Nonlin. Opt.*, 1990, *186*, 19.
68. S. Ariel and J. Trotter, *Acta Crystallogr.*, 1987, *C43*, 959.
69. S. Ariel and J. Trotter, *Acta Crystallogr.*, 1987, *C43*, 1100.
70. S. Ariel and J. Trotter, *Acta Crystallogr.*, 1987, *C43*, 1103.
71. S. Ariel, S. Askari, J. R. Scheffer and J. Trotter, *J. Org. Chem.*, 1989, *54*, 4324.
72. A. D. Gudmundsdottir and J. R. Scheffer, *Tetrahedron Lett.*, 1989, *30*, 419.
73. A. D. Gudmundsdottir and J. R. Scheffer, *Tetrahedron Lett.*, 1989, *30*, 423.
74. F. D. Lewis, R. W. Johnson and D. E. Johnson, *J. Am. Chem. Soc.*, 1974, *96*, 6090.
75. S. V. Evans, M. Garcia-Garibay, N. Omkaram, J. R. Scheffer, J. Trotter and F. Wireko, *J. Am. Chem. Soc.*, 1986, *108*, 5648.
76. M. Garcia-Garibay, J. R. Scheffer, J. Trotter and F. Wireko, *J. Am. Chem. Soc.*, 1989, *111*, 4985.
77. M. Garcia-Garibay, J. R. Scheffer, J. Trotter and F. Wireko, *Tetrahedron Lett.*, 1987, *28*, 4789.
78. J. Chen, M. Garcia-Garibay and J. R. Scheffer, *Tetrahedron Lett.*, 1989, *30*, 6125.
79. M. Garcia-Garibay, J. R. Scheffer, J. Trotter and F. Wireko, *Tetrahedron*, 1988, *29*, 2042.
80. J. R. Scheffer and J. Trotter, *Mol. Crystal. Liq. Crystal. Inc. Nonlin. Opt.*, 1988, *156*, 63.
81. M. Garcia-Garibay, J. R. Scheffer and D. G. Watson, *J. Chem. Soc., Chem. Commun.*, 1989, 600.
82. P. R. Pokkuluri, J. R. Scheffer and J. Trotter, *Tetrahedron Lett.*, 1989, *30*, 1601.
83. M. Garcia-Garibay, J. R. Scheffer, J. Trotter and F. Wireko, *Tetrahedron Lett.*, 1987, *28*, 1741.
84. P. R. Pokkuluri, J. R. Scheffer and J. Trotter, *J. Am. Chem. Soc.*, 1990, *112*, 3676.
85. T. Matsuura, J. B. Meng, Y. Ito, M. Irie and K. Fukuyama, *Tetrahedron*, 1987, *43*, 2451.
86. Y. Ito, H. Ito, M. Ino and T. Matsuura, *Tetrahedron Lett.*, 1988, *29*, 3091.
87. Y. Ito in *Photochemistry on Solid Surfaces*, M. Anpo and T. Matsuura, eds., Elsevier, Amsterdam, 1989, p. 469.
88. (a) C. Decoret, J. Vicens and J. Royer, *J. Mol. Structure (Theochem)*, 1985, *121*, 13.
    (b) J. Vicens, *Tetrahedron*, 1987, *43*, 1361.
89. J. Baro, D. Dudek, K. Luther and J. Troe, *Z. Phys. Chem.*, 1984, *140*, 167.
90. H. E. Zimmerman and M. J. Zuraw, *J. Am. Chem. Soc.*, 1989, *111*, 2358.
91. H. E. Zimmerman and M. J. Zuraw, *J. Am. Chem. Soc.*, 1989, *111*, 7974.
92. E. Ciganek, *J. Am. Chem. Soc.*, 1966, *88*, 2882.

93. H. E. Zimmerman in *Molecular Rearrangements in Ground and Excited States*, Vol. 3, P. de Mayo, ed., Academic, New York, 1980, Ch. 16.
94. J. Jacques, A. Collet and S. H. Wilen, *Enantiomers, Racemates and Resolutions*, Wiley Interscience, New York, 1983.
95. M. Lahav, B. S. Green and D. Rabinovich, *Acc. Chem. Res.*, 1979, *12*, 191.
96. L. Addadi and M. Lahav in *Origins of Optical Activity in Nature*, D. C. Walker, ed., Elsevier, Amsterdam, 1979, Ch. 14.
97. H. E. Zimmerman and A. P. Kamath, *J. Am. Chem. Soc.*, 1988, *110*, 900 and references cited therein.
98. See, however, L. A. Paquette and E. Bay, *J. Org. Chem.*, 1982, *47*, 4597.
99. P. W. Rabideau, J. B. Hamilton and L. Friedman, *J. Am. Chem. Soc.*, 1968, *90*, 4465.
100. W. Adam, O. De Lucchi, K. Peters, E. M. Peters and H. G. von Schnering, *J. Am. Chem. Soc.*, 1982, *104*, 5747.
101. H. E. Zimmerman, R. S. Givens and R. M. Pagni, *J. Am. Chem. Soc.*, 1968, *90*, 6096.
102. H. E. Zimmerman and C. O. Bender, *J. Am. Chem. Soc.*, 1970, *92*, 4366.
103. C. O. Bender and S. S. Shugarman, *J. Chem. Soc., Chem. Commun.*, 1974, 934.
104. C. O. Bender and D. W. Brooks, *Can. J. Chem.*, 1975, *53*, 1684.
105. Y. Ito and T. Matsuura, *J. Photochem. Photobiol. A*, 1989, *50*, 141.
106. T. Matsuura, Y. Sata, K. Ogura and M. Mori, *Tetrahedron Lett.*,1968, 4627.
107. H. E. Zimmerman and J.W. Wilson, *J. Am. Chem. Soc.*, 1964, *86*, 4036.
108. S.V. Evans and J. Trotter, *Acta Crystallogr.*, 1989, *B45*, 500.
109. S.V. Evans and J. Trotter, *Acta Crystallogr.*, 1989, *B45*, 159.
110. S. V. Evans, C. Hwang and J. Trotter, *Acta Crystallogr.*, *C44*, 1988, 1457.
111. S. Ariel, S.V. Evans, M. Garcia-Garibay, B. R. Harkness, N. Omkaram, J. R. Scheffer and J. Trotter, *J. Am. Chem. Soc.*, 1988, *110*, 5591.
112. S. V. Evans and J. Trotter, *Acta Crystallogr.*, 1988, *B44*, 63.
113. J. R. Scheffer and J. Trotter, *Rev. Chem. Int.*, 1988, *9*, 271.
114. S. Ariel, M. Garcia-Garibay, J. R. Scheffer and J. Trotter, *Acta Crystallogr.*, 1989, *B 45*, 153.
115. S. V. Evans and J. Trotter, *Acta Crystallogr.*, 1988, *C44*, 874.
116. S. V. Evans and J. Trotter, *Acta Crystallogr.*, 1988, *B44*, 533.
117. S. V. Evans and J. Trotter, *Acta Crystallogr.*, 1988, *C44*, 1459.
118. S. Ariel and J. Trotter, *Acta Crystallogr.*, 1986, *C42*, 71.
119. S. Ariel and J. Trotter, *Acta Crystallogr.*, 1986, *C42*, 485.
120. S. Ariel and J. Trotter, *Acta Crystallogr.*, 1986, *C42*, 1166.
121. T. J. Lewis, S. J. Rettig, J. R. Scheffer, J. Trotter and F. Wireko, *J. Am. Chem. Soc.*, 1990, *112*, 3679.
122. F. Toda, M. Yagi and S. Soda, *J. Chem. Soc., Chem. Commun.*, 1987, 1413.
123. F. Toda, *Mol. Crystal. Liq. Crystal. Inc. Nonlin. Opt.*, 1988, *161*, 355.
124. A. Sekine, K. Hori, Y. Ohashi, M. Yagi and F. Toda, *J. Am. Chem. Soc.*, 1989, *111*, 697.
125. T. Hasegawa, J. Moribe and M. Yoshioka, *Bull. Chem. Soc. Jpn.*, 1988, *61*, 1437.
126. P. J. Wagner and B. Zhou, *Tetrahedron Lett.*, 1989, *30*, 5389.
127. P. J. Wagner, *Acc. Chem. Res.*, 1989, *22*, 83.
128. Y. Ito and T. Matsuura, *Tetrahedron Lett.*, 1988, *29*, 3087.
129. Y. Ito, T. Matsuura, K. Tabata, M. Ji-Ben, K. Fukuyama, M. Sasaki and S. Okada, *Tetrahedron*, 1987, *43*, 1307.
130. E. J. Corey, J. D. Bass, R. LeMahieu and R. B. Mitra, *J. Am. Chem. Soc.*, 1964, *86*, 5570.
131. J. R. Scheffer, C. Fyfe and L. Randall, unpublished results.

132. R. Popovitz-Biro, C. P. Tang, H. C. Chang, M. Lahav and L. Leiserowitz, *J. Am. Chem. Soc.*, **1985**, *107*, 4043.
133. C. P. Tang, H. C. Chang, R. Popovitz-Biro, F. Frolow, M. Lahav, L. Leiserowitz and R. K. McMullan, *J. Am. Chem. Soc.*, **1985**, *107*, 4058.
134. A. Gavezzotti, *Tetrahedron*, **1987**, *43*, 1241.
135. S. K. Kearsley and J. M. McBride, *Mol. Crystal. Liq. Crystal. Inc. Nonlin. Opt.*, **1988**, *156*, 109.
136. M. D. Hollingsworth and J. M. McBride, *Mol. Crystal. Liq. Crystall. Inc. Nonlin. Opt.*, **1988**, *161*, 25.
137. M. D. Hollingsworth and J. M. McBride in *Advances in Photochemistry*, Vol. 15, D. Volman, G. S. Hammond and K. Gollnick, eds., Interscience, New York, 1990, p. 279.
138. X. W. Feng and J. M. McBride, *J. Am. Chem. Soc.*, **1990**, *112*, 6151.
139. R. Lomolder and H. J. Schafer, *Angew. Chem. Int. Ed. Engl.*, **1987**, *26*, 1253.
140. J. Baro, D. Dudek, K. Luther and J. Troe, *Ber. Bunsenges. Phys. Chem.*, **1983**, *87*, 1161.
141. J. Baro, D. Dudek, K. Luther and J. Troe, *Ber. Bunsenges. Phys. Chem.*, **1983**, *87*, 1155.
142. M. D. Pace, *Mol. Crystal. Liq. Crystal. Inc. Nonlin. Opt.*, **1988**, *156*, 167.
143. N. J. Karch, E .T. Koh, B. L. Whitsel and J. M. McBride, *J. Am. Chem. Soc.*, **1975**, *97*, 6729.
144. J. M. McBride and R. A. Merrill, *J. Am. Chem. Soc.*, **1980**, *102*, 1723.
145. A. Gavezzotti, *J. Am. Chem. Soc.*, **1983**, *105*, 5220.
146. J. M. McBride, B. E. Segmuller, M. D. Hollingsworth, D. E. Mills and B. A. Weber, *Science*, **1986**, *234*, 830.
147. M. Feldhues and H.J. Schafer, *Tetrahedron*, **1985**, *41*, 4195.
148. M. D. Pace and W. B. Moniz, *J. Magn. Reson.*, **1982**, *47*, 510.
149. S. M. Aldoshin, M.V. Alfimov, L.O. Atovmyan, V. F. Kaminsky, V. F. Razumov and A. G. Rachinsky, *Mol. Crystal. Liq. Crystal.*, **1984**, *108*, 1.
150. K. H. Pfoertner, G. Englert and P. Schoenholzer, *Tetrahedron*, **1987**, *43*, 1321.
151. G. Kaupp, H. Frey and G. Behmann, *Chem. Ber.*, **1988**, *121*, 2135.
152. E. Hadjoudis, M. Vittorakis and I. Moustakali-Mavridis, *Tetrahedron*, **1987**, *43*, 1345.
153. E. Hadjoudis, J. Argyroglou and I. Moustakali-Mavridis, *Mol. Crystal. Liq. Crystal. Inc. Nonlin. Opt.*, **1988**, *156*, 39.
154. E. Hadjoudis, M. Vitorakis and I. Moustakali-Mavridis, *Mol. Crystal. Liq. Crystal.*, **1986**, *137*, 1.
155. N. H. Cromwell and J. A. Caughlan, *J. Am. Chem. Soc.*, **1945**, *67*, 2235.
156. N. H. Cromwell and H. Hoeksema, *J. Am. Chem. Soc.*, **1949**, *71*, 708.
157. A. Padwa and L. Hamilton, *J. Heterocycl. Chem.*, **1967**, *4*, 118.
158. H. W. Heine, R. H. Weese, R. A. Cooper and A. J. Durbetaki, *J. Org. Chem.*, **1967**, *32*, 2708.
159. A. M. Trozzolo, T. M. Leslie, A. S. Sarpotdar, R. D. Small, G. J. Ferraudi, T. DoMinh and R. L. Hartless, *Pure Appl. Chem.*, **1979**, *51*, 261.
160. T. DoMinh, *Res. Chem. Int.*, **1989**, *12*, 125.
161. A. Lottermoser, *J. Prakt. Chem.*, **1896**, *54*, 113.
162. K. Maeda, N. Kihara and N. Ishimura, *J. Chem. Soc. Perkin Trans. II*, **1985**, 887.
163. Y. Mori, Y. Ohashi and K. Maeda, *Acta Crystallogr.*, **1988**, *C44*, 704.
164. Y. Mori, Y. Ohashi and K. Maeda, *Bull. Chem. Soc. Jpn.*, **1988**, *61*, 2487.
165. Y. Mori, Y. Ohashi and K. Maeda, *Bull. Chem. Soc. Jpn.*, **1989**, *62*, 3171.
166. J. Shibuya, M. Nabeshima, H. Nagano and K. Maeda, *J. Chem. Soc. Perkin Trans. II*, **1988**, 1607.

167. S. Nespurek, S. Bohm and J. Kuthan, *J. Mol. Struct. (Theochem)*, 1986, *136*, 261.
168. E. Hadjoudis and I. Pulima, *Mol. Crystal. Liq. Crystal.*, 1986, *137*, 29.
169. J. H. Golden, *J. Chem. Soc.*, 1961, 3741.
170. H. Kamogawa and T. Suzuki, *J. Chem. Soc., Chem. Commun.*, 1985, 525.
171. W. R. Cullen and M. Williams, *J. Fluorine Chem.*, 1979, *13*, 85.
172. H. Gusten, G. Heinrich and H. J. Ache, *J. Photochem.*, 1985, *28*, 309.
173. H. Durr, *Angew. Chem. Int. Ed. Engl.*, 1989, *28*, 413.
     Photochromism, H. Durr and H. Bous-Laurent, eds., Elsevier, Amsterdam, 1990.
174. A. Peres de Carvalho, *Ann. Chim. (France)*, 1935, *4*, 449.
175. J. M. Tien and I. M. Hunsberger, *Chem. Ind. (London)*, 1955, 199.
176. J. M. Tien and I. M. Hunsberger, *J. Am. Chem. Soc.*, 1955, *77*, 6604.
177. E. F. Ullman and W. A. Henderson, Jr., *J. Am. Chem. Soc.*, 1966, *88*, 4942.
178. Y. Ohashi, *Acc. Chem. Res.*, 1988, *21*, 268.
179. A. Uchida and J. D. Dunitz, *9th International Conference on the Chemistry of the Organic Solid State*, Como, Italy, July, 1989, Abstract PC 72.
180. Y. Ohashi, A. Sekine and A. Uchida, *9th International Conference on the Chemistry of the Organic Solid State*, Como, Italy, July, 1989, Abstract OC 11.
181. N. N. Majeed, G. S. MacDougall and A. L. Porte, *J. Chem. Soc. Perkin Trans. II*, 1988, 1027
182. N. N. Majeed and A. L. Porte, *J. Chem. Soc. Perkin Trans. II*, 1987, 1139.
183. J. Pacansky, H. C. Coufal and D. W. Brown, *J. Photochem.*, 1987, *37*, 293.
184. J. Pacansky, H. Coufal, R. J. Waltman, R. Cox and H. Chen, *Radiat. Phys. Chem.*, 1987, *29*, 219.
185. J. Reisch, N. Ekiz and T. Guneri, *Arch. Pharm. (Weinheim)*, 1986, *319*, 973.
186. B. Marciniec, *Pharmazie*, 1983, *38*, 848.
187. Y. Matsuda and R. Masahara, *J. Pharm. Sci.*, 1983, *72*, 1198.
188. G. Reisch and J. Reisch, *Pharmazie*, 1980, *35*, 402.
189. J. Reisch, N. Ekiz and M. Takacs, *Arch. Pharm. (Weinheim)*, 1989, *322*, 173.
190. K. Jochym, H. Barton and J. Bojarski, *Pharmazie*, , *43*, 621.
191. C. R. Theocharis, A. M. Clark, S. E. Hopkin, P. Jones, A. C. Perryman and F. Usanga, *Mol. Crystal. Liq. Crystal. Inc. Nonlin. Opt.*, 1988, *156*, 85.
192. H. Langhals, T. Potrawa, H. Noth and G. Linti, *Angew. Chem. Int. Ed. Engl.*, 1989, *28*, 478.
193. Y. Ohgo, K. Orisaku, E. Hasegawa and S. Takeuchi, *Chem. Lett.*, 1986, 27.
194. C. Gianotti, G. Merle and J.R. Bolton, *J. Organomet. Chem.*, 1975, *99*, 145.
195. D. K. MacAlpine, A. L. Porte and G. A. Sim, *J. Chem. Soc. Perkin Trans. I*, 1981, 2533.
196. A. A. Freer, D. K. MacAlpine, J. A. Peacock and A. L. Porte, *J. Chem. Soc. Perkin Trans. II*, 1985, 971.
197. G. Ferguson, C. J. Fritchie, J. M. Robertson and G. A. Sim, *J. Chem. Soc.*, 1961, 1976.
198. P. J. Roberts, R. C. Petterson, G. M. Sheldrick, N. W. Isaacs and O. Kennard, *J. Chem. Soc. Perkin Trans. II*, 1973, 1978.
199. J. Hajdu, P. A. Machin, J. W. Campbell, T. J. Greenhough, I. J. Clifton, S. Zurek, S. Gover, L. N. Johnson and M. Elder, *Nature (London)*, 1987, *329*, 178.
200. T. Hayashi and K. Maeda, *Bull. Chem. Soc. Jpn.*, 1960, *33*, 565.
201. H. Zimmerman, H. Baumgartel and F. Bakke, *Angew. Chem.*, 1961, *73*, 808.
202. D. M. White and J. Sonnenberg, *J. Am. Chem. Soc.*, 1966, *88*, 3825.
203. K. Maeda and T. Hayashi, *Bull. Chem. Soc. Jpn.*, 1970, *43*, 429.
204. M. Tajima, H. Inoue and M. Hida, *Denki Kagaku*, 1989, *57*, 1225.

Chapter 6

Probing Reaction Pathways via Asymmetric Transformations in Chiral and Centrosymmetric Crystals

M. Vaida, R. Popovitz-Biro, L. Leiserowitz and M. Lahav

Department of Structural Chemistry,
The Weizmann Institute of Science,
Rehovot, Israel.

Contents

1. Introduction

Organized assemblies, and crystals in particular, provide matrices for the performance of reactions with a stereo-, regio-, and enantiospecificity generally found in enzymes. Moreover, the crystalline state is unique in the sense that in properly designed systems, an ensemble of nonchiral molecules can be spontaneously transformed by reaction into chiral molecules in the absence of any outside chiral influence. This is reflected by the fact that in recent years, asymmetric transformations with a high enantiomeric excess (ee) have been achieved by using the chiral environment of the molecule in the crystal.[1] Such experiments cannot be performed in solution, where asymmetric transformations can only occur from one chiral system into another. Therefore, reactions in the solid phase may serve as model systems related to the fundamental question on the origin of chirality in molecules of living matter.

A further advantage of solid-state reactions is that one is able to determine, via x-ray or neutron diffraction, the orientation of the constituent molecules prior to reaction and, in some favorable systems, even after its completion. The potential for reactivity and the elucidation of its pathway in the solid requires very particular packing arrangements. Therefore, the field is closely related to crystal engineering, which involves the design of molecular packing motifs (see Chapter 4). At present, it is hardly feasible to engage in such a design by computational methods using atom–atom potential energy criteria, since general computational algorithms for *ab-initio* construction of packing motifs are not yet available. Moreover, weak inter- and intramolecular forces are not known, with the required reliability, in many systems. Nevertheless it is possible to use rule of thumb methods for the design of packing motifs, and certainly to vary known types of crystal structures.

The use of chiral crystals for reaction pathway studies has an additional advantage; one may compare the absolute structure of the crystal before reaction with the absolute configuration of the reaction product. Of particular interest are transformations that occur in single crystals which preserve their integrity during reaction. In such systems one may determine the orientation and conformation of the molecular product inside the reacting matrix before the dissolution step required for extraction and recrystallization of the product.

Ever since the pioneering studies of Farina et al.[2] and Penzien and Schmidt,[3] asymmetric transformations have been obtained from nonchiral molecules crystallizing in chiral arrangements. Paradoxically one is not confined to chiral crystals; "centrosymmetric" crystals may also be used, for it is possible (within limits) to obtain chiral solid solutions via occlusion of guest molecules into the centrosymmetric host during growth. This type of crystal engineering provides an extension for the performance of spontaneous asymmetric synthesis.

The purpose of this review is not to examine a wide list of asymmetric reactions but to outline general concepts by selecting representative examples. We shall focus on those systems where detailed structural information on the reaction pathways is available or, if not, can at least be subjected to mechanistic analysis.

Finally, we shall review the way in which the symmetry of a centrosymmetric crystal is reduced, followed by a description of several asymmetric photoreactions inside such crystals.

1.1. Chiral and Centrosymmetric Space Groups

We first present a brief outline of those features of three-dimensional space groups essential for an understanding of asymmetric synthesis in the solid state. There are 230 space groups, which represent the different symmetry arrangements in which molecules may crystallize.[4] They can be divided into two classes: One comprises chiral space groups, of which there are 65. These contain symmetry elements only of the first kind, which are translations, rotations, and combinations thereof, i.e., the corresponding screw operations. The most common space groups for organic molecules are monoclinic $P2_1$, C2, and orthorhombic $P2_12_12_1$, although we shall in this review encounter structures with triclinic P1 or tetragonal $P4_12_12$ symmetry. Chiral space groups are enantiomorphous; i.e., the crystal structures of the two enantiomorphs are related by inversion symmetry. The second class comprises 165 space groups, which contains symmetry elements of both the first and the second kind, such as centers of inversion, mirror, and glide planes. Common space groups for molecular crystals of this class are triclinic $P\bar{1}$, monoclinic $P2_1/c$ and C2/c and orthorhombic Pbca and $Pna2_1$. All these nonchiral space groups are centrosymmetric, except for $Pna2_1$ which belongs to a subclass containing a polar axis. By comparison, all the chiral space groups are polar, some of which contain polar axes. Crystals containing polar axes display important physical properties, such as piezo- and pyroelectricity and optical nonlinearity leading to second harmonic generation.

Chiral-resolved molecules must crystallize into chiral space groups, but a racemic mixture in solution may either aggregate to form a nonchiral racemic compound or undergo a spontaneous resolution where the two enantiomers segregate into a conglomerate of chiral crystals. Nonchiral molecules may crystallize into either a nonchiral or a chiral space group. If they crystallize into the latter, however, the nonchiral molecules reside in a chiral environment imposed upon by the lattice. Among the most common examples are quartz, sodium chlorate, urea inclusion complexes, and the γ–form of glycine.

2. Asymmetric Transformations in Chiral Crystals
2.1. Chiral Molecules in Chiral Crystals

There are several ways to induce the formation of chiral crystals. As already mentioned, one way is to make use of chiral-resolved molecules. In such a crystal, not only the chiral center of the molecule resides in a chiral environment, but the entire molecule senses a chiral field. Therefore in contrast to solution, where an asymmetric induction would be exerted solely by the chiral handle, in the solid the chiral environment of the lattice may play a dominant role. Depending upon the type of reaction, the two effects may operate in unison or not. In many solid photochemical transformations where the crystallinity is preserved, the asymmetric induction is exerted primarily by the lattice; the overall effect of the chiral handle is simply to force the molecule into a chiral lattice. In heterogeneous solid-state transformations as in gas–solid or liquid–solid reactions, which are generally associated with pronounced disruption of the lattice during the progress of the reaction, the asymmetric induction is primarily due, as in solution, to the chiral influence exerted by the chiral center of the molecule. The asymmetric induction of the lattice is sensed only at the early stages of the conversion.

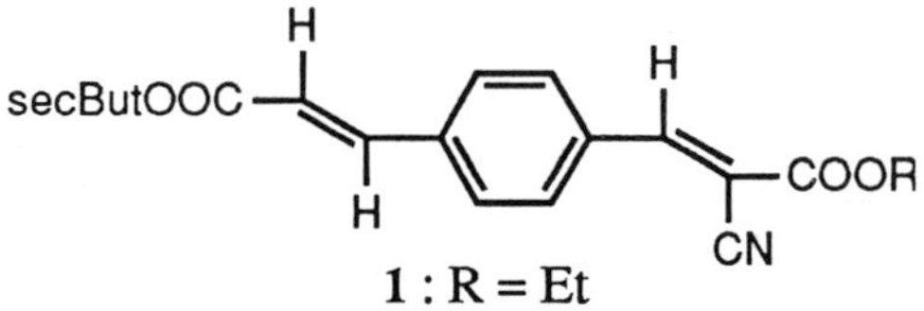

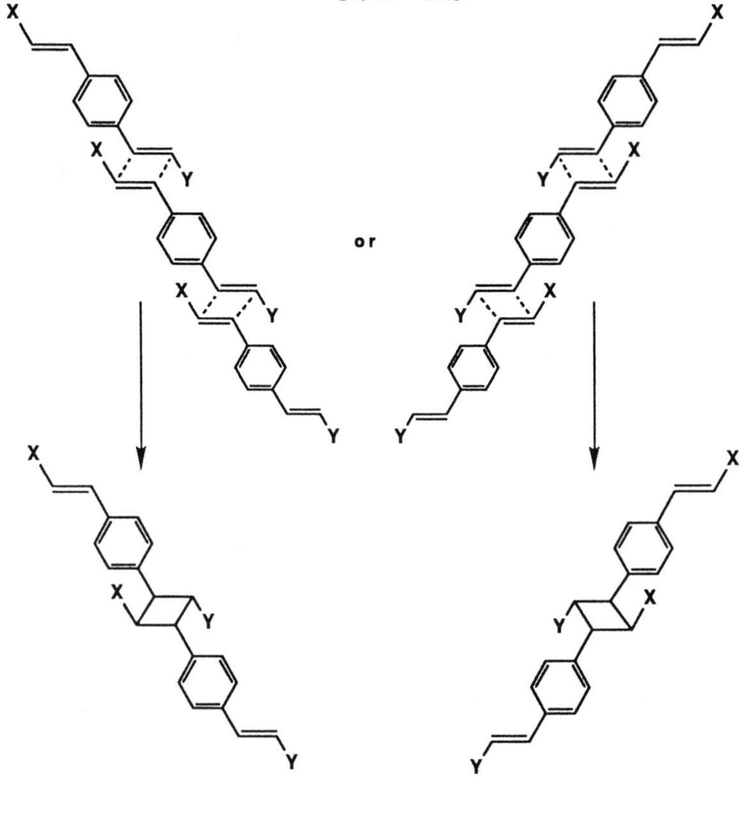

Scheme 1

Some mechanistic studies have been directed toward deconvoluting these effects, for example, the asymmetric photopolymerization of the chiral resolved diene **1** by Addadi and Lahav.[5,6,7] This molecule has been designed following some rules of crystal engineering. It crystallizes in a triclinic space group P1 with one molecule in the unit cell, so that the molecules in the crystal are related only by translation. A schematic representation of its packing arrangement is shown in Scheme 1.

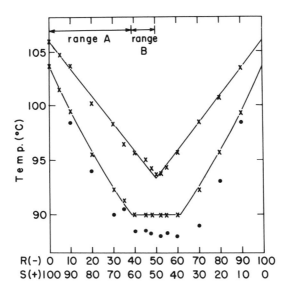

Figure 1. Phase diagram of **1** revealing the eutecticum at 40R:60S.

Irradiation of polycrystalline samples of the optically pure monomer at 5°C, $\lambda > 310$ nm, yields chiral dimers, trimers, and oligomers. The reaction occurs with a quantitative enantiomeric yield. The monomer with the absolute configuration (S) yielded the chiral cyclobutanes of absolute configuration (SSSS), in keeping with the molecular packing arrangement in the crystal. In this reaction it has been demonstrated that the asymmetric induction is completely due to the chiral environment of the crystal lattice and not due to that of the chiral handle; the role of the handle is just to induce a chiral arrangement. This conclusion follows from three different studies. First, the phase diagram of the two enantiomers showed that they form almost an ideal solid solution in the range up to 40:60 and with an eutecticum between 40R:60S and 60R:40S (Figure 1). A nice correlation of the optical yield of the photodimerization as a function of the composition of the two enantiomers is shown in Figure 2. In the region of a composition of 60R:40S, where full miscibility of the two enantiomers takes place, the ee of the photoproduct is quantitative. Another example is illustrated with the photobehavior of compound 2. This material is dimorphic; form α crystallizes in the triclinic space group P1, Z = 1. Irradiation of a polycrystalline sample of the R enantiomer yields dimers and oligomers with a quantitative diastereomeric yield. The absolute configuration of the product at the four chiral centers is (RRRR). The same enantiomer crystallizes in a different polymorph, space group $P2_1$, Z = 2. The orientation of the dienes in this lattice is almost enantiomeric to that of the same dienes in the P1 form. Irradiation of this crystalline lattice yields diastereoisomeric dimers which, after removal of the *sec*-butyl groups, become enantiomeric to the product obtained from the P1 crystal after a similar treatment.[8]

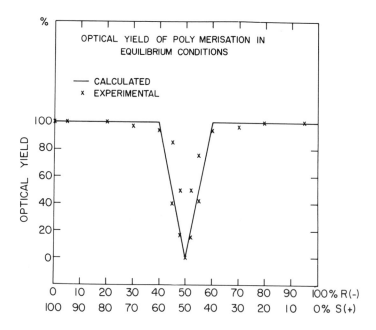

Figure 2. Expected and measured dependence of optical purity of dimers of **1** upon composition of the initial monomer.

Additional support is provided by the photochemistry of (3). This compound, although resolved, crystallizes in a quasi centrosymmetric space group. Irradiation of a solid sample of this compound results in the formation of two diastereoisomers that, after removal of the *sec*-butyl group, yields a racemic mixture of the two dimers, implying that the *sec*-butyl group does not exert a chiral induction in the reaction (Scheme 2). This intrinsic property of chiral disorder in molecules bearing the *sec*-butyl group follows on geometrical considerations.[5,7]

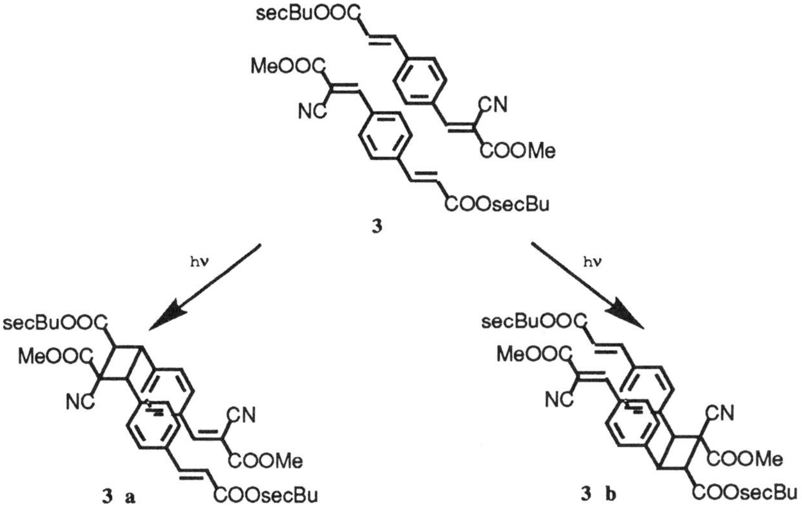

Scheme 2

Systematic studies of the packing arrangements of a large variety of molecules containing this handle in chiral-resolved, racemic, and diastereoisomeric crystals have demonstrated that the *sec*-butyl moiety may assume two different conformations inside the crystalline lattice, gauche and trans. For example, from the four isomers of the two diastereoisomers of racemic isoleucine,[9] it was shown that the sec-butyl group adopts a gauche conformation (Scheme 3). Moreover, these groups are disordered in the solid state by virtue of having almost the same surface envelope, as shown in Scheme 3.

This enantiomeric disorder also appears in crystals of *sec*-butyl phthalates (**4**) and phthalamides (**5**); within each compound the racemate and enantiomeric counterparts are isostructural[10]; the *sec*-butyl groups assume an extended trans conformation. A crystal structure analysis of the 1:1 R:S mixture of the phthalamide shows how the disorder is expressed; the R and S *sec*-butyl groups occupy the same volume by a slight lateral adjustment of their chiral C* atoms, as shown in Scheme 4. An x-ray analysis of a 2:1 R:S mixture of the phthalamide (**5**) demonstrated how an R molecule replaces the site of an S molecule in the lattice by the same mechanism.

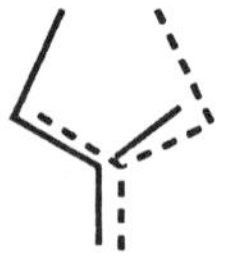

Scheme 3

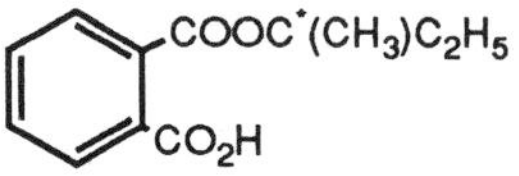

C*⌐⌐⌐*C   ⟨benzene⟩COOC*(CH₃)C₂H₅ / CO₂H   ⟨benzene⟩CONHC*(CH₃)C₂H₅ / CO₂H

                    4 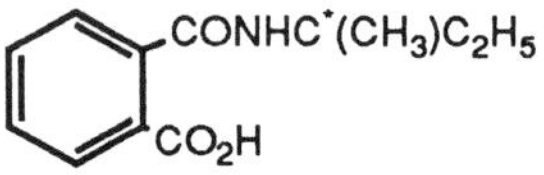   **5**

Scheme 4

Another example of an asymmetric transformation using a *sec*-butyl handle has been studied recently in the unimolecular di-π-methane rearrangement by Scheffer et al.[11] (See also Section 3.1. of this Chapter and 3.3. of Chapter 5).

2.2. Nonchiral Guest Molecules in Chiral Inclusion Complexes

Another efficient method by which a nonchiral molecule can be induced to reside in a chiral environment is to complex it with a host molecule that has a tendency to form chiral inclusion complexes. A variety of such systems has been reported in which the host may be either chiral or nonchiral; for example, urea, triorthothymotide, cyclodextrins, deoxycholic acid, and apocholic acid. The structures of some of these complexes have been studied by x-ray analysis (see Chapter 7).

The first asymmetric transformation reported was by Farina et al.[2] on the polymerization of 1,3-*trans*-pentadiene complexed with resolved perhydro-triphenylene (**6**). A polycrystalline sample of this complex subjected to γ-irradiation yielded a highly stereospecific 1,4-trans isotactic polymerization, albeit with low ee. More recently, polymers with higher optical activity were obtained from polymerization of *cis*-1,3-pentadiene inside the channels of deoxycholic acid (DCA) (**7**) and apocholic acid.[12,13] The polymers assume a predominantly 1,4-trans structure (90%), but the tacticity is not maintained, especially when the trans monomer is used. Even higher optical yields were obtained with 2-methyl 1,3-pentadienes.[14] The authors suggested that the high ee seems to depend on the packing of the monomers inside the channels; the larger the constraint of the molecular motion of the guest molecules inside the channel, the higher the ee. However, in the absence of knowledge of the detailed packing arrangement of the guest molecules, such an assumption has to be taken with caution.

2.3. Photoaddition of Ketones to Choleic Acids

The channel inclusion complexes of deoxycholic acid (DCA) (7) are an ideal vehicle to monitor photoreaction pathways for the following reasons. The chirality of DCA specifies the absolute structure of the crystal and thus of the host and guest at the reaction site. From x-ray studies of a variety of such complexes, it has been demonstrated that the arrangement of the guest molecules along the channel can be varied by crystal engineering.[15,16,17] Furthermore, the crystal maintains its integrity on photoreaction, so the solid solution containing host, guest, and the photoproduct in various ratios can be determined by x-ray diffraction. These properties were taken advantage of to glean information on the reaction pathway involving photoaddition of guest ketones to the host DCA molecules.

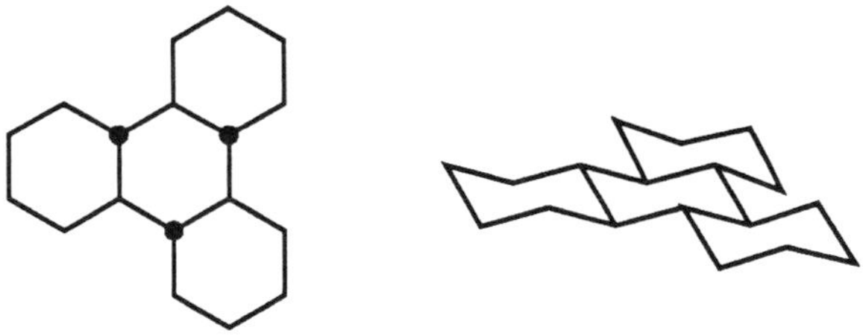

6

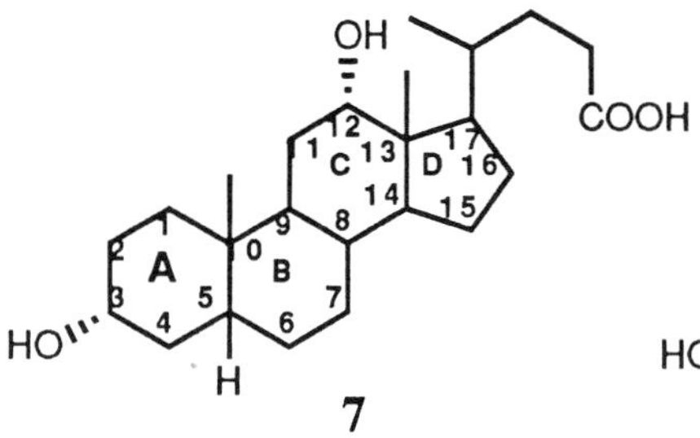

7

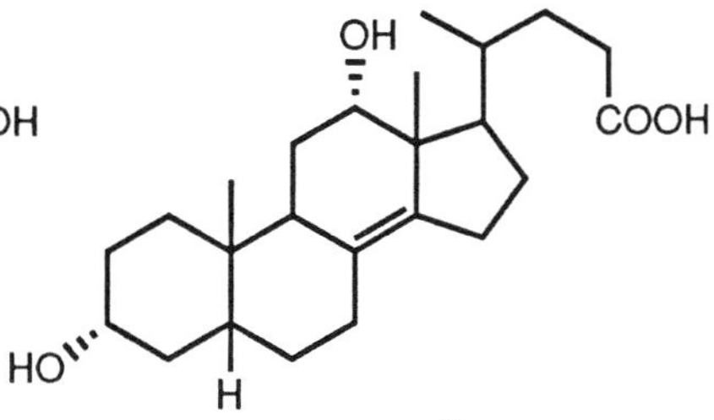

8

Deoxycholic acid generally crystallizes in one of three different forms, orthorhombic, tetragonal, or hexagonal (see also Section 2.1.7. of Chapter 7). The photochemical reactions were performed on complexes of the orthorhombic form, where the crystal structure embodies a two-dimensional bilayer motif, with axial dimensions b = 13.6, c = 7.1 Å. The molecules along the 13.6 Å axis are interlinked by O—H···O bonds to form chains. These molecules are further joined by hydrogen bonds about twofold screw axes parallel to b and spaced along c, thus generating the *bc* bilayer. Because these bilayers contain grooves parallel to the c axis, DCA is induced to form channel inclusion complexes (Figure 3).

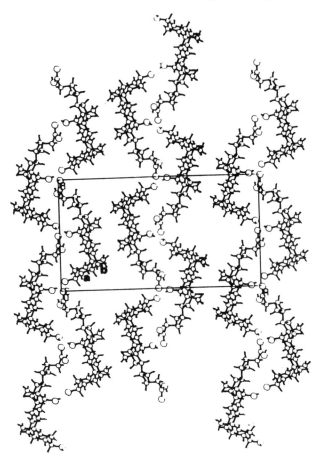

Figure 3. Packing motif of the host DCA (**7**) . View along the channel axis *c* . The channels formed by juxtaposition of neighboring bilayers are shown. The atoms of rings **a** and **b** exposed on the channel wall are specified.

These adjacent bilayers may adjust their relative positions, but only within limits, to form channels that best fit the guest molecules. The topochemical nature of the photoaddition of DCA was demonstrated by the photochemistry and host–guest arrangements of occluded aliphatic ketones, such as acetone (**9**), diethylketone (**10**), ethylmethyl-ketone (**11**) and cyclohexanone (**12**) (Scheme 5). Photoaddition took place at different sites on the channel steroid wall, depending on the geometric arrangement of the ketone group C'=O' vis-a-vis the steroid C—H moiety which is abstracted by the excited carbonyl prior to addition. Reaction takes place if the C—H bond tends to be perpendicular to the C'=O' moiety, with O'⋯H and C'⋯C distances ranging from 2.9 to 3.9 Å and from 3.7 to 4.2 Å, respectively. The aliphatic ketones generally induce a channel packing of the steroid which only permits hydrogen abstraction from rings A and B, so that photoaddition takes place at C5 and C6 for the aliphatic ketones. In order to functionalize ring D by a guest ketone, it is necessary to generate a channel of the type shown in Figure 4 by a bulky guest, in this case cyclohexanone. Here the guest carbonyl group C'=O' is in close proximity to ring *D*. Irradiation yields the addition product at C16ax (**12f**) (Scheme 5); C16ax being the most eligible candidate for addition in terms of both O'⋯H and C'⋯C contacts. Prochiral ketones $R_1R_2C'=O'$ were used to probe the reaction pathway since their photoaddition to steroids leads to formation of a new chiral center, whose absolute configuration may be compared with the prochiral configuration about the plane of the guest carbonyl. The crystalline complex of ethylmethylketone proved to be unsuitable for this purpose because of its packing in the channel. The guest molecules form centrosymmetric pairs (Figure 5a) in such a way that leads to the formation of both diastereoisomers, the reactive centers of the steroid at sites C5 and C6 being equally well exposed to the two opposite faces of the ketone molecule (Figure 5b). To preclude this type of packing we chose acetophenone, whose complex[17e] gave only one addition product (**13**), at position C5 on UV irradiation (Scheme 6).

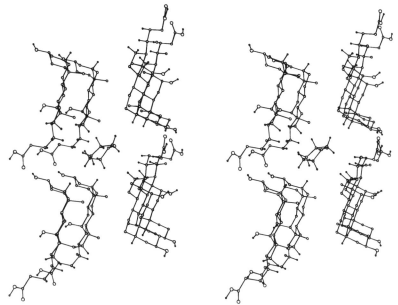

Figure 4. Stereoscopic view of the packing motif of DCA with guest cyclohexanone. The D ring is exposed to the guest in the channel.

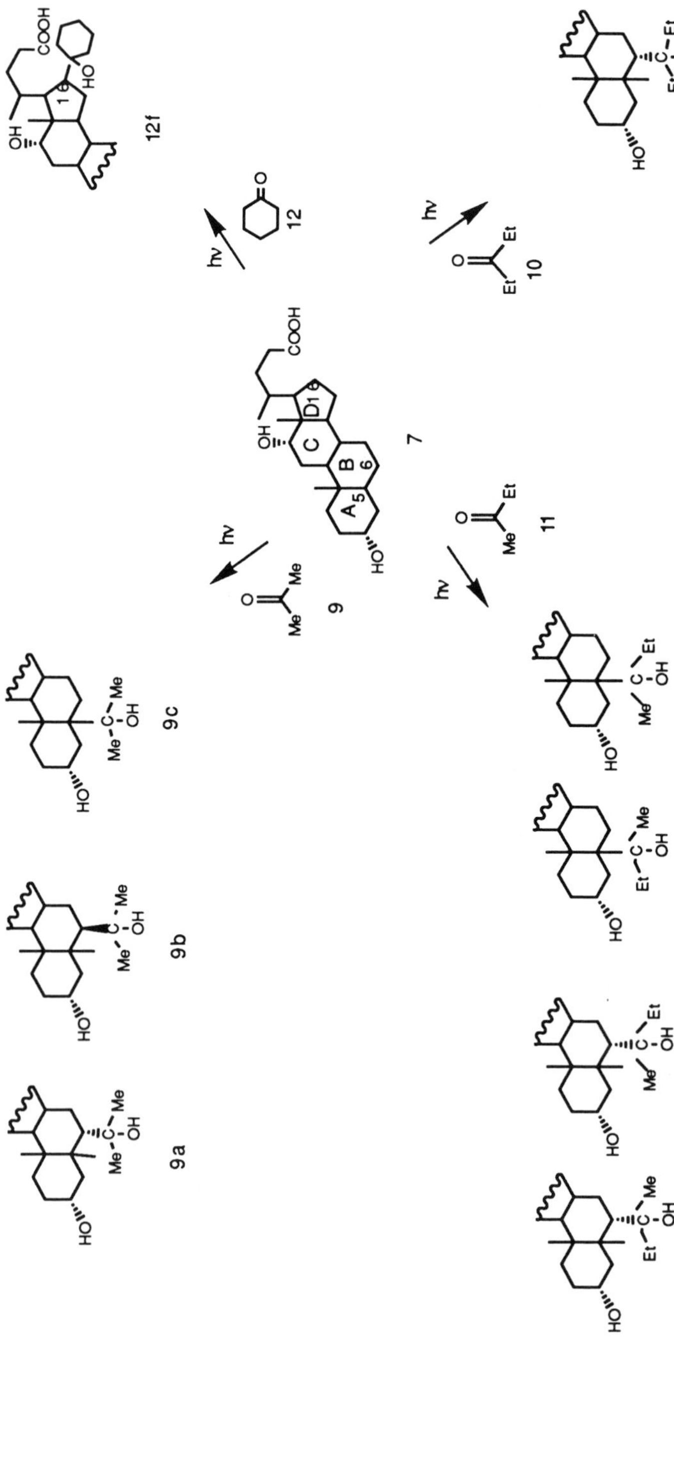

Scheme 5

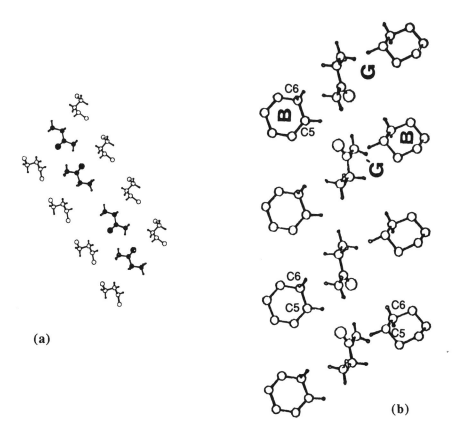

Figure 5. Packing arrangement of DCA-methylethyl ketone showing: (a) Centro-symmetric pairs of guest inside the channel. (b) Atoms C5 and C6 being equally well exposed to the two opposite faces of the ketone molecule.

13 : X=H
14 : X =F

Scheme 6

The host–guest arrangement of 5:2 DCA–acetophenone is depicted in Figure 6. There are two independent guest molecules, G and G', which form a chain of closely packed pairs in the channel. The molecules in a pair are separated by 8 Å along the channel axis of length 7.2 Å.

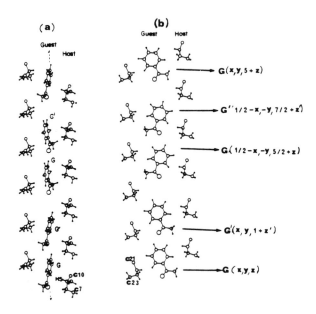

Figure 6. DCA-acetophenone. Packing of guest acetophenone molecules in the channel showing arrangement of the close-packed pairs G and G'. (a) View edge on to the plane of the guest molecules. (b) View perpendicular to the plane of the guest molecules.

The orientation of the acetophenone molecules in proximity to the (abstracted) H5 atom of the steroid is such that the guest ketone exposes its *re* face to the steroid C5—H bond (Figure 7b). A comparison of this arrangement with that of the acetophenone addition product in its own crystal structure (Figure 7d) shows that the ketone adds from that face of acetophenone which is the more distant from the steroid in the starting structure, to generate a new chiral center of absolute configuration S. This stereochemical relationship, exemplified in Scheme 7, implies that an unusual motion of the guest acetyl group is required for reaction. The reaction course was further elucidated by a determination of the crystal structure after reaction. One such structure of a solid solution containing host, guest, and 41% of the addition product was determined.[17f] Both molecules G and G' reacted. With regard to the attained degree of 41% conversion, the structure indicates that it would be impossible for both nearest neighbors G and G' molecules to react. The resulting arrangement would require too close a distance between the reacted G and G' molecules. Therefore, 50% would appear to be the maximum achievable conversion. The molecular structure of the photoproduct in the solid solution shows that the phenyl ring underwent minimal positional and orientational change on reaction (Figure 7c). Thus it was inferred that the ketyl function undergoes a net rotation of 180° prior to addition. Upon isolation and recrystallization, the product (Figure 7d) adopts a conformation different from that found in the partially reacted crystal. To test whether a net 180° rotation of the acetyl group in the channel is feasible, atom–atom potential energy calculations were performed on DCA–acetophenone. The results indicated that it is indeed possible for the guest acetyl group, with a sp^3 configuration about the carbon "radical" atom, to perform a net rotation of 180° without incurring impossibly short host–guest contacts, provided minor positional adjustments of the neighboring steroid molecules occur. One way to rationalize the stereochemistry of the photoproduct in DCA–acetophenone, or DCA–*p*-fluoroacetophenone is depicted in Scheme 7.

Here we invoke the electrophilic nature of the P_y orbital of the excited oxygen. In several studies on hydrogen abstraction by ketones in solution,[18] and in rigid molecules in the solid,[19] it has been reasoned that the ketone abstracts the hydrogen through the P_y orbital of the excited oxygen on n-π* photoexcitation. We observed that for the guest acetophenone molecules the P_y orbitals are orthogonal to the C5—H (abstracted) bond. Assuming the above abstraction pathway and in the light of our experimental results we propose that the abstraction involves a reorientation of the carbonyl bond to bring its P_y orbital along the line of sight of the C5—H bond (Scheme 7). Moreover, after abstraction, it is the α lobe (rather than the β lobe) of the π* orbital on atom C' which appears to be closer to the steroid atom C5. This leads to a net rotation of the ketyl radical by 180° prior to coupling to the steroid.

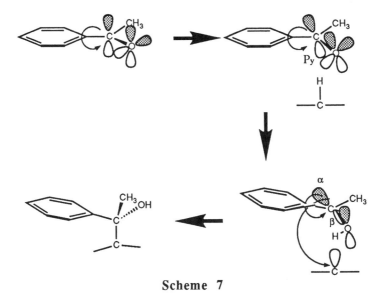

Scheme 7

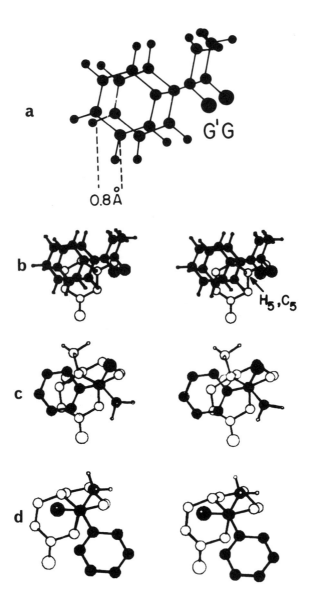

Figure 7. (a) The two independent acetophenone guest molecules G and G' shown occupying the same crystallographic site. (Actually they are separated by 8 Å along the channel wall, i.e., 7.2 + 0.8 Å, where 7.2 is the length of the channel c axis. (b) Stereoscopic view of the orientation of the two acetophenone guest molecules G and G' in proximity to the steroid C5—H bond. Only the A ring of the steroid moiety is shown. (c) Stereoscopic view of the DCA-acetophenone addition product inside the partially reacted crystal. Only the A ring of the steroid is shown. (d) Stereoscopic view of the addition product after isolation in its own crystal structure.

In order to determine whether the rotation is orbital steered, two experimental approaches were adopted. One involved the design of a host guest arrangement in which the guest ketone exposes its *si* face to the steroid C5—H bond instead of the opposite face *re*, which, if the photoreaction is orbital steered, should then yield the diastereomeric product of absolute configuration R at the newly generated chiral carbon center. The other involved inhibition of the 180° rotation of the ketone moiety by attaching a flexible bulky ketone group to the phenyl ring. We discuss these two approaches in turn.

The host-guest arrangement in which the guest ketone exposes its si face to the steroid C5—H bond was engineered by modifying the arrangement of guest acetophenone molecules through atomic substitution. The host–guest arrangement of DCA–acetophenone at the site of the reaction is depicted schematically in Figure 8, showing guest molecules G (original) and G' separated by 8 Å.

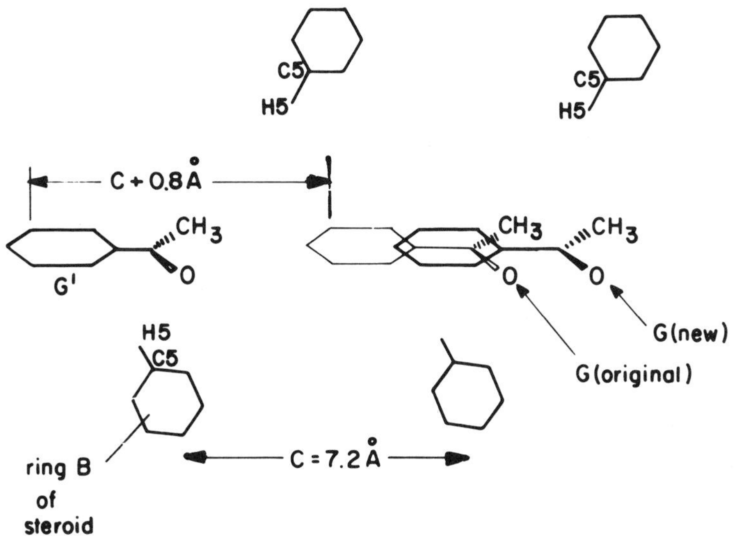

Figure 8. A schematic arrangement of host DCA and G' guest molecules in which G (original) has been shifted along the channel *c* axis from its original position by a distance sufficient to expose the *si* side of its acetyl group to the C5—H bond.

This arrangement may be modified by inducing a para-substituted acetophenone molecule G to occupy a new position G (new), approximately 9.2 Å removed from G' along the -c direction. In this way the *si* face of a substituted acetophenone may be exposed to the steroid C5—H bond. *p*-Fluoroacetophenone was chosen as an appropriate guest. The host–guest arrangement in 3:8 DCA–*p*-fluoroacetophenone contains two independent molecules G and G' which form close-packed triplets G'GG'.....G'GG'..... in the channel (Figure 9).

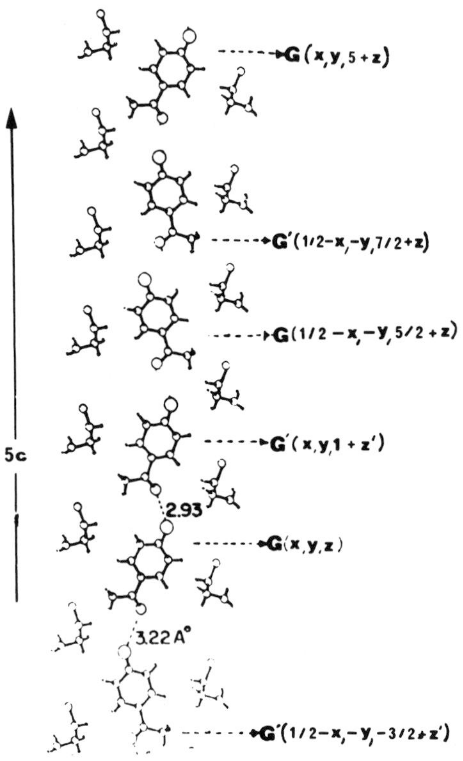

Figure 9. Packing motif of DCA–fluoroacetophenone. View perpendicular to the plane of the guest molecules, showing close-packed triplets G'GG'.

The *p*-fluoroacetophenone guest molecule G' makes the same contact with the potentially reactive C5—H bond of DCA (Figure 10), exposing the *re* face of its acetyl group to C5—H as in DCA–acetophenone (Figure 8). On the other hand the fluoroacetophenone G molecule exposes mainly the si face of its acetyl group to a steroid C5—H bond. Consequently purely on the basis of the host–guest packing, photoirradiation of DCA–*p*-fluoroacetophenone should yield two diastereomeric products (one from G and the other from G'). However uv irradiation yields only one diastereomeric photoaddition product (**14**) with the same chirality (S) about the newly generated chiral carbon atom as the photoproduct of DCA–acetophenone. This apparent inconsistency was resolved by determining the structure of a single crystal of photoirradiated DCA-*p*-fluoroacetophenone; it was found that only the G' molecules reacted, G remaining unaffected.[17f] This result was explained on steric grounds according to Figure 11 because, were it to react, the G molecule that is sandwiched between two G' molecules in the channel would make impossibly short contacts with a neighboring G' molecule.

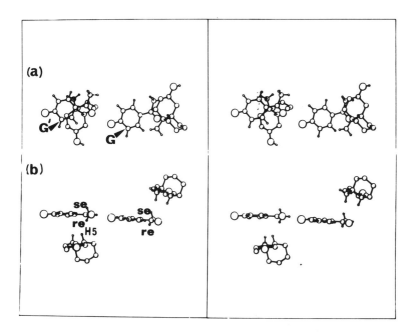

Figure 10. DCA-fluoroacetophenone. Stereoscopic view of host-guest packing at site of reaction. The two guest molecules G and G' and ring A of the steroid are shown. (a) View along the steroid H-C5 bond. (b) View perpendicular to the H—C5 bond.

2.4. DCA-Substituted Propiophenones

In order to inhibit the 180° rotation of the ketone moiety, we made use of substituted propiophenones as guests.[17g] Ultraviolet-irradiation of DCA–propiophenone gave two diastereomeric photoproducts **15a** and **15b** at site C5 and product **15c** at site C6 (Scheme 8, X = H). Irradiation of DCA–*p*-fluoropropiophenone yielded the analogous photoproducts **16a**, **16b** and **16c**, the major product being **16b** (Scheme 8 X = F). DCA–*p*-chloropropiophenone yielded the photoproduct **17b**. The absolute configurations about the newly generated chiral center bound to site C5 are S for product **a** and R for product **b**.

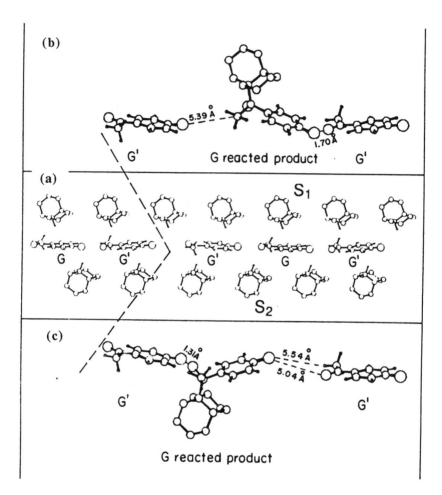

Figure 11. (a) Central picture showing fluoroacetophenone guest triplet G'GG' spanned by steroid molecules. (b) Hypothetical arrangement were steroid S_1 to react with G. (c) Hypothetical arrangement were steroid S_2 to react with G.

The absolute configurations of product **c** were not unambigously assigned for lack of suitable single crystals. According to x-ray diffraction analysis of both crystalline complexes, the channels contain two crystallographically independent molecules, G and $\bar{G}$, with host-guest molecular ratios very close to 1:3. There are essentially two different chain motifs in which the guest molecules pack in the channels. In one (Figure 12a) the guest molecules G and $\bar{G}$ form a chain $G\bar{G}G\bar{G}G\bar{G}$ ..., G and $\bar{G}$ being related by a pseudo-center of inversion. In the other, the G molecules form chains $GGGG$, etc., (Figure 12b).

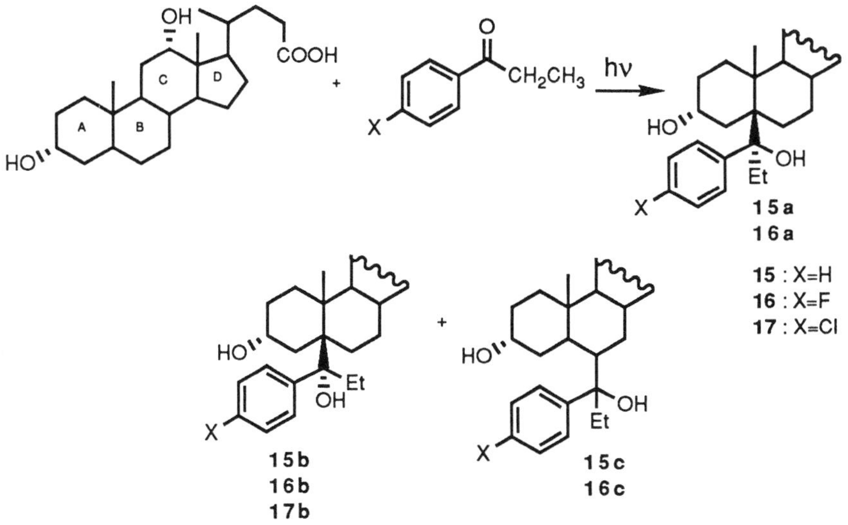

Scheme 8

The host–guest geometries at the sites of reaction are depicted in Figures 13a,b and Scheme 9. Each G and $\bar{G}$ molecule exposes its *re* face to a potentially reactive C5—H center. Therefore the formation of the products **15a** and **16a** necessitates rotation of the propionyl group by a net 180° prior to photoaddition, as was found in the DCA (substituted) acetophenone series. On the other hand the products **15b** and **16b** are formed without rotation of the propionyl group. These two results may be explained in terms of the guest packing. The propionyl groups in the arrangements shown in Figure 12a cannot easily undergo a rotation of 180° because that could eventually lead to unfavorable short contacts between neighboring guest methyl groups. Therefore were reaction involving molecules G or $\bar{G}$ to take place at such a site in the chain the propionyl group would have to bind to the steroid without rotation. In contrast the propionyl group in the arrangements shown in Figure 12b could undergo a net rotation of 180° without inducing prohibitively short contacts with the nearest-neighbor guest molecules.

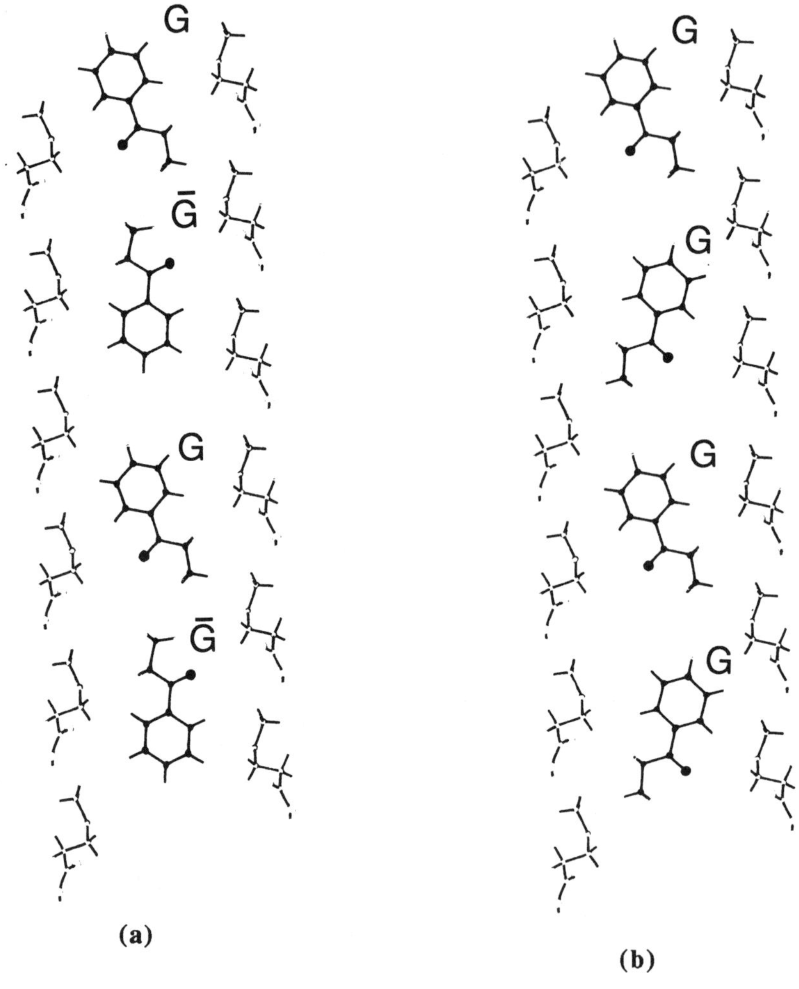

(a)

(b)

Figure 12. DCA–propiophenone packing arrangement of the guest inside the channel, showing the two different motifs. (a) Pseudo-centrosymmetrically related dimers of G(H) and Ḡ(H) molecules. The distances between the methyl C atoms of G and Ḡ are 4.1Å and between the para-H atoms, 3.9 Å. (b) String of G(H) molecules related by twofold screw symmetry. The intermolecular distance between the (methyl) C and C4 is 4.9 Å.

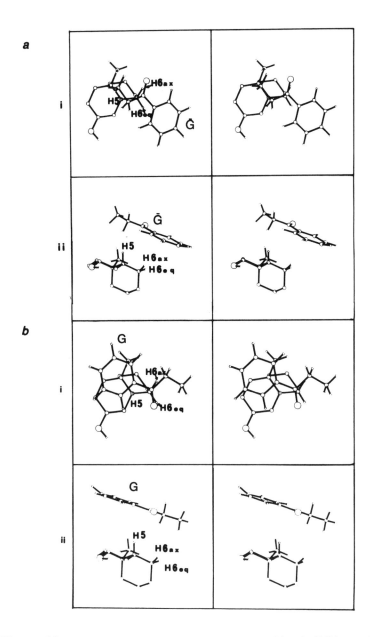

Figure 13. Stereoscopic view of the host–guest packing in DCA–propiophenone at the sites of the reaction for: (a) guest molecule $\bar{G}$(H): (b) Guest molecules G(H). (i) View along the steroid C5—H bond; (ii) View perpendicular to the C5—H bond.

The crystal structure of DCA–*p*-chloropropiophenone, including the host–guest packing at the C5 and C6 sites of reaction, was determined by x-ray diffraction[17g] (Figure 14). However, the absolute configuration about the newly generated chiral center of the photoaddition product **17c** was not unambigously determined. Therefore little can be said about the reaction pathway at site C6.

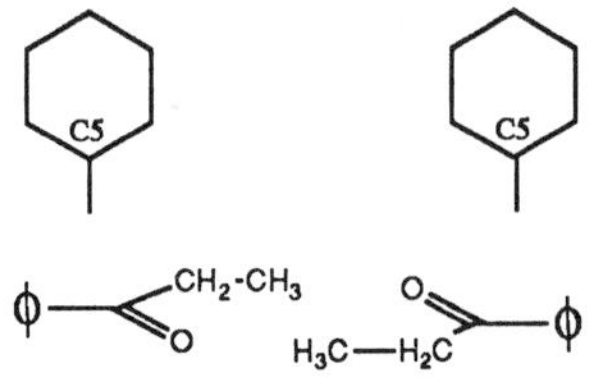

Scheme 9

The crystallographic and photochemical results on the channel inclusion complexes of DCA with aromatic ketones indicate that photoaddition of the ketone may take place with or without a 180° rotation of the ketone group. This rotation appears to be orbital controlled for it occurs in those systems where in terms of intermolecular steric contacts, reaction may easily occur without rotation of the ketone. In those systems where contacts between the guest molecules appear to strongly inhibit rotation, photoaddition of the ketone still takes place but without rotation.

Recently Ayoma et al.[20] reported another example of nonchiral guest molecules in chiral inclusion complexes; the asymmetric transformations of *N,N'*-dialkyl pyruvamides (**18**) in a polycrystalline solid phase of DCA or cyclodextrin to yield a chiral β-lactam (**19**) (Scheme 10). Although the product obtained in the solid phase was different from those obtained in solution, the ee generally ranged from 3 to 43% depending again upon the substituents. The authors propose that the molecules probably move inside the channels. Although this explanation is reasonable one cannot exclude other possibilities, for example that the guest molecules occupy more than one independent crystallographic site, as found for some of the aliphatic and aromatic ketones discussed previously.

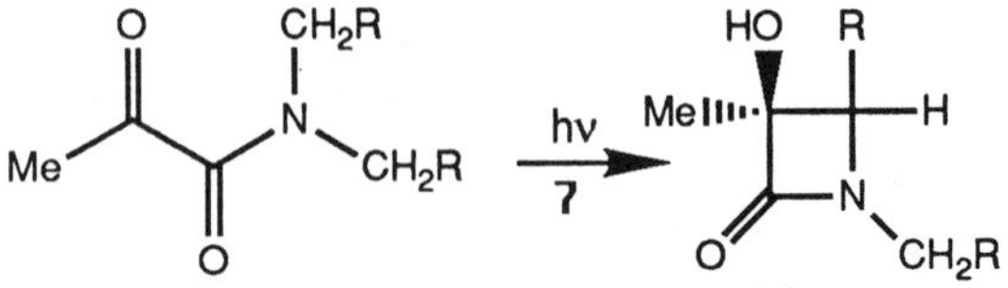

Scheme 10

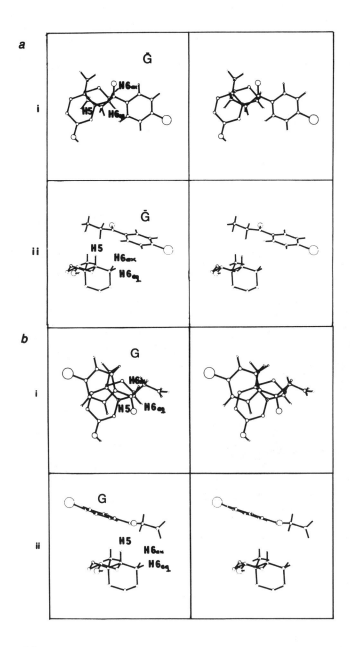

Figure 14. Stereoscopic view of host–guest packing in DCA–p-fluoro-propiophenone for (a) guest molecule $\bar{G}$(F); (b) guest molecule G(F). (i) View along the steroid C5-H bond. (ii) View perpendicular to the C5—H bond.

Toda reported that the resolved diol (**20**) has a strong tendency to form inclusion complexes with a large variety of molecules bearing polar functional groups which can form hydrogen bonds with the hydroxy groups of the diol (see also Section 3.3 of Chapter 4 and Section 2.1 of Chapter 7 for more details).[21] A number of monomolecular asymmetric transformations of nonchiral molecules complexed with this host were reported and are listed in Scheme 11. Among these are the pyruvamide (**21a**) to the β-lactam (**22a**). The asymmetric transformation is almost quantitative in contrast to results obtained within the inclusion complex of DCA.

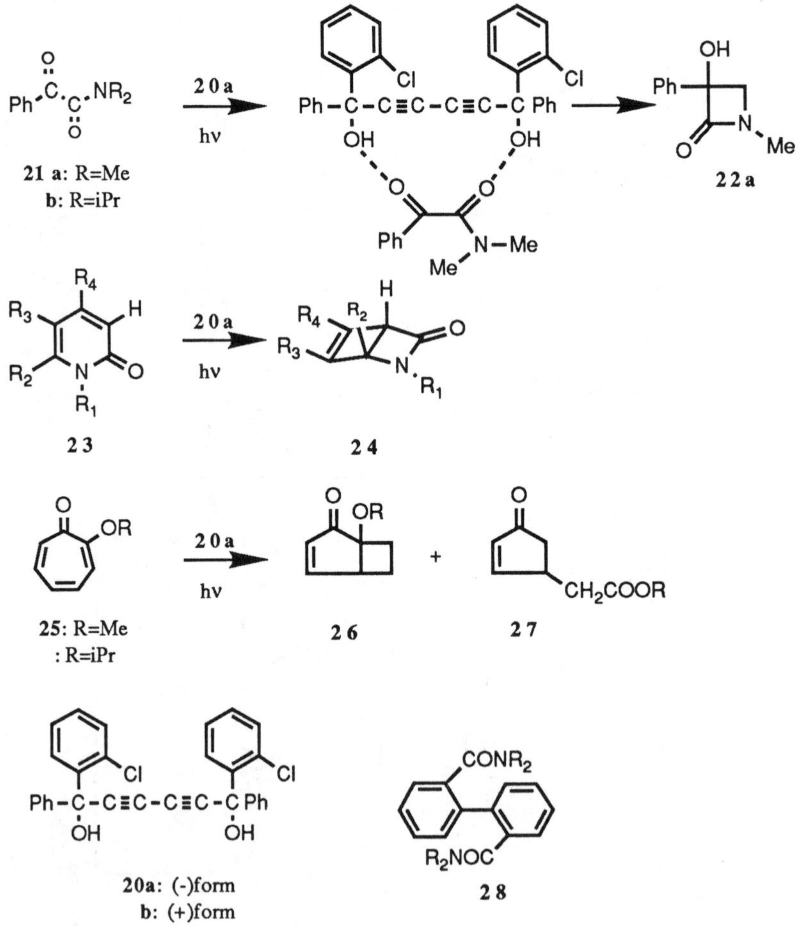

Scheme 11

This important transformation of pyruvamides to β-lactams has been studied with a variety of other complexing agents,[21] such as (**28**). Another enantioselective transformation is the electrocyclic transformation of the α-tropolone alkyl ethers (**25**) to product (**26**) with 100% ee.[21a] In the complex this product photorearranges further to (**27**) (72%-91%). The quantitative ee implies that this disrotatory photocyclization reaction occurs in one direction. An understanding of the rotation modes must await the crystal structure determination of the complex. In the second system, (**23**) was transformed with a quantitative ee to (**24**). x-Ray structure determinations of some of these complexes have been reported.[22] All these systems are ideal vehicles for studying reaction pathways. It will be rewarding to find systems in this class of compounds that undergo single-crystal to single-crystal transformations and so allow trapping of the intermediate conformations along the reaction coordinate.

3. "Absolute" Asymmetric Transformations: Nonchiral Molecules in Chiral Crystals

When a nonchiral molecule crystallizes in a chiral space group, it has equal probability of crystallizing in each of the two enantiomorphs. Thus, to assure the homochirality of the crystalline phase, large single crystals have to be grown in conditions close to equilibrium, unlike the case for chiral-resolved molecules, whose handedness fixes that of the crystalline phase even in polycrystalline samples. In well-designed chiral matrices composed of nonchiral molecules, the chiral environment can exert an asymmetric induction in a lattice-controlled solid-state transformation. As mentioned in Section 1, these "absolute" asymmetric syntheses have attracted wide interest since they may occur spontaneously in the absence of any outside chiral influence. Owing to the strict requirements, however, the discovery of new systems appropriate for such syntheses has been slow. Less than a dozen examples have been described since the pioneering report on the asymmetric gas–solid bromination of *p,p*-dimethylchalcone by Penzien and Schmidt,[3] 20 years ago.

Packing arrangements have been analyzed of nonchiral disubstituted ethylenes in chiral crystals, which are required for the formation of chiral cyclobutanes through (2π–2π) photo-addition reactions. As a result, three possible arrangements have been proposed, two of which have been materialized during the years.[23] The first chiral system is composed of host and guest molecules where the latter are oriented in a translation stack of axial length of 4 Å (Scheme 12). Upon selective excitation the guest molecule X—CH:CH—Z changes its conformation. In the crystal the excited molecule will be distorted in an asymmetric mode and will have different probabilities of interacting with the molecule above or below in the stack. This will result in the formation of two different diastereoisomeric transition states, yielding two enantiomers in unequal amounts.

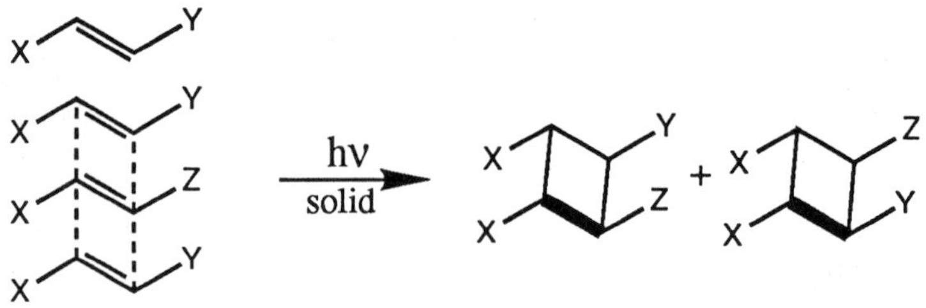

Scheme 12

This prediction was experimentally confirmed by Green et al.[24] with the mixed crystal of dichlorophenyl-4-phenyl–*trans,trans* -1,3-butadiene **29** with the 4-thienyl analogue **30**. The chlorine atoms were deliberately attached to the molecule to assure the short 4.0 Å axis needed for photodimerization. These two materials crystallize in two isostructural arrangements in the chiral space group $P2_12_12_1$. Large mixed crystals of the phenyl material **29** containing ~15% of the thienyl **30** as guest were prepared. The latter absorbs light at a longer wavelength. As a result of this selective excitation the thienyl reacts with a nearer phenyl neighbor in its ground state to form a mixed cyclobutane dimer **31**. This dimer has been isolated and demonstrated to be optically active with an enantiomeric excess of ~ 70%. Ludmer et al.[25] proposed the formation of an exciplex as a reaction intermediate. This was experimentally supported by a non-Gaussian shaped fluorescence emission with a resolvable shoulder at 475 nm. Warshel and Shakked[26] calculated the reaction pathway in the crystal and proposed a preferential interaction of the thienyl molecule with one of the two neighboring molecules. This theoretical model remains to be experimentally probed by correlating the absolute configuration of the reacting crystal with that of the absolute configuration of the enantiomer formed in excess.

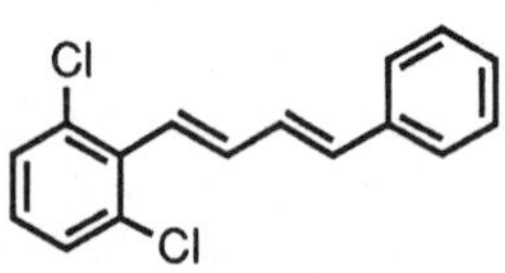

29 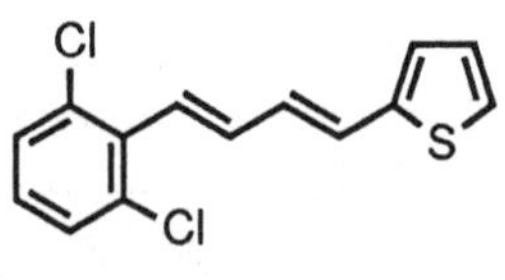   **30** 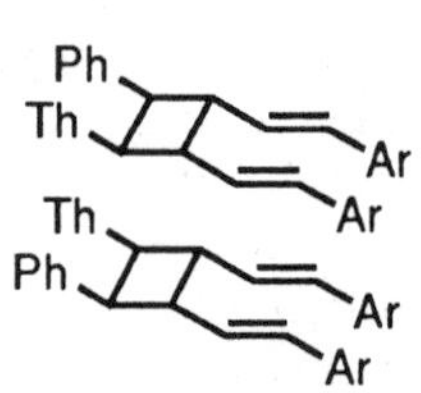   **31**

The second approach involves the design of a crystalline phase isomorphous with the chiral compound **1** but composed of nonchiral molecules. The rules of isomorphism state that isomorphous phases can be built from two similar systems if the overall volume occupied by the different groups is the same and the interactions are not changed.[6] From the packing arrangement of **1**, it can be seen that the *sec*-butyl groups are arranged in a chain (Scheme 13). A simple transfer of a methyl group along this chain replaces the chiral *sec*-butyl handles with two nonchiral isopropyl and 3-pentyl side chains, leaving the backbone of the molecules unchanged. The overall volume of the crystal should be maintained.[7] Addadi and Lahav found experimentally that while pure monomers **32** and **33** crystallize in a nonchiral space group, a 1:1 mixture precipitates in a chiral space group and the crystal is isomorphous to that of the *sec*-butyl diene **1**.

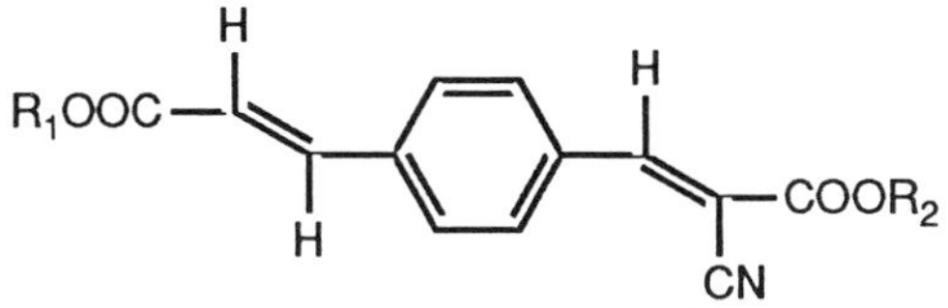

32: R_1= iPr , R_2=Et
33: R_1= 3Pentyl, R_2=Et
34: R_1=iPr, R_2=nPr
35: R_1=3Pentyl, R_2=Me

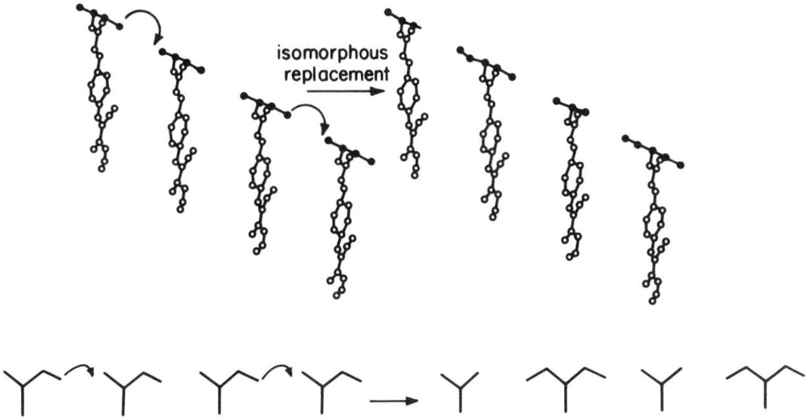

Scheme 13

Two other possible model motifs were envisaged in an analogous way. Figure 15 illustrates the close contacts between a *sec*-butyl groups residing at site 1, and two nearest-neighbor sites denoted as 2 and 3. A transfer of one methyl group from the *sec*-butyl by two different pathways may generate two possible isomorphous structures composed of nonchiral molecules. Path *a* (see Figure 15) generates compound **34**, where R_1 is isopropyl and R_2 *n*-propyl and path *b* generates **35** where R_1 is 3-pentyl and R_2 is methyl. Both compounds have been synthesized; although **34** packs in a non-chiral space group, **35** crystallizes in a chiral structure of space group $P2_1$, $Z = 2$, which, in spite of the difference in space group, is very close in structure to the parent material. Single crystals were prepared from the two systems; the 1:1 mixture of **32** and **33**, and of pure **35**. Compound **35** has a tendency to form perfect single crystals and indeed chiral dimers and oligomers with a quantitative enantiomeric yield were obtained, therefrom.

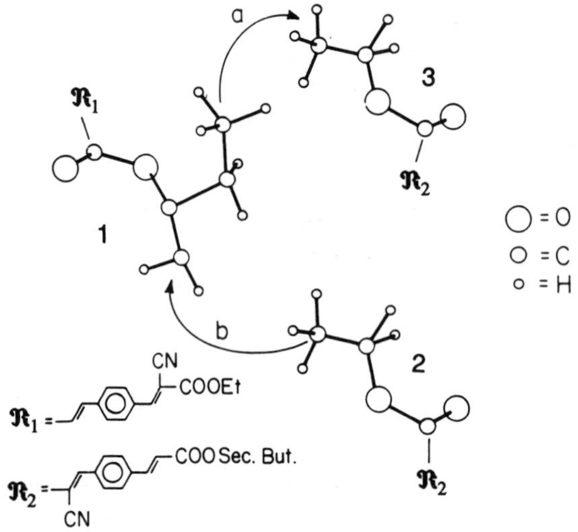

Figure. 15. Relative orientation of the *sec*-butyl group of the molecule *1* at site 1 with respect to the ethyl ester residues of the molecules at sites 2 and 3.

Finally, in these crystal engineering experiments advantage was taken of the tendency of molecules of racemates and resolved enantiomers bearing *sec*-butyl groups to crystallize in isomorphous structures. Indeed, crystals of racemic **1** are isomorphous with the crystals of the enantiomers so that the *sec*-butyl groups in the former are enantiomerically disordered but the polymerizing diene is arranged in the same fashion in both crystals. Unfortunately, it was not possible to grow homogeneous single crystals, since the system forms an eutectic mixture. The existence of this eutecticum implies that each crystal grown under conditions close to equilibrium is composed of domains of equal amounts of *d* and *l* crystallites. On the other hand, fast cooling provides a new metastable phase enriched with one of the enantiomorphs. Intermediate conditions of crystal growth of this phase yielded optically active products but with comparatively low optical yield. The structural reason for the generation of the immiscibility gap is due to diastereoisomeric interactions between the heterochiral *sec*-butyl groups in the chiral crystal. In order to remove these unfavorable interactions, the achiral isopropyl homologue was intercalated in racemic **1**. Introduction of 20% of isopropyl **32** already caused a substantial reduction of the eutectic range but it did not remove it completely.

Hasegawa et al.[27] reported another example of a $2\pi-2\pi$ asymmetric transformation in a chiral crystal. Ethyl 4-[2-(4-pyridyl)ethenyl] cinnamate crystallizes in a chiral space group $P2_12_12_1$ and upon irradiation yields a chiral dimer with 92% ee.

Recently, Scheffer et al. reported two elegant unimolecular "absolute" asymmetric transformations.[1e,28] This group demonstrated that the very well studied *di*-π-methane solution phase photorearrangement (Scheme 14) can also occur in the solid phase. The particular compounds investigated included dibenzobarrelene-11,12-diester derivatives (Scheme 15). The corresponding diisopropylester **36** is dimorphic and one of the forms grown from the melt is chiral (space group $P2_12_12_1$).

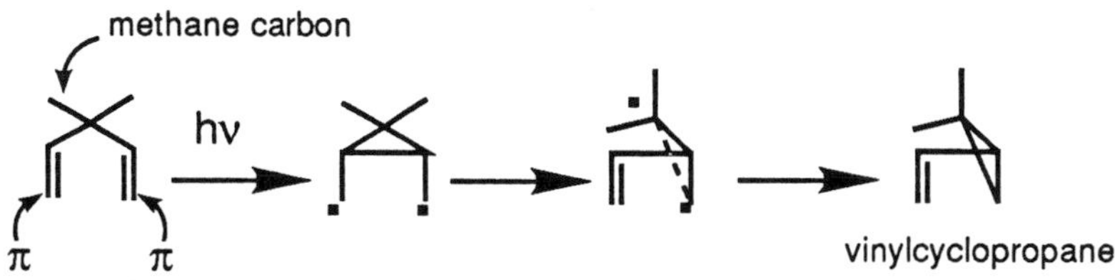

Scheme 14

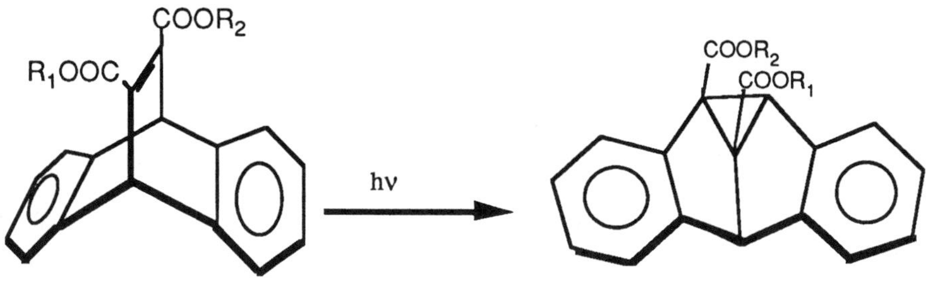

36 : $R_1 = R_2 = $ iPr
37 : $R_1 = $ iPr; $R_2 = $ *sec*-butyl

Scheme 15

Irradiation of single crystals resulted in the formation of semi-bullvalene derivatives with quantitative enantiomeric yield. The reaction may occur along four different pathways, as shown in Scheme 16. The absolute configuration of the reacting single crystal was assigned and correlated with that of the photoproduct formed in that single crystal. The correlation demonstrated that this photorearrangement proceeds only through paths 1 or 2 or a combination thereof.

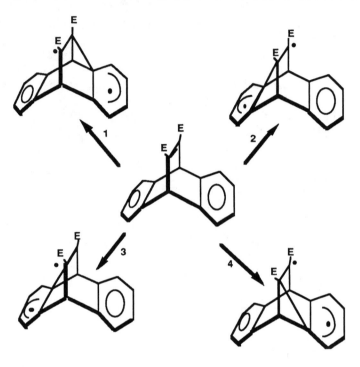

Scheme 16

Arguments supported by minimum conformational changes of the attached ester groups, and atom–atom potential calculations, are in agreement with path 2. An analysis of the pathway is presented in the chapter by Scheffer and Pokkuluri on unimolecular photorearrangements.[29] The racemic *sec*-butyl isopropyl ester **37** also crystallizes in a chiral space group and photo-rearranges to the corresponding chiral semi-bullvalene.

Two other unimolecular asymmetric transformations of the Norrish type II were reported. In the first system (Scheme 17), an adamantyl ketone derivative **38** crystallizes[28] in the chiral space group $P2_12_12_1$. Upon irradiation of single crystals of this ketone, an 1,4-biradical is formed that undergoes exclusive closure to a cyclobutanol derivative **39**. Deeper understanding of the reaction pathway of this system must await a crystallographic study akin to that carried out with the di-π-methane photorearrangement.

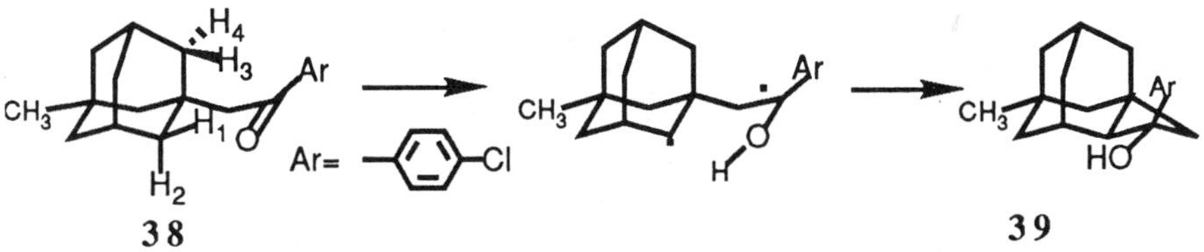

38   **39**

Scheme 17

In a second system, Toda et al.[30] demonstrated that N,N'-di-isopropylphenyl glyoxylamide, previously studied by Ayoma et al., crystallizes in a chiral space group $P2_12_12_1$. Irradiation of single homochiral crystals resulted in the formation of an optically active β-lactam (75%) with 93% ee. Using methods of seeding during recrystallization, both enantiomers could be obtained. Recently, the structure of the reacting crystal has been elucidated.[30b] According to the authors, there are two possible pathways, shown in Scheme 18. However, the results on the Norrish type II reactions inside the channels of DCA demonstrated that such simple stereochemical correlations might be misleading, since the acetyl group can undergo substantial deformation upon photoexcitation. Therefore, a Bijvoet analysis of the reacting crystal and correlation with the absolute configuration of β-lactam should provide an unambiguous answer to the possible pathway.

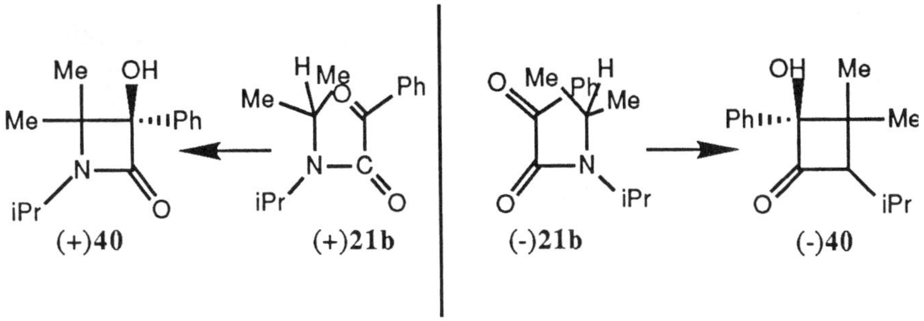

(+)40   (+)21b   (-)21b   (-)40

Scheme 18

An elegant use of reactivity in chiral crystals has recently been made by Feng and McBride[31] in the elucidation of the mechanism of hydrogen-atom transfer in dialkyl peroxide. Photolysis at a temperature of 100 K of 11-bromoundecanoyl peroxide 41a (spacegroup $P4_32_12$) yields a pair of 10-bromodecyl radicals (Scheme 19). One of these radicals abstract a hydrogen from the α-position of a neighboring intact molecule to form a secondary and primary pair of radicals. When α-hydrogens are replaced by deuterium **41b**, this abstraction is not observed. Instead, the two primary radicals abstract specific δ-hydrogens at 140 K to generate a different secondary-secondary radical pair. The esr studies suggest that only one equatorial hydrogen is abstracted in forming pair α. In order to differentiate between reactivity of the two hydrogens, di(13-bromo-2-deuteriotridecanoyl) peroxide **42**, was synthesized in an optically active form using a known chemical procedure. This material crystallizes in a chiral space group $P4_32_12$ or $P4_12_12$. The chiral peroxide, by virtue of H and D, has (almost) equal probability of crystallizing in one of the two enantiomorphs, since the crystal does not distinguish between protium and deuterium. The two hydrogens in solution are enantiotopic, but upon crystallization they assume different environments and thus become diastereotopic. When the chiral molecule crystallizes in space group $P4_32_12$, all protiums of the chiral center reside in the equatorial positions and are in a close proximity to the primary radical. When the molecule crystallizes in the "enantiomorphous" space group, hydrogen and deuterium atoms interchange their positions. In this enantiomorph, the deuterium is in the equatorial position and not transferrable to yield the α-pair. In this crystal, the product, as predicted, yields the formation of pair δ (Figure 16).

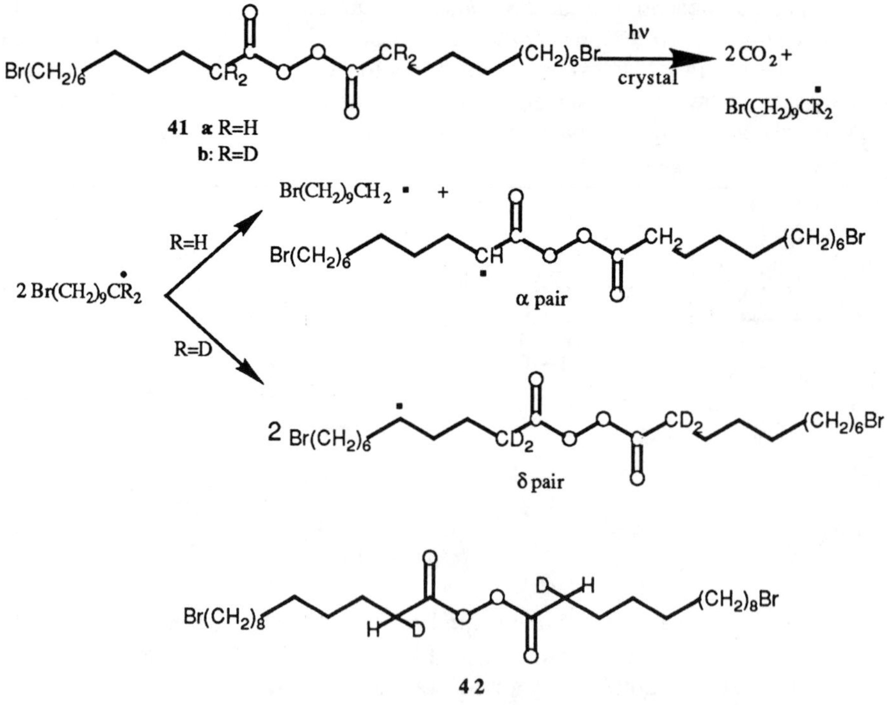

Scheme 19

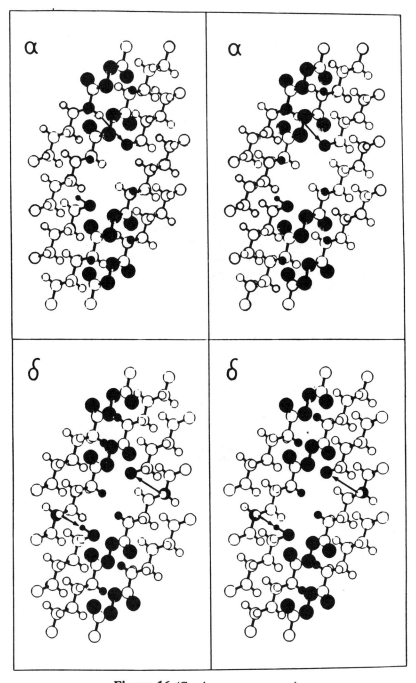

Figure 16 (Caption see next page)

Figure 16   Stereoscopic   plots   of   hydrogen   transfer   in   crystalline bromoundecanoyl peroxide viewed along (110), the twofold rotation axis of intact molecules; (α) space group $P4_32_12$; (δ) its enantiomorph, $P4_12_12$, related by inversion. Geminal hydrogens, classed as axial or equatorial with respect to the dyad, exchange roles between space groups. The central molecule has lost CO_2 and its radical carbons are shown without hydrogens as filled circles. The smaller filled circle indicates motion in the primary radical pair small shaded circles, their (pro-S)α-hydrogens. Arrows show the hydrogen transfers that are observed when the shaded hydrogens are replaced by deuterium. Carbons that donate hydrogen and become radical centers are shown by filled circles. In pair α the radical–radical vector is nearly vertical; in pair δ, nearly horizontal.

The success of the "absolute" asymmetric syntheses described in this chapter depends not only on the design of chiral crystals but also on the ability to prepare homochiral crystalline phases. The most commonly used method for the precipitation of the desired chiral phase is induction of crystallization by seeding. Another method described recently involves the use of "tailor-made" additives for the enantioselective inhibition of growth of the undesired enantiomorphs.[32] These chiral inhibitors resemble the substrate molecule and so can be adsorbed at specific faces during growth and retard crystal development. Since the inhibitor does not interact with the other enantiomorphs, its growth continues unperturbed (Scheme 20).

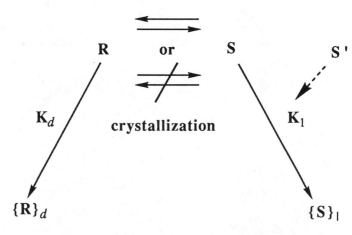

S' = stereochemically similar to S
{ }= crystalline phase

Scheme 20

4."Absolute" Asymmetric Synthesis in Nonchiral Crystals

The surface of a crystal generally exhibits symmetry lower than that of the bulk. Indeed, centrosymmetric crystals may be delineated by chiral faces. This property may be used for the performance of absolute asymmetric syntheses. Several years ago two groups[1a,33] proposed that an "absolute" asymmetric transformation might be performed not only in the medium of a chiral crystal, but also via the chiral surfaces of a centrosymmetric crystal. The first example illustrated here from work by Richardson and Holland,[33] does not involve a photochemical reaction but presents an important concept closely related to the subject of the present review. Tiglic acid **43** crystallizes in a centrosymmetric triclinic space group $P\bar{1}$. At the two faces of the crystal (Scheme 21) with Miller indices $(\bar{2}10)$ and $(2\bar{1}0)$, the molecules form coplanar hydrogen-bonded dimers oriented almost parallel to the face of the crystal. The exposed surfaces of the two molecules of a pair are essentially related by a twofold symmetry axis perpendicular to the crystal surface. Each molecule of tiglic acid has two enantiotopic faces labeled a and $\bar{a}$, a being exposed at the $(2\bar{1}0)$ face and $\bar{a}$ at the opposite $(\bar{2}10)$ face.

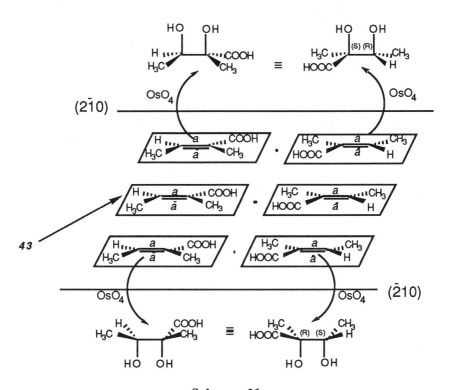

Scheme 21

It has been demonstrated that OsO_4 undergoes a heterogeneous reaction with crystalline tiglic acid to form a diol via a cis addition to the double bond. We expect the formation of a chiral resolved product at each of the two crystal faces provided the following criteria are met: the unreacted part of the crystal maintains its integrity and that the reaction occurs, layer by layer, from the exposed side of the same face and that the molecules thereon preserve their orientation during reaction. A large single crystal was glued at one of these faces and the other was exposed to the reagent. An asymmetric transformation with 16% ee was recorded. In centrosymmetric single crystals, the absolute orientation of the molecules in the crystal *vis-a-vis* the axes is fixed and therefore one can easily assign the absolute configuration of the chiral product formed in excess at each face. Thus one may anticipate an excess of enantiomer 2(S)–3(R) at the ($2\bar{1}0$) face and 2(R)–3(S) at the opposite face. More recently these authors have carried out experiments on asymmetric gas–solid bromination using other centrosymmetric crystals.[33c]

4.1. Reduction in Crystal Symmetry in Solid Solutions

We now describe another route by which the lower symmetry of a crystal surface may be used to perform an "absolute" asymmetric synthesis. It involves the loss of a center of inversion in a crystal when grown in the presence of specific additives. This work formed part of an overall study on the changes in crystal habit and composition during crystallization caused by the presence in solution of various concentrations (0.1–25%) of "tailor-made" additives with molecular structures similar to that of the host. The fraction eventually occluded (0.01–16%) depended on the ease with which an additive replaced a host molecule in the crystal. A stereochemical correlation was established between the molecular structure of the additive, the crystal structure of the host, and the faces affected on growth.[34] It could be inferred that the additive would be adsorbed only at the surface sites of those faces where the part of the adsorbate that differs from the host emerges from the crystal. In order to understand the reduction in crystal symmetry upon occlusion of a "tailor-made" additive, we shall present schematic examples before presenting the actual systems that were studied.

An important concept for the discussion that will follow is that crystal surfaces, although determined by the arrangement of the molecules within the crystal lattice, display surface structures that are different from each other and generally lower in symmetry than that of the bulk. For convenience we shall illustrate this aspect through an example of a schematic molecular arrangement in a motif of point group $2/m$ (the symbol $2/m$ specifies a twofold axis perpendicular to a mirror plane m, the combination of which generates a center of inversion).

The packing arrangement shown in Plate 1 exhibits space symmetry $P2_1/c$ (the c glide is a combination of the mirror symmetry and half a translation along the c axis; the twofold screw 2_1 along the b axis is a combination of a twofold rotation 2 and half a translation along b). In the arrangement shown, the *light blue* figures are related to the *pink* ones by twofold screw rotation, to the *green* by glide symmetry, and to the *yellow* by centers of inversion. Note the corresponding symmetry operations of the $2/m$ point group. The crystal is delineated by three different sets of symmetry-related faces, the two top and bottom faces, the four diagonal faces, and the two opposite side faces. We examine in turn the effect of additive on each of these sets. One can see easily that an additive molecule in *dark blue*, bearing an appropriately modified group, will be able to substitute for a substrate molecule at the top surface at the *light blue* and *pink* sites, but not at *green* and *yellow* sites because only at the former sites the modified group does not disturb the regular pattern of interactions at the crystal surface. Conversely, at the bottom face the additive can be adsorbed only at the *green* and *yellow* sites. Thus upon eventual occlusion in the crystal, the additives can occupy the *blue* and *pink* sites at the top half and the *green* and *yellow* sites at the bottom half (Plate 2). The additive molecules in the top and bottom halves of the crystal are related to each other by twofold screw symmetry but not by center of inversion or glide symmetry so there is an overall reduction in crystal symmetry in each half. Occlusion of additive through the four diagonal faces, as shown in Plate 3, will lead to a crystal composed of four sectors. The symmetry in each sector will be P1 since the occluded additive molecules therein are related by translation symmetry only. The effect of adsorption of additive on the two side faces would be a reduction in symmetry analogous to that depicted in Plate 2. However, instead of the loss of a center of inversion and glide symmetry, adsorption on the two side faces would lead to a loss of the twofold screw and center of inversion.

Consequently the additive molecule will be anisotropically distributed within the grown crystal, preferentially occluded through different subsets at surface sites on the various faces, leading to a mixed crystal composed of sectors coherently intertwined. Occlusion of the additive will thus lead to a reduction of the crystal symmetry to the symmetry of the surfaces through which it was adsorbed. This principle holds for each whole crystal sector, although the additive may occupy only a small fraction of all the unit cells. However it will not hold for the whole crystal, because the added effects in the different sectors would average out. In more exact terms the different crystal sectors will be related to each other by the symmetry elements of the original point group. This is in keeping with *Curie's law*, which states that the *overall* symmetry of a system cannot be reduced unless outside asymmetric forces have been applied to the system. Experimentally, reduction in crystal symmetry following the above principles was deduced in several host–guest systems by the techniques of optical second harmonic generation,[35] optical birefringence,[36] and "absolute" asymmetric photo-dimerization.[37] It was directly observed in a number of systems by x-ray and neutron diffraction.[38] We describe here some representative examples.

The α-form of glycine,[39] a prochiral molecule, packs in an arrangement of point symmetry $2/m$. The two crystal faces relevant to the discussion are analogous to the top and bottom faces as shown in Plate 2. They are of the type {010} shown in Figure 17.

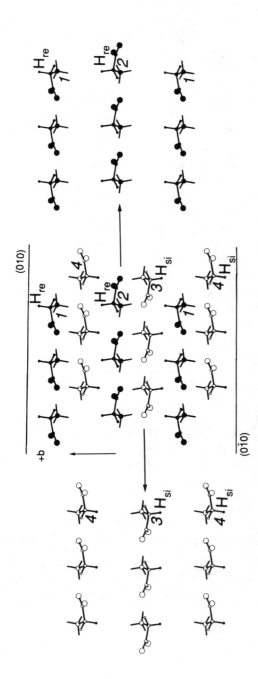

Figure. 17. Packing arrangement of α-glycine viewed along the *a* axis. The four symmetry-related molecules are labelled (*1-4*). The glycine moledules adopt a "chiral" conformation in the crystal. The crystal consists of chiral layers; Layers *1* and *2* are homochiral composed of say "d" molecules and "enantiomeric" to layers *3* and *4*, composed of "l" molecules. The crystal may be regarded as composed of two sublattices, each comprising "homochiral" molecules. The C-H$_{re}$ bonds of all "d"-glycine molecules point towards the (010) face and the C-H$_{si}$ bonds of the " l"-glycine molecules to the (0$\bar{1}$0) face.

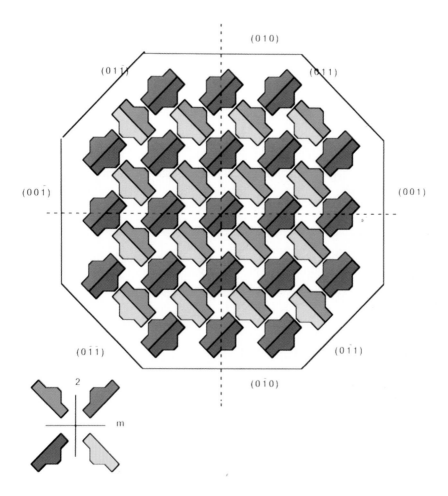

Plate 1. Schematic packing arrangement viewed along the *a* axis exhibiting space symmetry $P2_1/c$, belonging to pointgroup *2/m* shown at bottom left. The light blue figures are related to the pink ones by twofold screw rotation, to the green ones by glide symmetry, and to the yellow ones by centers of inversion. The "crystal" is delineated by three different sets of symmetry-related faces of types {010}, {011}, and {001}.

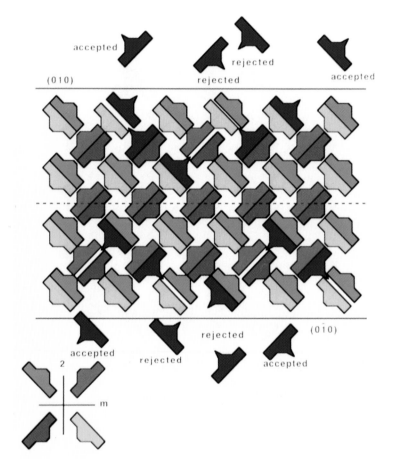

Plate 2. Schematic representation of adsorption and occlusion of dark blue additives through the top and bottom {010} faces in a crystal as shown in Plate 1. At the top (010) face the "additive" can replace the light blue and pink figures and at the bottom ($0\bar{1}0$) face the green and yellow figures, resulting in loss of the glide and inversion symmetry. The crystal symmetry as a consequence is reduced from $P2_1/c$ to $P2_1$ in the two crystal halves.

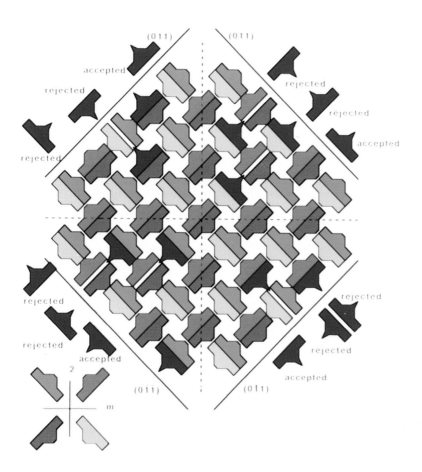

Plate 3. Schematic representation of adsorption and occlusion of dark blue additives through the four {011} slanted faces in the crystal depicted in Plate 1. In each of the four sectors only one of the four symmetry related figures can be replaced by the additive, which means loss of all symmetry elements. The crystal symmetry is reduced from $P2_1/c$ to P1.

Of the four symmetry-related molecules (*1, 2, 3, 4*), *1* and *2* are related by twofold screw symmetry and have their C—H$_{re}$ bonds emerging from the (010) face. By symmetry, molecules 3 and 4 are related to 1 and 2 by a center of inversion and have their C—H$_{si}$ bonds emerging from the (0$\bar{1}$0) face (Fig.17). Consequently, only R-α-amino acids can substitute glycine molecules at the *1* and *2* surface sites, and then only on face (010), while only S-α-amino acids can be adsorbed at sites *3* and *4* on face (0$\bar{1}$0). S-α-Amino acid additives induce the formation of pyramids with an (0$\bar{1}$0) basal plane. R-α-Amino acid additives induce the enantiomorphous morphology. Racemic additives cause the formation of {010} plates. These plates were found to contain 0.02–0.2% racemic additive occluded inside the crystal bulk with the two enantiomers totally segregated in the two crystal sectors at the +b and -b halves. As expected, the R enantiomers populate the +b and the S enantiomers the -b half of the crystal.[40] In terms of the argument given above, the crystal symmetry of each half of the crystal must be reduced from P2$_1$/n to P2$_1$, and the two sectors are enantiomorphous, as in Plate 2.

For second harmonic generation (SHG) to be active in a crystal, the material must be acentric. Thus SHG is an excellent diagnostic tool for detecting the loss of a crystallographic center of inversion. One requirement is that either the guest or the host molecules have large molecular hyperpolarizability tensors β, leading to large optical nonlinearity. Crystals of many promising molecules have proved unsuitable because they form centrosymmetric (i.e., anti-parallel) pairs. A site-selective replacement of one of the pair by a nonhyperpolarizable molecule would result in a crystal active to SHG. For the system of α-glycine we made use of guest amino acids which have high β-coefficients,[35] such as *p*-nitrophenyl derivatives of lysine and ornithine **44–47**. Another interesting demonstration is the example of the centrosymmetric host crystal of *p*-(*N*-dimethylamino) benzylidene-*p*-nitroaniline (**48**), which became acentric and SHG active upon site-selective occlusion of the guest molecule *p,p*-dinitrobenzylidene aniline **49**. In this system the host and guest molecules have, respectively ,large and negligible β coefficients.

McBride and Bertman have recently used the method of optical birefringence to demonstrate a reduction in crystal class.[36] Crystals that belong to a tetragonal, trigonal, or hexagonal system are optically uniaxial, whereas those that belong to a lower class, triclinic, monoclinic, or orthorhombic, are optically biaxial. The reduction of symmetry was illustrated for the crystal of di(11-bromoundecanol) peroxide (**50**), denoted Br···Br in the presence of guest Br···CH$_3$ (**51**), where a Br atom is replaced by a CH$_3$ group. The pure platelike crystals are not birefringent when viewed along the uniaxial tetragonal axis. Crystals of Br···Br containing 15% of Br···CH$_3$ proved to be birefringent, indicating unsymmetrical incorporation of the Br···CH$_3$ additives during crystal growth.

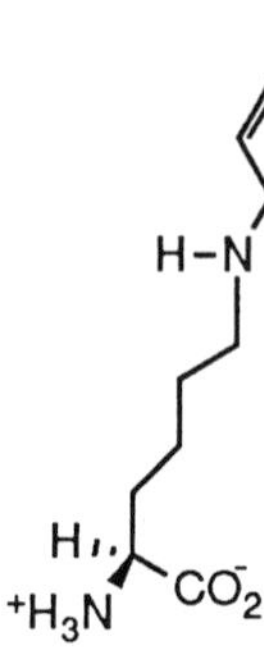

44 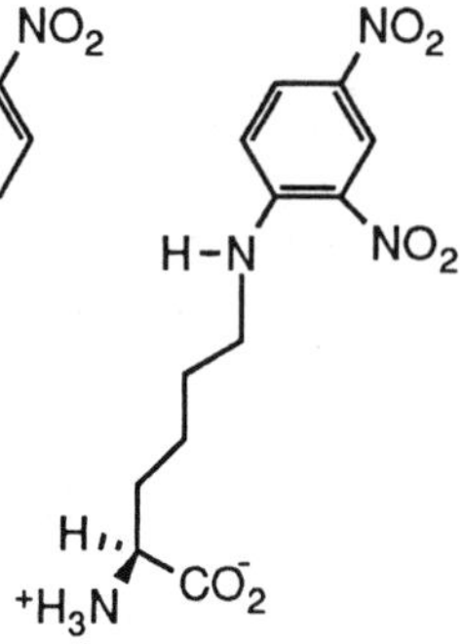 **45** 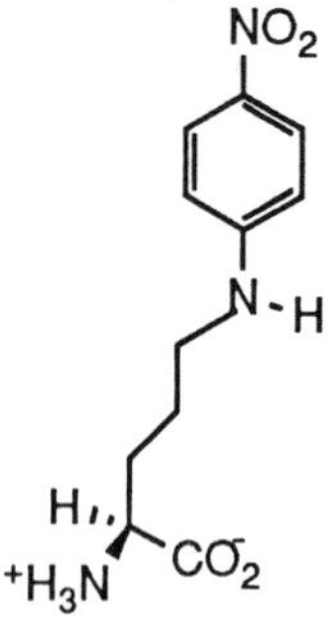 **46** 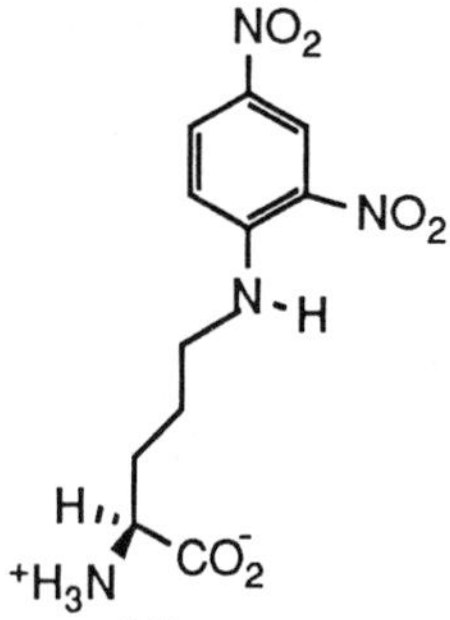 **47**

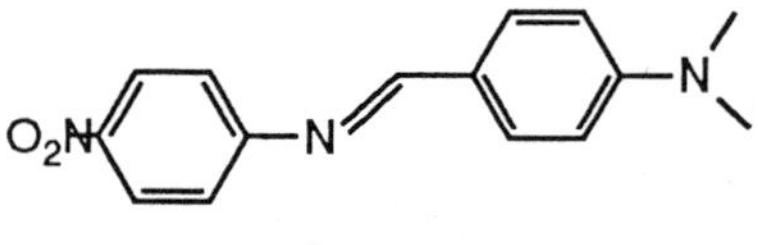

48 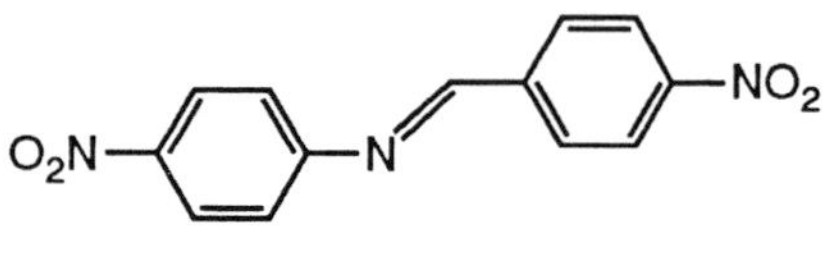 **49**

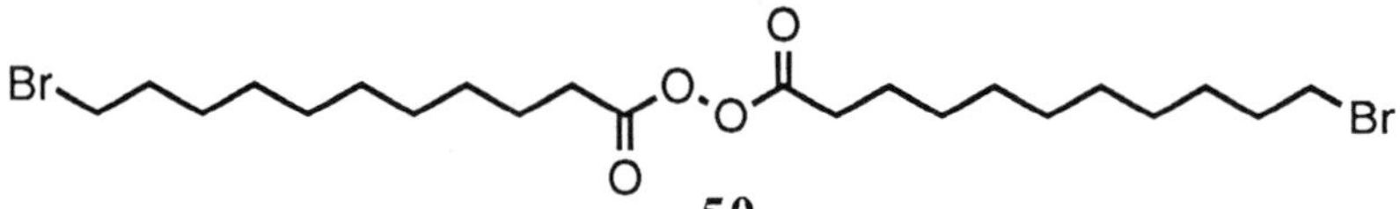

50

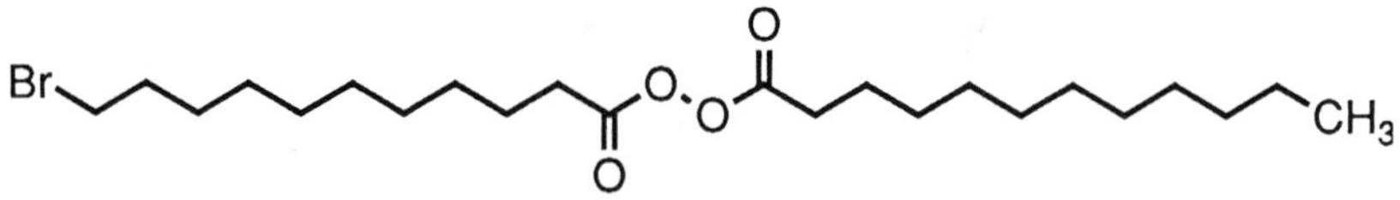

51

4.2. "Absolute" Asymmetric Synthesis in "Centrosymmetric" Crystals

In systems dealt with till now, a gross modification was introduced in the additive molecule. The question arose as to the minimal modification that would still be recognized and discriminated for by the growing crystal surface leading to reduction in symmetry. Indeed, solid solutions composed of host and additive molecules of similar structure and shape have been generally observed, or assumed, to exhibit the same symmetry as that of the host crystals.

We first discuss the case of solid solutions of carboxylic acids (XCO_2H) in primary amides ($XCONH_2$), where an NH_2 group is substituted by an OH moiety. In all the systems studied, the morphology of the host amide crystal undergoes a pronounced change when grown in the presence of the corresponding carboxylic acid.[37b,41] A strong inhibition of growth is invariably developed along the direction of the O=C—N—H_a···O=C hydrogen bond, H_a being the amide hydrogen atom in antiperiplanar conformation to the carbonyl (Scheme 22). Inhibition arises from repulsive O(hydroxyl)···O(carbonyl) interactions between the lone-pair electrons of an adsorbed carboxylic acid molecule and of an amide molecule at the site of an original N—H···O hydrogen bond (Scheme 22). The repulsive nature of such O···O interactions when the two oxygens are forced to be at a distance of 2.9–3.0 Å has been crystallographically confirmed by the hydrogen-bond arrangements of several monoacid/monoamide crystals[37b,41]; the repulsion is evaluated to be about 2 kcal mol^{-1} for a separation distance of 3.0 Å. This value may be compared with an attractive energy of -6 kcal mol^{-1} for the N—H···O bond.

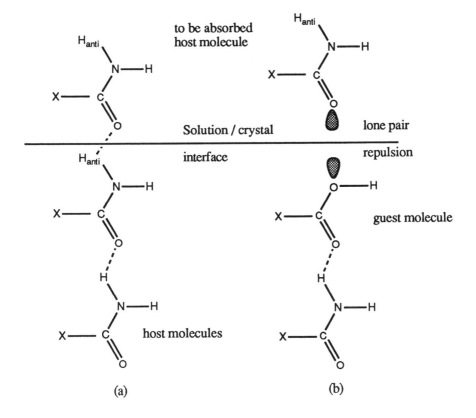

Scheme 22

The carboxylic acid additive molecule will thus be preferentially adsorbed at a surface site where the OH will emerge from the surface, rather than in the opposite orientation.

Reduction in crystal symmetry has been deduced in two amide–carboxylic acid systems. In (*E*)-cinnamamide–*E*-cinnamic acid, small amounts (<2%) of guest were occluded into the host crystal. (*E*)-cinnamamide **52**, packs in space group P2$_1$/c with four symmetry-related molecules, yielding prismatic crystals elongated in b (Figure 18). The basic structural motif consists of centrosymmetric hydrogen-bonded dimers interlinked by N—H$_a$···O bonds along b by translation to form ribbons. These ribbons make herringbone contact via twofold screw axes and plane-to-plane contact across C=C double bonds via centers of inversion along the a axis (Figure 19). Molecules of (*E*)-cinnamamide assume a chiral conformationin in the crystal. Thus each ribbon may be considered as composed of two "homochiral" stacks related to each other by a center of inversion. These stacks are given the same label (*1, 2, 3, 4*) as the corresponding symmetry-related molecule in its stack. At the surfaces at the +b end the N—H$_a$ bond points outwards for stacks *2* and *3* whereas for stacks *1* and *4* the corresponding N—H$_a$ bond points into the crystal. The opposite is true at the -b end.

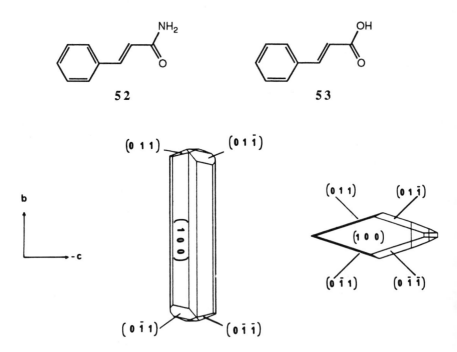

Figure 18. Morphology of cinnamamide: (left) pure; (right) grown in the presence of cinnamic acid.

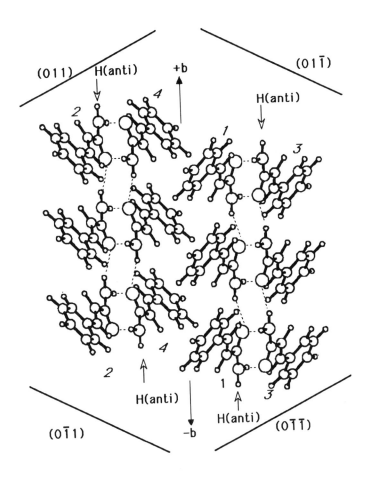

Figure 19. Packing arrangement of (E)-cinnamamide. View along the a axis. The {011} faces are shown and the four symmetry-related sites (*1–4*) are denoted.

Molecules of (*E*)-cinnamic acid **53**, will be attached preferentially from the +b side of the crystal at the end of stacks *2* and *3*, since the oxygen lone pair of the carboxyl group of the acid will bind to the N—H bond of the amide emerging at that side. Binding to the *1* and *4* type stacks from the same side would result in oxygen–oxygen repulsion. By symmetry, (*E*)-cinnamic acid is expected to bind and subsequently be occluded into the *1* and *4* sites from the -b side. Thus in keeping with Plate 2, the mixed crystal of cinnamamide-cinnamic acid is composed of two enantiomorphous halves each of symmetry $P2_1$. It is well documented that (*E*)-cinnamamide photodimerizes in the solid phase to yield the centrosymmetric dimer truxillamide. Such a reaction takes place between close-packed amide molecules of two "enantiomeric" stacks, *1* and *3* or *2* and *4* (Figure 21). It has also been established that solid solutions yield upon photodimerization the chiral mixed dimers with absolute configuration at the cyclobutyl carbons of (RRRR) (**54a**) and (SSSS) (**54b**) apart from α-truxillamide homodimers. Consequently, we expected the chiral mixed dimers (RRRR) and (SSSS) to form at the +b and -b sectors of the crystal, respectively (Figure 20). Large mixed crystals of (*E*)-cinnamamide and (*E*)-cinnamic acid were grown and the material removed from the opposite b poles and irradiated. The enantiomeric excess of the dimers at the two poles of the crystal was about 60%, depending upon the temperature at which the crystal was grown (Figure 22).

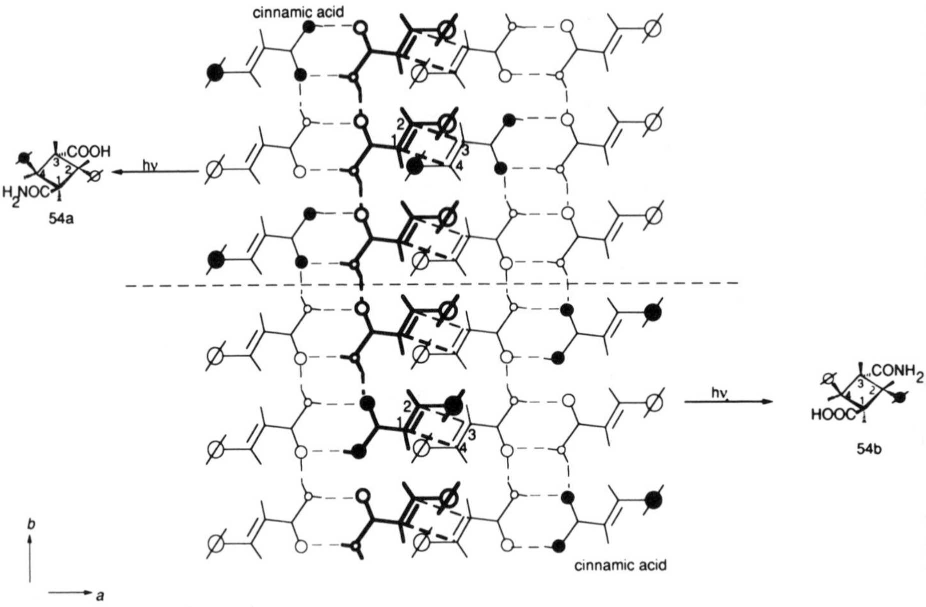

Figure 20. Packing arrangement of cinnamamide (open circles) containing occluded cinnamic acid (full circles). View along the c axis. The molecules are related by translation along b and a. Dimerization takes place between the molecules from the bold printed stack and those of the underlying "centrosymmetrically" related stack.

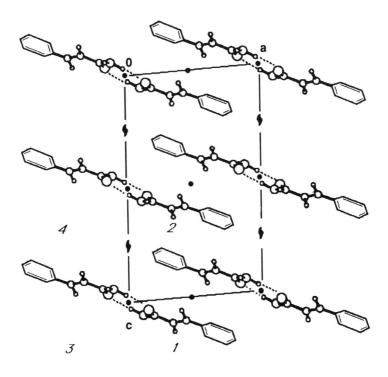

Figure 21. Packing arrangement of (*E*)-cinnamamide viewed along the b axis. The four symmetry related molecules are shown.

In the amide–acid system described above, the calculated energy difference in binding the guest molecule to a host site differs by as much as 8 kcal mol^{-1}, compared to binding a host molecule to the same site. We can now ask the question whether a reduction in crystal symmetry will occur also in systems where the difference in binding energy between the two molecules at the surface will be even smaller than that of an amide–acid, for example, in the solid solution between molecules bearing phenyl and thienyl groups. It is well documented that a phenyl ring can be easily replaced by a thienyl, whose atoms do not directly coincide with the atoms of the phenyl. A further advantage of the thienyl ring is that it contains sulfur, which is a relatively heavy x-ray scatterer.

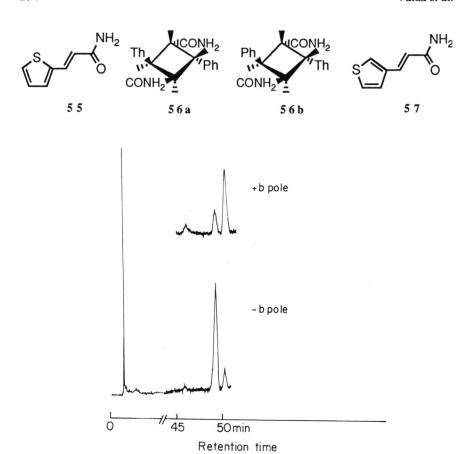

Figure 22. Gas chromatograms of the *O-i*-PR-*N*-TFA derivatives of compounds **54a** and **54b** isolated from the +b and -b poles of the crystal. (Chirasil–L-valine capillary column.)

When a crystal of (*E*)-cinnamamide (**52**) is grown in the presence of (*E*)-2-thienylacrylamide (**55**) the morphology of the crystal changes. The four {011} faces are more expressed, implying that the thienyl molecules are occluded primarily through these faces. From the space group of the crystal it follows that the four {011} faces are chiral. By virtue of the point symmetry 2/*m* the two faces (011) and (01$\bar{1}$) are homochiral and congruent and enantiomeric to the (0$\bar{1}$1) and (0$\bar{1}\bar{1}$) faces. Each of these faces exposed four independent surface sites, corresponding to the four symmetry-related sites in the bulk, as shown in Figure 19. This structural property has already been schematically depicted in Plate 1. Cinnamamide molecules at these faces display a herringbone arrangement. These contacts involve interactions between the aromatic C—H groups and the π electrons of the neighboring phenyl ring. To effect site-selective adsorption and occlusion, use was made of replacement of the herringbone contact involving C—H groups and π-electron clouds of neighboring phenyl rings by unfavorable contacts between sulfur lone-pair electrons and the π electron system (Figure 23). Consequently, as shown in Figure 24, 2-thienylacrylamide should easily be adsorbed at site *1* on face (0$\bar{1}$1), as the sulfur atom would emerge from the crystal surface. The guest cannot as easily occupy site *3* at which the sulfur would point into the crystal bulk. At sites *2* and *4*, the thienyl rings of a guest would lie almost parallel to the face and their relative ease of adsorption would depend only upon the tilt of the ring with respect to the face.

(a)   **(b)**

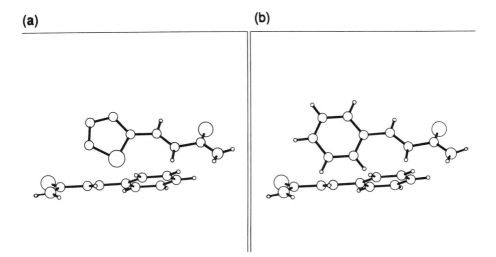

Figure 23. Herringbone contacts between (a) an absorbed 2-thienylacrylamide molecule and a cinnamamide molecule; (b) two cinnamamide molecules.

Any inequality in the average guest occupancy will result in reduction of symmetry; preferential occlusion at site *1* yielding space group P1, analogous to the representation in Plate 3. Adsorption of thienylacrylamide takes place also through faces of type {001}, as revealed by changes in morphology. A selective occlusion through faces of this type should lead to symmetry P*c*, given that the guest should be adsorbed more easily at sites *1, 2* than at *3, 4*. The mixed crystal should be thus divided in six sectors of reduced symmetry (Figure 25), with the structure of the sectors related to each other by the 2/*m* point symmetry of the host crystal: the two joined sectors of symmetry P1 of type A at one half of the crystal will be congruent and enantiomorphous to the two joined sectors of type $\bar{A}$ at the opposite half. The two P*c* sectors (type B), should have opposite polarity along the c axis.

Vaida et al.

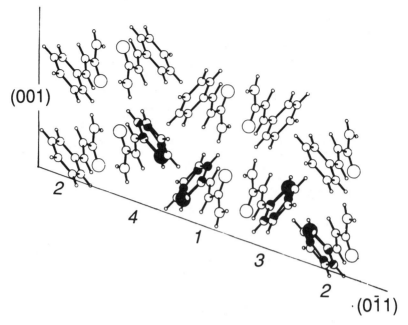

Figure 24. Packing arrangement of cinnamamide showing the four different surface sites at the $(0\bar{1}1)$ face. The filled atoms are those of 2-thienyl rings in the positions they would assume were they to replace cinnamamide molecules.

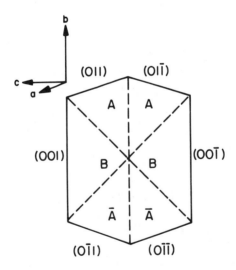

Figure 25. Morphological representation of a cinnamamide crystal with sectors of reduced symmetry.

Previously, it was demonstrated that irradiation with monochromatic light (λ > 350 nm) of such a solid solution grown from the melt led to the formation of racemic mixtures of the heterodimers.[42] In this mixed crystal there is no energy transfer from the host to the guest so that the photoproduct distribution reflects directly the molecular occupancy of thienylacrylamide inside the cinnamamide crystal.[43] Consequently, on the basis of the photochemical arrangements presented, we anticipated that irradiation of sectors A and $\overline{A}$ should yield optically active dimers **56a** and **56b** of opposite chirality.

In the $\overline{A}$ sector delineated by the ($0\overline{1}1$) face, thienylacrylamide molecules at the preferred sites 1 and 4 will photoreact with cinnamamide molecules at the two sites 3 and 2, respectively, to yield the chiral mixed dimer. This expectation was confirmed experimentally by circular dichroism and gas chromatography on a chiral column; the photodimers obtained from A and $\overline{A}$ type sectors show that the respective photoproducts are optically active with an ee in the range of 40–69% (Figure 26).[37a] As expected, irradiation of sector B of the crystal yields a racemic mixture of both enantiomers. Recently the absolute configuration of the two dimers obtained from sectors A and $\overline{A}$ was assigned by independent nmr studies and found to be in agreement with the present proposed mechanism. The reduction in symmetry of each given sector was directly demonstrated by x-ray and neutron diffraction. The proposed reduction in symmetry of sector B could not be demonstrated by an x-ray diffraction analysis. Yet a reduction in symmetry to Pc was shown from the analogous system of 3-thienylacrylamide (**57**) and cinnamamide: when a crystal of cinnamamide was grown with this additive a thin plate elongated in the b direction was observed. About 15% of additive was occluded inside the crystal of the host. According to morphological results, most of the guest molecules were occluded through the {001} faces. Therefore we expected reduction of the symmetry to a nonchiral Pc crystal, which was unambiguously confirmed by an x-ray analysis.[44]

The principle of reduction of symmetry is general and should be applicable to a large variety of crystals. Understanding the detailed mechanism at a molecular level allows one to plan the design of the new chiral phases provided the structure and morphology of the host crystals are known.

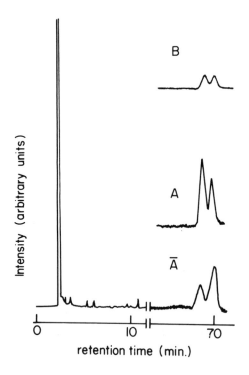

Figure 26. Gas chromatogram of the *O*-*i*Pr-*N*-TFA derivatives of compounds **56a** and **56b** of segments A, Ā, and B. The thiophene ring was reduced to an *n*-butyl group. (Chirasil–L–valine capillary column.)

5. Concluding Remarks

Research on reactivity in the crystalline state has been extended in recent years to a variety of new systems such that it can now be regarded as an important branch of organic chemistry. In the present review we have limited ourselves to two aspects that are unique to the crystalline state.

First, several new spontaneous asymmetric syntheses with very high enantiomeric yields were discovered. Furthermore centrosymmetric single crystals can be "transformed" in chiral solid solutions, thus greatly increasing the repertoire of matrices appropriate for such asymmetric transformations. The understanding of the detailed mechanism of the reduction in symmetry should permit the planning of new systems, incorporating molecules that may have played an important role in the early stages of evolution. Thus spontaneous generation of chiral materials on earth can be regarded as a probable event.

A second aspect covered is the possibility to monitor, albeit within limits, the progress of a chemical reaction in those crystals which preserve their integrity during the reaction process. The ability to determine the atomic coordinates of the reacting molecule before, during and after reaction by diffraction techniques provides a valuable tool to probe interactions which dictate the reaction pathway and the nature of the product. At the present time these analyses have been applied to a small number of systems because they are laborious and time consuming. With the advent of powerful synchrotron irradiation, which provides tunable x-ray sources, it should be possible to apply this methodology to more complex systems, such as transformations at the active sites of enzymes in biological crystals. Finally with such tools in hand it should become possible also in the not too distant future to monitor the structure of photoexcited molecules en route to product formation.[25]

Acknowledgments

We thank our colleagues Prof. L. Addadi, and Dr. I. Weissbuch for fruitful discussions. We also thank Prof. J.M. McBride for advice. We are grateful to the U.S. / Israel Binational Foundation, Jerusalem for financial support.

References

1. (a) B. S. Green, M. Lahav and D. Rabinovich, *Acc. Chem. Res.*, **1979**, *69*, 191.
    (b) L. Addadi, S. Ariel, M. Lahav, L. Leiserowitz, R. Popovitz-Biro and C. P. Tang in *Chemical Physics of Solids and Their Surfaces*, Specialist Periodical Reports, Vol. 8, M. W. Roberts and J. M. Thomas, eds., The Royal Society of Chemistry, London, 1980, Ch. 7.
    (c) V. Ramamurthy and K. Venkatesan, *Chem. Rev.*, **1987**, *87*, 433.
    (d) V. Ramamurthy, *Tetrahedron*, **1986**, *42*, 573.
    (e) J. R.Scheffer and M. Garcia-Garibay in *Photochemistry on Solid Surfaces*, Studies in Surface Science and Catalysis, Vol. 47, M. Anpo, T. Matsuura, eds., Elsevier, Amsterdam, 1989, p. 501.
    (f) *Organic Solid State Chemistry*, G. R. Desiraju, ed., Elsevier, Amsterdam, 1987.
    (g) *Organic Chemistry in Anisotropic Media*, Tetrahedron Symposia-in-Print, No. 29, V. Ramamurthy, J. R. Scheffer and N. J. Turro, eds., *Tetrahedron*, **1987**, *43*, issue 7.
    (h) B. S. Green, R. Arad-Yellin and M. D. Cohen in *Topics in Stereochemistry*, **1986**, *16*, 131.
    (i) Proceedings of various international symposia on organic solid-state chemistry, some of which have been published in Molecular Crystals and Liquid Crystals.
    (j) M. D. Hollingsworth and J. M. McBride, *Advances in Photochemistry*, **1990**, *15*, 279.
2. (a) M. Farina, G. Audisio, G. Natta, *J. Am. Chem. Soc.*, **1967**, *89*, 5071.
    (b) M. Farina, in *Inclusion compounds*, Vol. 2, J. L. Atwood, J. E. D. Davies, D. D. McNicol, eds., Academic, London, 1984, p. 69.
    (c) M. Farina in *Inclusion compounds*, Vol. 3, J. L. Atwood, J. E. D. Davies, D. D. McNicol, eds., Academic, London, 1984, p. 297.
    (d) M. Farina in *Topics in Stereochemistry*, **1987**, *17*, 1.
3. K. Penzien and G. M. J. Schmidt, *Angew Chem. Int. Ed. Engl.*, **1969**, *8*, 608.
4. For details, see *International Tables for Crystallography*, Vol. A, D. Reidel, Dordrecht, Holland, 1985.
5. L. Addadi and M. Lahav, *J. Am. Chem. Soc.*, **1978**, *100*, 2838.

6.   L. Addadi and M. Lahav, *J. Am. Chem. Soc.*, **1979**, *101*, 2152.
7.   L. Addadi and M. Lahav, *Pure Appl. Chem.*, **1979**, *51*, 1269.
8.   J. van Mil, L. Addadi , M. Lahav and L. Leiserowitz, *J. Chem. Soc., Chem. Commun.*, **1982**, 584.
9.   (a) V. Petraccone, P. Ganis, P. Corradini and G. Montagnoli, *Eur. Polym. J.*, **1972**, *8*, 99.
     (b) E. Benedetti, C. Pedone and A. Sirigu, *Acta Crystallogr.*, **1972**, *B 29*, 730.
10. S. Ariel. Ph.D. Thesis, The Feinberg Graduate School, The Weizmann Institute of Science, Rehovot, Israel, 1981.
11. M. Garcia-Garibay, J. R. Scheffer, J. Trotter and F. Wireko, *Tetrahedron Lett.*, **1988**, *29*, 2041.
12. (a) G. Audisio and A. Silvani, *J. Chem. Soc., Chem. Commun.*, **1976**, 481.
     (b) G. Audisio, A. Silvani and L. Zetta, *Macromolecules*, **1984**, *17*, 29.
13. M. Miyata, Y. Kitahara and K. Takemoto, *Polym. Bull.*, **1980**, *2*, 671.
14. (a) M. Miyata, Y. Kitahara and K. Takamoto, *Polym. J.*, **1981**, *13*, 111.
     (b) M. Miyata and K. Takemoto, *Polym. J.*, **1977**, *9*, 111.
15. B. M. Craven and G. T. DeTitta, *J. Chem. Soc., Chem. Commun.*, **1972**, 530.
16. S. C. De Sanctis, E. Giglio, V. Pavel and C. Quagliata, *Acta Crystallogr.*, **1972**, *B28*, 3656.
17. (a) R. Popovitz-Biro, H. C. Chang, C. P. Tang, N. Shochet, M. Lahav and L. Leiserowitz, *Pure Appl. Chem.*, **1980**, *52*, 2693.
     (b) R. Popovitz-Biro, H. C. Chang, C. P. Tang, N. Shochet, M. Lahav and L. Leiserowitz, in *Chemical Approaches to Understanding Enzyme Catalysis: Biomimetic Chemistry and Transition-State Analogues*, B. S. Green, Y. Ashani, D. Chipman, eds., Elsevier, Amsterdam, 1982, p. 88.
     (c) H. C. Chang, R. Popovitz-Biro, M. Lahav and L. Leiserowitz, *J. Am. Chem. Soc.*, **1982**, *104*, 614 .
     (d) R. Popovitz-Biro, C. P. Tang, H. C. Chang, M. Lahav and L. Leiserowitz, *J. Am. Chem. Soc.*, **1985**, *107*, 4043.
     (e) C. P. Tang, H. C. Chang, R. Popovitz-Biro, F. Frolow, M. Lahav, L. Leiserowitz and R. K. McMullan, *J. Am. Chem. Soc.*, **1985**, *107*, 4058.
     (f) H. C. Chang, R. Popovitz-Biro, M. Lahav and L. Leiserowitz, *J. Am. Chem. Soc.*, **1987**, *109*, 3883.
     (g) Y. Weisinger-Lewin, M. Vaida, R. Popovitz-Biro, H. C. Chang, F. Manning, F. Frolow, M. Lahav and L. Leiserowitz, *Tetrahedron*, **1987**, *43*, 1449.
18. P. J. Wagner, *Topic Curr. Chem.*, **1976**, *66*, 1.
19. P. J. Wagner in *Rearrangements in Ground and Excited States*, Vol. 3, P. de Mayo, ed., Academic Press, New York, 1980, p. 381.
20. H. Aoyama, K. Miyazaki, M. Sakamoto and Y. Omote, *Tetrahedron*, **1987**, *43*, 1513.
21. (a) F. Toda and K. Tanaka, *J. Chem. Soc., Chem. Comm.*, **1986**, 1429.
     (b) F. Toda and K. Tanaka, *Chem. Lett.*, **1987**, 2283.
     (c) F. Toda and K. Tanaka, *Tetrahedron Lett.*, **1988**, *29*, 4299.
     (d) F. Toda, *Mol. Cryst. Liq. Cryst.*, **1988**, *161*, 355.
     (e) F. Toda, *Top. Curr. Chem.*, **1988**, *149*, 211.
     (f) F. Toda, K. Tanaka and T. C. W. Mak, *Chem. Lett.*, **1989**, 1329.
     (g) F. Toda, K. Tanaka and M. Yagi, *Tetrahedron*, **1987**, *43*, 1495.
22. M. Kaftory, *Tetrahedron*, **1987**, *43*, 1503.
23. B. S. Green, M. Lahav and G. M. J. Schmidt, *Mol. Cryst. Liq. Cryst.*, **1975**, *29*, 187.
24. (a) A. Elgavi, B. S. Green and G. M. J. Schmidt, *J. Am. Chem. Soc.*, **1973**, *95*, 2058.
     (b) D. Rabinovich and Z. Shakked, *Acta Crystallogr.*, **1975**, *B 31*, 819.
25. Z. Ludmer, G.E. Berkovic and L. Zeiri, *Tetrahedron*, **1987**, *43*, 1579.

26. A. Warshel and Z. Shakked, *J. Am. Chem. Soc.*, **1975**, *97*, 5679.
27. M. Hasegawa, C. M. Chung, N. Murro and M. Maekawa, *J. Am. Chem. Soc.*, **1990**, *112*, 5676.
28. (a) S. V. Evans, M. Garcia-Garibay, M. Omkaram, J. R. Scheffer, J. Trotter and F. Wireko, *J. Am. Chem. Soc.*, **1986**, *108*, 5648.
   (b) J. R. Scheffer and J. Trotter, *Mol. Cryst. Liq. Cryst. inc. Nonlin Opt.*, **1988**, *156*, 63.
29. J. R. Scheffer and P. R. Pokkuluri, Chapter 5 of this volume and references therein.
30. (a) F. Toda and K. Mori, *J. Chem. Soc., Chem. Commun.*, **1989**, 1245.
   (b) A. Sekine, K. Mori, Y. Ohashi, M. Yagi and F. Toda, *J. Am. Chem. Soc.*, **1989**, *111*, 697.
31. X. W. Feng and J. M. McBride, *J. Am. Chem. Soc.*, **1990**, *112* , 6151.
32. (a) J. van Mil, E. Gati, L. Addadi and M. Lahav, *J. Am. Chem. Soc.*, **1981**, *103*, 1248.
   (b) J. van Mil, L. Addadi, E. Gati and M. Lahav, *J. Am. Chem. Soc.*, **1982**, *104*, 3429.
   (c) L. Addadi, S. Weinstein, E. Gati, I. Weissbuch and M. Lahav, *J. Am. Chem. Soc.*, **1982**, *104*, 4610.
   (d) D. Zbaida, I. Weissbuch, E. Shavit-Gati, L. Addadi, L. Leiserowitz and M. Lahav, *Reactive Polymers*, **1985**, *6*, 241 .
   (e) E. Staab, L. Addadi, L. Leiserowitz and M. Lahav, *Adv. Mater.*, **1990**, *2*, 40.
33. (a) H. L. Holland and M. F. Richardson, *Mol. Cryst. Liq. Cryst.*, **1980**, *58*, 311.
   (b) P. Ch. Chenchaiah, H. L. Holland amd M. F. Richardson, *J. Chem. Soc., Chem. Commun.*, **1982**, 436.
   (c) P. Ch. Chenchaiah, H. L. Holland, B. Munoz and M. F. Richardson, *J. Chem. Soc. Perkin Trans. II*, **1986**, 1775.
34. (a) L. Addadi, Z. Berkovitch-Yellin, N. Domb, E. Gati, M. Lahav and L. Leiserowitz, *Nature*, **1982**, *296*, 21.
   (b) Z. Berkovitch-Yellin, J. van Mil, L. Addadi, M. Idelson, M. Lahav and L. Leiserowitz, *J. Am. Chem. Soc.*, **1985**, *107*, 3111.
   (c) I. Weissbuch, L. J. W. Shimon, Z. Berkovitch-Yellin, L. Addadi, L. Leiserowitz and M. Lahav, *Isr. J. Chem.*, **1985**, *25*, 353.
   (d) I. Weissbuch, Z. Berkovitch-Yellin, L. Leiserowitz and M. Lahav, *Isr. J. Chem.*, **1985**, *25*, 362.
   (e) L. Addadi, Z. Berkovitch-Yellin, I. Weissbuch, J. van Mil, L. J.W. Shimon, M. Lahav and L. Leiserowitz, *Angew Chem. Int. Ed. Engl.*, **1985**, *24*, 466.
   (f) L. Addadi, Z. Berkovitch-Yellin, I. Weissbuch, M. Lahav and L. Leiserowitz, *Topics in Stereochemistry*, **1986**, *16*, 1.
35. I. Weissbuch, M. Lahav, L. Leiserowitz, G. R. Meredith and H. Vanherzeele, *Chem. Mater.*, **1989**, *1*, 14.
36. (a) J. M. McBride and S. B. Bertman, *Angew. Chem. Int. Ed. Engl.*, **1989**, *28*, 330.
   (b) J. M. McBride, Ibid, **1989**, *28*, 377.
37. (a) M. Vaida, L. J. W. Shimon, Y. Weisinger-Lewin, F. Frolow, M. Lahav, L. Leiserowitz and R. McMullan, *Science*, **1988**, *241*, 1475 .
   (b) M. Vaida, L. J. W. Shimon, J. van Mil, K. Ernst-Cabrera, L. Addadi, L. Leiserowitz and M. Lahav, *J. Am. Chem. Soc.*, **1989**, *111*, 1029.
38. Y. Weisinger-Lewin, F. Frolow, R. McMullan, T. F. Koetzle, M. Lahav and L. Leiserowitz, *J. Amer. Chem. Soc.*, **1989**, *111*, 1035.
39. S. P. Legros and A. Kvick, *Acta Crystallogr.*, **1980**, *B36*, 3052 and references therein.
40. (a) I. Weissbuch, L. Addadi, Z. Berkovitch-Yellin, E. Gati, S. Weinstein, M. Lahav and L. Leiserowitz, *J. Am. Chem. Soc.*, **1983**, *105*, 6613.

(b) I. Weissbuch, L. Addadi, Z. Berkovitch-Yellin, E. Gati, M. Lahav and L. Leiserowitz, *Nature*, **1984**, *310*, 161.

(c) I. Weissbuch, L. Addadi, M. Lahav and L. Leiserowitz, *J. Am. Chem. Soc.*, **1988**, *110*, 561.

41. M. Vaida, Ph.D. Thesis submitted to the Feinberg Graduate School, The Weizmann Institute of Science, Rehovot, Israel, November 1990.

42. J. D. Hung, M. Lahav, M. Luwish, G. M. J. Schmidt, *Isr. J. Chem.*, **1972**, *10*, 585.

43. M. D. Cohen, R. Cohen, M. Lahav and P. L. Nie, *J. Chem. Soc.*, **1973**, 1095 .

44. L. J. W. Shimon, Ph.D. Thesis, Feinberg Graduate School, The Weizmann Institute of Science, Rehovot, Israel, 1989.

Chapter 7

Photoprocesses of Host–Guest Complexes in the Solid State

V. Ramamurthy

Central Research and Development Department,
The Du Pont Company, Wilmington, DE, USA.

Contents

1. Introduction

When a host and a guest combine, the result is a chemical system known as a host–guest compound (complex). The term *host–guest compound (complex)* is the highest generic term that is used to express a host–guest relationship.[1] Over time numerous terms have been used to establish the relationship—supramolecular assembly, extramolecular assembly, addition compound, inclusion compound, occlusion compound, clathrate, intercalate, etc., to name a few. A single feature common to all these structures is that the forces holding the two components together are much weaker than the covalent bonds keeping a molecule's atoms together.[2] These are the weak intermolecular forces (e.g., hydrogen bonding, ion pairing, π-acid to π-base interactions, dipole–dipole interaction, and van der Waals attraction) which are also responsible for some molecules being liquids and some others being gases. Since these forces are distance dependent, close encapsulation of the guest by the host provides a stronger binding. Such close encapsulation can be achieved if the host and the guest have complementary binding sites and steric requirements.[3] Hosts in general can contain cavities/cages/channels that are rigid or that are developed by reorganization of the hosts during the process of complexation. Such complexation can occur both in solution and in the solid state. A special type of complexation that occurs only in the solid state but not in solution is clathration.[4] Under these conditions guests are retained by the hosts through crystal lattice forces. A few examples of such species are Dianin's compound, urea, and perhydrotriphenylene. Since at least one system (cyclodextrin) that is presented in this chapter can exhibit complexation both in solution and in solid state, we prefer to use the highest generic term "host–guest complex" for all the systems that are being described here. Guests will be only organic molecules.

Photochemical interest in host–guest complexes stems from the two approaches chemists have taken to achieve selectivity in photoreactions. One of these involves the use of highly organized and rigid organic crystals[5] and the other less organized and less rigid systems such as surfaces, liquid crystals, micelles, etc.[6] as media for reactions. Although organic crystals provide very high selectivity, their use has been limited since they curtail, almost completely, the freedom of motion of the reactant molecules. With other less organized media selectivity or the modification imposed has not been high. Such a difference can be attributed to the size and nature (flexibility) of the "reaction cavity" available for the reactant molecules in these media (for details on "reaction cavity" concept see Section 2.3 of Chapter 4). If one views crystals as an extreme example of host–guest complexes (the excited molecule is the guest in a host crystal consisting of mostly ground-state molecules), extension of the reaction cavity concept, originally invoked in the case of reactions in crystalline state by Cohen,[7,8] to real host–guest complexes is obvious. If one can control and provide a well-defined small and flexible cavity (a cavity that can accommodate the guest with some void space around) for the reactant molecule to confine itself, one may predict that a large number of molecules will undergo selective photo-transformations in the solid state. However, this cannot be done in the crystalline state of the reactants alone since the factors controlling molecular packing have not been understood— the main problem, namely, to be able to preorganize the molecules in the crystal lattice the way one would like, has not been fully surmounted (see Section 3 of Chapter 4).[9] It is in this context that host systems which crystallize with channels/cages/cavities have been valuable. There are a large number of organic and inorganic systems that exhibit an ability to include a variety of organic guest molecules in the solid state.[10,11] Organic guests have been accommodated into these well-defined and uniform host channels/cages/ cavities and their photochemical and photophysical behavior have been examined. Such studies are reviewed in this chapter. A brief description of the solid state structure of these host-guest complexes is also provided. An appreciation of these structures is essential to understand the photobehavior of guest molecules imprisoned in these host structures.

2. Structure and Packing Arrangement of Host–Guest Complexes in the Solid State

In this section a brief description of the solid-state structure of several organic host systems is presented. Many of these in the absence of guests do not exhibit the presence of a channel or a cage. Except for cyclodextrin all other host systems presented here do not have the ability to include guests in solution. Only through cooperative effects do they display an ability to include organic guest molecules in the solid state. While x-ray structure solution and precise assignment of the atomic positions in the host are not that difficult, those of guests often meet with difficulty. Since the forces holding them are very weak, guests are often disordered in the channel/cage created by the host. We have chosen to provide a brief description only for systems that have been used as hosts in at least one photochemical or photophysical study. For structural details of the systems briefly described here and of many other hosts that are not covered here readers are advised to consult the literature.[10] *J. Inclusion Phenomena and Molecular Recognition* is a journal to consult for new host systems.

2.1. Urea and Thiourea[12]

In 1940, Bengen discovered by accident that urea formed addition compounds with a great variety of aliphatic linear alcohols.[13] Since then urea has been shown to form inclusion complexes with a large number of aldehydes, ketones, acids, aromatics, etc., provided that their main chain consists of at least six carbons. The ability of thiourea to form inclusion compounds with a variety of organic compounds was first reported by Angla in 1947.[14] It was van Bekkum who showed in 1967 that selenourea also forms inclusion compounds with a large number of hydrocarbons.[15]

Free urea crystallizes as tetragonal prisms. The urea inclusion compounds generally crystallize in long, hexagonal prisms or occasionally as hexagonal plates.

x-Ray structure analysis shows clearly the existence of a *central channel* in the urea–*n*-hydrocarbon complex (Figure 1). The diameter of the channel is about 5.3 Å. As illustrated in Figure 1 the *chiral environment* about the guest is created by the *helical packing* of the urea host. Although urea molecules in each single crystal are arranged in only one enantiomeric helix, when a polycrystalline sample is precipitated, a racemic mixture of both crystal forms will be obtained. The arrangement of thiourea molecules in the rombohedral crystals is similar to that of urea in the urea inclusion compounds. The channel size is about 6.1 Å.

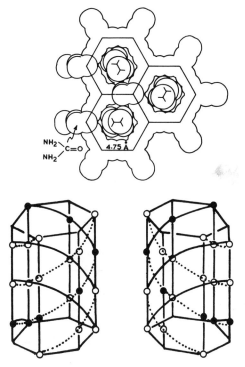

Figure 1. (above) End view cross-section of the urea–n-paraffin complexes; (below) idealized arrangement of urea representing the chiral forms.

Urea and thiourea channel-type inclusion complexes are generally stabilized by van der Waals forces between the host and the guest and by hydrogen bonding between the host molecules. Crystalline inclusion compounds of urea as well as thiourea with the guest organic compounds can be obtained either by direct mixing or by dissolving both the host and the guest and crystallizing from methanol solution.

Molecules such as dialkyl ketones, dialkyl peroxides, and *n*-alkanes are known to form complexes with urea. On the other hand, larger sized molecules such as ferrocene, nickelocene, and alkyl substituted dienes and aromatics such as naphthalene form complexes with thiourea.

2.2. Deoxycholic Acid, Apocholic Acid and Cholic Acid[16]

The ability of deoxycholic acid to form inclusion compounds was reported as early as 1916.[17] Since then typical steroidal bile acids such as deoxycholic acid (DCA), apocholic acid (ACA), and cholic acid (CA) have been shown to form channel type inclusion compounds with a wide variety of organic molecules. Of these DCA has been extensively investigated.

From the x-ray analyses of several DCA complexes, it has become clear that they can be grouped into three crystal forms: orthorhombic, tetragonal, and hexagonal. Orthorhombic form is more generally observed. In the orthorhombic structures one observes a two-dimensional bilayer motif with axial dimensions of b ~ 13.6 Å and c ~ 7.2 Å. DCA molecules form chains by translation along the b axis, being interlinked front to end by O(hydroxyl)—H- - -O(carbonyl) hydrogen bonds. Rows of DCA molecules developed along the b axis are further joined by hydrogen bonds about the 2_1 axes, which are parallel to the b axis and spaced along the c axis, so generating the bilayer (see Figure 3 in Chapter 6). These bilayers contain grooves parallel to the c axis which induce DCA to form channel inclusion complexes. These channels have a variable size and shape depending on the mutual positions of two adjacent bilayers along the b axis as illustrated in Figure 2. This accounts for the ability of the DCA host lattice to accommodate guest molecules of very different dimensions. The channel dimension often varies between ~2.6 x 6.0 Å and ~5 x 7 Å depending on the size of the guest. The length of the channel is clearly dependent on the guest.

Hexagonal crystals of DCA inclusion complex are characterized by the packing helicies of DCA molecules, generated by 6_5 axes held together by van der Waals forces. The channel created by the helix has a diameter of ~4 Å. The interior surface of the helix is covered mainly by polar groups, which give rise to spiral hydrogen bonds. This is in contrast to the situation in orthorhombic crystals, wherein the guest molecules included in the channels are surrounded by nonpolar groups. Therefore, the hexagonal DCA crystal can be utilized to include polar guests and the orthorhombic ones to accommodate apolar guests. Tetragonal crystals are obtained only with very small solvent molecules and therefore are not of interest to us. *Since the host DCA itself is chiral the inclusion complex obtained with DCA always belongs to a chiral space group.*

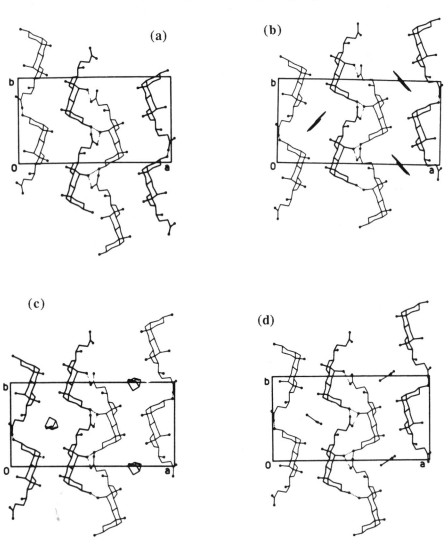

Figure 2. DCA packing illustrating the flexible size of the channel cross-section. Packing viewed along the *c*-axis: (a) no guest, (b) phenanthrene, (c) norbornadiene and (d) acetone as guests. The channels have variable size and shape depending on the mutual positions along the *b*-axis of two adjacent bilayers. In this Figure one can notice that the area of the channel cross section increases with the shift in the central bilayer towards more positive values of *b*. Reproduced with permission from Ref. 16, p.215.

DCA–guest interactions are usually van der Waals type. However, in the case of aromatics strong interaction with the π cloud could be important. DCA–guest inclusion compounds are often obtained by slow evaporation of the solvent from an ethanol or methanol solution containing both DCA and the guest in a well-defined stoichiometry or by crystallizing the complex from a solution of DCA in a liquid guest.

Guests of photochemical interest that form complexes with DCA are aromatics such as naphthalene, acenaphthene, 1,2-benzantharacene, and phenanthrene; azo dyes and dialkyl ketones; and arylalkyl ketones such as acetophenone.

Very recently, cholic acid (CA) itself has been shown to form an inclusion complex with acetophenone.[18] As illustrated in Figure 3 the channel space provided by CA molecules is much larger (~8 Å) than in DCA complexes. It is of interest to note that two molecules of acetophenone are accommodated side by side in the channels of CA.

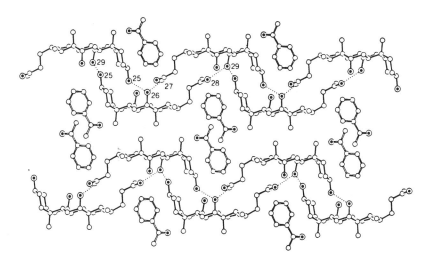

Figure 3. The crystal structure of the inclusion compound between cholic acid and acetophenone viewed down the *b*-axis. (Ref. 18).

2.3.   Perhydrotriphenylene[19]

Perhydrotriphenylene (PHTP) is a chiral molecule. Optically pure PHTP was accidently discovered by Farina in 1963 to form *chiral* inclusion complexes with a large number of organic guests.[20]   The equatorial isomer of PHTP gives rise to a wide variety of inclusion compounds with different kinds of molecules, ranging from those with a nearly spherical or planar shape to linear ones.   All the investigated adducts have a channel like structure with the PHTP molecules arranged in infinite stacks whose axes are parallel to the threefold axis of the molecule (Figure 4). The diameter of the channel is about ~5 Å and it is slightly flexible. Guest molecules are held within the hydrophobic channels through van der Waals interaction. The preparation of the inclusion complex of PHTP with the guest has been achieved through a variety of techniques. For volatile guests it is enough to expose the guest vapor to the PHTP crystal; for liquid guests the complex is readily prepared by direct contact of the liquid with the solid PHTP; for solid guests brief grinding of the two is sometimes enough. An often used procedure involves dissolving the host and the guest in methylethyl ketone and letting the complex crystallize. Guests of interest include alkenes and aromatics such as naphthalene and anthracene.

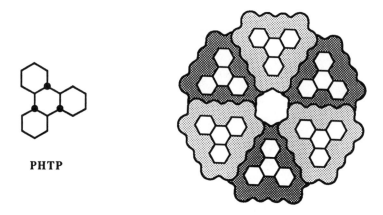

PHTP

Figure 4. View of the hexagonal PHTP inclusion compound in the *ab* plane. (Ref. 19). PHTP inclusion compounds are composed of infinite stacks of host molecules, repeating at about 4.78 Å, parallel to the molecular threefold axes. The regular packing of the stacks gives rise to parallel channels.

2.4. Dianin's Compound and Related Hosts[21]

4-*p*-Hydroxyphenyl-2,2,4-trimethylchroman, widely known as Dianin's compound, was first prepared by Dianin in 1914.[22] He reported the remarkable ability of this compound to retain tightly certain organic solvents. This host belongs to the class of hydroquinones and phenols reported even earlier (late nineteenth century) to form inclusion compounds with several gaseous molecules.[23] Since the cavity sizes of hydroquinone and phenol are small (~4.5 Å) they will not be discussed.

The general crystal structure of the Dianin's compound lattice consists of hexamers of Dianin's compound held together by a ring of hydrogen bonds involving the phenolic hydroxy groups. The monomeric units of the hexamers form chains of hourglass-shaped cages. As illustrated in Figure 5, the upper half of each cage consists of three molecules of Dianin's compound from one hexamer, and the lower half is three molecules of Dianin's compound from another hexamer.

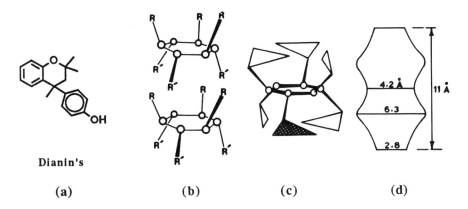

Dianin's

(a)   (b)   (c)   (d)

Figure 5. The shape and size of the cavity/channel of Dianin's compound. (a) molecular structure; (b) the cage formed by two sets of hexamers; (c) the top or bottom half of the cage, hexamers are held together by hydrogen bonds and (d) schematic illustration of the cage/channel. The stacking is along the *c*- axis. (Ref. 21).

The cage is held together at the roof and floor by the hydrogen bond network and at the waist by van der Waals forces between molecules. The cages are quite large and can accommodate even bulky molecules. Figure 5 also shows a view of the cage normal to the c axis. Each one of these hourglass-shaped cages connected on both the top and the bottom to yet another cage and thus this arrangement gives rise to a column consisting of bulged channels. Each particular column is ideally infinite in extent and runs parallel to the c axis. Each column is surrounded by, but not interconnected with six other identical columns related by a threefold screw axis. Although Dianin's compound possesses a chiral center, *only the racemic compound has the ability to include organic molecules; optically pure S-(-)-Dianin's compound has no ability to form inclusion complexes.*

Based on careful analysis of the structure and understanding of the features controlling the structure, several modifications on the molecular framework of Dianin's compound have been carried out. In Figure 6 we show how slight modification of the structure results in cages of different dimensions. Readers are referred to recent reviews for details.[21]

Inclusion complexes of Dianin's compound with organic molecules are prepared by slow crystallization of unsolvated Dianin's compound from the liquid guests. Large-sized molecules such as decahydronaphthalene and arylalkyl ketones form inclusion complexes with Dianin's compound.

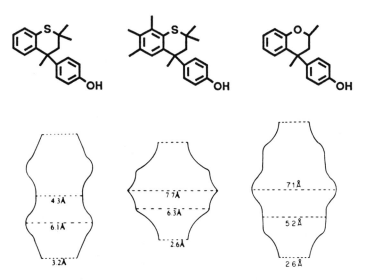

Figure 6. Variation in the channel size with a slight alteration in the structure of Dianin's compound. (Adopted from ref. 21). Compare the channel/cage size and shape with Figure 5.

2.5.   Cyclophosphazenes[24]

An accidental discovery by Allcock in 1964 brought cyclophosphazene (CPZ) to the list of organic host compounds.[25] Tris(*o*-phenylenedioxy)cyclophosphazene exists in a hexagonal modification when acting as a host. As shown in Figure 7 each molecule of tris(*o*-phenylenedioxy)cyclophosphazene has a paddle and wheel shape. Superimposition of the two layers leads to the arrangement in which a channel (~5 Å) is created in the middle. Replacement of the phenyl group by naphthyl leads to a host structure in which the channel size is considerably larger, ~ 9–10 Å. This is true in the case of tris(2,3-naphthalene-dioxy)cyclophosphazene. Depending on the organic spiro side group, cyclophosphazene can give rise to either *channel* or *cage* structures. For example, tris(1,8-naphthalenedioxy)-cyclophosphazene yields a cage structure. However, the same host has not been shown to exhibit both cage and channel structures as occurs with other hosts, such as triorthothymotide (see Section 2.8). Since the initial discovery by Allcock, several related cyclophosphazenes have been reported to behave in ways that suggest inclusion phenomena.[24]

In the case of channel-forming hosts the complex can be prepared by contacting the hosts with the guest in either a vapor or a liquid state. Inclusion into a cage–forming host can be achieved only by crystallizing the complex from the guest liquid or by dissolving and crystallizing both the guest and the host from a sterically bulky solvent. Similar to PHTP the channels in CPZ are hydrophobic and the guests are held within these channels by van der Waal's forces.

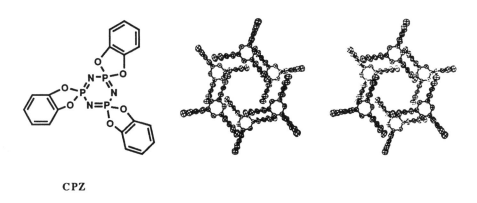

CPZ

Figure 7. View of the channel formed by tris(*o*-phenylenedioxy)-cyclophosphazene. (Ref. 24 and 25). The location of the channel ($d = 4.5–5$ Å) is clearly evident in this figure.

2.6. Hexahosts and Related Structures

Close examination of the structures of hydroquinone, phenol, Dianin's compound, and related structures led MacNicol to the discovery of a new class of hosts, termed hexahosts, in 1976.[27] Of the various derivatives that have been examined, the one of interest is hexakis(*p-t*-butylphenylthiomethyl)benzene (**1**), which forms inclusion compounds with long as well as flat organic molecules such as squalene, naphthalene, and pyrene. A channel is clearly seen in the structure of the complex of **1** with squalene (Figure 8). These complexes are readily prepared by crystallizing hexahost from the liquid guest or by dissolving and crystallizing the host and the guest from a hydrocarbon solvent such as mesitylene. For details on other related structures readers are referred to a recent review by MacNicol.[26]

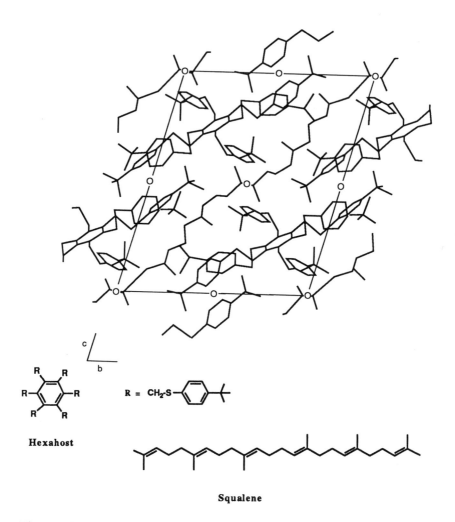

Hexahost

R = CH₂-S-⟨⟩-⊣

Squalene

Figure 8. A view onto the *bc* plane of the adduct hexakis(*p*-*t*-butylphenyl-thiomethyl)benzene with squalene. Squalene is accomodated in continous channel running through the crystal (Ref. 26).

2.7. Wheel and Axle-Type Hosts[28]

Based on the concept that compounds with one long molecular axis (axle) and with large and relatively rigid groups at each end (wheels) would pack loosely in the solid and thus function as hosts, Toda prepared several new host systems in 1968.[29,30] Examples of such hosts are characterized by bulky ends and thin centers. The detailed structures of several clathrates have been characterized and a certain degree of selectivity in complexation has been noticed. Most of these complexes are of the channel type.

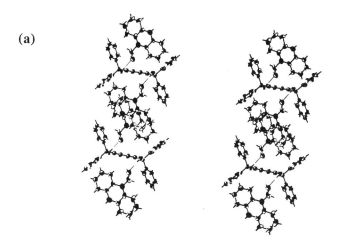

2   3

Of the various hosts reported by Toda, 1,1,6,6-tetraphenylhexa-2,4-dyne-1,6-diol (2) has attracted considerable attention in photochemical studies. One of the advantages of using this host is the *flexible packing* it provides. When the size of the guest molecule is on the order of the size of the host molecule, the host molecules accommodate themselves to form a channel. When the guest molecule is smaller than the host, the latter forms a channel with a cross section enclosing the guest molecules in an antiparallel fashion. Two such structures are shown in Figure 9. In both these structures hydrogen bonding between the host and the guest plays an important role in guest inclusion. In the context of chiral hosts, 1,6-bis(*o*-chlorophenyl)-1,6-diphenyl-2,4-dyine-1,6-diol (3) is of interest. The ability of chiral hosts to form channel inclusion complexes has been utilized to resolve optical isomers and to conduct asymmetric photochemical transformations.

(a)

Figure 9 (a). Caption on next page.

(b)

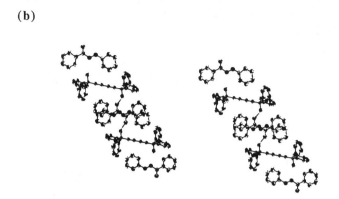

Figure 9. A stereoscopic view of the packing in the inclusion compound of 1,1,6,6-tetraphenylhexa-2,4-dyne-1,6-diol with   (a) 9-anthraldehyde and   (b) chalcone. 9-Anthraldehyde gives cage type and chalcone gives channel type structures. (Ref. 28).

Host–guest complexes of **2** and **3** with liquid guests can be prepared by mixing them and with solid samples by grinding them together. It is surprising to note that Toda claims that with solid guests inclusion complexes can be obtained by just mixing the host **2** and the guest together even without grinding.

2.8. Tris-orthothymotide and Related Hosts[31]

As early as 1909 tris-orthothymotide (TOT) was synthesized and isolated as an adduct with solvent benzene.[32]   Only in 1952 was its inclusion ability established by Baker, Gilbert, and Ollis.[33]   TOT forms two types of crystalline inclusion complexes: in one the guest molecules are enclosed in *discrete closed cavities*; in the other the guest molecules are accommodated in continuous *linear channels* running through the crystal along a crystallographic axis.

While uncomplexed TOT crystallizes in a racemic orthorhombic form, TOT complexes generally crystallize in a chiral space group. TOT exists in solution as an equilibrium mixture of several chiral conformers and crystallizes with spontaneous resolution while forming an inclusion complex with a guest. Thus under appropriate conditions, a single enantiomer may crystallize selectively. Analysis of the unit cell dimensions of a large number of complexes by Lawton and Powell revealed that cage-type complexes are formed with guests of length less than ~9 Å and that channel-type structures are formed with long chainlike molecules.[34]   The cage in the former type of structures is comprised of eight TOT molecules related pairwise about a crystallographic twofold axis. The average diameter of the ellipsoid cage is about 12 Å (Figure 10). These cages deform to a limited extent to accommodate molecules of different dimensions.

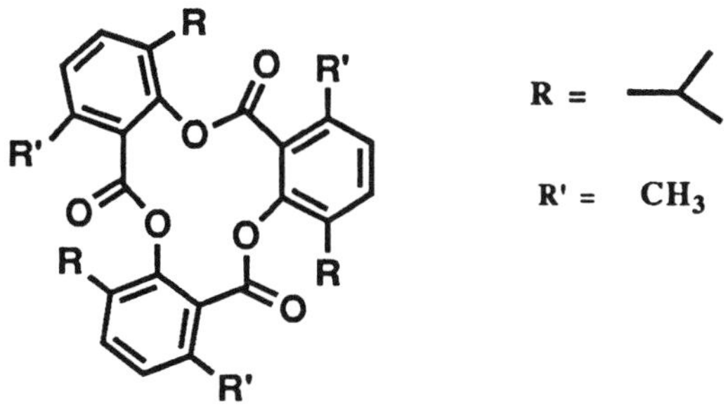

R = ─⟨

R' = CH₃

TOT

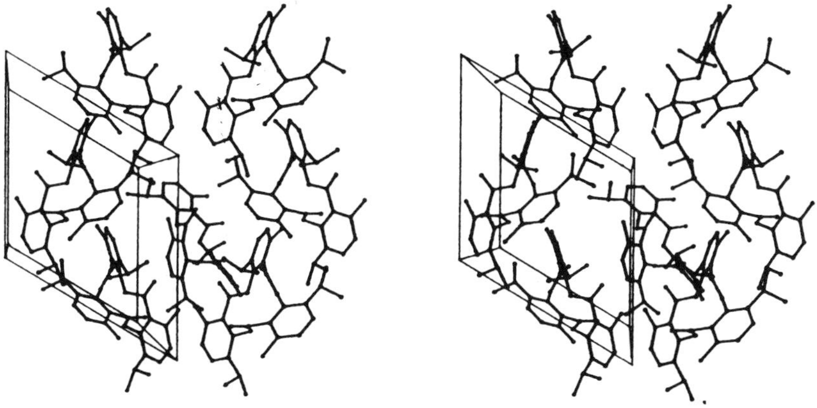

Figure 10. Stereoview of the cage in the TOT complex. (ref.31).

Channel structures are formed for example in the case of *trans*-stilbene and benzene as guests. In these cases at least two independent channels are present; one along the a and the other along the b axis as illustrated in Figure 11. The channel along the a axis is of fairly uniform cross section, whereas the one along the b axis presents a succession of bulges and constrictions. As pointed out earlier the channel dimensions vary slightly depending on the guest.

TOT cages and channels lack any specific binding sites and the host–guest complexation therefore is essentially controlled by van der Waals's forces. It is obvious from the discussion above that TOT possesses cages and channels that are fairly large compared to the other host systems presented above. Consistent with this, TOT serves as a host for a large number of organic molecules. Olefins such as *trans-* and *cis-*stilbenes, carbonyl compounds, and aromatics such as naphthalene, pyrene, etc., form complexes with TOT. TOT complexes are easily prepared by dissolving the TOT in the liquid guest and letting it crystallize. When the guest is a solid both TOT and the guests are dissolved in an appropriate solvent such as methanol or trimethylpentane and slowly cooled to crystallize the adduct.

(a)

(b)

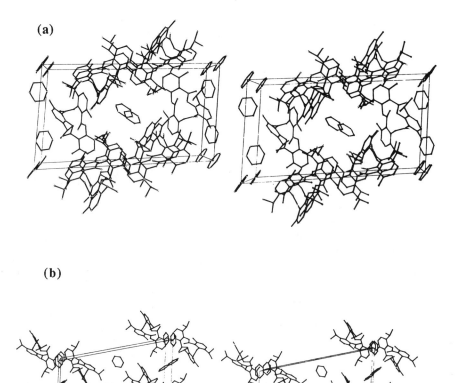

Figure 11. Stereoview of the two channels present in TOT–benzene complex. (a) view along the a-axis and (b) viewed down the b-axis. One channel runs along the a-axis and the other along the b-axis. (Ref. 31).

Several modifications on the structural framework of TOT have been carried out. Several trianthranilides, a nitrogen analogue of TOT, have been established to be good hosts.[35] These form helical chiral channels into which guests are accommodated. So far they have not been used as hosts in photochemical studies.

2.9. Cyclodextrins[36]

Cyclodextrins (CD), one of the most commonly used host systems, possess hydrophobic cavities that are able to include, in aqueous solution, a variety of organic molecules whose character can vary from hydrophobic to ionic. Internal diameters and depths of cyclohexaamylose (α-CD), cycloheptaamylose (β-CD), and cyclooctaamylose (γ-CD) provide cavities of different sizes (Figure 12). The oligosaccharide ring forms a torus, with the primary hydroxyl groups of the glucose residues lying on the narrow end of the torus. The secondary glucopyranose hydroxyl groups are located on the wider end.

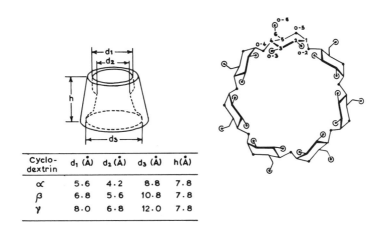

Cyclo-dextrin	d_1 (Å)	d_2 (Å)	d_3 (Å)	h(Å)
α	5.6	4.2	8.8	7.8
β	6.8	5.6	10.8	7.8
γ	8.0	6.8	12.0	7.8

Figure 12. Shape and structure of cyclodextrin cavity.

Inclusion complexes of known ratio can be precipitated from aqueous solutions of CD when an excess of guest is added. Such precipitates contain the guest accommodated within the cavities of CD. In addition to the local structure, the global structure of the solid is determined by how these individual complexes are arranged in the solid state. Based on the overall appearance, these are described as cage- or channel-type structures. In channel-type structures, CD molecules are stacked one on top of each other like coins in a roll, the new linearly aligned cavities producing channels in which the guest molecules are embedded. The channel structures may be further divided into two types, namely, one in which head to tail packing occurs and the other in which head to head packing occurs. In the cage type of crystals, the cavity of one CD molecule is blocked off on both sides by an adjacent CD, thereby leading to isolated cavities. Once again two types of structures are encountered: brick type and herring bone type. These are illustrated in Figure 13. In the case of α-CD, small molecular guests form cage packing, whereas long and ionic guests prefer channel structures. However, such distinctions cannot be readily made in the case of β- and γ-CD complexes. Another important property of CD is its chirality: β-CD is dextrorotatory with $[\alpha]_D$ +162°. Therefore CD can show selectivity in the inclusion of optical isomers.

(a)   (b)

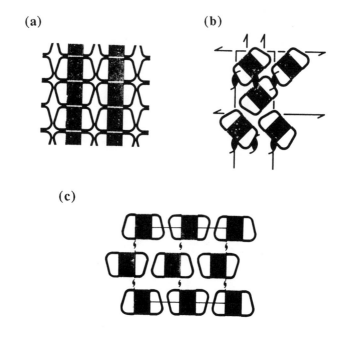

(c)

Figure 13. Schematic representation of the packing arrangement in cyclodextrin complexes: (a) channel type, (b) cage or herringbone type and (c) brick type. (Ref. 36).

The number of organic molecules known to form complexes with CD is large.[36] Most of these studies have been carried out in aqueous solution. The potential of CD inclusion complexes in the solid state is yet to be fully explored.

2.10. Miscellaneous Host Systems

In addition to the above well-known and popular host systems, several lesser known organic host systems have been reported, many of them without much crystallographic detail. Recent reviews by Davies, Finocchiaro, and Herbstein[37] and by Goldberg[28b] are good sources of information on some of these. Cyclotriveratrylene and its derivatives have been studied in some detail and they are established to form crystalline inclusion compounds with small sized organic molecules.[38]

3. Choice of a Host for Photophysical and Photochemical Investigations[39]

In choosing a host system for a particular study the following aspects should be considered:

1. There should be a size matching between the host and the guest; the guest should fit into the cavities or channels created by the packing of the host molecules.

2. The guest should possess a distinctly different electronic absorption from that of the host. Also the host should not possess excited singlet and triplet energies lower than the guest.

3. The host emission, if any, should not interfere with the emission from the included guest.

4. Included guests upon excitation should not undergo reaction with the host.

The first criterion can be fulfilled by knowing the molecular size of the guest and the sizes of the host crystal cavity or the cage. The second criterion can be fulfilled by examining the absorption spectra (diffuse reflectance spectra) of the host and the guest. PHTP, CD, DCA, and urea do not absorb significantly above 250 nm. TOT, hexahost, and thiourea show strong absorption up to 350 nm and CPZ up to 300 nm and therefore one should be cautious in choosing them as hosts. Examination of the emission spectra of the above hosts recorded at 77 K revealed that CD, DCA, PHTP and urea do not show any emission upon excitation in the region 240–300 nm. TOT, hexahost and thiourea showed a weak emission in the region 400–600 nm upon excitation between 250–300 nm. CPZ exhibits a moderate emission with a max at 320 nm upon excitation between 250–290 nm.

One of the earliest photoreactions investigated in the solid state in fact involves the reaction between the host and the guest. Friedman et al. isolated three products upon photolysis (or thermolysis) of di-*tert*-butyl peroxide included in the channels of DCA (Scheme 1).[40] All three products contained the DCA framework, indicating that the main reaction is the attack of the peroxy radical on the host DCA. Selectivity in the position of attack is attributed to the fact that only a limited number of tertiary hydrogens of the steroid on the inner wall of the channel are exposed to the guest and therefore available for abstraction. Another example is also provided by the same group led by Lahav and Leiserowitz. The acetone complex of DCA on exposure to uv radiation gave several products, all containing the DCA framework (Scheme 1).[41] This reaction has been studied in detail with a large number of ketones. The details of enantioselective product formation by this process of host functionalization are presented in Chapter 6. At this juncture it is sufficient to point out that the reaction between the guest ketones and the host DCA has been established in several systems.

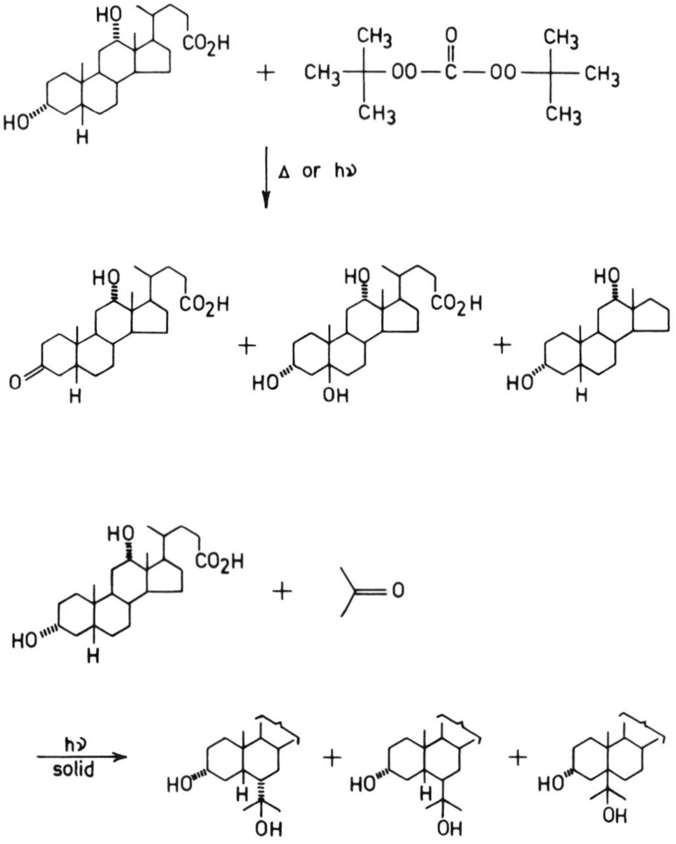

Scheme 1

Reaction between the guest and the host has also been reported with CD as the host. Two examples are provided in Scheme 2. Abelt and Pleier isolated a carbene insertion product of CD when phenylmethyldiazirine was photolyzed or thermolyzed as a complex with CD.[42] Veglia et al. have shown that the CD framework can serve as a hydrogen donor.[43] During the solid-state irradiation of phenyl acetate in CD they isolated phenol. The authors believe that this resulted from the abstraction of hydrogen from CD by the phenoxy radical generated by the photofragmentation of phenylacetate. No mention of the fate of the CD radical is made. In all the above examples the reactive guest molecules (or the intermediates derived therefrom) attack the host structure at activated/exposed centers. The reverse, namely the activated host reacting with the unactivated guests, is also known. For example, γ or x, irradiation of crystalline empty PHTP gives radical centers having a long lifetime.[44] These centers are known to initiate polymerization of monomers included in the PHTP channels. From the examples presented in this section it is clear that hosts most often contain active hydrogens and other chromophoric centers, similar to solvents, that are capable of participating in photoreactions. Therefore, one should carefully consider these possibilities in choosing a host during photochemical investigations.

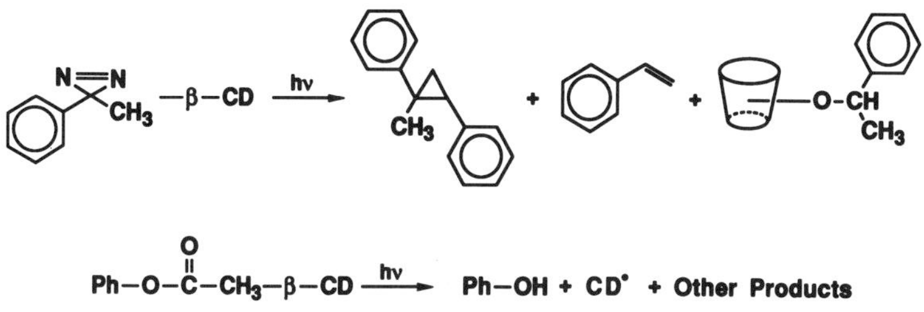

Scheme 2

4. Photophysical Studies

In spite of considerable interest in photophysical studies of organic molecules incorporated in various organized assemblies, host–guest systems in the solid state have not received much attention. Considering the literature that exists on the photophysics of the related assembly, mixed crystals of aromatic molecules, this lack of interest is surprising.

Room temperature phosphorescence of organic molecules in micellar media, on silica surfaces, and in aqueous CD solution has received much attention in recent years (see Chapter 2). Along the same line, emission studies of several organic molecules (e.g., phenanthrene and other aromatics, aminobenzoic acids, phenols, benzoquinoline) included in solid mixtures of α- and β-CD and NaCl have been carried out by Hurtubise and co-workers.[45] In general, enhancement of phosphorescence has been observed both at low and at room temperatures. Reduction in radiationless deactivation from S_1 to S_0 and enhancement of intersystem crossing from S_1 to T_1 are suggested to be responsible for the ready observation of phosphorescence from organic guests included in α- and β-CD. In this context observations by Eaton are of interest.[46] Phosphorescence was recorded at room temperature from naphthalene and dimethylnaphthalenes included in TOT and thiourea crystals. This is attributed to reduction in quenching of the triplet state by oxygen. The host is suggested to serve as a barrier to oxygen penetration. Kook and co-workers have observed phosphorescence from 4-methyl-benzophenone included in deoxycholic acid.[47]

Both the band position and structure of the phosphorescence spectra of the guest 4-methylbenzophenone are strongly dependent on the occupancy level of the guest within the host matrix (Figure 14). At concentrations below 14% (by weight) the spectra are different from that of the stoichiometric crystal (DCA:guest = 2:1); at ~ 16% loading the spectrum resembles that of the stoichiometric crystal. It is suggested that due to percolation, this high occupancy (16%) may be sufficient to have a neighboring guest site occupied by another 4-methylbenzophenone molecule and this may be enough to provide an environment similar to that of a stoichiometric crystal. No energy transfer between 4-methylbenzophenone molecules was noted in DCA complex, although such transfer has been reported earlier in polystyrene matrices for the same system (spectrum shifts to longer wavelength due to energy transfer to lower energy sites). The low rate of energy transfer is attributed to the large intermolecular distance between guest molecules. It is to be noted however that energy transfer is expected on the basis that there will be low-energy defect sites in the DCA–4-methylbenzophenone complex. However, this need not be the case.

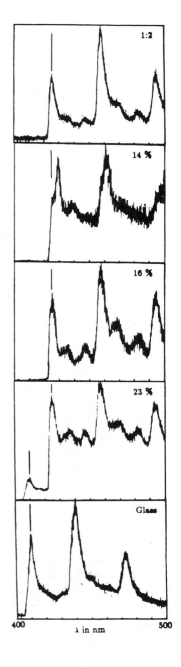

Figure 14 (Caption next page)

Figure 14. Steady state phosphorescence spectra of 4-methyl-benzophenone included in DCA at 4.2 K. (a) Stoichiometric complex: 4-methyl-benzophenone : DCA = 1:2; (b), (c) and (d): various % of 4-methyl-benzophenone in DCA and (e) 4-methyl-benzophenone alone in glass. Note the difference in spectral position between glassy and DCA complexed states and also the slight difference between various loading levels.

The absence of face to face intermolecular contacts between guest molecules even at high occupancy levels has also been inferred by Eaton and co-workers from their luminescence studies of aromatics in TOT, PHTP, and thiourea.[46] Pyrene, which is known to form excimers in solution and exists essentially in excimeric orientation in crystals, shows only monomer fluorescence when included in TOT (host:guest = 1:1) and PHTP (3:1). Similar observations are also made in the cases of methylnaphthalenes and 9-methylanthracene. Such absence of strong face to face intermolecular contact permitted Eaton to probe the feasibility of energy transfer between naphthalene and anthracene included in PHTP matrix.[46] Fluorescence from both naphthalene and anthracene was seen from a solid inclusion complex of naphthalene and anthracene with PHTP in the ratio PHTP:aromatics 75:25 (aromatics consist of anthracene and naphthalene in the ratio 1:500). At the excitation wavelength used naphthalene absorbs >99% of the light. Lifetime measurements reveal that the lifetime of anthracene that emits in the mixed guest system is different from that of the anthracene present alone in the PHTP channel. The S_1 of anthracene in the PHTP–anthracene complex decays with a lifetime of 0.4 ns, whereas that in the PHTP–anthracene–naphthalene complex described above shows a double exponential behavior with lifetimes of 4.4 ns (90%) and 37 ns (10%). These observations suggested that naphthalene, upon population of the excited singlet state, transfers energy to the anthracene present in the PHTP matrix in an environment different from that of anthracene molecules present alone in the PHTP–anthracene complex. Further work is desirable to extend these observations. In this context of energy transfer the theory being developed by Guarino should be noted.[48] In particular, Guarino has emphasized the importance of symmetry matching between the host and the guests in energy transfer processes that occur in inclusion complexes.

The above studies clearly point out that host–guest complexes can be an interesting medium for photophysical investigations. Unequivocal interpretation of the results in many of these cases has been prevented by the lack of structural details on these complexes. While x-ray structural investigation often provides the structure of the host matrix, the location of the guests has been difficult to establish in most cases due to disorder in the guest arrangement. Solid-state nmr investigations have been useful in this context.[49] Photophysical studies can provide information concerning the location and the freedom of the guests within the host matrix. We believe that such studies will attract attention in the future.

5. Photochemical Studies
5.1. Prealignment Effect (Crystal Engineering)

The efficiency of a solid-state photochemical reaction can be enhanced by having the reactants 'prealigned" for the reaction. It is clear from Chapters 4, 5 and 6 that prealignment is a must for the reaction to take place in the crystalline state. If the molecules are not prealigned no reaction occurs in the crystalline state, since very little molecular motion is tolerated by the crystalline environment. One can also utilize a host medium to align or prearrange the molecules toward a particular reaction. Such prealignment may result from specific (geometry-dependent) interaction between the host and the guest or from the limited space available for the guest(s) to accommodate themselves. In Chapter 4 (Section 3.3) this aspect was briefly discussed as it related to bimolecular reactions and crystal engineering. This chapter provides additional examples of dimerization, polymerization, and intramolecular hydrogen abstraction reactions of guests in host–guest solid-state systems.

5.1.1. Dimerization

Chalcone dimerizes inefficiently to give several products in the solid state. However, when it is included in the channels created by the host 1,1,6,6-tetraphenylhexa-2,4-dyne-1,6-diol, (2, wheel and axle-type host) smooth dimerization occurs to yield a single dimer, the syn-head-to-tail isomer.[50] This is true of other derivatives of chalcone as well (see Scheme 17 of Chapter 4). The reason for the well-controlled reaction is evident when the crystal structure of the complex is analyzed. From the packing diagram provided in Figure 9 it is seen that hydrogen bonding between the hydroxyl groups of the host and the carbonyl group of the guest plays an important role in packing chalcone molecules close together in the complex.[51] The double bonds are parallel and the distance between them is short (3.86 Å). Other examples of this effect include [4+4] dimerization of 9-anthraldehyde and 2-pyridone[52] and [2+2] dimerization of coumarins.[53] The prealignment is achieved through the hydrogen bonding interaction between the host and the guest in these cases also. Photodimerization of benzylideneacetone in 2,6-diphenylhydroquinone is another example of such an effect (see Scheme 18 of Chapter 4).[52]

β- and γ-cyclodextrins possess a cavity that can include two molecules. Enclosing two molecules in a small cavity can bring the molecules within reactive distance (<4.2 Å). Specific interaction between the host rim hydroxyls and the guest may also provide a selective orientation of the reactant guest molecules. Such effects have been utilized in the dimerization of coumarins[53,54] (for details see Chapter 4) and cinnamic acids[55] included in β- and γ-cyclodextrins. p-Methoxy-cis-cinnamic acid, but not p-methoxy-trans-cinnamic acid, is reported to dimerize when included in β-cyclodextrin; the nature of the dimer has not been characterized.

5.1.2. Polymerization

Some of the best examples of the "prealignment" concept are provided by γ- or x-irradiation-induced polymerization of olefins included in the channels of urea, thiourea, cyclophosphazines, cyclodextrins, deoxycholic acid, and apocholic acid. We briefly touch upon these examples to illustrate the concepts involved as these may be valuable in designing photochemical studies. For details readers are referred to an excellent review by Farina which critically summarizes the extensive literature that exists in this area.[56] In the narrow channels of these host systems monomers are aligned in such a way that γ irradiation results in radical centers which initiate polymerization (Figure 15). This is often termed *one-dimensional polymerization*.

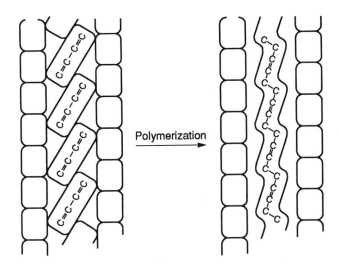

Polymerization

Figure 15. Polymerization of 2,3-dimethylbutadiene in the thiourea channel.

The brief summary provided below demonstrates the extent of selectivity that one can obtain during polymerization in host–guest included systems.

1. When butadiene and 2,3-dimethylbutadiene are included in the channels of urea and thiourea, respectively 1,4 addition invariably results to yield polymers with chemical and stereo regularities (Scheme 3).[57] Note addition could also occur in the 1,2 fashion but this is prevented by the narrow channel. Similarly, high selectivity was obtained when butadiene, vinyl chloride, and styrenes were polymerized in the channels of cyclophosphazenes.[58]

2. Syndiotactic polymer alone is obtained from vinyl chloride included in urea; this is apparently the first example of inclusion polymerization of a

vinyl polymer in which control is exerted over the steric configuration of the tetrahedral carbon atom (Scheme 3).[57]

3. Highly isotactic polymer is obtained from 1,3-pentadiene when included in perhydrotriphenylene matrix (Scheme 3).[59] Note that addition could occur at either end (i.e., C_1 to C_1, C_1 to C_4 and C_4 to C_4) but the channel directs the addition selectively to only one, namely C_1 to C_4.

4. Polymerization of dienes takes place in DCA and ACA channels but with less selectivity.[60]

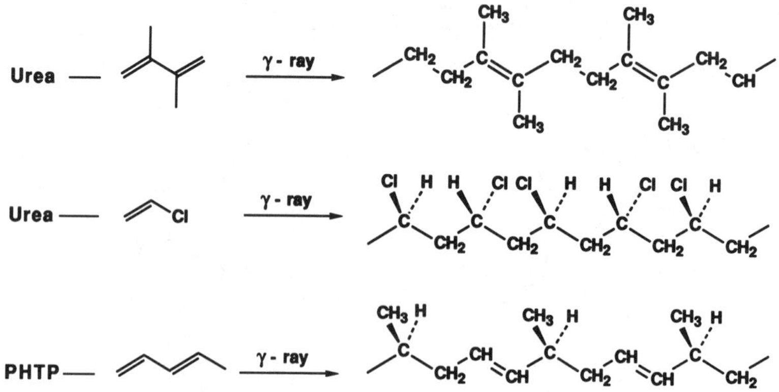

Scheme 3

The selectivities noted above correlate well with the channel size. The lower selectivity obtained in DCA and ACA channels compared to other channels is a reflection of the freedom experienced by the included monomers. It is clear that an ordered arrangement of monomers in the channel is of fundamental importance. Also, the arrangement should favor interaction with neighboring guest molecules.

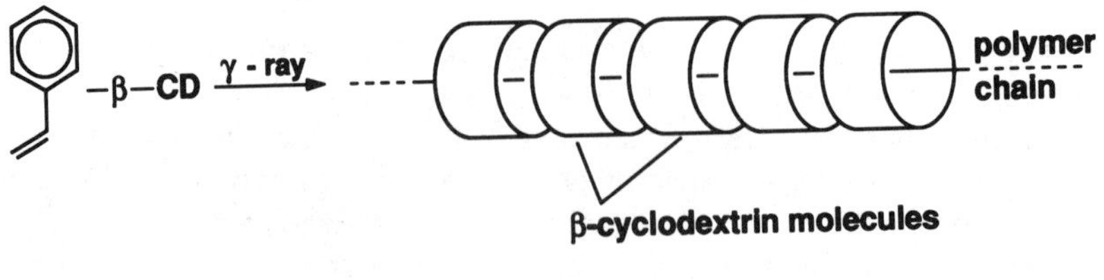

Scheme 4

Before concluding this section we wish to point out an elegant example provided by Maciejewski et al.[61] Solid-state γ-ray-induced polymerization of styrene, methylmethacrylate, and vinylidene chloride included in cyclodextrin gave a polymer in which the host cyclodextrin is appended to the polymer chain like a "string of beads" (cyclodextrins are not supported by any chemical linkage; Scheme 4). This is probably the result of channel-type packing present in the CD–monomer complex.

5.1.3. Norrish II Reactions

The success of the Norrish type II reaction depends on the proximity and on the geometry of the γ hydrogen with respect to the carbonyl chromophore (see Section 3.4. of Chapter 5).[62] Therefore, favorable prealignment of the γ hydrogen with respect to the carbonyl chromophore should favor the Norrish type II reaction. In flexible systems conformational freedom lowers the efficiency of the type II process (Scheme 5). In this context, application of the inclusion strategy has been valuable. By specifically including the reactive conformer in a host system one can prealign the molecule for the type II reaction. Therefore, in the case of γ-hydrogen abstraction the "prealignement" essentially refers to 'conformational control of the reactive ketone. We illustrate below with several examples how CD can selectively include a particular conformer of some ketones within its cavity.

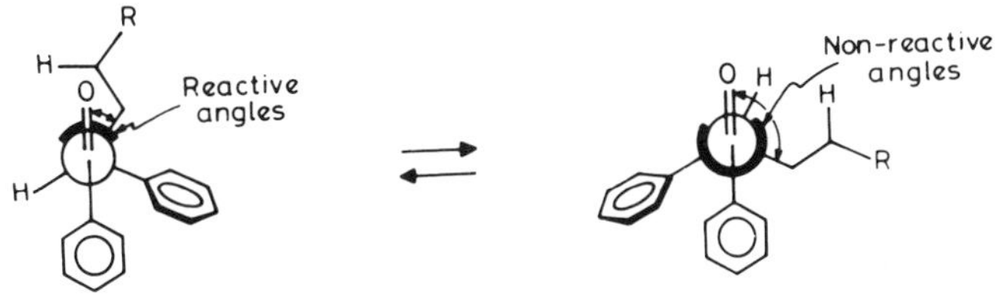

Scheme 5

A remarkable effect was observed by Reddy et al. on the photoreactivity of benzoin alkyl ethers **4**, and α-alkyldeoxybenzoins **5**, upon inclusion in β-cyclodextrin.[63-65] Benzoin alkyl ethers are known to undergo Norrish type I reaction as the only photoprocess in organic solvents (Scheme 6). The competing type II reaction, although feasible in these substances, is not observed at all in organic solvents. Quite interestingly, the solid β-CD complexes of several benzoin alkyl ethers upon irradiation yielded only the type II products in nearly quantitative yields.[63] In Scheme 7 we provide the results on benzoin methyl ether. An examination of Figure 16 reveals that of the two possible representative conformations **A** and **B** available for these substrates, only **B** is capable of undergoing the type II process. The preferable complexation of this conformer in the CD cavity would account for the substantial difference. The observation of the type II process from conformation **B**, trapped inside the CD cavity, is possible only when the competing type I reaction is suppressed by the large cage effect of the cavity (see below). This is indeed the case as shown by the results of the

photolysis of solid complexes in an aerated atmosphere. Under this condition
oxygen-trapped products of the initially formed type I radicals were isolated.

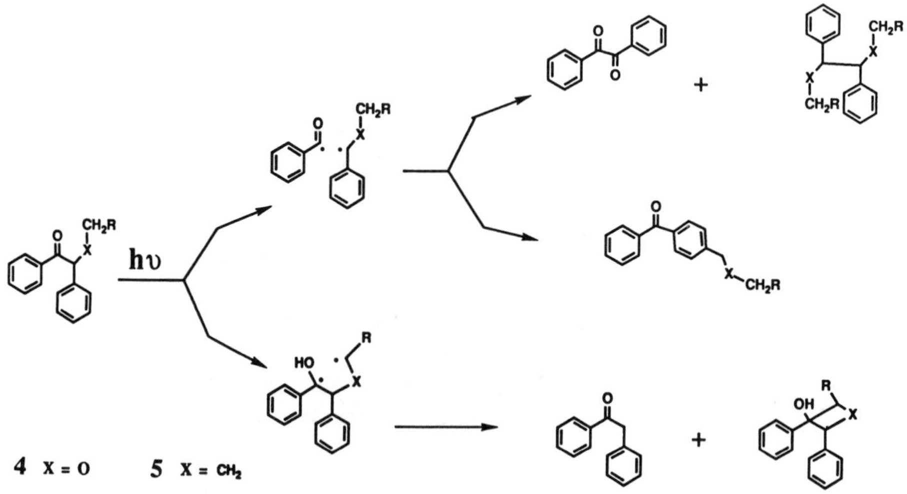

4 X = O   5 X = CH₂

a = H   a = CH₂CH₃
b = CH₃   b = CH₂CH₂CH₃
c = (CH₂)₅CH₃   c = (CH₂)₆CH₃
d = (CH₂)₈CH₃

Scheme 6

The yield of the type II products reaches >60%, suggesting that the majority
of the molecules are trapped within CD in a favorable conformation (above 60%
conversion the reaction slows down since one of the products absorbs the light).
The idea that one can utilize CD to control the conformation of benzoin ethers is
supported by additional examples. On the basis of the behavior of the short-chain
alkyl benzoin ethers, it was anticipated that long-chain alkyl ethers would exhibit
different behavior.[64] It was speculated that a longer alkyl chain would prefer to
reside inside the cavity and therefore conformation A would be preferred instead of
B, preferred by short chains (Figure 17).

In such a case even in the presence of cage control there is no possibility of
observing the type II process. Results obtained with benzoin hexyl ether, benzoin
octyl ether, and benzoin decyl ether support this role of conformational control by
the CD cavity. Photolysis of the complexes of the above three ethers, both in
aqueous solution and in the solid state, gave essentially the type I products—
benzil, benzaldehyde, and pinacol ethers (Scheme 8).

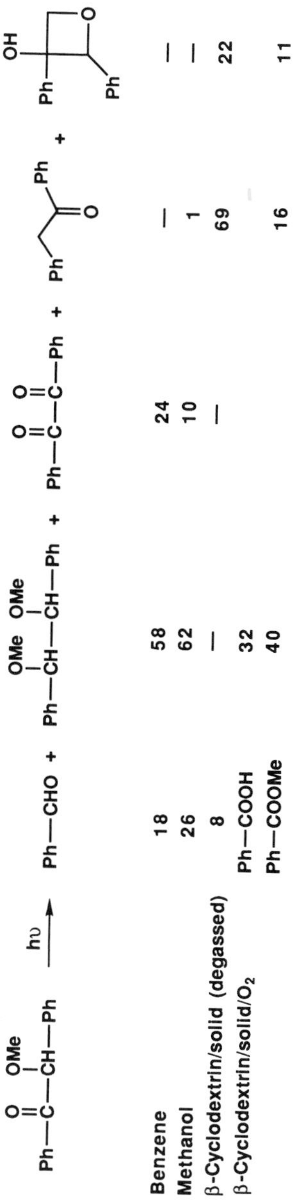

Scheme 7

$$\underset{\text{OMe}}{\underset{|}{\text{Ph}-\text{C}-\text{CH}-\text{Ph}}}\ \ \ \overset{h\upsilon}{\longrightarrow}\ \ \ \text{Ph}-\text{CHO}\ +\ \underset{\text{OMe }\ \text{OMe}}{\text{Ph}-\text{CH}-\text{CH}-\text{Ph}}\ +\ \text{O}=\text{C}-\text{C}-\text{Ph}\ +\ \text{Ph}-\text{CO}-\text{CH}_2-\text{Ph}\ +\ \text{(oxetane: OH, Ph, Ph)}$$

	Ph–CHO	Ph–CH(OMe)–CH(OMe)–Ph	O=C–C–Ph	Ph–CO–CH₂–Ph	oxetane
Benzene	18	58	24	—	—
Methanol	26	62	10	1	22
β-Cyclodextrin/solid (degassed)	8	—	—	69	11
β-Cyclodextrin/solid/O₂				16	
Ph—COOH		32			
Ph—COOMe		40			

Scheme 8

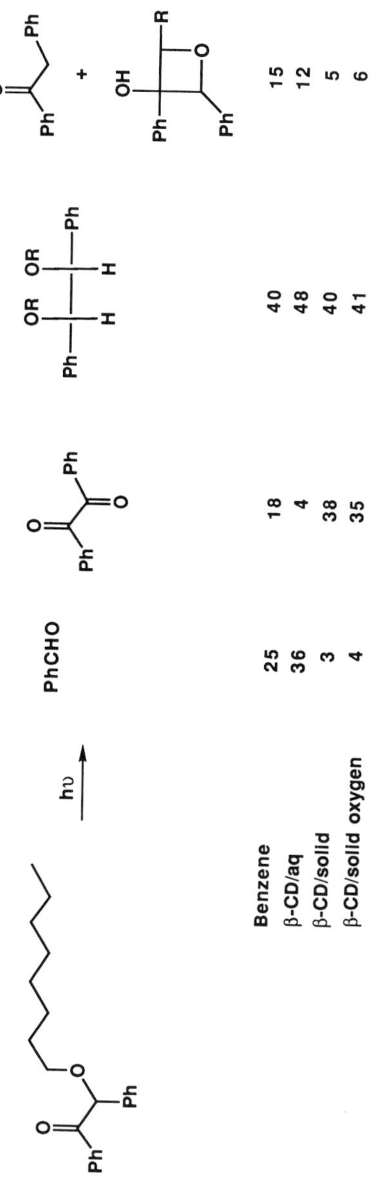

$$\text{(long-chain alkyl)}-\text{O}-\underset{\text{Ph}}{\overset{}{\text{CH}}}-\text{CO}-\text{Ph}\ \ \ \overset{h\upsilon}{\longrightarrow}\ \ \ \text{PhCHO}\ +\ \text{O}=\text{C}-\text{C}=\text{O (Ph, Ph)}\ +\ \underset{\text{Ph}}{\overset{\text{OR}\ \ \text{OR}}{\text{C}-\text{C}}}\ +\ \text{(oxetane: OH/R, Ph, Ph)}$$

	PhCHO	benzil	OR–CH(Ph)–CH(Ph)–OR	oxetane
Benzene	25	18	40	15
β-CD/aq	36	4	48	12
β-CD/solid	3	38	40	5
β-CD/solid oxygen	4	35	41	6

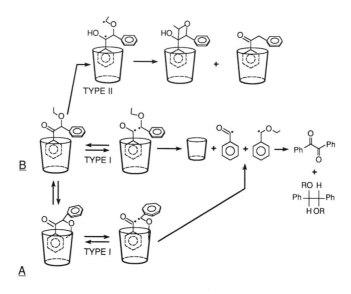

Figure 16. Inclusion of benzoin methyl ether into β-CD

Figure 17. Dependence of conformational preference on the α−alkyl chain length. Hydrophobic effect is speculated play a role in controlling whether the phenyl or the alkyl group is accommodated within the cavity.   Smaller chain prefer to stay outside with respect to the phenyl where as longer chain prefers to enter the cavity.

Another example of the prealignment effect is provided by alkyldeoxybenzoins.[65] Several α-alkyldeoxybenzoins **5**, (α-ethyl-, α-propyl-, α-butyl-, α-octyl-, and α-dodecyldeoxybenzoins, Scheme 6) formed complexes with CD. The type II pathway, generally the major reaction pathway in benzene, was further enhanced both in aqueous CD solution and in the solid CD complex and their behavior resembled that of benzoin alkyl ethers, a closely analogous system (Scheme 9, see page 334). Predictable behavior observed with several alkyldeoxybenzoins on the basis of the understanding of the benzoin alkyl ethers provides further support for the ability of the CD cavity to prealign the reactive centers in ketones.

It is conceivable that the conformation preferred in the CD cavity may vary from one system to another depending on the structure of the guest. A comparison between the α-alkyldeoxybenzoins and the α-alkyldibenzyl ketones, **6**, highlights the differences in the conformational preferences between 1,2-diphenyl and 1,3-diphenyl systems in the cavity of CD (Scheme 10).

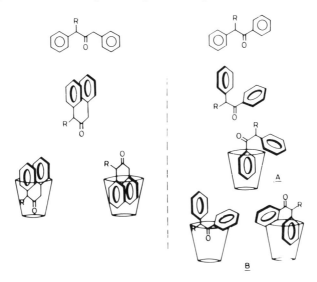

Scheme 10

Once again, molecules capable of both type I and type II reactions were chosen for examination by Rao et al.[66] α-Alkyldibenzyl ketones (Scheme 11) gave products resulting from the Norrish type I and type II reactions upon photolysis in benzene and methanol. Photolysis of the solid CD complex gave only **AB**. The results obtained with a few of **6** are provided in Scheme 12. The absence of type II products both in aqueous solution and in the solid state suggests that CD imposes a conformation on α-alkyldibenzyl ketones that is not suitable for γ-hydrogen abstraction. Probable structures for the CD complexes of α-alkyldibenzyl ketone and α-alkyldeoxybenzoin are shown in Scheme 10.

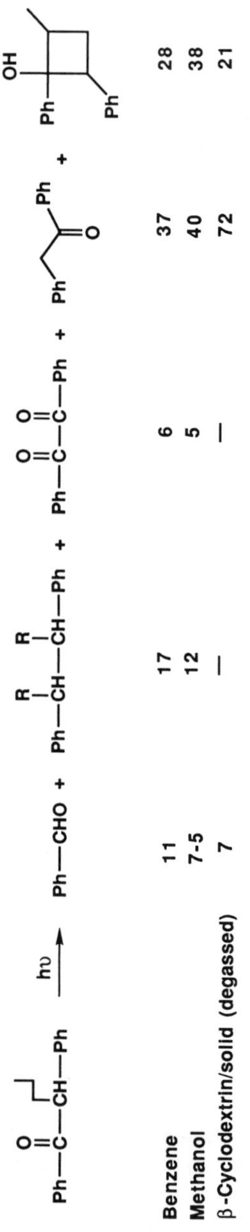

$$\text{Ph—C(=O)—CH—Ph} \xrightarrow{h\nu} \text{Ph—CHO} + \text{Ph—CH(R)—CH(R)—Ph} + \text{Ph—C(=O)—C(=O)—Ph} + \text{Ph—CH}_2\text{—C(=O)—Ph} + \text{(cyclobutanol: Ph, Ph, OH, R)}$$

	Ph—CHO	Ph—CH(R)—CH(R)—Ph	Ph—CO—CO—Ph	Ph—CH₂—CO—Ph	cyclobutanol
Benzene	11	17	6	37	28
Methanol	7·5	12	5	40	38
β-Cyclodextrin/solid (degassed)	7	—	—	72	21

Scheme 9

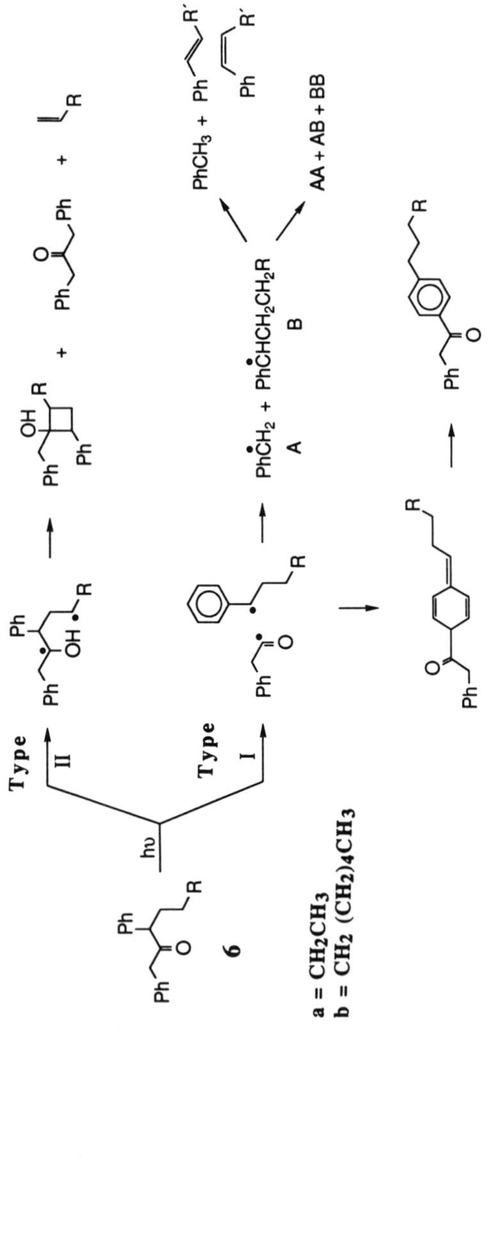

Scheme 11

a = CH₂CH₃
b = CH₂ (CH₂)₄CH₃

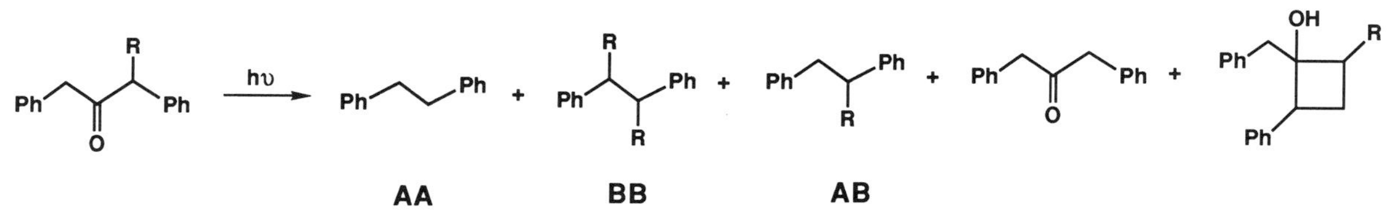

$$R\text{-}CH(Ph)\text{-}CO\text{-}CH_2\text{-}Ph \xrightarrow{\ h\upsilon\ } AA + BB + AB + (Ph\text{-}CH_2\text{-}CO\text{-}CH_2\text{-}Ph) + \text{cyclobutanol}$$

	AA	BB	AB		
R = CH₂—CH₃					
Benzene	21	21	45	13	—
Methanol	23	21	44	13	—
β-Cyclodextrin (solid)	<0.5	<0.5	99	·	—
R = CH₂—CH₂—CH₃					
Benzene	23	22	45	4	7
Methanol	21	19	39	4	8
β-Cyclodextrin (solid)	<0.5	<0.5	99	—	—
R = CH₂—CH₂—CH₂—CH₃					
Benzene	17	17	38	14	19
Methanol	17	17	35	15	19
β-Cyclodextrin (solid)	<0.5	<0.5	99	—	—

Scheme 12

Goswami et al. measured the type I to type II ratio for several ketones in Dianin's complex.[67] Results on two of these are summarized in Scheme 13. The type I to type II ratio increases significantly in proceeding from liquid media to the environment of the solid Dianin's complex. This could once again be the result of selective inclusion of a particular conformer in the complex. It is likely that the unfavorable conformer (for type II) is preferred over the favorable one in the Dianin's complex.

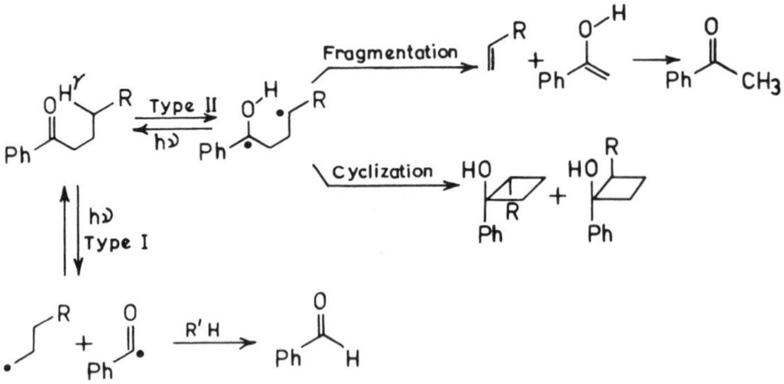

Guest Ketone	Medium	F/C	t/c	Type I/ Type II
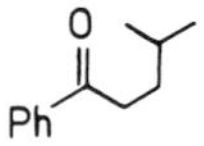	benzene	7.3		
	tert-butyl alcohol	15.7		
	complex	27.6		
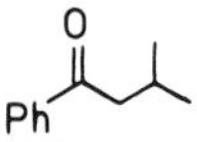	benzene	7.3		
	tert-butyl alcohol	9.5		
	complex	6.1		
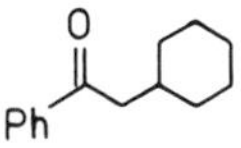	benzene	0.5	•2.2	
	tert-butyl alcohol	0.5	1.3	
	complex	0.9	1.9	
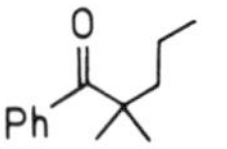	benzene	0.2	1.6	<0.1
	tert-butyl alcohol	0.5	1.1	0.1
	complex	0.9	1.7	0.6
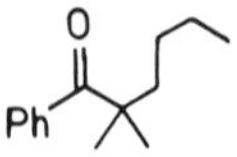	benzene	0.3	1.6	<0.1
	tert-butyl alcohol	0.7	1.1	<0.1
	complex	0.7	1.7	0.7

Scheme 13

Vincens has attempted to utilize this prealignment concept with **2** as the host.[68]   He has investigated the Norrish type II reaction of butyrophenone and valerophenone complexed to **2**. Surprisingly no reaction was observed with butyrophenone, while valerophenone gives both elimination and cyclization products. He attributes this to conformational control by the host matrix. It is not clear why the matrix should favor a nonreactive conformer in the case of butyrophenone and the reactive conformer in the case of valerophenone.

5.1.4. Photo-Fries and Related Rearrangements

Of the many different ways in which CD can prealign the molecules for a reaction, one may be by sterically blocking certain potential sites of the substrate from intermolecular attack by their very mode of complexation. Hence it can be expected that unimolecular rearrangements that involve an initial cleavage, followed by reorganization of the fragments, when conducted in CD, will give rise to one particular isomer specifically. Syamala et al. have succeeded in exploiting this unique feature of CD complexation in the photo-Fries, photo-Claisen, and related rearrangements.[69,70]   In this approach the prealignment is achieved through selectively blocking certain reactive sites and thus channeling the reagents to a particular site.

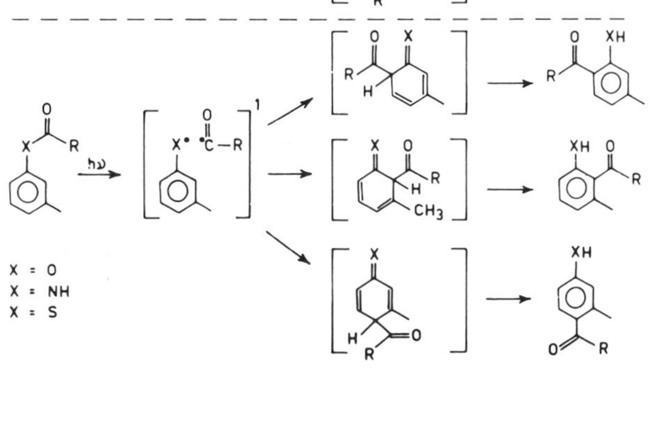

Scheme 14

   The photo-Fries rearrangement of phenyl esters is thought to involve an initial cleavage of the phenyl ester to produce a phenoxy and an acyl radical entrapped in a solvent cage which may recombine in three different ways (Scheme 14).[71] Any escape of the phenoxy radical from the solvent cage with subsequent hydrogen abstraction from the solvent would result in the formation of phenol. When the phenyl ester has a meta substituent, the photo-Fries rearrangement can then produce two ortho-rearranged products along with the para and the deacylated products (Scheme 14). The corresponding photorearrangement of either the unsubstituted or the meta-substituted anilides is expected to follow a pathway similar to that of the esters. The most striking observation was made by Syamala et al. with the solid cyclodextrin complexes of several phenylesters and anilides, which upon photolysis yielded only the ortho-rearranged product (results on one such system is summarized in Scheme 15).[69,43]

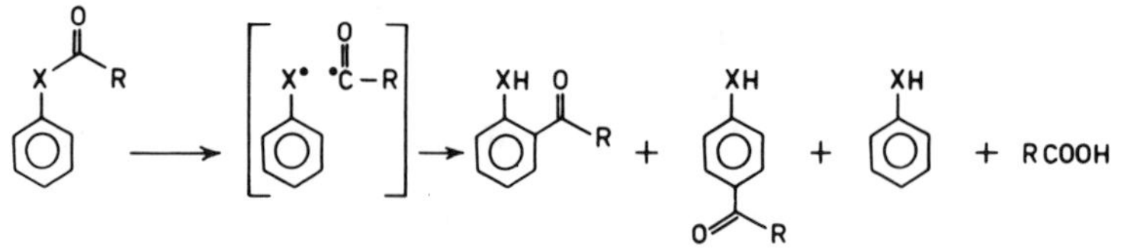

X = O; R = Ph

MeOH	48	30	14	8
C_6H_6	55	30	15	—
β-CD-H_2O (1:10)	~99	≃1	—	—
β-CD solid	~99	≃1	—	—

X = NH; R = Ph

EtOH	65	35	—	—
β-CD-H_2O (1:10)	96	4	—	—
β-CD solid	100	—	—	—

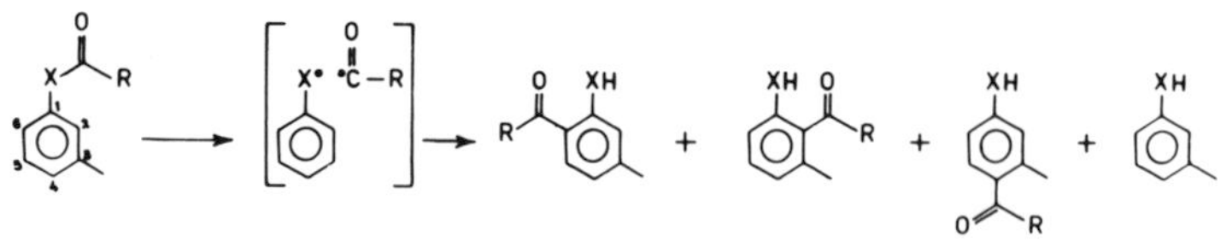

X = O; R = CH_3

MeOH	31	20	18	41
C_6H_6	36	37	13	14
β-CD-H_2O (1:10)	43	45	0.4	12
β-CD solid	94	5	—	—

X = NH; R = CH_3

EtOH	43	26	30	—
β-CD-H_2O (1:10)	75	18	6	—
β-CD solid	91	9	—	—

Scheme 15

The remarkable observation of "ortho-selectivity" in the solid β-cyclodextrin complexes is in accordance with the formation of inclusion complexes with structures shown in Scheme 16. It can be seen that in such an orientation, the β-cyclodextrin sleeve protects the para position of the aromatic ring from attack by the acyl radical exposing the two ortho positions for the reaction. Further, the tight packing of the surrounding molecules, in a solid complex, forces the acyl radical either to recombine with the phenoxy radical to regenerate the starting material, or to attack the exposed ortho positions (Scheme 16). Scheme 16 also shows the most probable orientation in the β-cyclodextrin complex for meta-substituted derivatives. It may be seen that in such an orientation, the cyclodextrin sleeve encircles the molecule in such a way that only one of the two ortho position is exposed for attack. Thus the acyl radical, once formed, is compelled to move toward the only accessible 6-ortho position in the solid β-cyclodextrin complexes (Scheme 16).

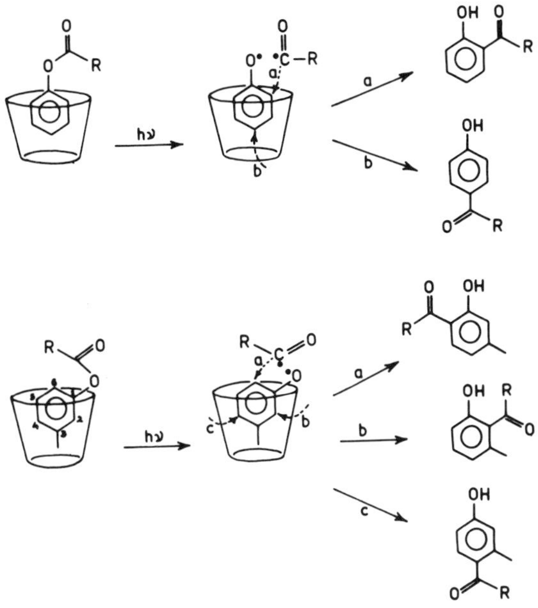

Scheme 16

A tight fit between the host and the guest molecules would be expected to be necessary to bring about maximum selectivity in photoreactions. To fulfill this criterion a suitable choice of the host system and the guest would be needed. Results obtained in the photo-Claisen rearrangement of *m*-alkoxyphenyl allyl ethers highlight the importance of this criterion (Scheme 17).[70] Out of the two possible ortho isomers and the para isomer that are formed during photolyses in organic solvents, only one ortho isomer was obtained, with a remarkable selectivity, upon irradiation of the α-cyclodextrin complexes of *m*-alkoxyphenyl allyl ethers. At the same time, the β-cyclodextrin complexes of *m*-alkoxyphenyl allyl ethers did not yield any significant selectivity (Scheme 18).

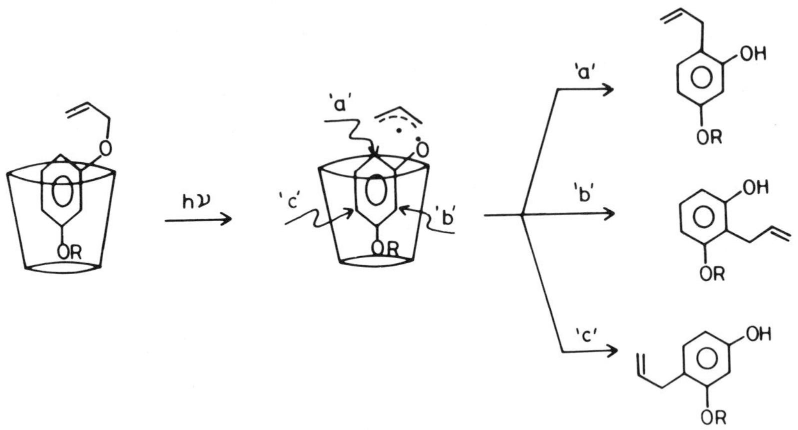

Scheme 17

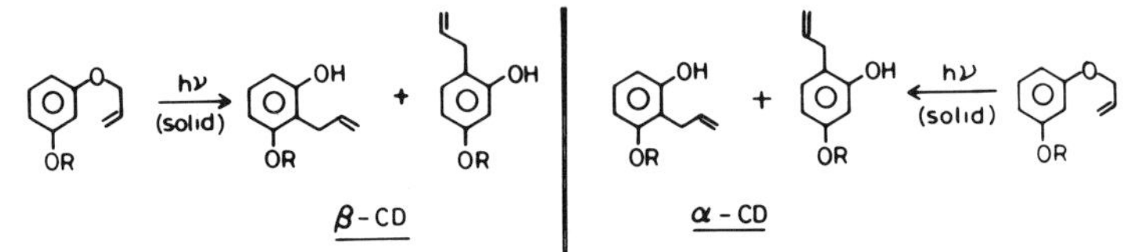

	β-CD		α-CD	
R = Methyl	62	30	0	100
Propyl	51	49	5	95
Hexyl	24	76	12	88
Octyl	10	90	12	88
Dodecyl	25	75	36	64

Scheme 18

While α-cyclodextrin, with a smaller cavity was able to bring about selectivity, the larger cavity of β-cyclodextrin probably failed to hold the molecule tightly. The tight fit necessary for achieving selectivity in β-cyclodextrin can apparently be provided simply by increasing the space filling capacity of the substrate, e.g. by adding a long alkyl chain (Scheme 19). Thus, with increasing chain length of the substituent the selectivity observed in β-cyclodextrin increased (Scheme 18). At the same time, too long an alkyl chain might be expected to occupy most of the cavity and push the aromatic ring outside to expose the ortho positions to attack. This would lead to no selectivity at all in the cases of both α- and β-cyclodextrin complexation. This was indeed the case with *m*-dodecyloxyphenylallyl ether which did not give rise to significant selectivity upon photolysis as CD complex. Thus it is possible to manipulate the substituents to achieve tight complexation to bring about selectivity. Proposed variation in the complex structure with the chain length is shown in Scheme 19.

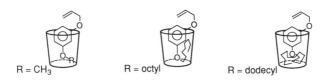

R = CH₃   R = octyl   R = dodecyl

Scheme 19

In this section we have illustrated, with several examples belonging to different classes, that one can pre-align reactive molecules with the help of the inclusion strategy. The strategy required for each reaction is different; yet with the judicious choice of the host system one can engineer the reactants toward a particular reaction.

5.2. Restriction of Motions

Guest molecules enclosed in restricted spaces such as cages and channels are expected to experience less freedom of motion (translational and rotational) when compared to that in isotropic solvents. Solid-state nmr studies of several inclusion complexes have shown that such is indeed the case.[72] Restriction of translational and rotational motions of the reactant molecules as well as that of the reactive intermediates (derived from the reactant molecules) often tends to be reflected in the product distribution. There are several such examples in the literature and these are highlighted below.

5.2.1. Translational Motion

Photolysis of dibenzyl ketones and benzylphenyl acetates in isotropic solvents yields products resulting from the coupling of the two radicals derived by the elimination of carbon monoxide and carbon dioxide, respectively.[73] In general, unsymmetrical starting ketones and esters yield all three possible products namely **AA**, **AB** and **BB** (Scheme 11). Lack of translational motion for the secondary radical pair became evident from the selective formation of the product **AB** from several dibenzyl ketones and benzylphenyl acetates included in DCA, Dianin's compound and β-cyclodextrin.[74] Further, the absence of any rearrangement product 7 (Scheme 11) in any of these cases clearly shows that both the primary and the secondary radical pairs do not undergo translational motion. Such a restriction of translational motion is termed the *cage effect*.

5.2.2. Conformational Motion

In this section, *rotational motion* refers to rotation about a C—C bond and not to the rotation of a molecule as a whole. An example of the restriction of the rotation around the C—C bond is provided by the behavior of the 5-hexenyl radical generated by the photolysis of 6-heptenoyl peroxide (Scheme 20) in the channels of urea.[75] In this case >80% of the radicals failed to undergo cyclization, the favored process in solution. It has been suggested by Griller et al. that the folding process is inhibited by the channel, as the diameter of the cyclized radical is too close to that of the urea channel.

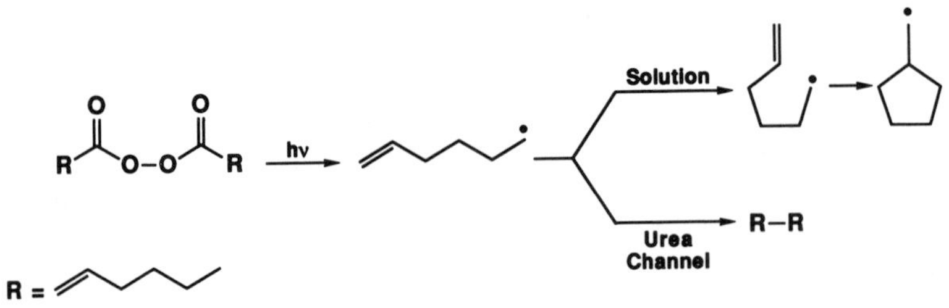

Scheme 20

Another such example is the photocyclization of 5-nonanone in the channels of urea.[76] On photolysis 5-nonanone included in urea underwent the Norrish type II process to yield 2-hexanone (the fragmentation product) and the cis isomer of the cyclobutanol (Scheme 21). Both 2-undecanone and 2-hexanone show the same type of behavior; i.e., only one isomer of the cyclobutanol product is obtained.

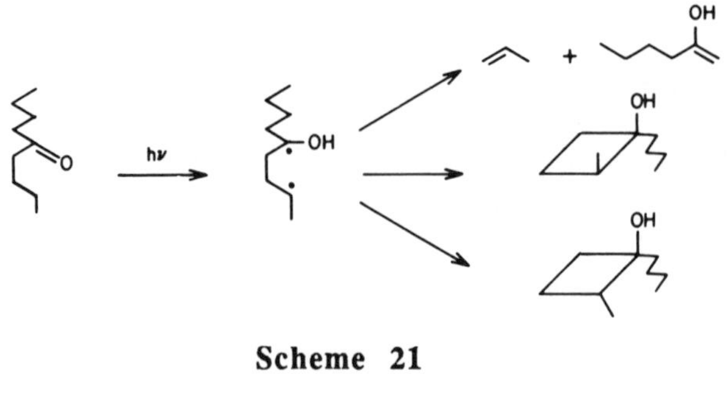

Scheme 21

It is important to note that no *trans* cyclobutanol was obtained in any of these cases. de Mayo, Scaiano, and co-workers believe that between the two isomers trans and cis, the formation of the latter requires minimal rotational motion and only that much motion is tolerated by the medium. Although such a remarkable effect was not obtained with arylalkyl ketones in Dianin's complex, a modest increase in fragmentation to cyclization ratio was recorded in going from solution media to Dianin's complex (Scheme 13).[67] This was attributed to the restriction brought on by the medium on the rotation around the C—C bond in the 1,4-diradical. The lower selectivity noted here compared to the urea as the host is understandable on the basis of the size of the channel (Dianin's > urea's).

Another example of rotational restriction of intermediate diradicals is provided by the extensive studies carried out by Reddy et al. on CD complexes of arylalkyl ketones and alkyldeoxybenzoins.[65, 77] The elimination to cyclization ratio (E/C; the diradical decays by an elimination process to give acetophenone and the corresponding olefin and by cyclization to yield cyclobutanol) was measured for several arylalkyl ketones and alkyldeoxybenzoins included in CD in the solid state and in aqueous solution (Scheme 22). The results are summarized below: (a) the E/C ratio is always lower in CD than in *tert*-butanol; (b) the extent of decrease depends on the length of the alkyl chain; and (c) the influence of CD is greater in the solid state than in aqueous CD.

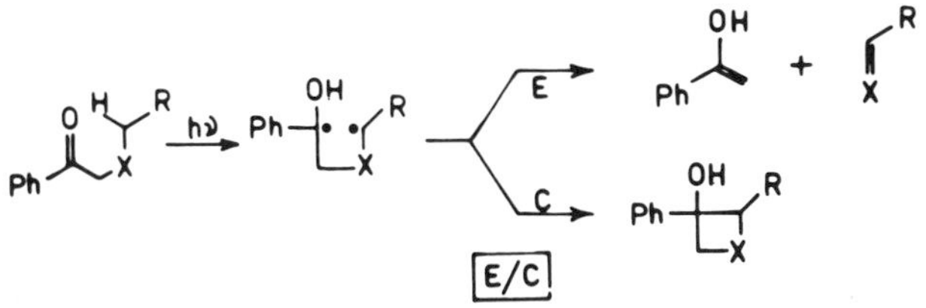

	Benzene	t-Butanol	β-CD soln (1 : 1)	β-CD solid
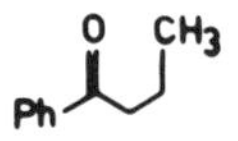 Ph—CO—CH_3	6.56	8.55	3.88	3.5
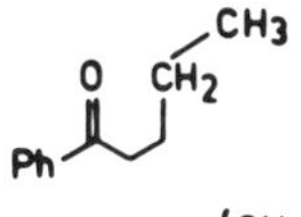 Ph—CO—CH_2—CH_3	3.03	4.22	3.82	2.65
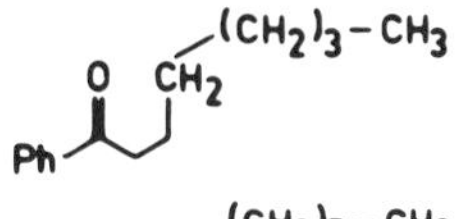 Ph—CO—CH_2—$(CH_2)_3$—CH_3	1.2	2.55	1.76	0.79
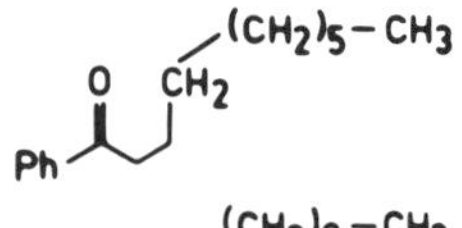 Ph—CO—CH_2—$(CH_2)_5$—CH_3	2.48	3.28	1.64	0.69
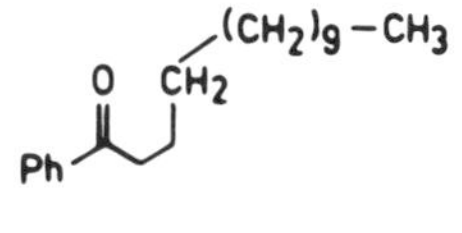 Ph—CO—CH_2—$(CH_2)_9$—CH_3	1.58	2.88	1.42	0.35

Scheme 22

The authors rationalize their observations on the basis of the influence of CD on the rotational freedom of the central C—C bond in the 1,4-diradical derived via the γ-hydrogen process. We only briefly refer to the mechanism of Norrish type II reaction here (refer to Section 3.4 of Chapter 5 and Section 8 Chapter 14 for details) to illustrate the effect of CD on the 1,4-diradical decay. As illustrated in Scheme 23, excitation of the aryl alkyl ketone included in the CD cavity would result in a 1,4-diradical in which the two singly occupied p orbitals are perpendicular to each other. The triplet diradical that is generated in the skew form readily equlibrates in isotropic solvents between the cisoid and the transoid forms, the transoid being more favored in protic solvents such as *tert*-butanol. It may be noted that the transoid form can undergo only fragmentation while the cisoid form can cyclize in addition to the fragmentation. In the CD complex, the phenyl group is locked inside the host cavity, so the rotations required for skewed–cisoid–transoid interconversion have to occur only by the rotation of the alkyl portion of the intermediate. It is speculated that the conversion of the skewed to the transoid conformer would require larger motion than conversion of skewed to the cisoid form. Such a large motion would be restricted by the CD matrix; the longer the alkyl chain, the greater the restriction. Therefore, conversion of skewed to cisoid would be favored in the CD cavity over to the transoid form. On this basis, more cyclization would be expected in CD complex than in solvent media such as *tert*-butanol and it is indeed the case.

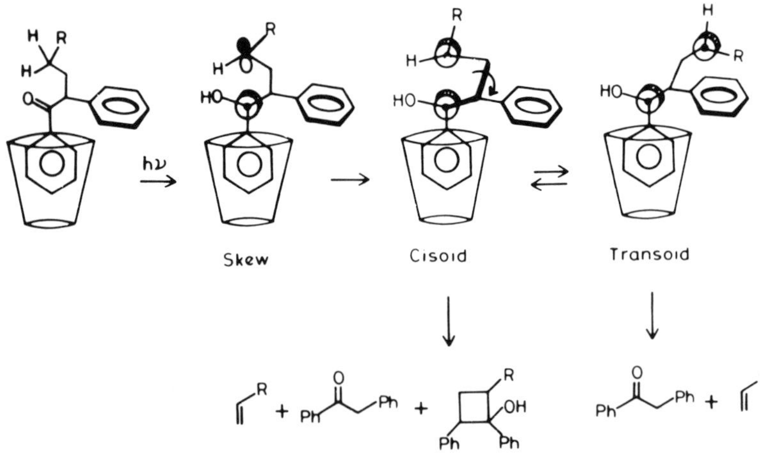

Scheme 23

Such a situation is further illustrated by the photobehavior of 2-phenylcycloalkanones included in the cavities of CD (Scheme 24).[78] These ketones undergo type I cleavage to yield a diradical. This diradical follows two pathways, namely disproportionation and coupling with rearrangement. CD favors the disproportionation reaction (to yield enal) over the radical coupling process and that the extent of the influence depends both on the size of the ketone and on the cavity.

The coupling reaction (to yield *p*-cyclophane) requires larger motion of the radical center than does disproportionation. Rao, Han and Turro proposed that such a large motion required for coupling (with rearrangement) is restricted by the CD cavity. The relationship between the selectivity and the sizes of the guest and the cavity (α-, β- and γ-CD) suggests that the extent of restriction depends on the tightness of inclusion.

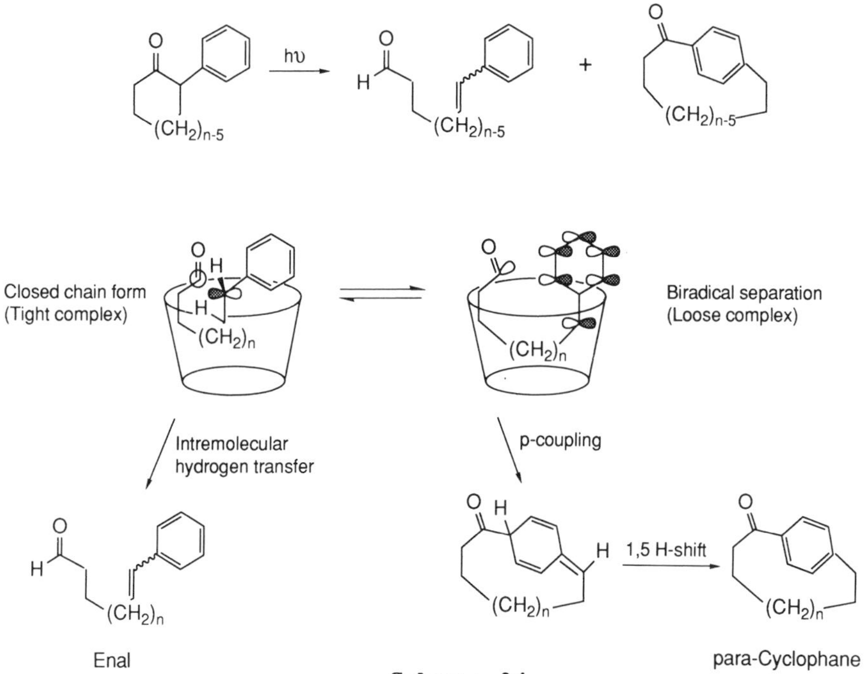

Closed chain form (Tight complex)

Biradical separation (Loose complex)

Intremolecular hydrogen transfer

p-coupling

Enal

para-Cyclophane

1,5 H-shift

Scheme 24

Other examples of restricted rotation involve geometric isomerization of stilbenes and methyl cinnamates as TOT [79] and DCA [80] complexes. While isomerization of *cis*-stilbene to *trans*-stilbene occurs readily both in TOT and in DCA matrix the corresponding trans to cis isomerization does not take place. On the other hand, photointerconversion between *trans* and *cis* methyl cinnamates takes place readily in the above two complexes. Similar observations with stilbenes and alkyl cinnamates have been made with CD complexes in aqueous solution but this will not be discussed here as it does not pertain to the solid state.[81]

Photoisomerization of stilbene and methyl cinnamate in TOT has been subjected to extensive investigation by Arad-Yellin et al.[79] The crystal structure of *trans*-stilbene–TOT complex has two channels in which *trans*-stilbene molecules are accommodated (Figure 18). The void volumes available for guest accommodation are different in these two channels: ~570 Å^3 (a- axis) and ~830 Å^3 (b- axis). In the case of the *cis*-stilbene–TOT complex, the guest molecules could not be located. However, the authors suggest that the *cis*-stilbene complex has a structure similar to that of the trans isomer although the exact position of the guest might be slightly different.

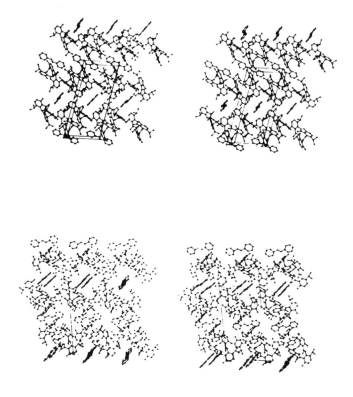

Figure 18. Stereo-packing arrangement of *trans*-stilbene–TOT complex illustrating the presence of channels along two directions. (a) Viewed along the a-axis and (b) viewed along the b-axis. (Ref. 79).

The occupancy level of *cis*-stilbene in the TOT complex is lower than that of *trans*-stilbene in its TOT complex, leaving more void space within the *cis*-stilbene–TOT structure. Another point Arad-Yellin et al. would like us to note is the symmetry property of the guests: except for *trans*-stilbene, other guests, namely *cis*-stilbene and methyl cinnamates, do not possess a center of symmetry. Due to symmetry mismatching between the TOT complex (P$\bar{1}$) and the guests (except for *trans*-stilbene), the guest is expected to be disordered. The facile cis to trans isomerization of stilbene is suggested to be favored by the low cage occupancy in the TOT–*cis*-stilbene complex and by the coincidence of the *trans* guest symmetry with that of the cavity. This symmetry matching is considered to be a major element of control over the reaction since the non-centrosymmetric *cis*-

stilbene never develops the same complementarity with the centrosymmetric cavity as the ordered product; hence the marked preference for *trans*-stilbene over *cis*-stilbene. The stabilizing, directive influence of guest–cage symmetry coincidence is further supported by the reaction pattern of the methyl cinnamate isomers which are disordered in their respective cavities. Both undergo photoisomerization in the TOT complexes.

Guarino reports that the *trans*-stilbene–DCA complex is photoinert.[80] On the other hand, *cis*-stilbene in *cis*-stilbene–DCA complex isomerizes to the trans, but unlike the TOT complex it does not convert to 100% trans isomer. Isomerization proceeds only up to 48%. The similarity in photobehavior between methyl cinnamates–DCA complexes and methyl cinnamates–TOT complexes is obvious. Both *cis* and *trans* isomers undergo geometric isomerization upon photolysis and establish a stationary state of ~30% trans in DCA and ~50 % in TOT. No structural details are available. Gurino attributes these differences in behavior to the symmetry mismatching between the host and the guest.

It is clear from the above photoisomerization studies of DCA and TOT complexes that rotation about the C=C bond in the excited state in these media is restricted to some degree. One should keep in mind that the space required for positional change of the end groups during isomerization in cinnamates (COOMe) is smaller than in stilbenes (phenyl). It is quite likely that this difference in spatial requirement accounts for the difference in restriction experienced by the two molecules. Further, between the *trans*- and *cis*-stilbenes the latter, being bent and bulky, would be expected to be loosely packed, thus allowing space for geometric isomerization upon photoexcitation.

5.3. Enantioselective Reactions

Enantioselective reactions, otherwise known as asymmetric transformations, in host–guest assemblies tend to be of the "show and tell" variety. Several examples in the literature need to be examined in detail before any order can be brought to this area. As discussed in Section 2 there are a large number of host systems that are either inherently chiral or attain chirality in the crystalline state. These host systems have attracted considerable attention as a medium to conduct chiral transformations. As opposed to liquid crystalline media, where attempts to use them as the media for chiral transformations have most often been met with failure (see Chapter 14), there are several successful examples in host–guest assemblies in the solid state. One can choose a chiral host to drive the guest molecules to pack in the cavity/channel in an asymmetric fashion. Since disorder is common in guest packing in the host matrix, predictability with respect to the feasibility of enantioselective reaction has not been high.

γ-Ray-induced polymerization of 1,3-*trans*-pentadiene included in PHTP and DCA gave optically active polymer.[82] Although selectivity was not high, these were the first example that clearly established the feasibility of obtaining enantioselectivity in the photoproducts with the use of the host–guest strategy in the solid state. Since the initial report, PHTP has not attracted further interest as a medium for asymmetric transformations. However, DCA has been utilized as the matrix for such reactions by Lahav, Leiserowitz, and co-workers.[83] Their detailed and elegant studies are summarized in Chapter 6 and will not be repeated here. Aoyama et al. succeeded in obtaining enantioselective photoreactions of N,N-dialkylpyruvamides as DCA and cyclodextrin complexes (see Scheme 10 in Chapter 6).[84] Enantioselectivity was not high in CD (enantiomeric excess <10%). In this context a recent report by Rao and Turro is of interest.[85] Photolysis of β-CD complexes of benzaldehyde gave benzoin as the major product with an enantiomeric excess of ~15% (Scheme 25). The absence of enantioselectivity in γ-CD prompted them to suggest that tight complexation is essential for optical induction.

An elegant example of chirality transfer during singlet oxygen oxygenation of Z-2-methoxybut-2-ene in TOT complex is provided by Gerdil et al.[86] Complexes of Z-2-methoxybut-2-ene with TOT can be separated into crystals of opposite handedness by visual examination. Thus separated and characterized (by polarimetry and by x-ray analysis), the crystals were photolyzed in the presence of Rose Bengal to give the hydroperoxide in 100% yield (Scheme 26).

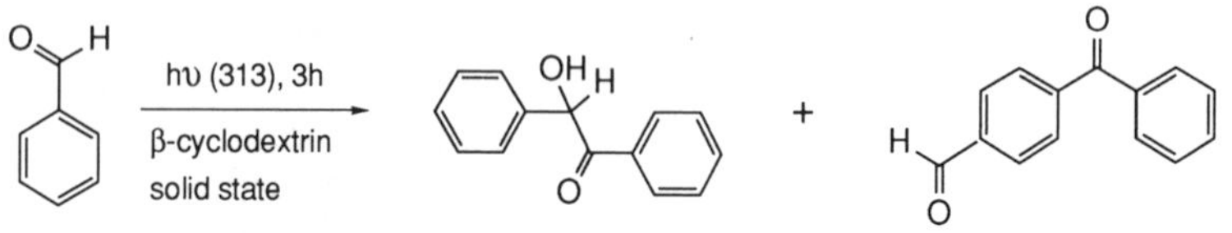

hυ (313), 3h
β-cyclodextrin
solid state

56%   24%

R-(-)-Benzoin
$[\alpha]^{25}$ -18.8 ± 0.4 (c = 1.5, acetone)
% e.e. = 15 ± 1%

Scheme 25

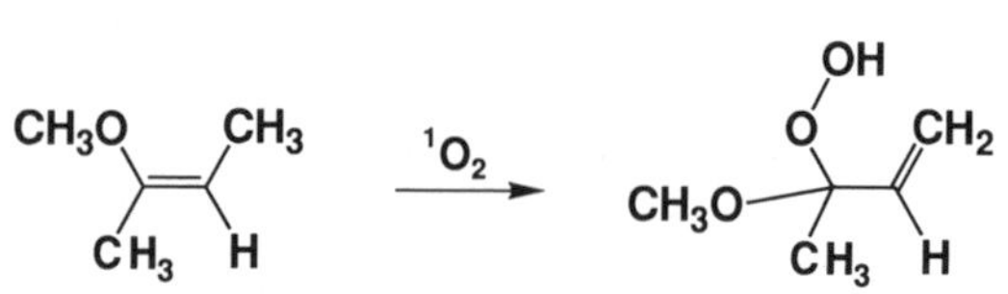

Scheme 26

Surprisingly, the peroxide obtained from the two opposite-handed crystals possessed opposite optical rotation (+0.060 and -0.051), corresponding to a substantial enantiomeric excess. x-Ray structural analysis of the complex reveal that the olefin is disordered and s-cis conformations with torsion angles C=C—O—C of +60° and -60° are equally populated (Scheme 26). This suggested that the chirality transfer must occur not at the inclusion stage but in the transition state of the singlet oxygen addition to the olefin. Unfortunately, no followup work has been carried out either by the same group or by others. This therefore stands as a lone example of asymmetric phototransformation in a TOT complex.

A large number of asymmetric transformations with 1,6-bis(o-chlorophenyl)-1,6-diphenyl-2,4-dyine-1,6-diol (3), as the host have been reported by Toda and Kaftory and their co-workers.[87] A list of such reactions is provided in Scheme 11 of Chapter 6. We examine some of them in detail below. Most of these examples involve trapping of a single chiral conformer in the host matrix. The first example involves intramolecular γ-hydrogen abstraction in α-oxoamides.[88] During the photoreaction two asymmetric carbon centers are generated and thus may result in four different stereomeric products as illustrated in Scheme 27.

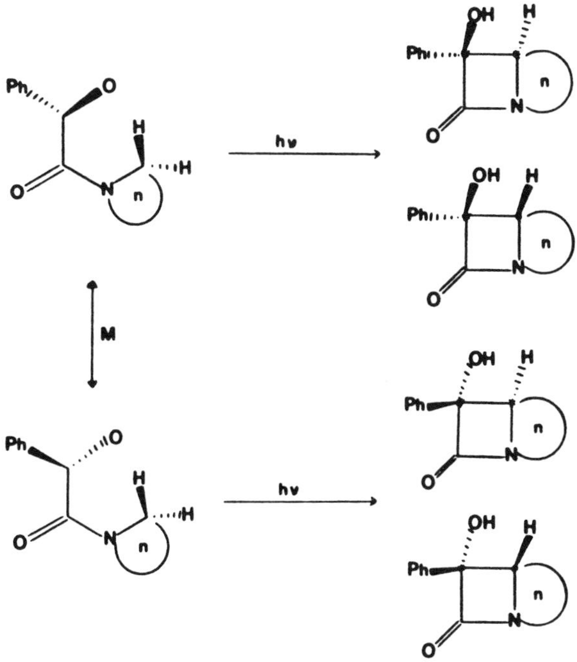

Scheme 27

Toda et al. argued that by controlling the conformation of the reacting α-oxoamide it might be possible to control, to some extent, the enantioselectivity of the reaction. Indeed when 1:1 complexes of several α-oxoamides with 3 were irradiated, optically active products were obtained (Scheme 28). In the case of N, N-dimethyl α-oxobenzenacetamide even 100% enantioselectivity was claimed. The x-ray crystal structure of the complex of *N,N*-dimethyl α-oxobenzenacetamide with 3 indicates that only a single conformer of the guest is included in the complex. The enantioselectivity is controlled by the conformation about the

O=C—C=O single bond. The observed torsion angle is +110° and in the absence of a mirror-symmetry-related molecule (torsion angle -110°), a single enantiomer is obtained. Consistent with this rationale when a 1:2 (host:guest) complex was irradiated no selectivity was obtained. In the 1:2 complex two mirror-symmetry-related conformers are present in the crystal lattice. The results obtained in 3 as a host should be compared with that in TOT as a host with Z-2-methoxybut-2-ene as the guest. In TOT no conformer preference was observed, while 3 showed a preference for one of the two optical (conformationally dependent) isomers of the guest. Although the guests involved are different, it is likely that stronger interaction between 3 and the guest (hydrogen bonding vs. van der Waals forces) might be responsible for the selectivity.

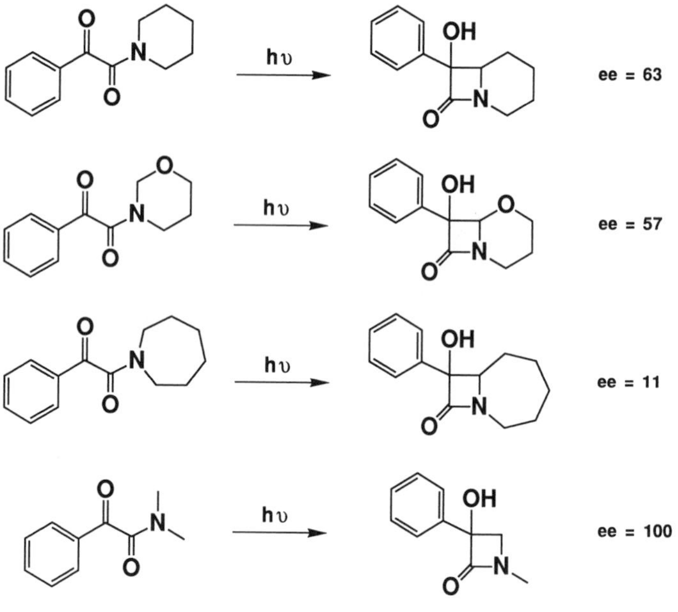

Scheme 28

Results with other molecules, however, are not always as clearcut as the above example. Cycloocta-2,4,6-trien-1-one exists as an equilibrium mixture of two optical isomers; i.e., it has an average plan of symmetry. Photolyses of these in solution give racemic bicyclo[4.2.0]octa-4,7-dien-2-one (Scheme 29). However, when a 2:1 solid complex of cycloocta-2,4,6-trien-1-one and 3 was irradiated, optically active product bicyclo[4.2.0]octa-4,7-dien-2-one was obtained; the optical purity was not established.[89] x-Ray crystal structure analysis of the complex revealed that both conformers were present. The authors attribute the enantioselectivity to the difference in thermal motions of the two conformers in the crystalline matrix. However, it is not obvious why such differences should result in enantioselective cyclization. Based on the behavior of cycloocta-2,4,6-trien-1-

one it is difficult to understand why **3** prefers to complex only with one optical isomer of cycloocta-2,4-dien-1-one. Irradiation of this complex gives optically active dimers. The distance between the reactive double bonds is reported to be > 7.7 Å. In order to rationalize the occurrence of dimerization in spite of the large distance of separation between the reactive double bonds and the enantioselectivity in the dimer obtained Fujiwara et al. suggest several rotational motions and ring flipping (boat to chair form) motions in the crystal.[89] However, it is to be noted that the distance between the reactive double bonds reported by the authors is not correct as unit cell transformation has not been performed. If it is carried out the distance between the reactive double bonds would be well within 4 Å. We hope it is clear from the examples discussed above that it is not possible to predict at this stage whether **3** would include only one of the many conformationally dependent optical isomers in a given case.

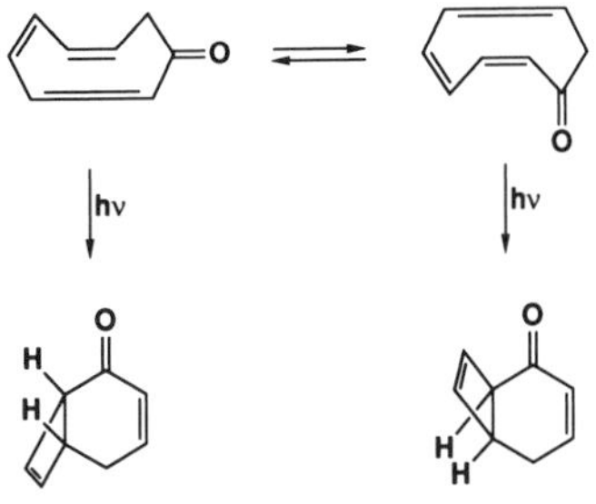

Scheme 29

Another set of examples consists of enantioselective photocyclizations. α-Tropolone alkyl ethers and pyridones upon excitation undergo intramolecular cyclization to the corresponding bicyclic products by disrotatory ring closure.[90] Depending on the direction of disrotation, opposite optical isomers will be formed. Racemic products are obtained in solution, but irradiation of crystalline inclusion complexes of α-tropolone alkyl ethers and pyridones with **3** gave cyclic products of 100% optical purity (Scheme 30). The x-ray structure of the 1:1 α-tropolone ethyl ether complex with **3** shows that the guest molecule is held by hydrogen bonds to two host molecules as illustrated in Scheme 31. On the basis of crystal packing, Kaftory argues that the enantiomeric control results from the chiral environment provided by the host and from the differences in space available at both sides of the planar molecule for rotation (bending) to the product.[88] Similar conclusions have been reached by Fujiwara et al. in the case of pyridones on the basis of the crystal structure of the 1:1 complex of 4-methoxy-1-methylpyridone and **3**.[91]

Another example of enantioselective photoreaction consists of oxaziridine formation from nitrones.[92] Inclusion complexes of several nitrones with 1,6-bis(*o*-chlorophenyl)-1,6-diphenyl-2,4-dyne-1,6-diol gave the corresponding oxaziridines in high optical yields. Enantiomeric excesses varied between 10 and 100% (Scheme 32). No structural details are available.

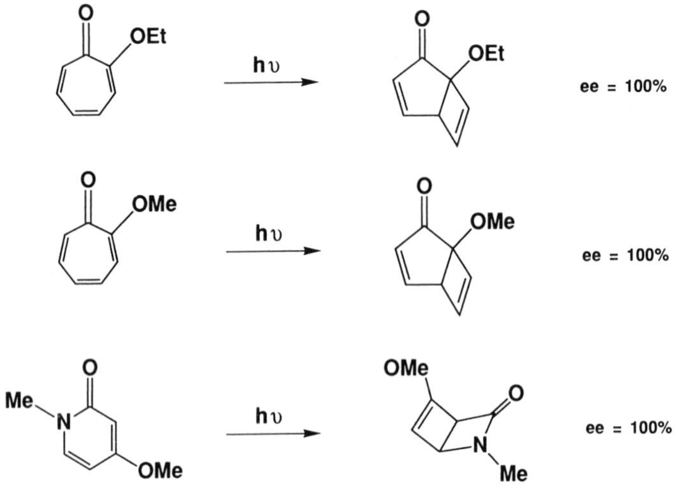

ee = 100%

ee = 100%

ee = 100%

Scheme 30

(1*S*, 5*R*) - (-)   (1*R*, 5*S*) - (+)

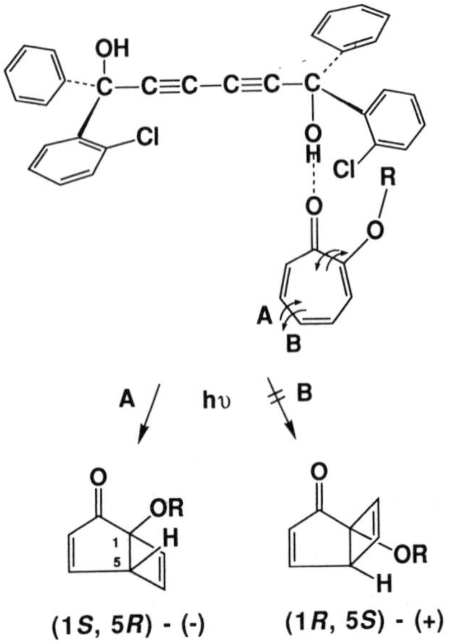

Scheme 31

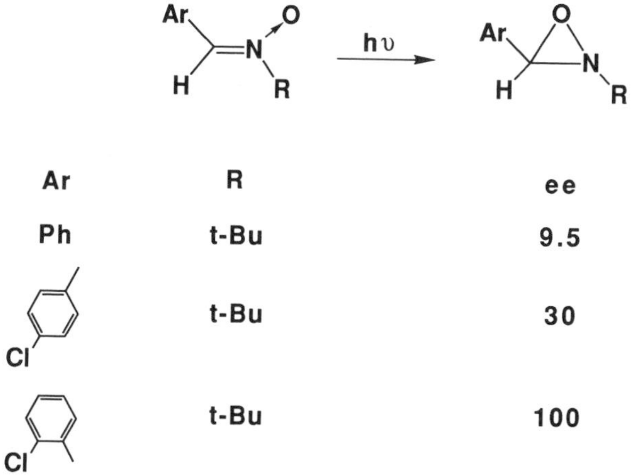

Ar	R	ee
Ph	t-Bu	9.5
(4-chloro, methyl-substituted aryl)	t-Bu	30
(2-chloro, methyl-substituted aryl)	t-Bu	100

Scheme 32

From the above examples it is clear that among the various host systems DCA and **3** have attracted considerable attention as media for asymmetric transformations. Structural details are often utilized to understand the origin of optical selectivity. The origin of enantioselectivity appears to vary with the guest systems, thus making general conclusions difficult.

5.4. Miscellaneous Studies

There are a few reports that do not fit into the categories discussed above and these are presented below. Conversion of norbornadiene to quadricyclene has attracted considerable attention for solar energy storage. Some of the problems in using this reaction as a viable energy source involve its unpleasant odor, flammability, and side reactions. Inclusion complexes of norbornadiene with cyclodextrin and DCA have been examined to alleviate some of these problems.[93] Odor seems to have been eliminated, and to a large extent, side reactions (formation of cyclohepta-1,3,5-triene, 6-methylfulvene, and polymers) on complexation. While the complexes would still be flammable, the flash point would presumably be higher than that of the liquid norbornadiene or quadricyclene. An important observation to note is that in both of these systems the triplet energy transfer technique has been used to bring about efficient rearrangement. In CD complex, acetophenone, and in DCA complex, 3,3-dimethyl-2-butanone have been used as the triplet sensitizers. Details are not yet available.

Photolysis of α-cyclodextrin complexes of 10-methyl-2-octalone has also been investigated.[94] Only a small influence of the cavity is felt on the rearrangement process. Padmanabhan et al. have investigated the oxidation of di-*tert*-butyl thione as a DCA complex.[95] While in solution it is readily oxidized to the corresponding ketone and S-oxide via a singlet oxygen pathway, in DCA complex it is stable.

Stability results from the absence of space within the channel for oxygen to diffuse.

Abelt and Pleier have investigated the photobehavior of β-CD complexes of phenylmethyldiazirine in the solid state.[42] The products obtained in the neat liquid phase and as a CD complex are summarized in Scheme 2. In addition to CD insertion products, upon photolysis of the CD complex several products common to the neat liquid were obtained, although in different yields. The formation of styrene in substantial yield is a reflection of the cage effect offered by the CD cavity. A similar increase in the yield of *trans*-diphenylmethylcyclopropane is noteworthy. The nature of the packing in the solid state is suggested to be responsible for this selectivity.

6. Summary

At the end of this Chapter one comes to an inescapable conclusion that the large number of host systems that are available as matrices for photochemical reactions have not yet attracted considerable attention. The field is still in its infancy. Uptill now, it has not been obvious how simple the preparation of the host-guest complex can be. But it is clear that a few host systems such as DCA, cyclodextrin and wheel and axel type hosts **2** and **3** have been popular and where investigated have yielded novel results.

The reason for an interest in host-guest photochemistry is worth emphasizing. Reactions that take place in the solid-state (crystals, silica and clay surfaces, and host-guest complexes) can be visualized to occur in a reaction cavity as pointed out in the introduction section. Host-guest complex strategy alone allows us to control the size of this reaction cavity. While crystals provide a very tight cavity, the surfaces of silica and clay provide a large cavity accounting for the poor selectivity (see Chapters 8 and 9). Therefore, both very small and very large cavity can work against our wishes of controlling the behavior of the guest molecules. The reacting molecule require small room around it to exert motions required for a reaction. If there is no room as is the case when the reaction cavity is tight, no reaction will occur; this is often the case with organic crystals which tend to pack tightly. However, if one provides a very large room around the reacting molecule no selectivity will be achieved. Relative sizes of the guest and the free space around it is an important factor to be considered while choosing a host system. With the proper choice of the guest and the host one can control the relative sizes of the reaction cavity and the free space around the guest and thus will have a handle on the photobehavior of the included guest.

In order to predict the behavior of the guest molecule one would like to know the structure and the nature of the microenvironment around the guest. Excepting for a few systems unequivocal structure of the complex has not been known partly due to the fact that the guests tend to experience disorder thus making the x-ray crystal structure determination non-trivial. Basic intuition and a knowledge of the nature of the interaction that may be involved in the complexation can guide us in predicting to a certain degree the microenvironment of the guest molecule. Among the various choices available as the microenviromnet for a reaction, host-guest complexes offer a better predictability on the structure and location of the reactant guest molecules. Therefore, the photochemistry and the photophysics of host-complexes need to be investigated further in order to bring this field to a state of maturity.

References

1. E.Weber and H.P. Josel, *J. Inclusion Phenomena*, **1983**, *1*, 79.
2. H. Colquhoum, F. Stoddart and D. Williams, *New Scientist*, **1986**, *May 1*, 44.
3. C. J. Pederson, *Science*, **1988**, *241*, 536.
    D. J. Cram, *J. Inclusion Phenomena*, **1988**, *6*, 397.
    J. M. Lehn, *Angew. Chem. Int. Ed. Engl.*, **1988**, *27*, 89.
4. J. E. D. Davies, W. Kemula, H. M. Powell and N. O. Smith, *J. Inclusion Phenomena*, **1983**, *1*, 3.
    L. Mandelcorn, *Chem. Rev.*, **1959**, *59*, 827.
    H. M. Powell, *Recueil*, **1956**, *75*, 885.
    G. Tsoucaris in *Organic Solid State Chemistry*, G. R. Desiraju, ed., Elsevier, Amsterdam, 1987, p. 207.
5. V. Ramamurthy and K. Venkatesan, *Chem. Rev.*, **1987**, *87*, 433 and Chapters 4, 5 and 6 of this monograph.
6. V. Ramamurthy, *Tetrahedron*, **1986**, *42*, 5753.
    K. Kalyanasundaram, *Photochemistry in Microheterogeneous Systems*, Academic, New York, 1987.
7. M. D. Cohen, *Angew. Chem. Int. Ed. Engl.*, **1975**, *14*, 386.
8. G. S. Murthy, P. Arjunan, K.Venkatesan, and V.Ramamurthy,*Tetrahedron*, **1987**, *43*, 1225.
    S. K. Kearsley in *Organic Solid State Chemistry*, G. R. Desiraju, ed., Elsevier, Amsterdam, 1987, p. 69.
9. G. R. Desiraju in *Crystal Engineering: The Design of Organic Solids*, Elsevier, Amsterdam, 1987.
10. J. L. Atwood, J. E. D. Davies and D. D. MacNicol, eds., *Inclusion Compounds*, Vol. 1-3, Academic Press, New York, 1984.
11. S. M. Hagan, *Clathrate Inclusion Compounds*, Reinhold Pub., New York, 1962.
    L. Mandelcorn, ed., *Non-Stoichiometric Compounds*, Academic , New York, 1964.
12. K. Takemoto and N. Sonoda in *Inclusion Compounds*, Vol. 2, J. L. Atwood, J. E. D. Davies and D. D. MacNicol, eds., Academic Press, New York, 1984, p. 47.
    K. D. M. Harris and J. M. Thomas, *J. Chem. Soc., Trans. Faraday*, **1990**, *86*, 2985.
13. M. F. Bengen and W. Schlenk, *Experientia*, **1949**, *5*, 200.
14. B. Angla, *Compt. Ren.*, **1947**, *224*, 402.
15. H. van Bekkum, J. D. Remijnse and B. M. Wepster, *J. Chem. Soc., Chem. Commun.*, **1967**, 67.
16. E. Giglio in *Inclusion Compounds*, Vol. 2, J. L. Atwood, J. E. D. Davies and D. D. MacNicol, eds., Academic Press, New York, 1984, p. 207
    W. C. Herondon, *J. Chem. Educ.*, **1967**, *44*, 724.
17. H. Wieland and H. Sorge, *Z. Physiol.*, **1916**, *97*, 1.
18. K. Miki, A. Matsui, N. Kasai, M. Miyata, M. Shibakami and K. Takemoto, *J. Am. Chem. Soc.*, **1988**, *110*, 6594.
19. M. Farina in *Inclusion Compounds*, Vol. 2, J. L. Atwood, J. E. D. Davies and D. D. MacNicol, eds., Academic Press, New York, 1984, p. 69.

20. M. Farina, *Tetrahedron Letters*, **1963**, 2097.
    M. Farina, G. Allegra and G. Natta, *J. Am. Chem. Soc.*, **1964**, *86*, 516.
21. D. D. MacNicol in *Inclusion Compounds*, Vol. 2, J. L. Atwood, J. E. D. Davies and
    D. D. MacNicol, eds., Academic Press, New York, 1984, p. 1.
    D. D. MacNicol, J. J. McKendrick and D. R. Wilson, *Chem. Soc. Rev.*, **1978**, 7,
    65.
22. A. P. Dianin, *J. Russ. Phys. Chem. Soc.*, **1914**, *46*, 1310.
23. F. Wohler, *Justus Liebigs Ann. Chem.*, **1849**, *69*, 297.
    A. Clemm, *Justus Liebigs Ann. Chem.*, **1859**, *110*, 357.
    F. Mylius, *Chem. Ber.*, **1886**, *19*, 999.
24. H. R. Allcock in *Inclusion Compounds*, Vol. 1, J. L. Atwood, J. E. D. Davies and
    D. D. MacNicol, eds., Academic Press, New York, 1984. p.351.
    H. R. Allcock, *Acc. Chem. Res.*, **1978**, *11*, 81.
25. H. R. Allcock and L. A. Siegal, *J. Am. Chem. Soc.*, **1964**, *86*, 5140.
26. D. D. MacNicol in *Inclusion Compounds*, Vol. 2, J. L. Atwood, J. E. D. Davies and
    D. D. MacNicol, eds., Academic Press, New York, 1984, p. 123.
27. D. D. MacNicol and D. R. Wilson, *J. Chem. Soc. Chem. Commun.*, **1976**, 494.
    A. D. U. Hardy, D. D. MacNicol and D. R. Wilson, *J. Chem. Soc., Perkin II.*,
    1979, 1011.
28. a) F. Toda , *Topics Curr. Chem.*, **1987**, *140*, 43.
    b) I. Goldberg, *Topics Curr. Chem.*, **1988**, *149*, 1.
29. F. Toda and K. Akagi, *Tetrahedron Letters*, **1968**, 3695.
30. E. Weber, M. Hecker, I. Csorech and M. Czugler, *Mol. Crys. Liq. Crys.* **1990**,
    187, 165.
    H. Hart, L. T. W. Lin and D. L. Ward, *J. Am. Chem. Soc.*, **1984**, *106*, 4043.
31. R. Gerdil, *Topics Curr. Chem.*, **1987**, *140*, 71.
32. R. Spallino and G. Provenzal, *Gazz. Chim. Ital.*, **1909**, *39*, 11.
33. W. Baker, B. Gilbert and W. D. Ollis, *J. Chem. Soc.*, **1952**, 1443.
34. D. Lawton and H. M. Powell, *J. Chem. Soc.*, **1958**, 2339.
35. W. D. Ollis and J. F. Stoddart in *Inclusion Compounds*, Vol. 2, J. L. Atwood, J. E.
    D. Davies and D. D. MacNicol, eds., Academic Press, New York, 1984, p. 169.
36. W. Sanger in *Inclusion Compounds*, Vol. 2, J. L. Atwood, J. E. D. Davies and D. D.
    MacNicol, eds., Academic Press, New York, 1984, p. 231.
    M. L. Bender and M. Komiyama, *Cyclodextrin Chemistry*, Springer, Berlin, 1978.
37. J. E. D. Davies, P. Finocchiaro and F. H. Herbstein in *Inclusion Compounds*, Vol.
    2, J. L. Atwood, J. E. D. Davies and D. D. MacNicol, eds., Academic Press, New
    York, 1984, p. 407.
    E. Weber, *Topics Curr. Chem.* **1987**, *140*, 1.
38. A. Collet, *Tetrahedron*, **1987**, *43*, 5725.
    A. Collet in *Inclusion Compounds*, Vol. 2, J. L. Atwood, J. E. D. Davies and D. D.
    MacNicol, eds., Academic Press, New York, 1984, p. 97.
39. For photochemical and photophysical studies of host-guest complexes in solution
    see Chapter 16 of this volume.
40. N. Friedman, M. Lahav, L. Leiserowitz, R. Popovitz-Biro, C. P. Tang and Z.
    Zaretzkii, *J. Chem. Soc. Chem. Commun.* **1975**, 864.
41. M. Lahav, L. Leiserowitz, R. Popovitz-Biro and C. P. Tang, *J. Am. Chem. Soc.*,
    1978, *100*, 2544.
42. C. J. Abelt and J. M. Pleier, *J. Org. Chem.*, **1988**, *53*, 2159.
43. A. V. Veglia, A. M. Sanchez and R. H. de Rossi, *J. Org. Chem.*, **1990**, *55*, 4083.
44. M. Farina, U. Pediretti, M. T. Gramegna and G. Audisio, *Macromolecules*, **1970**,
    3, 475.
    P. Sozzani, R. Scotti and F. Morazzoni, *J. Chem. Soc., Farady Trans. I.*, **1989**,
    85, 2581.
45. M. D. Richmond and R. J. Hurtubise, *Applied Spect.*, **1989**, *43*, 810.

J. M. Bello and R. J. Hurtubise, *Applied Spect.*, **1988**, *42*, 619.
J. M. Bello and R. J. Hurtubise, *Anal. Chem.*, **1988**, *60*, 1291 and 1285.
M. D. Richmond and R. J. Hurtubise, *Anal. Chem.*, **1989**, *61*, 2643.
J. R. Bello and R. J. Hurtubise, *Anal. Chem.*, **1987**, *59*, 2395.

46. D. F. Eaton, J. V. Caspar and W. Tam in *Photochemical Energy Conversion*, J. R. Norris and D. Meisel, eds., Elsevier, New York, 1989, p. 122.
47. S. K. Kook, D. Y. Kim and D. M. Hanson, *Chem. Phys. Lett.*, **1989**, *164*, 409.
48. A. Guarino in *Inclusion Compounds*, Vol. 3, J. L. Atwood, J. E. D. Davies and D. D. MacNicol, eds., Academic Press, New York, 1984, p. 147.
    A. Guarino, *Chem. Phys. Lett.*, **1984**, *86*, 445.
    A. Guarino, *J. Photochem.*, **1986**, *35*, 1.
49. B. F. Chemelka and A. Pines, *Science*, **1989**, *246*, 71.
50. K. Tanaka and F. Toda, *J. Chem. Soc., Chem. Commun.*, **1983**, 593.
51. M. Kaftory, K. Tanaka and F. Toda, *J. Org. Chem.*, **1985**, *50*, 2154.
52. M. Kaftory, *Tetrahedron*, **1987**, *43*, 1503.
    K. Tanaka and F. Toda, *J. Chem. Soc. Japan., Chem. Ind. Chem.*, **1984**, 141.
53. N. Narasimha Moorthy and K. Venkatesan, Unpublished results.
54. Y. Tanaka, S. Sasaki and A. Kobayashi, *J. Inclusion Phenomena*, **1984**, *2*, 851.
55. E. Hadjoudis, I. Moustakali-Mavridis, G. Tsoucaris, F. Villain and G. Le Bas, *Mol. Crys. Liq. Crys. Inc. Nonli. Opt.*, **1988**, *156*, 405.
56. M. Farina in *Inclusion Compounds*, Vol. 3, J. L. Atwood, J. E. D. Davies and D. D. MacNicol, Eds., Academic Press, New York, 1984, 297.
57. J. F. Brown and D. M. White, *J. Am. Chem. Soc.*, **1960**, *82*, 5671.
58. H. R. Allcock, W. T. Ferrar and M. L. Levin, *Macromolecules*, **1982**, *15*, 697.
59. M. Farina, G. Audisio, G. Allegra and M. Loffelholz, *J. Poly. Sci.*, **1967**, *C16*, 2517.
60. M. Miyata and K. Takemoto, *J. Polym. Sci., Polym. Lett. Ed.*, **1975**, *13*, 221.
    K. Takemoto and M. Miyata, *Mol. Cryst. Liq. Cryst.*, **1990**, *186*, 189.
61. M. Maciejewski and Z. Durski, *J. Macromol. Sci. Chem.*, **1981**, *A16*, 441.
    M. Maciejewski, *J. Macromol. Sci. Chem.*, **1979**, *A13*, 77.
62. P. J. Wagner in *Molecular Rearrangements in Ground and Excited States*, Vol. 3, P. de Mayo, ed., Wiley Science, New York, 1980, p. 381.
63. G. Dasaratha Reddy, G. Usha, K. V. Ramanathan and V. Ramamurthy, *J. Org. Chem.*, **1986**, *51*, 3085.
64. G. Dasaratha Reddy and V. Ramamurthy, *J. Org. Chem.*, **1987**, *52*, 3952.
65. G. Dasaratha Reddy and V. Ramamurthy, *J. Org. Chem.*, **1987**, *52*, 5521.
66. B. Nageswer Rao, M. S. Syamala, N. J. Turro and V. Ramamurthy, *J. Org. Chem.*, **1987**, *52*, 5517.
67. P. C. Goswami, P. de Mayo, N. Ramanth, G. Bernard, N. Omkaram, J. R. Scheffer and Y. F. Wong, *Can. J. Chem.*, **1985**, *63*, 2719.
68. J. Vincens, *Mol. Cryst. Liq. Cryst.*, **1990**, *187*, 115.
69. M. S. Syamala, B. Nageswer Rao and V. Ramamurthy, *Tetrahedron*, **1988**, *44*, 7234.
70. M. S. Syamala and V. Ramamurthy, *Tetrahedron*, **1988**, *44*, 7223.
71. D. Bellus, *Adv. Photochem.*, **1971**, *8*, 109.
72. For a review: H. Lechert and W. D. Basler, *J. Phys. Chem. Solids.*, **1989**, *50*, 497.
73. P. Engel, *J. Am. Chem. Soc.*, **1970**, *92*, 6074.
    W. K. Robbins and R. H. Eastman, *J. Am. Chem. Soc.*, **1970**, *92*, 6076.
    R. S. Givens and W. F. Oettle, *J. Am. Chem. Soc.*, **1972**, *37*, 4325.
74. B. Nagerwer Rao, N. J. Turro and V. Ramamurthy, *J. Org. Chem.*, **1986**, *51*, 460.
75. H. L. Casal, D. Griller, R. J. Kolt, F. W. Hartstock, D. M. Northcott, J. M. Park and D. D. M. Wayner, *J. Phys. Chem.*, **1989**, *93*, 1666.
    T. Ichikawa, *J. Phys. Chem.*, **1979**, *83*, 1358.

76. H. L. Casal, P. de Mayo, J. F. Miranda and J. C. Scaiano, *J. Am. Chem. Soc.*, 1983, *105*, 5155.
77. G. Dasaratha Reddy, B. Jayasree and V. Ramamurthy, *J. Org. Chem.*, 1987, *52*, 3107.
78. V. Pushkara Rao, N. Han and N. J. Turro, *Tetrahedron Letters*, 1990, *31*, 835.
79. R. Arad-Yellin, S. Brunie, B. S. Green, M. Knossow and G. Tsoucaris, *J. Am. Chem. Soc.*, 1979, *101*, 7529.
    R. Arad-Yellin, B. S. Green, M. Knossow, N. Rysanek and G. Tsoucaris, *J. Inclusion Phenomena*, 1985, *3*, 317.
80. A. Guarino, E. Possagno and R. Bassanelli, *Tetrahedron*, 1987, *43*, 1541.
81. M. S. Syamala, S. Devanathan and V. Ramamurthy, *J. Photochem.*, 1986, *34*, 219.
    G. L. Duveneck, E. V. Sitzman, K. B. Eisenthal and N. J. Turro, *J. Phys. Chem.*, 1989, *93*, 7166.
82. M. Farina, G. Audisso and G. Natta, *J. Am. Chem. Soc.*, 1967, *89*, 5071.
    G. Audiso and A. Silvani, *J. Chem. Soc., Chem. Commun.*, 1976, 481.
    M. Miyata and K. Takemoto, *Polym. Lett.*, 1975, *13*, 221.
83. Y. Weisinger-Lewin, M. Vaida, R. Popovitz-Biro, H. C. Chang, F. Mannig, F. Frolow, M. Lahav and L. Leiserowitz, *Tetrahedron*, 1987, *43*, 1449.
    H. C. Chang, R. Popovitz-Biro, M. Lahav and L. Leiserowitz, *J. Am. Chem. Soc.*, 1987, *109*, 3883.
    R. Popovitz-Biro, C. P. Thang, H. C. Chang, M. Lahav and L. Leiserovitz, *J. Am. Chem. Soc.*, 1985, *107*, 4043 and 4058.
84. H. Aoyama, K. Miyazaki, M. Sakamoto and Y. Omote, *J. Chem. Soc., Chem. Commun.*, 1983, 333.
    H. Aoyama, K. Miyazaki, M. Sakamoto and Y. Omote, *Tetrahedron*, 1987, *43*, 1513.
85. V. Pushkara Rao and N. J. Turro, *Tetrahedron Letters*, 1989, *30*, 4641.
86. R. Gerdil, G. Barchietto and C. W. Jefford, *J. Am. Chem. Soc.*, 1984, *106*, 8004.
87. M. Kaftory, F. Toda, K. Tanaka and M. Yagi, *Mol. Cryst. Liq. Cryst.*, 1990, *186*, 167.
    F. Toda, *Mol. Cryst. Liq. Cryst.*, 1990, *187*, 41.
    F. Toda, *J. Incl. Phenom. Mol. Recog. Chem.*, 1989, *7*, 24.
    F. Toda, *Topics Curr. Chem.*, 1988, *149*, 211.
    F. Toda, *Mol. Cryst. Liq. Cryst. Inc. Nonli. Opt.*, 1988, *161*, 355.
88. M. Kaftory, M. Yagi, K. Tanaka and F. Toda, *J. Org. Chem.*, 1988, *53*, 4391.
89. F. Toda, K. Tanaka and M. Oda, *Tetrahedron Letters*, 1988, *29*, 653.
    T. Fujiwara, N. Nanba, K. Hamada, F. Toda and K. Tanaka, *J. Org. Chem.*, 1990, *55*, 4532.
90. F. Toda and K. Tanaka, *J. Chem. Soc. Chem. Commun.*, 1986, 1429.
    F. Toda, K. Tanaka and M. Yagi, *Tetrahedron*, 1987, *43*, 1495.
    F. Toda and K. Tanaka, *Tetrahedron Letters*, 1988, *29*, 4299.
91. T. Fujiwara, N. Tanaka, K. Tanaka and F. Toda, *J. Chem. Soc., Perkin Trans. I.*, 1989, 663.
92. F. Toda and K. Tanaka, *Chemistry Letters*, 1987, 2283.
93. T. Yumoto, K. Hayakawa, K. Kawase, H. Yamakita and H. Taoda, *Chemistry Letters*, 1985, 1021.
    A. Guarino, E. Possagno and R. Bassanelli, *J. Inclusion Phenomena*, 1987, *5*, 563.
94. V. Wintgens, B. Guerin, H. Lennholm, J. R. Brisson and J. C. Scaiano, *J. Photochem. Photobiol. A: Chemistry*, 1988, *44*, 367.
95. K. Padmanabhan, K. Venkatesan and V. Ramamurthy, *Can. J. Chem.*, 1984, *62*, 2025.

Chapter 8

Phototransformations of Organic Molecules Adsorbed on Silica and Alumina#

L. J. Johnston

**Steacie Institute for Molecular Sciences,
National Research Council of Canada,
Ottawa, Ontario, Canada.**

Contents

1. Introduction

The use of heterogeneous systems such as micelles, solid supports, polymers, and membranes for controlling chemical reactions is currently an area of considerable activity. Much of the interest in solid supports such as metals, metal oxides, zeolites, and semiconductors has resulted from the catalytic effects of these materials and their importance in industrial processes. During the last two decades

---

Issued as NRCC 32821.

there has been a marked increase in the use of photochemical techniques to study the behavior of a variety of surface–adsorbate systems.[1,2] These investigations of the photochemistry and photophysics of adsorbed molecules have the potential to yield several sorts of information. First, an appropriate probe molecule can provide information about the surface properties and adsorption sites. Second, the restrictions on both the rotational and the diffusional mobility of an adsorbed species may substantially modify its normal solution reactivity. Since excited states and photochemically produced intermediates with a wide range of lifetimes can be studied, these techniques may provide evidence for differing degrees of mobility and changes in surface environment as a function of the time scale examined.

The coverage of this chapter will be limited primarily to the photochemistry of organic molecules adsorbed on dry silica and alumina. These supports are relatively inert and are not usually directly involved in the chemistry of the adsorbate, in contrast to results for semiconductor or metal surfaces. Examples dealing with aqueous colloidal silica systems and other adsorbent–solvent systems are also discussed, although these account for a relatively small fraction of the available data. Silica- or alumina-supported catalysts are not included. Although the emphasis is mainly on the photochemistry of adsorbed species, a brief description of the surface properties of the adsorbents and the photophysics of molecules adsorbed on them is also presented (for details see Chapters 12 and 13).

2. Description of Silica and Alumina Surfaces

Both amorphous and crystalline silicas have the composition $SiO_2 \cdot x\ H_2O$ and are constructed from SiO_4 tetrahedra.[3-5] Since the crystalline silicas have low surface areas, only the amorphous silicas (porous silica gels and powders, colloidal silica, and porous glass) are commonly used as adsorbents. Silica gel is a rigid three-dimensional network of silica particles produced by condensation of silicic acids or by aggregation of silica particles. Its porous spongelike structure results in large surface areas ($100–600\ m^2\ g^{-1}$), which correspond primarily to areas on the internal pore walls. The surface areas are usually determined by adsorption isotherms for nitrogen or other small molecules and do not necessarily reflect the area available for larger organic molecules. Pore sizes range from micropores of <20 Å to macropores of >2000 Å, although values from 20 to 150 Å are typical of chromatographic silicas. The sizes of even the smallest pores are considerably larger than the dimensions of most adsorbed molecules. As a result, restrictions on molecular mobilities are not expected to be as pronounced as in the case of zeolites, which have much smaller channels or pores (≤ 13 Å; see Chapter 10).

Silica aerogels and porous silica powders are structurally similar to silica gel and differ only in their method of preparation. Porous glass (commonly called porous Vycor) is also structurally similar to the other porous silicas but has a composition of $\sim97\%\ SiO_2$ and $\sim3\%\ B_2O_3$. It is prepared by heating a homogeneous borosilicate glass until the two phases separate as mutually interpenetrating networks and then cooling and dissolving the boron-rich phase with strong acid. The resulting glass has 40 Å pores, 200 $m^2 g^{-1}$ surface areas and surface properties similar to those of other silicas. Colloidal silicas are stable dispersions of spherical particles of amorphous silica (5–50 nm diameters) which are stabilized against interparticle bonding by an ionic charge on the surface in the presence of base. The surface consists of SiO– groups with adsorbed OH– ions and exchangeable Na^+ counterions and is therefore useful for binding charged species. The surface areas of colloidal silicas refer to external surfaces since the particles are nonporous.

The surface of porous silicas consists of a network of siloxane (Si—O—Si) and silanol (Si—OH) linkages and physically adsorbed water molecules (Figure 1). There are generally 4–5 silanols nm^{-2} and these may be isolated, geminal, or vicinal (hydrogen bonded to water or to each other), as illustrated in Figure 1. Both physically adsorbed water and adjacent hydroxyls can be removed by heating the silica and at sufficiently high temperatures a relatively hydrophobic surface is produced. Surface adsorption can occur via (1) dispersion forces arising from induced dipole interactions, (2) induction forces, and (3) charge-transfer interactions such as hydrogen bonding. Adsorption of most polarizable organic molecules is generally accepted to involve interaction with the surface hydroxyls and it has been suggested that the vicinal silanols provide active sites for adsorption. However, the relatively large size of many adsorbed molecules will usually permit their interaction with several surface silanols.

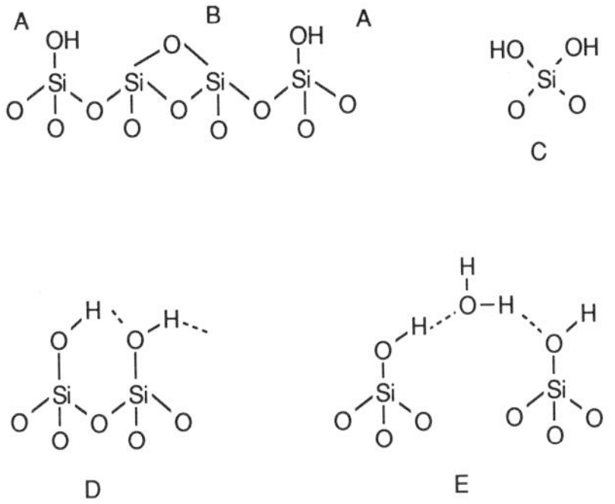

Figure 1. Schematic representation of silica surfaces: (A) isolated silanols; (B) siloxane bonds; (C) geminal silanols; (D) hydrogen-bonded silanols; (E) hydrogen-bonded water.

Alumina (Al_2O_3) occurs in various amorphous and crystalline forms; for the latter, the α, η, and γ phases are the most important.[6-8] Of these γ-alumina is the least acidic and catalytically active. The aluminum cations are found in both tetrahedral and octahedral positions, with the latter being the less favorable. In contrast to the porous silicas, the typical (<200 m^2 g^{-1}) surface areas of alumina correspond to external surfaces. The surface has both hydroxyl groups, of which five different types have been identified, and physically adsorbed water molecules. There are 12–13 hydroxyls per nm^2 and their properties are determined by their coordination and net charge. Dehydroxylation of the surface leads to neighboring coordinatively unsaturated aluminum and oxygen sites. The latter acceptor sites are thought to be responsible for the charge-transfer interactions with adsorbed aromatics and for the generally "active" nature of alumina adsorbents.

3. Photophysics of Adsorbed Molecules

Any discussion of the photochemistry of adsorbed molecules should begin with a brief discussion of the photophysical behavior of systems that do not undergo any net chemistry. This provides much valuable background data on how singlet and triplet states are affected by adsorption. Although a detailed discussion of all the available data on the photophysics of adsorbed molecules would be beyond the scope of the present chapter, some of the key findings are summarized below, along with illustrative examples. The reader is referred to several reviews on various aspects of the subject for further information.[1, 9-13]

Some of the earlier studies in this area examined the shifts in uv absorption spectra for a variety of molecules in silica gel–solvent slurries; the observed shifts are consistent with a polar environment for the adsorbed molecules.[14-16] Although in these experiments the substrates were distributed between the solvent and support, similar results have been obtained for dry silica systems.[17-19] A variety of steady-state and time-resolved luminescence techniques have proved to be very sensitive probes for the nature of the environment of the excited state as well as studies of the mobility (both rotational and diffusional) and distribution of the substrate on the surface. In particular, the fluorescent singlet states of aromatic hydrocarbons have been extensively studied. In general, the fluorescence lifetimes are similar to those in solution except for cases in which the surface prevents twisting motions necessary for deactivation or for which there is direct complexation with the surface. However, the fluorescence decays are generally nonexponential, reflecting the presence of multiple lifetimes arising from multiple adsorption sites. Pyrene has been the probe of choice for many of these investigations, since the relative intensities of the vibrational bands of its fluorescence are sensitive to the polarity of the environment (Ham effect) and since it has a long singlet lifetime and exhibits excimer emission. Pyrene adsorbed on silica shows monomer fluorescence which is indicative of a polar environment and which does not show single exponential decay kinetics.[20] Excimer emission arises, in part, from ground-state complexes, indicating a nonhomogeneous distribution of molecules on the surface.[20,21] However, modification of the silica surface by coadsorption of polar molecules such as 1-decanol or by chemical bonding of a hydrocarbon layer did result in dynamic excimer formation with both monomer and excimer emissions typical of a solution environment.[22,23] Fluorescence polarization experiments have indicated that on silica gel there is rearrangement of the pyrene ground-state complex on the time scale of the excimer emission.[24] Quenching of pyrene monomer fluorescence by halonaphthalenes has also been demonstrated to occur via static and dynamic mechanisms and has been used to estimate activation energies for diffusion of 4 and 2 kcal mol^{-1} for dry and decanol-covered silicas, respectively.[25,26] Similar results to those on silica have been obtained for pyrene on alumina, with the exception of the fact that single exponential decays were observed.[12]

In contrast to the above results, there are also examples in which more substantial modifications of the singlet properties of adsorbed aromatic hydrocarbons have been observed. For example, the fluorescence spectra of aminopyrene and acridine on silica both show evidence for protonation.[27,28] In the former case only geminal silanols appear to result in protonation. A study of the fluorescence of 9,9'-bianthryl adsorbed on porous glass demonstrated that there was a lack of rotational reorientation on the time scale of the singlet lifetime,[29] although similar studies for 1,1'-binaphthyl[25] and 1-(*N*,*N*-dimethylamino)-4-benzonitrile[30] on silica indicated that rotational motion was relatively unrestricted.

Phosphorescence from triplet states of a variety of adsorbed organics is readily observed at room temperature, in contrast to solution results.[2,31,32] A number of triplets have also been examined by diffuse reflectance laser flash photolysis of solid samples.[33-40] Spectra of triplets on silica are similar, although frequently broader, than those recorded in polar solvents. On the other hand triplet–triplet absorption spectra for some aromatic hydrocarbons on alumina show shifts that have been attributed to transitions to charge-transfer states of surface–adsorbate systems.[41] Enhanced triplet lifetimes and nonexponential decays are typical.[34,36,38] In one case, the study of dynamic quenching of triplet benzophenone by naphthalene on a silica surface has provided an estimate of 7.3 x 10^{15} dm^2 mol^{-1} s^{-1} for the rate constant for triplet energy transfer.[35] In a related study the rate constants for quenching of triplet benzophenone adsorbed on silica by both gas phase and adsorbed oxygen were shown to be a function of the silica pore size.[42] Singlet energy transfer between Rhodamine 6G and malachite green on silica has also been shown to be sensitive to the pore dimensions.[43] On large pore silicas (1000 Å) the energy-transfer process probes only the local environment, whereas for small-pore silicas it is sensitive to the pore network rather than the local details of the surface.

The above discussion indicates that multiexponential decays are common for singlet and triplet decays of adsorbed molecules. Further, the distribution of adsorbates is frequently inhomogeneous. In many cases attempts have been made to fit the data with a combination of several exponentials and to associate these exponentials with the number of surface sites. However, it has been demonstrated that the single and double exponential fluorescence decays frequently reported for adsorbed molecules may fit equally well a wide distribution of rate constants (see Chapter 13).[44] Obviously new approaches are needed to deal with the problems of multiexponential fluorescent decays and the typical triplet decays which extend over many orders of magnitude for many adsorbed molecules. In this regard it has recently been suggested that the fractal approach to surfaces (see Chapter 12) provides a useful method for dealing with the irregularities of a variety of supports, including silicas.[45-47] However, a more rigorous examination of some porous silica gels has indicated that not all of these materials have fractal surfaces.[48]

4. Radical Reactions

The photochemical generation of radical pairs and/or biradicals from molecules adsorbed on silica has been used extensively as a probe for the occurrence of rotational and diffusional mobility of adsorbed species, and there have also been some investigations of alumina and porous Vycor. Such studies have provided considerable information on the degree of control that a surface may exert on the reactivities of adsorbed molecules. These reactions have been studied in more detail than most other photochemical reactions of adsorbed species. As a result data on effects of surface coverage, pore size, surface additives, temperature, and magnetic field on the reactivity of radical pairs are available and lead to a more complete picture of the modifications of normal solution behavior that may be expected.

One of the earlier examinations of radical pair behavior on surfaces claimed on the basis of product analysis that cyanopropyl radicals generated by photolysis of bis(azoisobutyronitrile) (1, reaction 1) in silica gel–benzene matrices were not free to rotate to produce the unsymmetrical coupling product (3).[49] In contrast, rotational motion of the acyl–phenoxy radical pair generated in the Photo-Fries rearrangement on silica gel (reaction 2) occurred readily to give both ortho and para products, in some cases more efficiently than in solution.[50] A similar reaction for aromatic amides was shown to be intramolecular in nature, indicating that recombination of the geminate radical pair does not compete with its diffusional separation.[51]

$$(1)$$

These apparently conflicting results concerning the rotational mobility of adsorbed radicals were reconciled by a reinvestigation of reaction (1). In this case substantial amounts of both 2 and 3 (and its hydrolysis product) were produced, even on dry silica gel, indicating that the radicals were not anchored to the surface.[52] The amounts of geminate recombination for cyanopropyl radicals produced by direct and triplet-sensitized irradiation of mixtures of deuterated and undeuterated **1** were also examined.[52] In both cases the results indicated some diffusional mobility of the adsorbed radicals. For example, there was approximately 31% geminate recombination for the triplet-sensitized photolysis as compared to 84% for the direct singlet reaction, indicating a more efficient escape of radicals from the longer lived triplet radical pair.

$$(2)$$

In agreement with the above results several esr experiments have demonstrated that radicals such as cumyl[53] and benzyl[54] are rotationally mobile on silica surfaces. The behavior of the benzyl radical varied dramatically from large-pore silicas on which rapid equilibration of radicals between sites occurred to small pore (20 Å) silicas on which the radicals were isolated, motionally restricted, and relatively unreactive. The results also indicated two distinct binding sites.[54]

The photolysis of a variety of substituted dibenzyl ketones has been one of the most extensively studied reactions on silica surfaces.[52, 54-61] This reaction is outlined in Scheme 1 and illustrates the various reaction pathways available to a geminate radical pair. These include (1) recombination to regenerate starting material, (2) recombination or disproportionation to give products, (3) transformation to a new radical pair, and (4) diffusional separation to give free radicals. Whether or not diffusional separation will be able to compete with product formation within the geminate pair will be determined by factors such as the multiplicity of the radical pair (since triplet pairs must first intersystem cross to the singlet before reacting) and the viscosity of the medium. Thus, the measurement of cage effects as a probe for the amount of geminate recombination has proved a very useful tool for evaluating the diffusional mobility of adsorbed radicals.

As shown in Scheme 1 and discussed in more detail in Chapter 1, the photolysis of dibenzyl ketone proceeds via cleavage of a short-lived triplet to give a triplet phenylacetyl–benzyl radical pair. The latter decays by a competition between loss of carbon monoxide to generate a triplet benzyl radical pair, and intersystem crossing (ISC), to give a singlet pair which combines to regenerate starting material or rearranged ketone. In solution the rate constant for decarbonylation of the phenylacetyl radical is 5.2×10^6 s^{-1}, although the effect of surface adsorption on this rate constant has not been determined.[62] The benzyl radical pair decays by a competition between intersystem crossing followed by geminate reaction and cage escape, presumably to give free radical coupling products. The amount of geminate coupling for an asymmetric ketone thus provides a probe for the amount of translational mobility of the adsorbed radicals.

The efficiency of geminate coupling is usually expressed as a cage effect, which for photolysis of the substituted dibenzyl ketone shown in Scheme 1, is defined according to equation (3).

$$\text{Cage effect} \quad = \quad \frac{AB - AA - BB}{AB + AA + BB} \qquad\qquad (3)$$

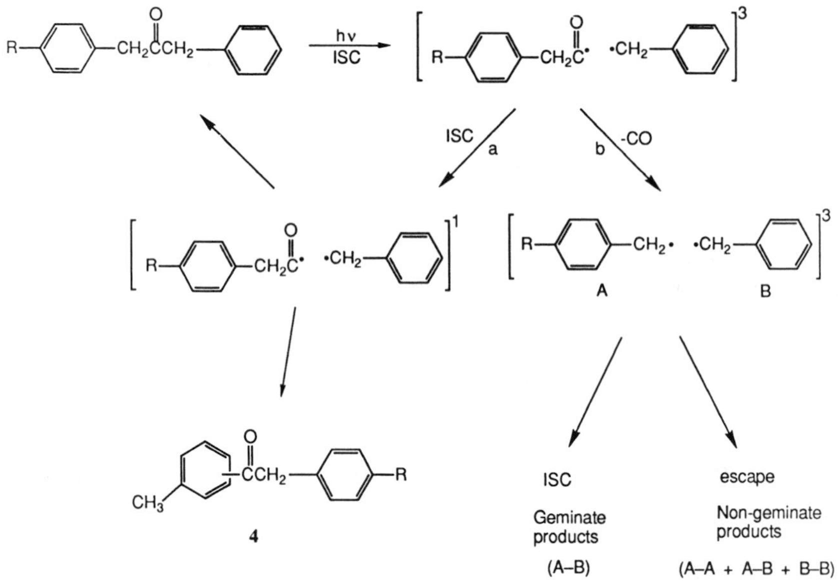

Scheme 1

Cage effects for benzyl radicals produced by photolysis of substituted dibenzyl ketones under a variety of conditions are listed in Table 1. The results show that cage effects of 20% are typical at room temperature over a range of coverages (1–50%) for silica gel. Considerably larger cage effects were measured at lower temperatures (~50% at -55°C).[57] Dehydration of the silica leads to slightly more mobile radicals whereas coadsorption of 1,4-cyclohexanediol gives only minor changes in the cage effect, even at -55°C.

The behavior of a singlet benzyl radical pair generated from benzyl phenyl acetate **5** (reaction 4) has also been examined.[57] In this case the singlet nature of the radical pair allows recombination to compete more favorably with diffusion and the cage effects at room temperature (Table 1) are approximately twice those obtained for the same benzyl radical pair produced from a triplet ketone precursor under the same conditions. At -55°C only geminate coupling products were observed. Similar results for a dibenzyl sulfone suggest that the cleavage in this case occurs primarily from the singlet.[57]

Table 1. Cage Effects for Singlet and Triplet Benzyl Radical Pairs Produced from Dibenzyl Ketones and Benzyl Phenylacetates, Respectively, on Various Types of Silica.

Radical pair	Support	Coverage (%)	Temp. (°C)	% Cage	Ref.
4-CH$_3$-C$_6$H$_4$ĊH$_2$ /	Silica gel	1	20	25	57
4-CH$_3$O-C$_6$H$_5$ĊH$_2$[a]		10	20	23	57
		50	20	21	57
		2	-55	60	57
		50	-55	51	57
	Silica gel + diol[b]	2	-55	69	57
4-CH$_3$-C$_6$H$_4$ĊH$_2$ /	Silica gel	1	20	51	57
4-CH$_3$O-C$_6$H$_4$ĊH$_2$[c]		10	20	39	57
		50	20	32	57
		10	-55	>96	57
C$_6$H$_5$ĊH$_2$ /	TLC[d]	—	27	9	63
C$_6$H$_5$ĊH(CH$_3$)[a]	Porous glass	—	27	38	63
	RPTLC[e]	—	27	15	63
C$_6$H$_5$ĊH$_2$ /	Porous glass	—	27	51	63
C$_6$H$_5$ĊH(CH$_2$C$_6$H$_5$)[a]	Porous glass	—	-77	56	63
	Silica		-77	79	63

[a.] Triplet precursor.
[b.] 0.45 mmol 1,4-cylcohexanediol per gram silica gel.
[c.] Singlet precursor.
[d.] Thin-layer chromatography plates.
[e.] Reverse-phase thin-layer chromatography plates.

$$CH_3 - \langle\!\!\!\!\bigcirc\!\!\!\!\rangle - CH_2CO_2CH_2 - \langle\!\!\!\!\bigcirc\!\!\!\!\rangle - OCH_3 \xrightarrow[-CO_2]{h\nu} CH_3 - \langle\!\!\!\!\bigcirc\!\!\!\!\rangle - \cdot \; + \; CH_3O - \langle\!\!\!\!\bigcirc\!\!\!\!\rangle - \cdot \quad (4)$$

5

A detailed investigation of the effect of the silica pore size on benzyl radical coupling has also been reported (Table 2).[58,61] In this case the cage effect at higher coverages (>5%) increases by ~50% on going from 95 Å to 22 Å pore silicas, and for both samples the cage effect is substantially increased at low coverages (<4%). For each pore size there is clearly a high coverage region (>5%), where the cage effect is more or less independent of coverage, as well as a second region (<5%), where the cage effect increases sharply with decreasing coverage. Again, enhanced effects are observed at lower temperatures. The cage effects in various systems are modified by photolysis in a magnetic field (2 kG) but only at low coverages for silicas of small pore diameters.[61] It has been suggested that the observed variations in cage effects for the different pore size silicas are due to two distinct silanol sites, the proportions of which depend on the pore dimensions.[61] The more strongly binding site is assigned to adjacent hydrogen-bonded silanols, while the weakly binding site is thought to be an isolated silanol group. Further, the cage effects for small-pore silicas were unaffected by coadsorbed water, whereas large-pore silicas gave decreased cage effects in the presence of water. These contrasts in the behavior of the various silicas were thought to be related to structural differences between the large and small pores.

Table 2. Cage Effects for Triplet 4-methylbenzyl–benzyl Radical Pairs on Silicas with Various Pore Sizes.[a]

Pore size	Coverage (%)	Temp. (°C)	% Cage
22 Å	16	r t	18
	3	r t	23
	1	r t	28
	—	-70	46
95 Å	25	r t	6
	4	r t	14
	1	r t	21
	0.5	r t	32
	—	-70	74

[a.] Data taken from ref. 61.

Although the above results for benzyl radical pairs show that restrictions to mobility are not severe in most cases, somewhat different results are obtained for the shorter lived phenylacetyl–benzyl pair (Scheme 1). In this case the competition between pathways a and b (Scheme 1) results in several modifications of the normal solution behavior. First, the presence of magnetic isotopes such as ^{13}C and ^{17}O increases the rate of intersystem crossing in the radical and thus leads to more recombination to give starting ketone.[56,59,60,64] Thus, if partially labeled starting material is photolyzed, there is an enhancement of the magnetic isotope in the recovered ketone. This isotopic enrichment is a direct consequence of the slower diffusional separation of the radical pair on a silica surface as compared to solution. The amounts of enrichment observed for silica photolysis are comparable to those reported for dibenzyl ketones in micellar systems. The amount of recombination of the initial phenylacetyl–benzyl radical pair has been estimated for *meso*-2,4-diphenylpentan-3-one (**6**, reaction 5). This was accomplished by measuring the relative amounts of *meso*- and *dl*-ketone isomers recovered after irradiation of the meso isomer as a function of the amount of ketone converted to products. This leads to the conclusion that on porous glass and TLC plates ~20% of the ketones that react undergo photoisomerization[61,63] and so give a direct measure of the fraction of initial radical pairs that return to starting material. A second modification in the usual solution behavior of the phenylacetyl–benzyl radical pair is the observation of products resulting from ortho or para coupling of the initial radicals to give the isomeric ketones (**4**, Scheme 1).[57] Similarly, recombination of the initial radical pair from dibenzyl sulfone produces the rearranged sulfinate.[57] The observation of these alternate coupling products is a result of the restricted translational motion for the adsorbed radicals.

$$\text{(5)}$$

6 (meso)

meso + d,l

+ PhCH=CH$_2$

In addition to the above results, several α-substituted benzyl radical pairs have also been examined. Substantial cage effects (10–50% at room temperature, Table 1) were observed for benzyl–α-methylbenzyl and benzyl–α-benzylbenzyl radical pairs on porous Vycor and a variety of silicas.[63] There is some indication that the cage effect is greater for larger radicals, although there are only a limited number of direct comparisons. Cumyl radicals have also been generated by the photolysis of azocumene.[53]

$$\text{(6)}$$

7

The amount of geminate combination was estimated at ~80–90% based on product studies for deuterium-labeled azocumene and the trapping of radicals which escape their initial partners. Further, ortho and para coupling of the cumyl radicals followed by rearomatization gave head-to-tail dimers (**7**, reaction 6) which accounted for ~8% of the products and were produced only from geminate radical pairs. This led to the suggestion that there are preferential adsorption sites which act as catalysts for aromatization of the initial head-to-tail adducts. Minor changes in products were also observed for varying degrees of dehydroxylation of the silica.

Diphenylmethyl radicals have been generated from 1,1,3,3-tetraphenylacetone on silica and zeolite surfaces. The results indicate that the radicals are very long lived in these environments and that their decays do not fit any of the usual single or double exponential kinetic expressions.[65] Although ~80% of the radicals decay within <100 μs on silica gel there are some that have lifetimes on the order of minutes. The radicals are efficiently quenched by oxygen on the silica surface.

Another example of modification of the behavior of a radical pair by adsorption on a surface is found in the variations in the relative rates of disproportionation versus coupling (k_d/k_c). For example, values of k_d/k_c of 0.055 and 0.28 were measured for cumyl radicals (reaction 7) in benzene and on silica, respectively.[53] The results have been rationalized on the basis of constraints on the motion of the radicals on the surface. Since there are more orientations suitable for disproportionation than for coupling, a more constrained radical pair has less chance of coupling. Similarly, increases in the yields of disproportionation products have been observed for α-methylbenzyl radical pairs at low temperature on porous glass and several types of silica.[63] For example, k_d/k_c ratios of 0.054 and 1.33 were obtained in pentane and on porous glass at -77°C. Constraints to the mobility of adsorbed radicals are also indicated by the fact that phenyl migration in the neophyl radical is inhibited on silica.[66]

$$2 \quad \begin{array}{c} CH_3 \\ | \\ -C\cdot \\ | \\ CH_3 \end{array} \quad \xrightarrow{k_d} \quad \begin{array}{c} CH_3 \\ | \\ -CH \\ | \\ CH_3 \end{array} \quad + \quad \begin{array}{c} CH_3 \\ | \\ -C=CH_2 \\ | \\ CH_3 \end{array}$$

$$\searrow k_c \qquad (7)$$

$$\left(\begin{array}{c} CH_3 \\ | \\ -C- \\ | \\ CH_3 \end{array} \right)_2$$

$$R \overset{O}{\underset{O}{\rVert}} \overset{CH_2R'}{\underset{}{\overset{|}{N}}}_{CH_2R'} \quad \xrightarrow{h\nu} \quad \begin{array}{c} R' \\ O \\ R \end{array} \overset{}{\underset{O}{N}}_{CH_2R'} \quad + \quad R \overset{OH}{\underset{O}{\rVert}} \overset{R'}{\underset{N}{\rVert}}_{CH_2R'} \qquad (8)$$

In addition to the above results for radical pairs generated via type I and related cleavages, there have been a number of investigations of type II hydrogen abstraction reactions and of the competition between type I and type II reactions for both alkyl and aryl ketones. The relevant reactions are outlined in Scheme 2. For example, the type II reaction of α-oxoamides (reaction 8) on silica and alumina gives the same two products as in solution.[67] The product ratios are consistent with a surface polarity similar to that of acidic methanol. The related δ-hydrogen abstraction of β-oxoamides also occurs facilely on a silica gel surface, with enhanced yields relative to solution that are probably due to the greater keto content of the amide on the surface.[68]

The photolysis of valerophenone and several related ketones adsorbed on silica gives exclusively type II products as in solution.[61] The cyclization–elimination ratios for the biradical increased slightly at room temperature as compared to solution results and at -125°C cyclization was the only observed reaction. The lifetime of the adsorbed valerophenone triplet has been measured using diffuse reflectance laser flash photolysis and is 300 ns as compared to <5 ns in solution. This has led to the suggestion that on the silica surface the kinetics of triplet decay are controlled by conformational dynamics (i.e., the molecule must achieve the appropriate conformation for hydrogen abstraction) rather than by hydrogen abstraction as in solution.[36] The biradical was not observed in the transient experiments which implies that its lifetime is shorter than that of the triplet, again in contrast to solution results.

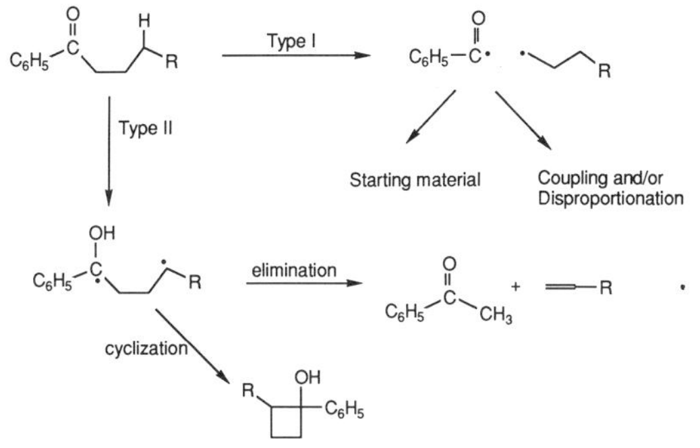

Scheme 2

Adsorption on a silica surface has been shown to substantially affect the competition between type I and type II reactions in a number of systems. Anpo and co-workers have carried out an extensive examination of these effects for 2-pentanone and related dialkyl ketones on porous Vycor.[69-74] The results in selected systems are shown in Table 3. In all cases there was a marked increase in the amount of Type I reaction as compared to either solution or gas phase results. The fraction of type I products was maximum at low surface coverages and varied with the prior heat treatment of the surface. The larger type I selectivity for bulkier ketones is consistent with the enhanced steric requirements for type II hydrogen abstraction.[73] Similarly, Turro and co-workers have reported a significantly larger type I/type II ratio for ketones **8** on silica as compared to *t*-butanol (Table 4), again consistent with adsorption restricting the ability of the ketone to acquire the necessary conformation for hydrogen abstraction during its triplet lifetime.[61]   However, the biradical once formed is already in the cis conformation necessary for cyclization and the product mixture reflects this in a lower elimination to cyclization ratio than in solution (Table 4).

Table 3. Ratios of Type I/Type II reaction for Photolysis of Alkyl Ketones in Solution, in the Gas Phase, and Adsorbed on Porous Vycor

Ketone	Medium	Type I/Type II	Ref.
2-Pentanone	Methanol	0.25	72
	Eater	0.40	72
	Porous Vycor	0.89 (4×10^{-6} mol g^{-1})	72
		0.65 (2.2×10^{-5} mol g^{-1})	72
3-Methyl-2-pentanone	Gas phase	0.35	74
	Porous Vycor	5.1 (5×10^{-6} mol g^{-1})	73
		4.0 (4.3×10^{-5} mol g^{-1})	73

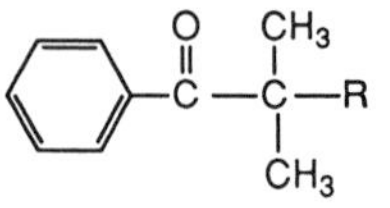

R = (CH$_2$)$_n$CH$_3$; n = 2,8,15

8

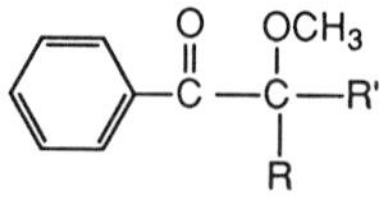

1 0

a: R=R'=H;   b: R=H, R'=CH$_3$
c: R=R'=CH$_3$; d: R=H; R'=Ph
e: R=CH$_3$, R'=Ph

In contrast to these results ketone **9** (Scheme 3) gives a much higher yield of type II reaction on a silica surface than in methanol (Table 4). For example, type II cleavage does not occur for **9** (R = CH$_3$) at -80°C in methanol but on silica it accounts for 32% of the products.[75] These results may suggest that in this case the adsorbed ketone is already in a favorable confomation for hydrogen abstraction. Enhancements in the amount of type II cleavage also occur for α-methoxy-acetophenones such as **10**.[76] It is not clear whether the observed results are due to changes in the ability of the ketone to achieve the necessary conformation for hydrogen abstraction on the surface, an enhanced stabilization of the adsorbed biradical or a decrease in the efficiency of α-cleavage products. For both **9** and **10** the type I cleavage products at low temperature resulted primarily from addition of the acyl radical to the aromatic ring of the initial alkyl radical (Scheme 3), demonstrating once again that the diffusional mobility of the geminate radical pair is restricted by adsorption.[75,76]

Table 4. Ratios of Type I/Type II Reaction and Biradical Elimination/Cyclization (Elim./Cycl.) for Photolysis of Ketones in Solution and Adsorbed on Silicas

Ketone	Medium (°C)[a]	Type I/Type II	Elim./Cycl.	Ref.
8, $n = 2$	*t*-butanol	0.04	0.91	61
	TLC[b]	0.28	0.50	61
	RPTLC[c]	0.38	0.50	61
8, $n = 15$	*t*-butanol	0.07	0.50	61
	TLC	0.48	0.78	61
	RPTLC	0.22	0.27	61
9, R = CH$_3$	Methanol (56)	13.6	2.1	75
	Methanol (-80)	>100	—	75
	Silica gel (56)	3.6	2.6	75
	Silica gel (-78)	2.1	0.66	75

[a] Room temperature unless otherwise specified.
[b] Thin-layer chromatography plates.
[c] Reverse-phase thin-layer chromatography plates.

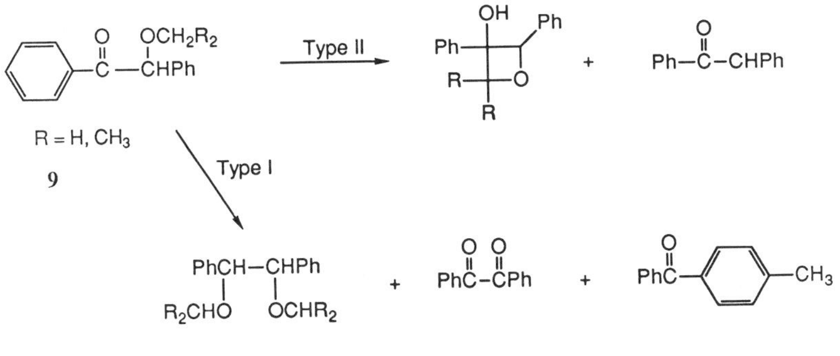

Scheme 3

5. Cycloadditions and Cyclodimerizations

There are a number of examples in which the possibility of controlling the stereochemistry of cycloaddition and dimerization reactions of adsorbed molecules has been examined. Two early studies reported changes in product ratios for cyclobutane dimers of 3-methyl-4-nitro-5-styrylisoxazole (11)[77] and 2-methyl-1,4-naphthoquinone (12).[78] In the former case the selectivity (>90% of a single isomer) was much higher than in solution and approached that observed for photodimerization in the solid state.[77]

11   12

13   14

Another example has shown that the efficiency of dimerization of dibenzotropone (13) on silica is modified by surface treatment, although the product formed is the same as in solution. Dry silica surfaces on which the ketone is more tightly bound gave less efficient dimerization than either wet or fluorinated silica.[17] The results suggest that aggregate formation is not important in this case. It has also been reported that silica adsorption prevents the intermolecular [2+2] cycloaddition of 1,5-bis(4-dimethylamino)phenyl-1,4-pentadien-3-one (14), either by changing the excited-state energy levels or by restricting the molecular mobility.[79]

One of the more detailed studies has looked at the photodimerization of acenaphthylene (reaction 9). In solution dimerization occurs from both singlet and triplet excited states, with the singlet pathway giving only the cis dimer, while the triplet gives a mixture of cis and trans products. On the surface of silica gel both dimers are observed; the singlet reaction was shown to occur only from nearest-neighbor interactions and to increase with increasing surface coverage, indicating a nonuniform distribution of acenaphthylene on the silica surface.[25,80] A combination of sensitization (Rose Bengal) and ferrocene quenching experiments indicated that translational movement of both the monomer and the dimer occurs on the time scale of the triplet lifetime. From this data it was estimated that acenaphthylene can migrate ~300 Å during its triplet lifetime, which is probably at least as long as that which has been measured in solution (2 μs).[25] Further, a rate constant of 7.0 x 10^{15} dm^2 mol^{-1} s^{-1} was estimated for quenching of the acenaphthylene triplet by ferrocene. The dimerization of 9-cyanophenanthrene on silica gel was also examined.[25] In contrast to the results for singlet acenaphthylene ($\tau \sim 1$ ns), the longer singlet lifetime of 9-cyanophenanthrene ($\tau \sim 20$ ns) allows molecules to diffuse together and dimerize.

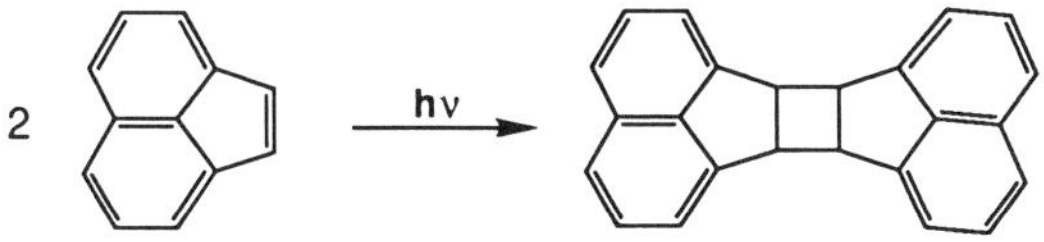

(9)

(cis and trans)

A number of photocycloadditions of alkenes and allenes to steroidal enones such as **15** have been examined on both silica and alumina.[81,82] Adsorption of the enone on the less hindered face directs the cycloaddition to the more hindered β face (reaction 10, **17**), in contrast to the usual preference for α attack (reaction 10, **16**) in solution. Although mixtures of products are typical, in the example shown in reaction (10) a complete reversal of the solution stereochemistry was observed for the addition of allene to enone **15**.[82] Adsorption on silica also reduces the amount of trans-fused cycloadducts, presumably by restricting the conformational inversion of the biradical required for their formation.

The intramolecular cycloaddition of the tricyclic dienedione (**18**) adsorbed on silica gel has also been studied.[83] The reaction occurs with a quantum yield of 1 and has been proposed as a suitable actinometer for determinations of quantum yields on this support.

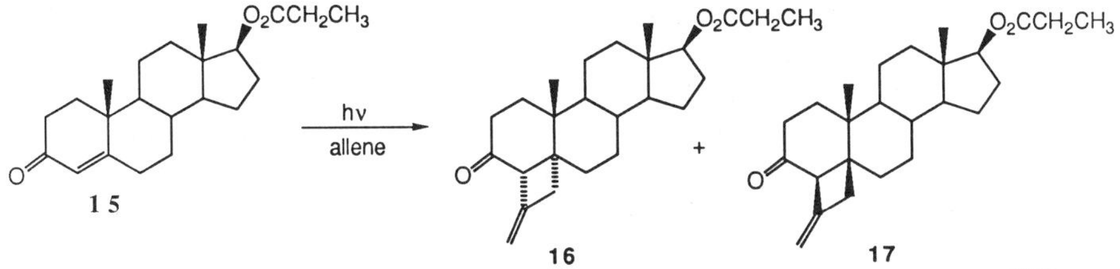

(10)

Methanol	83	17
Silica gel	47	53
Alumina	20	80

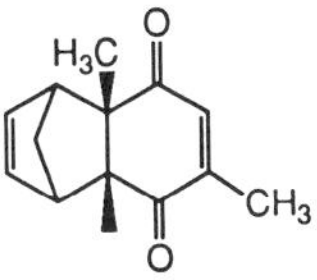

6. Isomerizations

A variety of photoisomerizations of molecules adsorbed on silica and alumina have been examined. One of the more interesting is the photochromism of spiropyrans such as **19**, which upon uv irradiation in solution undergo C—O ring heterolysis to produce the colored merocyanine **20**, which reverts to **19** both thermally and photochemically (reaction 11). In contrast, adsorption of the spiropyran on silica gel (either dry or as a solvent slurry) results in "reversed photochromism" in which the merocyanine is the stable form; irradiation now results in conversion to the closed spiropyran, which thermally reverts to the open form.[16,84-86]

1 9 (colorless)   2 0 (colored)   (11)

Preferential adsorption or stabilization of one isomer has also been suggested to be important in other surface-mediated isomerizations. For example, preferential adsorption of trans versus cis isomers of α,β-unsaturated carbonyl compounds and related compounds (e.g., **21, 22**) on aluminosilicates increases the proportion of cis isomers in the photostationary-state mixtures.[87] Further, stabilization of *cis*-thioindigo (**23**) on an alumina surface has been shown to prevent its isomerization to the trans isomer although the reverse reaction (trans–cis) occurs facilely.[88] In contrast to this result, reversible cis–trans isomerization of perinaphthothioindigo occurs on a silica surface, although with lower quantum yields than in solution.[89] In this case the thermal cis–trans conversion occurs ca. 100 times faster on silica than in solution. The failure to observe cis–trans conversion for thioindigo on alumina may also indicate a rapid thermal back reaction for the trans isomer.

21   22   23

Changes in the photostationary-state ratios for cis–trans isomerizations of retinal isomers[90] and piperylenes[91] were rationalized on the basis of changes in the excited state energy levels for the adsorbed molecules. Stilbene in a silica gel–solvent matrix also gave a different photostationary state from that in solution,[16] whereas on alumina a largely reversible photochromism was observed and was postulated to involve an intermediate with substantial bonding to an active surface site.[92, 93] Similarly, surface interactions have been invoked to explain the double bond migration that occurs in addition to cis–trans isomerization for irradiation of 2-butenes adsorbed on porous Vycor (reaction 12).[94-96] The photoisomerizations are postulated to occur at Lewis acid sites on the surface.

$$\diagdown\!\!=\!\!\diagup \quad \xrightarrow{\;h\nu\;} \quad \diagdown\!\!=\!\!\diagup \quad + \quad \diagdown\!\!\diagup\!\!\diagdown \qquad (12)$$

In contrast to these variations in isomerizations of adsorbed molecules, the isomerization of *trans*-azobenzene (**24**) is not modified by adsorption on alumina.[11] It was suggested that for azobenzene the isomerization occurred via an inversion mechanism which did not require desorption from the surface. For both stilbene and azobenzene cis–trans isomerization was accompanied by minor amounts of cyclization to dihydroaromatics, as is also observed in solution.[11] Finally, the photoisomerization of 1-(9-anthryl)-4,4-diphenyl-2,3-diazabutadiene (**25**) on silica gel has been suggested as an actinometer, although the technique is applicable only to transparent silica gel solvent slurries.[97]

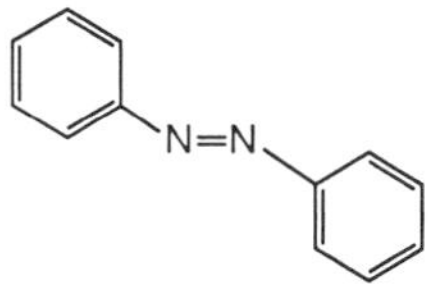

24

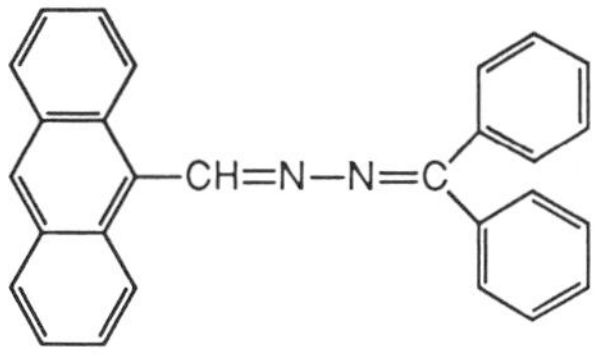

25

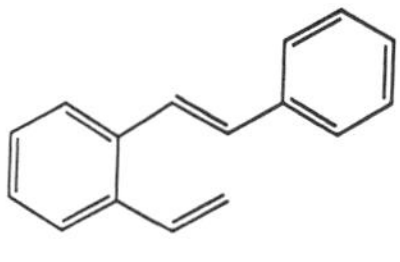

26

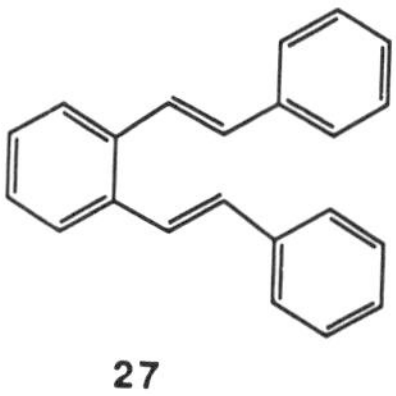

27

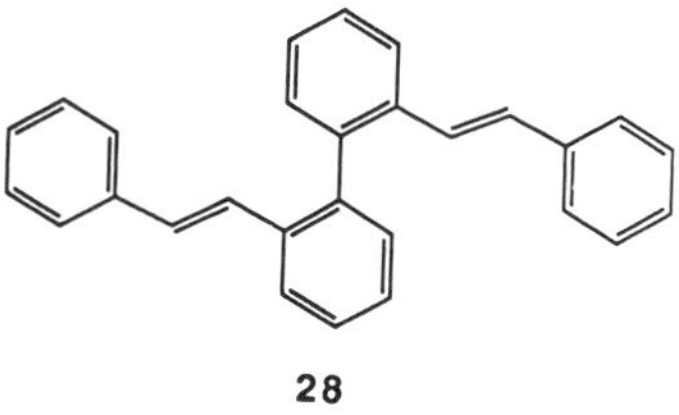

28

The photochemistry of 2-vinylstilbene (26) is substantially modified by adsorption on silica gel. In addition to the formation of 5-phenylbenzobicyclo-[2.1.1]hex-2-enes which are the only products formed in solution, a variety of naphthalene and indene derivatives are also produced.[98] The additional products are ascribed to variations in the reactivities of the various possible conformers as well as to the decreased mobility of the intermediate biradicals on the silica surface.

Although the irradiation of 1,2-distyrylbenzene (27) on silica gel also gives a much more complex product mixture than is obtained in solution, irradiation of the related 2,2'-distyrylbiphenyl (28) gives the same two products in similar ratios in solution and on silica gel.[98] These results again point out the difficulty in predicting the manner in which surface adsorption will modify the normal solution behavior for any particular molecule.

7. Electron Transfer Reactions

These reactions may be divided into two distinct categories: photoionization of adsorbed species and electron transfer between two adsorbed species (or between an adsorbed species and a donor or acceptor in solution). In the former, the support itself is the acceptor and these reactions have a number of similarities to the chemistry of species adsorbed on semiconductor surfaces. However, the photoionizations that have been observed on silica and alumina are always induced by excitation of the adsorbed molecule, not the support.

The irradiation of benzene adsorbed on silica gel and alumina at 77 K has been examined by electron spin resonance.[99] On silica spectra due to phenyl radical and benzene radical cation and dimer radical cation were identified, whereas on alumina an additional species assigned to the cyclohexadienyl radical was also observed. Two-photon ionization was suggested to result in formation of the radical cation, which then reacted with benzene to produce the dimer radical cation. The photoionization of triphenylamine adsorbed on porous Vycor at 77 K has also been studied by esr.[100] Signals due to the amine radical cation were observed only when an electron acceptor such as methyl bromide was available to trap the photoejected electron. The observation of methyl radicals demonstrated that trapping of the electron followed by loss of bromide had occurred. Evidence in favor of electron migration on the Vycor surface was presented, although this conclusion assumes a uniform distribution of adsorbed species. More recently, the photoinduced disproportionation of $Ru(bpy)_3^{2+}$ exchanged onto the surface of porous Vycor has been examined.[101] Photolysis of the adsorbed complex leads to disproportionation, which occurs via a sequential biphotonic ionization followed by capture of the photoejected electron by another adsorbed complex [reactions (13) and (14)]. Analysis of the temperature dependence of the quantum yield led to an estimated barrier to electron transport on the glass surface of 6.87 kcal mol^{-1}. A similar mechanism has been postulated for the production of $MV^{+\bullet}$ via trapping of an electron photoejected from $Ru(bpy)_3^{2+}$ adsorbed on porous Vycor.[102]

$$Ru(bpy)_3^{2+}(ads) \quad \xrightarrow{2h\nu} \quad Ru(bpy)_3^{3+}(ads) \quad + \quad e^- \tag{13}$$

$$e^- + Ru(bpy)_3^{2+}(ads) \quad \longrightarrow \quad [Ru(bpy)_2(bpy)^-]^+(ads) \tag{14}$$

Recently diffuse reflectance laser flash photolysis techniques have been applied to monitor the radical cations on solid supports. For example, radical cations of diphenylpolyenes adsorbed on alumina (**29, 30, 31**) and on silica (**31**) have been observed.[38] The radical cations showed very nonexponential decays extending over six orders of magnitude and decayed almost exclusively by radical cation–electron recombination. The mechanism for radical–cation formation on alumina is not well understood. It has been suggested that the concentration of surface defects such as aluminum ions with coordinatively unsaturated sites that can act as electron acceptors is too low to allow formation of charge-transfer complexes with polyenes.[38] This is further supported by the similarity of the ground-state absorption spectra for the polyenes on alumina and in solution. The ejected electron may eventually populate these sites or it may be taken up by the bulk of the support.

The behavior of Rose Bengal on silica, alumina, and titanium dioxide has also been examined using transient techniques.[40] Only the triplet was observed on silica, whereas excitation of the dye on alumina generated both triplet and radical cation. The latter was produced in a biphotonic process via an upper singlet or triplet state, in contrast to the results for titanium dioxide, where excitation produced radical cation via a direct monophotonic process. The latter reaction has been shown to occur via charge injection from the excited dye into the conduction band of the support and is largely an irreversible process. The amount of degradation of the dye on the three supports was proportional to the ease of formation of the radical cation. A similar biphotonic photoionization has been reported for pyrene on alumina; in this case the radical cation eventually decays by back electron transfer.[34] These results, as well as those above for the diphenylpolyenes, highlight the greater activity of alumina as compared to silica supports in terms of promoting electron-transfer reactions.

There are a number of reports on the use of aqueous solutions of negatively charged colloidal silica particles to enhance the efficiency of electron-transfer reactions by preventing the energy-wasting back electron transfer reaction.[103-105] For example, excitation of $Ru(bpy)_3^{2+}$ electrostatically adsorbed to the colloid surface in the presence of a neutral zwitterionic viologen (propyl viologen sulfonate, PVS) as acceptor leads to quantum yields of reduction that are substantially higher than those in homogeneous solution (0.033 versus 0.005). The enhancement results from the much slower back electron transfer reaction in the colloid system due to electrostatic repulsion of the acceptor radical anion by the negatively charged interface (reaction 15).

$$Ru(bpy)_3^{2+*} + PVS \longrightarrow Ru(bpy)_3^{2+} + PVS^- \qquad (15)$$

$$\underset{\text{back electron transfer}}{\swarrow} \qquad \underset{\text{free ions}}{\searrow}$$

Subsequent investigations of similar systems have identified other zwitterionic viologens that have suitable redox properties for producing hydrogen from water at the basic pHs necessary for stabilizing colloidal silica.[106,107] A related report has shown that photoinduced electron transfer between an iridium(III) complex trapped in a porous silica glass and 1,4-dimethoxybenzene dissolved in the water phase of the glass pores occurs readily.[108] Further, the back electron transfer reaction is retarded by four orders of magnitude relative to the same system in solution and at acidic pH the reduced iridium(II) species is able to catalyze the formation of hydrogen. Colloidal silica systems have also been useful in enhancing electron-transfer reactions in other systems. For example, large enhancements in the efficiency of $Ru(bpy)_3^{2+}$-sensitized cis–trans isomerizations for styrylpyridinium ions in colloidal silica systems have been reported.[109] Electron-transfer reactions have also been examined on positively charged colloidal alumina-coated silica particles.[110] Reductive quenching of an excited anionic ruthenium complex by a coadsorbed anionic electron donor was shown to be greatly enhanced over the rate in homogeneous solution; electron transfer occurred via a static quenching process as a result of the high local concentrations of the two species on the particle surface.

8. Oxidations and Reductions

Relatively few oxidations and reductions have been examined on solid supports, which makes it difficult to draw any general conclusions as to the effects to be expected. The photochemical oxidation of anthracene to anthraquinone has been shown to be strongly enhanced by adsorption on the surface of either silica or alumina.[111] Irradiation of 1,1-diphenylethylene adsorbed on silica gel, alumina, and florisil in an oxygen atmosphere leads to oxidative cleavage to benzophenone and similar behavior is observed for other olefins.[112] The efficiencies are much higher than those observed in solution and competing dimerization of the radical cations is not observed. The reaction has been suggested to involve electron transfer within an olefin–oxygen charge-transfer complex, followed by reaction of the resulting olefin radical cation with either oxygen or the superoxide radical anion.

The 1,4-dicyanoanthracene-sensitized photooxidation of dienes adsorbed on silica gel has also been examined.[113] Depending on the particular diene used either addition of singlet oxygen or electron-transfer photooxidation products are produced. For 1,4-diphenyl-1,3-butadiene only cleavage products resulting from superoxide addition are observed (reaction 16), whereas in solution both cleavage products and the endoperoxide resulting from singlet oxygen addition are formed. For 1,4-diphenyl-1,3-cyclohexadiene the endoperoxide is isolated (reaction 17) both in solution and on silica.[113,114] It has been suggested that adsorption of 1,4-diphenyl-1,3-butadiene on the surface prevents the necessary cis conformation required for singlet oxygen addition and thus modifies its normal solution behavior.

$$C_6H_5 \diagup\!\!\!\diagdown\!\!\!\diagup C_6H_5 \xrightarrow[\text{O}_2,\text{ silica}]{\text{Sens, h}\nu} C_6H_5CHO + C_6H_5CH{=}CHCHO \qquad (16)$$

$$C_6H_5{-}\!\!\!\bigcirc\!\!\!{-}C_6H_5 \xrightarrow[\text{O}_2,\text{ silica}]{\text{Sens, h}\nu} C_6H_5{-}\!\!\!\bigcirc\!\!\!{-}C_6H_5 \qquad (17)$$

Several examples of surface-mediated photoreductions have also been reported. For example, the photoreduction of methylene blue adsorbed on silica in the presence of water is reversible but eventually leads to demethylation of the dye.[115] This is in contrast to the stability of the dye in dilute aqueous solutions. The photosensitized reduction of heterocyclic N-oxides by eosin adsorbed on alumina has been suggested to occur from the triplet state of the N-oxide:[116] The latter is produced either via triplet energy transfer from eosin or by trapping of the eosin radical cation by the N-oxide, followed by back electron transfer to generate the singlet excited state of the N-oxide. The photodecomposition of chlorinated compounds (e.g., 32) adsorbed on silica gel has been examined as a model system to simulate environmental conditions.[117] Both oxidation and reductive dechlorination products are produced with higher reaction efficiencies on the silica surface as compared to solution.

$$\text{isoquinoline-}N\text{-oxide} \xrightarrow[\text{eosin}]{\text{h}\nu} \text{isoquinoline} \qquad (18)$$

$$\text{32} \qquad R = CH_2Cl, CH_3$$

9. Miscellaneous Reactions

Several reactions that do not fit any of the above categories have also been examined on silica and alumina supports. For example, the photochlorination of stearic acid at monolayer coverage on alumina resulted in selective chlorination (>90%) at the ω and ω-1 positions (reaction 19), indicating that only the terminal groups were exposed to reaction.[118] In a related study the photochlorination of aromatic hydrocarbons (reaction 20) occurred readily on alumina in the presence of ferric chloride, presumably via the intermediacy of a radical cation.[119] However, the reaction could not be carried out on silica gel.

$$CH_3(CH_2)_{16}CO_2H \xrightarrow{\ h\nu\ } CH_2Cl(CH_2)_{16}CO_2H \ + \ CH_3CHCl(CH_2)_{15}CO_2H \qquad (19)$$

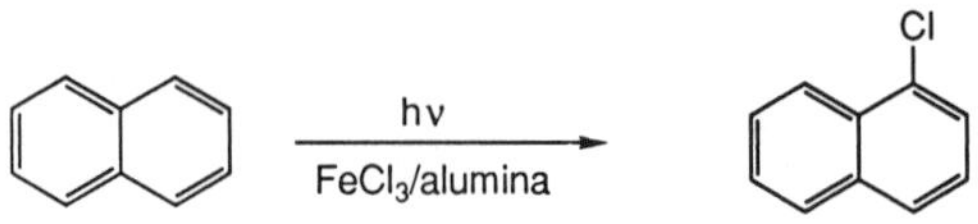

(20)

There are also several reports of cation generation on solid supports. Aryl cations produced by photolysis of diazonium salts (reaction 21) have been detected by esr at low temperatures on silica and their reactivity toward donors has been examined.[120] The 9-phenylxanthenyl cation (**33**) has also been shown to be relatively stable on a silica surface and its luminescence properties have been investigated.[121]

$$Ar\overset{+}{-}N\!\equiv\!N\ X^- \xrightarrow{\ h\nu\ } \overset{+}{Ar} \ + \ N_2 \ + \ X^- \qquad (21)$$

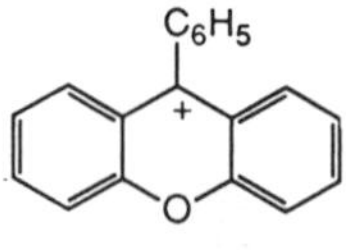

33

10. Conclusions

The photochemical reactions of molecules adsorbed on silica and alumina usually show some deviations from their solution behavior. These are generally reflected in modifications of product ratios and in a few cases in the occurrence of reaction pathways that are not otherwise observed. These effects can be attributed primarily to restrictions on the diffusional mobility of the adsorbed molecules, their excited states or various intermediates along the reaction pathway. In general, the effects of surface adsorption are greatest for short-lived intermediates, although most species are sufficiently long-lived that their rotational movement is not affected. In addition conformational effects are important in determining product ratios in some isomerizations and biradical reactions. Many of the photophysical and photochemical studies have also provided useful information on the surface polarity and the distributions of adsorbed molecules. Further, in those cases where the effects have been examined, there is generally a trend toward more solution-like behavior on the modified surface. In general, modifications in reaction pathways are often more pronounced upon inclusion of molecules in zeolites as compared to adsorption on metal oxide surfaces. Despite the changes that can be observed for photochemical reactions of adsorbed molecules on silica and alumina, it is still not particularly easy to predict beforehand the direction of change in any given case. Additional experiments are obviously required to understand the complex range of factors that control the behavior of adsorbed species.

References

1. P. de Mayo and L. J. Johnston in *Preparative Chemistry Using Supported Reagents*, P. Laszlo, ed., Academic Press, San Diego, CA, 1987, Ch. 4.
2. *Photochemistry on Solid Surfaces*, Studies in Surface Science and Catalysis, Vol. 47, M. Anpo and T. Matsuura, eds., Elsevier, Amsterdam, 1989.
3. A. V. Kiselev and V. I. Lygin, *Infrared Spectra of Surface Compounds*, Keterpress Enterprises, Jerusalem, 1975.
4. R. K. Iler, *The Chemistry of Silica*, John Wiley, New York, 1979, Chps. 4 - 6.
5. K. K. Unger, *Porous Silica*, J. Chromatographic Library, Vol. 16, Elsevier, Amsterdam, 1979.
6. H. P. Boehm, *Adv. Catalysis*, **1966**, *16*, 179.
7. J. B. Peri, *J. Phys. Chem.*, **1965**, *69*, 211.
8. H. Knozinger and P. Ratnasamy, *Catal. Rev. Sci. Eng.*, **1978**, *17*, 31.
9. P. de Mayo, *Pure Appl. Chem.*, **1982**, *54*, 1623.
10. P. de Mayo, L. V. Natarajan and W. R. Ware in *Organic Phototransformations in Microheterogeneous Media*, ACS Symposium Series, M. A. Fox, ed., American Chemical Society, Washington, DC, **1985**, p.1.
11. D. Oelkrug, W. Flemming, R. Fullemann, R. Gunther, W. Honnen, G. Krabichler, M. Schafer and S. Uhl, *Pure Appl. Chem.*, **1986**, *58*, 1207.
12. J. K. Thomas, *J. Phys. Chem.*, **1987**, *91*, 267.
13. K. A. Zachariasse in *Photochemistry on Solid Surfaces*, Studies in Surface Science and Catalysis, Vol. 47, M. Anpo and T. Matsuura, eds., Elsevier, Amsterdam, 1989, Ch. 2.3.
14. M. Robin and K. N. Trueblood, *J. Am. Chem. Soc.*, **1957**, *79*, 5138.
15. P. A. Leermakers, H. T. Thomas, L. D. Weis and F. C. James, *J. Am. Chem. Soc.*, **1966**, *88*, 5075.
16. L. D. Weis, T. R. Evans and P. A. Leermakers, *J. Am. Chem. Soc.*, **1968**, *90*, 6109.
17. Z. Grauer, H. Daniel and D. Avnir, *J. Colloid Interface Sci.*, **1983**, *96*, 411.
18. D. Fassler, M. Raddatz and D. Baezold, *Z. Chem.*, **1984**, *24*, 411.

19. P. de Mayo, A. Safarzadeh-Amiri and S. K. Wong, *Can. J. Chem.*, **1984**, *62*, 1001.
20. R. K. Bauer, P. de Mayo, W. R. Ware and K. C. Wu, *J. Phys. Chem.*, **1982**, *86*, 3781.
21. C. Francis, J. Lin and L.A. Singer, *Chem. Phys. Lett.*, **1983**, *94*, 162.
22. R. K. Bauer, P. de Mayo, L. V. Natarajan and W. R. Ware, *Can. J. Chem.*, **1984**, *62*, 1279.
23. D. Avnir, R. Busse, M. Ottolenghi, E. Wellner and K. A. Zachariasse, *J. Phys. Chem.*, **1985**, *89*, 3521.
24. T. Fujii, E. Shimizu and S. Suzuki, *J. Chem. Soc. Faraday Trans. I*, **1988**, *84*, 4387.
25. R. K. Bauer, R. Borenstein, P. de Mayo, K. Okada, M. Rafalska, W. R. Ware and K. C. Wu, *J. Am. Chem. Soc.*, **1982**, *104*, 4635.
26. P. de Mayo, L. V. Natarajan and W. R. Ware, *J. Phys. Chem.*, **1985**, *89*, 3526.
27. P. Hite, R. Krasnansky and J. K. Thomas, *J. Phys. Chem.*, **1986**, *90*, 5795.
28. S. Suzuki and T. Fujii in *Photochemistry on Solid Surfaces*, Studies in Surface Science and Catalysis, Vol. 47, M. Anpo and T. Matsuura, eds., Elsevier, Amsterdam, 1989, Ch. 2.4.
29. N. Nakashima and D. Phillips, *Chem. Phys. Lett.*, **1983**, *97*, 337.
30. A. Levy, D. Avnir and M. Ottolenghi, *Chem. Phys. Lett.*, **1985**, *121*, 233.
31. E. M. Schulman and C. Walling, *Science*, **1972**, *178*, 53.
32. W. Honnen, G. Krabichler, S. Uhl and D. Oelkrug, *J. Phys. Chem.*, **1983**, *87*, 4872.
33. P. L. Piciulo and J. W. Sutherland, *J. Am. Chem. Soc.*, **1979**, *101*, 3123.
34. G. Beck and J. K. Thomas, *Chem. Phys. Lett.*, **1983**, *94*, 553.
35. N. J. Turro, M. B. Zimmt and I. R. Gould, *J. Am. Chem. Soc.*, **1985**, *107*, 5826.
36. N. J. Turro, I. R. Gould, M. B. Zimmt and C. C. Cheng, *Chem. Phys. Lett.*, **1985**, *119*, 484.
37. D. Oelkrug, W. Honnen, F. Wilkinson and C. J. Willsher, *J. Chem. Soc., Faraday Trans. 2*, **1987**, *83*, 2081.
38. D. Oelkrug, G. Krabichler, W. Honnen, F. Wilkinson and C. J. Willsher, *J. Phys. Chem.*, **1988**, *92*, 3589.
39. F. Wilkinson and L. F. V. Ferreira, *J. Lumin.*, **1988**, *40 and 41*, 704.
40. K. R. Gopidas and P. V. Kamat, *J. Phys. Chem.*, **1989**, *93*, 6248.
41. R. W. Kessler and F. Wilkinson, *J. Chem. Soc. Faraday Trans. I*, **1981**, *77*, 309.
42. J. M. Drake, P. Levitz, N. J. Turro, K. S. Nitsche and K. F. Cassidy, *J. Phys. Chem.*, **1988**, *92*, 4680.
43. P. Levitz, J. M. Drake and J. Klafter in *Molecular Dynamics in Restricted Geometries*, J. Klafter and J. M. Drake, eds., John Wiley, New York, 1989, Ch. 7.
44. D. R. James, Y. S. Liu, P. de Mayo and W. R. Ware, *Chem. Phys. Lett.*, **1985**, *120*, 460.
45. D. Avnir, D. Farin and P. Pfeifer, *Nature*, **1984**, *308*, 261.
46. D. Avnir, *J. Am. Chem. Soc.*, **1987**, *109*, 2931.
47. D. Pines, D. Huppert and D. Avnir, *J. Chem. Phys.*, **1988**, *89*, 1177.
48. J. M. Drake, P. Levitz and J. Klafter, *New. J. Chem.*, **1990**, *14*, 77.
49. P. A. Leermakers, L. D. Weis and H. T. Thomas, *J. Am. Chem. Soc.*, **1965**, *87*, 4403.
50. D. Avnir, P. de Mayo and I. Ono, *J. Chem. Soc., Chem. Commun.*, **1978**, 1109.
51. M. M. Abdel-Malik and P. de Mayo, *Can. J. Chem.*, **1984**, *62*, 1275.
52. L. J. Johnston, P. de Mayo and S. K. Wong, *J. Org. Chem.*, **1984**, *49*, 20.
53. J. E. Leffler and J. J. Zupancic, *J. Am. Chem. Soc.*, **1980**, *102*, 259.

54. N. J. Turro, K. C. Waterman, K. M. Welsh, M. A. Paczkowski, M. B. Zimmt and C. C. Cheng, *Langmuir* , **1988**, *4*, 677.
55. D. Avnir, L. J. Johnston, P. de Mayo and S. K. Wong, *J. Chem. Soc. Chem. Commun.*, **1981**, 958.
56. G. A. Epling and E. Florio, *J. Am. Chem. Soc.*, **1981**, *103*, 1237.
57. B. Frederick, L. J. Johnston, P. de Mayo and S. K. Wong, *Can. J. Chem.*, **1984**, *62*, 403.
58. N. J. Turro, C. C. Cheng and W. Mahler, *J. Am. Chem. Soc.*, **1984**, *106*, 5022.
59. N. J. Turro, C. C. Cheng, P. Wan, C. J. Chung and W. Mahler, *J. Phys. Chem.*, **1985**, *89*, 1567.
60. N. J. Turro, M. A. Paczkowski and P. Wan, *J. Org. Chem.*, **1985**, *50*, 1399.
61. N. J. Turro, *Tetrahedron*, **1987**, *43*, 1589.
62. L. Lunazzi, K. U. Ingold and J. C. Scaiano, *J. Phys. Chem.*, **1983**, *87*, 529.
63. B. H. Baretz and N. J. Turro, *J. Am. Chem. Soc.*, **1983**, *105*, 1309.
64. L. J. Johnston and S. K. Wong, *Can. J. Chem.*, **1984**, *62*, 1999.
65. G. Kelly, C. J. Willsher, F. Wilkinson, J. C. Netto-Ferreira, A. Olea, D. Weir, L. J. Johnston and J. C. Scaiano, *Can. J. Chem.*, **1990**, *68*, 812.
66. J. E. Leffler and J. T. Barbas, *J. Am. Chem. Soc.*, **1981**, *103*, 7768.
67. H. Aoyama, K. Miyazaki, M. Sakamoto and Y. Omote, *Chem. Lett.*, **1983**, 1583.
68. T. Hasegawa, J. Moribe and M. Yoshioka, *Bull. Chem. Soc. Jpn.*, **1988**, *61*, 1437.
69. Y. Kubokawa and M. Anpo, *J. Phys. Chem.*, **1974**, *78*, 2442.
70. M. Anpo, S. Hirohashi and Y. Kubokawa, *Bull. Chem. Soc. Jpn.*, **1975**, *48*, 985.
71. M. Anpo, T. Wada and Y. Kubokawa, *Bull. Chem. Soc. Jpn.*, **1975**, *48*, 2663.
72. M. Anpo and Y. Kubokawa, *Bull. Chem. Soc. Jpn.*, **1975**, *48*, 3085.
73. M. Anpo and Y. Kubokawa, *Bull. Chem. Soc. Jpn.*, **1976**, *49*, 2623.
74. M. Anpo, T. Wada and Y. Kubokawa, *Bull. Chem. Soc. Jpn.*, **1977**, *50*, 31.
75. P. de Mayo, A. Nakamura, P. W. K. Tsang and S. K. Wong, *J. Am. Chem. Soc.*, **1982**, *104*, 6824.
76. P. de Mayo and N. Ramanth, *Can. J. Chem.*, **1986**, *64*, 1293.
77. D. Donati, M. Fiorenza and P. Sarti-Fantoni, *J. Heterocyclic Chem.*, **1979**, *16*, 253.
78. H. Werbin and E.T. Strom, *J. Am. Chem. Soc.*, **1968**, *90*, 7926.
79. J. M. Eisenhart and A. B. Ellis, *J. Org. Chem.*, **1985**, *50*, 4108.
80. P. de Mayo, K. Okada, M. Rafalska, A. C. Weedon and G. S. K. Wong, *J. Chem. Soc., Chem. Commun.*, **1981**, 820.
81. R. Farwaha, P. de Mayo and Y. C. Toong, *J. Chem. Soc. Chem. Commun.*, **1983**, 739.
82. R. Farwaha, P. de Mayo, J. H. Schauble and Y. C. Toong, *J. Org. Chem.*, **1985**, *50*, 245.
83. S. Lazare, P. de Mayo and W. R. Ware, *Photochem. Photobiol.*, **1981**, *34*, 187.
84. T. R. Edwards, A. F. Toth and P. A. Leermakers, *J. Am. Chem. Soc.*, **1967**, *89*, 5060.
85. C. Balny and P. Douzou, *C. R. Acad. Sci. Paris, Ser. C*, **1967**, *264*, 477.
86. D. Levy and D. Avnir, *J. Phys. Chem.*, **1988**, *92*, 4734.
87. R. F. Childs, B. Duffey and A. Mika-Gibala, *J. Org. Chem.*, **1984**, *49*, 4352.
88. H. D. Breuer and H. Jacob, *Chem. Phys. Lett.*, **1980**, *73*, 172.
89. K. Fukunishi, M. Kobayashi, A. Morimoto, M. Kuwabara, H. Yamanaka and M. Nomura, *Bull. Chem. Soc. Jpn.*, **1989**, *62*, 3733.
90. M. E. Zawadzki and A. B. Ellis, *J. Org. Chem.*, **1983**, *48*, 3156.
91. L. D. Weis, B. W. Bowen and P. A. Leermakers, *J. Am. Chem. Soc.*, **1966**, *88*, 3176.

92.   H. G. Hecht and J. L. Jensen, *J. Photochem.*, **1978**, *9*, 33.
93.   H. G. Hecht and R. L. Crackel, *J. Photochem.*, **1981**, *15*, 263.
94.   A. Morikawa, M. Hattori, K. Yagi and K. Otsuka, *Z. Phys. Chem.*, **1977**, *104*, 309.
95.   K. Otsuka and A. Morikawa, *Bull. Chem. Soc. Jpn.*, **1975**, *48*, 3025.
96.   K. Otsuka, M. Fukaya and A. Morikawa, *Bull. Chem. Soc. Jpn.*, **1978**, *51*, 367.
97.   D. Fassler, R. Gade and W. Guenther, *J. Photochem.*, **1980**, *13*, 49.
98.   A. J. W. Tol and W. H. Laarhoven, *J. Org. Chem.*, **1985**, *51*, 1663.
99.   T. Tanei, *Bull. Chem. Soc. Jpn.*, **1968**, *41*, 833.
100. P. K. Wong, *Photochem. Photobiol.*, **1974**, *19*, 391.
101. J. Fan, W. Shi, S. Tysoe, T. C. Strekas and H. D. Gafney, *J. Phys. Chem.*, **1989**, *93*, 373.
102. W. Shi and H. D. Gafney, *J. Am. Chem. Soc.*, **1987**, *109*, 1582.
103. M. Calvin, I. Willner, C. Laane and J. W. Otvos, *J. Photochem.*, **1981**, *17*, 195.
104. C. Laane, I. Willner, J. W. Otvos and M. Calvin, *Proc. Natl. Acad. Sci U.S.A.*, **1981**, *78*, 5928.
105. I. Willner, J. W. Otvos and M. Calvin, *J. Am. Chem. Soc.*, **1981**, *103*, 3203.
106. D. N. Furlong, O. Johansen, A. Launikonis, J. W. Loder, A. W. H. Mau and W. H. F. Sasse, *Aust. J. Chem.*, **1985**, *38*, 363.
107. J. L. Bourdelande, J. Camps, J. Font, P. D. March and E. Brillas, *J. Photochem.*, **1985**, *30*, 437.
108. A. Slama-Schwok, D. Avnir and M. Ottolenghi, *J. Phys. Chem.*, **1989**, *93*, 7544.
109. K. Takagi, K. Aoshima and Y. Sawaki, *J. Chem. Soc. Perkin Trans. II*, **1986**, 1771.
110. P. V. Kamat and W. E. Ford, *J. Phys. Chem.*, **1989**, *93*, 1405.
111. V. G. Kortum and W. Braun, *Liebigs Ann.*, **1960**, *632*, 104.
112. C. Aronovitch and Y. Mazur, *J. Org. Chem.*, **1985**, *50*, 149.
113. Y. Cao, B. W. Zhang, Y. F. Ming and J. X. Chen, *Studies Org. Chem.*, **1988**, *33*, 57.
114. Z. Bao-Wen, C. Jian-Xin and C. Yi, *Acta Chim. Sinica*, **1989**, *47*, 502.
115. N. I. Litsov, V. I. Nikolaevskaya and A. A. Kachan, *High Energy Chem.*, **1981**, *15*, 178.
116. N. Hata, *Chem. Lett.*, **1975**, 401.
117. H. Parlar, *Chemosphere*, **1988**, *17*, 2141.
118. C. Eden and Z. Shaked, *Isr. J. Chem.*, **1975**, *13*, 1.
119. M. Hasebe, C. Lazare, P. de Mayo and A. C. Weedon, *Tetrahedron. Lett.*, **1981**, *22*, 5149.
120. D. Bazold, D. Fassler and R. Kunert, *J. Prakt. Chem.*, **1982**, *324*, 209.
121. R. M. Berger and D. Weir, *Chem. Phys. Lett.*, **1990**, *169*, 213.

Chapter 9

Photochemistry and Photophysics in Clays and Other Layered Solids

W. Jones

Department of Chemistry,
University of Cambridge, Cambridge, U.K.

Contents

1. Introductory Remarks

When interacting with a layered host, guest molecules may either be adsorbed onto the external surfaces of the material or they may be accommodated—intercalated—within the bulk of the guest matrix. Intercalation, the insertion of guest molecules between the sheets of a layered host matrix, has been extensively studied for a variety of reasons.[1] Among the hosts that have been investigated are graphite, clays, transition metal dichalcogenides, phosphates, halides, and oxyhalides.[1] From a crystallographic viewpoint, the arrangement of the guest molecules with respect to the host lattice is known to be considerably dependent upon such factors as stoichiometry, temperature, and preparation procedures. Very subtle two-dimensional phase transformations are known to take place. The macroscopic stoichiometry of host and guest, as opposed to the microscopic domain-like composition, is almost infinitely variable. Whereas in the graphite intercalates such variation in stoichiometry is generally accommodated by the process of "staging," for sheet silicates interstratification and the insertion of one, two, three, and up to 10 or more sheets of guest between each pair of host layers gives considerable scope for controlling local reactant (i.e., guest) concentrations and geometric arrangement, see Figure 1.[2]

A great variety of layered materials has been studied. These include clays, graphite, transition metal dichalcogenides, double hydroxides, phosphates, and phosphonates.[1] This review does not aim to cover all the possible range of materials. It concentrates upon three: clays, layered double hydroxides (LDH), and transition metal oxides. We begin by concentrating on the extensive studies that have been reported on clay complexes.[3] Such studies highlight the important attributes of layered solids.

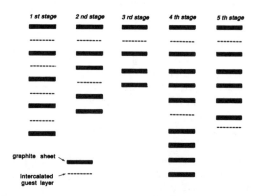

Figure 1 (a)   Staging in graphite intercalates results in the formation of well-defined compounds. It is rare in such materials to find more than one layer of guest molecules in any single gallery. The driving force for intercalation is principally electronic and a large number of graphite intercalates are air-sensitive.

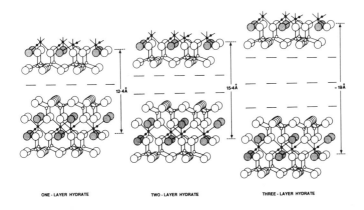

ONE - LAYER HYDRATE   TWO - LAYER HYDRATE   THREE - LAYER HYDRATE

Figure 1 (b) For certain sheet silicates, for example montmorillonite, several layers of guest may be intercalated in a gallery region. The spacings illustrate the values obtained with water as the guest molecules. The principal driving force for the production of these intercalates appears to be a direct interaction between the guest and the charge-balancing cations. A detailed explanation of the structure of the aluminosilicate clay is given in the text and in Figure 2.

2. Sheet Silicates

A great variety of natural sheet silicates are known. Although the basic structure of each is the same, variation in chemical composition has given rise to a family of materials with very subtle changes in properties. While such variation has been helpful, the development of synthetic procedures for producing clays of controlled composition has been equally important. Furthermore, extensive commercial utilization of both natural and synthetic clays has led to a detailed understanding of compositional variations. Very pure materials in large quantities have become available.

2.1. Structural Characteristics of Sheet Silicates

There are four classes of silicate minerals: amorphous silicates (allophanes), chain silicates (pyroxenes and amphiboles), sheet silicates (clays and micas), and infinite three-dimensional frameworks (feldspars and zeolites). The silicates of interest here are the clays and micas although some reference to zeolites will also be made (see Chapter 10 for details on zeolites). The term clay is variously defined[4] but may generally be taken to describe sheet aluminosilicates with particle sizes within the micron range.[4,5] They are composed of two distinct types of layers consisting of [SiO4] tetrahedra and [M(O,OH)6] octahedra with M being, in general, Al^{3+}, Mg^{2+}, or $Fe^{2+/3+}$. The silicates of principal interest here result when one octahedral layer is sandwiched by two tetrahedral layers (see Figure 2).

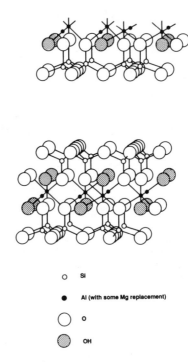

○	SI
●	Al (with some Mg replacement)
◯	o
◎	OH

Figure 2. Schematic illustration of the structure of a montmorillonite-like clay. Two layers of silica-oxygen tetrahedra and joined together by an octahedrally co-ordinated aluminium-oxygen/hydroxide layer to give an individual clay sheet. The ideal composition of the composite sheet would be $Si_8Al_4O_{20}(OH)_4$ but as more and more magnesium replaces aluminium so a greater number of charge-balancing cations are found in the interlayer region. These charge-balancing cations are generally hydrated. The replacement of either the charge-balancing cation or the water or both generates the intercalation compound.

When the octahedral cation is divalent (e.g., Mg^{2+}) then all the possible octahedral sites within the structure will be occupied and the silicate will be of the trioctahedral type. For trivalent ions (e.g., Al^{3+}) only two thirds of the available sites are occupied leading to a dioctahedral silicate. In addition to substitution in the octahedral layer, the tetrahedral layers may also be partially substituted by ions of lower valency. This produces a negative charge on the layers which is balanced by interlamellar exchangeable cations. In naturally occurring clays these are frequently Na^+ or Ca^{2+}.

Scheme 1 illustrates the principal members of the dioctahedral family where the major octahedral cation is Al^{3+}. Scheme 2 illustrates the principal members of the trioctahedral family. The subtle interplay of the location, type, and amount of isomorphous replacement that is possible can be understood from examination of these two schemes. Thus, the neutral end members pyrophyllite (based on Al) and talc (based on Mg) carry no residual charge and therefore no significant cation-exchange capacity. As a result, they do not readily expand and are generally of little interest.

Important clay formulae - dioctahedral

Pyrophyllite
$[Si_8](Al_4)O_{20}(OH)_4$
Neutral end member

$\Downarrow$

Montmorillonite
$M_x[Si_8](Al_{4-x}Mg_x)O_{20}(OH)_4 \cdot nH_2O$

Beidellite
$M_x[Si_{8-x}Al_x](Al_4)O_{20}(OH)_4 \cdot nH_2O$

Mica - muscovite
$K_2[Al_2Si_6](Al_4)O_{20}(OH)_4 \cdot nH_2O$
$(x=2)$

Scheme 1

Montmorillonite, on the other hand, is an extensively studied host mineral whose octahedral substitution pattern is primarily the replacement of Al^{3+} by Mg^{2+}. The extent of replacement is of the order of 80 meq per 100 g. This appears to be ideal for a great variety of applications, particularly for lamellar Brønsted acid catalysts.[6] Scheme 3 gives the important compositional details concerning a montmorillonite-type clay. The chemical composition of natural montmorillonites depends very much on their location. Table 1 gives analytical details for a variety of sheet silicates, including four montmorillonites.[7]

Important clay formulae - trioctahedral

Talc
$$[Si_8](Mg_6)O_{20}(OH)_4$$
Neutral end member

$$\Downarrow$$

Hectorite
$$M_x[Si_8](Mg_{6-x}Li_x)O_{20}(OH)_4 \cdot nH_2O$$

Saponite
$$M_x[Si_{8-x}Al_x](Mg_6)O_{20}(OH)_4 \cdot nH_2O$$

Mica - phlogopite
$$K_2[Al_2Si_6](Mg_6)O_{20}(OH)_4 \cdot nH_2O$$
$$(x=2)$$

Scheme 2

Montmorillonite
$$M_x[Si_8](Al_{4-x}Mg_x)O_{20}(OH)_4$$

In montmorillonite *principal* substitution is Mg^{2+} for Al^{3+} in the octahedral layer. The extent varies slightly but is in the region $x = \sim 0.5$ to 0.8

We have, therefore, as variables:

(i) The number of exchangeable cations - the cation exchange capacity - determined by the value of x.

(ii) The nature of the exchangeable cation i.e. the identity of M. The choice of M is important in controlling the Bronsted acidity of the clay.

Scheme 3

More recent studies have utilized the commercial material Laponite. This is a synthetic hectorite mineral (supplied by Laporte plc) where the octahedral cation is primarily Mg^{2+} and residual charge results from the substitutional incorporation of Li^+. Table 1 gives a typical set of analytical data for a natural hectorite with a cation-exchange capacity of 92 meq per 100 g.[7] An important attribute of Laponite is that it is available in a pure and reproducible form. This is valuable when, for example, the absence of paramagnetic or colored impurities is sought.

Table 1. Compositional Variations and Cation Exchange Capacities for a Variety of Clays

	Tetrahedral layer	Octahedral layer	CEC[a]
Montmorillonite "Gelwhite L"	$Si_{7.65}Al_{0.35}$	$Al_{3.20}Fe^{III}_{0.09}Mg_{0.74}$	100
Montmorillonite Chambers, Arizona (Cheto)	$Si_{7.7}Al_{0.3}$	$Al_{2.67}Fe_{0.44}Mg_{0.90}$	150
Montmorillonite Belle Fourche, South Dakota	$Si_{7.88}Al_{0.14}$	$Al_{3.03}Fe_{0.43}Mg_{0.45}$	110
Montmorillonite Camp Berteau, Morocco	$Si_{7.75}Ti_{0.03}(Al,Fe^{III})_{0.22}$	$(Al,Fe^{III})_{3.22}Mg_{0.69}$	115
Montmorillonite Wyoming ("Volclay")	$Si_{7.84}Al_{0.16}$	$Al_{3.10}Fe^{III}_{0.38}Mg_{0.52}$	87
Beidellite Black Jack mine	$Si_{6.92}Al_{1.08}$	$Al_{3.92}Fe^{III}_{0.08}$	142
Beidellite Rupsroth	$Si_{7.35}Ti_{0.03}(Al,Fe^{III})_{0.64}$	$(Al,Fe^{III})_{3.54}Mg_{0.45}$	148
Hectorite	$Si_{8.0}$	$Mg_{5.45}Li_{0.55}$	92
Kozakov saponite	$Si_{6.6}Al_{1.36}Fe_{0.04}$	$Mg_{5.0}Fe^{III}_{0.48}Fe^{II}_{0.52}$	120

[a] In mequiv per100 g

Laponite material as generally used, however, does not form such well-ordered intercalates and frequently, therefore, crystallographic studies are directed to montmorillonite-like hosts.[7,8]

In beidellite (trioctahedral) and saponite (dioctahedral) the isomorphous substitution occurs in the tetrahedral sites, with some Al^{3+} replacing Si^{4+}. Such replacement distinguishes, for example, montmorillonite-like materials from beidellitic materials, in terms of the coordination of the substituted ion as well as its proximity to the interlayer region.

These descriptions, as the data in Table 1 indicate, however, are very idealized. In the majority of silicate structures a considerable range of substitutions (both in identity and location) are possible. This is particularly true of naturally occurring specimens.

2.2. Principle Techniques of Investigation

It is not possible to cover all the experimental techniques available for studying intercalates and layered solids. For clays the following are important.[4,9]

1. Cation-exchange capacity (cec). Alongside the determination by chemical analysis, cation-exchange measurements are important in indicating the number of exchangeable species that may be inserted into the gallery regions. The area of a single six-ring site in a clay is of the order of 24 $Å^2$. Each of these rings is sufficiently large to allow cations to be partly or wholly keyed into them when the clay is outgassed and dehydrated.[5] If the cation-exchange capacity is high (e.g., in micas, at 250 meq per 100 g) all six-ring sites will be occupied. In montmorillonite, on the other hand, only some of the six-ring sites will be occupied. In the hydrated states, when expanded, the hydrated cations may be "floating" in the interlamellar water layers.[5]

2. The water content of the clay (see Figure 3) is important since it controls many important properties of the clay, e.g., the acidity (and hence extent of protonation of guest molecules) within the interlayer region. In addition, the hydrophobic nature of the interlayer will be controlled by the amount of water present and hence the interaction, for example, with organic guests. Thermogravimetric analysis is important for determining the water (and other guest) content. The absorption of the water is theoretically reversible and the separation between the layers of the crystals returns to the normal value when the water is removed. However, probably as a result of unsymmetrical drying, there is generally a permanent increase in the overall volume after drying.[4]

3. Infrared spectroscopy using thin self-supporting films of the appropriate purified clay will lead to some indication as to the composition of the material as well as the number and nature of intercalated molecules. For amines, for example, both neutral and protonated guest species may exist, with the relative proportions being controlled by the charge-balancing cations. The intensity of interlamellar water vibrations, for example, decrease on the uptake of γ-butyrolactone, this decrease being associated with the

displacement of the water by the organic guest. Uv–visible spectroscopy allows electronic modifications and perturbations to be studied.

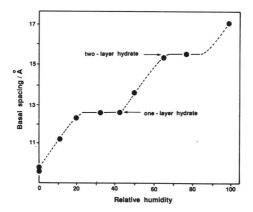

Figure 3. The control of the relative humidity above a sodium exchanged clay controls the amount of guest taken up and the accompanying gallery height. The values obtained correspond with those indicated in Figure 1(b). Note that the interconversion is gradual and that by careful preparation it is possible to produce quite well-defined compositional (and hence structural) materials.

4. x-Ray diffraction measurements (and in particular d_{00l}) indicate the expansion of the interlayer region that may accompany uptake of guest molecules. One-dimensional Fourier projection maps may be extracted from the intensity of $(00l)$ deflections when highly oriented films (and hence large values of l) of the intercalate are formed.
5. Nuclear magnetic resonance (NMR), both conventional and with magic angle spinning (MAS), may yield information about the nature of the intercalated species and changes in structure of the sheet silicate.[10] Aluminum-27 MASNMR has become an important tool since it readily allows octahedral and tetrahedral substitution to be distinguished. By techniques established for other solids, in particular zeolites, ^{29}Si NMR (Scheme 4),[10] allows a quantitative determination of the Si:Al ratio. Whenever possible, samples low in impurities such as iron and manganese are used: otherwise these paramagnetic centers will lead to line broadening and an increase in the contribution from spinning sidebands.[10] Both esr and Mössbauer spectroscopy have also been used to probe both the clay and the guest structures.
6. Luminescence and electronic spectra have contributed greatly to our understanding of the photophysical properties of the adsorbed and intercalated molecules. Changes in fluorescence spectra, for example, allow changes in

local aggregation to be established by monitoring both ground-state interactions (through absorption spectra) and excited-state perturbations (through excimeric emission).

Typical values for common sheet silicates obtained from MASNMR

Pyrophyllite $[Si_8](Al_4)$		-95.1 ppm
Talc $[Si_8](Mg_6)$		-98.1 ppm
Montmorillonite $M_x[Si_8](Al_{4-x}Mg_x)$		-93.1 ppm
Beidellite	(0Al)	-93.0 ppm
$M_x[Si_{8-x}Al_x](Al_4)$	(1Al)	-88.0 ppm

Scheme 4

2.3. Intercalation of Water and Other Guest Molecules and Species
2.3.1. Water

The fact that clays swell in the presence of water results from the increased uptake of interlamellar water. This uptake can take place in a very controlled manner, giving rise to so-called one-layer, two-layer, three-layer, and so on hydrates.[5] A typical water uptake curve is shown in Figure 3 for a Na-montmorillonite. While the exact arrangement of water for particular intercalates may be unclear, the existence of the well-defined "hydrates" is unambiguous. The neutral end members talc and pyrophyllite do not take up water. The major contribution to the driving force for water uptake is the change in free energy when interlayer cations are hydrated rather than solvated by the lattice oxygen. If there are too many cations then the positive free energy change of sheet separation exceeds the decrease due to the penetration of the water. If there are no exchangeable cations (e.g., talc, pyrophyllite) then water penetration becomes energetically unfavourable.[5] In some Na-montmorillonites complete swelling in water occurs with an almost infinite separation of the clay sheets.

The properties of the interlayer water depend very much upon the identity of the exchangeable cation, the amount of water present and the magnitude of the cation-exchange capacity. For basic molecules, for example, the extent of protonation within the interlayer region is substantially different for an Al^{3+}-exchanged sample compared with a Na^+-exchanged sample (Figure 4).[8] Proton nmr also reveals this difference: the water associated with the Na^+ cation is readily displaced by D_2O. Such is not the case for the water around an Al^{3+} cation.[11] Clearly, therefore, it is expected that a rich variety of properties may be engineered for the interlayer water (e.g., catalytic activity) depending upon the exact structure and composition of the sheets and the identity of the interlayer cation.

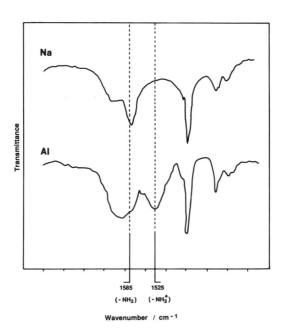

Figure 4. NH and CH absorptions in the infrared spectra of cyclohexylamine intercalated into a Na^+ and Al^{3+} exchanged montmorillonite. The uptake of guest in this instance was from the vapor and resulted in partial displacement of the interlayer water. The extent of polarization of the residual interlayer water is, however, controlled by the charge-balancing cations. In the case of the Na-clay the vast majority of cyclohexylamine molecules are in a neutral state. For the Al-clay a substantial proportion of the molecules are protonated ($R-NH_2$ at 1585 cm^{-1}; $R-NH_3^+$ at 1525 cm^{-1}).

2.3.2. Organic Guests

To a considerable extent the interlayer water may be replaced by other (neutral) guest molecules. Amines, for example, readily intercalate with the elimination of much (if not necessarily all) of the interlayer water. A considerable variety of neutral organic molecules have been studied. The hydrated nature of the interlayer region tends to make the uptake of nonpolar molecules more difficult. However, once an organic "surface" is generated the sorption properties will change. Pyrene, for example, is absorbed little by a clay until detergent molecules are also sorbed (see Section 2.4). Such so-called organoclays find uses as thickners in paints, in inks, as lubricants, in ointments, and in gas–solid chromatography.[3,5]

Very early work[3] demonstrated that cationic organic molecules, e.g., alkylammonium cations, may be exchanged into the interlayer region. A considerable amount of information is available concerning the structure and arrangement of such molecules within the layers. Such materials (either as prepared or modified—for example, by partial solvent extraction) give rise to a family of so-called organoclays.[5]

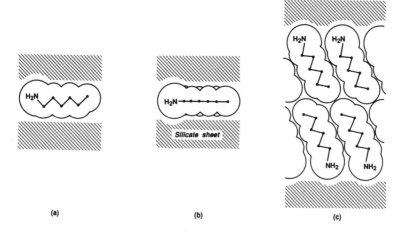

(a)   (b)   (c)

Figure 5. Idealised diagram illustrating the three possible arrangements of intercalated n-hexylamine molecules inside a sheet silicate. In the so-called α_1-configuration (a) the molecular chain lies parallel to the sheets. In the α_{11}-configuration (b) a single layer is again present with the chain parallel to the sheets but rotated with respect to (a) by $90°$. In (c) the β-configuration the chains are inclined to the sheets and form a bi-layer. The notation used is after that of G.W. Brindley and R.W. Hoffmann, Clays Clay Miner., **9**, 546 (1962).

Figure 5 illustrates some of the proposed structures for intercalated *n*-hexylamine. This molecule is readily intercalated into the interlayer region where its strong basic character results in protonation. The extent of protonation reflects the acidity of the interlayer region. In both the α_1 and α_{11} configurations the guest molecules lie parallel to the sheets. In the β configuration the chains are inclined to the sheets.[3,5] Such steep orientations relative to the sheets are frequently observed.[5] In the case of a long-chain alkylammonium salts sorption of the organic may be by an ion-exchange mechanism or by intercalation of whole salt molecules.[5] It will be shown later that the exact orientation and agglomeration of the guests within the gallery region may be controlled by several factors including the presence of co-adsorbed molecules.

2.3.3. Polyoxocations and Metal Oxide Pillared Clays

Figure 5 suggests that quite large repeat separations are possible with long-chain alkylammonium cations, and linear correlations between chain length and repeat distances have, indeed, been observed.[3] However, in such materials the available free pore volume can be quite low; much will depend upon the cation-exchange capacity and the amount of guest accommodated between the layers.

It was appreciated that the appropriate choice of organic guests could allow the possible generation of a permanently expanded porous solid[12,13] with adsorption properties possibly similar to that of a zeolite. The absence of an induction period in the adsorption isotherm of a clay "pillared" by the molecule DABCO (1,4-diazabicyclo[2,2]octane) demonstrated that a controlled porosity had been introduced. Such "permanently expanded" materials were also able to intercalate many nonpolar molecules, as well as permanent gases that were not intercalated in the parent clay. This approach was more recently extended to the intercalation of quite large polyoxocations of the type $[Al_{13}O_4OH_{24}(H_2O)_{12}]^{7+}$ and $[Zr_4(OH)_8(H_2O)_{16}]^{2+}$. These types of pillared materials generated considerable interest as zeolite-like solids possessing greater thermal and hydrothermal stability compared with their organic counterparts.[14] Figure 6 illustrates, schematically, the type of material envisaged.[15]

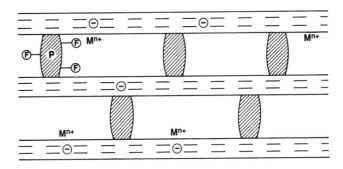

Figure 6. In order that permanent three-dimensional porosity might be introduced into these two-dimensional hosts a variety of robust inorganic complexes have been exchanged. Subsequent calcination results in the generation of metal-oxide "pillars" - labelled P. Shape-selective effects are observed with these types of solids. The diagram illustrates the several concepts which have been applied to these materials including: (i) possible functionalisation of the pillars (F); co-exchange with other cations (M^+) for possible dual-function applications; and (c) the variation in inter-pillar separation (and hence average pore diameter) by controlling *inter alia* the original charge on the sheets.

The various types of expanded layer silicates that have been produced have been recently reviewed.[13] The fact that both the layers and the pillars are composed of metal oxide material imparts considerable thermal stability to the pillared clay. The data given in Table 2[16] compares the surface area and pore size distribution of a clay, a pillared clay, and a zeolite (zeolite Y). Therefore, processes taking place within a zeolite-like solid are also expected to occur within a pillared clay.

Table 2. Characteristics of a Clay, a Pillared Clay and a Zeolite Y

	Clay	Pillared Clay	Zeolite Y
BET area, m^2/g	46.6	270.0	508.0
average pore radius, Å	35.7	13.4	12.7
pore volume, cm^3/g	0.08	0.18	0.32
pore volume distribution, area %			
$R < 10$ Å	25.2	87.1	94.9
$10 < R < 15$ Å	24.8	7.4	1.8
$15 < R < 20$ Å	11.3	2.1	0.5
$20 < R < 100$ Å	29.1	2.4	1.4
$100 < R < 300$ Å	9.6	0.5	0.1
	100.0	100.0	100.0

2.4. Adsorption of Fluorescent Probes

A variety of probes have been used to identify the manner in which organic molecules are held by layered materials. Thomas[17] has highlighted three factors that will be important in controlling processes on clay surfaces: the local environment, the proximity of the adsorbed species to one another and to other adsorbed species and the geometric arrangement of the molecules at the surface. Such studies have additionally aimed at monitoring any changes in molecular geometry or rotational freedom which occur as a result of uptake onto the surface of the material. In the case of clays, tris(2,2′-bipyridyl)ruthenium ($[Ru(bpy)_3]^{2+}$) and cationic pyrene derivatives have received attention.[18,19] Such charged molecules as these, which are located onto the clay surface by the process of ion exchange, are expected to adhere quite strongly to the clay surface. The influence of coadsorbed species has also been studied. When the coadsorbent is a detergent molecule then the clay surface is sufficiently modified that neutral molecules (e.g., pyrene) may also be adsorbed and studied, the neutral molecule being solubilized in domain-like clusters of detergent molecules on the clay surface.[17]

2.4.1. Pyrene derivatives

Viaene et al.[19] have studied the adsorption of [3-(1-pyrenyl) propyl]trimethyl-ammonium bromide (P3N), [8-(1-pyrenyl)octyl]ammonium chloride (P8N), and (1-pyrenyl)trimethyl ammonium chloride (PN) (Scheme 5). In addition to studying the fluorescent properties of the probes when the clay–organic composite was colloidally dispersed in an aqueous medium the fluorescence characteristics were also studied in organic dispersants. Such a comparison has allowed the hydrophobic character of the clay and the adsorbed guest to be studied.[17,19] Pyrene itself, for example, is poorly sorbed by a conventional clay but is readily absorbed by organoclays.[17]

$$Py - (CH_2)_n{-}NR_3{}^+ \quad X^-$$

P3N   [3(1-pyrenyl)propyl]trimethylammonium bromide

       $Py = pyrene; \ n = 3; \ R = CH_3; \ X^- = Br^-$

PN   (1-pyrenyl)trimethylammonium chloride

       $Py = pyrene; \ n = 1; \ R = CH_3; \ X^- = Cl^-$

P8N   (8-(1-pyrenyl)octyl)ammonium chloride

       $Py = pyrene; \ n = 8; \ R = H; \ X^- = Cl^-$

Scheme 5

Spectral Properties of the Adsorbed Species: Viaene et al.[19] point to a ground-state interaction between molecules of P3N and the very efficient excimer formation that is observed in the fluorescence spectra for this probe on the clay surface. Aggregation into clusters on the surface is assumed to have taken place. However, although there was evidence for a ground-state interaction between adsorbed PN molecules, no excimer emission was observed (see Figure 7). The generation of an excimeric state clearly is impeded for PN, suggesting that the movement of excited PN molecules on the clay surface is restricted. It is likely that a significant requirement for molecular movement to an excimeric state in this case is the presence of the alkyl chain, which is available in P3N but not PN.

402 *Jones*

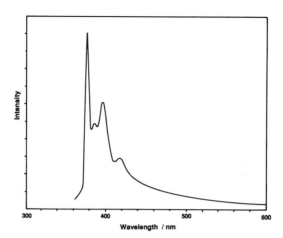

Figure 7. Fluorescence spectrum obtained for PN adsorbed on a clay surface which is dispersed in an aqueous medium. Despite evidence for a ground state interaction between adsorbed molecules (as revealed in excitation spectra) no excimeric emission around 480 nm is observed. The absence of a flexible chain between the pyrenyl head group and the nitrogen end group does not allow the required movement of the pyrene groups for excimer formation. (After Viane *et al.*, reference 19).

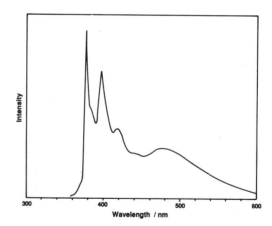

Figure 8. Fluorescence spectrum for P8N adsorbed on a clay surface which is dispersed in an aqueous medium. It is suggested that a folding of the relatively long alkyl chain allows an interaction between the pyrene head group and the nitrogen group. This folding and interaction is favoured by the repulsion between the aromatic head group and the aqueous medium. (After Viane *et al.*, reference 19).

For P8N very low intensity fluorescence spectra are obtained from the adsorbed species (Figure 8). This low intensity is accompanied by a bathochromic shift and change in vibrational structure compared to P3N and PN. The pyrene end group is likely able to interact with the ammonium head group via folding of the longer alkyl chain with consequential and efficient quenching. In P3N and PN the alkyl chain is not sufficiently long to allow this interaction to occur. When cationic detergent molecules are present (e.g., cetyltrimethylammonium bromide, CTAB) the properties of P8N become similar to P3N and PN seemingly to support this folding mechanism; the alkyl chain in P8N adjusts to interact with adjacent detergent molecules, thereby removing the chain-folding characteristics.[19]

Cluster Formation: The heterogeneous adsorption of probes to form clusters has been well established. Charge inhomogeneity in the sheet structure has been proposed as the cause of this clustering. Viaene et al.[19] however, suggest that the driving force for cluster formation is the minimization of the contact surface between the hydrophobic pyrene derivative and the surrounding water phase. Evidence for this is suggested to come from quite pronounced fluorescence changes when the dispersing medium is changed from water to toluene or methanol. Figure 9 illustrates the changes that occur when the fluorescence spectra for P8N adsorbed on a clay are measured with the sample suspended in water and in methanol. In nonaqueous media no excimeric emission is observed, the initial clusters of molecules "dispersing" over the clay surface. The intramolecular interaction between the pyrene chromophore and the ammonium head group described above for P8N is also lost in an organic dispersant. Such changes in spectral properties clearly indicate mobility of the molecules on the clay surface.

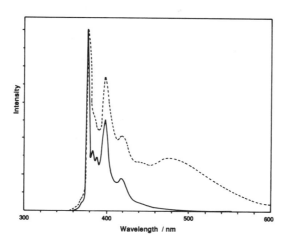

Figure 9. Fluorescence spectra for P8N absorbed on a clay surface which is dispersed in water (----) and in methanol (—). The interaction between the head-group and the N-tail is lost and the molecules disperse when the suspension medium is methanol. The excimeric emission is also lost. (After Viane *et al.*, reference 19).

Coadsorbed Detergent Molecules: As already mentioned, several studies have investigated the influence of coadsorbed detergent molecules on the properties of clay surfaces and the associated spectral features of adsorbed probes.[17,19] In terms of organization and movement of the pyrene derivatives, a dependence on chain length of coadsorbed detergent molecule and length of alkyl chain within the probe is obtained. The dominant factor appears to be a requirement for the coadsorbed species to shield the pyrene chromophore from the aqueous dispersant phase. If this shielding is not possible, separate aggregates of detergent and probe will form. The geometric relationship between probe and clay appears to be such that in aqueous media the probe molecules lie nearly perpendicular to the surface (with the pyrene chromophore close to the surface), whereas in organic dispersants the absence of any strong hydrophobic interactions allows the probes to lie parallel to the clay surface.

2.4.2. Tris(1,2'-bipyridyl)ruthenium

The fluorescence properties of $[Ru(bpy)_3]^{2+}$ have been used as a probe for a range of sheet silicates including both natural and synthetic specimens.[17,18,20,21] The use of nonreactive materials such as Laponite is useful since quenching of excited $[Ru(bpy)_3]^{2+}$ by lattice ions, e.g., iron, may be avoided.[17] The probe is strongly adsorbed. An important distinction between $Ru(bpy)_3^{2+}$ adsorbed on the exterior surfaces of the clay particles and $Ru(bpy)_3^{2+}$ intercalated between the layers emerges in terms of, for example, motional freedom. (Careful choice of the silicate allows this experimental distinction to be made - kaolinite, for example, adsorbs only on the exterior surface of the particles whilst montmorillonite under appropriate conditions will adsorb the probe molecules within the clay layers.[17])

The absorption spectrum of $[Ru(bpy)_3]^{2+}$ in aqueous solutions has a metal-ligand (d–π) charge-transfer band around 460 nm and a π–π^* transition for the ligands around 300 nm. When intercalated into a montmorillonite there is a red shift in the charge-transfer band and splitting of the π–π^* absorption. Adsorption on exterior surfaces alone does not result in the π–π^* splitting with a spectrum very similar to that obtained from aqueous solutions.[17] The transition is gradual; as more and more probe molecules are intercalated rather than simply adsorbed, so the extent of splitting increases.

Ghosh and Bard,[21] as part of a study dealing with clay-modified electrodes, have studied the absorption, emission, and resonance Raman spectra of adsorbed $[Ru(bpy)_3]^{2+}$. The Raman spectra confirm that the molecules remain intact. Intercalation results in very high local concentrations and in a very efficient excited-state self-quenching process (60% of the intensity from excited complexes decays with an average lifetime of 55 ns when as little as 7% of the molecules are initially excited). A series of spectra has been obtained for various loadings of the complex onto the clay particles. No significant changes were observed as the concentration was increased and the suggestion was made that this could be attributed to the preferential uptake of the guest molecules into galleries, i.e., individual galleries are saturated or completely empty. Such segregation may occur because of intrinsic charge heterogeneity or because of energetic reasons associated with differences in size and solvation energies between the Ru complex and, for instance, solvated sodium ions. Loading of the $[Ru(bpy)_3]^{2+}$-impregnated material with methylviologen (MV^{2+}) also leads to a low quenching efficiency, again, it is suggested, because of the segregation of guest species into separate layers. When neutral molecules, e.g., a neutral viologen (propylviologen sulfonate, PVS) are added then efficient quenching is observed. The importance of charge–charge interactions is clear from these observations. Interestingly, in contrast to the segregation observed with Na^+ and MV^{2+}, $[Ru(bpy)_3]^{2+}$ and $[Zn(bpy)_3]^{2+}$ do not segregate when coadsorbed. These two species appear to mix readily within the interlayer region with a progressive decrease of excited-state self-quenching rate as the $[Zn(bpy)_3]^{2+} : [Ru(bpy)_3]^{2+}$ ratio increases.

2.5. Photophysics of Dye Molecules, Pigments, Phthalocyanines, and Porphyrins

Endo et al.[22] have studied the fluorescence of the dye coumarin-1, chemical formula $C_{14}H_{17}NO_2$, intercalated in a synthetic clay, saponite. This clay has a high transparency in the visible region. By confining the dye between the sheets of an expandable layered silicate transparent to uv–visible radiation they expected to make the dye more thermally stable and to show an increased efficiency of fluorescence. The clay–dye composites for this study were prepared by direct interaction between a dispersed clay in water and a solution of the dye molecule. Both neutral and protonated molecules could be intercalated. Two interlayer spacings were observed: at 13.0 and 18.5 Å (compared to the 11.9 Å spacing of the original Na^+ saponite clay). Figure 10 illustrates the three possible arrangements that have been proposed for the dye molecules within the layers for the 13.0 and 18.5 Å forms, with the arrangement of the dye molecules depended significantly upon the amount of coumarin intercalated. In the 13.0 Å intercalate, designated *dl*, the coumarin molecules are arranged parallel to the aluminosilicate surfaces of the clay. In the 18.5 Å intercalate, designated *dh* - two arrangements satisfy the gallery height of 8.9 Å. In model 1 the molecules are positioned parallel to the sheets in a parallel structure, whereas in model 2 the molecules lie perpendicular to the sheets. Preference is given by the authors to model 2. The quantum efficiency of fluorescence was also affected by the separation of the molecules. The variation in the observed emission intensities could be explained by conventional concentration quenching phenomena.[22]

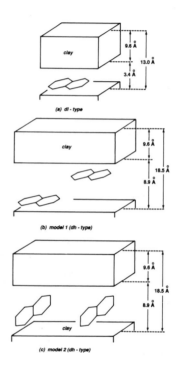

Figure 10. Schematic illustration of the arrangement of coumarin molecules intercalated into a magnesium rich (saponite) clay. The arrangement of the guests depends significantly upon the amount of coumarin present in the gallery region. For the 13.0 Å intercalate the available height is 3.4 Å - the metal-oxide/hydroxide sheet being approximately 9.6 Å thick - and corresponds to a single coumarin layer. For the 18.5 Å intercalate two models are proposed - either as a double layer, model 1, with the molecules parallel to the layers or as a single layer but with the molecules perpendicular to the sheets. (After Endo *et al*.,reference 22).

The intercalation of Rhodamine 590,[23] which occurs at room temperature, was accompanied by expansion of the lattice up to the thickness of the dye molecules. The fluorescence spectra of the dye–clay complex when compared with the corresponding solution spectra showed a reduced intensity and the values of λ_{max} were shifted toward higher wavelengths as the quantity of intercalated dye increased. Correlations were observed between changes in d spacing, relative intensity, and λ_{max} (Figure 11). The red shifts in λ_{max} were associated with an electrostatic interaction between the xanthene dye in its excited state and the negatively charged aluminosilicate clay layers. The aluminosilicate layers were subjected to considerable mechanical strain by the intercalation of the dye (Figure 12).

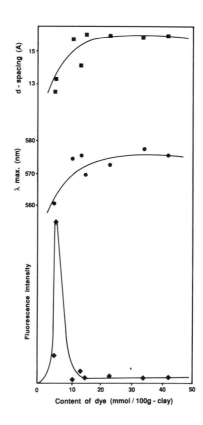

Figure 11. Variation of fluorescence intensity, position of λ_{max} and d-spacing as a function of the amount of rhodamine 590 incorporated into saponite. (After Endo *et al.*, reference 23).

Methylviologen intercalates[24] readily into expandable smectites giving a repeat separation of 12.6 Å. The species present was believed to be MV^{2+}. Time-resolved fluorescence spectra as well as fluorescence intensity variations with changing MV^{2+} : clay ratio suggest two different binding sites for MV^{2+} adsorbed on the clay colloidal particles. A significant increase in intensity was observed for MV^{2+} in a colloidal suspension compared to that of an aqueous solution of the same concentration. The MV^{2+} adopts a planar conformation when intercalated, with the gallery height being 2.95 Å. The observation of enhanced fluorescence, therefore, for an intercalated molecule probably results from a decrease in the nonradiative deactivation processes when it is in this geometry. Structural iron serves as a quencher and therefore no fluorescence has been observed for nontronite, a sheet silicate rich in iron.

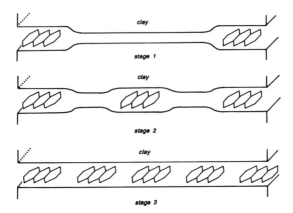

Figure 12. From the variation in spectra and repeat distances described in Figure 11, Endo *et al.* propose a three-stage process for intercalation of rhodamine and other dyes into saponite. In stage 1 dye molecules begin to intercalate and deform the clay sheets. The individual dye molecules are too isolated at this concentration for any spectroscopic interactions between them to be observed. In the second stage further molecules are intercalated with the penetration of the dye governed by a balance between the concentration gradient of the dye and the coulombic interaction between the dye cations and the negatively charged layers. There is still residual strain within the matrix. In the third stage strain-free galleries are constructed with concentration quenching observed in the fluorescence spectra. (After Endo *et al.*, reference 23).

The electronic absorption spectrum of *meso*-tetraphenylporphyrin (TPPH$_2$) as a function of the nature of the interlamellar hydrated cations have been reported; the porphyrin, its dication, and its metallo complex have approximate surface areas of 300 Å^2.[25] Under the conditions used and with the loading levels chosen, approximately 2000 Å^2 of basal surface per porphyrin moiety was available, or an interporphyrin distance of approximately 45 Å. For VO^{2+} and Fe^{3+} as the initial interlayer cation, uptake is rapid with the loss of the four free-base porphyrin bands at 510, 545, 588, and 645 nm. The electronic absorption spectra (Figure 13) suggest demetallation has occurred and that the intercalated species is the dication (TPPH$_4$$^{2+}$), where the protons result from hydration of the interlamellar cations, i.e., the following scheme:[25]

$$2\, M(H_2O)_x^{n+} + TPPH_2 \rightarrow 2\, M(H_2O)_{x-1}(OH)^{(n-1)+} + TPPH_4^{2+}$$

where M = VO^{2+}; Fe^{3+}.

x-Ray data support the idea that the porphyrin rings are parallel to the silicate sheets (Figure 14). The basal spacing is 14.2 Å, corresponding to a thickness of 4.6 Å, the approximate thickness of the porphyrin nucleus.

For Cu^{2+}, Zn^{2+}, and Co^{2+} as the exchangeable cations a different type of reaction occurs and is shown schematically as

$$M(H_2O)_x^{2+} + TPPH_2 \rightarrow 2\, H_3O^+ + M(TPP)$$

The absorption bonds of the free base ($TPPH_2$) are replaced by those of the corresponding Cu^{2+}, Zn^{2+}, or Co^{2+} complex, i.e., $Cu(TPP)$, $Zn(TPP)$, and $Co(TPP)$. The results indicate that surface metallation had occurred with desorption of the complex. For Co, however, the desorption rate was slower than for Cu or Zn. With Na^+ or Mg^{2+} as the interlayer cation no decrease in the absorption for the free base was observed. In conclusion, therefore, two distinct types of reaction take place. One is a Bronsted acid–base reaction between the hydrated cations and the free base porphyrin, which affords the diprotonated porphyrin dication. The second reaction involves metallation of the porphyrin ring and subsequent desorption of the metalloporphyrin from the silicate surface.

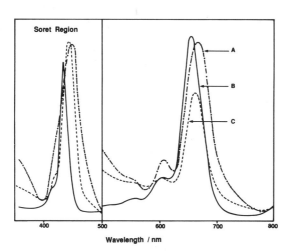

Figure 13. Electronic absorption spectra for (A) $TPPH_2$ bound to Fe^{3+}-montmorillonite, (B) $TPPH_4^{2+}$ in glacial acetic acid and (C) a $TPPH_4^{2+}$ intercalated montmorillonite. Comparison of these spectra indicate that the porphyrin molecule binds to Fe^{3+} (and VO^{3+}) exchanged clays as the diprotonated cation $TPPH_4^{2+}$. (After Cady and Pinnavaia, reference 25).

The particular metalloporphyrin, Co(II) *meso*-tetrakis(1-methyl-4-pyridyl)porphyrin (CoTMPyP), has been reported to intercalate intact between the clay sheets by ion exchange in acid solution.[26] Initially adsorption on the external surface takes place but this is subsequently followed by intercalation. When intercalated a gallery height of 3.7 Å was obtained; the molecules again lie parallel to the sheet layers.

2.6. Photochemical Processes
2.6.1. Photooxidation and Photodegradation

The photooxidation of tryptophan, photosensitized by methylene blue (MB) exchanged on a variety of clays has been investigated.[27] Three factors affect the reaction (1) Fe(III) in the structure, which quenches the excited state of MB; (2) the adsorption site of MB (photooxidation only takes place when MB is located on the external surface of the clay), and (3) dye aggregation, which decreases the yield of the reaction.

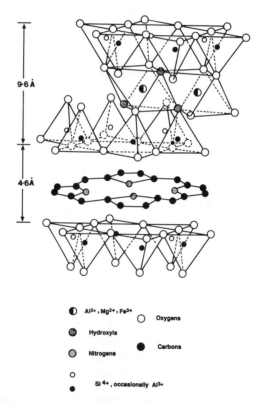

Figure 14. Schematic illustration of the $TPPH_4^{2+}$ porphyrin nucleus within the gallery region of a montmorillonite. Demetallation appears to have accompanied intercalation (for example, when the initial exchangeable cation is Fe^{3+} or VO^{2+}) with the incorporation of the diprotonated cation. An (001) basal reflection of 14.2 Å is observed giving a gallery height of 4.6 Å. This approximates to the thickness of the porphyrin nucleus and suggests that the guests lie parallel to the sheets. (After Cady and Pinnavaia, reference 25).

2.6.2. Photooxidation and Photoreduction of Water

A variety of reports have appeared on this topic.[28] Early attempts to photooxidize water with a clay-supported catalyst used hectorite-supported ReO_2 dispersed in water.[29,30] The clay was also exchanged with the hexamine ruthenium(III) cation with $[Co(NH_3)_5Cl]^{2+}$ also exchanged as a sacrificial acceptor. No significant activity was observed, however, it was believed because the ruthenium and cobalt complexes were present within the gallery regions, whereas the ReO_2 particles were located on the external surfaces.

For sepiolite, a fibrous mineral, this difficulty does not arise and an active system was generated. The position concerning the development of clay-supported reduction catalysts has also been reviewed by Van Damme et al.[28] and others.[30,31-34]

2.6.3. Photodimerization and Photochromism

The spatially controlled photocycloaddition of stilbazolium cations inside a saponite clay has been reported:[35] uv irradiation of the intercalate resulted in the generation of the syn head-to-tail dimer as the predominant dimer. There was also a sharp decrease in the cis–trans isomerization compared with homogeneous photolysis. Such differences in reactivity are ascribed to the restricted molecular arrangement in the clay interlayers. x-Ray analysis indicated an expanded gallery height of 6.2–6.8 Å and an arrangement compatible with the experimental results is shown in Figure 15. In this arrangement alkene molecules are packed alternately in an antiparallel alignment.

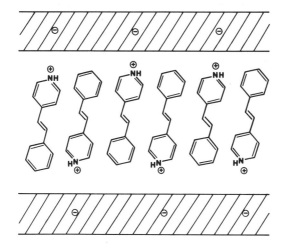

Figure 15. Schematic illustration of the packing of stilbazolium cations inside a magnesium-rich saponite clay. Stereoselective dimerization takes place to give the *syn* head-to-tail dimer. The reaction takes place, therefore, in a restricted reaction cavity. (After Takagi *et al.*, reference 35).

Work by Wang and Jones[36] has demonstrated that acridizinium cations can also be intercalated within the clay sheets. Again efficient dimerization takes place. For such planar aromatic cations it has been proposed that bilayers are formed with a slight increase in gallery height accompanying the dimerization.

Photochromic effects for intercalated systems have also been described[37] and several patents have appeared in this area. In a recent study[38] of the interaction of various ion-exchanged clays with the photochromic fulgide α-2,5-dimethyl-3-furylethylidene (isopropylidene) succinic anhydride (see Scheme 6), Adams and Gabbutt have described interesting bathochromic shifts of the fulgide and the photochrome of 20 and 80 mm, respectively.

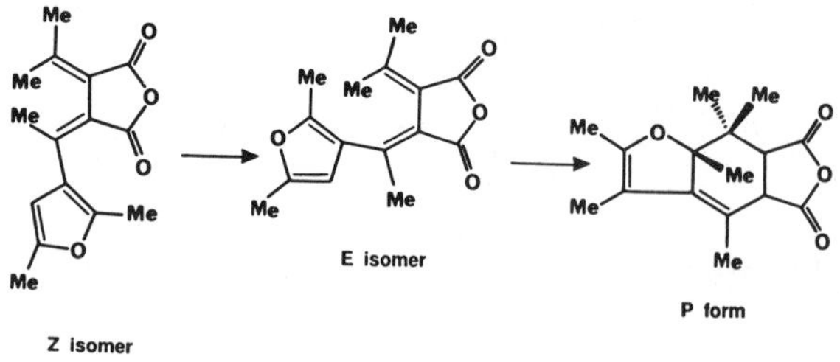

Scheme 6

They report that conversions that are normally accomplished with light are obtained over clays in the dark and are thermally activated. The photochrome generated by the clay is chemically identical to that generated photochemically. Bleaching, however, i.e., the light-induced reversible process, is difficult on the clay surface. In addition, clay tends to introduce side reactions, leading to unwanted decomposition products. The possible presence of geometric constraints on the reaction which result from the fulgide being present in the interlayer region is discussed.[38]

2.7. Photocatalytic Activity of Intercalated Nanosized Particles of Metal Oxides

The photochemical and photocatalytic properties of microcrystalline TiO_2 incorporated into the interlayer space of montmorillonite has been investigated and the properties of this composite compared with those of TiO_2 powder.[39] The pillared TiO_2 particles (see section 2.3.3) contained within a 15 Å high gallery region showed a ca. 0.58 eV blue shift in its absorption and emission spectra. The excited electronic states of the pillared TiO_2 were determined to be 0.36 V more negative than that of the TiO_2 powder particles. An enhanced activity for the photodecomposition of 2-propanol to acetone and hydrogen was observed[39]

(Table 3). Similar enhanced activity was observed for the photocatalytic conversion of a range of carboxylic acids to the corresponding alkane and carbon dioxide. A variety of procedures have been described for incorporating TiO_2 with the use of titanium tetraisopropoxide appearing to be a particularly useful procedure. Such TiO_2 pillared samples have a significantly increased surface area ($347 \ m^2 \ g^{-1}$) compared to the original clay ($72 \ m^2 \ g^{-1}$). This increase in surface area, however, is not considered to be the major contribution to the high photocatalytic efficiency. The perturbation of the electronic energy levels of the TiO_2 particles is also being beneficial for hydrogen evolution in the decomposition of 2-propanol. In a study of dichlormethane photodegradation, Tanguay et al.[40] observed that bentonite clays pillared by titanium oxide were more catalytically active than straight conventionally exchanged titanium clays. The efficiency of clays pillared with aluminum oxide as well as mixed titanium–aluminum oxide pillars were also investigated, although such materials were found to be less active than their titania counterpart. Fan et al.[41] have designed integrated chemical systems in which TiO_2 incorporated into Nafion (a polymeric ion-exchange resin) or clay films have been studied. Both systems showed photocatalytic activity in the reduction of methylviologen with oxidation of triethanolamine. For Fe_2O_3 oxide microcrystallites incorporated into a Na-montmorillonite clay[42] the oxide particles exhibited a ca. 0.28 eV greater bandgap than α-Fe_2O_3 powder. The Fermi level of the incorporated oxide was 0.23 eV more negative than that of free α-Fe_2O_3. Again enhanced activity for decomposition of acetic acid was observed and attributed to both an increase in surface area as well as electronic perturbations. The efficiency of uranyl-exchanged clays in the oxidation of alcohols to alkanes has also been investigated.[43,44]

Table 3. Photocatalytic Decomposition of 2-Propanol

| catalyst | Production rate—μmol per hour | | | | |
	H_2	CH_4	CO	CO_2	$(CH_3)_2CO$
TiO_2	0.26	0.07	0.07	trace	0.28
TiO_2/clay	1.34	0.35	0.12	0.15	1.40
Pt/TiO_2	183.0	trace	trace	trace	172.0
Pt/TiO_2/clay	697.0	4.0	trace	trace	661.0

2.8. Semiconductor Particles Incorporated into Clays

The properties of CdS and mixtures of CdS and ZnS included into colloidal suspensions of clays have been examined.[45] While particulate and colloidal semiconductor suspensions supported on silica or in vesicles have been studied the advantages of clays (intercalation, high surface area, ion-exchange properties and thermal stability) suggested they may be particularly advantageous in the preparation of semiconductor "integrated chemical systems." The photoredox properties of the CdS and CdS + ZnS particles were modified following incorporation into the clay colloid. The various experimental observations (including amounts of hydrogen photoproduced and pH dependence) could be understood if most of the CdS + ZnS mixed particles were intercalated within the clay layers. x-Ray measurements suggest that the particles are extremely small with catalytic efficiency appearing to be strongly dependent on particle size.

2.9. Concept of Anchoring Molecules to Clay Surfaces

Casal et al.[46] have demonstrated that Os-bearing catalysts may be derived from clay catalysts. The strategy employed was to graft vinyl groups onto sepiolite (a high surface area fibrous clay) followed by addition of osmium tetroxide to the carbon–carbon double bond. Mild reduction in hydrogen at 473 K gave the active catalyst. The existence of a large number of surface Si—OH groups on the sepiolite mineral were reacted with, for example, methylvinyldichlorosilane. The effectiveness of the Os-sepiolite (OsS) catalyst was tested in the photooxidation of water in a classical three-components system [sensitizer; $[Ru(bpy)_3]^{2+}$; sacrificial acceptor; $[Co(NH_3)_5Cl]^{2+}$; catalyst, OsS]. Although not particularly active for the reaction the catalyst was the first example of a molecularly dispersed water–oxidation catalyst grafted onto a colloidal support.

3. Layered Double Hydroxides
3.1. Structure of Layered Double Hydroxides (LDHs)

The structures of LDHs are very similar to that of brucite, $Mg(OH)_2$. Magnesium is octahedrally surrounded by six oxygens in the form of hydroxide; the octahedral units, then, through edge sharing, form infinite sheets. The sheets are stacked on top of each other through hydrogen bonding (Figure 16). When some of the magnesium in the lattice is replaced by a higher charged cation, the resulting overall single layer (e.g., Mg^{2+}–Al^{3+}–OH) gains a positive charge. Sorption of an equivalent amount of hydrated anions renders the structure electrically neutral. In nature the anion is frequently found to be the carbonate anion, although OH⁻ and Cl⁻ are occasionally found.

LDHs may be represented by the general formula;

$$M_a^{2+}M_b^{3+}(OH)_{2a+2b}(X^-)_b, x \, H_2O$$

where M^{2+} may be Mg^{2+}, Fe^{2+}, Co^{2+}, Ni^{2+}, or Zn^{2+} and M^{3+} may be Al^{3+}, Cr^{3+}, or Fe^{3+}. When M^{2+} is Mg^{2+}, M^{3+} is Al^{3+} and with X⁻ being carbonate, the material may be either hydrotalcite (the 3R structural variant) or mannaseite (the 2H polytype). The two forms are indistinguishable physically and can be distinguished only by powder x-ray analysis. Rhombohedral hydrotalcite has cell parameters of a = 3.1 Å and c = 23.1 Å, while hexagonal mannaseite has a = 3.1 Å and c = 15.3 Å. $M^{2+}:M^{3+}$ ratios of between 1 and 5 are possible. There are definite limits to a and b, the nature of M^{2+} and M^{3+}, the X^- anion, and the value of x.

(a)

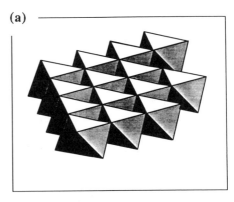

(b)

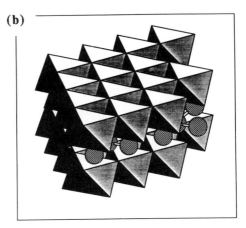

Figure 16. (a) Schematic illustration of a single brucite $(Mg(OH)_2)$ sheet. (b) Following isomorphous replacement of some of the octahedral Mg^{2+} by Al^{3+} the need for charge-balancing anions arises. In this particular case carbonate anions are shown. Water molecules are also present along with the carbonate anions. (After Reichle, reference 48).

Synthetically[47,48] there are a wide range of variables giving rise to the possibility of producing tailor-made materials. These possible variables are :

1. Different M^{2+}s and mixtures
2. Different M^{3+}s and mixtures
3. Possibility of M^+ incorporation, e.g., Li^+
4. Different charge balancing anions
5. Amount of interlayer water
6. Crystal morphology and size

The form normally obtained synthetically is the three-layer polytype (i.e., rhombohedral) type. Taylor has reviewed the structure of various anionic materials.[49]

3.2.   Exchange Properties of LDHs

The number of exchangeable anions within the structure depends upon the charge density carried on the host layer. In nature the most common anion is the carbonate anion (which is held tenaciously) and occasionally the hydroxyl ion (see Table 4).

While cation exchange for cationic clays may be carried out with relative ease, anion exchange for the LDHs is not easily achievable. There are two principal reasons. First, the charge density carried on the LDHs may be quite large [the exchange capacity for $Mg_6Al_2(OH)_{16}CO_3.4 H_2O$ is in the range 4.1 to 2.4 meq g^{-1} depending upon exact composition—values much higher than for montmorillonites which exhibit exchange capacities typically between 0.7 and 1.0 meq g^{-1}]. Second, the readily incorporated carbonate anion is held tenaciously within the layers and is difficult to exchange. Anion-exchange properties of LDHs have been examined by Miyata,[50] who demonstrated that the ion selectivities of monovalent anions are in the order of $OH^- > F^- > Cl^- Br^- > NO_3^- > I^-$, and that divalent anions have higher ion selectivities than monovalent anions. Of the divalent ions, CO_3^{2-} was found to be the most selective. The presence of CO_2 during synthesis is, therefore, highly undesirable when preparing non-carbonate LDHs.[51]

Table 4. Composition and formulae for some hydrotalcite-like minerals

M^{2+}	M^{3+}	Structure	Name
Mg	Al	$Mg_6Al_2(OH)_{16}(CO_3^{2-}).4H_2O$	Hydrotalcite
Mg	Fe	$Mg_6Fe_2(OH)_{16}(CO_3^{2-}).4H_2O$	Pyroaurite or sjogrenite
Mg	Cr	$Mg_6Cr_2(OH)_{16}(CO_3^{2-}).4H_2O$	Stichtit
Ni	Fe	$Ni_6Fe_2(OH)_{16}(CO_3^{2-}).4H_2O$	Reevesit
Ni,Zn	Al	$(Ni,Zn)_6(OH)_{16}(CO_3^{2-}).H_2O$	Eardlegit
Ni	Al	$Ni_6Al_2(OH)_{16}(CO_3^{2-}).4H_2O$	Takovite
Mg	Al	$Mg_6Al_2(OH)_{16}(OH^-)_2.4H_2O$	Meixnereit

One method of anion exchange has been described by Bish[52] using the reaction of dilute mineral acids with the carbonate form of the LDH. The expulsion of carbon dioxide results in anion exchange according to the equation:

$$LDH.CO_3 + 2\ HCl \rightarrow LDHCl_2 + CO_2 + H_2O$$

This method is restricted, however, to acids stronger than carbonic acid. A more recent method makes use of the reversible decomposition of the LDH (Figure 17) to yield a calcined precursor such that a variety of anions may be introduced (Scheme 7) during the rehydration process.[51,53]

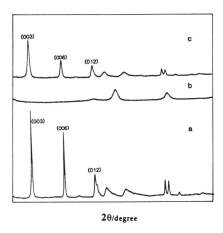

2θ/degree

Figure 17. Calcination of the parent LDH results in expulsion of the interlayer carbonate anions (as CO_2) and water along with decomposition of the brucite-like sheet. The process is reversible, however, and the semi-amorphous solid may be reconverted to the LDH on exposure to a carbonated aqueous solution. Exposure of the calcinated LDH to dilute solution of other anions (both organic and inorganic) leads to their incorporation and the formation of ordered intercalation compounds - see Scheme 7. The Figure shows the X-ray powder patterns for (a) Mg-Al-CO_3 LDH as prepared; (b) after heat-treatment at $450^\circ C$ for 18 hr in air; and (c) after exposure, at room temperature, to a 0.05 M sodium carbonate solution.

Swelling Properties: Swelling of LDHs is possible and has been demonstrated by Ross and Kodama,[54] who observed a dependence of basal spacing for hydrotalcite on levels of humidity. They concluded that a 7.9 Å spacing characterized the fully dehydrated phase. Observations by Bish[52] indicated that the LDH sulfate and LDH chloride could be solvated with glycols and glycerol. Swelling of LDHs depends on several factors similar to those observed for cationic clays including: (a) dependence on the nature of the exchangeable anion (i.e., charge, mass, structure, etc.); (b) nature of the solvent (polarity, molecular dimensions, etc.); (c) the layer charge. Compared to cationic clays, the available data on the swelling properties of LDHs is limited.

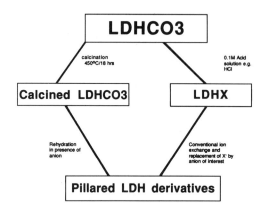

Scheme 7

3.3. Pillared LDH Derivatives and Photophysics of the Pillars

Polyoxometalate pillared LDHs are desirable since they are suitable for developing materials with unique two-dimensional (2D) galleries and zeolitic porosites. Such porous solids could be useful as catalysts and photocatalysts. The strategies described in the literature make use of the fact that Cl^- or NO_3^- anions are significantly easier to displace than, for example, CO_3^{2-}. Kwon et al.[55] made use of the Cl- form of an LDH to exchange for $[V_{10}O_{28}]^{6-}$, whereas Woltermann[56] used the NO_3^- form to exchange for numerous polyoxometalate anions, including $[Ta_6O_{18}OH]^{7-}$, $[Nb_6O_{18}OH]^{7-}$, $[V_{10}O_{28}]^{6-}$, $[PMo_6V_6O_{40}]^{3-}$, $[PMo_6W_6O_{40}]^{3-}$, $[P_3O_{10}]^{5-}$, $[PMO_{12}O_{40}]^{3-}$, $[NiW_6O_{24}H_6]^{4-}$, and $[H_3Ru_4(CO)_{12}]^-$. Another approach has relied upon the use of LDH initially synthesized with a large organic anion (typically the terephthalate dianion) as the intercalated species. The organic anion is then subsequently displaced by the polyoxometalate species.[57] Acidifying the Mg-Al-terephthalate in the presence of $NaVO_3$, for instance, leads simultaneously to polymerization of the monovanadate and migration of terephthalate anion and results in the smooth intercalation (see Scheme 8) of the polyoxovanadate species.[51,58]

A recent report by Kwon and Pinnavaia[59] has investigated the possibility of pillaring LDHs with α-$[XM_{12}O_{40}]^{n-}$ polyoxometalates of the Keggin-type structure. It was found that the reactivity of the $[XM_{12}O_{40}]^{n-}$ species toward intercalative ion exchange depended strongly on both the net charge and polyhedral form of the ion. For example, an LDH was found to undergo facile and complete intercalative ion-exchange reaction with α-$[H_2W_{12}O_{40}]^{6-}$ and α-$[SiV_3W_9O_{40}]^{7-}$ Keggin ions but no reaction was observed for the Keggin ions α-$[PW_{12}O_{40}]^{3-}$ and α-$[SiW_{12}O_{40}]^{4-}$. Furthermore, only partial intercalation was observed under equivalent conditions for the Keggin-like species $[PCuW_{11}O_{39}(H_2O)]^{5-}$. It is therefore concluded that the accessibility of the LDH galleries depends upon both the charge on the polyoxometalate and on geometric and symmetry considerations. $Zn_2Al[\alpha$-$H_2W_{12}O_{40}]$ and $Zn_2Al[\alpha$-$SiV_3W_9O_{40}]$ exhibited N_2 BET surfaces areas of 63 and 155 $m^2\ g^{-1}$, respectively. The respective pore volumes were 0.023 and 0.061 ml g^{-1}. Compared with the unpillared LDH, which possessed a surface area of 26 $m^2\ g^{-1}$, these materials may be considered to be pillared with "props" of the intercalated Keggin ions holding the layers apart. This contrast quite sharply with the values of 30 m^2/g observed for the polyvanadate intercalated LDHs for which it could be said that the intercalated species holding the layers apart were occupying all the available gallery space.[59]

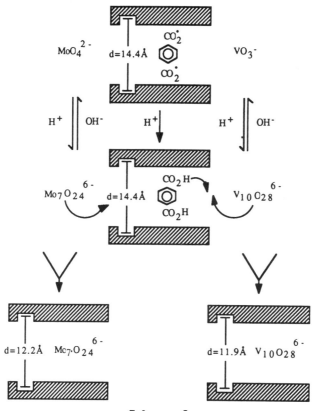

Scheme 8

3.4.   Insertion of Photoactive Species

Giannelis et al.[60] have immobilized complex anions within the layered structure. $[Ru(BPS)_3]^{4-}$ (BPS = 4,7-diphenyl-1,10-phenanthroline disulfonate) produces a hydrotalcite-like layered structure with a basal spacing of 22 Å. The complex was believed to be oriented with the C_3 axis normal to the double hydroxide layers. Various spectroscopic measurements (i.e., electronic absorption, emission, and vibrational studies) indicated that the complex was accommodated intact. Self-quenching reaction rate values were possible by cointercalation of $[Ru(BPS)_3]^{4-}$ with $[Zn(BPS)_3]^{4-}$. The influence of guest–host interactions on the excited state properties of *trans*-$[ReO_2(CN)_4]^{3-}$ has been investigated by Newsham et al.[61] Electronic and absorption spectra confirm that a structurally unperturbed oxo complex is present in the intracrystalline region of a hydroxide. The formation of intercalates is confirmed by powder x-ray diffraction (Figure 18).

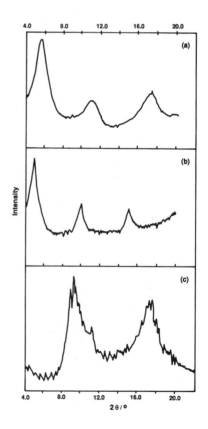

Figure 18. X-ray powder patterns for three metal-oxide exchanged intercalates. (a) ReO_2-$(py)_4$ intercalated hectorite; (b) ReO_2-$(py)_4$ intercalated fluorohectorite; and (c) $ReO_2(CN)_4$ intercalated hydrotalcite. The basal spacings demonstrate occupation of the gallery regions by the dioxorhenium (v) species. The respective gallery heights are (a) 6.7, (b) 9.0 and (c) 4.3 Å suggesting that the actual orientation of the guests varies from one intercalate to another depending upon the identity of the host matrix. (After Newsham *et al.*, reference 61).

In this study, comparisons were also made with cationic counterparts (e.g., *trans*-[ReO$_2$(Py)$_4$]$^+$) immobilized in hectorite. Quite distinct emission spectra are observed, depending very much upon the nature of the host. Figure 19 illustrates the emission spectra (at 9 K) for a hectorite and fluorohectorite host. An important factor controlling the luminescence spectra is the "keying" of the [ReO$_2$(Py)$_4$]$^+$ ions into the oxide/hydroxide layers. For the LDH materials the structural matching between the host and the guest is extremely poor and luminescence is not observed.The possible utility of polyoxometallate-exchanged materials in photocatalysis has been described by Kwon et al.[61] who indicated that for [V$_{10}$O$_{28}$]$^{6-}$ guests the photoinduced conversion of isopropyl alcohol to acetone proceeded with greater activity than with the homogeneous catalyst, despite the accompanying scattering by the host particles. More recently, Chibwe et al.[62] have studied the [2+2] photodimerization of anions within LDHs. Their results are compared with the chemistry of the pure cinnamate salts. Such studies are proposed as a means of probing the geometric relationship between intercalated guest species.

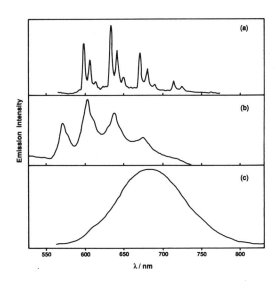

Figure 19. Low temperature (9 K) emission spectra for (a) solid [ReO$_2$(py)$_4$]; (b) ReO$_2$(py)$_4$-hectorite intercalate; and (c) a ReO$_2$(py)$_4$-fluorohectorite intercalate. While the emission band for the fluorohectorite intercalate is featureless the emission of the hectorite sample is quite structured consisting of a distinct progression of 900 cm^{-1} subdivided by a less pronounced progression of 200 cm^{-1} - quite similar to that of the pure solid. An important parameter emerging from this work is the concept of "keying" of the guest species into the host matrix. This keying is very host-guest sensitive. (After Newsham *et al.*, reference 61).

3.5.   Photophysics of Porphyrins inside LDHs

The intercalate with 5,10,15,20-tetra(4-sulfonatophenyl)porphyrin (TSPP) was prepared by direct synthesis.[63]   The basal spacing increases to 22.4 Å from 8.0 Å when the Cl⁻ was replaced by the porphyrin group. The ir and visible spectra for the TSPP were similar to those in either solution or the solid state, suggesting the molecule had been intercalated intact. The size of the TSPP molecular plane is estimated to be 18 x 18 Å and this, along with other evidence, suggests that the molecular planes are perpendicular to the basal sheets of the aluminosilicate. The amount of material intercalated fitted with this geometric model. Figure 20 illustrates the configuration postulated by the authors.

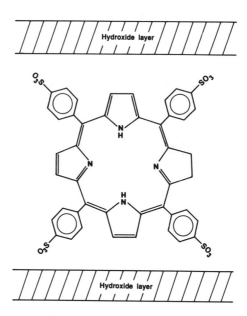

Figure 20. Proposed arrangement for the tetraphenylporphyrin inside the LDH. The proposal is based on observations concerning the size of the molecule (estimated a 18 x 18 Å), the gallery height which was observed (17.6 Å), the layer charge and the amount of guest from chemical analysis. (After Park *et al.*, reference 63).

4.   Layered Alkali Metal/Transition Metal Oxides

An example within this family of layered oxides is provided by KTiNbO$_5$. In the structure alkali ions lie between layers composed of zigzag chains of edge-sharing MO$_6$ octahedra (Figure 21).[64] The material undergoes topotactic exchange reactions; for example, protons may stoichiometrically displace K^+ ions when the material is treated with strong acid. As with other cation-exchangeable materials long-chain organic molecules such as alkyl amines may be intercalated.[65] Such expanded structures then serve as precursors for the incorporation of even larger, more complex molecules.

The protonated form $HTiNbO_5$ (titaniobic acid) has been intercalated with methylviologen and the photochemical behavior of the composite investigated.[66] In this particular case exchange directing into $HTiNbO_5$ was not possible and an intermediate propylammonium–$HTiNbO_5$ intercalate was required. Irradiation of the methylviologen intercalate by uv light under a N_2 atmosphere or under vacuum resulted in the compounds turning blue. This was interpreted in terms of the generation of a blue radical cation by a one-electron reduction process. Electron transfer was believed to occur from the oxygen-deficient layered oxide to the viologen. The blue color was stable under vacuum conditions and relatively stable in air.

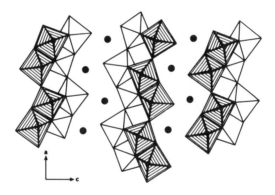

Figure 21. Structure of the layered oxide $KTiNbO_5$. The metal oxide octahedra are shown with potassium as circles. K^+ replacement by H^+ results from acid treatment (1N HCl) of the $KTiNbO_5$ material. Direct intercalation of methylviologen into $HTiNbO_5$ was not possible and an intermediate compound with propylamine was first formed. (After Nakato *et al.*, reference 66).

When tetratitanic acid (ideal formula $H_2Ti_4O_9H_2O$) is intercalated with methylviologen and irradiated with uv light, electron transfer from the oxide to viologen also takes place.[67] However, in this case x-ray diffraction measurements suggest that the guest molecules lie parallel to the sheets, an increase of 3.6 Å in spacing being observed. For $HTiNbO_5$ the situation is less clear; a gallery height of 5.9 Å is observed. Possible arrangements include a transverse orientation of the viologen molecules with respect to the layer. Bimolecular layers of viologen are not likely to be found, nor, indeed, guests perpendicular to the sheets. Given the potential application of such materials in display devices or in optical memories further work on such materials is likely.

The intercalation of methylviologen into $K_4Nb_6O_{17}$ has also been investigated.[68] $K_4Nb_6O_{17}$ is a particularly interesting solid in that it contains two alternating interlayer spaces; interlayer I containing hydrated K^+ ions and interlayer II containing unhydrated K^+ ions. $K_4Nb_6O_{17}$ has been shown useful as a photocatalyst in water splitting. The properties of a methylviologen intercalate of $K_4Nb_6O_{17}$ have also been studied. Photoreduction again occurs following uv irradiation (Figure 22), but, unlike the case of $HTiNbO_5$ and $KTiNbO_5$, photoreduction also takes place in air. The radicals that are generated are also much more stable in air than was the case for $HTiNbO_5$ and $KTiNbO_5$.

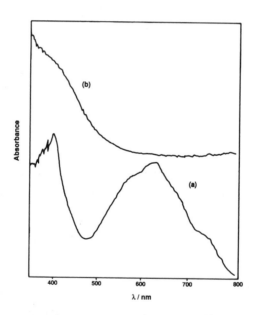

Figure 22. Visible spectra of methylviologen - $K_4Nb_6O_{17}$ intercalation compound (a) after irradiation in air and (b) before irradiation - photoreduction of the methylviologen cations has taken place with electron transfer from the transition metal oxide host to the guest. (After Nakato *et al.*, reference 68).

5. Concluding Remarks

Layered solids present significant opportunity for the development of new types of materials. Through the process of intercalation it is possible to engineer solids with either desired physical properties, e.g., pore size distributions akin to those of zeolites, or prepared materials with subtlely modified chemical properties, e.g., as catalysts. Part of the recent studies on such intercalated materials has been targeted toward the generation of pillared layered solids (PLSs). A recent compilation of the attractive properties of such materials has appeared. Promising areas of interest that were identified for such pillared materials included:[69]

- Photochemistry and photocatalysis: The low absorption coefficient of most PLS host materials in the visible and near-uv range makes them important and interesting host materials for carrying out pillar-induced photochemical processes in a constrained shape-selective environment.

- Photophysical and nonlinear optical properties: Again the transparencies of most PLS materials and their ability to adsorb and/or intercalate a wide range of organic, inorganic, and organometallic species suggest that a wide range of extremely flexible luminescent materials in which complex energy-transfer processes are possible may be developed. With semiconductors this provides an opportunity for bandgap engineering and the formation of quantum well structures.

- As sensors: via color or luminescent changes or by changes in ionic conductivity.

It is clear, however, that for the majority of these materials structural characterization is not straightforward. Some of the experimental techniques described earlier in this chapter have provided considerable insight into the arrangement and mobility of guest species. Frequently, however, the structure of the host matrix itself is unknown or unclear. Standard techniques such as ir and x-ray diffraction will continue to be important tools, as will neutron diffraction and neutron inelastic scattering. Nuclear magnetic resonance will also continue to be an important technique for structure elucidation as well as monitoring the motion of guest molecules.

Acknowledgment

Support by VAL-ERO under contract DAJA45-90-C-0013 is appreciated.

References

1. M. S. Whittingham and A. J. Jacobson, eds., *Intercalation Chemistry*, Academic Press, New York, 1982.
2. W. Jones, J. M. Thomas, D. T. B. Tennakoon, R. Schlogl and P. Diddams in *The New Surface Science in Catalysis*, M. L. Delviney and J. L. Gland, eds., American Chemical Society, Washington, D. C., 1985, p. 472.

3.   B. K. G. Theng, *The Chemistry and Clay of Organic Reactions*, John Wiley, New York, 1974.
4.   R. W. Grimshaw, ed., *The Chemistry and Physics of Clays*, Ernest Benn, London, 1971.
5.   R. N. Barrer, *Zeolites and Clay Minerals as Sorbants and Molecular Sieves*, Academic Press, London, 1978.
6.   J. M. Thomas in reference 1, p. 55.
7.   J. W. Stucki and W. L. Banwart, eds., *Advanced Chemical Methods for Soil and Clay Minerals Research*, D. Reidel, Dordrecht, Holland, 1980.
8.   W. Jones, D. T. B. Tennakoon, J. M. Thomas, L. J. Williamson, J. A. Ballantine and J. H. Purnell, *Proc. Indian Acad. Sci. (Chem. Sci.)*, **1983**, *92*, 27.
9.   D. T. B. Tennakoon, J. M. Thomas, W. Jones, T. A. Carpenter and S. Ramdas, *J. Chem. Soc,. Faraday Trans. 1*, **1986**, *82*, 545.
10. D. T. B. Tennakoon, W. Jones, J. M. Thomas, *J. Chem. Soc., Faraday Trans. 1*, **1986**, *82*, 3081.
11. D. T. B. Tennakoon, R. Schlogl, T. Rayment, J. Klinowski, W. Jones and J. M. Thomas, *Clay Minerals*, **1983**, *18*, 357.
12. R. M. Barrer and D. M. MacLeod, *Faraday Trans.*, **1954**, *50*, 980.
13. R. M. Barrer in *Pillared Layered Structures: Current Trends and Applications*, I. V. Mitchell, ed., Elsevier Applied Science, London, 1990, p. 55.
14. R. Burch, ed., *Pillared Clays, Catalysis Today*, Vol. 2, Elsevier, Amsterdam, 1988.
15. J. Shabtai, F. E. Massoth, M. Tokarz, G. M. Tsai and J. M. McCauley in *Proc. 8th Int. Congress Catal., Berlin, 1984*, Vol. IV, Verlag Chemie, Weinheim, 1984, p. 735.
16. M. L. Occelli and J. E. Lester, *Ind. Eng. Chem. Prod. Res. Dev.*, **1985**, *24*, 27.
17. J. K. Thomas, *Acc. Chem. Res.*, **1988**, *21*, 275.
18. V. G. Kuykendall and J. K. Thomas, *J. Phys. Chem.*, **1990**, *94*, 4224.
19. K. Viaene, R. A. Schoonheydt, M. Crutzen, B. Kunyima and F. C. De Schryver, *Langmuir*, **1988**, *4*, 749.
20. A. Habti, D. Keravis, P. Levitz and H. Van Damme, *J. Chem. Soc., Faraday Trans. 2*, **1984**, *80*, 67.
21. P. K. Ghosh and A. J. Bard, *J. Phys. Chem.*, **1984**, *88*, 5519.
22. T. Endo, N. Nakada, T. Sato and M. Shimada, *J. Phys. Chem. Solids*, **1989**, *50*, 133.
23. T. Endo, N. Nakada, T. Sato and M. Shimada, *J. Phys. Chem. Solids*, **1988**, *49*, 1423.
24. G. Villemure, C. Detellier and A. G. Szabo, *J. Am. Chem. Soc.*, **1986**, *108*, 4658.
25. S. S. Cady and T. J. Pinnavaia, *Inorg. Chem.*, **1978**, *17*, 1501.
26. H. Kameyama, H. Suzuki and A. Amano, *Chem. Lett.*, **1988**, 1117.
27. J. Cenens and R. A. Schoonheydt, *Clay Minerals*, **1988**, *23*, 205.
28. H. van Damme, F. Bergaya and B. Challal in *Homogeneous and Heterogeneous Photocatalysis*, E. Pelizzetti and N. Serpone, eds., D. Reidel, Dordrecht, Holland, 1986.
29. H. Nijs, M. I. Cruz, J. J. Fripiat and H. van Damme, *Nouv. J. Chim.*, **1982**, *6*, 551.
30. H. Nijs, H. Van Damme, F. Bergaya, A. Habtai and J. J. Fripiat, *J. Mol. Catal.*, **1983**, *21*, 223.
31. H. Nijs, M. Cruz, J. J. Fripiat and H. van Damme, *J. Chem. Soc., Chem. Commun.*, **1981**, 1026.
32. G. Villemure, H. Kodama and C. Detellier, *Can. J. Chem.*, **1985**, *63*, 1139.
33. C. Detellier and G. Villemure, *Inorg. Chim. Acta*, **1984**, *85*, L19.

34. G. Villemure, G. Bazan, H. Kodama, S. Hideomi, A. G. Szabo and C. Detellier, *Appl. Clay Sci.*, **1987**, *2*, 241.
35. K. Takagi, H. Usami, H. Fukaya and Y. Sawaki, *J. Chem. Soc., Chem. Commun.*, **1989**, 1174.
36. W. N. Wang and W. Jones, unpublished results.
37. T. Seki and K. Ichimura, *J. Photopolym. Sci. Technol.*, **1989**, *2*, 147.
38. J. M. Adams and A. J. Gabbutt, *J. Incl. Phen.*, **1990**, *9*, 63.
39. H. Yoneyama, S. Haga and S. Yamanaka, *J. Phys. Chem.*, **1989**, *93*, 4833.
40. J. F. Tanguay, S. L. Suib and R. W. Coughlin, *J. Catal.*, **1989**, *117*, 335.
41. F. R. F. Fan, H. Y. Liu and A. J. Bard, *J. Phys. Chem.*, **1985**, *89*, 4418.
42. H. Miyosi and H. Yoneyama, *J. Chem. Soc., Faraday Trans. 1*, **1989**, *85*, 1873.
43. S. L. Suib and K. A. Carrado, *Inorg. Chem.*, **1985**, *24*, 863.
44. S. L. Suib, J. F. Tanguay and M. L. Occelli, *J. Am. Chem. Soc.*, **1986**, *108*, 6972.
45. O. Enea and A. J. Bard, *J. Phys. Chem.*, **1986**, *90*, 301.
46. B. Casal, E. Ruiz-Hitzky, F. Bergaya, D. Challal, J. Fripiat and H. van Damme, *J. Molecular Catal.*, **1985**, *33*, 83.
47. W. Feitknecht and G. Fischer, *Helv. Chim. Acta*, **1935**, *18*, 555.
48. W. T. Reichle, *Chemtech.*, **1986**, January, 58.
49. R. M. Taylor, *Clay Minerals.*, **1984**, *19*, 591.
50. S. Miyata, *Clays Clay Minerals*, **1983**, *31*, 305.
51. W. Jones and K. Chibwe in *Pillared Layered Structures*, I. V. Mitchell, ed., Elsevier Applied Sciences, London, 1990, p. 67.
52. D. L. Bish, *Bull. Mineral.*, **1980**, *103*, 170.
53. K. Chibwe and W. Jones, *J. Chem. Soc., Chem. Commun.*, **1989**, 926.
54. G. J. Ross and H. Kodama, *Am. Mineral.*, **1967**, *52*, 1036.
55. T. Kwon, G. A. Tsigdinos and T. J. Pinnavaia, *J. Am. Chem. Soc.*, **1988**, *110*, 3653.
56. G. M. Woltermann, U. S. Patent 4,454,244 (June 12, 1984; Ashland Oil, Inc. Patent).
57. M. A. Drezdon, *Inorg. Chem.*, **1988**, *27*, 4628.
58. K. Chibwe and W. Jones, *Chem. Mater.*, **1989**, *1*, 489.
59. T. Kwon and T. J. Pinnavaia, *Chem. Mater.*, **1989**, *1*, 381.
60. E. P. Giannelis, D. G. Nocera, T. J. Pinnavaia, *Inorg. Chem.*, **1987**, *26*, 203.
61. M. D. Newsham, E. P. Giannelis, T. J. Pinnavaia and D. G. Nocera, *J. Am. Chem. Soc.*, **1988**, *110*, 3885.
62. M. Chibwe, J. B. Valim and W. Jones, in preparation.
63. I. Y. Park, K. Kuroda and C. Kato, *Chem. Lett.*, **1989**, 2057.
64. A. D. Wadsley, *Acta Crystallogr.*, **1964**, *17*, 623.
65. J. F. Lambert, Z. Deng, J. B. d'Espinose and J. J. Fripiat, *J. Colloid Interface Sci.*, **1989**, *132*, 337.
66. T. Nakato, H. Miyata, K. Kuroda and C. Kato, *Reactivity of Solids*, **1988**, *6*, 231.
67. H. Miyata, Y. Sugahara, K. Kuroda and C. Kato, *J. Chem. Soc., Faraday Trans. 1*, **1988**, *84*, 2677.
68. T. Nakato, K. Kuroda, C. Kato, *J. Chem. Soc., Chem. Commun.*, **1989**, 1144.
69. I. V. Mitchell, ed., *Pillared Layered Structures*, Elsevier, London, 1990.

Chapter 10

Photoprocesses of Organic Molecules Included in Zeolites

V. Ramamurthy

Central Research and Development Department,
The Du Pont Company, Wilmington, DE, USA.

Contents

1. Introduction

In the case of non-porous inorganic oxides such as silica and alumina only a small fraction of the Si—O and Al—O atoms are exposed to the adsorbents. Only the surface atoms are accessible. These surfaces can be considered to provide an one-dimensional organized medium for the reaction. Therefore, the constrained environment experienced by the reactive molecules are not expected to be very high (Chapter 8). On the other hand, clays offer both internal and external surfaces as the medium for a reaction. Most clays possess layered structure and therefore, the microcavity wherein the guest reactants are accommodated, can be considered to be two dimensionally organized (Chapter 9). While the restriction experienced by the guest molecule in clays is expected to be higher than in uni-dimensional silica/alumina surface, it may not reach the level of organic host–guest systems discussed in Chapter 7. Quite different is the situation with zeolites. Via the entire internal micropore surface all atoms are accessible for the adsorbents. Most of the surface available for adsorption is internal in character. Further, the pore dimension is of molecular size. This unique arrangement provides a three dimensionally restricted arrangement for a reactant guest molecule. Therefore, silica, clays and zeolites provide an increasingly organized, complex and constrained environment for a reaction. In this Chapter, a summary of photochemical studies carried out utilizing zeolites as the medium is provided. A brief introduction to the structure of zeolites is also included.

2. Zeolites as Hosts

Zeolites are crystalline aluminosilicate materials with open framework structures. Commercially, they have found application in catalysis (e.g., hydrocarbon cracking), separations, drying, and in detergent formulations.[1] There are approximately 40 naturally occurring zeolites and over 100 synthetic forms. The primary building blocks of the zeolites are the $[SiO_4]^{4-}$ and $[AlO_4]^{5-}$ tetrahedra (see Figure 1).

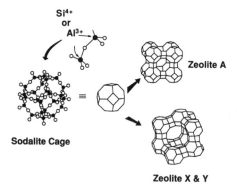

Figure 1. Illustrations of the $[SiO_4]^{4-}$ and $[AlO_4]^{5-}$ tetrahedra that are the primary building blocks of zeolites. Also shown are representations of the sodalite cage and zeolites A, X and Y.

These tetrahedra are linked by all their corners to form channels and cages or cavities of discrete size with no two aluminum atoms sharing the same oxygen. Substitution of framework $[SiO_4]^{4-}$ and/or $[AlO_4]^{5-}$ with other components such as those listed in Table 1 also give materials with zeolite-like properties. As such the term "molecular sieve" is generally used to describe any three-dimensional framework of oxygen ions generally containing tetrahedral sites. The term zeolite is reserved for molecular sieves containing only Si^{4+} and Al^{3+}.

Table 1. Cations that may form molecular sieve framework structures and the metal oxide charge possible [from R. Szostak, "Molecular Sieves. Principles of Synthesis and Identification", Van Nostrand: New York, 1989, p. 3].

$(M^{2+}O_2)^{2-}$	Be, Mg, Zn, Co, Fe, Mn
$(M^{3+}O_2)^{1-}$	Al, B, Ga, Fe, Cr
$(M^{4+}O_2)^{0}$	Si, Ge, Mn, Ti
$(M^{5+}O_2)^{1+}$	P, As

As a result of the difference in charge between the $[SiO_4]^{4-}$ and $[AlO_4]^{5-}$ tetrahedra, the total framework charge of an aluminum-containing zeolite is negative and hence must be balanced by a cation, typically an alkali or alkaline earth metal cation. As such, zeolites can be represented by the empirical formula $M_{2/n} \cdot Al_2O_3 \cdot x\ SiO_2 \cdot y\ H_2O$, where M is the cation of valence n (typically Na, Ca, Mg, etc), $x = 2-\infty$, and y varies from 0 to approximately 10. These cations can generally be exchanged by conventional methods. The cations and water molecules present are located in the cages, cavities, and channels of the zeolites. The position, size, and numbers of cations as well as the position and numbers of water molecules can significantly alter the properties of the zeolite. The aluminosilicate backbone of a zeolite can be represented in a number of ways, two of which are illustrated in Figure 1 for the beta- or sodalite-cage, a 'ball and stick' model and the 'tetrahedral array' in which 'single bonds' join the silicon and aluminum atoms. This cage structure is constructed of openings containing 4- and 6-membered rings of $[SiO_4]^{4-}$ and $[AlO_4]^{5-}$ polyhedra. In faujasite zeolites, the sodalite cages surround an even larger cage, the supercage.

The numerous framework topologies of the molecular sieves offer various systems of channels and cavities resulting in one-, two-, or three-dimensional diffusion for included guest molecules.[2] There are two types of structures: one provides an internal pore system comprised of interconnected cage structures; the second provides a system of uniform channels. For example, molecular sieves where the channels are parallel to one another and there are no connecting channels large enough for guest molecules to cross from one channel to the next are considered one-dimensional channel systems. The preferred type has two- or three

432 Ramamurthy

dimensional channels to provide rapid intracrystalline diffusion in adsorption and catalytic apllications. Access to these channels, cages or cavities is through a pore or window which can be of the same size or smaller than the size of the channels, cages or cavities. It is this pore dimension which determines the size of molecules that can be adsorbed into these structures.

The variety in internal structure types also results in a range of pore dimensions (3–8 Å; Table 2) from the small-pore materials, with the largest pore opening consisting of eight-membered rings of the tetrahedra, to the medium-pore zeolites with 10-membered ring openings, and the large-pore materials with 12-membered ring openings. Larger pore openings have been reported for the aluminum phosphate molecular sieves, such as the extralarge-pore AlPO$_4$-8 [3] and the very large pore VPI-5.[4] Here, 14-membered and 18-membered rings, respectively, are observed. As illustrated in Figure 2, for the ideal planar configuration, the O—O distance across the eight-membered ring would be 7.1 Å (4.4 Å assuming[5] O with an ionic radius[6] of 1.35 Å). Similarly, for the 10-membered ring 8.7 Å (6.0 Å assuming ionic O), and the 12-membered ring 10.4 Å (7.7 Å assuming ionic O).

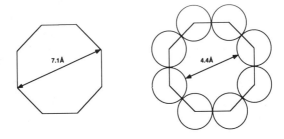

Figure 2. Illustration of a planar configuration of 8 oxygen atoms (and interatomic distance across the ring) and the estimated diameter of the opening assuming component oxygens to have radii of 1.35Å.

For the larger pore materials, these distances are 12.4 Å (9.7 Å assuming ionic O) for the 14-ring and 15.1 Å (12.4 Å assuming ionic O) for the 18-ring sieves. The actual openings vary significantly from these values as shown in Table 2. These variations are the result of differences in the structure generating the pore. The pore size can be further modified by framework composition[7], ion exchange[8], framework flexibility,[9] sorbents[10], and external surface modifications.[11] In ion exchange zeolites, the ions site in the pore opening or window as determined by the charge requirements of the framework as well as by the spatial and charge requirements of the cation. The size of the cation also contributes to the modification of the effective size of the opening (as illustrated in Figure 3 for the eight-ring opening of zeolite A).[8]

Table 2. Size of pore openings and dimensionality of the pore system for selected medium-pore and large pore molecular sieves. [a]

Molecular Sieve Name	Pore (window) Size (Å)	Channel/Cage Size
Medium Pore Zeolites:		
AlPO$_4$-11	3.9 x 6.3	Single Channel
ZSM-23	4.5 x 5.2	"
Laumontite	4.0 x 5.3	"
Partheite	3.5 x 6.9	"
Theta-1	4.4 x 5.5	"
Dachiardite	3.4 x 5.3 and (3.7 x 4.8)	Two interconnected channels
Stilbite	4.9 x 6.1 and (2.7 x 5.6)	"
Epistilbite	3.4 x 5.6 and (3.7 x 5.2)	"
Ferrierite	4.2 x 5.4 and (3.5 x 4.8)	"
ZSM-11	5.3 x 5.4	"
ZSM-5	5.3 x 5.6 and 5.1 x 5.5	"
Eu-1	4.1 x 5.7	Channel with side pockets
Heulandite	3.0 x 7.6 and (3.0 x 7.6 and 3.3 x 4.6)	Three interconnected channels
Large Pore Zeolites:		
AlPO$_4$-5	7.3	Single Channel
Cancrinite	5.9	"
ZSM-12	5.5 x 5.9	"
Linde Type L	7.1	Single channel with a lobe (d = 7.5)
Omega	7.4 (3.4 x 5.6)	Two noninter-connected channels
MAPSO-46	6.3 and (4.0)	Two interconnected channels
Gmelinite	7.0 and (3.6 x 3.9)	"
Mordenite	6.5 x 7.0 and (2.6 x 5.7)	"
Offretite	6.7 and (3.6 x 4.9)	"
Beta	7.5 x 5.7 and 6.5 x 5.6	Three dimensional channel
Faujasite (X and Y type)	7.4	Three dimensional channel with a cage (d = 12)
Extra Large and Very Large Pore Zeolites:[c]		
AlPO$_4$-8	7.9 x 8.7	Single Channel
VPI-5	12.1	"

[a] The numbers in paranthesis correspond to the smaller pore.

[b] For details see refs. 2, 7, 14 and 26.

[c] See refs. 3 and 4.

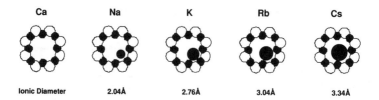

Figure 3. Illustration of the narrowing of the pore opening of zeolite A resulting from ion-exchange.

Here, only monovalent ions site in the eight-membered ring so in the calcium form, no constriction of the pore is observed. For the monovalent ions, the degree of constriction is directly related to the ionic diameter of the exchangable cation. Several zeolites, such as zeolite rho, exhibit exceptional framework flexibility on dehydration[12], thermal changes[13], and cation exchange.[9] Although useful for separations, ion-exchange, and catalytic applications, the pore sizes characteristic of the small-pore materials are significantly smaller than the molecular dimensions of most photochemically active probes of interest, so the remainder of this introduction to zeolites is devoted to the structure and properties of medium- and large-pore zeolites and the cage and channel structures that characterize many of them.

Among the various zeolites, faujasite type zeolites have attracted considerable attention. The two synthetic forms of faujasite are referred to as zeolite X and Y and have the following typical unit cell composition:[14]

X type $M_{86}(AlO_2)_{86}(SiO_2)_{106} \cdot 264 \; H_2O$
Y type $M_{56}(AlO_2)_{56}(SiO_2)_{136} \cdot 253 \; H_2O$

where M is a monovalent cation.

Charge-compensating cations present in the internal structure of zeolites are known to occupy three different positions in zeolites X and Y. As illustrated in Figure 4, the first type (site I), with 16 per unit cell (both X and Y), is located on the hexagonal prism faces between the sodalite units. The second type (site II), with 32 in number per unit cell (both X and Y), is located in the open hexagonal faces. The third type (site III), with 38 per unit cell in the case of X type and only eight per unit cell in the case of Y type, is located on the walls of the larger cavity. Only cations of sites II and III are expected to be readily accessible to the adsorbed organic. The free volume available for the organic within the supercage depends on the number and nature of the cation. Supercages are large as evident from their known capacity to include 28 molecules of water, 5.4 molecules of benzene, or 2.1 molecules of perfluorodimethylcyclohexane per cage. The largest pore opening is a 12-ring with dimensions of about 7.4 Å. As the calculated supercage volumes[15] given in Table 3 show, the free volume decreases as the cation size increases from Li to Cs.[16] The supercages form a three-dimensional network with each supercage connected tetrahedrally to four other supercages through the 12-membered ring opening.

CATION LOCATION INSIDE FAUJASITE CAGES

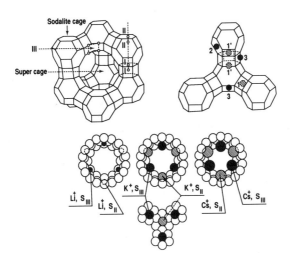

Figure 4. Supercage structure, cation location (I, II, III) within X and Y type zeolites. Bottom portion shows the reduction in available space (relative) within the supercage as the cation size increases.

Table 3. Effect of Ion Size on Estimated Supercage Volume for Zeolites X and Y

Cation	Ionic Radius (Å)	Supercage Volume (Å³)	
		Zeolite X	Zeolite Y
-		888	840
Li	0.76	873	834
Na	1.02	852	827
K	1.38	800	807
Rb	1.52	770	796
Cs	1.67	732	781
Tl	1.50	775	798

Among the medium-pore sized zeolites perhaps most studied are the pentasil zeolites, ZSM-5 (or silicalite)[17] and ZSM-11 (or silicalite-2)[18] (see Figure 5). These zeolites also have three-dimensional pore structures, in these cases comprised of pores containing 10 of the $[SiO_4]^{4-}$ and $[AlO_4]^{5-}$ tetrahedra. A major difference between the pentasil pore structures and those described above is the fact that the pores do not link cage structures as such. Instead, the pentasils are composed of two intersecting channel systems. This is perhaps best illustrated in the tubular representation given in Figure 5. For ZSM-5, one consists of straight channels with a free diameter of about 5.4 x 5.6 Å and the other consists of sinusoidal channels with a free diameter of about 5.1 x 5.5 Å. For ZSM-11, both are straight channels with dimensions of about 5.3 x 5.4 Å. The volume at the intersections of these channels is estimated to be 370 Å^3 for a free diameter of about 8.9 Å.[19]

Other large-pore zeolites of interest for photochemical studies include the large-pore zeolites L[20], mordenite[21], offretite[22], omega[2], and beta[24], as well as the aluminophosphate frameworks AlPO$_4$-5 (large-pore),[25] AlPO$_4$-8 (extra-large pore),[3] and VPI-5(very large-pore).[4] The structures of each of these are illustrated in Figures 6 and 7, respectively, and their respective pore openings are found in Table 2.

For further details on zeolites and molecular sieves, readers are referred to books by Breck,[14] Szostak,[7] Dyer,[26] and an edited volume by van Bekkum, Flanigen and Jansen[26] as well as various review articles.[1]

3. Guests within Zeolites
3.1. Inclusion

Two methods are employed to include organic and inorganic probes into zeolite structures. The method employed by physical chemists interested in studying adsorption and catalytic properties utilizes vacuum techniques. A typical procedure recorded in a paper reads as follows[27]:

"The zeolite material was activated for 48 h at 400°C. At this temperature the samples were kept in contact with a mercury diffusion pump until the pressure decreased to values less than 0.01 Pa. After cooling to room temperature, the adsorption was accomplished through the gas phase by freezing the adsorbate (guest) out of a vessel of known volume on the activated zeolite material. The amount adsorbed was checked gravimetrically. Afterwards the loaded zeolite material was transformed under vacuum into tubes and sealed."

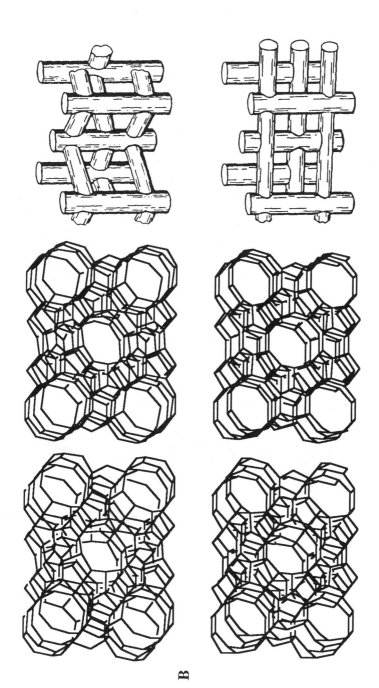

Figure 5. Stereo views of the framework structures and tubular representations of the channel systems for zeolites a). ZSM-5 and b). ZSM-11.

A

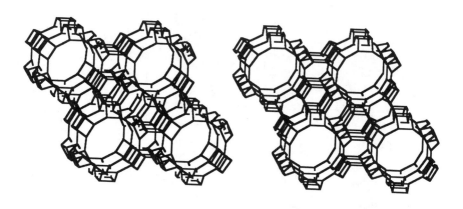

B

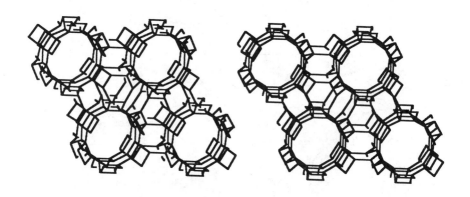

Figure 6 (Captions see page 440)

C

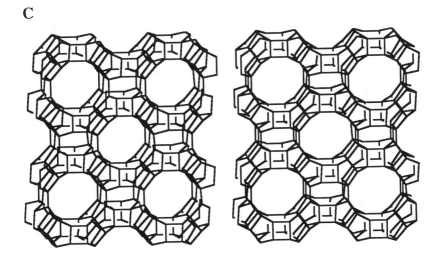

D

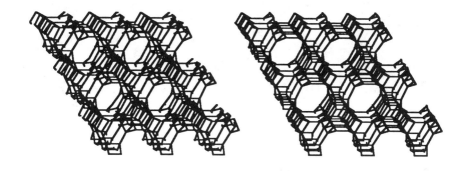

Figure 6 (Continued, Caption see page 440)

E

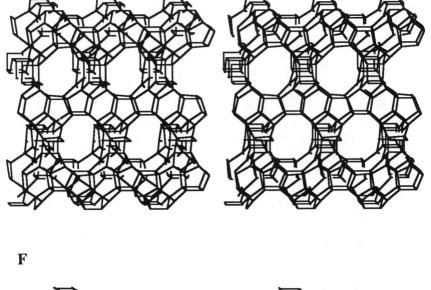

F

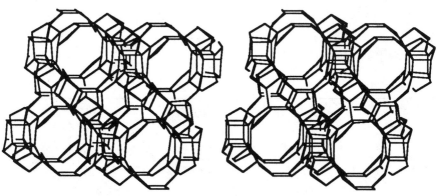

Figure 6. Stereo views of the framework structures of zeolites A). L; B). Mazzite (Omega); C). Mordenite; D). Offretite; E). Beta - Polymorph A; and F). Beta - Polymorph B. [D. H. Olson and coworkers at Mobil are acknowledged for access to their database of zeolite structural parameters.]

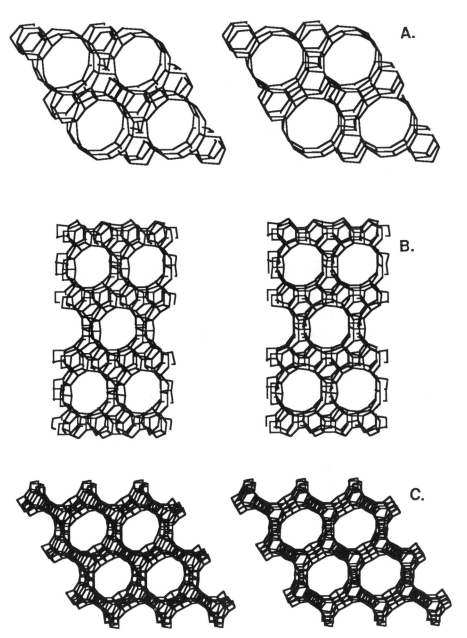

Figure 7. Stereo views of the framework structures of molecular sieves A). AlPO$_4$-5; B). AlPO$_4$-11; and C). VPI-5. [D. H. Olson and coworkers at Mobil are acknowledged for access to their database of zeolite structural parameters.]

A simpler solution technique is utilized by photochemists. A typical procedure is as follows[28]:

> "The zeolite was activated at 300-450° C for 3 h and then added to the guest-cyclohexane solution while stirring. Stirred suspension is kept in the dark for several hours (2–12 h) After filtration, the guest containing zeolite was washed with fresh cyclohexane three times. The filtrate was collected for guest analysis. The guest concentrations introduced by this procedure were calculated by subtraction of the guest left in the liquid from total amount of guest added."

While the solution method is simpler and relatively easier it gives rise to several ambiguities. Therefore, *a standard procedure must be employed to be able to compare the results from various laboratories.*

3.2. Location

To be able to understand and predict the photobehavior of a guest molecule within zeolites one should know the exact location of the guest within these structures. Since all guest molecules exert some motion within the channels/cages/cavities of zeolites, x-ray structural characterization has not been possible. However, considerable literature exists on the characterization of the location of guest molecules within zeolites based on other techniques. Most of these studies are concerned with small molecules such as benzene, *p*-xylene, and pyridine, molecules not of primary concern to photochemists. However, a brief summary of conclusions reached so far with these molecules is relevant.

The adsorption of benzene on X- and Y- type zeolites has been followed by several techniques: infrared spectroscopy,[29] Raman spectroscopy,[30] uv diffuse reflectance spectroscopy,[31] nmr spectroscopy,[32] neutron diffraction,[33] small-angle neutron scattering,[34] adsorption techniques,[35] and quantum chemical calculations.[36] These studies have shown that at high loadings there are three distinct types of benzene molecules, located within the supercages—one at the cation site (site II or III), one at the 12-ring window site, and the other corresponding to benzene clusters within the cage (Figure 8). At low loading levels the clustering can be avoided and the distribution between the window and the cation sites can be controlled by the loading level and by the nature of the cation. At the cation site, the benzene molecule is stabilized through the interaction between the cation and the π cloud. The binding strength depends on the cation charge density (acidity) or the electrostatic potential (e/r) of the cation. At the   12-ring window site, the interaction occurs through van der Waals forces and through acid–base interactions between the C—H bonds of benzene and the oxygens of the 12-ring window. The basicity of the 12-ring oxygens depends on the cation. For example in the case of Y-type zeolites, it has been estimated that the negative charge on the 12-ring oxygen increases with the decrease in cation acidity (Li–Y, -0.345; Na–Y, -0.351; and K–Y, -0.381).[37]   In the Y–type of zeolites, the binding energy between the cation and benzene decreases in the order Li > Na > K > Rb > Cs. The interaction energy with 12-ring oxygen follows the reverse trend. The 12-ring oxygens in Li– and Na–Y are not basic enough and the first benzene molecule in these zeolites prefers only the cation site.

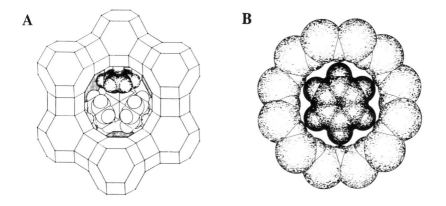

Figure 8. (a). The four benzene positions and the site II sodium ions within the supercage. (b). Benzene position at the window site. (Ref. 33)

In the case of pentasil zeolites (e.g., ZSM-5) similar studies have been carried out with *p*-xylene, benzene, and pyridine as guests.[38] All of these studies point to the intersection between channels as the preferred site at low loading levels. At higher levels of loading both sinusoidal and straight channels are also occupied by the guests.

The importance of cation sites has become evident in several other zeolites as well. In large pore zeolite N–β, similar to X and Y zeolites, two locations, namely the cation site and the 12-ring window site, have been identified for benzene through ir studies.[39] Neutron diffraction investigation of perdeutrobenzene in zeolite K L has identified the cation site as the location of the benzene guest.[40] Similar to other zeolites, benzene interacts with potassium in K–L through its π cloud. Quasielastic neutron scattering studies of benzene in Na-mordenite once again leads the investigator to the cation site as the preferred site for benzene.[41] Thus it is clear that the cation plays a determining role in the location of guests within zeolites.

3.3. Dynamics

A complete bibliography of the literature on the dynamics (diffusion and mobility) of guests included in zeolites with appropriate comments could easily fill a monograph. Large inconsistencies between reported diffusivities, often by several orders of magnitude, exist in the literature prior to 1985. However, in recent years attempts have been made to understand these discrepancies.[42] In the limited space available, we can only provide a very brief summary. A knowledge of the dynamics of guests within a zeolite structure is valuable for formulating a simple model to understand the photobehavior of the zeolite-included guests.

We pointed out in Section 3.2 that there are preferred sites for guests within zeolites. It has become clear through extensive nmr investigations that the molecules are not stationary at these sites.[43] They undergo rotational motions at the site itself and also jump from one site to the other. Therefore, the sites pointed out above should be considered as the places where the molecules reside between jumps. Deuterium nmr lineshape analyses have revealed that the motion of benzene adsorbed onto Na–X, Na–Y, Na M-5, ZSM-5, and K–L is temperature dependent.[44] At temperatures as low as 100 K, benzene molecules present at the cation site in X and Y zeolites and at intersections in ZSM-5 (low loadings) undergo anisotropic rotation around their C_6 axes. In the case of Na–mordenite-5, the narrow channel permits the molecule to reorient only around the sixfold symmetry axis. The correlation time for the above rotations must be shorter than 1 μs, the nmr time scale, as such motions are detected by nmr measurements. It is of interest to note that neutron scattering studies of benzene in ZSM-5 have shown that the correlation time for rotational motions is even less than 10^{-11} s.[45] The interaction between the π-electron density of benzene and the cation that is responsible for the stability of the adsorption site seems to be unaffected by such motions.

In addition to the molecular rotation at the site of adsorption, the guest molecules jump from one site to the other, leading to intracrystalline diffusion at temperatures above 200 K. The diffusion processes in zeolites have been extensively studied and several terms have been developed to describe this process—mean residence time (the time spent at a site between two successive jumps), mean jump length (the average distance of travel in one jump), and diffusion coefficient. These parameters depend on the type of zeolite, the nature of the guest molecule, loading level, the cation, and the temperature. Recently agreed numbers for the mean jump length and diffusion coefficient for benzene in Na–X are 0.24 nm and 1.9×10^{-10} m^2 s^{-1}, respectively, at 458 K at an occupancy level of two molecules per cage.[46] The mean residence time is of the order of nanoseconds at room temperature.[47] Analogous values for benzene in ZSM-5 at a loading of 1.5 molecule per channel intersection are: mean residence time, 10 μs; mean jump length, ~ 0.1 nm, and diffusion coefficient, ~1 x 10^{-14} m^2 s^{-1}.[47,48] Similar intracrystalline migration of benzene in Na–mordenite-5 and K–L has been monitored by ^2H nmr.[40,41] However, no rates have been measured. In addition to intracrystalline diffusion the guest molecule may undergo intercrystalline migration but the rate of diffusion is extremely slow and therefore need not be of serious concern.

It is clear from the above brief discussion that the guest molecules are not stationary at the site of preferred adsorption within zeolites. They undergo local as well as global motion within the crystallite. The time scale for such motions varies with the zeolite and the guest molecule. Although our discussion is limited to benzene, we hope that this provides a feeling for what one should expect for other molecules of interest to photochemists.

4. Unimolecular Photophysical Processes
4.1. Dependence on Microenvironment

The luminescence spectra of the zeolite-included organic guests are generally similar to those in an organic solution or glass. This is evident when the fluorescence spectra of phenanthrene and chrysene are compared in the crystalline state, in methylcyclohexane glass (MCH) and in Na–X zeolite (Figure 9).[49] A close similarity in fluorescence spectra between MCH glass and Na–X zeolite is obvious and this suggests that the aromatic molecules included in the cages of zeolites have a microenvironment that is very much like that in an organic solution and do not resemble that in the crystalline state. The detailed photophysical properties of the guest molecule depend very much on the method of sample preparation and handling. This is illustrated with a few examples below.

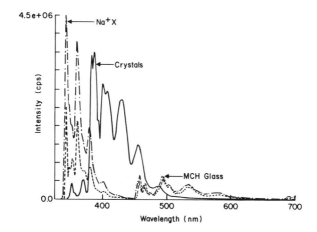

Figure 9. Fluorescence spectra of phenanthrene in various media. Notice the similarity between the MCH glass and zeolite included samples.

Pyrene has been employed as a probe by Suib and Kostapapas[50] and by Liu et al.[28] The Ham effect (intensities of bands I and III in fluorescence emission, see Chapters 2 and 3) and the ratio of the monomer to the excimer emission intensities were measured. These numbers depend critically on the solvent employed to load the pyrene into X- and Y- type zeolites (I_I/I_{III} in Na–X: cyclohexane, 0.22; carbon tetrachloride, 0.21; acetonitrile, 0.57; ethanol, 0.89). From the data obtained, the authors conclude that polar solvents such as alcohols are not suitable for loading guests into zeolite. They tend to occupy the zeolite cages, blocking the entry of guests. In general, under such conditions, pyrene molecules aggregate and adsorb on the exterior surface of the zeolite. Therefore, nonpolar solvents such as hydrocarbons should be employed to include organic guests into zeolites. While preparing complexes by the solution method, the zeolite interior is occupied by both the guest and the solvent molecules. Volatile solvent molecules can be desorbed by degassing the zeolite samples for several hours. The photophysical properties of a guest have been shown to depend upon the presence or absence of solvent in the interior of the zeolite. This is illustrated by the example of anthracene in Na–Y investigated by Dinesenko (Figure 10).[51] The fluorescence spectrum of anthracene in Na–Y recorded immediately after inclusion from heptane (without degassing) is similar to that in heptane solution except that it is slightly red shifted (250 cm^{-1}). This sample is believed to contain both anthracene and solvent heptane in the cages of Na–Y. However, the emission of the evacuated sample showed a significant blue shift. This is attributed to a strong interaction between the cation and the aromatic molecule; the presence of solvent in the nondegassed sample probably prevented such a close interaction.

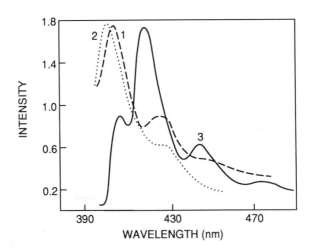

Figure 10. Fluorescence spectra of anthracene adsorbed on Na–Y zeolite showing the effect of sample handling on the λ_{max}. (ref. 51). (1): recorded immediately after adsorption, n-heptane evacuated; (2): sample (1) kept in air for ~50 h; (3): recorded after ~ 100 h of adsorption, n-heptane not evacuated.

Degassing of the sample prepared by the solution method leaves the interior of the zeolite filled only with the guest molecule. Samples thus prepared are highly hygroscopic and, if handled under laboratory conditions, depending on the humidity, water molecules may occupy the cages along with the guest. Under extreme conditions, water may even displace the guest from the interior to the exterior of the zeolite. Differences in photophysical properties of the guests present in cages coadsorbed with and without water have been reported. In this section the term *hydrated zeolites* means that the guests have been included into dry activated zeolites and then were exposed to water after degassing and removal of solvent. Dinesenko and co-workers have shown that aromatics such as anthracene, phenanthrene, and naphthalene included in Na–Y undergo slow oxidation to the corresponding quinones when the zeolite cages are hydrated.[52] Therefore, some caution should be exercised in interpreting the spectra obtained under such conditions. Liu et al. have recorded the emission spectra of pyrene in Na–Y under hydrated and dehydrated conditions.[28] The monomer to excimer emission ratio depended on the water content in the cage. The absence of water favors excimer ($I_E/I_M \sim 3$) while its presence facilitates monomer emission ($I_E/I_M \sim 0.8$). Excimer formation is inhibited in zeolite cages as water molecules tend to occupy the positions between pyrene molecules. Iu and Thomas[53] have also observed that spectral (emission and excitation) resolution depends on the water content in the cages (Figure 11). In the case of pyrene, better spectral resolution is obtained when the cages were occupied with both pyrene and water molecules. They attribute this to the presence of a highly organized water structure, which limits the motion of pyrene molecules. The quenching ability of oxygen is also curtailed by the presence of water in the interior of zeolites.

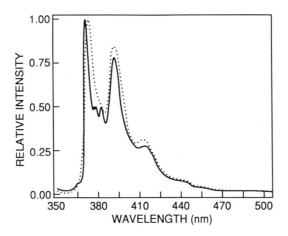

Figure 11. Effect of co-adsorbed water on the fluorescence spectra of pyrene adsorbed onto Na–X. (Ref. 53). Note the difference in spectral resolution between (——) hydrated and (- - - -) dehydrtaed samples.

Incavo and Dutta[54] have recorded an interesting dependence of the photophysical properties of tris(bipyridine)ruthenium(II) (Rubp), included in the cages of Na–Y, on water content. In the hydrated zeolite, Rubp exhibits an emission maximum at 621 nm (same as in aqueous solution), whereas the maximum in the dehydrated zeolite is at 586 nm (Figure 12). Furthermore, the degree of fluorescence polarization (p) shows an increasing trend from hydrated to dehydrated zeolite ($p = 0.05$ for hydrated and 0.15 for dehydrated zeolites). These suggest that there is stronger interaction between Rubp and the zeolite framework in the absence of water. Based on extensive studies involving resonance Raman spectroscopy, these authors have shown that such differences are to the influence of the zeolite framework on the ^3CT state of Rubp and not due to its influence on the ground-state structure.

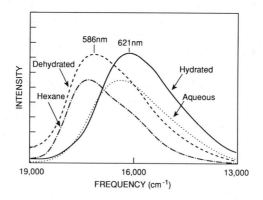

Figure 12. Emission spectra of Rubp in hydrated (—), dehydrated (- - -) and hexane–exposed (- · - · -) samples of Na–Y. Spectrum in aqueous solution (· · · ·) is also shown. (Ref. 54).

From the above studies, it is clear that sample preparation and handling are important aspects of zeolite photochemistry. If the solution method is employed, it is essential to use a nonpolar solvent to load the organic guest molecule into the zeolite. The solvent should be removed from the interior of the zeolite by a degassing procedure. Once the solvent is removed, it is advisable to keep the sample away from water, preferably in a sealed condition. The use of unactivated zeolites should be avoided as the interior of such zeolites is filled with water and, unless water is removed, no organic molecule can be included; adsorption may occur only on the exterior surface.

4.1.1. Micropolarity

Pyrene and pyrenealdehyde have been used as probes to monitor the micropolarity of the zeolite interior (see Section 4.1 of Chapter 2).[28,55] The emission maximum of pyrenealdehyde depends on the polarity of the medium.[56] Based on the recorded emission spectra of pyrenealdehyde on zeolites Na–Y and H– Y*, Baretz and Turro concluded that the interior of these zeolites is polar.[55] In the case of pyrene, fluorescence band intensities at 391 and 372 nm were measured (the paper denotes the ratio as III/I but the band at 391 nm is in fact either IV or V [57]). Based on the measured ratio Liu et al.[28] conclude that the cages of Na–X (III/I: 0.45) are more polar than those of Na–Y (III/I: 0.74). The III/I ratio depends also on the loading level of pyrene in zeolite. A higher loading level leads to an increase in the III/I ratio indicating that the presence of additional pyrene molecules in the cage decreases the internal micropolarity. The conclusions drawn in this paper do not agree with the earlier ones drawn by Suib and Kostapapas.[50] The III/I (I_{394}/I_{374}) ratios measured by these authors in Na–X and Na–Y are considerably higher (~1.5). Liu et al. attribute this difference to poor sample preparation (presence of microcrystals) by earlier workers and to poor spectral resolution of the recorded spectra.

We have measured the I_{380}/I_{370} (III/I) values for pyrene emission in various cation-exchanged X- and Y- type zeolites.[58] The ratio is dependent on the cation (M X zeolites: Li, 0.67; Na, 0.72; K, 0.52; Rb, 0.42; and Cs, 0.42). On the basis of these values one would infer that the micropolarity of the zeolite cage to increase as the cation size increases. On the other hand, the use of pyrenealdehyde as the probe leads us to conclude differently. The λ_{max} of monomer fluorescence shows a blue shift with the cation size: Li, 467; Na, 462; K, 455; Rb, 450; and Cs, 450 nm. Based on the known dependence of the emission λ_{max} of pyrenaldehyde on polarity,[56] the above trend would suggest that the micropolarity of the zeolite decreases with the cation size. Therefore, it is clear that the meaning of micropolarity in zeolite cages and the factors controlling the observed parameters are not fully understood. It is important to note that some controversy exists in interpreting the measured micropolarity. As presented in Section 3.2., aromatic molecules present in zeolite cages are expected to interact strongly with the framework cation. The consequence of such interactions on spectral band intensities and on λ_{max} is not known. Without this knowledge, attributing the variation in III/I band intensities and λ_{max} to micropolarity is to be viewed with caution.

An interesting observation was made by us during the photophysical investigation of α,ω-diphenylpolyenes and arylalkyl ketones in pentasil zeolites.[59] The resolution of the emission spectra was remarkably dependent on the Si/Al ratio of the ZSM-5 used. The fluorescence spectra of *trans*-stilbene and *trans, trans*-1,4-diphenylbutadiene and the phosphorescence spectra of octanophenone in ZSM-5 with varying Si/Al ratios are shown in Figure 13. In these cases at high Si/Al ratio, highly resolved spectra were obtained, but the spectral resolution decreased when the ratio was lower than ~40. The reason for such a dependence is not clear. One possibility is the change in micropolarity with the Si/Al ratio. Polarity would be expected to correlate directly with the aluminum content; i.e., lower aluminum content would mean lower polarity and higher aluminum content would imply higher polarity.

A
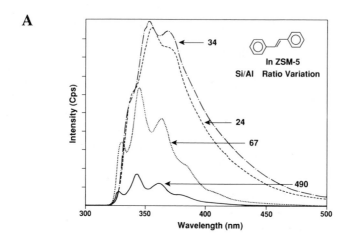

B
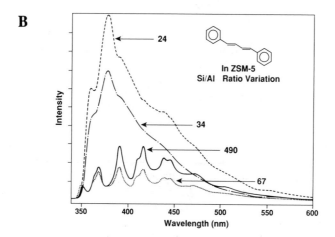

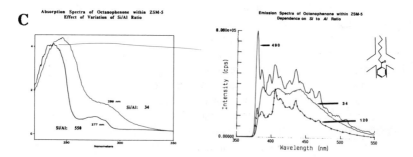

Figure 13. Dependence of spectral resolution on the Si/Al ratio: Fluorescence spectra of (a) *trans*-stilbene and (b) *trans,trans*-1,4-diphenylbutadiene and (c) phosphorescence spectra of octanophenone in ZSM-5.

We are unable to rule out the alternate possibility, namely that a higher aluminum content gives rise to inhomogenity within the porous structure and offers several sites for adsorption. This heterogeneous distribution of the guest molecule might also lead to broadening and loss of fine structure.

4.1.2. Site Inhomogeneity and Multiple Occupancy

The distribution of organic molecules within zeolites may or may not be uniform. This also means that the microenvironment around the guest molecules within a zeolite may not be uniform. Inhomogeneity in the microenvironment around a guest can arise for two reasons: variation in the occupancy number within a cage and the presence of sites of varying microenvironment. Even at low loading levels the cages may not be uniformly occupied; i.e., some may be singly occupied or multiply occupied, while others not occupied at all. A similar situation also arises in channel-forming pentasil zeolites. *The factors controlling the distribution pattern of guests within zeolites is an important problem yet to be addressed.* Several studies have clearly brought out the existence of site inhomogeneity in zeolites. Site inhomogeneity on solid surfaces has become the rule rather than an exception.

Time-resolved emission studies in the case of pyrene included in Na–X and Na–Y (< 0.1 molecule per cage) indicate that the pyrene excimers are formed within 10^{-9} s of excitation. This suggests that the pyrene complex is already formed in the zeolite cage prior to excitation.[28] Consistent with this postulate is the fact that the excitation spectra of the emission corresponding to the monomer and the excimer differ (Figure 14).

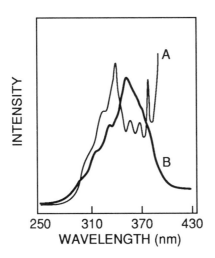

Figure 14. Excitation spectra of pyrene in Na–X: (A) excitation spectrum of pyrene monomer and (B) of pyrene excimer emission. (Ref. 28).

452 Ramamurthy

The latter shows a strong absorption at 350 nm which is not present in the monomer. These results imply that even at very low loading levels multiple occupancy occurs. Similar conclusions have been reached with pyrenealdehyde as the probe.[55]   In the above two cases, site inhomogeneity arises due to the differences in the occupation number between cages. Based on steady-state and time-resolved emission studies, Scaiano and co-workers have concluded that silicalite (a pentasil zeolite) provides at least two types of sites for guest molecules.[60] The triplet states of several arylalkyl ketones and diaryl ketones— benzophenone, xanthone, and benzil— have been used as probes. Phosphorescence from all these systems included in silicalite was observed. However, the decay of this luminescence was multiexponential. With the help of time-resolved diffuse reflectance spectroscopy, they were able to show that the triplet decay in the above systems follows complex kinetics and extends over a long period of time. Oxygen quenching experiments with benzophenone and arylalkyl ketones demonstrate that some sites are more easily accessible than others to oxygen. A more dramatic effect was observed in the decay of diphenylmethyl radical on Na–X and on silicalite.[61] Diffuse reflectance studies of diphenylmethyl radical decay on Na–X showed that the radicals decay over a time period of seven orders of magnitude (τ varies between 20 μs and 30 min). Again on silicalite, multiexponential decay was observed.

Caspar et al. have also observed multiexponential decay for naphthalene triplets in M–X (M = K, Rb, and Cs) zeolites.[62] Interestingly at temperatures below 150 K the decay was single exponential. However, at higher temperatures the lifetime was determined by at least two independent first-order decays (Figure 15).

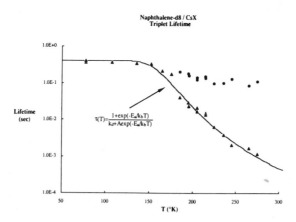

Figure 15. Temperature dependent decay of triplet naphthalene included in Cs–X. Note single exponential decay at temperatures below 150 K and double exponential decay above 150 K. (Ref. 62).

This suggested the presence of at least two independent sites for naphthalene in M–X zeolites. Consistent with this conclusion time-resolved emission spectra of naphthalene in Cs–X differed slightly for the slow and long-decaying components. The above studies conclusively demonstrate that there is a wide range of possible sites or environments for guest molecules within the zeolite interior.

4.1.3. Surface Acidity

The catalytic activity of zeolites depends critically on the number and strength of active acidic/basic sites present in the interior of these porous structures. Fluorescence techniques have been popular in monitoring the surface acidity of zeolites.[63] Fluorescence probes such as 8-quinolinol, 8-hydroxyquinoline, quinoline, α- and β-naphthols and acridine are used to measure the Bronsted surface acid strengths. Surface acidity is estimated from the emission intensities of neutral and acidic or basic forms. Lewis acid sites on the surfaces of zeolites are also monitored with aromatic probes. Most of these centers undergo full or partial electron transfer to give either radical cations or charge-transfer complexes (CTC) with the guest molecules. Such CTC formation has been observed on X- and Y-type zeolites with acridine, anthracene, naphthalene, and α-naphthylamines.[64]

4.2. Influence of Exchangeable Cations
4.2.1. Emission from the Framework Cations

The luminescence of rare earth and transition metal ion-exchanged zeolites has been extensively investigated.[65] Energy transfer between different framework cations (e.g., Cu^+/Co^{2+}, Cu^+/Cu^{2+}, UO_2^{2+}/Eu^{3+}) has also received attention.[66,67] Details of such studies are beyond the scope of this article. It is sufficient to point out that one should be careful in using zeolites containing photoemissive cations as they can interfere with the emission from zeolite-included organic molecules.

4.2.2. Interaction between the Organic Guest Molecule and the Framework Cation

Interaction between the excited state of the zeolite-trapped guest molecule and the framework cations may result in quenching of the excited state of the guest molecule. The excited singlet state of pyrene included in Y zeolites is reported to be quenched by framework cations such as Cs^+, Ag^+, Cu^{2+}, and Tl^+.[28,50,53] While no mechanism has been proposed in the first two cases,[28] quenching by Cu^{2+}, and Tl^+ is attributed to an electron-transfer process.[50,53] We wish to point out here that the electron-transfer process proposed by these authors may not the major pathway in the case of Tl^+ cation (see Section 4.2.3). An elegant example of energy transfer from the guest to the framework cation, Eu^{3+}, has been provided by Benedict and Ellis.[68] When tetramethyl dioxetane (TMD) included in Eu^{3+} Y was heated to 65°C intense emission in the 550–750 nm region characteristic of

Eu^{3+} was seen. The excited state of Eu^{3+} is generated by the energy transfer process from acetone[*], which in turn is produced by the chemiluminescent thermal decomposition of TMD. Much weaker interaction between the organic molecule and the framework cation is demonstrated to provide dramatic changes in the photophysical behavior of the aromatic guest molecule included in X- and Y-type zeolites. Such studies are highlighted below.

4.2.3. Heavy–Atom (Cation) Effect

The earliest report of a heavy-atom effect in zeolites is to be found in a paper by Bobonich.[69] In his studies on α-naphthol included in Y-type zeolite Bobonich noticed a new emission band attributable to phosphorescence from α-naphthol in Ba^{2+} and Cs^+ Y. Unfortunately, this work has not received any attention. Very recently, Ramamurthy et al. have established the utility of heavy cations in observing phosphorescence at room temperature from guest molecules included in zeolites.[70] They were able to utilize this technique to record phosphorescence from polyenes for which phosphorescence had not been seen before.

As shown in Figure 16, the emission spectrum of naphthalene is profoundly affected by inclusion in faujasites. For low-mass cations such as Li^+ and Na^+, the emission spectra show the typical naphthalene blue fluorescence. However, as the mass of the cation increases (e.g., from Rb^+ to Cs^+ to Tl^+), there is a dramatic decrease in fluorescence intensity and a simultaneous appearance of a new vibronically structured low-energy emission band that is readily identified as the phosphorescence of naphthalene. This effect is found to be general.

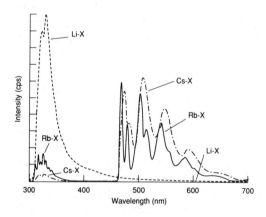

Figure 16. Emission spectra of naphthalene included in various cation–exchanged X zeolites. Note the ratio of fluorescence to phosphorescence emission depends on the cation. (Ref. 70).

Intense phosphorescence *alone* is observed for a wide range of different organic guests such as anthracene, acenaphthene, phenanthrene, chrysene, fluoranthene, pyrene, and 1,2,3,6,7,8-hexahydropyrene when included in Tl^+-exchanged faujasites. The only set of examples of guests for which phosphorescence is not observed are fused aromatics, which are too large in diameter to fit through the 8 Å windows of the X- and Y-type zeolites (e.g., coronene and triphenylene). In these cases the observed emission spectrum closely resembles that for the crystalline guest with no evidence of heavy-atom perturbation.

The correlation of the appearance of the phosphorescence with cation mass clearly suggests that the effect is due to an external heavy-atom perturbation. It is well known that the effect of external heavy-atom perturbation scales with the square of the perturbers spin–orbit coupling constant, ξ^2 and that a log–log plot of τ^{-1} vs. ξ^2 should be linear with a maximum predicted slope of unity.[71] As shown in Figure 17, the expected dependence is observed.[62] The magnitude of the heavy-atom effect observed in zeolites is significantly larger than that observed for the 1,5-naphtho-22-crown-6 exchanged with heavy-atom cations where the cation is rigidly held over the naphthalene π face.[72]

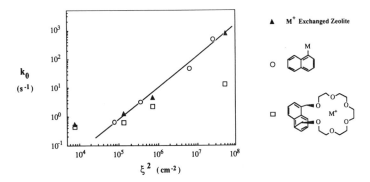

Figure 17. Correlation between the triplet decay and the spin–orbit parameter: for naphthalene included in cation exchanged zeolites. For comparison, results on other related systems are also shown. (Ref. 62).

In fact, the zeolite samples show heavy-atom effects nearly as large as for a series of 1-halonaphthalenes where the perturbers are covalently attached to the chromophore.[73] The unusually high external heavy-atom effect observed here is attributed both to the close approach between naphthalene and the heavy atom (cation), which is enforced by the zeolite supercage, and to the presence of more than one heavy atom (cation) per supercage, which leads to high effective concentrations of the heavy atom in the vicinity of the naphthalene. In order to obtain a picture of the geometry of the cation–chromophore complexes in X- and Y-type faujasites, Ramamurthy et al. took advantage of the heavy-atom-induced phosphorescence, which allows the use of optical detection of magnetic resonance (ODMR) in zero applied magnetic field.[74]   The sublevel specific dynamics for adsorbed naphthalene show a distinct increase in relative radiative character and in total rate constant of decay of the out-of-plane x-sublevel with increasing mass of the cation perturber. Interpretation of the kinetic results on the basis of a well-established model, which will not be described here, suggests that the naphthalene is adsorbed through its π cloud at a cation site.

It is easy to appreciate the potential of this unusual environment when one realizes that even olefins, systems that under normal conditions do not show phosphorescence, emit from their triplet states when included in Tl^+-exchanged zeolites. Excitation of *trans*-stilbene included in Tl^+-exchanged faujasites (X and Y type) and pentasils (ZSM-5, -8, and -11) emits intense phosphorescence both at room temperature and at 77 K (Figure 18).[75]

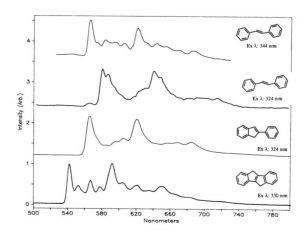

Figure 18. Phosphorescence spectra of *trans*-stilbene and related systems included in Tl^+-exchanged zeolites. (Ref. 75 and 82).

Emission in the 550–750 nm region having a sub-msec lifetime is quenched by oxygen. Furthermore, the emission band positions are in the same region reported earlier for phosphorescence (EPA glass λ_{max}: 580, 636, and 698 nm). More importantly, the excitation spectra consist of both S_0 to S_1 and S_0 to T_1 transitions, the latter being in the same region recorded by the oxygen perturbation technique. These factors strongly indicate that the long-wavelength emission is indeed phosphorescence. This observation is significant as only very weak phosphorescence from *trans*-stilbene and several substituted *trans*-stilbenes has been recorded at 77 K in organic glasses containing ethyl iodide as the heavy-atom perturber.

In an effort to expand the utility of these zeolite hosts for the observation of phosphorescence from triplet states that have not heretofore been observable, Ramamurthy et al. have investigated the spectroscopy of the *all-trans-*α,ω-diphenylpolyenes 1,4-diphenyl-1,3-butadiene (DPBD), 1,6-diphenyl-1,3,5-hexatriene (DPHT), and 1,8-diphenyl-1,3,5,7-octatetraene (DPOT) included in heavy-atom-exchanged zeolites.[74] These polyenes exhibit very low intersystem crossing efficiencies and efficient fluorescence and are expected to phosphoresce at low energies where detection with conventional photomultipliers is impractical. All these considerations conspire to make the detection of phosphorescence from these species difficult and to our knowledge no authentic phosphorescence spectra from them have been reported. The authors' approach to this problem has been to combine the use of the zeolite hosts and their prodigious heavy atom effects with the use of a sensitive germanium detector to enable the detection of the phosphorescence of the α,ω-diphenylpolyenes. Figure 19 shows the observed phosphorescence of the α,ω-diphenylpolyenes included in Tl$^+$-exchanged X-type faujasite. At 77 K, a well-resolved structured emission for each of the polyenes with a prominent vibronic spacing of 1200–1400 cm^{-1} as expected for triplet phosphorescence was observed. Empty zeolites showed no emission in this region. The singlet–triplet energy gaps ($\Delta T_1 \rightarrow S_0$) obtained from the observed zero–zero lines are in excellent agreement with literature predictions from $S_0 \rightarrow T_1$ absorption spectra obtained by the O$_2$ perturbation method and from energy-transfer studies.[76] These results demonstrate the utility of zeolite hosts as spectroscopic matrices for the investigation of organic triplet states.

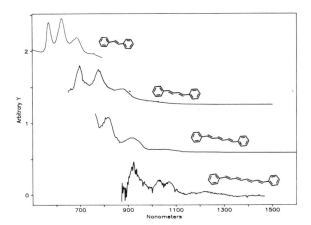

Figure 19. Phosphorescence spectra of several α,ω-diphenyl polyenes included in Tl$^+$-exchanged zeolites.

4.2.4. Light–Atom (Cation) Effect

In Section 3.2, the importance of the electronic interaction between the cation and the guest molecules in stabilizing the guests within zeolites was summarized. Based on the ^{2}H nmr of phenanthrene in M^+ X zeolites we have been able to show that the strength of binding interaction between the cation and the guest molecule is directly dependent on the charge density/electrostatic potential of the cation. The higher the charge density (i.e., charge per unit volume of the cation), the stronger the binding (ΔH binding: Na^+, 14.9; K^+, 11.0; Cs^+, 7.9 kcal mole^{-1}).[77] If such electronic interactions also influence the photophysical properties of the guest, the light cations are expected to have more influence than the heavier and the larger ones—thus the name "light atom effect." This aspect of zeolite photochemistry has not received much attention. However, recently a brief study has been initiated in our laboratory. A summary of the preliminary results is provided below.

The role of the medium on the λ_{max}, the spectral resolution, and the band intensities of fluorescence emission of organic molecules is well established.[78] Dependence of the above three emission characteristics on the cation in the case of guests included within zeolites has been recently observed by us.[49] Spectral resolution of the fluorescence emission from olefins, arylalkyl ketones, and aromatics included in zeolites varied with the nature of the alkali metal cation — the smaller the size of the cation, the poorer the spectral resolution. The variation in vibrational band intensities and a shift in the λ_{max} with respect to cation is also noticed with a few aromatic molecules. In all the systems examined, the largest effect was observed with smaller cations such as Li^+ and Na^+. A clear illustration of the light-cation influence is seen in the (electronic absorption) diffuse reflectance spectra of pyrene and anthracene (Figure 20).

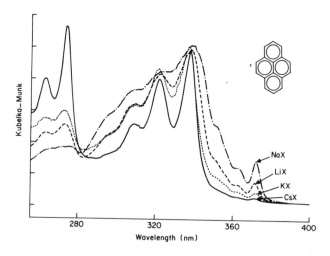

Figure 20. Diffuse reflectance spectra of pyrene included in cation-exchanged X zeolites. Note the intensity dependence of λ_{max} at ~370 nm on the cation.

In both cases, the intensity of a forbidden band depends on the cation. All of the above changes, we believe, are a consequence of the strong electronic interaction between the included guest and the cation.

We have also observed that the lifetimes of the excited singlet and triplet states of guests are influenced by the cations.[49] A consistently shorter lifetime was recorded in Li–X than in Na–X for the excited singlet state of aromatic molecules (e.g., naphthalene in Li–X , 33 and Na–X, 36 ns; phenanthrene in Li–X, 28 and Na–X, 37 ns; pyrene in Li–X, 92 and Na–X, 115 ns). This is temporarily attributed to the light-cation-induced enhancement of radiative and/or radiationless processes. Detailed mechanistic studies are yet to be carried out. A similar cation-dependent variation in the triplet lifetime of valerophenone has been recorded by Johnston and Ramamurthy (Li–X, 4; Na–X, 0.4; and K–X, 0.9 μs).[79] This is clearly a consequence of the cation influence on the triplet decay. Interaction between the cation and the carbonyl chromophore prohibits the γ-hydrogen from approaching the n orbital. The cation reduces the rate of γ-hydrogen abstraction and thus prolongs the triplet lifetime. Details of this process are presented in Section 6.2.

The results presented above illustrate that one will have to consider carefully the nature of the cation present along with the guest in a cage. Either a strong electronic interaction or a weak spin–orbit coupling may be present. These can have significant effects on the photophysical properties of the cage-included guests.

5. Bimolecular Photophysical Processes
5.1 Energy Transfer

Energy transfer (singlet–singlet and triplet–triplet) between donors and acceptors included within zeolites have not received much attention. However, there are a few literature observations that seem to indicate the feasibility of such a process between zeolite-exchanged cations and the included guests (Section 4.2.2). In this context, a report of energy transfer from acetone triplet to the framework Eu^{3+} in Y-type zeolite is noteworthy.[68] Although no mechanistic explanation has been offered for the oxygen quenching of the emission from the framework cation (Eu^{3+})[80] and from the included guests such as pyrene,[28] benzil, and arylalkyl ketones[60] (in X-, Y-, and pentasil-type zeolites), this process most likely involves an energy-transfer process. Pettit and Fox have established the occurrence of an energy transfer between $[Ru(bpy)_3]^{2+}$ and oxygen in zeolite Y.[81] In this study singlet oxygen oxidation of tetramethylethylene and 1-methylcyclohexene has been conducted with $[Ru(bpy)_3]^{2+}$ exchanged Y zeolite as the sensitizer. We have recorded emission from the singlet oxygen upon excitation of α,ω-diphenyl-polyenes included in faujasite- and pentasil- type zeolites.[82] This we believe is the result of energy transfer either from the singlet or the triplet states of polyenes to oxygen.

5.2. Electron Transfer

Reactions within zeolites involving electron-transfer process as the primary step have received considerable attention in recent times. In one set of studies, a radical-ion pair is generated by the γ-radiation of the guest-included zeolites. Long-lived cation radicals are formed when linear alkanes (hexane and octane)[83] and cyclic alkanes (hexamethyl Dewarbenzene and tetramethylcyclopropane)[84] included in ZSM-5 and Y- type zeolites, respectively, are subjected to γ radiation. A working mechanism, which needs further confirmation, is shown in Scheme 1.

$$
\text{Zeolite} \xrightarrow{\ \gamma\text{-ray}\ } \text{Zeolite}^+ + \text{e}^-
$$

$$
\text{Zeolite}^+ + \text{Guest} \longrightarrow \text{Guest}^{+\bullet} + \text{Zeolite}
$$

$$
\text{M}^+ + \text{e}^- \longrightarrow \text{M}
$$

$$
\text{M} + [\text{M}_3]^{3+} \longrightarrow [\text{M}_4]^{3+}
$$

where M = Na, K, Rb, and Cs

Scheme 1

Support for this mechanism comes from spectral identification of $[Na_4]^{3+}$ and other $[(\text{alkali metal})_4]^{3+}$ clusters within zeolite cavities.[85] Very recently, radical cations from α,ω-diphenylpolyenes and from thiophene oligomers have been generated even without the use of γ-, x-, and uv-radiation and stabilized within ZSM-5 zeolites.[86] Mechanistic details are awaited.

Electron transfer between $[Ru(bpy)_3]^{2+}$ and acceptors such as $[\text{methylviologen}]^{2+}$ incorporated within zeolites has attracted the attention of two groups. Dutta and Incavo[87] have investigated the electron transfer between $[Ru(bpy)_3]^{2+}$ and $[MV]^{2+}$ in Na–Y zeolite. Laser illumination (413.1 or 457.9 nm) of orange pellets of $[Ru(bpy)_3]^{2+}$–$[MV]^{2+}$–Na–Y resulted in a color change and the resulting blue color persisted for several hours. This blue color lasted for days when the above pellet was impregnated with EDTA before illumination. The resonance Raman spectrum characteristic of $[MV]^{+\bullet}$ was recorded with these samples. The mechanistic sequence can be visualized as follows: photoexcitation of $[Ru(bpy)_3]^{2+}$ brings about electron transfer to $[MV]^{2+}$ held in adjacent cages. The $[MV]^{+\bullet}$ formed is stabilized within the supercages of Y zeolite through a reduction in the rate of back electron transfer from $[MV]^{+\bullet}$ to $[Ru(bpy)_3]^{3+}$. Kruger et al.[88] have devised a zeolite-based molecular assembly in which radical ions generated by electron-transfer process are stabilized. The model used by this group consists of covalently linked donor – acceptor system **1**. In fluid solution, the forward and reverse electron transfer processes occur in about 300 ps. Kruger et al. immobilized **1** onto zeolite L in the manner shown in Figure 21.

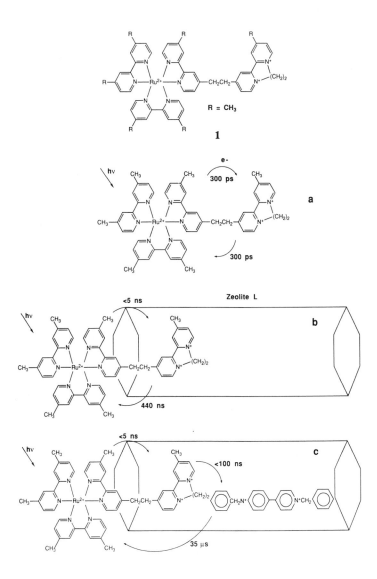

Figure 21. Light induced electron transfer reactions of a donor–acceptor assembly in : (a) acetonitrile solution; (b) zeolite L and (c) in zeolite L exchanged with benzyl-viologen[2+]. (Ref. 88).

Flash photolysis study of this assembly in an aqueous medium revealed that the charge- separated state $[Ru(bpy)_3]^{3+}$–diquat]$^{+•}$ decays with a lifetime of 440 ns. This dramatic increase in lifetime with respect to that in solution is attributed to slow reverse electron transfer between the rigidly held donor and acceptor ends of the molecule in the zeolite assembly. Inclusion of a secondary acceptor such as [benzylviologen]$^{2+}$ and $[BV]^{2+}$ into the assembly (Figure 21) results in a rapid light-induced $[Ru(bpy)_3]^{3+}$–$[BV]^{+•}$ formation. This state lasts for at least 35 μs both in L- and Y-type zeolites. It is clear from the two examples above that the rate of electron transfer can be controlled with a judicious choice and assembly of donors and acceptors within zeolites. A comparison between the works of Dutta and Incavo and Kruger et al. reveals that the charge-separated state in the Dutta and Incavo study lives for a much longer time than that in the case of Kruger et al. study. Although the donor and the acceptor systems are slightly different in the two cases, the reasons for the large difference in the lifetime of the charge-separated state need to be addressed.

Yoon and Kochi[89] have very recently utilized $[MV]^{2+}$- and $[DQ]^{2+}$- exchanged Y-type zeolites to form stable charge transfer complexes with a number of arenes. In continuation of their elegant study, Sankararaman et al.[90] revealed that the back electron transfer between radical ion pairs generated by photolysis of the charge-transfer band of the acceptor – arene complex, stabilized in the supercages of Y- type zeolites, is reduced by a factor of 10^5 –10^6 relative to that in solution (for the proposed structure of the complex see Figure 22).

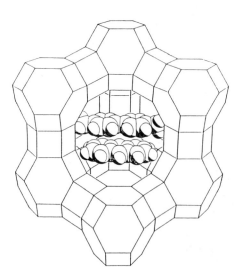

Figure 22. Proposed structure of the methylviologen dication and anthracene CTC included within zeolite Y. (Ref. 89).

Generality of this process has been established with a number of examples (naphthalenes and anthracenes as donors, methyl viologen^{2+}, tropylium$^+$, pyridinium$^+$ ions as acceptors). This observation is in line with the results discussed above in the case of the $[Ru(bpy)_3]^{3+}$–$[MV]^{2+}$ system.

Electron-transfer-initiated photoreactions have also received some attention. For example, photolysis of 1,3-cyclohexadiene–Na–Y–methylene chloride slurry yields two dimers (Scheme 2).[91] This reaction is presumed to proceed via initial electron transfer from an excited state of 1,3-cyclohexadiene to Na–Y. A similar electron-transfer process is proposed to occur during the photo-Cope rearrangement of 1,3,4-triphenyl-1,5-hexadiene in Na–Y (Scheme 2).[92]

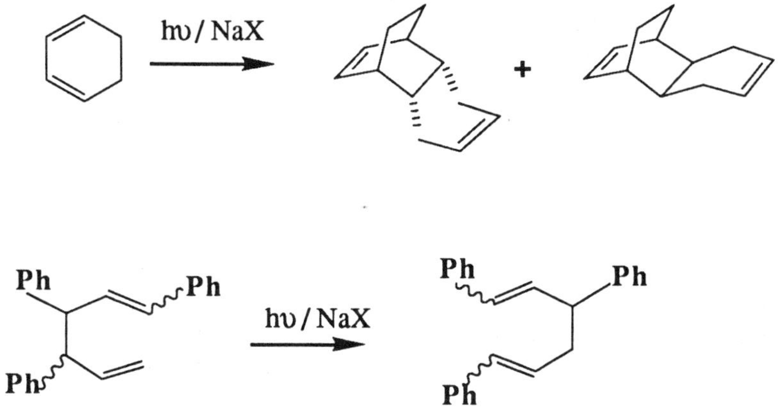

Scheme 2

One is not sure in either of these examples where the reaction occurs, i.e., inside or outside of the zeolite. In none of these cases has evidence been presented for the formation of radical cations. In 1,3,4-triphenyl-1,5-hexadiene, a conventional electron transfer method (photosensitized single-electron transfer using dicyanobenzene) failed to produce the Cope rearrangement. Therefore, at this stage these reactions should be viewed as interesting but less well-established examples of electron-transfer-initiated photoreactions within zeolites.

Decomposition of water into hydrogen and oxygen also has been achieved with zeolite as catalyst supports. Jacobs et al.[93] as well as Leutwyler and Schumacher[94] have shown that water splitting can be achieved by visible light irradiation of silver ion-exchanged Y zeolite suspended in water. The initial step involves reduction of Ag^+ to Ag^0. The reaction sequence is shown Scheme 3.

$$2 \, Ag^+ \; + \; 2 \, ZO^- \; + \; H_2O \; \xrightarrow{\; h\upsilon \;} \; 2 \, Ag^0 \; + \; 2 \, ZOH \; + \; 1/2 \, O_2$$

$$Ag^0 \; + \; ZOH \; \xrightarrow{\; \Delta \;} \; Ag^+ \; + \; ZO^- \; + \; 1/2 \, H_2$$

where ZO = Zeolite.

Scheme 3

Hydrogen is evolved only when the reduced zeolite is heated to 600°C. Photocatalytic hydrogen evolution has also been achieved with zeolite-supported CdS particles.[95] The efficiency of hydrogen evolution upon irradiation of CdS-containing zeolite depends on the nature of the sacrificial reagent such as Na_2S and Na_2SO_3 and on the availability of a cocatalyst such as ZnS or platinum. However, not much difference in efficiency was noticed between zeolite-supported and silica-supported CdS. Persaud et al.[96] has achieved water-splitting with an elegant trimolecular assembly of [zinc tetra(N-methyl-4-pyridyl)porphrin]$^{4+}$, [MV]$^{2+}$ and EDTA (Figure 23). This assembly is put together with the help of platinized zeolite L. Under visible light illumination, this integrated system produces hydrogen from water via singlet state electron transfer quenching of the [zinc tetra(N-methyl-4-pyridyl)porphrin]$^{4+}$ by [MV]$^{2+}$ as the primary step. The reduced acceptor molecule [MV]$^{+\bullet}$ is stabilized through a sacrificial electron donor EDTA present in the aqueous phase. EDTA transfers an electron to [zinc tetra(N-methyl-4-pyridyl)porphrin]$^{5+}$ and thus frees [MV]$^{+\bullet}$ to liberate hydrogen from water ($2 \, [MV]^{+\bullet} \; + \; 2 \, H^+ \rightarrow H_2 \; + \; [MV]^{2+}$).

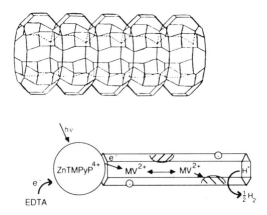

Figure 23. Trimolecular assembly of [zinc tetra(N-methyl-4-pyridyl)porphrin]$^{4+}$, [MV]$^{2+}$ and EDTA in zeolite L. Water splitting is achieved with this setup by the proposed mechanism shown in this Figure. Structure of zeolite-L on the top.

Ozin et al.[97] have established that the dimerization of alkanes can be effected upon photolysis of Ag^+ Y in an atmosphere of alkanes. For example, they were able to effect the dimerization of methane to ethane, ethane to butane, and propane to hexane. As with water splitting discussed in the previous paragraph, the dimerization is proposed to be initiated by the creation of an electron-deficient center, which the authors term the V center. The photoejected electron that is lost to the zeolite matrix is trapped by the Ag^+ and this results in the formation of a reduced silver cluster. A simple mechanism proposed by Ozin et al., shown in Scheme 4, has all the features of the mechanism proposed in the case of water-splitting.

$$T—O—Si \xrightarrow{h\upsilon} T—O^{\bullet}—Si \;+\; e^-$$

$$T—O^{\bullet}—Si \;+\; R_3C—H \longrightarrow \underset{\underset{H}{|}}{T—O—Si} \;+\; R_3C^{\bullet}$$

$$2\,R_3C^{\bullet} \longrightarrow R_3C—CR_3$$

where T = Al or Si;   R = alkyl

Scheme 4

Photooxidation of ethanol and diethyl ether with uranyl $[(UO_2)^{2+}]$-exchanged Y zeolite has been conducted.[98] Reaction involves an initial electron transfer from the ethanol and ether to uranyl ion.

6. Photochemical Process

In this section photoreactions conducted within zeolites with the view of either achieving selectivity in a reaction course or obtaining information regarding the microenvironment of the zeolite interior are presented.

6.1. Role of Cavity/Channel Size

With several examples below, we illustrate the importance of the critical matching of the channel/cavity dimensions of the host and the guest to achieve maximum influence of the host on the guest's behavior.

6.1.1. Tight vs. Loose Fit

Polyenes (stilbene, 1,4-diphenylbutadiene and 1,6-diphenylhexatriene) were incorporated into either faujasites (cation exchanged forms of zeolites X and Y) or pentasil zeolites (ZSM-5, -8, and -11).[99]   While both *cis*- and *trans*-stilbene can be included into faujasites, only the latter was accommodated by pentasils.   A similar difference in inclusion was noticed between *trans,trans*- and *trans,cis*-1,4-diphenylbutadienes. This is not surprising considering the channel size of pentasils and the molecular size and shape of the cis isomers.   Selectivity in inclusion is also reflected in the photobehavior of the included polyenes (Schemes 5 and 6).

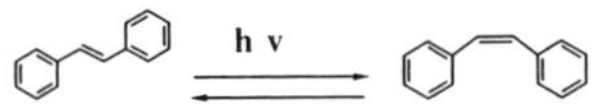

Medium	Initial	Photostationary state mixture	
		trans	*cis*
Benzene	trans	28	72
	cis	26	74
Li-X	trans	56	44
	cis	12	88
Cs-X	trans	73	27
	cis	34	66
ZSM-5	trans	100	- -
ZSM-8	trans	100	- -
ZSM-11	trans	100	- -

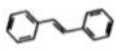

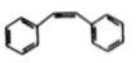

Scheme 5

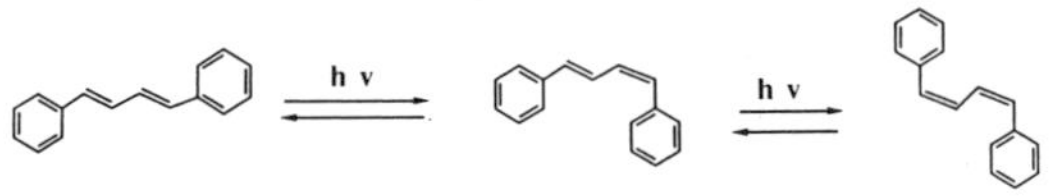

Medium	Initial	Photostationary state		
		trans,trans	*trans,cis*	*cis,cis*
Benzene	trans,trans	18	75	6
	trans,cis	17	76	7
Li-X	trans,trans	76	20	3
	trans,cis	41	45	12
Cs-X	trans,trans	73	17	10
	trans,cis	35	44	20
ZSM-5	trans,trans	100	.	.
ZSM-8	trans,trans	100	.	.
ZSM-11	trans,trans	100	.	.

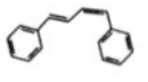

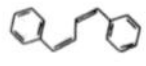

Scheme 6

Direct excitation of the trans (and all-trans) isomers of the above three olefins incorporated in pentasils resulted in no change suggesting that their inclusion in pentasils fully arrested the rotation of "π" bonds (Scheme 7).   However, both the trans and the cis isomers underwent geometric isomerization inside the supercages of faujasites. The above restriction of molecular motion is also indicated by drastic changes in the photophysical properties of the included guests.

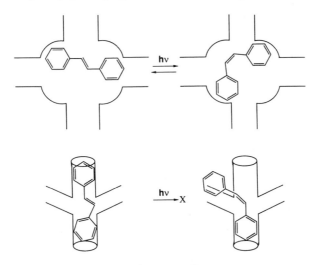

Scheme 7

Table 4: Consequence of rotational restriction on excited singlet state lifetime at room temperature

Medium	Lifetime (nano sec)
trans-Stilbene	
Methylcyclohexane	0.11
Na-X	0.21(96%); 4.6(4%)
Na-Y	0.20(98%); 4.7(2%)
ZSM-5	1.88
ZSM-8	1.87
ZSM-11	3.80
crystal	6.0
trans,trans-1,4-Diphenylbutadiene	
Methylcyclohexane	0.58
Na-X	0.60
Na-Y	0.67
ZSM-5	12.7
ZSM-8	15.1
ZSM-11	13.2
Molecular beam jet(4.2 K)	16.0

In the extremely confining space of the pentasil channels, the polyenes examined all exhibit enhanced fluorescence lifetimes (Table 4). Such lifetimes are significantly longer than in fluid solution or in the supercages of faujasites. For trans,trans-1,4-diphenylbutadiene the longest lifetime observed, 16 ns in ZSM-8, is identical within errors to the lifetime observed in a molecular beam experiment where the equivalent temperature is 4.2 K.[100a] This similarity emphasizes the rigidity of the zeolite environment. Likewise, for trans-stilbene in ZSMs the observed lifetimes approach those of rigid analogues.[100b] The enhanced lifetime is a direct reflection of the constraint provided by the host which prevents π-bond rotation.

Gessner et al.[101] have utilized the selective inclusion of the trans isomers into the pentasil channel and the photoinertness of trans isomers within the channels, to achieve one-way geometric isomerization of the cis- to the trans -stilbene.

Results of photolyses of the three of several alkanophenones investigated in isotropic solvents and in several zeolites are summarized in Table 5.[102] A dependence of the product distribution (cyclization to fragmentation ratio and cis-cyclobutanol to trans-cyclobutanol ratio) on the nature of zeolite host is evident from the table (for the size and shape of the zeolite cavity/cage/channel see Section 2). Details concerning the factors controlling the ratio of cyclization (cyclobutanol) to fragmentation (acetophenone) has been discussed in several Chapters of this volume and will not be repeated here (see Section 3.4 of Chapter 5, Section 5.1.3 of Chapter 7 and Section 8 of Chapter 14). Cyclization (C) of the type II derived 1,4-diradical to yield cyclobutanols is sterically more demanding than fragmentation (Scheme 8). Furthermore, cyclobutanol formation requires a large permanent displacement of the phenyl group.

Table 5. Photolysis of Ketones in Zeolites: Dependence of E/C Ratio on the zeolite Channel Size and Structure.

Zeolite	Butyrophenone	Valerophenone	
	E/C	E/C	CB_2/CB_1
Na,X	3.0	1.2	0.4
Na,Y	3.2	1.1	0.8
Na,L	4.5	2.1	0.8
Na,M-5	5.4	3.2	2.8
Na, Ω-5	1 1	4.6	1.9
Offeretite	5.4	2 6	2.6
ZSM-34	only E	only E	- -
ZSM-11	5 6	only E	- -
ZSM-5	7 3	only E	- -

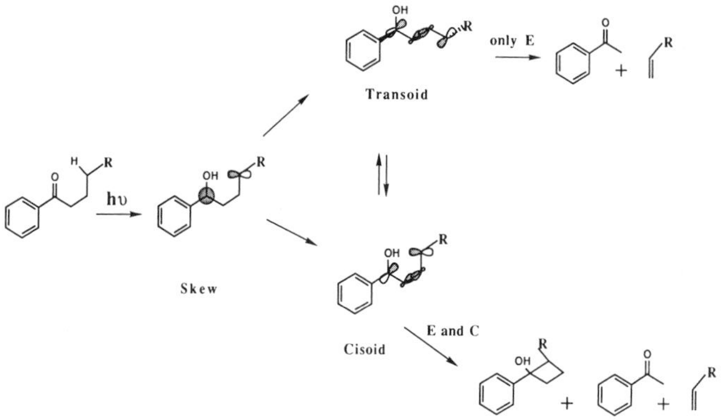

Scheme 8

Authors attribute the absence of cyclobutanols in medium-pore zeolites (ZSM-5, -11 and -34, Na–β and offeretite) to the restriction provided by the channel walls for the rotation of the central "σ" bond in the 1,4-diradical (Scheme 9) leaving elimination as the only mode of decay for the above diradical. Also, the size of cyclobutanols may be too large to fit within the channels of pentasils. In the case of smaller guests, e.g., butyrophenone, cyclization occurred to a small degree even in pentasils. As the cage/channel size increases both cyclization and fragmentation result from the 1,4-diradical.

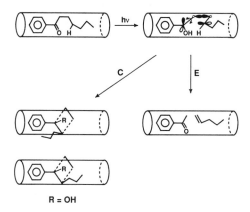

R = OH

Scheme 9

Although the hydrogen abstraction is not completely restricted within any of these zeolites, the lifetime of the reactive triplet depends on the nature of the zeolite.[103] This is believed to be a reflection of the effect of the channel/cage size and shape on the rate of hydrogen abstraction. For example the triplet lifetime of valerophenone in Na–X and Na–Y is ~ 1 μs, whereas that in ZSM-5 channel is about 15 μs. A similar effect of the nature of zeolite on the triplet lifetime and emission characteristics of β-phenylpropiophenone has been reported by Scaiano and co-workers.[60]   In this case, the triplet lifetime is controlled by intramolecular quenching of the carbonyl triplet by the β-phenyl group.   While intense phosphorescence was recorded in silicalite, none was seen in mordenite as the host. This is again the result of a medium-pore zeolite, silicalite (ZSM-5-like), restricting the molecule attaining the correct conformation for the intramolecular physical quenching (Figure 24).

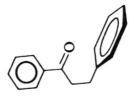

Figure 24. Conformation required for the intramolecular quenching of the triplet of β-phenylpropiophenone.   Attainment of such a conformation is restricted within silicalite.

In one of the earliest studies which in fact popularized the use of zeolites among photochemists, Turro and Wan[104] established the dependence of the cage effect, ^{13}C enrichment, and the yield of photorearrangement product from dibenzyl ketones on the nature of the zeolite.   The cage effect varies as follows: silicalite, ~75%; Na–X, ~35%; Na+ mordenite, ~45%; Na–Y, ~15%; and Na A, 0%. The yield of the photorearrangement product 1-(4'-methylphenyl)acetophenone also shows similar dependence: silicalite, 20±5%; Na–X, 2±2%; Na–mordenite, 1±0.5%; Na–Y, 0 ± 0.3%. These differences are attributed to the differences in molecular mobility of the photogenerated intermediate diradicals in these zeolites.

6.1.2. Inclusion and Exclusion Strategy[105]

Turro and co-workers have provided two elegant examples wherein they have used the zeolite either to selectively include (or imprison) a photoproduct within the cavity/cage or to selectively exclude a reactant to the outside surface.[106,107] Novelty comes from the fact that size and geometry of the reactive intermediates are sufficiently different from those of the reactant and the product to give rise to a unique photobehavior. These examples bring out the power of zeolite as a unique medium for a reaction.

The photolysis of small-ring 2-phenylcycloalkanones gives rise to products derived via α-cleavage. Details of this process was discussed in Chapter 7 (see section 5.2.2, Scheme 24). Irradiation results in α-cleavage. The radicals derived by this process yield a cyclophane as the major product. It was observed that the photolysis of 2-phenylcyclododecanone adsorbed on Na–X results in disappearance of the starting ketone but no significant amount of extractable products was observed. It was discovered, however, that dissolution of the zeolite framework followed by extraction of products resulted in the isolation of a high yield of the expected cyclophane (Scheme 24 of Chapter 7). This is attributed to the larger size of the cyclophane, which is not capable of being extracted.

Like a "ship-in-a-bottle" technique, an object that has a size and shape that is capable of passing through a window is converted to another object whose size and shape are sufficiently different to prevent reverse passage through the same window (Figure 25). When the starting ketone is too large it will adsorb only on the outside zeolite surface. Photolysis of the externally adsorbed 2-phenyl-pentadecanone resulted in the disappearance of the starting material, but not the appearance of significant amounts of extractable products. However, as before, dissolution and extraction of the zeolite on which the reactant was externally adsorbed gave the product cyclophane. Turro et al. conclude that photolysis of the externally adsorbed ketone produces an externally adsorbed biradical which rapidly reptates into the internal zeolite framework, enters a supercage, and then couples at the ends after rearrangement to produce a cyclophane structure whose size and shape prevent escape from the supercage, (Figure 25).

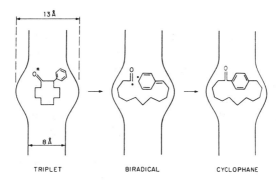

TRIPLET   BIRADICAL   CYCLOPHANE

Figure 25. Ship in a bottle strategy for encapsulation in a zeolite supercage. (Ref. 106).

Photolysis of *o*-methyldibenzyl ketone (*o*-ACOB) and *p*-methyldibenzyl ketone (*p*-ACOB) on pentasil leads to the exclusive formation of decarbonylation products (see Section 3 of Chapter 1 for the details concerning the mechanism of the cleavage of dibenzylketones).[107] However, in the former case AA and BB are the major products, whereas for the latter, AB is the major product. The qualitative difference in products is explained by molecular sieving of the radicals in the case of *o*-ACOB and internal hindrance to diffusion in the case of *p*-ACOB (Scheme 10). Photolysis of *o*-ACOB produces a primary radical pair on the external surface. This pair separates by diffusion and decarbonylates to produce a secondary radical pair, *o*-A• and B•.

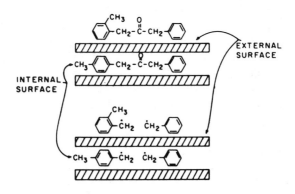

Scheme 10

At this point sieving occurs and the B• radicals enter the internal surface, while the *o*-A• radicals are constrained by size to the external surface. The B• radicals combine to form BB by diffusion within the internal surface and the *o*-A• radicals combine to form *o*-A—*o*-A by diffusion on the external surface. The photolysis of *p*-ACOB produces a primary radical pair on the internal surface. This pair undergoes decarbonylation but is constrained from diffusional separation out of the channel in which it is generated. As a result the *p*-A• and B• radicals combine to give only *p*-A—B. Turro et al. have provided support to their rationale through additional experiments involving trapping experiments (with 2,2,6,6-tetramethylpiperidin-1-oxyl) and blocking experiments (with water). A visualized mechanistic scheme in shown in Scheme 11.

The results observed with *o*-ACOB and 2-phenyl cycloalkanones demonstrate the innovative approaches to synthetic methodologies and mechanistic investigations that one could make using zeolites as the media for reactions. A similar strategy utilized by Turro et al. to achieve selective photochlorination of alkanes using pentasil zeolites is worth noting.[108]

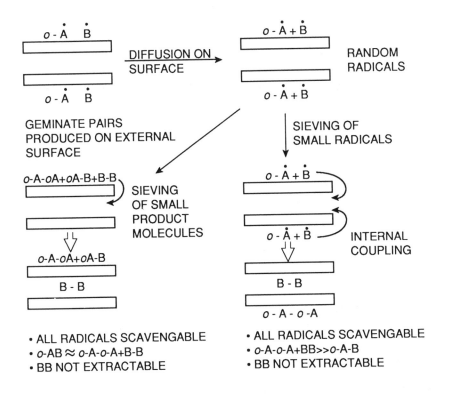

Scheme 11

6.2. Role of Cations

Based on the investigations of various cation-exchanged faujasites as the media for reaction, it has become clear that the cation can have a significant influence on the course of both thermal and photochemical reactions. The origin of the effect has been attributed to several factors: variation in cation size, variation in electrostatic field provided by the cation, or variation in the electrostatic potential of the cation and variation in the spin–orbit coupling parameter. Although the authors attribute the effect to a single factor it is quite likely that many of these effects operate cumulatively in several of the systems examined so far.

6.2.1. Variation in the Size of the Cations

Turro and co-workers have provided several examples related to dibenzylketones wherein the cations have significant influence on the reaction. In all of these cases the variation is attributed solely to variation in the free volume of the supercage with respect to the cation. These examples are discussed in Chapter 1 of this volume by Turro and Garcia-Garibay and will not be repeated here. However, three examples are provided from our own studies,[16,109] which illustrate the importance of the size of the cation (present within a cage along with the organic guest molecule) on the product distribution. Product distribution obtained upon photolysis of benzoin alkyl ethers, α-alkyldeoxybenzoins, and α-alkyldibenzyl ketones is dependent on the cation as summarized in Tables 6 and 7 with one example from each class. Structures of the products and the mechanism of product formation in each case have been discussed earlier (Chapter 7, Section 5.1.3). Scheme 6 of Chapter 7 is reproduced here (Scheme 12) for ready reference.

Table 6: Product distributions upon photolysis of benzoin methyl ether and α-propyldeoxybenzoin within zeolites[a]

Medium	Type I Products		Type II Products	
	Benzil/Pinacol ether	Rearrangement product	Deoxy benzoin	Cyclobutanol
benzoin methyl ether				
Benzene	26/67	1.0	1	7
Li–X	3	77	13	8
Na–X	4	72	10	14
KX	7	48	14	18
Rb–X	5	46	18	22
Cs–X	8	34	17	31
α-propyldeoxybenzoin				
Benzene	5/24	--	54	17
Li–X	--	95	4	1
Na–X	--	88	5	7
K–X	--	48	31	21
Rb–X	--	32	22	45
Cs–X	--	21	27	42

[a] see Scheme 12 for structures.

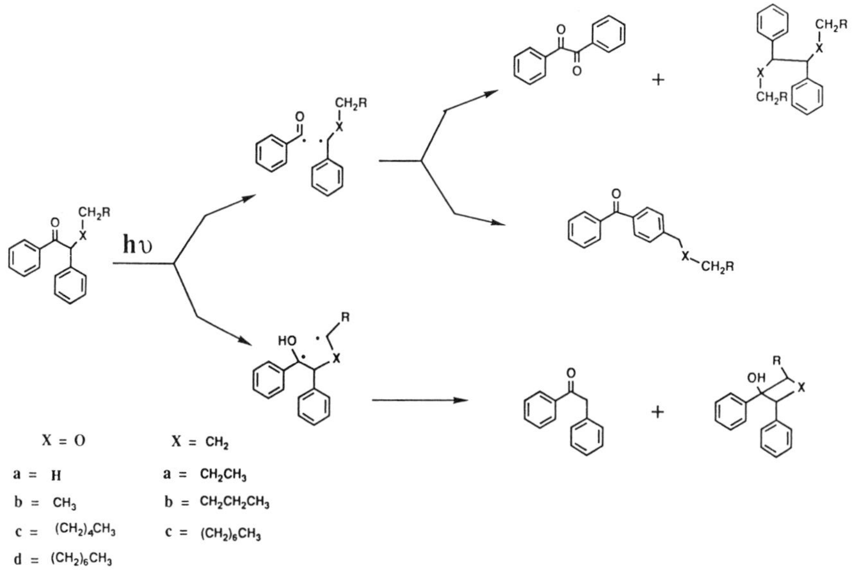

Scheme 12

In solution the termination process of the benzyl radicals derived from α-alkyldibenzyl ketones consists only of the coupling between the two benzylic radicals and results in diphenylalkanes AA, AB and BB in a statistical ratio of 1:2:1. Details of this process is discussed in section 5.1.2 of Chapter 7. Scheme 11 of Chapter 7 provides the structure of products and a mechanism for the formation of these products. This is reproduced in Scheme 13. Within supercages, on the other hand, termination proceeds by both coupling and disproportionation (Table 7). A schematic diagram for the termination processes between the benzylic radicals is shown in Scheme 14. Ramamurthy et al. interpret the preference for disproportionation within the supercage as follows: The association between benzylic radicals which would favor coupling, would be prohibited inside the cavity, especially in the presence of large cations, because of the reduction in free volume. Further, more drastic overall motion would be required to bring benzylic radicals together for head-to-head coupling than to move an alkyl group so that one of its methylene hydrogens would be in a position for abstraction by the benzylic carbon radical. It is logical to expect the radical pair to prefer the pathway of "least volume and motion" when the free space around it is small. Thus, as smaller cations are replaced with larger ones and as shorter alkyl chains are replaced with longer ones, one would indeed expect enhanced yields of olefins as observed in the reported study (Figure 26).

Table 7: Product distribution upon photolysis of α-hexyl dibenzyl ketone within zeolites[a]

Medium	Olefin	(AB)	Rearrangement Product
Li–X	39	17	37
Na–X	19	18	57
K–X	23	29	36
Rb–X	38	23	29
Cs–X	60	15	22

[a] see Scheme 13 for structures

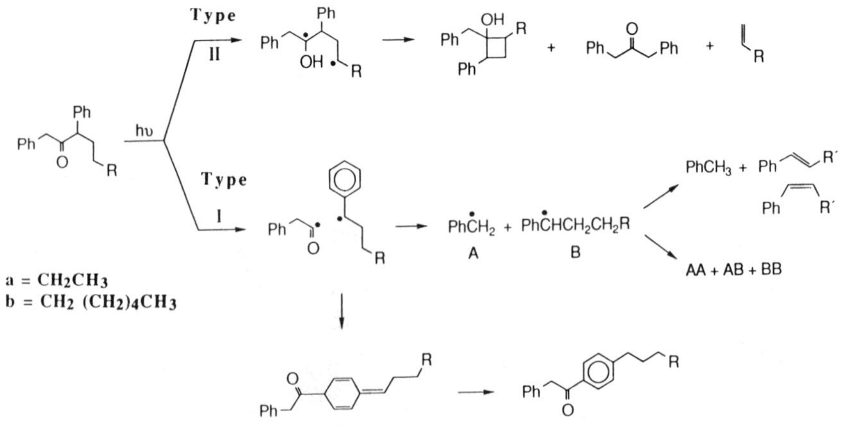

Scheme 13

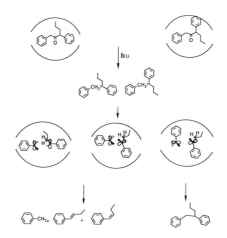

Scheme 14

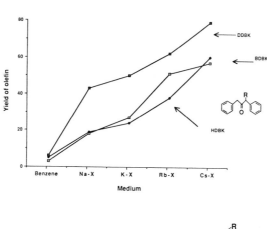

Figure 26. Yield of olefin (disproportionation product) on the cation

The above conclusion is also supported by the pathways undertaken by the primary triplet radical pair (Scheme 13) generated by the α-cleavage of the α-alkyl dibenzylketones and α-alkyl benzoin ethers and deoxybenzoins. Perusal of Tables 6 and 7 reveals that while the rearrangement takes place in all cation-exchanged X and Y zeolites, the yield of the rearrangement product varies depending on the cation (Figure 27). The yield decreases as the cation present in the supercage is changed from Li$^+$ to Cs$^+$. Such a trend is attributed to the decrease in the free space within the supercage.

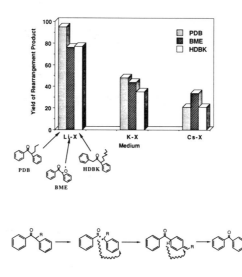

Figure 27. Dependence of rearrangement process on cations: as the cation size increases the yield of of the rearrangement product decreases.

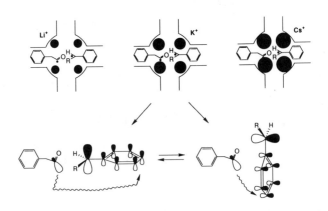

Scheme 15

As the available free space inside the supercage is decreased by the increase in the size of the cation, the translational and rotational motions required for the rearrangement process become increasingly hindered (Scheme 15). Under these conditions, competing paths, such as coupling to yield the starting ketone and decarbonylation, both of which require less motion, dominate.

Examination of Table 6 reveals that in the case of α–methylbenzoin ether as well as in α-propyldeoxybenzoin the C/E ratio, i.e., the ratio of the yield of cyclobutanol, the cyclization (C) product, to that of deoxybenzoin, the elimination product (E) resulting from the 1,4-diradical derived via the Norrish type II γ-hydrogen abstraction process, depends on the cation present in the supercage. Also, the C/E ratio increases as smaller cations are replaced with the larger ones; i.e., C/E increases from Li^+ to Cs^+. The above dependence of the product distribution on the cation can be understood on the basis of the well-understood mechanism of the type II reaction (see Section 3.4 of Chapter 5, Section 5.1 of Chapter 7 and Section 8 of Chapter 14). The enhancement of cyclization within the supercages of X and Y zeolites in the presence of larger cations is believed to reflect the rotational restriction brought on the skewed–transoid–cisoid 1,4-diradical inter-conversion (Scheme 8). Authors propose that as the cation size increases, the 1,4-diradical is forced to adopt a compact geometry due to reduction in the available supercage free volume. Thus, the skewed 1,4-diradical first formed would be encouraged to relax to the cisoid- rather than to the transoid- conformer. Severe constraints would be imposed by the supercage on the cisoid–transoid interconversion and the barrier for the cisoid to transoid conversion would be accentuated. These factors are expected to enhance the yield of cyclobutanol.

6.2.2. Variation in the Electrostatic Potential of the Cation

Adsorption and spectral studies have indicated that if a molecule with π-bond or polar functional groups is adsorbed on zeolites, it interacts strongly with the exchangeable cations. The interaction energy depends on factors such as the type, radius, and number density of the exchangeable cations; type of zeolite; and structure of adsorbed molecules. Singly charged cations such as lithium and sodium show stronger interaction than the larger potassium, rubidium, and cesium cations. Such differences in the strength of interaction has also been invoked to understand the cation-dependent photochemistry of a few organic compounds. These examples are briefly discussed below.

Photolysis of several α,α-dialkylphenyl ketones in benzene and as complexes of zeolite M^+ X and M^+ Y (M = Li, Na, K, Rb, and Cs) gave products resulting from both Norrish type I and type II processes.[110] Results with two typical examples are illustrated in Figure 28. It is clear from the figure that the yield of benzaldehyde, a product of the Norrish type I process, is enhanced significantly within zeolites with respect to benzene. It is also to be noted that Li^+ and Na^+ generally exert larger influences than other cations.

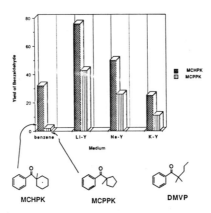

Figure 28. Dependence of α-cleavage products on the cation: Smaller cations yield higher yield of α-cleavage products

This phenomenon has been understood on the basis of the change in the binding ability of the cation with the carbonyl chromophore due to variation in the electrostatic potential of the cation. Although no clue to binding interaction between the cation and the ketones is obtained from the absorption spectra, thermogravimetric analyses indicated that the temperature required to desorb these ketones from Li–X is much higher than from Cs–X; the temperature of desorption decreases in the order Li > Na > K > Rb > Cs. The extent of cation influence on the photoproduct distribution also follows the same trend: the extent of influence decreases from Li to Cs as cations. It is proposed that the cations present within the supercage interact electronically with the carbonyl chromophore and thus, impede the hydrogen abstraction sterically (Scheme 16).

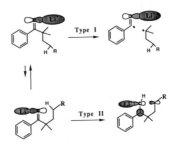

Type II Slowed due to Li⁺ Interaction with Oxygen 'n' Orbital

Scheme 16

The effect is expected to be more with Li⁺ because it binds very strongly. Such a binding would affect the rate of hydrogen abstraction but not the α -cleavage and so would be expected to decrease the efficiency of the type II process alone. Such a binding would not have any effect on the type I process. Therefore, the ratio of the type I to type II products would be expected to increase as the binding strength with the cation increases. This is indeed the case. As mentioned in Section 4.2.4, the triplet lifetime of valerophenone in Li–X is much longer than in other alkali cation-exchanged X zeolites. This is attributed to the reduced reactivity of the ketone towards the Norrish type II process in Li–X.

Another example is provided by the photobehavior of α-alkylbenzoin ethers and α-alkyldeoxybenzoins (Table 6).[111] The behavior of benzoin alkyl ethers and alkyldeoxybenzoins when viewed together provides important additional information (Figure 29). The zeolite cavity induces α-alkyl benzoin ethers to yield products derived via the type II pathway, a minor pathway in benzene.

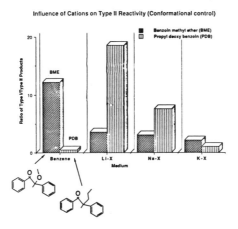

Figure 29. Dependence of type I to type II ratio on the cation: opposite influence on the benzoin ether and deoxybenzoin.

On the other hand, zeolites inhibit α-alkyldeoxybenzoins from proceeding via the type II pathway, which is favored in benzene. This is attributed to the ability of the cation present in the cavity to control the conformation of the included molecules (Scheme 17). The presence of an alkoxy chain in α-alkylbenzoin ethers most likely directs the chelation of the cation to a conformer that is favorable for the type II process. Similarly, in α-alkyldeoxybenzoins, the phenyl ring directs the conformational preference in the cavity. Such a hypothesis is supported by the results on dealuminated zeolite-Y, in which the Si to Al ratio is very high (>550). At very low levels of aluminum, the cation concentration is also low. Therefore, conformational control is expected to be minimal and, indeed, only the type I products dominate the product mixture in both the cases.

Scheme 17

Wada et al.[112] have investigated the photobehavior of iron pentacarbonyl included in a variety of alkali-exchanged Y zeolites. Photocatalytic isomerization of *cis*-but-2-enes was investigated. The activity decreased with the decrease in the electrostatic field of the cation: Li–Y, 27.3 ; Na–Y, 18.7; K–Y, 1.22; Rb–Y, 0.45 and Cs–Y, 0.71 x 10^{-6} mol min^{-1}. This dependence, although the details of the mechanism are yet to be understood, is attributed to the change in the electrostatic field with the cation.   Earlier, a similar study was conducted by Suib et al.[113] with an iron pentacarbonyl and 1-pentene system, and no cation effect was observed. This discrepancy, according to Wada et al., is due to the presence of solvent molecules within the cage, which may weaken the interaction between the cation and the guest iron pentacarbonyl.

6.2.3. Variation in the Spin–Orbit Coupling Effect of the Cation

As discussed at length in Section 4.2.3 cations can be utilized to control the efficiency of the triplet generation from an organic molecule included in the cavities of zeolites.   One example provided below illustrates how such an effect can also be utilized to control product distribution in a photoreaction.

The photobehavior of acenaphthylene is unique in that it has been extensively studied in various constrained media and has been subjected to one of the largest heavy-atom effects on its dimerization.[114] The irradiation of acenaphthylene in solution yields the cis and the trans dimers; the singlet gives predominantly cis dimer, whereas the triplet gives both cis and trans dimers in comparable amounts (Scheme 18). Photolyses of dry solid inclusion complexes of acenaphthylene in various cation (Li, Na, K, Rb)-exchanged Y zeolites gave the cis and trans dimers.[115] Cis to trans dimer ratio, relative efficiency of dimerization, relative triplet yields, and triplet lifetimes of acenaphthylene are dependent on the cation as summarized in Table 8.

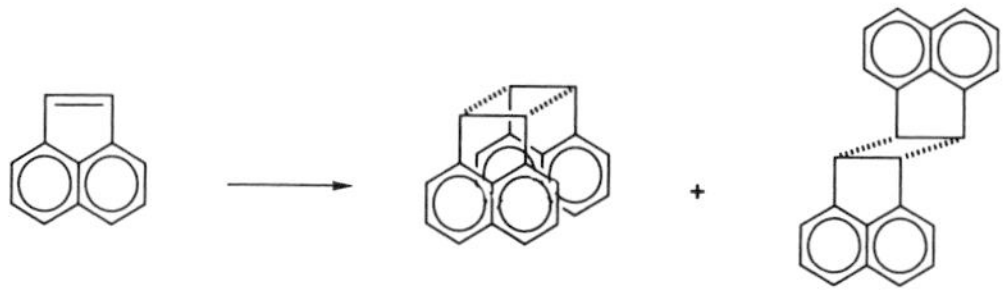

- Excited singlet (S_1) gives essentially *cis* dimer (t/c ~ 0.03)

- Excited Triplet (T_1) gives *trans* and *cis* dimers (t/c ~1.7); slightly solvent polarity dependent

- Poor intersystem crossing from S_1 to T_1 in the absence of external perturbations

Scheme 18

Table 8. Cation Dependent Photodimerization of Acenaphthylene Included in M^+Y Type Zeolites (<S> = 0.5)

Zeolite	Cis/Trans dimer	Relative Efficiency of Dimerization	Relative Triplet Yield	Triplet Lifetime
Li–Y	25	0.2	-	--
Na–Y	25	0.2	-	--
K–Y	2.3	0.4	0.2	9.6 µs
Rb–Y	1.5	1.0	0.5	5.7 µs
Cs–Y	4.2	0.8	0.7	2.1 µs

The absence of triplet formation in Li–Y and Na–Y is consistent with the solution behavior in which the intersystem crossing yield from S_1 to T_1 is reported to be near zero. This lead Ramamurthy et al. to conclude that the dimerization in the supercages of Li–Y and Na–Y is from the excited singlet state. Preferential formation of the cis dimer also supports this conclusion. The high triplet yield in K and Rb–Y is thought to be a consequence of a "heavy-atom effect" caused by the cations present within the supercage. The trends observed in the variation of the triplet yield and the triplet lifetime with the increasing mass of the cation is consistent with the expected spin–orbit-coupling-induced triplet formation. Formation of the trans-dimer (the triplet-derived product) in the cages of K and Rb–Y is in agreement with triplet generation.

The use of cations to effect triplet reactions within the internal structure of zeolite is yet to be fully investigated.

6. Zeolite-Solvent Slurry Photolysis

Photobehavior of organic molecules included in zeolites is routinely investigated only in the dry state. In this section, we illustrate with two examples that the product distribution obtained upon u.v. irradiation of organic molecules included in zeolite-solvent slurries is distinctly different from conventional dry powder photolysis. For more examples, readers are referred to the original article.[116] A typical experiment consisted of the following: Zeolites were activated at 500° C overnight before use. To the guest organic molecule in hexane (or other solvent), the activated zeolite, cooled to room temperature, was added and stirred for about an hour. In general, hexane was found to be the best solvent for zeolite-solvent slurry preparations since total adsorption of the organic molecules on zeolites occurred in hexane slurries. In the other solvents, some amount of guest remained in the bulk solvent. Magnetically stirred translucent zeolite-hexane slurries were photolyzed under a helium atmosphere (to avoid quenching by atmospheric oxygen). Gas chromotagraphic analysis of the hexane portion at the end of irradiation of the slurry revealed no products in hexane layer indicating that the products also remain within zeolites. However, products and unreacted guests were readily extracted (material balance ~90%) into the ether layer by stirring the reaction mixture in diethyl ether for about 10 h.

Product distributions upon photolyses of valerophenone and octanophenone included in Na–X and K–X zeolites in dry state and as zeolite-hexane slurries are provided in terms of a bar graph in Figure 30. Examination of the figure reveals that in zeolite-hexane slurry cyclobutanol formation is significantly enhanced with respect to dry zeolites.

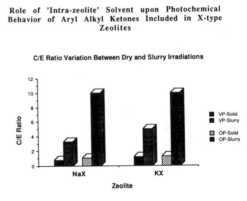

Figure 30. Dependence of cyclization to elimination product ratio on the nature of irradiation: slurry vs. solid state.

Such a trend can be correlated with the preference of the cisoid geometry for the 1,4-diradical with the change in the medium. Preference for the cisoid geometry would be anticipated with the decrease in the reaction cavity volume. Such a preference would be reflected as an increase in cyclization process. One would predict that the type II 1,4-diradical generated within the supercage of zeolite-solvent slurry would be forced to maintain a compact geometry due to reduction in the available supercage free volume. Thus, the first formed skewed 1,4 diradical would be encouraged to relax to the cisoid rather than to the transoid conformer. Severe constraints would be imposed by the "intra-crystalline hexane" on the cisoid-transoid interconversion and the barrier for the cisoid to transoid conversion would be accentuated. Thus the enhanced participation of the cisoid conformer in the transoid-cisoid equilibrium would be expected to give higher yield of cyclobutanol in zeolite-hexane slurry. Results observed in Na–X and K–X - hexane slurries are indeed consistent with the above predictions.

Figure 31 provides the difference in cis to trans dimer ratio between slurry and solid state irradiation of acenaphthylene included in zeolites. Enhanced yield of the cis dimer during slurry irradiation is attributed to the reduced mobility of the acenaphthylene molecules from one cage to the other. The yield of the *trans*-dimer which is believed to be formed by the encounter of acenaphthylene molecules present in singly occupied cages, is expected to be affected much more than that of the cis dimer by the solvent present in cages.

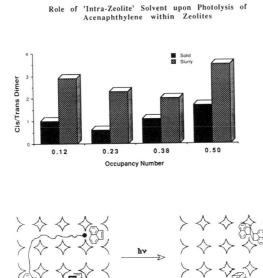

Figure 31. A comparison of the product dimer ratio between slurry and solid state irradiations.

7. Final Remarks

The approaches taken thus far in "Photochemistry in organized media" can be considered under two categories – one in which the photochemical and photophysical tools are utilized to understand the media itself and in the second the media is utilized to modify the photochemical and photophysical behavior of the included guest molecule. In using photochemistry as a tool fairly well understood and generally well established probes are utilized. There is a tremendous need to understand the physical and chemical characteristics of the internal pore structure of zeolites. In the past, use of photochemistry in this area has not been significant. However, one might predict that in the coming years photochemistry will play a significant role as a tool to understand the physical characteristics of zeolites – location, aggregation, mobility and diffusion of guests within zeolites. Competing techniques such as solid state nmr, laser Raman, x-ray- and neutron-diffraction etc., are more complicated, time consuming and less routine.

Attempts to modify the photochemical and photophysical behavior of a guest molecule have involved several strategies. In one of them the rotational and translational motions of a molecule are restricted utilizing a constrained medium. A comparison of the extent of restriction offered by silica and alumina surfaces, clays and zeolites clearly suggest that zeolites will play an increasingly important role in the future as a medium for a reaction. A simplified model of the environment provided by these three media are illustrated in Figure 32. Examination of this figure indicates that only in zeolite matrix restriction on all three dimensions is felt by the guest molecule.

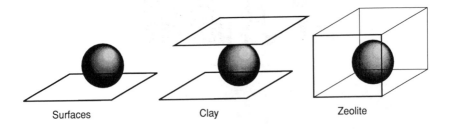

Figure 32. Restriction enforced on the guest by surfaces of silica, clay and zeolites

Selectivity can be enhanced if only one can pick out or convert all the reactive molecules, at the very beginning, toward a particular reactive conformer or geometry. By this process, one will avoid the quantum loss through unproductive processes by the unwanted conformers (those normally present as an equilibrium mixture with the wanted conformer or geometric isomer). Zeolites whose internal pore dimensions are of molecular sizes have been utilized to control the conformer or geometry of an adsorbed molecule. Efforts in this direction which is expected to be scientifically rewarding, will bring out many such examples. In addition to driving the reactive molecule toward a particular equilibrium conformer, it should be possible to hold the molecule in a geometry suitable for a reaction. Such is achieved in solution through complicated setups. Cations present within the zeolite structure can serve as 'pins' or 'coat hangers' to hold a guest molecule in the required geometry (Figure 33). Furthermore, zeolite cavity can become a medium for asymmetric reactions when cations are exchanged with optically active ammonium ions. Concerted efforts are required to bring out these features of zeolites. Advantages offered by zeolites are many when compared to silica and alumina surfaces and therefore it is not a surprise to note that zeolites have been receiving increasing attention recently.

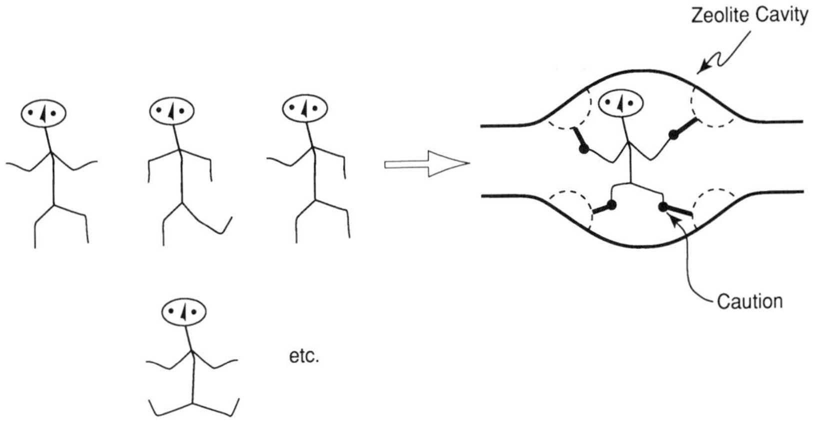

Figure 33. A cartoon representation of the conformational restriction brought on the guest by the zeolite cavity. Cations help to hold the molecule in the correct geometry for the reaction.

When compared to organic hosts described in Chapter 7, zeolites offer several advantages. Zeolites do not absorb in the region where most organic molecules do.   Generally, they do not undergo reaction with the guest molecules. Zeolites with a number of sizes and shapes are commercially available and therefore one has a choice. The internal properties of zeolites can also be fine tuned by the variation of silicon to aluminum ratio, by co-inclusion of solvents or other adsorbents and by the variation of cations. It is hard to be less optimistic about zeolites being an excellent medium for investigating the photoprocesses of both organic and inorganic molecules.

Acknowledgment

It is indeed a pleasure to thank Dr. D. R. Corbin for his help in the preparation of this article. His critical reading, sound advice and useful suggestions are reflected in the article. My own research in this area has benefitted immensely from my association and collaboration with Drs. J. V. Caspar and D. F. Eaton. I have been fortunate to be able to interact periodically with Professor Turro whose global vision has helped me to shape my thoughts in this area.

References

1.   J. Turkevich, *J. Catal. Rev.*, **1967**, *1*, 1.
       P. B. Weisz, *Pure Appl. Chem.*, **1980**, *52*, 2091.
       E. M. Flanigen, *Pure Appl. Chem.*, **1980**, *52*, 2191.
       J. V. Smith, *Chem. Rev.*, **1988**, *88*, 149
       W. Holderich, M. Hesse, F. Naumann, *Angew. Chem., Int. Ed. Engl.*, **1988**, *27*, 226.
       G. Ozin, A. Kuperman, A. Stein, *Angew. Chem., Int. Ed. Engl.*, **1989**, *28*, 359.
       J. Dwyer and A. Dyer, *Chemistry and Industry*, **1984**, 237.
       J. Dwyer, *Chemistry and Industry*, **1984**, 258.
       G. T. Kerr, *Scientific American*, **1989**, 100.
2.   R. M. Barrer in *Inclusion Compounds I*, J. L. Atwood, J. E. D. Davies, and D. D. MacNicol, eds., 1984, Academic Press: London, pp. 191-248.
       W. M. Meier and D. H. Olson in *Atlas of Zeolite Structure Types*, Butterworths: Cambridge, 1987 (Second Revised Edition).
3.   R. M. Dessau, J. L. Schlenker, J. B. Higgins, *Zeolites*, **1990**, *10*, 522.
4.   P. R. Rudolf and C. E. Crowder, *Zeolites*, **1990**, *10*, 163.
       M. E. Davis, C. Montes, P. E. Hathaway, D. L. Hasha, J. M. Garces, *J. Am. Chem. Soc.*, **1989**, *111*, 3919.
5.   The degree of ionicity versus covalency of silicates has been debated and probably should be for zeolites. [see F. Liebau, *Structural Chemistry of Silicates*, Springer-Verlag, Germany, 1985, p 32].
6.   R. D. Shannon, *Acta Crystallogr.*, **1976**, *A 32*, 751.
7.   For example, dealumination of zeolite Y leads to replacement of trivalent aluminum in the framework with the smaller tetravalent silicon. Similarly, replacement of boron for aluminum in ZSM-5 or incorporation of boron into the silicalite framework leads to a significant decrease in the unit cell volume. In both examples, a change in the pore and channel dimensions would be expected. [see R. Szostak, *Molecular Sieves. Principles of Synthesis and Identification*, Van Nostrand, New York, 1989, pp. 223-228, 291-294]
8.   D. R. Corbin, L. Abrams, C. Bonifaz, *J. Catal.*, **1989**, *115*, 420.

9.   D. R. Corbin, L. Abrams, G. A. Jones, M. M. Eddy, W. T. A. Harrison, G. D. Stucky, D. E .Cox, *J. Am. Chem. Soc.,* 1990, *112*, 4821.

10.   D. R. Corbin, W. C. Seidel, L. Abrams, N. Herron, G. D. Stucky, C. A. Tolman., *Inorg. Chem.,* 1985, *24*, 1800.

11.   H. E. Bergna, M. Keane, Jr., D. H. Ralston, G. C. Sonnichsen, L. Abrams, R. D. Shannon, *J. Catal.,* 1989, *115*, 148.

12.   J. B. Parise, T. E. Gier, D. R. Corbin, D. E. Cox, *J. Phys. Chem.,* 1984, *88*, 1635.
    L. B. McCusker, C. Baerlocher in *Proceedings of the Sixth International Zeolite Conference,* D. Olson and A. Bisio, eds., Butterworths, Guildford, U. K., 1984, p. 812.

13.   J. B. Parise, L. Abrams, T. E. Gier, D. R. Corbin, J. D. Jorgensen, E. Prince, *J. Phys. Chem.,* 1984, *88*, 2203.

14.   D. W. Breck, *Zeolite Molecular Sieves: Structure, Chemistry, and Use,* John Wiley and Sons, New York, 1974.

15.   Calculations of polyhedral volumes were performed using a modification of the POLYVOL Program [D. Swanson and R. C. Peterson, *The Canadian Mineralogist,* 1980, *18*(2), 153; D. K. Swanson and R. C. Peterson, "POLYVOL Program Documentation", Virginia Polytechnic Institute, Blacksburg, VA] assuming the radius of the TO_2 unit to be 2.08Å (equivalent to that of quartz).

16.   V. Ramamurthy, D. R. Corbin, D. F. Eaton, *J. Org. Chem.,* 1990, *55*, 5269.

17.   G. T. Kokotailo, S. L. Lawton, D. H. Olson, and W. M. Meier, *Nature 272,* 1978, 437.

18.   C. A. Fyfe, H. Gies, G. T. Kokotailo, C. Pasztor, H. Strobl, D. E. Cox, *J. Am. Chem. Soc.,* 1989, *111*, 2470.

19.   This volume was estimated using a framework model obtained from Beevers assuming a sphere at the intersection and the diameter of a TO2 group to be 2.08 Å

20.   R. M. Barrer, H. Villiger, *Z. Kristallogr.,* 1969, *128*, 352.

21.   W. M. Meier, *Z. Kristallogr.,* 1961, *115*, 439.

22.   J. A. Gard, J. M. Tait, *Acta Cryst.,* B, 1972, 28, 825.

23.   E. Galli, *Rend. Soc. Ital. Mineral. Petrol.,* 1975, 31, 599.

24.   J. M. Newsam, M. M. J. Treacy, W. T. Koetsier, C. B. de Gruyter, *Proc. R. Lond.,* A, 1988, 420, 375.
    J. B. Higgins, R. B. LaPierre, J. L. Schlenker, A. C. Rohrman, J. D. Wood, G. T. Kerr, W. J. Rohrbaugh, *Zeolites,* 1988, *8*, 446.

25.   S. Qiu, W. Pang, Kessler, H., J. -L. Guth, *Zeolites,* 1989, *9*, 440.

26.   A. Dyer, *An Introduction to Zeolite Molecular Sieves,* John Wiley and Sons, Bath, 1988.
    H. van Bekkum, E. M. Flanigen and J. C. Jansen, eds., *Introduction to Zeolite Science and Practice,* Elsevier, Amsterdam, 1991.

27.   A. Germanus, J. Karger, H. Pfeier, N. N. Samulevic, S. P. Zdanov, *Zeolites,* 1985, *5*, 91.

28.   X. Liu, K. K. Iu and J. K. Thomas, *J. Phys. Chem.,* 1989, *93*, 4120.

29.   V. N. Abramov, A. V. Kiselev and V. I. Lygin, *Russ. J. Phys. Chem. Eng. Trans.,* 1963, *37*, 613.
    K. T. Geodakyan, A. V. Kiselev and V. I. Lygin, *Russ. J. Phys. Chem. Eng. Trans.,* 1967, *41*, 227 and 476.
    C. L. Angell and M. V. Howell, *J. Colloid. Sur. Sci.,* 1968, *28*, 279.
    B. Coughlan, W. M. Carroll, P. O'Malley and J. Nunan, *J. Chem. Soc., Faraday Trans. I,* 1981, *77*, 3037.
    A. de Mallman and D. Barthomeuf, Proc. 7th International Zeolite Conference, Y. Murakami, A. Iijima and J. W. Ward, eds., Tokyo, 1986, p. 609.

A. de Mallman and D. Barthomeuf, *Zeolites*, **1988**, *8*, 292.
A. de Mallman and D. Barthomeuf, *J. Chem. Soc. Chem. Commun.*, **1989**, 129.
A. de Mallman, S. Dzwigaj and D. Barthomeuf in *Zeolites: Facts, Figures, Future*, P. A. Jacobs and R. A. van Santen, eds., 1989, Elsevier, Amsterdam, 935.
P. J. O'Malley, *Chem. Phys. Lett.*, **1990**, *166*, 340.
30. J. J. Freeman and M. L. Unland, *J. Catalysis.*, **1978**, *54*, 183.
31. M. L. Unland and J. J. Freeman, *J. Phys. Chem.*, **1978**, *82*, 1036.
M. Primet, E. Garbowski, M. V. Mathieu and B. Imelik, *J. Chem. Soc., Faraday Trans. I*, **1980**, *76*, 1942.
32. H. Lechert and K. P. Wittern, *Ber. Bunsenges. Phys. Chem.*, **1978**, *82*, 1054.
V. Yu. Borovkov, W. K. Hall and V. B. Kazanski, *J. Catalysis.*, **1978**, *51*, 437.
R. Ryoo, S. B. Liu, L. C. de Menorval, K. Takegoshi, B. Chmelka, M. Trecoske and A. Pines, *J. Phys. Chem.*, **1987**, *91*, 6575.
L. C. de Menorval, D. Raftery, S. B. Liu, K. Takegoshi, R. Ryoo and A. Pines, *J. Phys. Chem.*, **1990**, *94*, 27.
33. A. N. Fitch, H. Jobic and A. Renouprez, *J. Chem. Soc., Chem. Commun.*, **1985**, 284.
A. N. Fitch, H. Jobic and A. Renouprez, *J. Phys. Chem.*, **1986**, *90*, 1311.
M. Czjek, T. Vogt and H. Fuess, *Angew. Chem. Int. Ed. Engl.*, **1989**, *28*, 770.
34. A. Renouprez, H. Jobic and R. C. Oberthur, *Zeolites*, **1985**, *5*, 222.
H. Jobic, A. Renouprez, A. N. Fitch and H. J. Lauter, *J. Chem. Soc., Faraday Trans. I*, **1987**, *83*, 3199.
35. G. V. Tsitsishvili and T. G. Andronikashvili in *Moleculr Sieve Zeolites II*, E. M. Flanigen and L. B. Sand, American Chemical Society, Washinton, 1971, p. 216.
W. Basler and H. Lechert, *Ber. Bunsenges. Phys. Chem.*, **1974**, *78*, 667.
A. de Mallmann and D. Barthomeuf, *J. Phys. Chem.*, **1989**, *93*, 5636.
36. J. Sauer and D. Deininger, *Zeolites*, **1982**, *2*, 114.
P. Demontis, S. Yashonath and M. L. Klein, *J. Phys. Chem.*, **1989**, *93*, 5016.
37. D. Barthomeuf, *J. Phys. Chem.*, **1984**, *88*, 42.
38. A. Jentys and J. A. Lercher in *Zeolites as Catalysts, Sorbents and Detergent Builders*, H. G. Karge and J. Weitkamp, eds., Elsevier, Amsterdam, 1989, p. 585.
H. van Koningsveld, F. Tuinstra, H. van Bekkum and J. C. Jansen, *Acta. Cryst.*, **1989**, *B45*, 423.
B. F. Mentzen, *J. Appl. Cryst.*, **1989**, *22*, 100.
H. Thamm, H. G. Jerschkewitz and H. Stach, *Zeolites*, **1988**, *8*, 151.
P. T. Reischman, K. D.Schmitt and D. H. Olson, *J. Phys. Chem.*, **1988**, *92*, 5165.
H. Thamm, *J. Phys. Chem.*, **1988**, *92*, 193.
B. F.Mentzen, *Mat. Res. Bull.*, **1987**, *22*, 337 and 489.
J. C. Taylor, *J. Chem. Soc., Chem. Commun.*, **1987**, 1186.
J. C. Taylor, *Zeolites*, **1987**, *7*, 311.
B. F. Mentzen and F. Vigne-Maeder, *Mat. Res.Bull.*, **1987**, *22*, 309.
39. S. Dzwigaj, A. de Mallmann and D. Barthomeuf, *J. Chem. Soc. Faraday Trans.*, **1990**, *86*, 431.
40. J. M. Newsam, B. G. Silbernagal, A. R. Garcia and R. Hulme, *J. Chem. Soc. Chem. Commun.*, **1987**, 664.
41. H. Jobic, M. Bee and A. Renouprez, *Sur. Sci.*, **1984**, *140*, 307.
42. S. F. Garcia and P. B. Weisz, *J. Catalysis.*, **1990**, *121*, 294.
J. Karger and D. M. Ruthven, *Zeolites*, **1989**, *9*, 267.
43. For a few selected papers:
M. D. Sefcik, *J. Am. Chem. Soc.*, **1979**, *101*, 2164.
W. D. Hoffmann, *Z. Phys. Chemie, Leipzig.*, **1976**, *257*, 315.
H. Lechert, W. Haupt and K. P. Wittern, *J. Catalysis*, **1976**, *43*, 356.
H. Lechert and H. J. Hennig, *Z. Natuurforsch.*, **1974**, *29 A*, 1065.

M. Nagel, D. Michel and D. Geschke, *J. Colloid. Interfac. Sci.*, **1971**, *36*, 254.

J. B. Nagy, E. G. Deroune, H. A. Resing and G. R. Miller, *J. Phys. Chem.*, **1983**, *87*, 833.

44. For recent papers:

B. G. Silbernagel, A. R. Garcia, J. M. Newsam and R. Hulme, *J. Phys. Chem.*, **1989**, *93*, 6506.

R. Eckman and A. Vega, *J. Am. Chem. Soc.*, **1983**, *105*, 4841.

J. Phys. Chem., **1986**, *90*, 4679.

B. Boddenberg and R. Burmeister, *Zeolites*, **1988**, *8*, 488.

R. Burmeister, H. Schwarz and B. Boddenberg, *Ber. Bunsenges. Phys. Chem.*, **1989**, *93*, 1309.

B. Boddenberg, R. Burmeister and G. Spaeth in *Zeolites, as Catalysts, Sorbents and Detergent Builders*, H. G. Karge and J. Weitkamp, eds., Elsevier, Amsterdam, 1989, p. 533.

I. Kustanovich, D. Fraenkel, Z. Luz, S. Vega and H. Zimmermann, *J. Phys. Chem.*, **1988**, *92*, 4134.

I. Kustanovich, H. M. Vieth, Z. Luz and S. Vega, *J. Phys. Chem.*, **1989**, *93*, 7427.

45. H. Jobic, M. Bee and A. J. Dianoux, *J. Chem. Soc., Faraday Trans. I*, **1989**, *85*, 2525.

46. H. Jobic, M. Bee, J. Karger, H. Pfeifer and J. Caro, *J. Chem. Soc., Chem. Commun.*, **1990**, 341.

C. Forste, J. Karger and H. Pfeifer, *J. Am. Chem. Soc.*, **1990**, *112*, 7.

47. B. Zibrowius, J. Caro and H. Pfeifer, *J. Chem. Soc., Faraday Trans. I*, **1988**, *84*, 2347.

48. M. Bulow, J. Caro, B. Rohl-Kuhn and B. Zibrowius in *Zeolites, as Catalysts, Sorbents and Detergent Builders*, H. G. Karge and J. Weitkamp, eds., Elsevier: Amsterdam, 1989, p. 505.

49. V. Ramamurthy and D. F. Eaton, unpublished results.

50. S. L. Suib and A. Kostapapas, *J. Am. Chem. Soc.*, **1984**, *106*, 7705.

51. G. I. Dinesenko, *Zh. Prikl. Spektrosk.*, **1968**, *9*, 307.

52. G. I. Dinesenko and V. A. Lisovenko, *Zh. Prikl. Spektrosk.*, **1971**, *14*, 702.

M. V. Kost, D. T. Taraschenko and A. M. Erimenko, *Zh. Prikl. Spektrosk.*, **1971**, *15*, 333.

M. V. Kost, A. M. Erimenko, M. A. Piontkovskaya and I. E. Neimark, *Teor. Eksp. Khim.*, **1972**, *8*, 396.

G. I. Denesenko, M. A. Piontkovskaya and I. E. Neimark, *Ukr. Khim. Zh.*, **1970**, *36*, 260.

53. K. K. Iu and J. K. Thomas, *Langnuir*, **1990**, *6*, 471.

54. J. A. Incavo and P. K. Dutta, *J. Phys. Chem.*, **1990**, *94*, 3075.

55. B. H. Baretz and N. J. Turro, *J. Photochem.*, **1984**, *24*, 201

56. K. Kalyanasundaram and J. K. Thomas, *J. Phys. Chem.*, **1977**, *81*, 2176.

57. K. Kalyanasundaram and J. K. Thomas, *J. Am. Chem. Soc.*, **1977**, *99*, 2039.

58. V. Ramamurthy, unpublished results.

59. V. Ramamurthy in *Inclusion Phenomena and Molecular Recognition*, J. Atwood, ed., Plenum, New York, 1990, p. 351.

60. H. L. Casal and J. C. Scaiano, *Can. J. Chem.*, **1985**, *63*, 1308.

J. C. Scaiano, H. L. Casal and J. C. Netto-Ferreria, ACS Symposium Series, **1985**, *278*, 211.

F. Wilkinson, C. J. Willsher, H. L. Casal, L. B. Johnston and J. C. Scaiano, *Can. J. Chem.*, **1986**, *64*, 539.

61. G. Kelly, C. J. Willsher, F. Wilkinson, J. C. Netto-Ferreira, A. Olea, D. Weir, L. J. Johnston and J. C. Scaiano, *Can. J. Chem.*, **1990**, *68*, 812.

62.   J. V. Caspar, V. Ramamurthy and D. R. Corbin, *Coordination Chem. Rev.*, **1990**, *97*, 225.
63.   For a review: A. M. Eremenko, *Adsorbtsiya Adsorbenty*, **1980**, *8*, 48.
64.   A. A. Eremenko, F. M. Bobonitz, M. V. Kost, M. A. Pionkovskaya, M. Yu. Sakhnovski and I. E. Niemark, *Opt. Spectrosc.*, **1973**, *35*, 131.
       M. V. Kost, A. M. Eremenko, M. A. Piontkovskaya, I. E. Neimark, *Zh. Prikl. Spektosk.*, **1971**, *14*, 106.
65.   For reviews: J. F. Tanguay and S. L. Suib, *Catal. Rev. Sci. Eng.*, **1987**, *29*, 1.
       G. T. Pott and W. H. J. Stork, *Catal. Rev. Sci. Eng.*, **1975**, *12*, 163.
66.   D. H. Strome and K. Klier in *Adsorption and Catalysis on Oxide Surfaces*, M. Che and G. C. Bond, Eds., Elsevier, Amsterdam, 1985, p. 41.
       D. H. Strome and K. Klier, *ACS Symposium Series*, **1980**, *135*, 155.
67.   S. L. Suib and K. A. Carrado, *Inorg. Chem.*, **1985**, *24*, 200.
68.   B. L. Benedict and A. B. Ellis, *Tetrahedron*, **1987**, *43*, 1625.
69.   F. M. Bobonich, *Zh. Prikl. Spektosk.*, **1978**, *28*, 145.
70.   V. Ramamurthy, J. V. Caspar, D. R. Corbin and D. F. Eaton, *J. Photochem. Photobiol., A: Chemistry*, **1989**, *50*, 157.
71.   S. P. McGlynn, T. Azumi and M. Kinoshita, *Molecular Spectroscopy of the Triplet State*, 1969, Prentice Hall, Englewood Cliffs, N. J.
72.   J. M. Larson and L. R. Sousa, *J. Am. Chem. Soc.*, **1978**, *100*, 1942.
73.   D. S. McClure, *J. Chem. Phys.*, **1949**, *17*, 905.
74.   V. Ramamurthy, J. V. Caspar, D. R. Corbin, B. D. Schlyer and A. H. Maki, *J. Phys. Chem.*, **1990**, *94*, 3391.
75.   V. Ramamurthy, J. V. Caspar and D. R. Corbin, *Tetrahedron Letters*, **1990**, *31*, 1097.
76.   B. Hudson and B. Kohler, *Ann. Rev. Phys. Chem.*, **1974**, *25*, 437.
77.   M. Hepp, V. Ramamurthy, D. R. Corbin and C. Dybowski, unpublished results.
78.   P. Suppan, *J. Photochem. Photobiol. A: Chemistry*, **1990**, *50*, 293.
79.   L. Johnston and V. Ramamurthy, unpublished results.
80.   M. D. Baker, M. M. Olken and G. A. Ozin, *J. Am. Chem. Soc.*, **1988**, *110*, 5709.
81.   T. L. Pettit and M. A. Fox, *J. Phys. Chem.*, **1990**, *90*, 1353.
82.   J. V. Caspar and V. Ramamurthy, unpublished results.
83.   K. Toriyama, K. Nunome and M. Iwasaki, *J. Am. Chem. Soc.*, **1987**, *109*, 4496.
84.   X. Z. Qin and A. D. Trifunac, *J. Phys. Chem.*, **1990**, *94*, 4751.
85.   P. H. Kasai and R. J. Bishop, *J. Phys. Chem.*, **1973**, *77*, 2308.
       M. R. Harrison, P. P. Edwards, J. Klinowski and J. Thomas, *J. Solid State Chem.*, **1984**, *54*, 330.
       K. B. Yoon and J. K. Kochi, *J. Chem. Soc., Chem. Commun.*, **1988**, 510.
86.   V. Ramaurthy, J. V. Caspar and D. R. Corbin, *J. Am. Chem. Soc.*, **1991**, *113*, 594.
       J. V. Caspar, V. Ramaurthy and D. R. Corbin, *J. Am. Chem. Soc.*, **1991**, *113*, 600.
87.   P. K. Dutta and J. A. Incavo, *J. Phys. Chem.*, **1987**, *91*, 4443.
88.   J. S. Kruger, J. A. Mayer and T. E. Mallouk, *J. Am. Chem. Soc.*, **1988**, *110*, 8232.
       J. S. Kruger, C. Lai, Z. Li, J. A. Mayer and T. E. Mallouk in *Inclusion Phenomena and Molecular Recognition*, J. Atwood, ed., 1990, Plenum, New York, p. 365.
89.   K. B. Yoon and J. K. Kochi, *J. Am. Chem. Soc.*, **1989**, *111*, 1128.
90.   S. Sanakararaman, K. B. Yoon, T. Yabe and J. K. Kochi, *J. Am. Chem. Soc.*, **1991**, *113*, 1419.
91.   S. Ghosh and N. L. Bauld, *J. Catal.*, **1985**, *95*, 300.
92.   K. Lorenz and N. L. Bauld, *J. Catal.*, **1985**, *95*, 613.

93. P. A. Jacobs, J. B. Uytterhoven and H. K. Beyer, *J. Chem. Soc., Chem. Commun.*, **1977**, 128.
94. S. Leutwyler and E. Schumacher, *Chimia*, **1977**, *31*, 475.
95. M. A. Fox and T. L. Pettit, *Langmuir*, **1989**, *5*, 1056.
96. L. Persaud. A. J. Bard, A. Campion, M. A. Fox, T. E. Mallouk, S. E. Webber and J. M. White, *J. Am. Chem. Soc.*, **1987**, *109*, 7309.
97. G. A. Ozin, F. Hugues, S. M. Mattar and D. F. McIntosh, *J. Phys. Chem.*, **1983**, *87*, 3445.
   G. A. Ozin and F. Hugues, *J. Phys. Chem.*, **1982**, *86*, 5174.
98. S. L. Suib, J. F. Tanguay and M. L. Occelli, *J. Am. Chem. Soc.*, **1986**, *108*, 6972.
99. V. Ramamurthy, J. V. Caspar, D. R. Corbin, D. F. Eaton, J. S. Kauffman and C. Dybowski, *J. Photochem. Photobiol. A, Chem.*, **1990**, *51*, 259.
100. (a) L. A. Heimbrook, B. E. Kohler and T. A. Spiglanin, *Proc. Nat. Acad. Sci. U.S.A.*, **1983**, *80*, 4580.
   (b) C. D. DeBoer and R. H. Schlessinger, *J. Am. Chem. Soc.*, **1968**, *90*, 803.
   J. Saltiel, O. C. Zafiriou, E. D. Megarity and A. A. Lamola, *J. Am. Chem. Soc.*, **1968**, *90*, 4759.
101. F. Gessner, A. Olea, J. H. Lobaugh, L. J. Johnston and J. C. Scaiano, *J. Org. Chem.*, **1989**, *54*, 259.
102. N. J. Turro and P. Wan, *Tetrahedron Lett.*, **1984**, *25*, 3655.
   V. Ramamurthy, D. R. Corbin and D. F. Eaton, *J. Chem. Soc., Chem. Commun.*, **1989**, 1213.
103. V. Ramamurthy and L. J. Johnston, unpublished results.
104. N. J. Turro and P. Wan, *J. Am. Chem. Soc.*, **1985**, *107*, 678.
105. N. J. Turro, *Pure Appl. Chem.*, **1986**, *58*, 1219.
   N. J. Turro in *Molecular Dynamics in Restricted Geometries*, J. Klafter and J. M. Drake, ed., 1989, John Wiley, New York, p. 387.
106. X.G. Lei, C. E. Doubleday, M. B. Zimmt and N. J. Turro, *J. Am. Chem. Soc.*, **1986**, *108*, 2444.
107. N. J. Turro, X. G. Lei, C. C. Cheng, D. R. Corbin and L. Abrams, *J. Am. Chem. Soc.*, **1985**, *107*, 5824.
   N. J. Turro, C. C. Cheng, L. Abrams and D. R. Corbin, *J. Am. Chem. Soc.*, **1987**, *109*, 2449.
108. N. J. Turro, J. R. Fehlner, D. P. Hessler, K. M. Welsh, W. Ruderman, D. Firnberg and A. M. Braun, *J. Org. Chem.*, **1988**, *53*, 3731.
109. D. R. Corbin, D. F. Eaton and V. Ramamurthy, *J. Org. Chem.*, **1988**, *53*, 5384.
   V. Ramamurthy, D. R. Corbin, D. F. Eaton and N. J. Turro, *Terahedron Lett.*, **1989**, *30*, 5833.
110. V. Ramamurthy, D. R. Corbin, N. J. Turro and Y. Sato, *Tetrahedron Lett.*, **1989**, *30*, 5829.
111. D. R. Corbin, D. F. Eaton and V. Ramamurthy, *J. Am. Chem. Soc.*, **1988**, *110*, 4848.
112. Y. Wada, Y. Yoshizawa and A. Morikawa, *J. Chem. Soc., Chem. Commun.*, **1990**, 319.
113 S. L. Suib, A. Kostapapas, K. C. McMohan, J. C. Baxter and A. M. Winiecki, *Inorg. Chem.*, **1985**, *24*, 858.
114. D. O. Cowan and R. L. Drisko, *Elements of Organic Photochemistry*, 1976, Plenum, New York, p. 435.
115. V. Ramamurthy, D. R. Corbin, C. V. Kumar and N. J. Turro, *Tetrahedron Lett.*, **1990**, *31*, 47.
116. V. Ramamurthy, D. R. Corbin, N. J. Turro, Z. Zhang and M. A. Garcia-Garibay, *J. Org. Chem.*, **1991**, *56*, 255.

Chapter 11

Phototransformations on Reactive Surfaces: Semiconductor Photoinduced Organic Reactions

Hussain Al-Ekabi

Nulite, A Division of Nutech Energy Systems Inc., London, Ontario, Canada.

Contents

1. Introduction
1.1. Photosensitized Electron-Transfer Organic Reactions

Photosensitized electron-transfer organic reactions in homogeneous solutions have received considerable attention over the past two decades.[1] Such electron-transfer processes may be induced by exciting the donor, the acceptor, or the donor–acceptor charge-transfer complex. The feasibility of photochemical electron transfer in polar solvents, shown in equation (1), can be predicted on the basis of the well-known Weller equation (eq. 2):[2,3]

$$A + D \xrightarrow{ h\upsilon } A^{-\bullet} + B^{+\bullet} \qquad (1)$$

$$\Delta G^{\circ} = e\,\{E^{\circ}(D+) - E^{\circ}(A)\} - U^{*} - e^{2}/(4\pi\varepsilon_{o}\varepsilon a) \qquad (2)$$

where ΔG° is the standard Gibbs energy change, $E^{\circ}(D^{+})$ and $E^{\circ}(A)$ are the standard reduction potentials of the oxidized donor D^{+} and the acceptor A, and $e^{2}/(4\pi\varepsilon_{o}\varepsilon a)$ is the energy gained by bringing the two radical ions to the encounter distance a in a solvent of dielectric constant ε. U^{*} is the singlet–singlet excitation energy of the chromophore.

The energy stored in the radical ion pair is given by the difference between the oxidation potential of the donor and the reduction potential of the acceptor. When the excitation energy (singlet or triplet, depending on the multiplicity of the reacting species) of the sensitizer (donor, acceptor, or donor–acceptor complex) exceeds the energy stored in the pair by a few kcal mol^{-1}, the electron-transfer reaction occurs at a diffusion-controlled rate. The reaction proceeds thereafter via a radical-ion mechanism. This is, however, not always the case. When the energy stored in the radical-ion pair is higher than the triplet energy of one of the reactants, recombination can lead to the triplet state of this reactant.[4-6] Consequently, the reaction will not be different from the triplet sensitization of the reactant. In addition, in some photosensitized electron-transfer reactions the sensitizer and the substrate form an addition product.[7-9] Most of these reactions proceed by proton transfer from the donor radical cation to the acceptor radical anion followed by coupling of the donor radical to the acceptor radical. Both of these complications can be avoided by the use of semiconductor photocatalysts which act only to form the radical ion of the substrate. It is the aim of this chapter to focus on semiconductor-photoinduced organic reactions.

1.2. Semiconductor Photocatalysts

A catalyst, by definition, is a substance that accelerates the reaction rate or changes the reaction path without being consumed as a reactant. This definition is applicable for both a thermal catalyst and a photocatalyst even though they are activated by different energies. A thermal catalyst is activated by heat, while a photocatalyst is activated by photons. The thermal energy (kT) required for activation of a thermal catalyst ranges from 0.03 to 0.1 eV. Accordingly, only reactions with negative free energy change (ΔG) can be catalyzed. In contrast, the photon energy required to activate a photocatalyst ranges from 1 to 4 eV and thus not only reactions with a negative ΔG value can be catalyzed but also uphill reactions ($\Delta G > 0$) can be driven under mild conditions (see Section 2.6). This chapter deals with semiconductors as photocatalysts for organic reactions. Organic photochemical reactions, in conventional photochemistry, are usually initiated either by direct excitation or indirectly by sensitization of the organic reactant. In both cases, excitation of the organic molecule follows promotion of an electron from a filled (HOMO) to an unfilled (LUMO) molecular orbital. The excited states so produced are more powerful oxidants or reductants than the ground states of the molecules. In other words, an electron is more easily removed from the excited state since the excited electron has the greatest energy, and it is easier to add an electron to the excited state since the "hole" generated in the highest occupied molecular orbital is the lowest in energy.

Generally, the same is true of excitation of semiconductors. Here, the filled and empty molecular orbitals are well separated by the bandgap of the semiconductor, which consists of a set of closely spaced, filled molecular orbitals termed the valence band and a set of closely spaced, empty molecular orbitals termed the conduction band. Accordingly, light of energy equal to or greater than the bandgap energy is required to promote an electron from the valence band to the conduction band (e^-_{CB}) leaving an electron deficiency or "hole" in the valence band (h^+_{VB}). Electrons promoted into the conduction band are considered mobile. Both e^-_{CB} and h^+_{VB} are able to participate in redox reactions. By convention, for semiconductors where the majority charge carriers are negative electrons, the material is referred to as "n-type," and where the majority charge carriers are positive "holes" the material is referred to as "p-type."

Just as the free energy of a redox couple in solution is described by its redox potential (E_{redox}) so is the free energy of the electrons in a semiconductor described by the Fermi level (E_F°). The Fermi level of an intrinsic semiconductor lies at exactly the median potential between the highest lying orbital of the valence band and the lowest lying orbital of the conduction band and can be adjusted by doping. In doped n-type semiconductors the Fermi level will move from the median position of an undoped semiconductor to a value just positive of the conduction band edge.

One of the major practical problems in the use of semiconductors as photocatalysts is the rapid electron–hole recombination process. The kinetic

barrier for this process is very low and the overall result of its occurrence is merely
the generation of heat. However, when a semiconductor is in contact with a redox
couple, an energy equilibrium is set up in which the Fermi level ($E_F°$) becomes
equal to the potential of the redox couple (E_{redox}). Equilibrium is reached by
electron-transfer from the phase of higher potential to that at the lower potential.
The electron-transfer produces an electric field described as "band bending." One
can consider a situation, where the electrons are directed toward the bulk of the
semiconductor while the "holes" are directed toward the surface, where oxidation
can occur. This phenomenon has been described by Gerischer,[10,11] Bard,[12-14] and
Wrighton[15] and many others. The reader is referred to these contributions for
detailed descriptions of semiconductor band structure.

Band bending thus provides a route for electron–hole pair separation, since it
could retard the back transfer of the electron from the conduction band to the
valence band. There is, therefore, a limited amount of time for a redox reaction to
occur, but this time is short. Typically, the lifetimes of excited semiconductors
vary between picoseconds and nanoseconds. A redox reaction will occur if it can
compete effectively with electron–hole pair recombination. Charge carrier
recombination can be diminished, however, by scavenging either the e^-_{CB} by
suitable surface-adsorbed electron acceptors or the h^+_{VB} by suitable surface-adsorbed
electron donors. This diminishes the extent of electron–hole recombination,
thereby increasing the potential photoactivity of the semiconductor toward redox
reactions. Figure 1 shows the energetics for an n-type semiconductor before and
after immersion in a redox solution: E_{VB} denotes the top of the valence band,
which can be considered the position of the highest filled molecular orbitals, and
E_{CB} denotes the bottom of the conduction band, which is considered the lowest
unfilled molecular orbitals. The valence band and the conduction band are separated
by the bandgap, E_{BG}, and light equal to or greater than the bandgap energy is
required to excite electrons from the valence band to the conduction band.

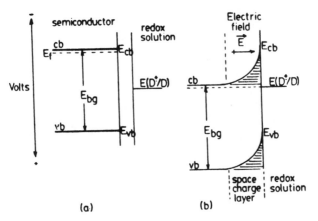

Figure 1. Band structure in an n-type semiconductor, (a) before contact with a redox
solution; (b) after equilibrium with a redox solution.

The potential that a semiconductor has for oxidation of an organic substrate is determined by the position of the valence band and the oxidation potential of the organic substrate with respect to a standard electrode. Likewise, reduction of an organic substrate is determined by the position of the conduction band and the reduction potential of the organic substrate with respect to a standard electrode. The redox reactions will occur when the electron-transfer processes are thermodynamically possible. Figure 2 shows the oxidation of a substrate D and the reduction of a substrate A by excited cadmium sulfide. The bottom of the conduction band, where the photogenerated electron is to be found, is at -0.8 V and the top of the valence band, where the photoinduced "hole" is to be found, is at +1.6 V, with respect to a standard calomel electrode (SCE). If the oxidation potential of the organic substrate D is more negative than the valence band it may transfer an electron to the valence band, giving the cation radical $D^{+\bullet}$. Similarly, if the reduction potential of the organic substrate A is more positive than the conduction band it may accept an electron from the conduction band, giving the anion radical $A^{-\bullet}$. If the subsequent reactions of the cation radical $D^{+\bullet}$ and the anion radical $A^{-\bullet}$ are faster than back electron-transfer, products are formed. Typical values for the band positions in a number of commonly used semiconductors are listed in Table 1. The oxidation potentials of most of the organic substrates treated in this chapter are collected in Table 2.

Table 1. Band Positions for the Most Commonly Used n-Type Semiconductors in Acetonitrile[a]

Semiconductor	Valence Band (V vs. SCE)	Conduction Band (V vs. SCE
CdS^{b}	1.6	-0.80
ZnO^{b}	2.4	-0.80
$TiO_2^{c,d}$	2.2	-1.20

[a] The band positions for most of semiconductors have been determined in aqueous solutions at various pH values. Since this chapter deals with organic reactions in organic solvents, only semiconductors whose band positions have been determined in acetonitrile are presented in this table. The actual positions of the valence and conduction bands in methylene dichloride are not known; however many reactions discussed in this chapter were conducted in CH_2Cl_2, and so it is assumed here that the positions of these bands in CH_2Cl_2 and in CH_3CN are the same.

[b] Ref. 16.

[c] Ref. 17.

[d] Anatase.

Table 2. Oxidation Potentials of Some Organic Compounds Used in This Chapter

Compound	Oxidation Potential, $E_{1/2}$ V vs. SCE (CH_3CN)	Ref.
cis-Cinnamonitrile	2.21	18
trans-Cinnamonitrile	2.21	18
cis-1,2-Di-*p*-Anisylcyclopropane	0.95	18
trans,1,2-Di-*p*-Anisylcyclopropane	0.85	18
1,1-Di-*p*-Anisylethylene	0.78[a]	19
Dianthracene	1.55	20
1,4-Dimethoxybenzene	1.35	21
1,1-Diphenylethylene	1.80	19
Hexamethylbenzene	1.62	22
Hexamethyl(Dewar)benzene	1.58	22
cis-Methylcinnamate	2.21	18
trans-Methylcinnamate	2.21	18
Norbornadiene	1.49	23
Phenyl vinyl ether	1.28[a]	24
Quadricyclane	0.91	23
cis-Stilbene	1.63	25
trans-Stilbene	1.49	26
1,2,4,5-Tetramethoxybenzene	0.81	27
1,2,4-Trimethoxybenzene	1.12	21
1,3,5-Trimethoxybenzene	1.49	21
N-Vinyl carbazole	0.08[a]	28

[a] vs. Ag/Ag$^+$.

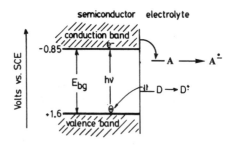

Figure 2. Donation of an electron from an organic substrate D to the valence band and the acceptance of an electron by an organic substrate from the conduction band of CdS.

2. Semiconductor-Photoinduced Organic Reactions

Various organic reactions have been photocatalyzed by semiconductors. Semiconductors have been used to initiate organic reactions in both gaseous and liquid phases. Gas-phase photooxidation of various organic compounds mediated by illuminated semiconductors has been reviewed recently.[29] This chapter is restricted to organic reactions conducted in solutions.

The major events, currently accepted, of the semiconductor-photocatalyzed reaction are outlined in Scheme 1. This involves exciton generation (SC^*), followed by hole–electron separation.

$$SC + h\nu \longrightarrow SC^*$$

$$SC^* \longrightarrow SC(h^+, e^-)$$

$$SC(h^+, e^-) + A \longrightarrow SC(h^+) + A^{-\bullet}$$

$$SC(h^+, e^-) + O_2 \longrightarrow SC(h^+) + O_2^{-\bullet}$$

$$SC(h^+) + D \longrightarrow SC + D^{+\bullet}$$

Scheme 1

After migration of these charge centers to the surface of the irradiated semiconductor, capture of the photogenerated electron by suitable electron acceptor A (or O_2) adsorbed on the surface and/or the photogenerated "hole" by suitable electron donor D adsorbed on the surface can occur. The radical ions so produced then undergo fast chemical reaction. The processes outlined in Scheme 1 appear applicable to all the semiconductor-photocatalyzed organic reactions.

2.1. Geometrical Isomerization

Semiconductors have been used in catalyzing various geometrical isomerization reactions.[30-40] Six substituted styrenes (**1** to **6**) have been shown to achieve thermodynamic equilibrium on irradiated CdS.[30,31] Table 3 summarizes these results.

The CdS-photoinduced isomerization of 4-substituted *cis*-stilbenes has been examined.[33,34] The Hammett plot with donor and acceptor substituents gave a sharp break at $\sigma^+ = -0.19$, indicating a change in the rate-determining step with a change of the electron-donating ability of the substituents. Exergonic and Endergonic electron-transfer processes were proposed to rationalize these results. While the electron-transfer to the "hole" on the CdS valence band must be endergonic in *cis*-stilbenes with electron acceptor substituents, the process is expected to be exergonic with electron donor substituents.

Olefin	R^1	R^2	R^3
1	H	H	Ph
2	H	H	COMe
3	H	CN	Ph
4	OMe	H	Me
5	H	H	COOMe
6	H	H	CN

Table 3. Isomeric Composition of the Thermodynamic Cis–Trans Equilibria of **1 – 6** Initiated by CdS, Iodine[a] or Heat[b]

	Cis %			Trans %		
Compound	CdS	Iodine	Thermal	CdS	Iodine	Thermal
1	<1	<1	4	>99	>99	96[c]
2[d]	<1	<1			>99	>99
3	<1	2		98	98	
4	3	3		97	97	
5	<1	<1		>99	>99	
6	25		37[e]	75		63

[a] Equimolar solutions (0.01 M) of olefin and iodine in methylene dichloride were irradiated at $\lambda > 430$ nm with a 1-kW Xe lamp.

[b] Reproduced with permission from Ref. 31.

[c] the thermodynamic equilibrium of *cis*- and *trans*-1 has been estimated in the liquid phase at 200°C (ref. 41).

[d] Irradiation at $\lambda > 460$nm.

[e] The thermodynamic equilibrium of *cis*- and *trans*-6 has been estimated in the gas phase at 352°C (ref. 42).

The cis–trans isomerization of simple olefins can also be induced by CdS or ZnS dispersions.[32] The reactions occur efficiently and the photostationary states, in most cases, are identical with the thermodynamic equilibria achieved by diphenyl sulfide (Table 4).

The cis–trans isomerization of olefins occurs via two possible mechanisms: (1) isomerization via the radical cation of the olefins and/or (2) isomerization via a sulfur radical originated from sulfur surface states on CdS or ZnS. The isomerization via cation radical is possible when the electron-transfer from the olefin to the photogenerated hole is thermodynamically favorable. For CdS this means that the oxidation potential of the olefin should be less positive than 1.6 eV, the potential of the valence band. This is, at least, the case for stilbene (where the oxidation potentials for *cis*- and *trans*-stilbene are 1.57 and 1.43 V vs. SCE, respectively) and anithole 4 (where the oxidation potentials for both *cis*- and *trans*-anithole are 1.26 V vs. SCE).[18]

Table 4. Thermodynamic Equilibria of Olefins Induced by Semiconductors and Diphenyl Sulfide[a]

Olefin[b]	ZnS (Sol)[c]	CdS (Sol)[c]	PhSPh[d]
2-Pentene	0.29	0.29	0.29
2-Hexene	0.29	0.29	0.29
3-Hexen-1-ol	0.22	0.39	0.23
3 Hexene	0.17	0.35	0.22
2-Octene	0.29	0.30	0.30
Methyloleate	0.20	0.44	0.22

[a] Reproduced with permission from Ref. 32.

[b] Concentration 0.05 M.

[c] For 2 mL of methanol solution in the presence of 0.04 mM of the solutions irradiated at λ 313 nm.

[d] Irradiation of a cyclohexane or benzene solution at λ 313 nm.

One mode of demonstrating the involvement of cation-radical intermediates in a chemical reaction is by quenching the reaction with methoxybenzenes which are good electron donors. 1,2,4-Trimethoxybenzene ($E^{ox}_{1/2}$ = 1.12 V vs. SCE) quenches the CdS-mediated *cis*-stilbene isomerization reaction very efficiently even at very low concentrations[31,34,35] by donating an electron to the cation-radical of *cis*-stilbene. By contrast, similar concentrations of 1,2,4-trimethoxybenzene quenches the isomerization of cinnamonitrile inefficiently.[35] More direct evidence for the involvement of a cation-radical intermediate can be obtained using a flash spectroscopic technique. When a colloidal suspension of TiO_2 or CdS is employed the optically transparent medium allows for direct characterization of transient intermediates on the microsecond–nanosecond time scale. The formation of *trans*-stilbene cation radical has been demonstrated when a colloidal suspension of TiO_2 containing *trans*-stilbene is flashed with a neodymium YAG laser.[43]

The second route for isomerization of olefins becomes important when the electron-transfer from the olefin to the photogenerated "hole" is unfavorable on the CdS or ZnS particle surface. Here, the reaction may proceed via the mechanism suggested in Figure 3 in which the photoformed radical, by an addition–elimination process, permits isomerization at the double bond.[32,36,38-40] Based on XPS examination, recent study has suggested, that the supersulfide radical ion ($S_2^{-\bullet}$) produced by "hole" trapping of the disulfide ion is the active species.[35] The isomerization of **5** and **6** and the simple olefins presented in Table 4 may proceed via this route.

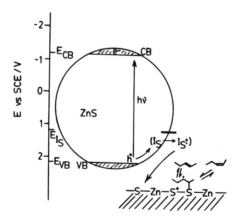

Figure 3. The cis-trans isomerization of simple olefins mediated by a sulphur radical originated from sulphur surface states on CdS or ZnS. (Reproduced with permission from Ref. 32).

The cis–trans isomerization of 1,2-diarylcyclopropane **7** to **8** can also be induced by illuminated CdS.[37]

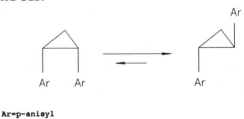

Ar=p-anisyl

      7   8

Approached from both sides, the equilibrium composition for the photocatalytic isomerization of 1,2-di-*p*-anisylcyclopropane is 95% trans, 5% cis. This composition resembles that expected for a thermodynamic equilibrium between the cis and trans closed-shell isomers. Based on quenching studies with methoxy benzenes, a cation-radical mechanism is proposed for this reaction.

2.2. Valence Isomerization

The rearrangement of highly strained hydrocarbons to their more stable valence isomers can be induced by illuminated semiconductor suspensions.[44,45] For example, the CdS- or ZnO- catalyzed valence isomerization of the strained cage molecule **9** gave the corresponding diene **10**. The reaction is very inefficient in the presence of 1,2,4,5-tetramethoxybenzene. A cation-radical mechanism is proposed to rationalize this observation.[44] The reaction is also inefficient in the absence of oxygen.

9   10

The CdS, TiO$_2$, and ZnO mediated valence isomerizations of hexamethyl-(Dewar)benzene (**11**) to hexamethylbenzene (**12**) has also been explored.[46,47] As in **9**, the presence of oxygen is necessary for the reaction to proceed efficiently.

Table 5. Quantum Yields of Valence Isomerization of **11** to **12** photoinduced by CdS, TiO$_2$, and ZnO at two different concentrations of **11**.[a,b]

Semiconductor	[11] M	Quantum yield ($\pm 10\%$)
CdS	0.13	0.29
	0.57	0.68
TiO$_2$	0.13	0.84
	0.57	1.23
ZnO	0.13	0.48
	0.57	1.23

[a] Solutions of **11** in CH$_2$Cl$_2$ were irradiated for 30 min at λ 320–390 nm (Corning filter CS,7-60) in the presence of 50 mg of the semiconductor, in air.
[b] Reproduced with permission from Ref. 47.

11   12

Also 1,2,4-trimethoxybenzene and 1,2,4,5-tetramethoxybenzene quench the reaction efficiently. A cation-radical chain mechanism, in which the electron-transfer from **11** to the cation radical **12$^{+\cdot}$** is the key step, is implicated since quantum yields greater than one are obtained[47] in the TiO_2- and ZnO-mediated reactions (Table 5).

Quadricyclane (**13**) rearranges to norbornadiene (**14**) over illuminated semiconductors.[48,49] The isomerization of **13** to **14**, via a radical-cation mechanism is well known in homogeneous solution.[23,50-55] It has been established that **14$^{+\cdot}$** is of lower energy than **13$^{+\cdot}$** and the radical-cation rearrangement proceeds as in Scheme 2.

13   13$^{+\cdot}$ 14$^{+\cdot}$ 14

It should be mentioned that the valence isomerizations of **9** to **10**, **11** to **12**, and **13** to **14** can also be regarded as cycloreversions.

2.3. Dimerization

The [2+2]-cyclodimerization of phenyl vinyl ether (**15**) to *trans*-cyclobutane (**16**) and *cis*-cyclobutane (**17**) on irradiated semiconductors such as ZnO, TiO_2, and CdS has been demonstrated.[56,57] The ratio of **16:17** (ca. O.52) at the concentration of **15** used (ca. O.22 M) is the same regardless of the semiconductor used. The reaction is efficiently quenched by 1,3,5-trimethoxybenzene and 1,2,4,5-tetramethoxybenzene.

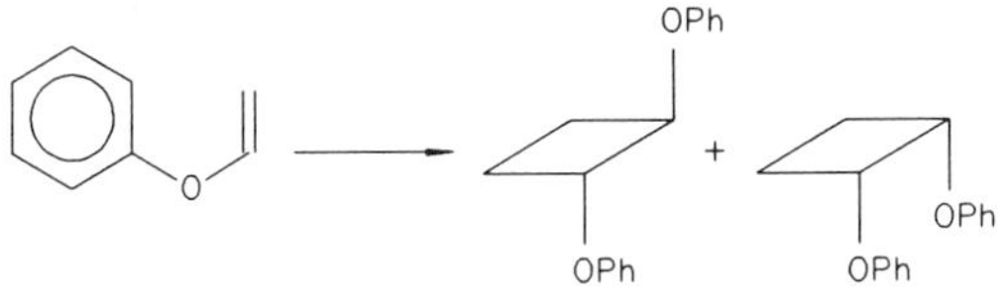

15	**16**	**17**

The thermodynamic stability of the two dimers determines their isomeric composition. The rates of ring opening and ring closure of the two isomers can be influenced by their thermodynamic stabilities. Since the trans dimer **16** is thermodynamically more stable than the cis dimer **17**, ring opening of **17** should be more efficient than that of **16**.[1] The stability of dimers **16** and **17** has been tested by irradiating a mixture (**16:17** = 0.59) in methylene dichloride over the concentration range 6 x 10^{-4} – 10^{-1} M in the presence of CdS or ZnO.[56] While no change occurs in the total amount of dimers after 12 h of irradiation, at low concentration (<10^{-3} M), some conversion of **17** to **16** has been observed with an increase in the **16:17** ratio to 0.77. Scheme 3 demonstrates these processes.

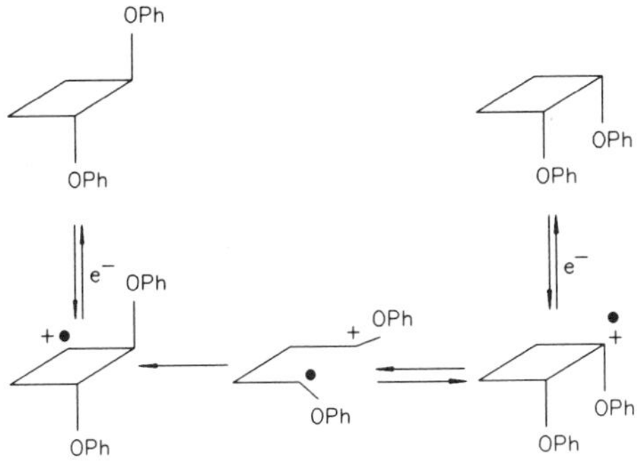

Scheme 3

In contrast to the dimerization of **15**, the CdS-mediated dimerization of *N*-vinyl carbazole **18** leads exclusively to the *trans*-cyclobutane dimer **19**.[58] Both 1,2,4-trimethoxybenzene and 1,2,4,5-tetramethoxybenzene quench the reaction efficiently. Dimerization in homogeneous solution by electron-transfer photosensitization has been reported for **15**[59-63] and **18**.[28,64-67] Scheme 4 summarizes the most recent view of reaction mechanism in both homogeneous and heterogeneous systems.

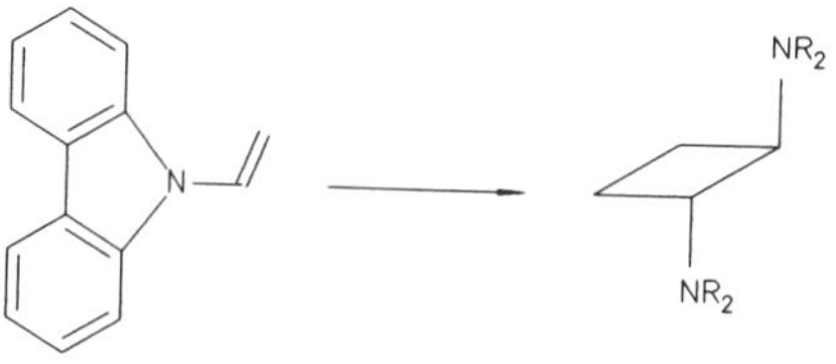

18   **19**

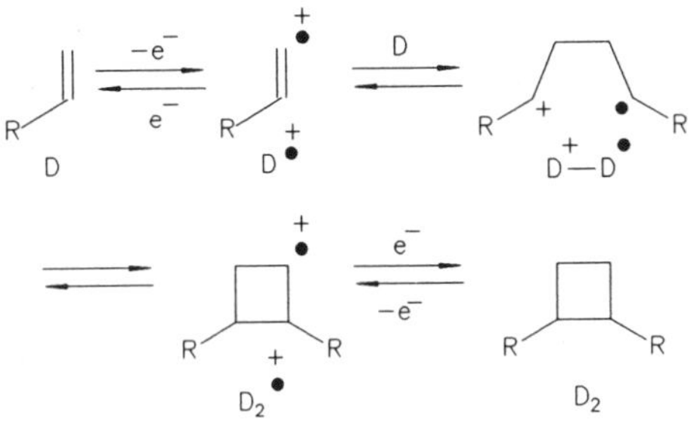

Scheme 4

The essential feature of this mechanism is the interconvertability of the cation radicals $D^+-D^\bullet$ and $D_2^{+\bullet}$.[64] Experiments have confirmed the reversibility of the step $D^+-D^\bullet \rightleftharpoons D_2^{+\bullet}$. Thus, when the dimer **19** is irradiated in the presence of CdS under the same conditions used to induce the dimerization of **18**, about 2.5% of the dimer is converted to **18**.[58] The inefficient cleavage of dimer **19** suggests that the rate of electron-transfer to $D_2^{+\bullet}$ is much faster than the rate of conversion to $D^+-D^\bullet$.[63]

The CdS-mediated dimerization of 1,1-di-*p*-anisylethylene (**20**) to [4+2] - cyclized dimers **21** and **22** and open-chain dimers **23** and **24** has also been reported.[68] The relative amounts of the cyclized dimers **21** and **22** vs the open chain dimers **23** and **24** are in favor of the open-chain dimers. Depending on the nature of the CdS sample the ratio of the cyclized to open-chain dimers is 1:3–6. All the products are derived from the cation radical $20^{+\bullet}$; their formation is efficiently quenched by 1,2,4,5 tetramethoxybenzene.

The addition of the radical cation $20^{+\bullet}$ to the neutral molecule **20** gives the 1,4-dimeric radical cation, which undergoes both cyclization and loss of a proton. The 1,6-cyclization product after loss of a proton and subsequent disproportionation, or electron-transfer followed by proton transfer, gives **23** and **24**. On the other hand, loss of a proton from the dimeric radical cation gives a radical that upon disproportionation or electron-transfer followed by proton transfer gives **21** and **22**.

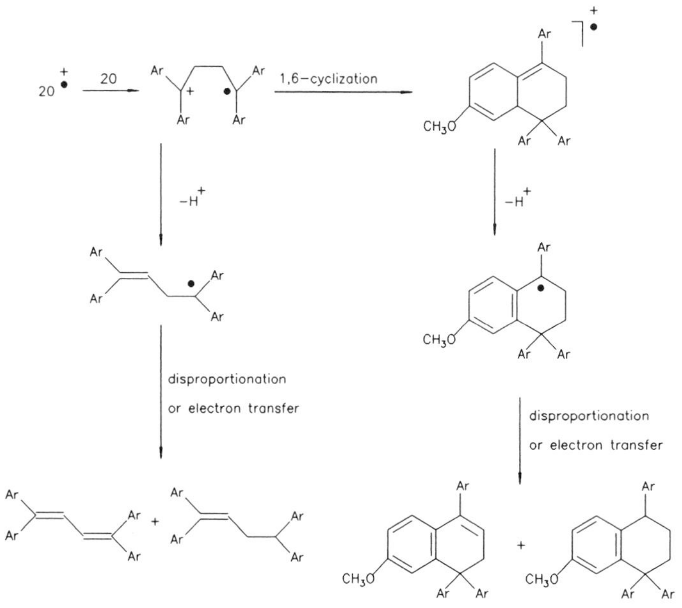

21 22 23 24

Scheme 5

2.4. Cycloreversion

The cycloreversion of the dianthracene **25** to anthracene **26** can be induced by illuminated CdS.[20]

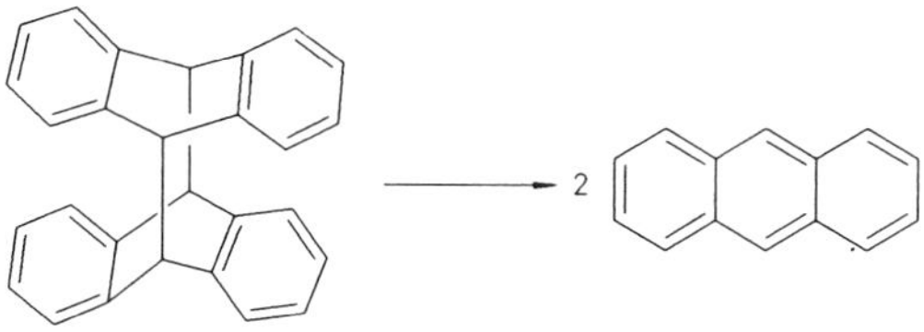

25 26

The view that the reaction proceeds via a cation-radical mechanism is supported by the fact that addition of *p*-dimethoxybenzene ($E^{ox}_{1/2}$ = 1.35 V vs. SCE) to the reaction reduced the conversion by half. The reaction was also reduced to 10–20% of its value by purging of the system with nitrogen.

The cycloreversion of the *N*-methylquinoline dimer (**27**) to *N*-methylquinoline (**28**) can also be induced by illuminated semiconductors.[46] Irradiation of suspensions of CdS, ZnO, or TiO$_2$ in methylene dichloride in the presence of **27** at the wavelengths where only the semiconductor absorbs the light resulted in formation of the corresponding monomer **28**.

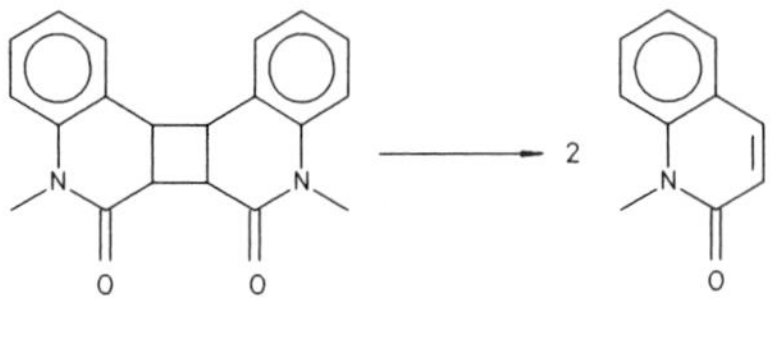

27   **28**

The yield, as in the cycloreversion of the dianthracene, is dramatically reduced when quenchers such as methoxybenzenes are added to the reaction mixture before irradiation, indicating that the reaction proceeds via a cation-radical mechanism. When the reaction mixture is purged with nitrogen prior to irradiation the yield is also reduced.

2.5. [1,3]-Sigmatropic Rearrangement

[1,3]-Sigmatropic rearrangement reactions can also be induced by irradiated semiconductors.[46,69,70] For instance, irradiation of suspensions of CdS or TiO$_2$ in argon-saturated acetonitrile containing 2,2-bis(4-methoxyphenyl)-1-dideuterio-methylene cyclopropane (**29**) resulted in the formation of 2,2-bis(4-methoxy-phenyl)-3,3-dideuterio-1-methylene cyclopropane (**30**).

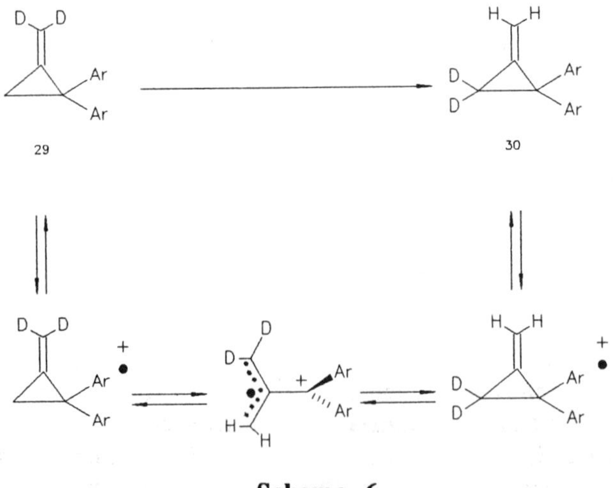

29   **30**

Scheme 6

The rearrangement was not efficient, however, probably because the initially formed radical cation **29**[+•] is quenched by the back electron-transfer much faster than the bond cleavage of **29**[+•] to give the trimethylenemethane cation radical, which serves as a key intermediate in this degenerate rearrangement. The rearrangement mechanism is presented in Scheme 6.

The suprafacial [1,3]-sigmatropic hydrogen shift in diphenylcyclobutene (**31**), takes place via cation radicals in irradiated CdS suspensions.[69,70] This is an interesting finding since the suprafacial pathway is symmetry forbidden in closed-shell systems. The reaction is inefficient, however, and its inefficiency has been attributed, in part, to a fast back reaction (**32** → **31**). Indeed, irradiation of **32** in the presence of CdS affords **31**.

2.6. Oxidation

The semiconductor-mediated oxidations of various organic compounds have been reported.[71,72] These reactions have attracted considerable attention in the past two decades. Alcohols were among the first organic compounds oxidized by irradiated semiconductors. In most cases alcohols were used as a sacrificial reagent for the photocatalytic production of hydrogen. Methanol and ethanol are more efficiently oxidized than other primary alcohols and the photocatalytic oxidation of tertiary alcohols was much more difficult than that of primary or secondary alcohols. While primary and secondary alcohols oxidize to the corresponding carbonyl compounds, tertiary alcohols give carbon–carbon bond cleavage products.[73]

Generally, oxidation of alcohols proceeds more efficiently in platinized than in metal-free semiconductor suspensions.[74-76]

The oxidative cleavage of arylated olefins occurs upon irradiation of platinized or metal-free semiconductor suspensions.[77,78] For instance, it has been reported that 1,1-diphenylethylene, under optimized conditions, gives benzophenone almost quantitatively.[78]

A reaction mechanism based on cation-radical formation of 1,1-diphenylethylene was proposed. One mode of demonstration of the involvement of positively charged intermediate in a chemical reaction is by invoking the Hammett relationship: a negative slope should be found in the plot of the relative rate of the reaction and σ^+. This, in fact, has been demonstrated.[79]

The photocatalytic oxidation of five- and six-membered lactams and *N*-acylamines gives the corresponding imides upon irradiation in the presence of oxygenated aqueous suspensions of TiO_2.[80]

However, a different oxidation product was observed when Cu(II) rather than oxygen was used as an oxidant.

The oxidative cleavage of the double bonds of conjugated dienes has been induced by uv irradiation of oxygenated CH_3CN suspensions of TiO_2.[81]

Different oxidation products of toluene have been observed when a suspension of TiO_2 in toluene was irradiated in the presence or absence of oxygen.[82]   In the presence of oxygen, benzaldehyde was selectively formed. By contrast, bibenzyl was formed when the reaction was conducted in the absence of oxygen.

Depending on the initial concentration of the amine, two pathways can be observed in the photocatalytic oxidation of a primary amine by irradiated TiO_2 powders suspended in oxygenated acetonitrile.[83,84]

Azo products were formed in the photocatalytic oxidation of toluidines in oxygenated aqueous suspensions of TiO_2.[85,86]

Apparently N—H deprotonation to form a nitrogen radical is more favorable than α-deprotonation of the pendant alkyl group.

The photocatalytic oxidation of diphenylmethane, in an oxygen saturated TiO_2 suspension, to benzophenone has also been reported.[87]   The yield is found to increase with increasing solvent polarity. An electron-transfer mechanism is proposed in which the diphenylmethane donates an electron to the photogenerated "hole" to produce a cation radical and oxygen accepts an electron from the conduction band to form a superoxide ion. The subsequent reaction of the diphenylmethane cation radical with the superoxide ion produces benzophenone with trace amounts of benzohydrol.

Photocatalytic carbon–carbon bond formation by illuminated semiconductors has also been reported.[88-90] Dehydrodimers of cyclic ethers, triethylamine, or methanol can be formed with concurrent hydrogen evolution by irradiation of aqueous colloidal solutions of zinc sulfide containing the organic substrate.

The formation of the dehydrodimers has been rationalized on the basis that the cation radical, produced via the electron-transfer from the organic substrate to the photogenerated "hole," loses a proton to give free radicals which recombines to give the dehydrodimer.

The photocatalytic decarboxylation of simple carboxylic acids to hydrocarbons and CO_2 is well documented in the literature.[17,91-95] This reaction type resembles the electrochemical-Kolbe reaction and is thus referred to as the photo-Kolbe reaction. For example, irradiation of an aqueous suspension of a platinized TiO_2 in the presence of acetic acid gives methane, ethane, propionic acid, and CO_2. Methane is the major product.

$$CH_3COOH \longrightarrow CH_4 + C_2H_6 + CH_3CH_2COOH + CO_2$$

The concentration of the methyl radicals produced by decarboxylation of the acetate radical determines the reaction pathway. Mechanistic studies suggest that there are two routes for methane formation:

$$^{\bullet}CH_3 + H^{\bullet} \longrightarrow CH_4$$

$$^{\bullet}CH_3 + CH_3COOH \longrightarrow CH_4 + {^{\bullet}CH_2COOH}$$

The hydrogen atom is produced from reduction of H^+ on Pt sites:

$$H^+ + e^- \longrightarrow {^{\bullet}H_{ad}}$$

The dimerization of $^{\bullet}CH_3$ radicals forms ethane, while reaction between $^{\bullet}CH_3$ and $^{\bullet}CH_2COOH$ radicals produces propionic acid:

$$^{\bullet}CH_3 \ + \ ^{\bullet}CH_3 \longrightarrow CH_3CH_3$$

$$^{\bullet}CH_3 \ + \ ^{\bullet}CH_2COOH \longrightarrow CH_3CH_2COOH$$

The formation of low yields of various amino acids from methane, ammonia, and water mixture by uv irradiation of Pt/TiO_2 slurries has also been reported.[96,97] Since the overall reaction is uphill, such a reaction is photosynthetic rather than photocatalytic. Higher quantum efficiencies (up to 35%) of amino acids are achieved when the reaction of organic acids, ammonia, and water is photocatalyzed by semiconductor.[98,99] Ketocarboxylic acids or hydroxycarboxylic acids can be used for amino acid preparation. The conversion of a ketocarboxylic acid to the corresponding amino acid may proceed via the following steps:

$$RCOCOOH + NH_3 \longrightarrow RC(=NH)COOH + H_2O$$

$$RC(=NH)COOH + 2H^+ + 2e^- \longrightarrow RCH(NH_2)COOH$$

The thermal condensation of the ketocarboxylic acid with ammonia produces imino acid, which is photocatalytically reduced to the corresponding amino acid. However, the primary step in the conversion of hydroxy carboxylic acids to amino acids may be the photocatalytic dehydrogenation of the hydroxy acid to the corresponding keto acid. The reaction may proceed, thereafter, as in keto-carboxylic acid reactions. These reactions resemble the Knoop reaction where the imino acid in that reaction is reduced thermally to amino acid by hydrogen and a platinum catalyst. Therefore, by analogy with the Knoop reaction, these reactions may be referred to as "a photo-Knoop reaction."

Formation of diglycine to pentaglycine peptides from glycine could be induced on irradiated semiconductors.[100] The yield of the peptides was efficiently enhanced by platinization of the semiconductors.

$$\overset{+}{H_3N}\ CH_2CO_2^{-} \longrightarrow \overset{+}{H_3N}\ CH_2\overset{\overset{\displaystyle O}{\|}}{C}NHCH_2CO_2^{-}$$

The formation of hypoxanthine, a nucleic acid base, by irradiation of aqueous suspensions of TiO_2 or CdS containing KCN and NH_3 has been demonstrated.[101] Platinization caused a marked increase in the yield of hypoxanthine for both the semiconductors tested.

$$\text{KCN} + \text{NH}_3 + \text{H}_2\text{O} \longrightarrow$$

2.7. Reduction

In contrast to the many oxidation reactions induced by illuminated semiconductors, only a few reduction reactions mediated by semiconductors have been reported. The modestly negative potential of electrons at the conduction band of the conveniently accessible semiconductors and the ease of oxygen reduction by electrons account for the limited success of organic reductions by semiconductors. However, when oxygen is removed some organics can accept the photogenerated conduction band electron.[72]

The reduction of methylviologen (MV^{2+}) by illuminated semiconductors has been extensively investigated.[102-107] In some cases the reaction was chosen to examine the photocatalytic activity of the semiconductor under study, while in other cases it has been used as photoelectrochemical relay.[108]

The reduction of MV^{2+} by illuminated titanium dioxide or cadmium sulfide occurs readily in the absence of oxygen. However, introducing oxygen to the reduced system ($MV^{+\cdot}$) results in rapid oxidation of $MV^{+\cdot}$ back to MV^{2+} with the concomitant formation of superoxide ion.

The irreversible one- and two-electron reductions of halothane (2-bromo-2-chloro-1,1,1-trifluoroethane) by conduction band electrons, e^-_{CB}, photogenerated in aqueous colloidal suspensions of platinized titanium dioxide (Pt/TiO_2) have been reported.[109] Both bromide and fluoride ions (the respective products of the one-electron and two-electron-transfer to halothane) are formed with high quantum yields provided an efficient "hole" scavenger (e.g., methanol) is present.

Since relatively mild conditions are sufficient to catalyze the overall process, such systems could be employed to obtain substituted alkenes.

Nonmetallized CdS, in acetonitrile containing triethylamine as a sacrificial electron donor, reduces benzophenone derivatives **33** whose reduction potentials lie between -1.42 and -1.90 V vs. SCE, giving alcohols **34** and/or pinacols **35**.[110]

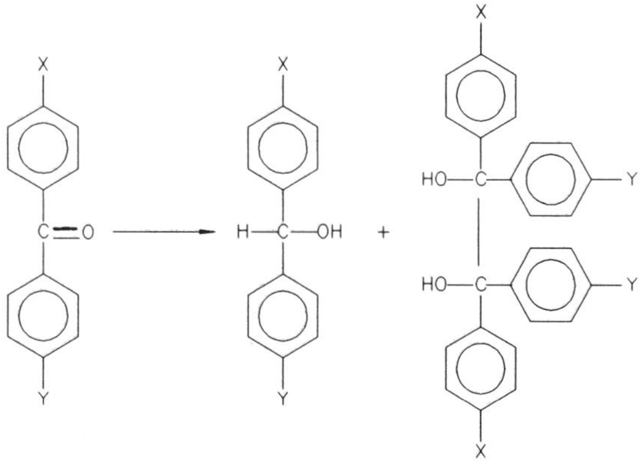

33   **34**   **35**

a: X = CN, Y = H
b: X = Y = Cl
c: X = Cl, Y = H
d: X = Y = H
e: X = MeO, Y = H

However, no reduction was observed with a platinized CdS. To explain these results it has been suggested[110,111] that the conduction band potential of CdS should be 1 V vs. SCE more negative than the generally accepted -0.85 V vs. SCE in CH_3CN value.

Two-electron reduction of an aldehyde by illuminated ZnS has also been reported.[112] Thus, the ZnS suspension prepared from cold oxygen-free aqueous $ZnSO_4$ and Na_2S solutions catalyzes photoredox reactions of acetaldehyde, giving ethanol as a two-electron reduction product and acetic acid, biacetyl, and acetoin as oxidation products.

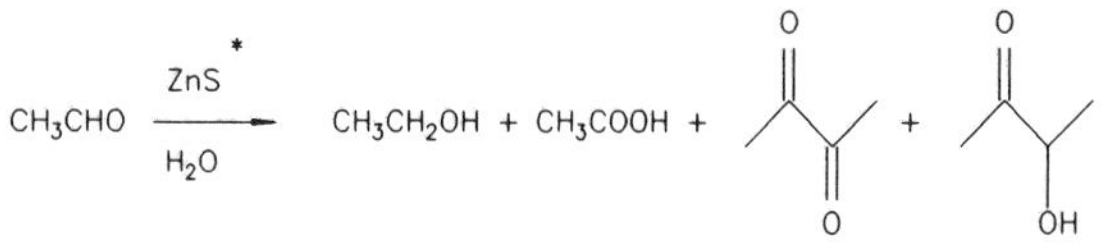

Since the Cannizzaro reaction of aliphatic aldehydes is generally impossible because of the ease of their aldol condensation in the presence of alkali, the present reaction, which is a photoreduced disproportionation of the acetaldehyde, may be referred to as "photo-Cannizzaro reaction."

Reduction of protons in oxygen-free suspensions of semiconductors by the photogenerated conduction band electrons produces hydrogen atoms. If these hydrogen atoms are formed on metallized semiconductor, catalytic hydrogenation can occur. It has been reported that hydrogenation of acetylene and ethylene in aqueous sulfide solution occurs during the illumination of CdS particles loaded with Pt or Rh catalysts.[113] The sulfide acts in these reactions as a sacrificial electron donor.

$$H-C \equiv C-H \quad \xrightarrow[S_2^-,H_2O]{Pt/CdS^*} \quad CH_3CH_3$$

The photocatalytic hydrogenation of olefins, vinyl ethers, and the double bond of α- and β-unsaturated enones over platinized TiO_2 with ethanol acting as electron donor has also been reported.[114]

The N=N double bond of the diaryl azo dye methyl orange can also be reduced by colloidal titanium dioxide.[115]

2.8. Alkylation

N-Alkylation of amines can be induced by irradiated semiconductors.[116–122] Primary amines, in the absence of oxygen, can be converted to secondary amines through Schiff base formation. It has been suggested that the positive "hole" (h^+) oxidizes the amine to imine:

$$RCH_2NH_2 \quad \xrightarrow[-H^+]{h^+} \quad RCH_2NH\bullet \quad \xrightarrow[-H^+]{h^+} \quad RCH=NH$$

The condensation of the imine and/or its hydrolyzed derivative aldehyde with the primary amine gives the Schiff base with liberation of ammonia:

$$RCH=NH + RCH_2NH_2 \longrightarrow RCH=NCH_2R + NH_3$$

The photocatalytic reduction of the Schiff base yields the corresponding secondary amine:

$$RCH=NCH_2R + 2e^- + 2H^+ \longrightarrow RCH_2NHCH_2R$$

N-Alkylation of primary amines can also be induced in alcohols. In this medium not only secondary but also tertiary amines can be produced[118]:

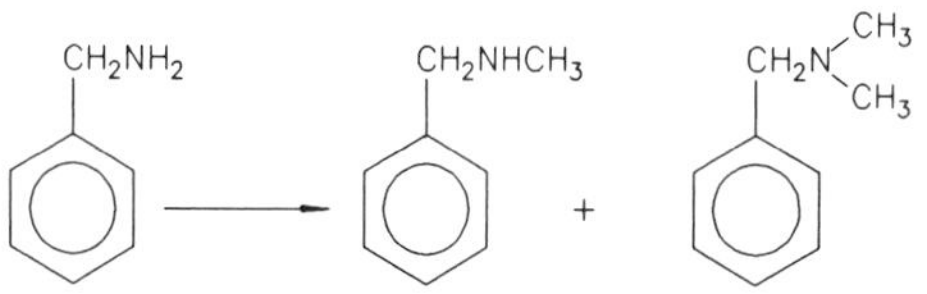

The primary step in the *N*-alkylation process in alcohol is the photocatalytic dehydrogenation of alcohol to the corresponding aldehyde followed by the condensation of aldehyde with the primary amine to give Schiff base intermediate. The photocatalytic reduction of Schiff base gives secondary amine.

Ammonia can also be *N*-alkylated into tertiary amines by irradiation of platinized titanium dioxide suspended in alcohols.[119] The photocatalytic dehydrogenation of alcohols to aldehydes, the condensation of the aldehyde with ammonia to give an imine and the thermal hydrogenation of the imine intermediate to primary amine account for the ammonia *N*-alkylation. Further *N*-alkylation leads to secondary and tertiary amines.

The photocatalytic cyclization of polymethylene-α,ω-diamine can be induced by irradiation of an aqueous suspension of Pt/TiO_2[116,120]:

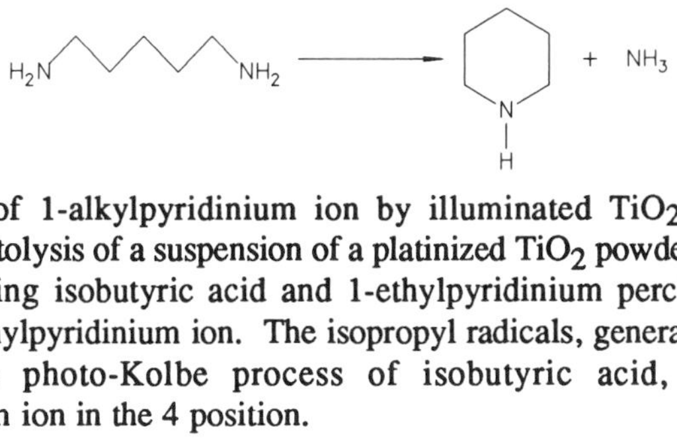

Alkylation of 1-alkylpyridinium ion by illuminated TiO_2 has also been reported.[123] Photolysis of a suspension of a platinized TiO_2 powder in an aqueous medium containing isobutyric acid and 1-ethylpyridinium perchlorate yielded 4-isopropyl-1-ethylpyridinium ion. The isopropyl radicals, generated at the TiO_2 surface by the photo-Kolbe process of isobutyric acid, alkylates the 1-ethylpyridinium ion in the 4 position.

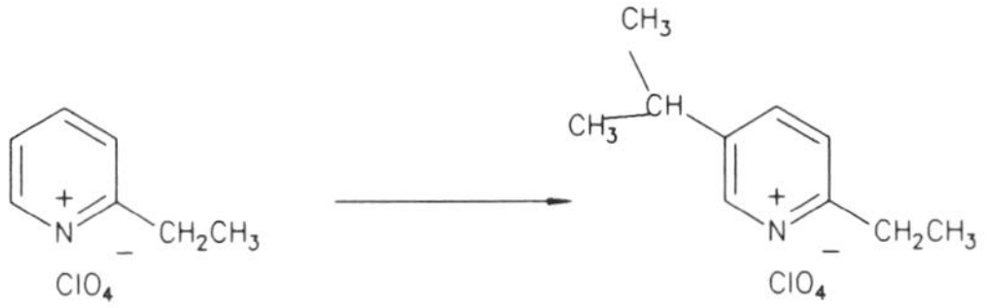

3. Selectivity

Cation radicals can be generated by several routes, including electrochemical, photochemical (both homogeneous and heterogeneous), thermal, and radiational activation. The ultimate chemical fate of these radical cations, however, varies with the conditions under which they are generated.[72] For instance, the poised metal electrode is a highly oxidizing medium that generates high local concentration of radical cations. This would not be the case when the cation radical is generated on the surface of an illuminated semiconductor. The cation radicals generated as adsorbed intermediates may also exhibit different chemistry than would be expected from fully solvated species which are generated in homogeneous solutions.

It has been reported that 1,1-diphenylethylene cation radicals generated in three different environments react via three chemical routes.[124] The electrochemical oxidation of 1,1-diphenylethylene at an inert metal electrode leads predominantly to dimerization of the cation radical to dicationic intermediates and products derived therefrom. In contrast to the highly oxidizing medium that an inert metal electrode provides, the surface of an illuminated semiconductor powder provides an environment for both oxidative and reductive electron-transfer to occur, simultaneously generating surface-bound radical ions (1,1-diphenylethylene cation and superoxide ion) that can interact with each other to produce benzophenone. Generation of the same radical cation via a homogeneously dispersed single-electron oxidant such as triarylaminium salts, where the cation radical is solvated, gives tetraphenyl cyclobutane as a third major product.

The oxidation of 1-methylnaphthalene is also governed by the environment in which the cation radical is generated.[124]

When 1-methylnaphthalene is photooxidized on titanium dioxide, the surface adsorbed cation radical reacts with its geminate reduction partner, superoxide ion, to form ring-cleaved product in very good chemical yield. In contrast, electron-transfer photosensitization by 1,4-dicyanobenzene in homogeneous solution leads to side-chain oxidation.

The oxidation of acetic acid to hydrocarbons and carbon dioxide represents another example in which the reaction environment plays a crucial role in the product distribution. The Pt/TiO$_2$ photocatalytic oxidation of acetic acid gives methane as the principal product. By contrast, the electrochemical oxidation of acetic acid yields ethane as the main product. The surface concentration of methyl radicals determines the product distribution. The high concentration of methyl radicals on the metal electrode allows dimer, namely ethane, formation while on the surface of the Pt/TiO$_2$ powder the concentration is so small that ethane is not easily produced.

Interesting differences were also observed when α-hydroxycarboxylic acids, such as lactic acid, are photooxidized by platinized CdS and TiO$_2$ powders.[125] Under irradiation, Pt/TiO$_2$ decomposes lactic acid to hydrogen, carbon dioxide, and acetaldehyde, while Pt/CdS gives hydrogen and pyruvic acid.

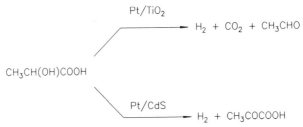

Electrochemical studies using metal electrode have shown that reaction products are independent of the electrode potential but do depend on the nature of the electrode materials. This suggests that the difference in selectivity is not caused by differences in the redox potential of the photogenerated "hole's" in the valence band of the two materials, but at least partially by differences in the adsorption on the two different surfaces.

4. Factors Affecting Semiconductor-Mediated Reactions
4.1. Electron Acceptors

The presence of oxygen proves useful in many semiconductor photoinduced organic reactions. As an electron acceptor, oxygen enhances remarkably the rate of the valence isomerization of 9[44] and 11;[46] the dimerization of 15,[56] 18[58] and 20;[68] and the cycloreversion of 25[19] and 27.[46] Oxygen traps the photogenerated electron at the conduction band as superoxide ion, O$_2^-\cdot$, thereby prolonging the lifetime of the photogenerated "hole" by delaying the collapse of the electron–hole pair.

That, indeed, a conduction band electron is trapped on CdS by adsorbed oxygen as superoxide ion has been confirmed by spin-trapping technique.[126,127] However, the disadvantage of the use of oxygen is that reaction may be accompanied by oxidation. While in certain cases oxidation may be a desired process, in many others it is not. Thus, replacement of oxygen by other electron acceptors not only may prove useful in preventing an undesired oxidation process, but also may lead to enhance the reaction rate. This has, indeed, been observed. For instance, replacement of oxygen by maleic anhydride ($E^{red}_{1/2}$ = –0.84 V vs. SCE) greatly improves the CdS-mediated dimerization of **15**.[56] Similarly, replacement of oxygen by methylviologen dication significantly increases the rate of valence isomerization of **11** to **12**.[49] However, these (including oxygen) are reversible electron acceptors. They, unless the reaction of the cation radical produced by the oxidation of the donor by the "hole" is fast, donate the electron back and thus quench the reaction.

Carbon tetrabromide is known to be a good electron acceptor ($E^{red}_{1/2}$ = -0.3 V vs. SCE)[128] and that the derived radical anion $CBr_4^{-\bullet}$ dissociates rapidly at room temperature into $\bullet CBr_3$ and bromide ion.[28,129-134] Thus CBr_4 could act as an irreversible electron acceptor if the rate of dissociation of $CBr_4^{-\bullet}$ were faster than the rate of reaction of the cation radical produced by the oxidation of the donor by the photogenerated "hole." Recent studies[135] show that using CBr_4 as electron acceptor considerably improves the rate of cis–trans isomerization of **1**, the valence isomerization of **11** and the cycloreversion of **25**. Table 6 summarizes the results of valence isomerization of **11** to **12** under various conditions.[135]

Table 6. Valence Isomerization of **11** to **12** under Different Conditions[a,b]

	% Chemical yield of **12** ($\pm$ 10%)			
CdS supplier	Argon[c]	Atm[d]	Atm + CBr$_4$[e]	Argon + CBr$_4$[e]
Strem (lot No. 16027-S1)	3	20	28	70
Aldrich (lot No. 1721 PJ)	4	10	18	38
Aldrich (lot No. 043087)	2	9	23	35

[a] Reproduced with permission from Ref. 135.

[b] Solution of **11** in CH_2Cl_2 (0.057 M; 6 mL) were irradiated for 15 min in the presence of CdS (50 mg), λ>400 nm, 150 W xenon lamp.

[c] Reaction mixtures were purged with argon (saturated with CH_2Cl_2) for 15 min prior to irradiation and during the irradiation.

[d] Reaction mixtures are open to atmosphere via a reflex condenser.

[e] [CBr$_4$] = 0.0026 M.

4.2. Electron Donors

Trapping of the photogenerated "hole" by suitable electron donors appears important if the organic reaction driven by the photogenerated electron is desired. For instance, addition of triethanolamine (10^{-2} M) to a CdS slurry containing heptylviologen (10^{-2} M) enhanced by a factor of 250 the rate of reduction of heptylviologen by the conduction band electrons.[136] Similarly, it has been reported that the presence of triethylamine is essential for CdS-catalyzed photoreduction of aromatic ketones to occur.[110] It has also been reported that the irreversible one- and two-electron reductions of halothane by conduction band electrons photogenerated in aqueous suspensions of platinized titanium dioxide (Pt/TiO_2) occur efficiently if "hole" scavengers such as methanol are present in the solution.[109] Evidently, trapping the photogenerated "hole" results in delaying the collapse of the electron–hole pairs and in enhancing the photocatalytic activity of the semiconductor toward redox reactions.

4.3. Light Intensity

It has been established that while at low light intensity the rate of semiconductor-mediated organic reactions usually increases linearly with light intensity, at high light intensity the rate is frequently proportional to the square root of light intensity. This behavior has been observed, at least, in the following photocatalytic reactions: (1) the oxidation of gaseous propan-2-ol on a thin uniform film of TiO_2,[137] (2) the gas-phase photo-Kolbe reaction of acetic acid over Pt/TiO_2,[94] (3) the oxidation of liquid propan-2-ol on suspended TiO_2,[138] (4) the photoelectrochemical cell system when a Pt electrode coated with TiO_2 powder is used as a photoanode,[139] (5) the isomerization of *cis*-stilbene on CdS,[31] and (6) the CdS-mediated dimerization of **18**.[58] This behavior has been rationalized on the basis that the competition between the electron–hole recombination process and the surface photo-oxidation of an organic substrate by the "hole" is light intensity dependent. The linear behavior, seen at very low light intensity, suggests that the electron–hole recombination process is negligible in that light intensity range. The quantum yield for electron–hole separation is considered unity at that range. However, with increasing light intensity, the quantum yield for electron–hole separation is no longer unity and the electron–hole recombination process becomes important. The fact that the electron–hole recombination process increases with light intensity has been demonstrated by measuring the quantum yield for the reduction of methylviologen on pulse-irradiated colloidal CdS particles as a function of incident light intensity.[103] The data clearly show that the quantum yield decreases with the increasing of the light intensity.

4.4. Semiconductor Samples

It has been found that different commercial samples of a semiconductor differ in their photocatalytic properties toward an organic reaction.[47,58,138,140,141] For instance, the rate of dimerization of **18** on CdS from different sources, with different crystal structures, different surface areas, and different purities vary by a factor of four[58] with no direct correlation with the physical properties (Table 7).

Table 7. Valence Isomerization of **11**[a] and Dimerization of **18**[b] Initiated by CdS of Different Properties[c]

CdS Supplier	Purity (%)	Surface area $(m^2 g^{-1})$[d]	Crystalline structure	Relative yields	
				12[e]	**19**[f]
Strem (Ultrapure) (lot #16693-S)	>99.99	7.1	α+β	1.00	0.31
Fisher (lot #792913)	99.00	12.0	β	0.81	0.32
Strem (luminescent) (lot #NATL)	99.99	0.5	β	0.78	0.74
Aldrich[g] (lot #053087)	98.8	7.0	β	0.75	0.24
Fluka (Puriss) (lot #249326-684)	99.999	1.4	β	0.71	0.59
Aldrich (gold label) (lot #1721PJ)	99.999	1.2	α	0.53	0.65
Strem (lot #NATL)	99.00	32.4	Polytype	0.50	1.00
Strem (Ultrapure)[h] (lot #16027-SI)	99.99	44	Polytype	0.43	0.80

[a] 0.062 M of **11** in CH_2Cl_2 (5 mL) was irradiated in the presence of CdS (50 mg at λ > 430 nm for 30 min with a 150 W xenon lamp.

[b] 0.22 M of **18** in acetone (6 mL) was irradiated in the presence of CdS (42 mg) at λ > 430 nm for 30 min with a 150 W xenon lamp.

[c] Reproduced with permission from ref. 58.

[d] see ref. 57.

[e] The highest **12** yield (0.055 M) was obtained with Strem (>99.99%).

[f] The highest dimer **19** yield (0.05 M) was obtained with Strem (99%.;

[g] Activation energy for dimerization of **18**: 4.2 kcal mole^{-1}

[h] Activation energy for dimerization of **18**: 1.7 kcal mole^{-1}

A different photocatalytic activity trend was also evident[47] when the various CdS samples were tested for the valence isomerization of **11** (Table 7). This behavior seems to be a common phenomenon. It is not unexpected since different surface states (vacancies, lattice defects, impurities) localized between the top of the valence band and the bottom of the conduction band, acting as electron donors or electron-acceptor centers, may well be present. The generation of these states depends[142-145] on the preparation method of CdS sample, the subsequent heat treatment, and the impurity level. In analogy with a similar observation found in the TiO_2-photoinduced oxidation of liquid propan-2-ol,[138] these states may trap the photogenerated electron at different depths below the conduction band of CdS.[58] Supporting evidence for this notion originated with the activation energies of two selected samples of CdS (Table 7). The difference in the activation energies of the CdS-mediated dimerization of **18** suggests that these energies are associated with processes within the semiconductor[138] rather than with the dimerization of **18**.[58] The low values of the activation energies are taken as indicative of the presence of traps at different depths but should be very close to the conduction band. The presence of traps at different depths (0.05, 0.14, 0.25, 0.41, 0.63, and 0.83 eV) has been reported in CdS single crystals.[146-151] Thus, the activation energies may reflect the promotion of the electron from these traps into the conduction band.

5. Kinetic Analysis
5.1. Rate vs. Concentration

   The kinetics of various semiconductor-photoinduced organic reactions have been reported in various publications.[31,34,47,56,58] The reader is referred to these publications for a detailed description of kinetics of these reactions. We therefore choose the CdS-photoinduced dimerization of *N*-vinyl carbazole to illustrate the kinetics of the semiconductor-photoinduced organic reactions in general. To simplify the discussion, Scheme 7 is invoked as a minimal representation of the processes involved.[58]

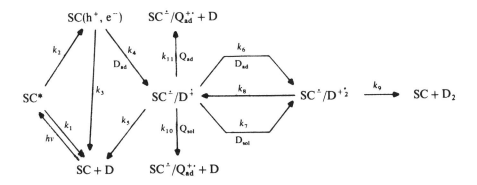

Scheme 7

where SC, SC^*, and $SC(h^+, e^-)$ are the semiconductor in the ground state, in the excited state (exciton), and after electron–hole separation, respectively; D and D_2 (and their respective ions) represent the substrate **18** and the dimer **19** (and their ions); while D_{ad} and D_{sol} denote **18** on the surface and in solution. Q_{ad} and Q_{sol} are the quencher on the surface and in solution, respectively.

The dimerization of **18**, being a bimolecular reaction, requires the cation radical **18**$^{+\cdot}$, produced by electron-transfer from **18** to the "hole" to react with a neutral molecule of **18**. To examine whether the neutral **18** molecule is preadsorbed on the surface or diffuses into the surface from the solution, the combination of the Langmuir-Hinshelwood (LH) and Eley-Rideal (ER) kinetics models has been adapted.

The LH kinetic model assumes that the reaction occurs between **18** cation radical on the surface and the neutral **18** molecule preadsorbed on the surface. The rate of reaction, at the photostationary-state concentration of **18** cation radical, when the reactant is significantly more strongly adsorbed on the surface than the product will be given by equation of the following form:

$$R_{LH} = k_{LH}\theta_D = \frac{k_{LH} K [D]}{1 + K [D] + K_s[S]} = \frac{k_{LH} K' [D]}{1 + K' [D]} \qquad (3)$$

where k_{LH} is the reaction rate constant, θ_D is the fraction of the surface covered by **18**, K and $[D]$ are the adsorption coefficient and the initial concentration of **18**, K_s and $[S]$ are the adsorption coefficient and the concentration of the solvent, and $K' = K/(1 + K_s[S])$. Since $[S] >> [D]$, however, $K_s[S]$ remains essentially constant at all concentrations of **18** used.

From Scheme 7, the concentration of the **18** cation radical at the photostationary state is

$$[SC^{-\cdot}/D^{+\cdot}] = \frac{k_2 k_4 I \theta_D}{k_5 (k_1 + k_2) (k_3 + k_4\theta_D)} \qquad (4)$$

where I is the flux, constant during the experiment. When the rate of electron-transfer from **18** to the "hole", $k_4\theta_D[SC(h^+,e^-)]$ is much greater than electron–hole recombination of the surface, $k_3[SC(h^+,e^-)]$, k_3 can be neglected comparing with $k_4\theta_D$;. then $[SC^{-\cdot}/D^{+\cdot}] \sim$ constant.

The ER kinetic model would imply that reaction occurs between **18** cation radical on the surface and the neutral **18** molecule diffusing from the solution. The rate of reaction, R_{ER}, at the photostationary state concentration of **18**$^{+\cdot}$ is given by equation 5.

$$R_{ER} = k_{ER} [D] \qquad (5)$$

where k_{ER} is the reaction rate constant.

If both LH and ER pathways are operating then the overall rate, R, should be given by equation (6):

$$R = k_{ER} [D] + \frac{k_{LH} K' [D]}{1 + K' [D]} \qquad (6)$$

Thus, in a plot of rate vs. concentration, as θ_D approaches a limiting value the overall increase in rate would be expected to approach linearity with a further increase in **18** concentration.

The effect of varying the concentration of **18** (0.022 – 0.22 M) on the rate of dimerization at constant CdS mass and constant light intensity has been examined. Figure 4 shows that at concentrations higher than 0.1 M, the overall rates of dimerization increase linearly with the concentration.

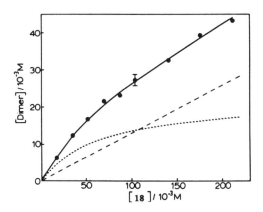

Figure 4. Effect of the concentration of **18** on the rates of dimer formation (•). The curve is calculated from R = 0.129[D] + 0.349[D]/(1 + 15.6[D]). The overall rates are separated into LH (....) and ER(---) components. Each sample was irradiated at > 430 nm for 30 minutes with a 150-W xenon lamp. (Reproduced with permission from Ref. 58)

Similar observations have been reported in, for instance, the CdS-, ZnO-, and TiO$_2$-photoinduced valence isomerization of hexamethyl(Dewar)benzene[47] and the gas-phase photooxidation of carbon monoxide over TiO$_2$[152] and of propan-2-ol over ZnO.[137] This behavior agrees with the kinetics suggested in equation (6) and indicates that both the LH and ER pathways contribute to the reaction of **18** cation radical with neutral molecules of **18**. The rate in Figure 4 has been fitted, using a non-linear least-squares method program, to a curve obtained by inserting appropriate values of the constants in Equation (7):

$$R = 0.129[D] + 0.349[D] / 1 + 15.6[D] \tag{7}$$

Having the values of k_{ER}, K', and $k_{LH}K'$, the ratio k_{ER}/k_{LH} is estimated to be 6, which clearly indicates that the ER pathway is favored. Similarly, using these values, it became possible to separate the overall rates of dimerization into LH and ER components.

5.2. Quenching Studies

One mode of demonstrating the involvement of the cation radical in a reaction scheme is the observation of its quenching by an electron donor. The CdS-photoinduced dimerization of **18** ($E^{ox}_{1/2}$ = 1.12 V vs. SCE) may be quenched, in principle, by (a) donation of an electron by the quencher, Q, to the photogenerated "hole" competitively with the reactant **18**, and (b) interception of the cation radical of **18** by the quencher. 1,3,5-Trimethoxybenzene (1,3,5-TMB, $E^{ox}_{1/2}$ = 1.49 V vs. SCE), 1,2,4-trimethoxybenzene (1,2,4-TMB, $E^{ox}_{1/2}$ = 1.12 V vs. SCE), and 1,2,4,5- tetramethoxybenzene (1,2,4,5-TMB, $E^{ox}_{1/2}$ = 0.81 V vs. SCE) have been tested as quenchers for the CdS-photoinduced dimerization of **18**. Plots of $(\phi°/\phi - 1)$ vs. [Q] for 1,3,5-TMB and 1,2,4-TMB are shown in Figure 5, which illustrates that, whereas 1,2,4-TMB quenches the reaction efficiently, the 1,3,5-isomer exhibits no quenching ability over the same concentration range (0.045 M).

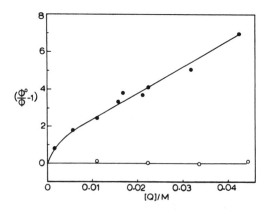

Figure 5. Effect of 1,2,4-trimethoxybenzene (O) and 1,3,5- trimethoxybenzene (•), as quenchers, on the rate of dimerization of **18**. (Reproduced with permission from Ref. 58).

This was surprising since 1,3,5-TMB was expected to quench the "hole" efficiently, although not the **18** cation radical. It therefore appears that "hole" quenching is not importantly involved in the overall quenching process. On the other hand, 1,2,4,5-TMB quenched the reaction dramatically (Figure 6) at a lower concentration range. Thus, for the same concentration (0.006 M), 1,2,4,5-TMB was found to quench the reaction about 10 times more efficiently than 1,2,4-TMB. To rationalize these results kinetically it has been assumed that only cation radical **18**$^{+\cdot}$ quenching is important and that the quenching involves surface (LH) and solution (ER) pathways. Application of the steady-state hypothesis to the excited and reactive intermediates (Scheme 7) in the presence and absence of quencher leads to equation (8), which is a Stern-Volmer equation for semiconductor-photomediated reactions:

$$\frac{\phi^0}{\phi} = 1 + A\,[Q] + \frac{B[Q]}{1 + C\,[Q]} \tag{8}$$

where $A = k_{11}/(k_5 + k_6\theta_D + k_7[D])$, $B = k_{11}K'_Q/(k_5 + k_6\theta_D^0 + k_7[D])$, and $C = K'$.

Figure 6 shows that at concentrations of 1,2,4,5-TMB higher than 0.001 M the rate of quenching increases linearly with the quencher concentration. Similar behavior was observed with 1,2,4-TMB. This behavior is similar to that discussed above for the relationship of dimerization rates to concentration of **18** (Figure 4). The data seem to arise from separate contributions from two quenching mechanisms to the overall quenching process.

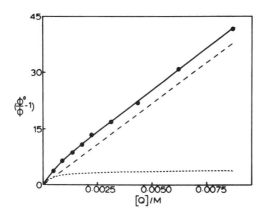

Figure 6. Quenching of the dimerization of **18** by 1,2,4,5- tetramethozybenzene (•). (Reproduced with permission from Ref. 58).

The data may be rationalized in terms of an LH mechanism, in which the cation radical of **18** is quenched by the preadsorbed quencher on the surface together with an ER mechanism, in which the **18** cation radical is quenched by diffusion of the quencher from the solution.

6. Summary

A variety of organic reactions can be photocatalyzed by semiconductors.[153] Semiconductor-photoinduced organic reactions such as: geometrical isomerization, dimerization, cycloreversion, [1,3]-sigmatropic shift, oxidation, reduction, and alkylation are discussed in this article. Some of these reactions have synthetic value. The photocatalytic activity of a semiconductor depends mostly, among other factors, on the presence of electron acceptors, electron donors, light intensity, and semiconductor samples. Controlling these factors optimizes the semiconductor performance toward organic phototransformation reactions. The mechanistic picture of photocatalytic activity is not very clear, even though several investigations have been carried out, and clearly more work is necessary to detail the mechanism path. However, delineating the Langmuir-Hinshelwood vs. Eley-Rideal approaches has shown that, even though based only on a preliminary result, analyzing photochemical reactions on semiconductor surfaces may help to clarify the reaction path for some processes. Finally, it may be pointed out that one of the disadvantages of the use of homogeneous electron-transfer photocatalysts is that the sensitizer and the substrate may form a mixed addition product.[7-9] In addition, back electron-transfer from the sensitizer radical anion to the substrate radical cation may result in formation of an excited state of the substrate.[4-6] Both of these complications can be avoided by the use of semiconductor photocatalysts. Furthermore, the radical cations generated as absorbed intermediates may also exhibit different chemistry than would be expected from solvated species which are generated in homogeneous solutions. A major disadvantage of the semiconductor photocatalysis is that the efficiency of the catalyzed reaction is low especially when a chain is not involved. The efficiency could be increased, however, by changing the environment of the semiconductor.[154] This appears to be a major challenge for those interested in this field.

Acknowledgment

I wish to acknowledge with great thanks the significant contribution of Paul de Mayo, who read the manuscript in detail and offered numerous useful suggestions, and J. Thomas Bolton, who helped me with typing and correcting.

References

1.   S. L. Mattes and S. Farid in *Organic Photochemistry*, Vol. 6, A. Padwa, ed., Marcel Dekker, New York, 1983, p. 233.

2.   A. Weller, *Fast React. Primary Processes Chem. Kinet., Proc. Nobel Sym., 5th*, 1967, 413.
3.   K. H. Grellman, A. R. Watkins and A. Weller, *J. Phys. Chem*, **1972**, *76*, 3132.
4.   K. A. Brown-Wensley, S. L. Mattes and S. Farid, *J. Am. Chem. Soc.*, **1978**, *100*, 4162.
5.   H. D. Roth and M. L. M. Schilling, *J. Am. Chem. Soc.*, **1979**, *101*, 1898.
6.   H. D. Roth and M. L. M. Schilling, *J. Am. Chem. Soc.*, **1980**, *102*, 4303.
7.   A. Albini, E. Fasani and M. Mella, *J. Am. Chem. Soc.*, **1986**, *108*, 4119.
8.   D. R. Arnold, P. C. Wong, A. J. Maroulis and T. S. Cameron, *Pure Appl. Chem.*, **1980**, *52*, 2609.
9.   F. D. Lewis and R. J. DeVoe, *Tetrahedron*, **1982**, *38*, 1069.
10.   H. Gerischer, *J. Electroanal. Chem.*, **1975**, *68*, 263.
11.   H. Gerischer, F. Willing, *Top. Curr. Chem.*, **1976**, *61*, 33.
12.   A. J. Bard, *J. Photochem.*, **1979**, *10*, 50.
13.   A. J. Bard, *Science*, **1980**, *207*, 139.
14.   A. J. Bard, *J. Phys. Chem.*, **1982**, *86*, 172.
15.   M. S. Wrighton, *Acc. Chem. Res.*, **1979**, *12*, 303.
16.   P. A. Kohl and A. J. Bard, *J. Am. Chem. Soc.*, **1977**, *99*, 7531.
17.   B. Kraeutler and A. J. Bard, *J. Am. Chem. Soc.*, **1978**, *100*, 5985.
18.   A. M. Draper, Ph.D. Thesis, University of Western Ontario, 1987.
19.   K. Mizuno, M. Ishii and Y. Otsuji, *J. Am. Chem. Soc.*, **1981**, *103*, 5570.
20.   R. A. Barber, P. de Mayo and K. Okada, *J. Chem. Soc., Chem. Commun.*, **1982**, 1072.
21.   A. Zweig, W. G. Hodson and W. H. Jura, *J. Am. Chem. Soc.*, **1964**, *86*, 4124.
22.   T. R. Evans, R. W. Wake and M. M. Sifain, *Tetrahedron Lett.*, **1973**, 701.
23.   P. G. Gassman, R. Yamaguchi and G. F. Koser, *J. Org. Chem.*, **1978**, *43*, 4392.
24.   K. Mizuno, H. Ueda and Y.Otsuji, *Chem Lett.*, **1981**, 1237.
25.   J. Eriksen and C. S. Foote, *J. Am. Chem. Soc.*, **1983**, *102*, 6083.
26.   L. T. Spada and C. S. Foote, *J. Am. Chem. Soc.*, **1980**, *102*, 391.
27.   H. Siegerman in *Techniques of Chemistry, Volume 5*, Part 2, N. L. Weinberg, ed., 1975, Wiley, New York, p. 667.
28.   Y. Shirota and H. Mikawa, *J. Macromol. Sci., Rev. Macromol. Chem.*, **1977-1978**, *C16(2)*, 129.
29.   S. J.Teichner and M. Formenti in *Photoelectrochemistry, Photocatalysis and Photoreactors*, M. Schiavello, ed., D. Reidel, 1985, p. 457.
30.   H. Al-Ekabi and P. de Mayo, *J. Chem. Soc., Chem. Commun.*, **1984**, 1231.
31.   H. Al-Ekabi and P. de Mayo, *J. Phys. Chem.*, **1985**, *89*, 5815.
32.   S. Yanagida, K. Mizumoto and C. Pac, *J. Am. Chem. Soc.*, **1986**, *108*, 647.
33.   T. Hasegawa and P. de Mayo, *J. Chem. Soc., Chem. Commun.*, **1985**, 1534.
34.   T. Hasegawa and P. de Mayo, *Langmuir*, **1986**, *2*, 362.
35.   G. M. Bancroft, A. M. Draper, M. M. Hyland and P. de Mayo, *New J. Chem.*, **1990**, *14*, 5.
36.   S. Kodama, A. Matsumoto, Y. Kubokawa and M. Anpo, *Bull. Chem. Soc. Jpn.*, **1986**, *59*, 3765.
37.   P. A. Carson and P. de Mayo, *Can. J. Chem.*, **1987**, *65*, 976.
38.   S. Kodama, M. Yabuta, M. Anpo and Y. Kubokawa, *Bull. Chem. Soc. Jpn*, **1985**, *58*, 2307.
39.   M. Anpo, M. Yabuta, S. Kodama and Y. Kubokawa, *Bull. Chem. Soc. Jpn.*, **1986**, *59*, 259.
40.   M. Anpo, A. Atsushi and S. Kodama, *J. Chem. Soc., Chem. Commun.*, **1987**, 1038.
41.   T. W. Taylor and A. R. Murray, *J. Chem. Soc.*, **1938**, 2078.
42.   G. B. Kistiakowsky and W. R. Smith, *J. Am. Chem. Soc.*, **1936**, *58*, 2428.

43. M. A. Fox, B. A. Lindig and C. C. Chen, *J. Am. Chem. Soc.*, **1982**, *104*, 5828.
44. K. Okada, K. Hisamitsu and T. Mukai, *J. Chem. Soc., Chem. Commun.*, **1980**, 941.
45. N. C. Barid, A. M. Draper and P. de Mayo, *Can. J. Chem.*, **1988**, *66*, 1579.
46. K. Okada, K. Hisamitsu, Y. Takahashi, T. Hanaoka, T. Miyashi and T. Mukai, *Tetrahedron Lett.*, **1984**, *25*, 5311.
47. H. Al-Ekabi and P. de Mayo, *J. Phys. Chem.*, **1986**, *90*, 4075.
48. A. M. Draper and P. de Mayo, *Tetrahedron Lett.*, **1986**, *27*, 6157.
49. H. Ikezawa and C. Kutal, *J. Org. Chem.*, **1987**, *52*, 3299.
50. P. G. Gassman and K. D. Olso, *Tetrahedron Lett.*, **1983**, *24*, 19.
51. E. Haselbach, T. Bally, Z. Lanyjova and P. Baertshi, *Helv. Chim. Acta*, **1974**, *62*, 583.
52. H. D. Martin, C. Heller, E. Haselbach and A. Lanyjova, *Helv. Chim. Acta*, **1974**, *57*, 465.
53. H. D. Roth, M. L. M. Schilling and G. Jones III, *J. Am. Chem. Soc.*, **1981**, *103*, 1246.
54. H. D. Roth and M. L. M. Schilling, *J. Am. Chem. Soc.*, **1981**, *103*, 7210.
55. K. Rajhavachari, R. C. Hadden and H. D. Roth, *J. Am. Chem. Soc.*, **1983**, *105*, 3110.
56. A. M. Draper, M. Ilyas, P. de Mayo and V. Ramamurthy, *J. Am. Chem. Soc.*, **1984**, *106*, 6222.
57. M. Ilyas and P. de Mayo, *J. Am. Chem. Soc.*, **1985**, *107*, 5093.
58. H. Al-Ekabi and P. de Mayo, *Tetrahedron*, **1986**, *42*, 6277.
59. S. Kurwata, Y. Shigemitsu and Y. Odaira, *J. Org. Chem.*, **1973**, *38*, 3803.
60. T. R. Evans, R. W. Wake and O. Jaenicke in *The Exciplex*, M. Gordon and W. R. Ware, eds., Academic Press, New York, **1975**, 345.
61. K. Mizuno, J. Ogawa, M. Kamura and Y. Ofsuji, *Chem. Lett.*, **1979**, 731.
62. K. Mizuno, H. Kagano, T. Kasuga and Y. Otsuji, *Chem. Lett.*, **1983**, 133.
63. S. L. Mattes, H. R. Luss and S. Farid, *J. Phys. Chem.*, **1983**, *87*, 4779.
64. P. Bereford, M. C. Lambert and A. Ledwith, *J. Chem. Soc., (C)* , **1970**, 2508.
65. R. A. Crellin, M. C. Lambert and A. Ledwith, *J. Chem. Soc., Chem. Commun.*, **1970**, 682.
66. A. Ledwith, *Acc. Chem. Res.*, **1972**, *5*, 133.
67. R. A. Crellin and A. Ledwith, *Maromolecules*, **1975**, *8*, 93.
68. H. Al-Ekabi and P. de Mayo, *J. Org. Chem.*, **1987**, *52*, 4756.
69. P. de Mayo and G. Wenska, *J. Chem. Soc., Chem. Commun.*, 1626, **1986**.
70. P. de Mayo and F. Wenska, *Tetrahedron*, **1987**, *43*, 1661.
71. M. A. Fox, *Acc. Chem. Res.*, **1983**, *16*, 314.
72. M. A. Fox, *Top. Curr. Chem.*, **1987**, *142*, 72.
73. S. Nishimoto, B. Ohtani, H. Shirai and T. Kagiya, *J. Chem. Soc., Perkin Trans. II*, **1986**, 661.
74. P. R. Harvey, R. Rudham and S. Ward, *J. Chem. Soc., Faraday Trans. I*, **1983**, *79*, 2975.
75. P. Pichat, M. N. Mozzanega, J. Disdier and J. M. Herrmari, *Nouv. J. Chim.*, **1982**, *6*, 559.
76. F. H. Hussein, G. Pattenden, R. Rudham and J. J. Russel, *Tetrahedron Lett.*, **1984**, *25*, 3363.
77. T. Kanno, T. Oguchi, H. Sakuragi and K. Tokumaru, *Tetrahedron Lett.*, **1980**, *21*, 467.
78. M. A. Fox and C. C. Chen, *J. Am. Chem. Soc.*, **1981**, *103*, 6757.
79. M. A. Fox and C. C. Chen, *Tetrahedron Lett.*, **1983**, *24*, 547.
80. J. W. Pavlik and S. Tantayanon, *J. Am. Chem. Soc.*, **1981**, *103*, 6755.
81. M. A. Fox, D. D. Sackett and J. N. Younathan, *Tetrahedron*, **1987**, *43*, 1643.

82. M. Fujihira, Y. Satoh and T. Osa, *J. Electroanal. Chem.*, **1981**, *126*, 277.
83. M. J. Chen and M. A. Fox, *J. Am. Chem. Soc.*, **1983**, *105*, 4497.
84. M. A. Fox and J. N. Younathan, *Tetrahedron*, **1986**, *42*, 6285.
85. M. A. Hema, V. Ramakrishnan and J. C. Kuriacose, *Indian J. Chem. Sect. B*, **1978**, *16* , 619.
86. H. Kasturirangan, V. Ramakrishnan and J. C. Kuriacose, *J. Catal.*, **1981**, 216.
87. J. J. Liang and T. J. Liu, *J. Chinese. Chem. Soc.*, **1986**, *33*, 133.
88. S. Yanagida, T. Azuma, H. Kawakami, H. Kizumoto and H. Sakurai, *J. Chem. Soc., Chem. Commun.*, **1984**, 21.
89. N. Zeug, J. Bucheler and H. Kisch, *J. Am. Chem. Soc.*, **1985**, *107*, 1459.
90. S. Yanagida, T. Azuma, Y. Midori and C. Pac, *J. Chem. Soc., Perkin Trans II*, **1985**, 1487.
91. B. Kraeutler and A. J. Bard, *J. Am. Chem. Soc.*, **1978**, *100*, 2239.
92. H. Keiche, W. W. Dunn, K. Wilbourn, F. R. F. Fan and A. J. Bard, *J. Phys. Chem.*, **1980**, *84*, 3207.
93. H. Yoneyama, Y. Takao, H. Tamura and A. J. Bard, *J. Phys. Chem.*, **1983**, *87*, 1417.
94. S. Sato, *J. Phys. Chem.*, **1983**, *87*, 3531.
95. B. Krauetler and A. J. Bard, *Nouv. J. Chim.*, **1979**, *3*, 31
96. H. Reiche and A. J. Bard, *J. Am. Chem. Soc.*, **1979**, *101*, 3127.
97. W.W. Dunn, Y. Aikawa and A. J. Bard, *J. Am. Chem. Soc.*, **1981**, *103*, 6893.
98. T. Sakata in *Homogeneous and Heterogeneous Photocatalysis*, Pellizetti, E., Serpone, N., eds., (NATO ASI Series C, *vol. 174*) D. Reidel, Dortrecht, The Netherlands, 1986, p. 311.
99. T. Sakata in *Photocatalysis*, N. Serpone, and E. Pellizetti, E., eds., John Wiley, New York, 1989, p. 397.
100. J. Onoe, T. Kawai and S. Kawai, *Chem. Lett.*, **1985**, 1667.
101. J. Onoe and T. Kawai, *J. Chem. Soc., Chem. Commun.*, **1988**, *681*.
102. Y. Nosaka, H. Miyama, M. Terauchi and T. Kobayashi, *J. Phys. Chem.*, **1988**, *92*, 255.
103. Y. Nosaka and M. A. Fox, *J. Phys. Chem.*, **1986**, *90*, 6521.
104. M. D. Ward, J. R. White and A. J. Bard, *J. Am. Chem. Soc.*, **1983**, *105*, 27.
105. R. Rossetti and L. E. Brus, *J. Phys. Chem.*, **1986**, *90*, 558.
106. N. Serpone, D. K. Sharma, M. A. Jamieson, M. Gratzel and J. Ramsden, *J. Chem. Phys. Lett.*, **1985**, *115*, 473.
107. J. Ramsden, *J. Proc. R. Soc. London, A*, **1987**, *410*, 89.
108. M. Gratzel in *Energy Resources Through Photochemistry and Catalysis*, M. Gratzel, ed., Academic Press, New York, 1983, p. 71.
109. D. W. Bahnemann, J. Monig and R. Chapman, *J. Phys. Chem.*, **1987**, *91*, 3782.
110. T. Shiragami, C. Pac and S. Yanagida, *J. Chem. Soc., Chem. Commun.*, **1989**, *831*.
111. D. Meissner, R. Memming and B. Kastening, *J. Phys. Chem.*, **1988**, *92*, 3476.
112. S. Yanagida, Y. Ishimaru, Y. Miyake, T. Shiragami, C. Pac, K. Hashimoto and T. Sakata, *J. Phys. Chem. Soc.*, **1989**, *93*, 2576.
113. A. J. Frank, Z. Foren and J. Willner, *J. Chem. Soc., Chem. Commun.*, **1985**, 1029.
114. H. Yamataka, N. Seto, J. Ichihara, T. Hanafusa and S. Teratani, *J. Chem. Soc., Chem. Commun.*, **1985**, 788.
115. G. T. Brown and J. R. Darwent, *J. Chem. Soc. Faraday Trans. 1*, **1984**, 1631.
116. S. I. Nishimoto, B. Ohtani, T. Yoshikawa and T. Kugiya, *J. Am. Chem. Soc.*, **1983**, *105*, 7180.
117. S. Yanagida, H. Kizumoto, Y. Ishimaru, C. Pac and H. Sakurai, *Chem. Lett.*, **1985**, 141.

118. B. Ohtani, H. Osaki, S. I. Nishimoto and T. Kagiya, *J. Am. Chem. Soc.*, **1986**, *108*, 308.
119. B. Ohtani, H. Osaki, S. I. Nichimoto and T. Kagiya, *Tetradedron Lett.*, **1986**, *27*, 2019.
120. F. G. Tang, H. Courbon and P. Pichat in *Catalysis and Fine Chemicals*, M. Guisnet, ed., Elsevier Science Publishers B.V., Amsterdam, 1988, p. 327 .
121. M. A. Fox and J. N. Younathan, *Tetrahedron*, **1986**, *42*, 6285.
122. B. Ohtani, H. Osaki, S. I. Nishimoto and T. Kagiya, *Chem. Lett.*, **1985**, 1075.
123. D. K. Ellison, D. C. Trulove and R. T. Iwamoto, *Tetrahedron*, **1986**, *42*, 6405.
124. M. A. Fox, *Nouv. J. Chim.*, **1987**, *11*, 129.
125. H. Harada, T. Veda and T. Sakata, *J. Phys. Chem.*, **1989**, *93*, 1542.
126. J. R. Harbour and M. L. Hair, *J. Phys. Chem.*, **1977**, *81*, 1791.
127. J. R. Harbour and M. L. Hair, *J. Phys. Chem.*, **1978**, *82*, 1397.
128. M. Stackelberg and W. Stracke, *Z. Electrochem.*, **1949**, *53*, 118.
129. T. Nyokong, Z. Gasyna and M. J. Stillman, *Am. Chem. Soc. Symp. Ser.*, *321*, American Chemical Society, Washington, D. C., 1986, p. 309.
130. T. Nyokong, Z. Gasyna, M. J. Stillman, *Inorg. Chem. Acta*, **1987**, *112*, 11.
131. T. Nyokong, Z. Gasyna, M. J. Stillman, *Inorg. Chem.*, **1987**, *26*, 548.
132. Z. Gasyna, W. R. Browett and M. J. Stillman, *Am. Chem. Soc. Symp. Ser.*, *321*, American Chemical Society., Washington, D. C., 1986, p. 298.
133. T. Nyokong, Z. M. Gasyna and J. Stillman, *Inorg. Chem.*, **1987**, *26*, 548.
134. L. Eberson and M. Ekstrom, *Acta. Chem. Scan.*, **1988**, *B42*, 101.
135. H. Al-Ekabi, A. M. Draper and P. de Mayo, *Can. J. Chem.*, **1989**, *67*, 1061.
136. F. D. Saeva, G. R. Olin and J. R. Harbour, *J. Chem. Soc., Chem. Commun.*, **1980**, 401.
137. J. Cunningham and B. K. Hodnett, *J. Chem Soc. Faraday Trans. I*, **1981**, *77*, 2777.
138. P. R. Harvey, R. Rudham and S. Ward, *J. Chem. Soc., Faraday Trans. I*, **1983**, *79*, 1391.
139. R. S. Davidson, R. M. Slater and R. R. Meck, *J. Chem. Soc., Faraday Trans. I*, **1979**, *75*, 2507.
140. A. W. H. Mau, C. B. Huang, N. Kakuta, A. J. Bard, A. Campion, A. Fox, J. M. White and S. E. Webber, *J. Am. Chem. Soc.*, **1984**, *106*, 6537.
141. S. Yanagida, H. Kawakami, K. Hashimoto, T. Sakata, C. Pac and H. Sakurai, *Chem. Lett.*, **1984**, 1449.
142. C. E. Reed and C. G. Scott, *Brit. J. Appl. Phys.*, **1965**, *16*, 471.
143. F. J. Bryant and A. F. J. Cox, *Brit. J. Appl. Phys.*, **1965**, *16*, 1065.
144. G. A. Marlor and J. Wood, *Brit. J. Appl. Phys.*, **1965**, *16* 1449.
145. G. Grill, G. Bastide, G. Sagnes and M. Rouzeyr, *J. Appl. Phys.*, **1979**, *50*, 1375.
146. K. H. Nicholas and J. Woods, *Brit. J. Appl. Phys.*, **1964**, *15*, 783.
147. K. H. Nicholas and J. Woods, *Brit. J. Appl. Phys.*, **1964**, *15*, 1361.
148. G. A. Marlor and J. Woods, *Brit. J. Appl. Phys.*, **1965**, *16*, 797.
149. D. S.Orr and L. Clark, J. Woods, *Brit. J. Appl. Phys.*, (*J. Phys. D.*), **1968**, *1*, 1609.
150. U. Buget and G. T. Wright, *Brit. J. Appl. Phys.*, **1965**, *16*, 1457.
151. P. Besomi and B. Wessels, *J. Appl. PhysI.*, **1980**, *51*, 4305.
152. A. Thevenet, F. Juillet and S. J. Teichner, *Proc. 2nd Int. Conf. Solid Surfaces*, *Jpn. J. Appl. Phys., Suppl.*, **1974**, *2*, 529.
153. H. Al-Ekabi and N. Serpone in *Photocatalysis*, N. Serpone and E. Pellizetti, eds., John Wiley, New York, 1989, p.457.
154. P. de Mayo, K. Muthuramu, private communication.

Chapter 12

Photoprocesses in Environments of Complex Geometry: Fractal and Porous Materials

D. Avnir and M. Ottolenghi

Institute of Chemistry, The Hebrew University of Jerusalem, Jerusalem 91904, Israel

Contents

1. Introduction
1.1. The Scope of the Review

This chapter reviews some recent activity in the authors' laboratories in elucidating the effects of adsorption interactions and of surface geometry on the course of reactions at interfaces with special emphasis on bimolecular, diffusion-controlled processes, and especially on the Eley-Rideal mechanism. In relation to the geometry details of the surface, the studies concentrated on two (linked) morphology parameters: the average pore size (aps) and the fractal dimension, D, of the surface accessible for molecular interactions. The research approach combined the verification of theoretical predictions with the identification of empirical trends, using random-walk simulation, numerical solutions of reaction/diffusion equations, detailed experimentation, and reanalyses and reinterpretations (according to our approach) of experimental data obtained in other laboratories.

Specific topics include the study of the effects of changes in the accessibility of a photoreactive surface on the kinetics of photoprocesses; of the effects that specific surface morphologies have on the pattern of reaction probability distribution on the surface; of the link between surface roughness and particle size effects in photocatalysis; of the effects that specific patterns of distribution of photoactive sites have on the efficiency of the reactions; of the effects that the pore-size distribution and the surface fractal dimension have on bimolecular photoprocesses (diffusional or long-range interactions) such as charge transfers and energy transfers with various reaction models and in various reaction configurations in which the excited state molecule is either adsorbed on the pore surface or dissolved and confined in the pore volume.

A brief review including preliminary observations on some of these topics appeared recently,[1] and here we extend it, update it and emphasize different aspects. Geometry effects on photoprocess have been studied in several other laboratories. Most of them have contributions in this volume and are therefore not treated here; others are cited below in the appropriate context. The interested reader is also referred to a recent edited volume,[2] where additional relevant review articles are collected.

The topics listed above are summarized in two main parts. In Section 2 we concentrate on theoretical aspects and their verifications via simulations, numerical solutions, and reanalyses of published experimental data. In Section 3 we concentrate on experimental studies and in particular on diffusion-controlled bimolecular processes. Two aspects will be dealt with in detail: Analysis of effects on diffusion pathways which determine the bimolecular encounter rate; and the analysis of the reaction mechanism following the primary encounter, i.e., the "intrinsic" reaction mechanism.

1.2. Some Basic Aspects of Fractal Analysis

The literature on applications of fractal geometry to the study of complex-geometry systems is at this stage quite voluminous. Although it is beyond the scope of this review to go into the details of fractal geometry, we provide here some basic elements of the underlying approach of that novel mathematical tool. This by no means can replace a systematic study of the field, and so, for the interested reader we suggest to start with some popular reviews collected in ref. 3, to browse through some books,[4] conference proceedings,[5] and more topic-oriented reviews[6] all in order to get a general impression of the scope of the applications of fractal geometry in the natural sciences. Then it is time to learn about fractal geometry in chemistry; this can be done through a comprehensive book on that topic,[7] through a very short review[8] tailored to the busy scientist who wishes to get only a quick glance at the field, or through a number of more detailed reviews.[9]

Fractal geometry has proved to be a useful tool in dealing with physical problems that involve nontrivial geometries. By "nontrivial" we mean those cases that can be treated only with difficulty by the classical Euclidean geometry. Actually, this situation is much more common in the physical world than the "well-behaved" geometry cases. For our purpose, we note here surfaces and porous structures in particular. Models dealing with these material characteristics have usually taken simple structures, such as cylinders with smooth walls. However, the awkwardness of using these simple structural units for amorphous and fractured objects has never escaped researchers' attention in this area.

Fractal geometry[10,11] has certainly not completely solved these problems, but it has made important conceptual progress in treating complex geometries: these are dealt with, not as a deviation from some simple Euclidean reference, but as autonomous cases that do not rely on outside reference for evaluation.

The basic approach of fractal analysis is the following: It is possible to describe quantitatively a complex geometry of an object if the object is symmetric to transformation of scale, which means that the same type of geometry features are seen either at different magnifications or by probes of various sizes. More generally, a power-law scaling relation characterizes one or more of the properties of an object or of a process carried out near the object:

$$\text{property} \propto \text{scale}^{\beta}$$

Examples for "property" are the surface area, the rate of a heterogeneous reaction, or the shape of an adsorption isotherm. The scales, or yardsticks, would be pore diameter, cross-sectional area of an adsorbate, particle size, or layer thickness. The exponent β is an empirical parameter that indicates how sensitive the property is to changes in scale, and depending on the case, it can be either negative (e.g., in length measurements) or positive (e.g., in measurements of mass distribution). In many instances theoretical considerations can be used to predict the limits of β. For instance, in length measurements, $-1 < \beta < 0$. In other cases the bounds of β are more difficult to predict, especially when several parameters dictate that general relation.

Originally, β was developed to deal with the relation between purely geometric parameters (length, area) and the size of the yardstick used to determine these parameters. For reasons explained in detail in refs. 10 and 11, the noninteger exponent β has, in these purely geometric cases, the meaning of dimension. Mandelbrot coined the term "fractal dimension" D for this type of dimension. Figure 1 shows some fractal objects that obey the relation:

$$\text{length} \propto \text{yardstick}^{1-D}$$

and Figure 2 shows an object that obeys:

$$\text{numbers of tips} \propto \text{yardstick}^{D}$$

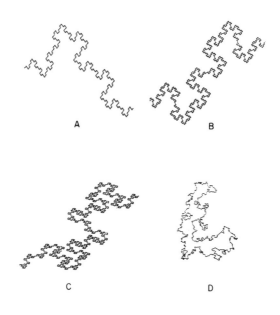

Figure 1. Some of the fractal objects used in this report. The D values are: A - 1.32; B - 1.45; C - 1.53; D - 1.33; and a straight line ($D = 1.00$). A, B and C are analytical fractals while D is a random fractal.

The extension of this concept to many "real" physical problems has become possible by recognizing that the *effective* geometries of various structures and for various processes can also be described in fractal geometry terms, resulting in an effective D_{property}. Examples include: D of the effective surface which is available for scattering of small angle x-rays[12] (here "property" is the intensity of the scattered rays and "scale" is the scattering angle); D of the surface which is available for a catalytic reaction[13] (here "property" is the rate of reaction and "scale" is the size of the catalyst particle); and D of the surface which is available for adsorption[14,15] (here "property" is the monolayer adsorption value and "scale" is the size of the adsorbate). Other examples for the application of this property/scale relation, appear in the following sections.

Figure 2. Illustration of a Devil-Staircase object. The results of the reaction with the tips of the stairs is shown in Figure 3. The fractal dimension of the collection of tips is 0.63.

2. Theoretical Studies
2.1. Surface Geometry Effects on the Eley-Rideal Mechanism—Random-Walk Simulation Studies

We analyze the basic case

$$B^* + S \rightarrow P \tag{1}$$

in which an excited-state molecule, B*, diffuses from the bulk and reacts with a (catalytic) surface, forming a product P (or is quenched to B) in a diffusion-limited process.[17-19] This is also known as the Eley-Rideal mechanism.[20]

Two approaches were used to analyze geometry effects on reaction (1). The first approach emphasizes global effects of irregularity and is based on the relation:[17,18]

$$B^*(t) = \text{const} \cdot t^{(3-D)/2} \tag{2}$$

in which B*(t) is the amount of B* molecules that have reacted after time t in the above mechanism, and D is the fractal dimension of the surface accessible to the interaction with B* molecules. Detailed random-walk simulations have indeed confirmed equation (2),[16,19] not only by retrieving the D value of the fractal object from the kinetic equation but also by retrieving the theoretical value of the prefactor in equation (2) (which is a function of the diffusion constant, the surface area, and the concentration). Interestingly, two opposite trends were found regarding the effects of increase in surface irregularity (increase in D): When the object size (particle size) is kept constant, the efficiency of the photocatalytic reaction increases with D (Figure 3); when the surface area is kept constant, the efficiency of the photocatalytic process increases as the surface becomes smoother (decreasing D; Figure 4). The difference between these two cases can be rationalized as follows: Keeping the total surface area fixed while changing D means that the proportion of screened surface sites increases with D; therefore, the $D = 2.0$ case is the most efficient. While the increase in screening with D occurs also for the fixed object size case, the increase in the number of easily accessible sites with D by virtue of the total increase in surface area is even faster under these conditions. (To understand the latter statement, notice that introducing even a mild irregularity can increase the surface area significantly, without affecting the exposure of the surface sites very much.)

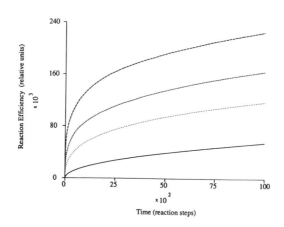

Figure 3. Increase in reaction efficiency with increase in D, for fixed object size. The D values from top to bottom are 1.53, 1.45, 1.32, and a line ($D = 1.00$). The three fractal objects are shown in Figure 1 (C, B, A, respectively). The reaction efficiency is given in units of the volume above the surface which has been depleted of starting material.

Common to many photochemical processes in confined environments is the situation in which the unimolecular self-decay of the excited state is on a similar time scale as its deactivation routes by collision with the surface. In such cases, the kinetics can be described not by equation (2), but by the product of probabilities of reaction routes. For a first-order decay of B* this will be $\sim B^*(t) \cdot e^{-t/\tau}$. The effect of geometry on the relative weight of the two decay routes is shown in Figure 5. As expected from the previous arguments, the relative weight of the self-decay of B* in the total decay rate is found to decrease with increase in D.

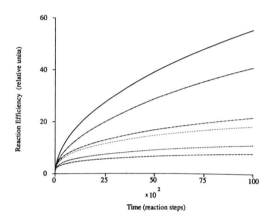

Figure 4. Increase in reaction efficiency with decrease of D, for fixed line length (fixed "surface area"). The D values from top to bottom are: 1.00 (a line); 1.08 (the hull of an Eden cluster)[4]; 1.31 (the random object shown in Figure 1-D); and the others are as in Figure 3. Reaction efficiency is determined as in Figure 3.

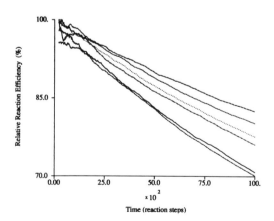

Figure 5. Effect of surface geometry on the relative contribution of self-decay ($k_r = 10^{-4}$) on scheme (1). Shown is the reacted volume with self-decay divided by the reacted volume without self decay. The D values from top to bottom are: 1.53, 1.45, 1.32, 1.31, 1.00 and 1.08. The objects are as in Figure 4.

A more detailed picture can be obtained by employing multifractal analysis.[21] This approach[22] analyzes the fractility of subsets in a given object. These subsets can be defined in various ways, one of which refers to collections of points that have the same reaction probability measures. As an example[21] let us look at reaction (1), performed on a devil's staircase[23] (Figure 2), which provides a realistic presentation of crystal defects.[24] Figure 6 shows the distribution of reaction probabilities along the surface, assuming that only the edges are active.

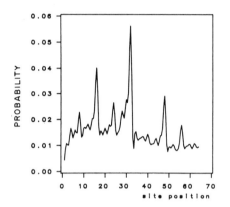

Figure 6. Distribution of reaction probabilities as a function of tip number from bottom to top in Figure 2.

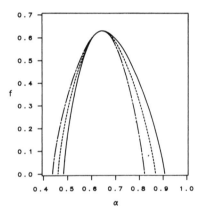

Figure 7. The $f(\alpha)$ spectrum for Figure 3. See text for explanation. (The different lines refer to different object sizes.)(ref. 21).

It is seen that the distribution function is definitely not smooth, with peaks of activity of different heights, reflecting the different accessibilities of each site. According to the multifractal formalism, one can present the data in Figure 6 in terms of a spectrum known as the $f(\alpha)$ spectrum:[22] Here, $f(\alpha)$ is the fractal dimension of the subset of all points that have a given probability measure α. This measure is defined through the relation:

(probability of reaction at site i) ~ (size of site i)$^{\alpha}$.

Figure 7 shows the $f(\alpha)$ spectrum corresponding to Figure 6, for which the distribution of the α-reaction probability measure is seen to be bell shaped. A detailed comparative analysis of these results is given in ref. 21.

2.2. The Effects of the Pattern of Distribution of Active Sites on the Eley-Rideal Mechanism—Numerical Solutions

The random-walk simulations described in the previous section are suitable for studying events on the single molecular level; on a larger scale, we study that problem by numerical solutions of diffusion/reaction equations. The Eley-Rideal mechanism (equation 1) has been chosen to explore the problem of how specific patterns of active-site distribution (fractal and nonfractal) affect the efficiency of the reaction. This situation mimics, e.g., the activity of specific sites on metal crystallites. The study was carried out by solving numerically (for technical details of the algorithms see refs. 25 and 26) in two dimension the following pair of coupled equations:

$$\partial[B^*]/\partial t = -k_r[B^*] + K_B\nabla^2[B^*] \qquad (3a)$$

$$\partial[P]/\partial t = k_r[B^*] + K_P\nabla^2[P] \qquad (3b)$$

where $[B^*]$ and $[P]$ are the concentrations of B^* and P (equation 1), k_r is the reaction constant and K_B, K_P are diffusion constants. B^* is homogeneously distributed above the surface in a finite reservoir. The active sites were placed on a (straight) line, and the following distributions were studied: equally spaced single sites; equally spaced pairs of active sites; and a Cantor-set fractal distribution with $D = 0.63$. (A Cantor set is a set of points spaced as the edge points of the devil's staircase in Figure 2.) In addition, we also studied the effects of randomizing the position of the active site, to mimic the dynamic nature of active-site location on the surface of metal catalysts. The results for two reaction rates are shown in Figure 8. It is seen that the least efficient distribution is the fractal one, that for a fixed pattern the equally spaced distribution is the most efficient one, that the efficiency increases with randomization rate, and that a slower reaction rate seems to blur the recognition of a pattern. These observations can be explained by the following picture[16]: A short while after reaction starts, the areas near the active sites are depleted of B^*, so that further reaction is very much dependent on the efficiency with which B^* can then reach the active zones. Thus, equal spacing or randomizations will be the most favorable conditions for such resupplying of unreacted B*. The decrease in the importance of the actual pattern as the reaction becomes more chemically controlled can be rationalized likewise: A slow reaction rate provides sufficient time for homogenization of the concentration profiles of the nonreacted B* molecules.

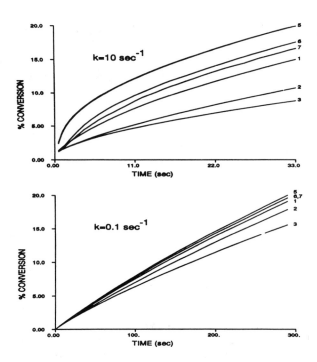

Figure 8. The effects of distributions of active sites on reaction efficiency: (1) Equally spaced single sites; (2) Equally spaced pairs; (3) Cantor fractal distribution ($D = 0.63$); (5) 10 randomizations/sec., any distribution; (6) 1 randomization/sec. of single sites; (7) 1 randomization/sec of pairs.

2.3. Effect of Surface Irregularity on Intermolecular Photoprocesses between Adsorbates: The Reaction Area[27]

We now change the configuration of the reaction and consider an intermolecular diffusional excited-state reaction between two adsorbates, B* and Q:

$$B^* + Q \rightarrow P \tag{4}$$

Some interesting phenomena emerge when one considers the effects of surface irregularity on that process, phenomena which originate from the different accessibilities of the surface toward B* and Q. For the sake of discussion, let us assume that B* is a smaller (subscript s) molecule than a large Q (subscript l). Obviously, the relation between the accessible areas, A, for these two molecules, is

$$A_S \geq A_l \tag{5}$$

where $A_S = A_l$ holds only for a smooth surface. In particular, it should be noted here that the surface area as measured with nitrogen, $A_{N_2} > A$ (of most organic molecules), from which it immediately follows that the N_2 surface area value *cannot* be used as a working number for calculating intermolecular distances between adsorbates (a practice that is used quite often in surface photochemistry studies). The only relevant area for that purpose is the area that is measured directly with either B or Q, or, if the fractal dimension of the area accessible for adsorption, D, is known, then:[28]

$$A = \text{const} \cdot \sigma^{(2-D)/2} \tag{6}$$

where σ is the cross-sectional area of the adsorbate. The preparation and use of calibration curves (equation 6) has been described.[29] The situation is further complicated by noticing that, in general, the ground state and excited state are isomers of each other, so that usually,

$$\sigma(B^*) \neq \sigma(B) \tag{7}$$

Current tools allow only for estimation of $\sigma(B^*)$ (which is still better than taking the N_2 surface area value).

If the surface is fractal, then one can estimate the fraction of area that is available to small molecules, but not to large ones, F, at any time:

$$F = \Delta A / A_S = 1 - (\sigma_l/\sigma_s)^{(2-D)/2} \tag{8}$$

where $\Delta A = A_s - A_l$. Notice that for a flat, $D = 2$ surface, $F = 0$, and that F grows with D. For instance, for $\sigma_l/\sigma_s = 8$ on a $D = 2.8$, $F = 50\%$.

Moving one step forward, we now recall that for any bimolecular process to take place, first an encounter complex, $(BQ)^*$, must form. By the same arguments presented above, $A_{BQ} \ll A_B$ or A_Q. A_{BQ} is the *only* area where the bimolecular reaction can take place, because only the features of that specific area are spacious enough to accommodate the encounter complex. We call this area, which is the only one relevant for the reaction, the *reaction area* ;[27] and again, it is smaller, not only than A_B or A_Q, but it is also much smaller than the standardly used

$A(\text{N}_2\text{-BET})$. To evaluate A_{BQ}, one has to estimate σ_{BQ}. We suggest that a good estimate is

$$\sigma_{BQ} \approx \sigma_B + \sigma_Q \tag{9a}$$

or

$$\sigma_{BQ} \approx \sigma_P \tag{9b}$$

if P is believed to be similar to $(BQ)^*$.

The realistic picture, therefore, is the following: B^* and Q diffuse in their own respective areas and there is more area for $B^*(s)$ to diffuse than for $Q(l)$. There is also a certain probability that their diffusional pathways will cross, but that is not sufficient: B^* and Q will react only if those crossings will occur in the small A_{BQ}. This, of course, is completely different from the situation in solution, where all of the volume is equally accessible to all starting materials and intermediates.

A related property that is unique to an irregular surface is the collapse of the classical meaning of the very concept of diffusional distance. Its value is now in the eye of the beholder: For the larger molecule the distance, d, to the smaller one is shorter than the opposite situation. For a fractal surface the relation is given by:

$$d(s{\rightarrow}l) \, / \, d(l{\rightarrow}s) = (\sigma_s/\sigma_l)^{(2-D)/2} \tag{10}$$

2.4. Fractal Analysis of Particle Size Effects in Photocatalysis

Size effects in fractal objects are revealed through the power law[10]

$$\text{property} \sim R^D \tag{11}$$

in which R is the object (particle) size. "Properties" that are found to obey equation (11) are the mass of aggregated objects,[30] the monolayer value in physisorptions,[8,28,31,32] the (catalytic or noncatalytic) reaction rate of particulate materials,[8,28,33] the activity of dispersed metals catalysts (structure–sensitivity relations),[34,35] and unsupported catalysts.[36] Although over 100 case analyses that obey equation (11) for surface interactions have been reported, the vast majority are ground-state reactions. Few cases of size effects in photocatalysis that obey equation (11) have been found and are they summarized in refs. 1 and 37. In these photochemical studies the relations

$$A \sim R^{D_R - 3} \tag{12}$$

$$A \sim R^{D_r - 3} \tag{13}$$

$$\Phi \sim A^{D_R - 3/D_r - 3} \tag{14}$$

were found, in which Φ is the quantum yield, D_R is the reaction dimension (the dimension of the collection of active sites), and D_r is the dimension of the surface available for physisorption. The areas determined by equations (12) and (13) may, but need not, coincide. The first is determined from the reaction rate (which is linearly proportional to the number of active sites), and the second is determined from an adsorption experiment. Examples for photochemical processes that obey equation (13) include hydrogenation and isomerization on binary TiO_2–Al_2O_3 photocatalyst particles, hydrogenation of acetylene on TiO_2, and water cleavage on Pt/TiO_2 and on $Pt/$(polyvinyl alcohol). The resulting D_R values in these analyses were interpreted as reflecting the collection of surface active sites which takes place in the process.

3. Experimental Studies
3.1. The Effects of Surface Morphology on Forster-Type Energy Transfer between Adsorbates

Fractal analysis is basically a resolution analysis: quantitative assessments of various properties are performed at various resolutions, i.e., with a set of yardsticks of various sizes. An interesting set of yardsticks is that of the intermolecular distances between two interacting molecules in the absence of diffusion, such as in nonradiative, one-step, Forster-type energy transfer between a donor and surrounding acceptors. Since the efficiency of this transfer process is distance dependent, then the specific profile of distribution of donor-to-acceptor distances, should show up in the details of the decay profile of the excited-state donor. Thus, the energy transfer in a three-dimensional solution environment behaves differently when limited to a two-dimensional plane; and for a D-dimensional environment it has been shown that:[38]

$$B^*(t) = e^{-\gamma(t/\tau^{(D/6)} + t/\tau)} \tag{15}$$

in which $B^*(t)$ is the survival probability of the excited donor, τ is its fluorescence lifetime, γ is a characteristic constant and D is the fractal dimension. γ and D are two adjustable parameters. For the adsorbed ionic donor–acceptor molecules, there is no diffusion. We showed that if the D-dimensional entity is a fractal surface (as indeed is the case if the interacting molecules are incorporated in that environment by adsorption) then γ becomes a simple linear function of D, and equation (15) reduces to a singly adjustable parameter equation (equation 16):[39]

$$\gamma = (N/A)(2\pi r^2/D) \; \Gamma(1 - D/6)(R_0/r)^D \tag{16}$$

where A is the surface area as measured with nitrogen (by the BET method), N is the number of adsorbed acceptor molecules per gram of absorbent, r is the radius of N_2, and R_0 is the Forster critical distance for a given donor–acceptor pair.

The validity of equation (15) was verified in a number of studies,[40] although it was argued that other models could be fitted to the observed decay profiles.[41] This problem, which is typical of virtually any attempt to fit a model to a decay profile, can be somewhat ameliorated by using the single adjustable parameter equation (15) + (16). Intensive experimental work has been performed by Huppert et al. and is recently reviewed.[42] Table 1 summarizes the results obtained for various SiO_2 materials. The D values from adsorption experiments are included for comparison purposes. There are two important aspects in that study: First, *different* donor–acceptor pairs, as well as donor–donor (fluorescence depolarization) and indirect donor–acceptor energy transfers were tested, resulting in virtually the same D values. Second, different surface coverages were tested, again resulting in unchanged D values. These two observations are in full agreement with what we expect to get from a fractal surface, namely, symmetry to dilation (i.e., invariance to magnification). The agreement with other molecule–surface interaction studies (Table 1) corroborates the fractal interpretation of these morphology-dependent energy-transfer results.

3.2. Surface Effects on Intrinsic Reactivities

In this section we shall consider surface geometry effects on the course of the reaction, which are associated with local, short-range, interactions between the reactants and the surrounding interface environment where the reaction takes place. We shall be mainly concerned with charge–transfer reactions, which are well known to be sensitive to environmental factors, such as the medium's polarity, as well as to the relative orientation of the interacting reactants. Both factors, as well as specific surface–adsorbate interactions, play a major role in determining the course of charge-transfer reactions at interfaces.

A model charge-transfer class of processes, which has been extensively studied in homogeneous solutions, is the reaction between excited organic electron acceptors (e.g., aromatics such as pyrene, Py) and electron donors (e.g., amines such as N,N'-diethylaniline, DEA). A major feature became clearly evident when such a reaction was carried out[42] on a series of silica surfaces with average pore size (aps) values ranging from 60 to 1000 Å (Figure 9).

Table 1. Determination of Surface Fractal Dimension of Various SiO_2 Materials from the Analysis of Energy-Transfer Kinetics.[39, 42]

Material	Pore Size (Å)/ surface area ($m^2\ gr^{-1}$)	Donor–Acceptor pair[a]	D	D from other adsorption studies
Silica	40/680	R6G/MG	2.96	
Silica	60/500	FL/DODCI	2.71	2.9–3.0[29,43,44]
		R6G/MG	2.71–2.78	
		RB/MG	2.82	
		RG6BG	2.71	
Silica	100/320	FL/DODCI	2.50	2.4[45]
		R6G/MG	2.51–2.57	
		RB/MG	2.57	
		R6G/R6b	2.50	
Silica	200/150	FL/DODCI	2.30	
		R6G/MG	2.32–2.35	
		RB/MG		
		R6G/R6G	2.30	
Silica	500/50	RB/MG	2.36	
Silica	1000/20	RB/MG	2.37	
Silica	2500/8	RB/MG	2.23	
Silica	5000/3	RB/MG	2.05	
Six controlled pore glasses	From 75/182 to 2000/13	R6G/MG	2.30–2.35 (all)	2.1–2.2[44,46] (2.20 from SAXS)[b]
Aerosil	nonporous /200	RB/MG	2.05	2.0–2.2[46,48]

[a] R6G, Rhodamine 6G; MG, Malachite Green; RB, Rhodamine B; FL, Fluorescein; DODCI, Diethyloxydicarbocyanineiodide,

[b] D(SAXS) is applicable in this case, but not for the silicas. See refs. 42 and 47 for a detailed explanation.

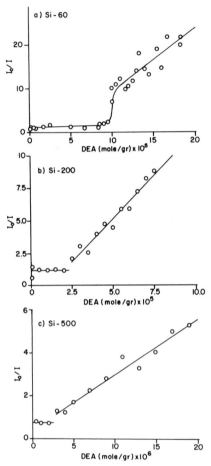

Figure 9. Stern-Volmer plots of the quenching of the pyrene fluorescence as a function of added (adsorbed) DEA on silicas with varying average pore size. Pyrene surface coverages: Si-60, $\theta_{Py} = 2.5 \times 10^{-3}$; Si-200 $\theta_{Py} = 5 \times 10^{-3}$; Si-500, $\theta_{Py} = 1.7 \times 10^{-3}$

In variance with the behavior in homogeneous solutions, quenching of the Py* fluorescence does not start from the very beginning of surface coverage by DEA, but only after an "induction" or "critical" amount of adsorbed DEA is reached. It turns out that in all of the above systems such critical amounts coincide with the DEA monolayer value as determined from the corresponding adsorption isotherms. Thus, no $^1Py^*$ quenching by DEA is observed as long as monolayer coverage is not attained. We have excluded inhibition of mutual surface diffusion of $^1Py^*$ and DEA as a possible reason for the lack of reactivity below $\theta_{DEA} = 1$ (where θ is the degree of coverage). Instead, it was shown that the latter phenomenon was due to specific adsorption interactions between DEA and the acidic hydrogens of the surface silanol groups. Stabilization of the nitrogen lone-pair electrons increases the DEA ionization potential to a level that completely inhibits the charge-transfer interaction with ^{1}Py*. It is only after the monolayer value is exceeded, or in the presence of coadsorbates that displace DEA from the surface, when "free" DEA molecules become available, that the quenching process is initiated.

To what extent does the above surface charge transfer (CT) quenching reaction resemble the analogous process in homogeneous solutions? In both cases, quenching of the ^{1}Py* fluorescence is accompanied by the appearance of a new red-shifted fluorescence band due to the excited complex (exciplex) formed between ^{1}Py* and DEA. A quantitative analysis of the surface data requires consideration of the general reaction scheme for the quenching of the fluorescence of an excited acceptor (^{1}B*, e.g., ^{1}Py*) by a ground-state donor (Q, e.g., DEA):

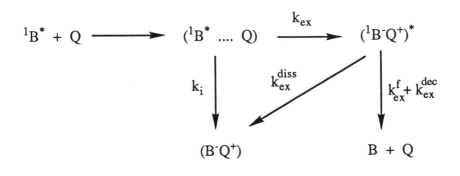

Scheme 1

in which ^{1}(B*... Q), ^{1}(B$^-$ Q$^+$)*, and (B$^-$ Q$^+$) represent the locally excited state, the fluorescent exciplex and the (geometrically correlated) ion pair, respectively. In nonpolar solutions $k_{ex} \gg k_i$ and $k_{ex}^f + k_{ex}^{dec} \gg k_{ex}^{diss}$ where k_{ex}^f and k_{ex}^{diss} are the rate constants for the fluorescent and nonfluorescent decay pathways and k_{ex}^{diss} measures the exciplex dissociation route). In polar solutions the situation is reversed, leading to a marked decrease in the relative exciplex fluorescence yield, Φ_e^f. The solvent polarity also effects the wavelength of maximum exciplex emission, λ_{ex}. Due to its polar nature, the exciplex is stabilized in polar solvents, resulting in a red-shifted emission.

Figure 10 compares these two polarity effects in homogeneous solutions with the corresponding phenomena as measured on silica (Si) surfaces. Rationalization of the data is best carried out by comparing the two extreme Si systems: the wide-pore Si-1000 (silica with aps of 1000 A), and the narrow-pore Si-60.

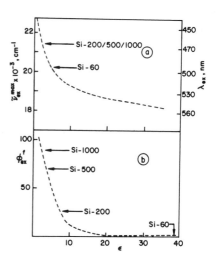

Figure 10. Py/DEA exciplex emission energies (a) and relative intensities, Φ^f_{ex}; (b) in homogeneous solutions as a function of the solvent dielectric constant (ε). Arrows for Si-60, Si-200, etc. denote the values of υ^{max}_{ex} (a) and Φ^f_{ex} (b) observed on the corresponding silica surfaces.

In the first case both the exciplex fluorescence yield and its emission frequency are comparable to the corresponding parameters observed in low-polarity homogeneous solutions such as toluene ($\varepsilon = 2.4$, where ε is the dipole moment). It was concluded that on the flat Si-1000 surface (covered by the ordered phenyl blanket formed by the DEA monolayer) the reaction between ^{1}Py* and DEA takes place in the (aromatic) low-polarity environment and is free from geometrical restrictions for generating the exciplex.

However, this solution-like behavior is not maintained in the case of Si-60. Thus, as shown in Figure 10 the Φ^f_{ex} value on Si-60 is much lower than on Si-1000, being comparable to that in a highly polar solvent such as ethanol ($\varepsilon = 23.9$). However, the value of λ_{ex} is similar to that of a relatively nonpolar environment ($\varepsilon = 5$). This implies that the low exciplex yield on Si-60 cannot be attributed to a high local surface polarity (i.e., to large k_i and k^{diss}_{ex}). Instead it was argued that on Si-60 the mutual orientation of ^{1}B* and Q is unfavorable for generation of the sandwich-like exciplex (small k_{ex}), leading to deactivation via the ground-state ion pair (B$^-$ Q$^+$). This conclusion is consistent with accumulated evidence obtained in homogeneous solutions showing that k_{ex} is much more sensitive than k_i to the relative orientation of ^{1}B* and Q. The basic argument is therefore that the surface geometry of the silicas plays an important role in dictating the ease with which the optimal exciplex alignment becomes possible: The extreme surface irregularity of Si-60 interferes with achieving the proper B–Q orientation for exciplex formation. This basic difference between the extreme cases of Si-1000 and Si-60 is also reflected by their different behavior with respect to

Eley-Rideal reaction mechanisms as well as with respect to their surface irregularity as measured by the fractal dimension.

In the above discussion we have seen that surface geometry and adsorption interactions affect the course of bimolecular, adsorbed-state Langmuir-Hinschelwood kinetics, charge-transfer reactions on surfaces by affecting the generation of sandwich-like exciplexes. Similar phenomena should also be observed in the case of Eley-Rideal reactions, e.g., in which B* is bound to the surface and Q diffuses from an adjacent liquid phase. Moreover, since chiral enantiomers may exhibit different adsorption interactions, such photoreactions between chiral B and Q species can be expected to lead to the photophysical "recognition" of chiral surfaces.

Experiments confirming these expectations were carried out in Si systems in which either of the two enantiomers R(-) and S(+) of the chiral acceptor 1,1-binaphthyl-2,2-dihydrogen phosphate were covalently bound to porous silica.[50] The excited bound enantiomer, R or S, was exposed to quenching by a chiral electron donor N,N-dimethyl-1-phenylamine diffusing from a nonpolar (cyclohexane) solution. A substantially higher (~ 30%) quenching rate constant of the R-surface by the S-quencher was observed relative to quenching by the R-quencher. A similar photophysical recognition of the chiral surface was observed when the S-surface was quenched by the R and S quenchers, respectively. The results were rationalized in terms of Scheme 1 by assuming that k_{ex}, which depends on the relative orientation of $^1B*$ and Q, is different for the two pairs of enantiomer combinations. On the other hand, the ion-pair route (k_i), associated with long-range electron transfer, is less sensitive to geometry and is thus comparable for both enantiomer pairs. This mechanistic interpretation is confirmed by the absence of any chiral discrimination when the cyclohexane liquid phase is replaced by a polar solvent such as methanol. As discussed above, in the latter case $k_i >> k_{ex}$ so that the exciplex route, responsible for chiral discrimination, is unavailable.

It is relevant to consider the factors that determine the effective polarity of the local microenvironment for a reaction carried out at a solid–liquid interface.[51] In the above naphthyl–amine system we have seen that the polarity is determined by that of the adjacent liquid phase. This is a case in which the reactants interact weakly with the surface so that the naphthyl moiety, which is covalently bound to the surface, is practically surrounded by solvent molecules. This is not the case when the reactant molecule is tightly bound to the surface. Relevant information in this respect was obtained by studying the effects of adsorption on the dual fluorescence of the probe molecule 1-(N,N-dimethylamino)-4-benzonitrile (DMABN) described by the scheme:

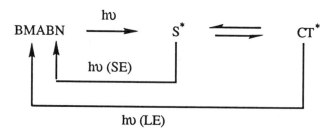

Scheme   2

where S* represents a planar nonpolar excited state characterized by a short wavelength emission (SE), and CT* is a twisted intramolecular charge-transfer state with a red-shifted emission (LE). Both energy and relative intensity (LE/SE ratio) of the LE emission are monitors of the local interface environment.

On plain silica surfaces the CT* state (originating in DMABN hydrogen bonded to surface silanols) undergoes effective solvation by orientational relaxation of neighboring silanols. The apparent surface polarity is comparable to that of a homogeneous ethanol solvent. It was also concluded that solvation of the CT* state and consequently the effective surface polarity is not affected by the presence of an adjacent cycloxane phase (although the latter does increase the rotational freedom about the C—N bond). This is in keeping with the strong adsorbant–adsorbate (amine–hydroxyl) interactions in the system. It is only when such interactions are disturbed, so as to displace the adsorbate from the vicinity of the silanols, that the effective polarity measured by DMABN is affected. This may be achieved either by coadsorption with alcohols or by covalently reversing the silica phase with aliphatic chains.

3.3.   The Diffusion Mechanism
3.3.1.   Diffusion on the Surface

The pyrene excimer formation process was studied by us on silica surfaces as a model for diffusion-controlled reaction. Steady-state and time-resolved fluorescence measurements, on a variety of unmodified silica surfaces, revealed that excimer generation was due to distinct nonequilibrium populations of close-lying ground-state molecules.[52] Excimer formation was found to be associated with intermolecular rearrangements occurring over time scales of up to several nanoseconds. Unfortunately, because the initial reactant distribution on the surface was not homogeneous, it was impossible to correlate the exciplex growing-in process with a surface diffusion mechanism.

The latter problem appears to be less serious on silica surfaces modified by chemical derivatization or by coadsorption with, e.g., alcohols. In the latter systems the pyrene surface mobility increases due to weaker adsorption interactions, allowing diffusion-limited interactions between nonneighboring molecules.[52] In such a case exciplex generation times are extended to the

10–100 ns range, yielding a value of 19 kJ mol^{-1} for the activation exciplex formation (on a C_{18} modified silica surface). However, even on such modified silicas, the contribution of "static" quenching mechanism (taking place over sub-nanosecond scales between ground-state aggregates) was substantial. This prevented any quantitative analysis of the diffusion-limited reaction in terms of specific surface parameters such as average pore size or surface irregularity (fractal dimension).

3.3.2. Diffusion from the Solvent Phase to the Surface

As mentioned above, a basic problem that has generated considerable interest is the effect of surface geometry on the rate at which a "marked" (e.g., excited) molecule diffuses from a solution to a reactive (e.g., quenching) surface, or one in which the "marked" (e.g., excited) molecule is situated at the solid–liquid interface and reacts with a molecule (quencher) diffusing from the intrapore liquid phase, in an Eley-Rideal type process. We have carried out an experimental study[53] of the latter system by applying a well-defined reaction between a static reactant (in this case an excited state, A*) adsorbed on the surface of a porous material and a mobile reactant (in this case a quencher, Q) diffusing to the surface from an intrapore liquid phase (i.e., an Eley-Rideal surface reaction). Characteristic results shown for the specific B = [Ru(bpy)$_3$]$^{2+}$, Q = anthracene pair are shown in Figure 11. The experiments indicate that the quenching reaction in the case of the narrow-pore glasses Si-100, Si-60, and Si-40 are associated with deviations from a single exponential behavior and are also accompanied by reduction of the effective quenching efficiency with respect to the wide-pore Si-1000 and controlled porous glass series.

Figure 11. Luminescence decay profiles of Ru(bpy)$_3$$^{2+}$ on various porous silica surfaces in the presence of anthracene (5x10^{-2}M) in an intrapore methylene chloride liquid phase. From top to bottom the decays are on Si-40, Si-60 and Si-100, respectively. The fractal dimensions of the surfaces, as determined by energy transfer experiments,[40,42] are $D = 2.96$, $D = 2.7$, and $D = 2.5$, respectively.

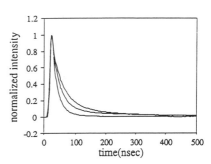

Since porous silicas and controlled porous glasses of comparable average pore size values behave differently, it was impossible to correlate the reduced reactivity in the Si-1000–Si-40 series to the reduction of the pore size. Other factors related to the irregularity of the surface were invoked. It was concluded that not all B*

molecules react with Q with an equal probability, as they do on the relatively "flat" controlled pore glass or wide-pore Si-1000 surfaces. Accordingly, the long-lived decay component was attributed to a fraction of B* molecules that are not easily accessible to Q, i.e., to a screening effect imposed by the tortuosity of the surface. The more accessible of the excited molecules react first, representing the fast initial decay component, followed by the less accessible ones, reflected by the longer-lived decay. Computer simulations suggested that the surface irregularity expressed by the above kinetic measurements might be correlated with the fractal nature of the porous silica surfaces. This is shown in Figure 12 which represents Monte Carlo simulations of a reaction involving a quencher diffusing from a two-dimensional space to a fractal line with adsorbed B* molecules. It is evident that the efficiency of the reaction decreases with the increase in the fractal dimension of the surface. The behavior is qualitatively similar to that of the experimental systems of Figure 11 in which the reactivity correlates with the fractal dimension of the surface as determined by energy-transfer experiments.

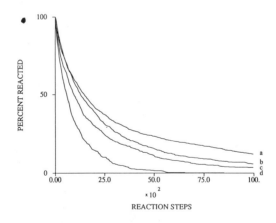

Figure 12. Computer simulations showing the decay of an excited molecule (B*) adsorbed at a solid-liquid interface, quenched by a molecule (Q) diffusing to the surface from an adjacent solution. Surface coverage by B* was 0.5% and $[Q] = 10^{-3}$ M/pixel^2. The fractal dimensions of the interfaces are (Figure 1): (a) $D = 1.53$; (b) $D = 1.45$; (c) $D = 1.32$; (d) $D = 1$.

3.4. Applications to Light-Energy Conversion

In the previous sections we have discussed the major parameters that affect the reactivity of surface-trapped molecules (adsorbed or covalently bound) with respect to reagents that are free to diffuse in the adjacent intrapore liquid phase. We now consider the application of such principles to the stabilization of photoinduced charge separation, which constitutes a key issue in the design of artificial photosynthetic systems.

Consider, for example, the following characteristic sequence of reactions:

$$B^* + Q \rightarrow B^- + Q^+ \tag{17}$$

$$B^* + Q^- \rightarrow B + Q \tag{18}$$

$$4Q+ + 2H_2O \xrightarrow{Cat} 4Q + O_2 + 4H^+ \tag{19}$$

$$2B^- + 2H_2 \xrightarrow{Cat} 2B + H_2 + 2OH^- \tag{20}$$

in which the back reaction (18) inhibits the catalyzed decomposition of, e.g., water into a useful fuel such as H_2. This goal is difficult to achieve in homogeneous solutions where the primary redox photoproduct Q^+ and B^- are free to diffuse. Thus, a wide variety of organized microenvironments have been applied for retardation of reaction. These are usually based on the compartmentalization of Q^+ and B^-, using boundaries of micelles or membranes, or through electrostatics or hydrophobic effects.

We have recently suggested[54] an alternative approach, based on the following reaction scheme, taking place at the solid–liquid interface of an inert porous solid:

$$A^*_{tr} + B_{sol} \longrightarrow A^-_{tr} + B^+_{sol} \tag{21}$$

$$A^-_{tr} + B^+_{sol} \longrightarrow A_{tr} + B_{sol} \tag{22}$$

$$B^+_{sol} \rightleftharpoons B^+_{tr} \tag{23}$$

$$B^+_{tr} + C_{tr} \longrightarrow B_{sol} + C^+_{tr} \tag{24}$$

where A^*_{tr} is an excited electron acceptor, trapped on the surface by adsorption or by covalent binding, while B_{sol} is a primary donor diffusing in the intrapore liquid phase. The back reaction (equation 22) between the charge-separated pair will obviously take place analogously to the homogeneous general process (equation 18). However, the solid–liquid heterogeneous microenvironment offers two potential routes for escaping charge recombination: First, B^+_{sol} may be removed from the intrapore liquid phase by trapping at the interface, e.g., by adsorption, forming B^+_{tr} via reaction (23). The strength of the respective adsorption interactions will determine the rate of the back electron transfer between B^+_{tr} and A^*_{tr} (via B^+_{sol}). Alternatively, B^+_{sol} may react with a secondary (trapped) donor, C_{tr} yielding a stable charge-separated pair $A^-_{tr} + C^-_{tr}$. In the second case B_{sol} acts as a diffusive charge carrier (shuttler) between the immobilized A^*_{tr} and C_{tr}. Stabilized species, such as A^-_{tr} and B^-_{tr} or C^-_{tr} may then react with the adjacent aqueous phase via catalyzed routes such as reactions (19) and (20), yielding molecular hydrogen and oxygen.

Implementation of the above principles has been recently achieved by immobilizing A_{tr} (and C_{tr}) in transparent porous oxide glasses, by the so-called "sol–gel" process. The method, which replaces conventional immobilization by adsorption or by covalent binding, is based on the addition of A and C to a polymerizing mixture of silanes. The product is a doped porous oxide glass that is both inert and transparent. At variance with the classical methods of preparing oxide glasses by high-temperature melting techniques, essentially any molecules, including organic and biological ones, may be incorporated in the porous matrix. Moreover, we have shown that a substantial fraction of such trapped molecules is accessible to reagents diffusing in the intrapore liquid phase.

Figure 13 shows kinetic data for the redox pair $A_{tr} = [Ir(bpy)_2(C^3,N')bpy]^{3+}$, an iridium tris-bipyridyl complex abbreviated Ir(III), and $B_{sol} = 1,4$-dimethoxy-benzene (DMB). Ir(III) is trapped in a porous silica glass generated by the "sol–gel" process, while DMB is dissolved in the aqueous intrapore phase. Pulsed laser excitation of Ir(III) is followed by a charge-transfer quenching reaction generating the $Ir(II)_{tr}, DMB^+_{sol}$ radical pair. After a relatively fast primary recombination, a retardation of four orders of magnitude of the back electron transfer process, with respect to homogeneous solutions, was observed at neutral pH for ~ 25% of the generated Ir(II). An analysis of pH and ionic strength effects indicated that the effect is due to the adsorption of DMB^+ on the pores' surface (yielding B^+_{tr}) along with the immobilization of Ir(II) by trapping in the glass matrix (i.e., A^-_{tr}). At acidic pH, retardation of the back reaction leads to catalyzed hydrogen generation from water via:

$$Ir(II) + H^+ \qquad \rightarrow \qquad [Ir(IV)H^-]$$

$$Ir(IV)H^+] + H^+ \qquad \rightarrow \qquad Ir(IV) + H_2$$

Work is in progress in our laboratory aiming at stabilizing charge separation in ternary donor–acceptor systems by implementing reaction (24) and by controlling the geometry features of the glasses, i.e., the surface area, the surface fractal dimension, and the average pore size.

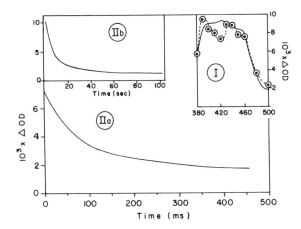

Figure 13. (I) Transient absorbance changes (ΔOD) following excitation of the trapped Ir(III) complex in the presence of DMB at neutral pH. Insert: -0-, absorbance changes recorded 200 μs after the laser pulse. The continuous line is the difference spectrum, due to the formation of Ir(II) and DMB^+ from Ir(III) and DMB, calculated from the absorption spectra of the respective species in aqueous solutions. (IIa,b) Decay profiles of Ir(II) generated by pulsed or continuous excitation in the presence of DMB at neutral pH. [DMB] = 7 mM; [Ir(III)] ≅ 5×10^{-7} mole/g. (a) Transient signal following N_2 laser pulse observed at λ = 290 nm; (b) Following continuous Hg lamp irradiation, measured with a diode-array spectrophotometer at λ = 400 nm.

4. Concluding Remarks

The aim of this report has been to draw attention to the importance of the geometry factor in heterogeneous photochemistry. Two geometrical parameters have been focussed on in this report: the surface fractal dimension and the average pore size. We used these to demonstrate through experiments, theory and simulation, how geometry affects the kinetics of arrival to a catalytic surface; how a specific pattern of distribution of active sites affects the efficiency of a photoprocess; we developed the concept of "reaction area" which accounts for the structural availability of an irregular surface to reactions between adsorbates; we showed how the analysis of particle size effect should incorporate surface roughness in it; we provided and analyzed experimental evidence to the effect of surface morphology on non-diffusional energy-transfer between adsorbates; we provided and analyzed experimental evidence for the effects of surface morphology on collisional electron transfer between adsorbates; we analyzed the effects of *local* geometry effects (in contradistinction with the globality of average pore size and surface fractal dimension) through the analysis of photo-interactions with chiral surfaces; we analyzed experimental data on the effects of pore-structure on diffusion from the bulk to an excited state surface; we gave an example for the experimental utilization of this new knowledge for practical applications of solar-light conversion; and there is still a lot left to do.

Acknowledgments

The authors acknowledge the sponsorships of the Fritz Haber Research Center for Molecular Dynamics and of the L. Farkas Center for Solar Energy Conversion (Minerva-Hebrew University Centers). They are deeply grateful to their collaborators: D. Farin, R. Gutfraind, J. Samuel, A. Seri-Levy, A. Slama-Schwock, M. Sheintuch, O. Citri, E. Wellner, H. Birenbaum, and R. Kosloff for their valuable contributions to the material reviewed here.

References

1.   A. Seri-Levy, J. Samuel, D. Farin and D. Avnir, in ref. 2, 353.
2.   *Studies in Surface Geometry and Catalysis Vol. 47. Photochemistry on Solid Surfaces*, M. Anpo and T. Matsuura, eds., Elsevier, Amsterdam, 1989.
3.   T. Waters, *Discover*, 1989, March, 26.
     J. Fricke, *Phys. Unserer Zeit*, 1986, *17*, 151 .
     K. Stein, *Omni*, 1983, Feb., 63.
     B. B. Mandelbrot, *Encylopedia Britanica Year Book of Science and Future*, 1981, 168.
     B. Schechter, *Discover*, 1982, June, 66.
     P. Engel, *The Sciences*, 1983, Sept./Oct., 63.
     J. Kappraff, *Comp. Math. Appl.* 1986, *12 B*, 655.
     B. Mandelbrot, *New Scientist*, 1990, Sept.*15*, 38.
4.   R. Jullien and R. Botet, *Aggregation and Fractal Aggregates*, World Scientific, Singpore, 1987.
     B. H. Kaye, *A Random Walk through Fractal Dimensions*, VCH, Weinheim, 1989.
     T. Vicsek, *Fractal Growth Phenomena*, World Scientific, Singapore 1989.
     J. Feder, *Fractals*, Plenum Press, N.Y., 1988.
     H. Takayasu, *Fractals in the Physical Sciences*, Manchester University Press, Manchester, 1989.
5.   *Fractals in Physics*, L. Pietronero and E. Tossatti, eds., North Holland, Amsterdam, 1986.
     On Growth and Form, H. E. Stanley and N. Ostrowski, eds., Nijhoff, Dordrecht, 1986.
     Random Fluctuations and Pattern Growth, H. E. Stanley and N. Ostrowski, eds., Kluwer, Dordrecht, 1988.
     Fractals, G. Cherbit, ed., Masson, Paris, 1987 (in French).
6.   H. E. Stanley, *Physica D*, 1989, *38*, 330.
     L. M. Sander, *Sci. Am.*, 1987, Jan., 82.
     B. J. West and A. L. Goldberger, *Am. Sci.* 1987,*75*, 354.
7.   *The Fractal Approach to Heterogeneous Chemistry: Surfaces, Colloids, Polymers*, D. Avnir, ed., J. Wiley, Chichester, 1989. (reprinted 1990)
8.   D. Avnir and D. Farin, *New J. Chem.*, 1990, *14*, 197.
9.   P. Meakin, *C. R. C. Solid State & Material Sci.*, 1987, *13*, 143.

P. Pfeifer, *Chimia*, **1985**, *39*, 120.

J. E. Martin and A. J. Hurd, *J. Appl. Crystallogr.*, **1987**, *20*, 61.

P. Pfeifer in *Preparative Chemistry using Supported Reagents*, P. Laszlo, ed., Academic Press, New York, **1987**.

10. B. B. Mandelbrot, *The Fractal Geometry of Nature*, Freeman, New York, **1982**.

11. P. Pfeifer and M. Obert, in ref. 7, chapter 1.2.

12. H. D. Bale and P. W. Schmidt, *Phy. Rev. Lett.*, **1984**, *53*, 596.

13. S. D. Jackson, *Reaction Kinet. Catal. Lett.*, **1989**, *39*, 223.

14. F. M. Gasparini and S. Mhlanga, *Phys. Rev.*, *B*, **1986**, *33*, 5066.

15. H. Spindler, P. Szargan and M. Kraft, *Z. Chem.*, **1984**, *27*, 230.

16. D. Avnir, O. Citri, D. Farin, M. Ottolenghi, J. Samuel and A. Seri-Levy, in *Optimal Structures in Heterogeneous Chemistry*", P. Plath, Ed., Springer, Berlin, *1989*, p. 65.

17. P. G. de Gennes, *C.R. Acad. Sci. Paris, Ser. II*, **1982**, *295*, 1061.

18. P. Pfeifer, D. Avnir and D. Farin, *J. Stat. Phys.*, **1984**, *36*, 699; **1985**, *39*, 263.

19. A. Seri-Levy and D. Avnir, *Surf. Sci.*, in press.

20. A.W. Adamson, *Physical Chemistry of Surfaces*, 5th ed., J. Wiley, New York, **1990**.

21. R. Gutfraind, M. Sheintuch and D. Avnir, *Chem. Phys. Lett.*, **1990**, *174*, 8.

22. S. Havlin and B. L. Trus, *J. Phys.*, **1988**, *A 21*, L731.

23. Ref. 10, 286.

24. S. E. Burkov, *J. Phys.*, **1985**, *46*, 317.

25. O. Citri, M. L. Kagan, R. Kosloff and D. Avnir, *Langmuir*, **1990**, *6*, 559.

26. D. Kosloff and R. Kosloff, *J. Comput. Chem.*, **1983**, *52*, 35.

27. D. Avnir, *J. Am. Chem. Soc.*, **1987**, *109*, 2931.

28. D. Farin and D. Avnir, in ref. 3, Ch. 4.1.2.

29. D. Farin, A. Volpert and D. Avnir, *J. Am. Chem. Soc.*, **1985**, *107*, 3368, 5319.
    A. Y. Meyer, D. Farin and D. Avnir, ibid., **1986**, *108*, 7879.
    D. Farin and D. Avnir, *J. Chromat.*, **1987**, *406*, 317.

30. M. Matsushita, in ref. 7, Ch. 3.1.3.

31. C. Fairbridge, H. S. Ng and A. D. Palmer, *Fuel*, **1986**, *65*, 1759.

32. D. Avnir, D. Farin and P. Pfeifer, *J. Coll. Interface* Sci., **1985**, *103*, 112.

33. D. Farin and D. Avnir, *J. Phys. Chem.*, **1987**, *91*, 5517.

34. C. Fairbridge, *Catal. Lett.*, **1989**, *2*, 191.

35. D. Farin and D. Avnir, *J. Am. Chem. Soc.*, **1988**, *110*, 2039.

36. D. Farin and D. Avnir, Preprints of the IUPAC Symp. Characterization of Porous Solids (COPS II), Alicante, Spain, May 1990, 172.

37. D. Farin, J. Kiwi and D. Avnir, *J. Phys. Chem.*, **1989**, *93*, 5851.

38. J. Klafter and A. Blumen, *J. Chem. Phys.*, **1985**, *80*, 874.

39. D. Pines-Rojanski, D. Huppert and D. Avnir, *Chem. Phys. Lett.*, **1987**, *139*, 109.
    D. Pines, D. Huppert and D. Avnir, *J. Chem. Phys.*, **1988**, *89*, 1177.

40. Y. Lin, M. C. Nelson and D. M. Hanson, *J. Chem. Phys.*, **1986**, *86*, 158.
    A. Takami and M. Mataga, *J. Phys. Chem.*, **1987**, *91*, 618.

41. P. Levitz and J. M. Drake, *Phys. Rev. Lett.*, **1987**, *58*, 686.

42. D. Pines and D. Huppert, *Isr. J. Chem.*, **1989**, *29*, 473.
    D. Pines and D. Huppert, *J. Chem. Phys.*, **1989**, *91*, 7291.

43. D. Avnir and P. Pfeifer, *Nouv. J. Chim.*, **1983**, *7*, 71.

44. S. V. Christensen and H. Topsoe, in ref. 7, p. 275.

45. H. Frank, H. Zwanziger and T. Welch, *Z. Anal. Chem.*, **1987**, *326*, 153.

46. L. D. Stremer, D. M. Smith and A. J. Hurd, *J. Coll. Interface Sci.*, **1989**, *131*, 592.

47. A. Hohr, H. B. Neuman, M. Steiner, P. W. Schmidt and D. Avnir, *Phys. Rev. B*, **1988**, *38*, 1462.

48. D. Avnir, D. Farin and P. Pfeifer, *J. Chem. Phys.*, **1983**, *79*, 3566.

49. H. Birenbaum, D. Avnir and M. Ottolenghi, *Langmuir*, **1989**, *5*, 48.

50. D. Avnir, E. Wellner and M. Ottolenghi, *J. Am. Chem. Soc.*, **1989**, *111*, 2001.
51. A. Levy, D. Avnir and M. Ottolenghi, *Chem. Phys. Lett.*, **1985**, *121*, 233.
52. D. Avnir, R. Busse, M. Ottolenghi, E. Wellner and K. A. Zachariasse, *J. Phys. Chem.*, **1985**, *89*, 3521.
53. J. Samuel, M. Ottolenghi and D. Avnir, *J. Phys. Chem.*, **1991**, *95*, 1890.
54. A. Slama-Schwok, D. Avnir and M. Ottolenghi, *J. Phys. Chem.*, **1989**, *93*, 7544.

Chapter 13

Recovery of Fluorescence Lifetime Distributions in Heterogeneous Systems

William R. Ware

Photochemistry Unit, Department of Chemistry, University of Western Ontario, London, Ontario, Canada.

Contents

1. Introduction

The exquisite sensitivity associated with fluorescence detection has over the last several decades given rise to the use of excited molecules as probes into a variety of heterogeneous systems. With the advent of the time-correlated single-photon technique for fluorescence decay measurements,[1-3] the experimenter has available a method for determining fluorescence decay curves with great sensitivity, precision, and accuracy. In addition, it has become possible to measure time-resolved fluorescence spectra, with nanosecond or subnanosecond resolution, as an adjunct to normal fluorescence spectra, that represent an average over time.[1]

Traditionally the single-photon technique has employed nanosecond hydrogen or nitrogen flashlamps, but more recently many laboratories have been using a mode-locked ion or Nd-Yag laser pumping a dye laser as an excitation source. Cavity dumping is used to control the time between pulses, and frequency doubling yields pulses in the uv. These laser sources provide excitation pulses of a few picoseconds width, and channel plate detectors typically have a response time of 50–75 ps. Thus, with appropriate deconvolution techniques, fluorescence decay times as short as 10 ps can now be measured. By way of comparison, using the conventional flashlamp with a width of about 1.6 ns, single exponential decays with lifetimes of approximately 200 ps can typically be recovered. Since the single-photon technique accumulates and stores the decay data in digital form in a multichannel pulse height analyzer (MCA), the potential exists for collecting data to very high precision.

There is an additional light source of significance that is used for time-correlated single-photon decay studies. This is the pulsed light that originates from a synchrotron storage ring loaded with one or more "bunches" of electrons.[4] As the electrons rotate they emit so-called synchrotron radiation with wavelengths from the x-ray region to the ir. The pulsewidth of the emitted light observed through a port is determined by the frequency of circulation, the width of the bunch in orbit, and the geometry of the sample excitation system. For example, the new Super-ACO (Anneau de Collision d'Orsay) ring at Orsay, France yields light pulses of approximately 500 ps width at repetition rates of 8.3 or 12.5 mHz when loaded with two or three bunches.[5] As with laser- or flashlamp-based systems, the detector is either a photomultiplier or a channel plate, and conventional time-correlated single-photon electronics are employed. Synchrotron radiation has the very great advantage of a light pulsewidth totally independent of wavelength, but the large repetition rate can be a disadvantage when measuring longer lifetimes, since the luminescent system is repumped before decay is complete, and numerical analysis must take this into account. Storage rings are of course large, expensive, multiuser installations, but there are several in operation with a port dedicated, at least partially, to fluorescence lifetime measurements.

The pulse method for fluorescence decay measurements has clearly evolved, such that today it represents a technique of considerable sophistication and power. It is only natural that it should be applied to systems of ever growing complexity and that the resultant decay data are not, in general, represented by a single exponential function. Such examples abound in the literature. Fluorescence probes have been employed extensively in systems consisting of micelles, vesicles, Langmuir-Blodgett films, and polymers in solution and in rigid media, and in systems of naturally occurring macromolecules, such as proteins and enzymes. Native fluorescent amino acids, such as tryptophan, on occasion serve as the probes in these biologically interesting macromolecules. Fluorescence molecules have been used to probe surfaces and cavities such as those found in cyclodextrins and zeolites. Immunofluorescence techniques depend for their sensitivity on the presence of a fluorescence probe and, in some protocols, the sensitivity is greatly enhanced by placing the probe in the protective environment of a micelle.

When fluorescence decay techniques are employed in these heterogeneous systems, it is quite common for the single exponential function to fail as a representation of the data. If this happens then two exponentials are tried. If two fail, then three are tried, and then four exponentials. This sequence of trial functions involves progressing from one to seven independent fitting parameters. It is not generally recognized how powerful the three- and four- exponential functions are in that they will fit virtually any decay data and yield random residuals and a flat autocorrelation plot. Herein lies the problem. If, for example, it is found that the data fail to fit a two-exponential function, but are represented well by a three exponential function, the temptation exists to consider simple models that predict just three exponentials. One is then tempted to attach physical significance to some or all of the five parameters and perhaps even to measure temperature coefficients and calculate activation parameters. The same scenario can be described for the case where three exponentials fail to fit the data properly and four exponentials are resorted to, with the associated seven fitting parameters.

In many heterogeneous systems a model may be considered that involves distributions of lifetimes. In the absence of methods for recovering such distributions this initial hypotheses may be rejected in favor of models that are more tractable, i.e., models that generate one to four exponentials for the predicted decay law. However, it has recently been demonstrated that the two-, three-, and four-exponential models fit decay data having their origins in *distributions of lifetimes*.[6] If the data are collected to only normal levels of precision, it has been shown that a variety of distributions can be present and appear in fact to fit a two- or three-exponential model.[6] For example, distributions having the shape of a Gaussian, a bimodal Gaussian, and a triangle and a rectangle can all be fitted very well indeed with two exponentials, provided the level of precision of the data is comparable to that commonly used, until very recently, for single-photon decay determinations.[6] Three exponentials give an even better fit, and four can fit almost any distribution!

The problem is therefore clear: How does one distinguish between distributions and discrete decays consisting of two or three or even four exponentials? How does one determine which is the best model given only the decay data? And finally, under what circumstances is it impossible to differentiate these two situations?

The discrete two- or three-component decay can of course occur in both simple and complex systems. Excimer or exciplex kinetics[7] in homogeneous solution yield a two-component decay for the monomer or initially excited species. Analysis with distribution recovery programs should yield two discrete peaks, the widths (at half maximum) of which are a manifestation of the quality of the data and the resolution of the algorithm. Likewise, mixtures of noninteracting emitters will, of course, give discrete components. Fluorescent probes situated in a homogeneous environment should give a single exponential decay, or a double exponential if there are two distinctly different, but homogeneous environments that influence the transition probability.

In heterogeneous systems, a multiplicity of interactions with different local environments is quite likely to give rise to a multiplicity of decay times and therefore, potentially, to a continuous distribution of lifetimes. This is in fact the most logical starting point for modeling heterogeneous systems, but one rarely witnesses its implementation because of the lack of distribution analysis programs in many laboratories where heterogeneous systems are studied with fluorescence probes.

There are also systems having decay curves that accurately follow complex, but nevertheless analytically well-defined, mathematical functions. Long-range energy transfer,[8] diffusion-controlled fluorescence quenching,[9] and quenching in micelles where the quencher is distributed according to Poisson statistics[10] provide examples. Here one has a choice. The decay data can be fitted with the complex decay law, or they can be fitted to a distribution that, in the cases mentioned above, is theoretically known exactly. Interestingly enough, one can also fit to three or four exponentials and obtain excellent results, but with meaningless parameters; whereas if one recovers the distribution or fits to the correct decay law, the resultant parameters of course carry the appropriate physical meaning.

If the potential presence of distributions is ignored and fluorescence decay data are fitted to the sum of two, three, or four exponentials, and then physical meaning is attached to the recovered parameters, there is the risk of entertaining interpretations that are pure nonsense. If the physical nature of the heterogeneous systems is such that distributions of lifetimes can be expected to be present, then one must attempt to differentiate the several potential models, i.e., two or three or even four discrete components vs. a continuous distribution or a combination of a continuous distribution and discrete components. Obviously, not all systems are well disposed to such differentiation.

Fluorescence lifetimes can also be measured by the so-called phase-shift method.[11] In this, a steady light source is modulated in the 1–250 mHz range and the phase shift and degree of modulation are measured. The high-frequency components of a pulsed light source can also be used. The modern approach to this classical method involves multifrequency phase and modulation determinations.[12] The resultant data can then be subjected to analysis with models that include Gaussian or Lorentzian distributions.[13,14] Proponents of this frequency-domain approach believe that it equals the pulse technique in power and versatility, but this contention has yet to be given a comprehensive test on difficult systems. Nevertheless, phase fluorimetry is now being used in a number of laboratories to collect data that are subjected to distribution analysis as well as to analysis with discrete component models. With assumed Gaussian or Lorentzian distributions, discrete components are recovered as narrow distributions, whereas broad distributions exhibit widths that exceed those predicted by simulation for discrete components. The limitation of preconceived distribution shapes, however, appears to be a serious problem, especially when the technique is applied to systems with broad and markedly skewed distributions. In addition, some experimenters prefer to see the actual shape of the decay curve and the residual associated with model fitting, and so are more favorably inclined toward pulse techniques. There can be no question, however, that both phase and pulse techniques are surviving the test of time.

This chapter considers the use of distribution analysis on both time-correlated single-photon data and phase and modulation data, although the emphasis is on the former. For pulse data, solutions to the distribution analysis problem based on the exponential series method (ESM) and the maximum entropy method (MEM) are discussed in considerable detail, as is the problem of differentiating the discrete from the continuous case through the interplay of simulated and real data analysis. Examples are drawn from systems with known distributions as well as from systems where at present one has only empirical knowledge as to the shapes of the fluorescence lifetime distributions. Included are systems consisting of fluorescent probes in micelles and cyclodextrins, molecules undergoing Forster energy transfer in rigid media, fluorescent probes in proteins, and fluorescent molecules adsorbed on surfaces such as silica gel. The importance of simulation as a tool for establishing confidence in the analysis of real data is emphasized in connection with both pulse and phase data.

Chapter 3 provides information regarding the experimental techniques used in this field.

2. The Recovery of Fluorescence Lifetime Distributions.

This section first considers the problem of recovering distributions from pulse data, and this subject is developed in some detail. This is followed by a discussion of the application of phase and modulation techniques (frequency-domain fluorimetry) to the same problem.

2.1. Pulse Methods

When the fluorescence decay originating from a system having a continuous distribution of lifetimes is measured, the measuring process may be viewed as equivalent to taking a Laplace transform of the distribution function in reciprocal lifetime space ($1/\tau = k$) or k space. This is clear from the definition of the transform

$$I(t) = \int_0^\infty F(k) \exp\{-kt\} \, dk \qquad (1)$$

and the notion of averaging a function over a distribution function. $I(t)$ is the δ-pulse response of the fluorescent system. The finite width of the excitation pulse and the temporal resolution of the photon detection system introduce distortions that must be taken into account in the analysis of real data, and this is normally done by introducing deconvolution[1-3] as an integral part of the analysis algorithm. Thus, the problem of recovering the distribution is formally equivalent to the problem of obtaining the inverse transform, i.e.,

$$F(k) = L^{-1}\{I(t)\} \tag{2}$$

This viewpoint is important because it is well known that the problem of numerically taking inverse Laplace transforms with data containing noise is intrinsically ill conditioned.[15,16] In practice, this means that the inversion problem does not possess a unique solution. Thus one is faced with a nontrivial problem in numerical analysis. Fluorescence lifetime data at the very least contain fluctuations derived from counting statistics. Since Poisson statistics prevail, the standard deviation of a given point is equal to the square root of the number of counts in that point (one channel in the MCA). The problem becomes one of selecting the best possible solution from a large set of acceptable solutions, and this can be approached in several different ways.

A straightforward approach to this problem involves assuming as a trial function a sum of exponentials consisting of 50–200 terms. Thus the δ-pulse response function of the luminescent system is represented by

$$I(t) = \sum a_k \exp\{-t/\tau_k\} \tag{3}$$

The lifetimes are fixed (in contrast with the one- to four-exponential fit, where the lifetimes are free parameters), and one then seeks a fit by varying the preexponentials. The fitting criterion is a minimum in Chi-square based on the differences between the fitting function and the actual decay curve. Since the decay curve is distorted at early times by the instrument response function (lamp- or laser-light pulse profile combined with the detector response), it is necessary to incorporate deconvolution[1,2] into the algorithm. With this procedure one selects a solution that minimizes Chi-square and expresses this solution as a set of amplitudes associated with the fixed lifetime exponentials. This is termed the exponential series method (ESM). The ESM was originally proposed a number of years ago[17] as a method for recovering the shapes of decay curves free of the distortion that arises from the instrument response function, with no assumptions as to the nature of the decay. Applications at that time consisted of analyzing the decay curves associated with diffusion-controlled reactions and the generation of time-resolved spectra that were free from instrument response function distortions.

The ESM has proved to be a remarkably successful tool for distribution recovery. The original algorithm of James and Ware[18] has been recently modified[19] to eliminate some instability, the result being a very robust procedure. Extensive tests have been conducted with both real and synthetic (simulated) data. These are discussed below.

An alternative to the ESM is the so called maximum entropy method (MEM).[20-22] This method has its origin in information theory.[20] Of all acceptable solutions this method selects the one that renders the Shannon-Jaynes entropy function

$$S = - \sum a_k \ln (a_k/a_{tot}) \tag{4}$$

a maximum, subject to a constraint on the value of Chi-square (normally set to be approximately unity). The factor a_{tot} is the sum of the preexponentials. Advocates of this method claim that it eliminates correlation problems between the recovered parameters. As with the ESM, the MEM uses a sum of exponentials with fixed lifetimes as a trial or probe function. Theoretical arguments can be advanced to show that the lifetimes should be fixed so that they are equally spaced in log τ space.[22]

Several groups have written software implementation for the MEM, following the general procedures outlined by Skilling and Bryant.[23] The algorithm used involves the method of conjugate gradients with the Chi-square constraint introduced via the Lagrange multiplier λ. Thus MEM maximizes the function

$$Q = S - \lambda C \tag{5}$$

where S is given above and

$$a_{tot} = \sum_{k=1}^{N} a_k \tag{6}$$

$$C = \sum_{n_1}^{n_2} \frac{(Y_i - \sum D_i^k a_k)^2}{Y_i} / (n_2 - n_1 - 1) \tag{7}$$

In these equations N is the number of exponents in the series with fixed lifetimes logarithmically spaced;[22] n_1 and n_2 are the initial and final channels containing the digital data; Y_i represents the number of counts in the i th channel, and

$$D_i^k = \int_0^{t_i} L (t_i - t') \exp \{-t'/\tau_k\} \, dt \tag{8}$$

a term that takes into account the need to deconvolute. In equation (8), L(t) is the excitation profile or the instrument response function.

The function Q (equation 5) is maximized subject to the constraint that $C \equiv 1.00$. Up to 200 terms can be used in the trial function. If one removes the condition of maximum entropy, the method reverts to the ESM. In both the MEM and ESM it is usual to start with all the coefficients of the trial function set equal to the same value. This eliminates any initial bias.

Tests of the success of distribution recovery are normally made with both simulated and real data. In the former, a decay curve is synthesized with a large number of components corresponding to the desired distribution, with added discrete components if required. This is convoluted with the instrument response function to yield a simulated decay curve which, after the addition of Poisson noise, is then subjected to distribution analysis. Realistic decays representing any shape or combination of continuous distributions and discrete components can be synthesized in this way and the success of the recovery method judged by the correspondence between the input and recovered distributions.

The possibility of differentiating between discrete and continuous distributions can be judged by the following procedure.[19] Let us assume that a continuous distribution has been recovered from the analysis of an experimental decay curve. It is also found that the data are well fit by a three-component model. A simulated decay curve is then prepared made up of the three components that have been found to fit the experimental data. This is subjected to distribution analysis. If three discrete components are recovered by distribution analysis, then it can be concluded that had the actual experimental data been from a system with these three discrete components, distribution analysis would have reveled their presence. It has been found, as one would expect, that if the three components found to fit the continuous distribution are close together, they will not resolve when subjected to distribution analysis and the two cases will not be differentiable. In other words, three or four *closely spaced* discrete lifetimes will in general behave like a continuous distribution and current methods appear to offer little hope of resolution. Only simulation studies or studies with real discrete data files will reveal the extent to which resolution is possible. Siemiarczuk et al.,[19] in fact, report that simulation with added Poisson noise is in general a highly reliable procedure for examining the potential resolution of either the MEM or the ESM.

For real data, there are at least two alternatives available when one wishes to conduct tests. One can make decay measurements on a system where the distribution shape is known a priori. A good example, which is discussed in greater detail below, is the case of Forster electronic energy transfer in rigid media. In the absence of a fluorescent system having a known distribution of the desired shape, real distributions can be synthesized by adding a large number of files (10–20) of single exponential decays. Since the amplitudes and lifetimes are known for each component, the shape of the resultant distribution is also known. The recovery procedure, of course, makes no attempt to match the lifetimes of the trial function with the input function. These tests based upon real data obviously incorporate Poisson noise, but the data also reflect other sources of error, such as instrument instability and rf interference, and so provide very realistic tests of the recovery methods.

A critical parameter in this context concerns the precision of the data. A commonly used measure is based on the number of counts in the peak channel (CPC). It is found that the ability of either the MEM or the ESM to resolve discrete components or recover the shape of a distribution in the presence of a nearby discrete component is strongly dependent on the CPC.[19] Likewise, the ability of these methods to recover the exact shape of a distribution depends on the precision of the data. When distribution analysis is contemplated, it is normal to collect between 10^5 and 10^6 CPC. This exceeds the CPC traditionally used in time-correlated single-photon counting, i.e., about 20,000 CPC. With weak samples it is frequently difficult to collect the required number of counts with a flashlamp-based fluorimeter, but this is rarely a problem with the laser- or synchrotron-based systems.

Recently Siemiarczuk et al.[19] have published a study of the relative merits of the MEM and ESM techniques as applied to a variety of distributions. In Figure 1 is shown a comparison of MEM and ESM for recovering a discrete component in the presence of a Gaussian distribution. Both methods work quite well, but the resolution is clearly dependent on the distance of separation in lifetime space of the discrete from the continuous component.

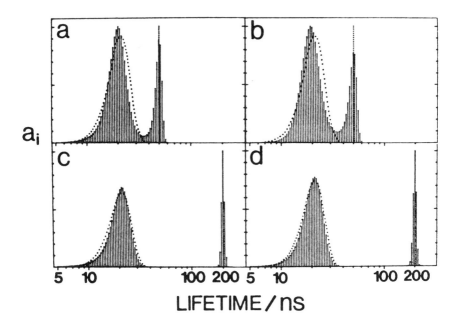

Figure 1. Gaussian distribution ($\langle \tau \rangle$ = 20 ns, σ = 5 ns) with added single lifetime component (dotted line) at 50 ns (a, b) and 200 ns (c, d) analyzed with MEM (a, c) and ESM (b,d). CPC = 8 x 10^5, 256 channels. Fitting range 3–60 ns with 80 exponentials, 3 ns/channel (a, b), and 3–240 ns with 120 exponentials, 6 ns/channel (c, d).

The extent to which these methods resolve a single component is illustrated in Figure 2 with both real and simulated data for a 383-ns lifetime; MEM is clearly superior. Figures 3 and 4 illustrate the recovery of Gaussians, including the bimodal case. The dotted lines represent the input distributions.

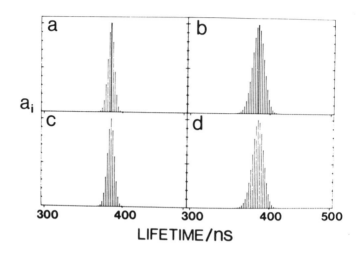

Figure 2. (a) MEM analysis of a simulated single-exponential decay at 383 ns (dotted line), CPC = 3 x 10^5. Probe function contained 80 terms over range of 300–500 ns. Fitting range of 460 channels with 6 ns/channel. (b) Same as (a) but with ESM. (C) MEM analysis of real data for pyrene decay in cyclohexane. Conditions the same as (a). (d) Same as (c) but with ESM.

While the recovery of the single Gaussian is essentially exact, the fidelity of the recovery in the bimodal case depends on the separation of the two maxima. The ESM and MEM are virtually identical. Figure 5 illustrates an attempt to differentiate the continuous and discrete cases. The Gaussian distribution shown in Figure 3 was fit to three components. A simulated decay based on these three components was then subjected to MEM and ESM analysis. It is clear that if three components had been present, ESM would have revealed their presence, but the resolution with MEM is not as good as ESM. In this case, the lifetimes were at 20.8, 47.4, and 78.6 ns, with the amplitudes indicated in the figure by the dotted lines. Figure 6 illustrates the successful recovery of an asymmetrical distribution.

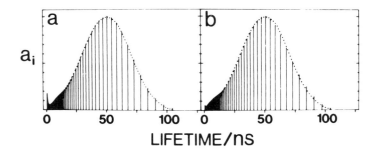

Figure 3. MEM (a) and ESM (b) analysis of simulated decays containing a Gaussian distribution (dotted line) of lifetimes. $<\tau>$ = 50 ns, σ = 20 ns, CPC = 2 x 10^5. Probe function of 80 terms over 0.5–150 ns range. Fitting range of 256 channels with 2.5 ns/channel.

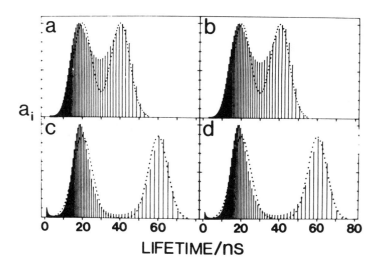

Figure 4. Analysis of simulated data with an underlying distribution composed of two Gaussians (dotted line). $<\tau_1>$ = 20 ns, $<\tau_2>$ = 40 ns, (a, b) or 60 ns (c, d), σ = 5 ns, CPC = 2 x 10^5, 256 channels. (a) MEM with 80-term probe function over 3–80 ns; (b) ESM with 80-term probe, same range; (c) MEM with 120-term probe function over range of 1–100 ns; (d) ESM with 120-term probe function over same range; 2 ns/channel (a, b) and 3 ns/channel (c, d).

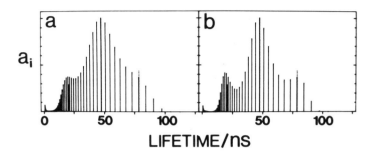

Figure 5. Analysis of a simulated three-exponential decay with lifetimes and amplitudes taken from the three-component fit (dotted lines) to the Gaussian in Figure 3. $a_1 = 0.17$ ns, $\tau_1 = 20.8$ ns; $a_2 = 0.58$, $\tau_2 = 47.4$ ns; $a_3 = 0.253$, $\tau_3 = 78.6$ ns. CPC $= 8 \times 10^5$. Probe function composed of 80 terms with range of 0.5–150 ns. (a) MEM and (b) ESM.

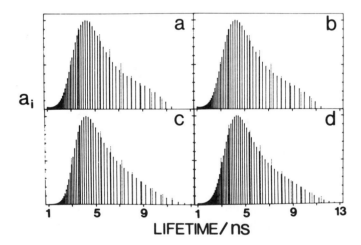

Figure 6. MEM and ESM analysis of a distribution composed of real and simulated data made up of the sum of 11 discrete decays of 3-methylindole in water at various temperatures. CPC $= 8 \times 10^5$, 153 ps/channel, 256 channels, $\tau_{max} = 4.6$ ns. Probe function contained 70 terms equally spaced in log τ. Solid lines, the MEM or ESM results; dotted lines, the input distribution. (a) Real data, MEM; (b) real data, ESM; (c) simulated data, ESM; (d) simulated data, ESM, time per channel doubled to 306 ps/channel.

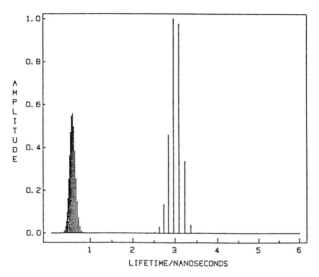

Figure 7. Lifetime distribution recovered with MEM for aqueous tryptophan at pH 5.2.

Tests of the MEM using simulated data have also been reported by Brochon and co-workers at Orsay.[24] The ability of the method to recover both continuous and discrete distributions was examined. In general, the resolution of discrete components and the ability to recover continuous distributions were comparable to those reported by Ware and co-workers.[19] Both groups have also measured tryptophan, but under different experimental conditions. However, if one compares the data of Ware and Wagner[25] shown in Figure 7 with those given in Figure 2 of ref. 22, it can be seen that similar resolution with MEM is obtained by both groups.

Long-range electronic energy transfer in rigid media provides a good test of distribution analysis and the ability of the ESM and MEM to differentiate continuous distributions from the discrete case. Dipole–dipole or so called Forster energy transfer in rigid media follows the well-known decay law[26]

$$I_F(t) = I^{\circ} \exp\left\{-(at + b^{1/2})\right\} \qquad (9)$$

where a is the unquenched lifetime and b is a function of the probability of energy transfer and the acceptor concentration.[26]

The inverse Laplace transform of $\exp\{-(at + bt^{1/2})\}$ is known and so the distribution in $k = 1/\tau$ space is known. The distribution function is

$$F(k) = 1/2b \, [\pi \, (k-a)^3]^{-1/2} \exp\{-b^2/4(k-a)\} \tag{10}$$

This distribution has a maximum at

$$k_{max} = a + b^2/6 \tag{11}$$

and the Forster critical transfer distance is related to k_{max} by

$$R_0 = \frac{3000 \, [6 \, (k_{max} + k_f)]^{1/2}}{4 \, (\pi^3 \, k_f)^{1/2} \, NC_A} \tag{12}$$

where k_f is the unquenched donor lifetime, C_A the acceptor concentration, and N Avogadro's number.

Wagner and Ware[27] recently reported a study of the system of phenanthrene quenched by acridine in polymethyl methacrylate matrix. It was first verified that Equation (9) was valid for this experimental system. Fluorescence decay data were collected to high precision with a laser-based system and the resultant decay analyzed according to

$$I_F(t) = I^0 \exp\{-(at + bt^n)\} \tag{13}$$

The value of n was varied and it was found that a sharp minimum in Chi-square was obtained at $n = 0.503$. This was taken as evidence that the Forster decay law was valid for this system with $n = 1/2$. The data were then fit to Equation (9) as well as with two, three, and four exponentials. Figure 8 illustrates the residuals. Clearly, both the three- and four-exponential fits are quite good, with four better than three. Four exponentials yield residuals that compare favorably with those obtained using Equation (9) as the fitting function. The data were then submitted to distribution analysis using both ESM and MEM. In Figure 9 are shown the resultant distributions along with the theoretical curve based on Equation (10) for both simulated and real data. It can be seen that ESM recovers the shape of the distribution for both the simulated and the real data with high fidelity, whereas there are minor discrepancies with MEM. Both methods accurately predict the location of the maximum, and the resultant R_0 calculated from k_{max} is in excellent agreement with that obtained from the b parameter in Equation (9) as well as with R_0 reported in the literature by others who have studied this system.

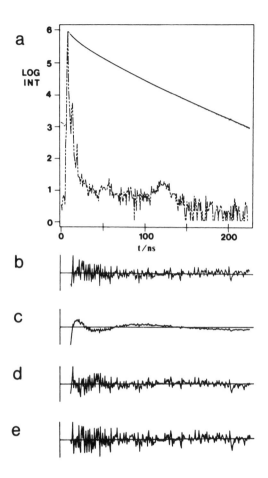

Figure 8. Fit of the decay of phenanthrene quenched by acridine in PMMA to various models: (a) (···) decay curve (-.-) laser profile, (—) fit to equation 9; (b) residuals for fit to equation 9; (c) residuals for two-exponential fit; (d) residuals for three-exponential fit; (e) residuals for four-exponential fit. [Ph] = 0.0028 M, [Ac] = 0.023 M, 9 x 10^5 CPC, 256 channels, 1100 ps per channel.

If one now takes the three- and four-component discrete fits to the experimental data, makes up simulated decay curves with added Poisson noise, and submits these to distribution analysis with ESM, the results obtained are shown in Figure 10. Shown also are the results of using the ESM distribution analysis on the experimental decay curve. Obviously, the distribution analysis would have recovered strong evidence of the presence of discrete components had this been the correct model for the experimental system. Since in fact the distribution is not discrete, the actual data gave correctly a continuous distribution. Note that in

Figure 9 the distributions are given in τ space, whereas in Figure 8 they are in k space.

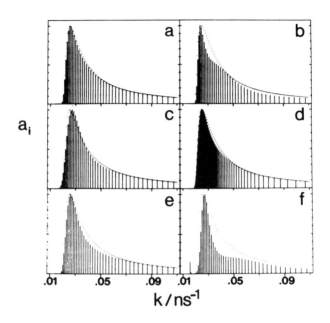

Figure 9. MEM and ESM recovery of rate constant distributions from a simulated Forster energy-transfer system (a = 0.0185 ns^{-1} and b = 0.2 ns^{-1}; eq. 9, 1100 ps/channel. (a) 5 x 10^6 CPC, 512 channels, log τ probe, ESM; (b) 5 x 10^6 CPC, 512 channels, log τ probe, MEM; (c) 5 x 10^6 CPC, 256 channels, log τ probe, ESM; (d) 5 x 10^6 CPC, 512 channels, linear τ probe, ESM; (e) 5 x 10^5 CPC, 512 channels, log τ probe, ESM; (f) 5 x 10^4 CPC, 512 channels, log τ probe, ESM. Probe function composed of 100 exponential terms.

The former makes it easier to visualize the three- and four-component results. Thus these experiments indicate that not only can one differentiate between two models which both give excellent residuals, namely the four-component model and the Forster model, but the shape of the Forster distribution is recovered with high fidelity. In the work cited,[27] distribution analysis was extended to the case where diffusion must be considered and R_0 was recovered from the k_{max} obtained from distribution analysis. Again good agreement was obtained with methods based upon determining R_0 from direct decay curve analysis.[28]

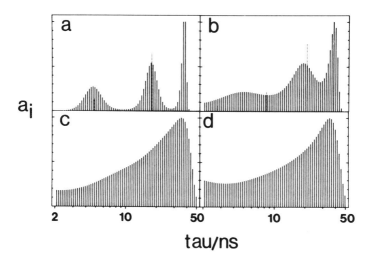

Figure 10. Lifetime distributions recovered by ESM analysis of real and simulated decay data for phenanthrene quenched by acridine in PMMA: (a) simulated three-exponential model; (b) simulated four-exponential model; (c) simulated equation 9; (d) experimental data ([Ph] = 0.0028 M, [Ac] = 0.0237 M). Discrete lifetimes in (a) and (c) are shown as ⋯. 9.99 x 10^5 CPC, 512 channels, 1036 ps per channel.

2.2. Frequency Domain Techniques

In 1987 the analysis of phase fluorimeter data using distributions of lifetimes was proposed by two groups.[13,14] This was a natural extension of studies involving the resolution of decays consisting of two or three components, a practice that became possible with the advent of multifrequency instruments, i.e., instruments where the modulation frequency could be varied over a wide range, for example, from 1 to 250 mHz. The δ-pulse response system is given by

$$I(t) = \sum \alpha_i \exp \{- t/\tau_i\}$$ (14)

where, if the amplitudes are described as continuous distributions, then

$$I(t) = \int_0^\infty \alpha(\tau) \exp \{-t/\tau\} \, dt$$ (15)

where

$$\int_0^\infty \alpha(\tau) \, d\tau = 1 \tag{16}$$

Two convenient distributions are then introduced arbitrarily.[13,14]

$$\text{Gaussian: } \alpha_G(\tau) = [1/\sigma \sqrt{2\pi}] \, \exp\{-1/2(\tau - \tau_m)^2/\sigma^2\} \tag{17}$$

$$\text{Lorentzian: } \alpha_G(\tau) = \Gamma/\{2\pi[(\tau - \tau_m)^2 + (\Gamma/2)^2]\} \tag{18}$$

In these equations, τ_m is the central value of the distribution, σ the standard deviation of the Gaussian, and Γ the full width at half maximum of the Lorentzian. A set of either Gaussian or Lorentzian distributions is now introduced such that

$$\alpha(\tau) = \Sigma_i \, g_i \, \alpha_i^o(\tau) = \Sigma_i \, a_i(\tau) \tag{19}$$

where the g_i are the amplitudes and the α_i^o are the shape factors for the distributions. The connection between the $\alpha_i(\tau)$ and the phase shifts ϕ_ω and modulations m_ω is established through the sine and cosine transforms of the impulse decay law, i.e.

$$N_\omega J = \int_0^\infty \int_0^\infty \alpha(\tau) \exp\{-t/\tau\} \sin \omega t \, d\tau \, dt \tag{20}$$

$$D_\omega J = \int_0^\infty \int_0^\infty \alpha(\tau) \exp\{-t/\tau\} \cos \omega t \, d\tau \, dt \tag{21}$$

Switching the order of integration yields

$$N_\omega J = \int_0^\infty \frac{\alpha(\tau) \, \omega \tau^2}{1 + \omega^2 \tau^2} \, d\tau \tag{22}$$

with

$$J = \int_0^\infty \alpha\,(\tau)\,\tau\,d\tau \qquad (24)$$

The phase and modulation values are related to these transforms through the equations

$$\phi_\omega = \text{arc tan } (N_\omega/D_\omega) \qquad (25)$$

$$m_\omega = (N_\omega^2 + D_\omega^2) \qquad (26)$$

Nonlinear least-squares techniques are then used to obtain the best fitting distributions as judged by the reduced Chi-square

$$\chi^2 = 1/\nu \sum_{\omega,\,\kappa} \frac{(\phi_\omega - \phi_{c\omega})^2}{\delta\phi^2} + 1/\nu \sum_{\omega,\,\kappa} \frac{(m_\omega - m_{c\omega})^2}{\delta m^2} \qquad (27)$$

where ν represents the number of degrees of freedom and $\delta\phi$ and δm are the uncertainties in the measured phase and modulations values. In the Chi-square calculation the sum extends over all frequencies employed in the measurements, and in some cases over multiple sets of data where measurements were made at multiple wavelengths. Typically 20 frequencies are employed, covering a range of two or more decades.

Tests of the distribution recovery capabilities of techniques based on the analysis of frequency-domain data do not appear to be as extensive as those available for the MEM or ESM analysis of pulse data. Both the groups of Lakowitz and of Gratton report simulations as well as tests of real data. The recovery of well-separated, discrete components appears comparable to the MEM or ESM results with time-correlated single-photon data. However, very little appears to have been reported concerning the recovery of non-Gaussian or non-Lorentzian distributions or combinations of discrete components and non-Gaussian or non-Lorentzian distributions. Some simulations have been reported for input distributions made up of one or more rectangles, or input distributions that are themselves Gaussian or Lorentzian.[14] The rectangular distributions are fit with one or two Gaussians or Lorentzians, so the fit is very approximate. Recovery of the widths or maximum values of input Gaussian or Lorentzian distributions appears reasonably satisfactory, although very shallow Chi-square minima are reported in some cases.

The fact that the method currently used to analyze frequency-domain data is based on Gaussian or Lorentzian distributions provides some serious limitations. For example, Gryczynski et al.[29] report the interesting variation of the decay characteristics of indole in cyclohexane–ethanol mixtures. Their distribution analysis recovers the discrete lifetimes in the pure solvents very well, but for the mixed solvents apparently good fits are obtained with either the single Gaussian or the single Lorentzian. The best fitting Gaussian had a width of almost three times that of the best fitting Lorentzian for the case of the mixed solvent system containing 1% ethanol. Since the distribution may be neither a Gaussian nor a Lorentzian, the information obtained from this type of analysis has obvious fundamental limitations. The extent to which these limitations can be overcome by using a large number of Gaussian or Lorentzian distributions appears to be incompletely investigated at this time, as does the power of the technique to differentiate continuous distributions from the discrete case with two, three, or four components.

3. Lifetime Distribution Analysis of Fluorescent Probes in Micelles and Cyclodextrin Cavities
3.1. Micelle Systems

Over the last several decades a vast literature on the physical properties of micelles has come into existence. While such techniques as light scattering and neutron scattering have been extensively employed, the use of fluorescent probes has without doubt played a dominant role in the investigation of aggregation numbers, micelle–micelle dynamics, polydispersity, the exchange of unbound molecules upon micelle–micelle collisions, sphere to rod transitions, and the polarity of core environments.[10,30-34] The micelle represents a classical heterogeneous system with the core and the solvent frequently at opposite ends of the polarity scale. The interfacial region is expected to be the most heterogeneous part of the micelle. The micelle size and shape are influenced by temperature, ionic strength, and, in the case of ionic micelles, the counterion.[10] The addition of polar molecules such as alcohols to ionic micelle systems in water can have a profound influence on both the size and shape of the micelles present in these systems, as well as on the degree of polydispersity.[35]

This is an obvious area for the application of fluorescence lifetime distribution analysis. As will be described below, distributions of lifetimes arise naturally because of the discrete nature of the quenching process. Micelles containing a fluorescent probe can potentially also contain one, two, three,... quencher molecules distributed theoretically according to Poisson statistics. In addition, the intrinsic heterogeneity of the micelle system suggests that polarity-sensitive probes would under many circumstances exhibit a broadening of the lifetime which in turn would be a function of factors that influence the micelle shape, size, aggregation number, and polydispersity. These ideas will be developed below and examples given to illustrate both the presence of distributions and the utility of distribution analysis in these interesting heterogeneous systems.

The commonly used scheme[10] for kinetic processes involving a fluorescent probe F^* in a micelle system in the presence of a bound or unbound quencher Q is as follows:

$$F_n^* \xrightarrow{\ k_f\ } F_n + h\upsilon_F$$

$$F_n^* \xrightarrow{\ nk_q\ } F_n$$

$$F_n^* + Q \xrightarrow{\ k^+\ } F_{n+1}^*$$

$$F_n^* \xrightarrow{\ nk^-\ } F_{n-1}^* + Q$$

$$F_n^* + M_j \xrightarrow{\ jk_e\ } F_{n-1}^* + M_{j-1}$$

$$F_n^* + M_j \xrightarrow{\ nk_e\ } F_{n-1}^* + M_{j+1}$$

where F_n^* designates an excited probe molecule in a micelle containing n quenchers, and M_j represents a micelle free of the fluorescent probe but containing j quencher molecules. If one now makes the assumption that the quenchers are distributed among the micelles according to Poisson statistics, it is easily shown[10,36] that the general decay law has the form

$$I(t) = A \exp \{-Bt - C\,[1 - \exp(-Dt)]\} \tag{28}$$

If, in addition, $k_q \gg k^- + k_c[M]$, then

$$I(t) = A \exp \{-k_o - <n>\,[1 - \exp(-k_q t)]\,\} \tag{29}$$

The above decay law is identical to

$$I(t) = A \sum_{j=0}^{\infty} P(j, <n>) \exp \{-(k_0 + jk_q)\,t\} \tag{30}$$

In these equations

$$k_o = k_F + S_2[Q] \tag{31}$$

$$P(j,<n>) = <n>j \exp(-<n>)/j! \tag{32}$$

$$S_2 = 1 + (k_e/k^-) [M] / (1/k^+ + [M]/k^-) \tag{33}$$

$$<n> = [Q] / (K^{-1} + [M]) \tag{34}$$

$$K = k^+/k^- \tag{35}$$

where $P(j,<n>)$ is the Poisson distribution equation. Equation (30) implies that the fluorescence decay can be represented by a relatively small discrete set of exponentials with Poisson distributed amplitudes given by $P(j,<n>)$ and lifetimes located at $(k_0 + jk_q)^{-1}$. Those micelles containing a probe molecule but no quencher contribute discrete terms at $\tau_0 = k_0^{-1}$.

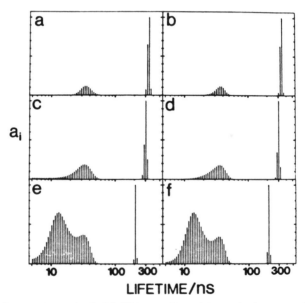

Figure 11. Pyrene decay in 0.1M SDS solution with cupric ion quencher analyzed with MEM. Quencher concentration: (a) 5×10^{-4} M; (c) 10^{-3} M, (e) 3×10^{-3} M. (b, d, f) represent results for MEM analysis of simulated Poisson distributions with $<n>$, k_q, and k_0 obtained from (a), (c), and (e), respectively. CPC = 3×10^5, 516 channels, 4.7 ns/channel.

The remaining lifetimes reflect the presence of one, two, three,... quenchers and the separation between components of the discrete distribution decreases rapidly with j for $j > 1$. Thus with a trial function of 50–200 exponentials one would predict the MEM or ESM to yield a spike at τ_0 and a continuous distribution at shorter lifetimes representing the remaining discrete components that are too closely spaced to resolve. Siemiarczuk and Ware[37] have recently reported studies of pyrene in the surfactant sodium dodecyl sulfate (SDS) in water quencher by Cu^{2+}. Figure 11 illustrates the distributions recovered as a function of $[Cu^{2+}]$ along with distributions recovered from simulated decays based on the recovered parameters using Equation (29) to generate the decay curve. As can be seen from Figure 11 the agreement between the predicted distributions and those recovered from the experimental decay curves is excellent.

Siemiarczuk and Ware[37] also show that there exists the following simple relationship between the average reciprocal lifetime within the distribution ($j \geq 1$) and k_q, provided the Poisson model for the distribution of quenchers is correct:

$$k_q = <n>^{-1} (<k> - k_0) [1 - \exp(-<n>)]^{-1} \tag{36}$$

where

$$<1/\tau> = <k> = \frac{\sum_{j=1}^{\infty} P(j, <n>)(k_0 + jk_q)}{\sum_{j=1}^{\infty} P(j, <n>)} \tag{37}$$

Use is also made of the integrated amplitude ratio between the terms at $j \geq 1$ and the spike at τ_0. This uniquely determines $<n>$ since if r is defined as

$$r = \sum_{j=1}^{\infty} P(j, <n>)/P(0, <n>) \tag{38}$$

then

$$<n> = \ln(r + 1) \tag{39}$$

$<k>$, k_0, and r are all measurable parameters, directly obtainable from the MEM or ESM results:

$$<k> = \sum a_i / \tau_i \tag{40}$$

$$k_0 = \Sigma \; a_m / \tau_m \tag{41}$$

$$r = \Sigma \; a_l / \Sigma \; a_m \tag{42}$$

where the summation indices l and m span the continuous distribution range and the τ_0 range, respectively.

Thus the analysis of distributions of the type shown in Figure 11 allows one to determine $<n>$ and k_q. Results show[37] that these parameters can be recovered with a precision of 2–3%.

It is also possible to verify the Poisson model directly by recovering the actual Poisson distribution of lifetimes. Siemiarczuk and Ware[37] used a modified MEM analysis where the lifetimes of the probe function are positioned at $\tau_j = (k_0 + jk_q)^{-1}$ and the number of exponents in the probe function reduced to 15. Since k_0 is known from the position of the spike or the tail of the decay curve, the only assumption made is that k_q is independent of the number of quenchers associated with a given micelle. *No distribution law is imposed.* The value of k_q is then varied, thus changing the positions of the τ_j in the probe trial function, until the best fit is found, based on Chi-square and the residuals. Thus, what are recovered are the individual terms in Equation (30). It is then possible to investigate the hypothesis that the recovered amplitudes are in agreement with the Poisson distribution. This is straightforward since $<n>$ can be obtained from an intrinsic property of the Poisson distribution, i.e.,

$$P_0 = <n> j / j! \tag{44}$$

and thus

$$<n> = \{(P_j / P_0) \, j!\}^{1/j} \tag{45}$$

For each quencher concentration $<n>$ can be calculated from Equation (45) through the use of amplitudes recovered with distribution analysis for several j values and the resultant $<n>$ then averaged. For the SDS system where the probe pyrene was quenched by Cu^{2+} ion, the values of $<n>$ obtained[37] compared favorably with those obtained from the continuous distribution analysis. Using these values of $<n>$ and the recovered value of k_q, one can calculate the expected Poisson amplitudes for comparison with the recovered amplitudes. As can be seen from Figure 12 the agreement is quite good except for the high quencher concentration where $<n>$ is very large (approximately 4). The values of $<n>$ and k_q recovered by these techniques agree well with those reported in the literature for this system.[38]

It is implicit in the above analysis that the fluorescent probe molecule has a single well-defined lifetime in the micelle in the absence of quencher. For SDS at a concentration of 0.1 mol L^{-1}, it is clear from Figure 11 that this is the case.

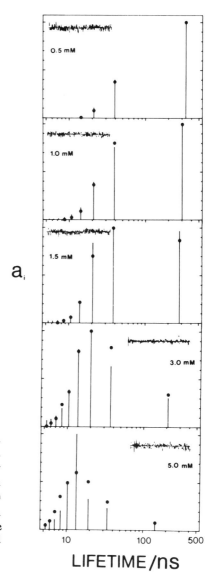

a_i

Figure 12. Distributions recovered from experimental decay with modified MEM with lifetimes positioned at $(k_0 + jk_q)^{-1}$ (vertical lines) and theoretical Poisson distributions (solid circles) calculated for $\bar{n}$ values determined from equation 45. [Cu^{2+}] are shown on the diagrams. Inserts represent weighted residuals.

LIFETIME /ns

A very narrow spike is recovered centered at τ_0. The fact that τ_0 decreases with increasing quencher concentration is a consequence of the kinetics,[10,36] as can be seen from Equation (31). However, it is perhaps surprising, considering the heterogeneous nature of the micelle, that a single lifetime should be observed. After all, the pyrene probe is polarity sensitive, and the fluorescence lifetime decreases by about a factor of 2 going from a hydrocarbon solvent to water. In fact, as has been shown by Siemiarczuk and Ware,[39] the narrow distribution found at 0.1 mol L^{-1} SDS is not found at either higher or lower surfactant concentrations. As can be seen from Figure 13 there is extensive broadening of the pyrene fluorescence lifetime, with the greatest width observed just above the critical micelle concentration (cmc), which is approximately 8×10^{-3}. There is a minimum in the width at about 0.1 mol L^{-1} SDS, after which the width increases again. Comparison of the recovered distributions with that obtained with pyrene in cyclohexane indicates that the width in 0.1 mol L^{-1} SDS is indistinguishable from that expected for a single exponential.[39] Ware and Siemiarczuk also found that the broadening is profoundly influenced by added salt.[39]

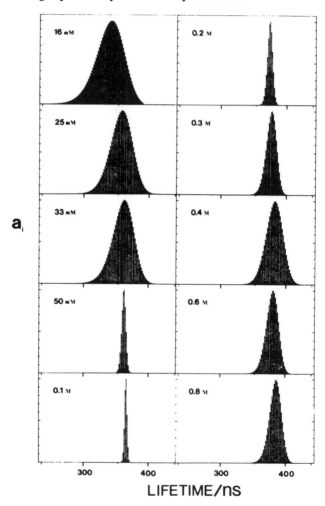

Figure 13. Lifetime distributions recovered by MEM from pyrene decays in SDS micelles. [SDS] indicated in the diagrams.

Simultaneous studies[39] of the Ham[40] effect for pyrene and the behavior of a water-sensitive probe in the above SDS system suggest that an interaction between pyrene and water is responsible for the lifetime distribution broadening. At high SDS concentrations the lifetime distributions probably also contain a contribution from increasing polydispersity of the micelles. It is of interest that many studies of SDS with pyrene have been done at 0.1 mo L^{-1}, where the broadening is absent. However, studies reported in the literature where the surfactant concentration is varied are suspect since in fact Equation 29 is being used when an average over the lifetime distribution is required.

The surfactant cetyl(hexadecyl)trimethylammonium chloride (CTAC) has also been studied extensively with probes such as pyrene and 1-methylpyrene.[10] Ware and co-workers[41] have found that both probes exhibit lifetime broadening at all concentrations above the CMC. This same group also finds profound lifetime broadening of pyrene in Triton-X and that for Triton-X the distribution of quenchers is non-Poissonian.[41] In this case, distribution analysis is extremely informative since there no question whatsoever as to the failure of the Poisson model. Comparison of Figure 14 with Figure 11 illustrates this point.

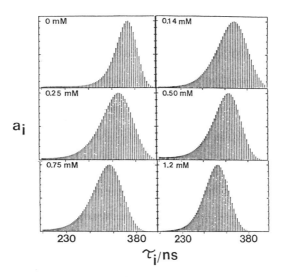

Figure 14. ESM-recovered distributions from experimental decays of 0.013 M pyrene in 0.024 M Triton-X with variable [I⁻] (shown on diagrams). The experimental decay curves were collected to 2 x 10^5 CPC.

When polydispersity in micelle systems is taken into account the kinetic formalism becomes quite complex.[42] Polydispersity also influences the determination of aggregation numbers by fluorescence quenching techniques. In the work described above,[37,39] at least for surfactant concentrations within an order of magnitude of the CMC for the ionic micelles, there is reason to believe that the systems are essentially monodisperse and that polydispersity is probably not responsible for the observed lifetime broadening.

Almgren et. al.[42] have developed a formalism for incorporating size distributions into the kinetics. However, this model does not explicitly take into account the phenomenon of lifetime broadening in the absence of quenchers described above when presumably the probe experiences a heterogeneous environment, nor does it represent a formalism for recovering size distributions.

It is clear that the determination of lifetime distributions present in micelle systems can be a powerful adjunct to the process of understanding micelle properties and the dynamics of quenching and micelle–micelle interactions. In particular, probes must be sought that do not suffer lifetime broadening over a wide range of surfactant concentration. Otherwise the kinetic schemes become unnecessarily cumbersome because of the addition of averaging over a distribution of unquenched lifetimes. The formalism is already somewhat complicated and contains a number of physically important parameters. In retrospect, pyrene appears to have been a very poor choice for a micelle probe for kinetic studies. Unfortunately, it and its 1-methyl derivative have been extensively used. That work must now be critically reexamined to determine the effect of lifetime broadening on the parameters determined and the conclusions drawn. At the same time there is potential benefit from this phenomenon since the broadening can provide evidence of heterogeneity that otherwise may be difficult to uncover.

3.2. Cyclodextrin Systems

Cyclodextrins (CD, α, β, γ) are toroidal polysaccharides consisting of six to eight D-glucose monomers. They have cavities with internal diameters from 4.7 to 8.3 Å (see Chapters 7, 16 and 19 for a detailed discussion). These cavities are considered to provide a hydrophobic environment. Small molecules can form inclusion complexes with CDs,[43,44] and CDs have been used extensively to model protein–ligand and enzyme–substrate interactions.[45] In addition, they have been used to effect separations and for selective synthetic procedures.[46] Fluorescence probe techniques have proved useful for studying the thermodynamics and kinetics associated with the formation of these inclusion complexes,[44,47-50] three types of which have been observed, i.e., 1:1, 1:2, and 2:2 (probe:CD).

Given the situation where a fluorescent probe is involved as a 1:1 inclusion complex, there naturally arises the question of the extent to which the probe's environment is homogeneous. This is an ideal situation for the application of fluorescence lifetime distribution analysis. Recently Bright et al.[51] have examined the system consisting of several anilinonaphthalene sulfonates (ANS) included into β-CD. Fluorescence lifetime data were collected with frequency-domain fluorimetry. For 2,6-ANS they found that discrete one- or two-component models failed to provide acceptable fits to their experimental data. The next level of complexity available consisted of a unimodal Gaussian or Lorentzian distribution of lifetimes. Both distributions gave superior fits as compared with the discrete model, but it was not possible to differentiate between the two types of distribution. These authors presented arguments[51] to support the view that this heterogeneity was not due to the presence of a mixture of 1:1 and 1:2 complexes and, in addition, probably not due to two distinct 1:1 complexes. They argued that a reasonable picture of the system involves the ANS probe included in β-CD with an ensemble of different conformations, all in coexistence with each other. Figure 15 illustrates a recovered distribution and the suggested range of conformations of the inclusion complex.

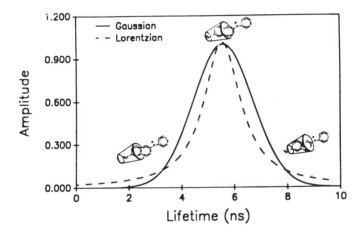

Figure 15. Recovered unimodal Gaussian (—) and Lorentzian (- - -) lifetime distributions for 10^{-5} M 2,6-ANS in 10 mM BCD at 25 °C. The inclusion complex structures represent possible conformations responsible for the lifetime distribution process. (Reproduced with permission from Ref. 51).

4. Lifetime Distributions of Molecules Adsorbed on Surfaces

The photochemistry and photophysics of molecules adsorbed on surfaces has received little attention until recently and, compared to their gas-phase and solution counterparts, are still very poorly understood. As soon as fluorescence lifetime measurements were made for systems composed of aromatic hydrocarbons adsorbed on such substrates as silica gel and porous glass, it was apparent that the resultant systems were very complex indeed.[52,53] Fluorescence decay data collected to a CPC of 20,000 were frequently very poorly fit by one exponential and the parameters for a two-exponential fit were in some cases wavelength dependent. This was clearly an ideal area for distribution analysis.

A common thread running through the early lifetime measurements, where quenching studies were carried out on the surface, was the approximate linearity of Stern-Volmer plots. These plots were constructed from $1/<\tau>$ vs. the estimated surface concentration of quencher; and $<\tau>$ was calculated from the best two-component fit as

$$<\tau> = \frac{\sum_{i=1}^{2} a_i \tau_i^2}{\sum_{i=1}^{2} a_i \tau_i} \tag{46}$$

That these observations were explicable by invoking distributions of decay times was demonstrated by James et al.[54] using either one or two distributions, initially Gaussian in the absence of quencher, and then simulating the effect of a quencher operating with either one or two quenching rate constants. In all cases examined, approximately linear Stern-Volmer plots were obtained for $1/<\tau>$ based on a two-component fit. Even for the intensity Stern-Volmers, i.e., I_0/I, approximately linear plots were obtained. Non-Gaussian distributions were not investigated. However, it is clear from these studies that the linear Stern-Volmer plots can conceal a multitude of complexities and are rather uninformative as to the detailed nature of the system under study. This work was preliminary to studies where actual distributions were recovered and examined as a function of the nature of the surface.

Another example based on assumed distributions is the recent work of Krasnansky et al.[55] Their study is concerned with the kinetic mechanism of quenching of pyrene and 9,10-diphenylanthracene by oxygen at the silica gel gas–solid interface. In order to avoid the assumption of a biexponential model, they employed a Gaussian model based on the approach of Albery et al.[56] and the work of Scott et al.[57] In this case the choice of a Gaussian was not influenced by the measurement technique since pulse methods were used rather than frequency-domain techniques. Once this somewhat arbitrary Gaussian assumption was in place, they developed models based on both the Langmuir-Rideal[58] and Langmuir-Hinshelwood[59] mechanisms for the surface reaction. The resultant decay laws and steady-state equations were derived and then compared with experimental data. This work simply underscores the need for recovering the actual distributions of lifetimes, both in the absence and presence of the quencher, in order to avoid the arbitrary assumption of the shape of the distribution for the unquenched lifetimes and their resultant distribution in the presence of quencher.

One reported application of photophysics of fluorophors adsorbed on surfaces involves the determination of the fractal dimension (see also Chapter 12 by Avnir in this volume for a discussion of fractals) of the adsorber from electronic energy-transfer parameters. For example Avnir and co-workers[60] have described experiments involving rhodamine-B as a donor on Fractosils 100 and 200. Single exponential lifetimes were measured for the unquenched donor of 3.8 and 4.0 ns when adsorbed on two different Fractosils. However, the lifetime measurements were carried out with a laser system and digitizer, and only 400 shots were averaged. Thus the data were far from ideal even for investigating deviations from single exponential decay. Ware et al.[61] have investigated this system using distribution analysis. While the values used by Avnir were close to the average values found, in all cases there was unequivocal evidence for the presence of distributions rather than a discrete unquenched lifetime. The distributions were smooth, slightly skewed in the long lifetime direction, and for Fractosil 200 at 290 K, a width of 2–2.5 ns was observed for a distribution with an average lifetime of about 4 ns. At 8 K the width doubled and the average lifetime increased slightly to about 4.5 ns. When a distribution was recovered, the decay curve was then also fit to three or four exponentials, and using these parameters, a simulated decay curve was prepared and subjected to the same distribution analysis. In all cases a discrete set rather than a distribution was recovered, indicating that had the actual data been properly represented by a discrete components, distribution analysis would have recovered the discrete components. Coverage of the silica was estimated to be about 0.003% of the available surface. Therefore the use of energy-transfer equations based on a single lifetime for the donor appears incorrect for this system, which in fact requires that the appropriate equations be averaged over the donor lifetimes.

Part of the problem with rhodamine-B may be that it can exist in three forms, at least in solution, i.e., the acid, base, and lactone, and that on the surface there may be heterogeneity ascribable to this as well as heterogeneity due to the surface itself. Rhodamine-6G, which is more attractive for this application,[62] does not have the complex acid–base character of rhodamine-B.

Energy transfer has also been used to obtain the fractal dimension for silica gel using samples that are suspended in a refractive index matching fluid.[62] For example, Levitz et al.[63] have used direct energy transfer from rhodamine-6G to Malachite Green in an index matching liquid composed of mixed alkanes (n = 1.4585). They employed a model for energy transfer that yielded the following decay law

$$I(t) = \exp \{-t/\tau_d - A_o \Gamma (1 - d/s) (t/\tau_d)^{d/s}\} \qquad (47)$$

where $\Gamma(x)$ is the gamma function, τ_d the donor lifetime, and d the fractal dimension; A_o is a time-independent factor and, for dipole–dipole transfer, $s = 6$. Levitz et al.[63] found that for the unquenched rhodamine-6G the decay appeared single exponential at a CPC of 20,000 counts. In the presence of Malachite Green Equation (47) was used as the fitting function and fractal dimensions of approximately 2 were observed, except for a silica preparation having a high surface area and low pore volume, where a value of approximately 3 was obtained.

Recently Drake et al.[64] have raised serious questions regarding the fractal concept as applied to the interfaces of porous materials such as silica gels. They employed electron microscopy, molecular size-dependent adsorption, small-angle x-ray scattering, and direct energy transfer. From these studies, they concluded that the materials under study did not exhibit fractal properties. This is a very important problem since the fractal concept allows one to characterize the degree of irregularity of a surface by one number, the fractal dimension d, where $2 < d < 3$. The applicability of the concept requires the material in question to exhibit morphological features that are scale invariant and self-similar. The higher the fractal dimension, the more space filling and wiggly the surface. The concept of the fractal dimension is used to determine the number of identical molecules needed to form a monolayer on a surface, and therefore the fractal nature of the surface enters critically into the problem of determining surface area seen by molecules of different cross section. In view of this recent work of Drake et al., one should proceed with caution when interpreting fractal dimensions recovered from energy-transfer studies, even when there is reason to believe that the energy-transfer model is valid and the unquenched lifetime represented by a discrete lifetime.

This and similar systems merit investigation with distribution analysis. In the case of $d = 3$, one has the mathematical form of the classical Forster transfer equation discussed above, and the system in the presence of both donor and acceptor should exhibit the characteristic distribution associated with this decay law (see Figure 9). In addition, when $d = 2$ there should be distortions to the Forster distribution. In addition, distribution analysis is needed to confirm that in the absence of acceptor, the donor decay is not broadened by adsorption on silica gel. Significant broadening might be difficult to identify at a CPC of only 20,000 counts.

As was mentioned above, when aromatic hydrocarbons are adsorbed on silica gel the fluorescence decay is in general not well fit by one and even two exponentials and in many cases the parameters recovered also vary with wavelength of emission and/or excitation. In general, distribution analysis reveals the presence of distributions, the width of which is a strong function of the surface treatment. In addition, the temperature at which the decay data is collected influences both the average lifetime and the width. For example, Liu et al.[65] have measured phenanthrene, chrysene, pyrene, benzoperylene, coronene, and perylene on both "wet" and "dry" surfaces. The wet surface is prepared by drying at 180°C for 12 h, followed by degassing in vacuum, whereas the dry surface preparation involves heating for 12 h at 180°C followed by heating in vacuo for 8 h at 800°C. It was observed that the average width ratio for wet to dry is about 13, whereas the k_{max} shifts such that the average ratio of wet to dry is about 5. The distributions for the wet surface are quite symmetrical in k space, but those associated with the dry surface are in general skewed toward longer k and in some cases there is a suggestion of bimodal character. A detailed mechanistic interpretation must await the completion of a study of quantum yields for this set of compounds. Nevertheless, there can be no doubt as to the dramatic changes brought about by dehydration of the surface to the point where isolated OH groups predominate.

It is interesting in this context to note that perylene does not exhibit the same changes as were observed with the above set of compounds.[65] Perylene has a ^1L_a lowest excited singlet state, whereas the remaining hydrocarbons are ^1L_b. Most of the compounds studied also show the Ham effect, which is absent in the case of perylene.

Ware et al.[61] have also examined the fluorescence decay of naphthalene at room temperature and at 8 K. The pretreatment of the silica surface involved heating for 12 h at 120°C. The silica gel was then introduced into a special cell, degassed, and combined with solid napthalene, which rapidly distributed on the surface. Broad lifetime distributions were recovered at both temperatures. At 290 K, the recovered distribution had a maximum at 14 ns and a width of 10 ns. At 8 K the maximum shifted dramatically to 50 ns and the width expanded to 60 ns. It is clear that the interactions with the surface responsible for the shortening of the lifetime are temperature dependent. However, the average lifetimes observed are much shorter than reported in solution, at room temperature,[66] in glasses at 77 K[66,67] in the gas phase, or in jets.[68]

5. Fluorescence Lifetime Distribution Analysis in Biological Systems

The use of fluorescent probes in biological systems has been common practice for a number of years. It is therefore perhaps surprising that the introduction of distribution analysis in this area has occurred only recently. Many of these systems would seem to be intrinsically heterogeneous. A probe, such as a tryptophan residue in a protein, will, for example, exhibit fluorescence lifetimes indicative of its exposure to solvent, the polarity and interactions associated with its local environment, and its accessibility to quenching (for a recent review see Beecham and Brand[69]). Distributions of lifetimes may arise from distributions of rates of interconversion of two or more conformations, as well as from a continuum of energy states in which the probe is situated. Since two or more conformations can experience a continuum of energy substates, the possibilities for distributions are manifold. These distributions should exhibit temperature sensitivity and change shape and position with added quencher. In biologically important macromolecules one can also have microhetrogeneity associated with combinations of specific and nonspecific binding sites. Heterogeneity of bilayers and vesicles can be revealed by fluorescence probes, and distributions are to be expected. In fact, when tests for the presence of distributions do reveal two or three discrete components, this is, in fact, highly informative, since it is implied that two or three homogeneous environments exist in an overall heterogeneous system. Early investigations in this area were done with frequency-domain techniques, but more recently work has also appeared that employs pulse techniques with both laser-based systems and synchrotron sources.

Vincent et al.[70] have used the MEM as well as nonlinear least-squares techniques to study the single tryptophan residue in horse heart apocytochrome c. They employed pulse techniques with the synchrotron ACO at Orsay as the source.

The MEM distribution analysis was consistent with a model based on the existence of four separate classes of lifetimes at 0.1–0.2, 1, 3, and 5 ns, in agreement with the results of a fit to four discrete exponentials. They studied this system as a function of temperature and in addition measured time-resolved fluorescence anisotropy to acquire information regarding the internal rotation of the tryptophan residue.

Merola et al.[71] have employed MEM for distribution analysis of fluorescence from thioredoxin from Escherichia *coli*, calf thymus, and yeast. A laser-based single photon system was employed with an instrument response time of 60 ps full width at half maximum. Distribution analysis revealed that the tryptophan fluorescence from these compounds was best represented by a small number of discrete components, and temperature studies suggested that these components do not exchange significantly on a nanosecond time scale. The MEM techniques employed were tested with simulations.

The effects of ligand binding on the conformation and internal dynamics in specific regions of porcine pancreatic phospholipase A_2 have recently been reported by Kuipers et al.[72] with tryptophan as a probe. The new Super-ACO synchrotron at Orsay was used as the source for time-correlated single-photon counting. This work contains a large number of examples of both narrow and broad distributions recovered with MEM analysis of single-photon decay data.

Phase techniques have been employed by Govindjee et al.[73] to examine changes associated with the opening or closing of the reaction center of photosystem II. Chlorophyll a was the fluorescent probe. Quite dramatic changes in the lifetime distribution were observed between the open and closed reaction center, the former extending from below 1 ns to beyond 20 ns, whereas the latter was confined to the region from below 1 ns to about 5 ns with a small tail extending to 10 ns. The disappearance of the slow components upon closure was interpreted as due to the absence of recombination reactions once pheophytin is reduced.

Analysis of frequency domain data with assumed Gaussian distributions has been reported by Barcellona and Gratton[74] for the DNA–4',6-diamidino-2-phenylindole (DAPI) complex. Rather broad distributions were recovered, which were interpreted as caused by different types of interactions between DAPI and DNA. Heterogeneity of binding sites was invoked because it is believed that the strong interaction between DAPI and DNA results in relatively long-lived complexes and thus the lifetime spread is interpreted as probably not attributable to dynamic effects.

Alcala et al.[75] have presented a theoretical discussion of fluorescence lifetime distributions in proteins which was then applied to several systems, including ribonuclease T_1, neurotoxin variant 3, and phospholipase A_2. Tryptophan, present in these molecules, was used as the fluorescent probe. Later Bismuto et al.[76] reported similar studies on apomyoglobin where the effect of unfolding on the tryptophanyl residues was used to study the microenvironmental characteristics of subconformations. Work along the same lines has been reported by Rosato et al.,[77] in which the fluorescent characteristics of horse spleen apoferritin and its subunits were studied by time-resolved fluorescent techniques. They sought to establish how the quaternary structure of the protein affects the decay pattern of the tryptophan emission and the mechanism of iron quenching.

Calmodulin, a calcium-binding protein, is of considerable interest because it is believed to regulate several different enzymes in diverse cells. Sanyal et al.[78] have recently reported a study of plant calmodulin where tryptophan fluorescence was used to study ligand binding surfaces. The protein does not contain tryptophan and the studies were made by binding tryptophan-containing amphiphilic peptides to wheat-germ calmodulin. Time domain fluorimetry was used and Lorentzian distributions assumed.

Distribution analysis has also been used in studies employing fluorescent probes in vesicles. Until recently, data obtained from time-resolved studies in such systems were subjected to one- or two-component fluorescence decay analysis and models constructed to be consistent with systems exhibiting discrete components. In many cases it is more reasonable to postulate distributions of lifetimes. James et al.[79] appear to have been the first to apply distribution analysis to membrane systems. They employed pulse techniques and ESM for distribution analysis. The fluorescence decay of the probe *cis*-parinaric acid (CPA) in dimyristoyl phosphatidylcholine (DMPC) vesicles was examined on either side of the liquid crystal–gel phase transition. Broad distributions are recovered with distinctly different maxima above and below the phase transition. When the two distributions are plotted with unitless parameters by scaling to the maximum amplitude and associated lifetime, the two distributions superimpose. This suggests that the underlying distributions is the same in both environments, with the phase transition causing simply a shift in the lifetime values.

Stubbs et al.[80] have also studied lifetime distributions in vesicles. Multifrequency phase fluorimetry was employed with assumed Lorentzian distributions. The probe diphenyl hexatriene was used with a number of lipid bilayer systems, and both narrow and broad distributions were recovered. Both environmental heterogeneity and environmental sampling by the fluorescent probe were considered in explaining the results obtained.

Membrane cholesterol heterogeneity and exchange have been studied with frequency domain techniques by Nemecz and Schroeder.[81] A fluorescent sterol was used to examine the sterol domains in small unilamellar vesicles as well as the exchange of sterol between such vesicles. A Lorentzian or Gaussian model was invoked. The results obtained were consistent with the presence of at least two pools of sterol in the unilamellar vesicles studied.

Pyrenedecanoyl labeled phospholipids have also been used for biomembrane studies. Prenner et al.[82] have studied the spontaneous transfer of phospholipids labeled with this probe from small unilamellar vesicles to human erythrocyte ghost membranes. Frequency-domain techniques were used. They observed monomer fluorescence lifetimes of the pyrene-labeled phospholipids to be much more heterogeneous in the biological membrane compared with the protein-free lipid bilayer systems, but the authors were unable to make a decision in favor of a specific decay mode.

As can be seen from the work quoted above, the recovery of discrete components in the course of distribution analysis is not uncommon. In an interesting recent study, Haydock et al.[83] have examined the tryptophan decay found in the variant 3 scorpion neurotoxin. Fluorescence decay was measured with the time-correlated single-photon technique where, in this case, the source was a mode-locked laser system. The MEM analysis very clearly revealed only three discrete components at 0.091, 0.45, and 2.05 ns. The authors then subjected their data to analysis with discrete models using the standard least-squares method,[1-3] the maximum likelihood method of Hall and Selinger,[84] and the recently proposed Pade-Laplace method of Yeramian and Claverie.[85] These three methods all recovered the discrete components with lifetimes and amplitudes in close agreement with each other and also in agreement with the MEM results. The least-squares method and the maximum likelihood method gave essentially identical results, and of the four methods the poorest agreement was between the Pade-Laplace method and the other three techniques.

While both frequency domain and pulse techniques generate data that can be subjected to distribution analysis, pulse techniques appear to offer a considerable advantage. In particular, the experimentalist has more control over the statistical quality of the data, and the distribution analysis procedures suited to pulse data appear to offer more flexibility since one is not constrained to assumed Gaussian- or Lorentzian-based distributions. Mode-locked lasers and synchrotron radiation both provide short duration pulses of excellent shape stability and high intensity, and the time-correlated single-photon technique provides data ideally suited for distribution analysis. Tests with simulated data also appear to be easier and more realistic with the single-photon technique. Software is commercially available for MEM and ESM analysis of pulse data.[86] Advocates of each method will no doubt continue to favor their chosen approach, but there is need for extensive comparative studies in the absence of bias where both phase and pulse techniques are applied to the same systems and especially to systems where the distributions present are known a priori.

While there do not appear to have been extensive and systematic studies aimed at comparing the pulse and phase methods as to their ability to recover and resolve distributions, Royer et al. have recently reported work where both methods were applied to the same systems.[87] They studied three *lac* repressor proteins, one of which contained two tryptophan residues per monomer, whereas the other two had only one tryptophan per monomer. Global analysis of multiwavelength frequency domain data from all three proteins was compared with single-photon data obtained with Super-ACO at Orsay. While in general there was quite good agreement, the MEM analysis of the single photon data produced components at short lifetimes that were not included in the frequency domain data. However, owing to the small contribution made by these additional components to the overall intensity, it was found difficult to establish whether they were significant. The authors do not report simulation studies, which might have clarified the situation with respect to the single-photon MEM results.

6. Final Remarks

It should be clear from the work surveyed above that in the last 4 years the level of sophistication of analysis of fluorescence decay data in heterogeneous systems has risen dramatically to where it now appears to be more or less consistent with the inherent complexity of the systems under study. In fact, one can now argue that the current challenge is associated with interpretation rather than with the problem of exposing the details of the inherent complexity.

Nevertheless, one must proceed with extreme care since distribution analysis has its limitations. It is important never to loose sight of the fact that the problem is ill-conditioned and that in general there will be a family of acceptable solutions from which one must choose the one best representing reality. In many cases this is not possible. The MEM is theoretically viewed as yielding the best possible solution from the family of acceptable solutions. However, as has been described above, the ESM frequently yields distributions that are virtually identical with those recovered by MEM. This suggests that the criterion of a minimum in Chi-square is, for single-photon decay data collected to high CPC, frequently a satisfactory tool for selecting the best solution. While simulation studies with synthetic data, or with data composed of combined files of real data, are essential and can justifiably instill considerable confidence in the ability of a given algorithm to recover distributions similar to those recovered from the experimental data under analysis, this by no means constitutes an absolute proof that the recovered distributions are real. Part of the problem lies in the difficulty or impossibility of exactly simulating real data which may contain scattered light, distortions due to drifts in the electronics, radiofrequency noise, variations in the excitation profile, etc.

A particularly difficult problem involves the resolution of close-lying discrete components and the differentiation of discrete cases from those best represented by continuous distributions. Analysis of simulated data is essential and yet appears to be somewhat less frequently used than would seem desirable in many studies.

While it is important to understand the risks inherent in attempts to recover distributions, it is also clear that the MEM and ESM have been very successful, not only in exposing the existence of distributions in single-photon decay data, but also in recovering distribution shapes with quite high fidelity.

The *comparative* success of the recovery of distributions from pulse and frequency domain data should be studied much more carefully, especially with respect to the question of the recovery of distributions of non-Gaussian or non-Lorentzian shape and distributions containing both discrete and continuous contributions. In addition, attention should be directed toward the problem of components and distributions of low relative intensity and the confidence we can place in their recovery. While the frequency-domain approach appears to have been used more extensively than the pulse technique to generate data destined for distribution analysis, it is by no means clear that the frequency-domain approach is indeed the method of choice. In fact, on the basis of published reports, the pulse technique with either the MEM or the ESM appears to have the edge in versatility, flexibility, and the fidelity of recovery of distributions.

Future developments will no doubt include other approaches to the numerical difficulties presented by the ill-conditioned or ill-posed nature of the problem of distribution recovery from decay data.[88]   For example, Kauffmann et al.[89] have recently proposed a novel ESM method that employs Tikhonov[90] regularization and has the potential for resolving negative as well as positive preexponentials.

References

1. D. V. O'Connor and D. Phillips, *Time Correlated Single Photon Counting,* Academic Press, Orlando, 1984.
2. J. N. Demas, *Excited State Lifetime Measurements,* Academic Press, New York, 1984.
3. W. R. Ware in *Time Resolved Fluorescence Spectroscopy in Biochemistry and Biology,* R. B. Cundall and R. E. Dale, eds., Plenum Press, New York, 1983.
4. I . H. Munro in *Time Resolved Fluorescence Spectroscopy in Biochemistry and Biology,* R. B. Cundall and R. E. Dale, eds., Plenum, New York, 1983.
5. O. Kuipers, M. Vincent, J. C. Brochon, B. Verheij, B. Hass and J. Gallay, *SPIE Int. Soc. Opt. Engr.,* 1990, *1024,* 620.
6. D. R. James and W. R. Ware, *Chem. Phys. Lett.,* 1985, *120,* 445.
7. W. R. Ware in *Time Resolved Fluorescence Spectroscopy in Biochemistry and Biology,* R. B. Cundall and R. E. Dale, eds., Plenum, New York, 1983.
8. J. B. Birks, *Photophysics of Aromatic Molecules,* Wiley-Interscience, London, 1970.
9. W. R. Ware and J. C. Andre in *Time Resolved Fluorescence Spectroscopy in Biochemistry and Biology,* R. B.Cundall and R. E. Dale, eds., Plenum, New York, 1983.
10. K. Kalyanasundaram, *Photochemistry in Microheterogeneous Systems,* Academic Press, Orlando, 1987.
11. J. R. Lakowicz, *Priniples of Fluorescence Spectroscopy,* Plenum, New York, 1984.
12. J. R. Lakowicz, G. Laczko and I. Gryczynski, *Rev. Sci. Inst.,* 1986, *57,* 2449. See also J. R. Lakowicz and B. P. Maliwal, *Biophys. Chem.,* 1985, *21,* 61.
13. J. P. Clcala, E. Gratton and G. Pendergast, *Biophys. J.,* 1987, *51,* 587; ibid, 597.
14. J. R. Lakowitz, H. Cherek, I. Gryczynski, N. Joshi and M. Johnson, *Biophys. Chem.,* 1987, *28,* 35.
15. M. Bertero, P. Boccacci and F. R. Pike, *Proc. Roy. Soc. Lon. A.,* 1982, *383,* 15.
16. J.G. McWhirter and E.R. Pike, *J. Phys. A. Math. Gen.,* 1978, *11,* 1729.
17. W.R. Ware, L. J. Doemeny and T. L. J. Nemzek, *J. Phys. Chem.,* 1973, *77,* 2038
18. D. R. James and W. R. Ware, *Chem. Phys. Lett.,*1986, *126,* 7.
19. A. Siemiarczuk, B. R. Wagner and W. R. Ware, *J. Phys. Chem.,* 1990, *94,* 1661.
20. C. P. Smith and W.T. Grady, Jr., *Maximum Entropy and Bayesian Methods in Inverse Problems,* Reidel, Boston, 1985.
21. A. K. Livesey and J. Skilling, *Acta Crystallog. Section A,* 1984, *41,* 113.
22. A. K. Livesey and J. C. Brochon, *Biophys. J.,* 1987, *52,* 693.
23. J. Skilling and R. K. Bryan, *Mon. Not. R. Astron. Soc.,* 1984, *221,* 111.

24. M. Vincent, J. C. Brochon, F. Merola, W. Jordi and J. Callar, *Biochem.*, 1988, *27*, 8752.
25. B. Wagner and W. R. Ware. Unpublished results.
26. T. Forster, *Naturwiss.*, 1946, *33*, 166; *Ann. Physik.*, 1948, *2*, 55; *Z. Natforsch.*,1949, *4a*, 321.
27. B. D. Wagner and W. R. Ware, *J. Phys. Chem.*, 1990, *94*, 34.
28. J. B. Birks and S. Georghiou, *J. Phys. B. Ser. 2.*, 1968, *1*, 958.
29. I. Gryczynski, W. Wiczk, M. L. Johnson and J. R. Lakowitz, *Biophys. Chem.*, 1988, *32*, 173
30. J. K. Thomas and M. Gratzel in *Modern Fluorescence Spectroscopy*, F. L. Wehry, ed., Plenum Press, New York, NY, 1972.
31. R. Malliaris, *Intern. Rev. Phys. Chem.*, 1988, *7*, 95.
32. N. J. Turro, M. Gratzel and A. M. Braun, *Angew. Chem. Int. Ed. Eng.*, 1980, *19*, 675.
33. J. K. Thomas, *Chem. Rev.*, 1980, *80*, 283.
34. N. J. Turro, B. H. Baretz and P. L. Kuo, *Macromolecules;* 1984, *17*, 132.
35. J. Lang, *J. Phys. Chem.*, 1990, *94*, 3734.
36. M. Tachiya, *Chem. Phys. Lett.*, 1975, *33*, 289.
37. A. Siemiarczuk and W. R. Ware, *Chem. Phys. Lett.*, 1989, *160*, 285.
38. J. C. Dederen, M. Van Der Auweraer and F. C. DeSchryver, *J. Phys. Chem.*, 1981, *85*, 1198.
39. A. Siemiarczuk and W. R. Ware, *Chem. Phys. Lett.*, 1990, *167*, 263.
40. K. Hara and W. R. Ware, *Chem. Phys.*, 1980, *51*, 61.
41. A. Siemiarczuk, K. LaPalm and W. R. Ware, unpublished results.
42. a) M. Almgren and J. E. Lofroth, *J. Chem. Phys.*, 1982, *67*, 2734.
    b) M. Almgren, J. Alsins, E. Mukhtar and J. van Stam, *J. Phys. Chem.*, 1988, *92*, 4479.
43. W. R. G. Baeyens, B. L. Ling, P. De Moerloose, B. Del Castillo and C. De Jonge, *Ann. Rev. Acad. Farm.*, 1988, *54*, 698.
44. W. Saenger, *Angew. Chem. Int. Ed. Engl.*, 1980, *19*, 344.
45. M. L. Bender and M. Komiyama, *Cyclodextrin Chemistry*, Springer-Verlag, Berlin, 1978.
46. Y. Ihara, E. Nakanishi, M. Nango and J. Koga, *Bull. Chem. Soc., Jpn.*, 1986, *59*, 1901.
47. G. C. Catena and F. V. Bright, *Anal. Chem.*, 1989, *61*, 905.
48. A. Hersey, B. H. Robinson and H. C. Kelly, *J. Chem. Soc. Faraday Trans., I*, 1986 *82*, 1271.
49. S. Kitamura, S. Matsumori and T. Kuge, *J. Inclu. Phenom.*, 1984, *2*, 725.
50. D. J. Jobe, R. E. Verrall, R. Palepu and V. C. Reinsborough, *J. Phys. Chem.*, 1988, *92*, 3582.
51. E. V. Bright, G. C. Catena and J. Haung, *J. Am. Chem. Soc.*, 1990, *112*, 1343.
52. P. de Mayo, L. V. Natarajan and W. R. Ware in *Organic Phtotransformations in Nonhomogeneous Media*, M. A. Fox, ed., ACS Symp. Series Vol. 278, American Chemical Society, Washington, DC, 1985, p. 1.
53. D. R. James, Y. S. Liu, A. Siemiarczuk, B. D. Wagner and W. R. Ware, *SPIE Int. Soc. Opt. Engr.*, 1985, *909*, 90.
54. D. R. James, Y. S. Liu, P. de Mayo and W. R. Ware, *Chem. Phys. Lett.*, 1985, *120*, 460.
55. R. Krasnansky, K. Koike and J. K. Thomas, *J. Phys. Chem.*, 1990, *94*, 4521.
56. W. J. Albery, P. N. Bartlett, C. P. Wilde and J . R. Darwert, *J. Am. Chem. Soc.*, 1985, *107*, 1854.
57. K. F. Scott, *J. Chem. Soc., Faraday Trans. I* , 1980, *76*, 2065.
58. E. K. Rideal, *Proc. Cambridge Philos. Soc.*, 1939, *35*, 130.

59. C. N. Hinshelwood, *Kinetics of Chemical Change*. Clarendon, Oxford, 1940.
60. D. Pines-Rojanski, D. Huppert and D. Avnir, *Chem. Phys. Lett.*, **1987**, *139*, 109.
61. W. R. Ware, R. Weersink and Y. S. Liu, unpublished results.
62. P. Levitz, J. M. Drake and J. Klafter, *Molecular Dynamics in Restricted Geometries* J. Klafter and J. M. Drake, eds., Wiley Interscience, New York, 1989. p. 165.
63. P. Levitz, J. M. Drake and J. Klafter, *J. Chem. Phys.*, **1988** *89*, 5224.
64. J. M. Drake, P, Levitz and J., Klafter, *New J. Chem.*, **1990**, *14*, 77.
65. Y. S. Liu, P. de Mayo and W. R. Ware, unpublished results.
66. B. Stevens and M. Thomaz, *Chem. Phys. Lett.*, **1968**, *1*, 549.
67. B. K. Selinger, *Aust. J. Chem.*, **1966**, *19*, 825.
68. N. Ohta and H. Buba, *J. Chem. Phys.*, **1982**, *76*, 1654.
69. J. M. Beechem and L. Brand, *Ann. Rev. Biochem.*, **1985**, *54*, 43.
70. M. Vincent, J. C. Brochon, F. Merola, W. Jordi and J. Gallay, *Biochem.*, **1988**, *27*, 8752.
71. F. Merola, R. Rigler and A. Holmgrent, J. C. Brochon, *Biochem.*, **1989**, *28*, 3783.
72. O. Kuipers, M.; Vincent, J. C. Brochon, B. Verheij, G de Hass and J. Gallay, *SPIE Intl. Soc. Opt. Engr.*, **1990**, *1024*, 1628.
73. M. Govindjee, M. van der Ven, C. Preston, M. Siebert and E. Gratton, *Biochim. Biophys. Acta*, **1990**,*1015*, 173.
74. M. Barcellona and E. Gratton, *Biochim. Biophys. Acta*, **1989**, *993*, 174.
75. J. R. Alcala, E. Gratton and F. G. Pendergast, *J. Biophys. Soc.*, **1987**, *5*, 925.
76. E. Bismuto, E. Gratton and G. Irace, *Biochem.*, **1988**, *27*, 2132.
77. N. Rosato, A. Finazzi-Agro, E. Gratton and S. Stefinini, *J. Biol. Chem.*, **1987**, *262*, 14487.
78. G. Sanyal, E. M. Thompson and D. Puett, *SPIE Intl. Soc. Opt. Engr.*, **1990**, *1024*, 611.
79. D. R. James, J. R. Turnbull, B. D. Wagner, W. R. Ware and N. O. Peterson, *Biochem.*, **1987**, *26*, 6272.
80. C. D. Stubbs, J. R. Williams and C. Ho, *SPIE Intl. Soc. Opt. Engr.*, **1990**, *1024*, 448.
81. G. Nemecz and F. Schroeder, *Biochem.*, **1988**, *27*, 7740.
82. G. Prenner, F. Paltauf and A. Hermetler, *SPIE Intl. Soc. Opt Engr.*, **1990**, *1024*, 604.
83. C. Haydoc, S. S. Sedaros, Z. Balzer and F. G. Pendergast, *SPIE Intl. Soc. Opt. Engr.*, **1990**, *1024*, 92.
84. P. Hall and B. K. Selinger, *Z. Physik. Chemie, N. F* ., **1984**, *141*, 77.
85. E. Yeramian and P. Claverie, *Nature* , **1987**, *326*, 169.
86. Photon Technology International, 1 Deerpark Dr., Suite F, South Brunswick, N. J., USA 08852.
87. C. A. Royer, J. A. Gardner, J. M. Beechem, J. C. Brochon and K. S. Matthews, *Biophys. J.*, **1990**, *58*, 363.
88. H. W. Engl and C. W. Groetsch, eds., *Inverse and Ill-Posed Problems*, Academic Press, Orlando, 1987.
89. H. F. Kauffmann, G. Landl, H. W. Engl in *Proc. NATO ASI., Large-Scale Molecular Systems: Quantum and Stochastic Aspects. Beyond the Simple Molecular Picture,"* W. Gans, A. Blumen, and A. Amann, eds., Plenum Press, New York, 1990.
90. A Tikhonov and V. Arsenin, *Solutions of Ill-Posed Problems*, John Wiley, New York, 1977.

Chapter 14

Photochemical Processes in Liquid Crystals

Richard G. Weiss

Department of Chemistry,
Georgetown University,
Washington, D.C., USA.

Contents

1. Introduction

Interest in employing liquid-crystalline media to explore the influence of anisotropic environments on thermal and photochemical reactions of dopant molecules has increased markedly during the last several years. The scope of reactions investigated in thermotropic liquid crystals has been reviewed recently.[1] This chapter focuses on photochemical and photophysical processes conducted in thermotropic and lyotropic liquid crystals and describes some of the criteria for conducting photochemical studies in liquid crystals.

Although lyotropic liquid crystals (i.e., those in which the mesomorphic[2] state is solvent induced, usually by water) have been used to make soap for at least 7000 years and thermotropic liquid-crystalline states (i.e., those which owe their mesomorphism to thermally induced processes) are estimated to exist in approximately 5% of all organic molecules,[3] the structural attributes of liquid-crystalline phases have been determined relatively recently and the discovery of thermotropic liquid crystals based upon optical microscopic and macroscopic properties is little more than a century old.[4] The first example of a solute reaction conducted in a liquid-crystalline solvent was reported in 1916 by Theodore Svedberg, who found an abrupt change in the rate of decomposition of picric acid at the mesophase/isotropic phase transition temperature of the solvent, p-azoxy-phenetole.[5]

2. Classes of Liquid Crystals

The structure and properties of liquid-crystalline phases have been reviewed extensively.[3,6-9] The structures and acronyms of many of the liquid crystals mentioned in this chapter are collected in the *Appendix*. Liquid crystals, as the name implies, are condensed phases in which molecules are neither isotropically oriented with respect to one another nor packed with a high degree of three-dimensional order like crystals: they can be made to flow like liquids but retain some of the intermolecular and intramolecular order of crystals (i.e., they are mesomorphic).[10]

A liquid-crystalline state exists within well-defined ranges of temperature, pressure, and composition. Outside these bounds, the phase may be isotropic (at higher temperatures), crystalline (at lower temperatures), or another type of liquid crystal. Liquid-crystalline phases can be grouped according to their microscopic organization into four major classes: nematics, smectics, cholesterics, and discotics. Macroscopically, liquid crystals are both fluid and anisotropic. They are different from plastic crystals, which have little or no orientational order and three degrees of translational order. Liquid-crystalline phases may be thermodynamically stable (enantiotropic) or unstable (monotropic). Monotropic phases can be formed only by cooling from a higher temperature phase. Due to their thermodynamic instability the period during which they retain their mesomorphic properties cannot be predicted accurately. For this reason it is advantageous to perform photochemical reactions in enantiotropic liquid crystals whenever possible.

The overall shape of molecules forming nematic, smectic, and cholesteric phases is rod like. They usually consist of at least one rigid group and at least one flexible chain at a molecular extremity. Plate like molecules (or molecular aggregates that adopt this shape) are found in discotic phases. They include a rigid central core from which several flexible chains emanate. A crude representation of the packing order in each class of liquid crystals is shown in Figure 1. The director (or common axis about which molecular order may be defined) is the long molecular axis.

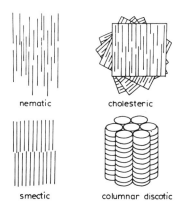

nematic   cholesteric

smectic   columnar discotic

Figure 1. Idealized cartoon representation of the molecular shapes and orientations of the major liquid-crystalline phase types.

2.1. Nematic Phases

Nematic phases exhibit one degree of orientational order. The directors of the constituent molecules are parallel to one another on average. Nematic phases can usually be oriented macroscopically by mechanical forces, electric fields, magnetic fields, and prepared surfaces.

2.2. Cholesteric Phases

Molecules that form cholesteric phases must be optically active or contain an optically active dopant. As the phase name implies, the constituent molecules are frequently steroids and most commonly are cholesteric esters or halides. Since cholesteric phases exhibit nematic-like order locally they are sometimes referred to as "twisted" nematics: molecules pack in a helical array along an axis perpendicular to their directors. A conceptual model of the cholesteric phase includes "layers" of molecules in nematic-like positions, each layer being twisted slightly with respect to the ones above and below it. When the phase consists only of optically active molecules, the angle of twist between layers is typically less than one degree.

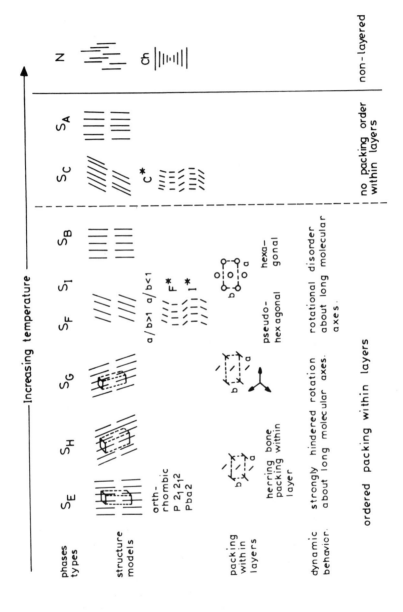

Figure 2. Phase types (especially smectic) according to their relative order. (Reproduced with permission from ref. 11).

2.3. Smectic Phases

Characteristically, smectic phases exhibit one degree of translational and one degree of orientational order. Layers of molecules are stacked on top of one another. The smectic phases can be differentiated further into a letter subclassification that depends upon the packing arrangement of molecules within a layer and the angle of the directors to the layer plane. Except for the S_D modification, which has units of cubic symmetry, the various packing arrangements of smectic subclasses are shown in Figure 2.[11] The star designates an optically active smectic phase.

2.4. Discotic Phases

Several subclasses of discotic phases exist, also.[12] In all, the molecular planes of the constituent molecules are parallel. However, the discs can pack in nematic-like arrangements (N_D) or in columns that are internally ordered (D_o) or disordered (D_d) and may be stacked vertically (as shown in Figure 1), tilted, or twisted.

3. Properties of Liquid Crystals

The anisotropic packing of liquid-crystalline molecules can result in altered conformations or specific orientations of intermolecular collisions for themselves or dopant molecules. As a consequence, liquid-crystalline media, on a microscopic scale, can potentially alter the rates and product distributions of reacting species.[1] In the extremes, the fluidity of the cybotactic regions may approach that of anisotropic liquid or the rigidity of a molecular crystal.[13] Many of the microscopic properties that make liquid crystals attractive as solvents to follow shape changes or collision trajectories along a reaction coordinate are responsible for macroscopic properties that complicate quantitative measurements and necessitate special analytical procedures.

3.1. Viscosities[14] and Diffusion

In isotropic media, rate constants for self-diffusion, k_{diff}, and for solute-solute collisions are calculated routinely from knowledge of viscosities (η) and temperature using the Debye equation.[15] In liquid crystals, rate constants calculated in this way are virtually meaningless due to the complex morphological changes induced by external forces during bulk viscosity measurements[14] on non-Newtonian liquids,[16,17] the nonnegligible influence of solutes on viscosity,[16-18] and the intrinsic solvent anisotropy.

At low shear rates, the enormous viscosities measured for cholesteric phases ($> 10^3$ poise) are significantly larger than for nematic phases.[19] However, flow activation energies (E_{FA}) in cholesteric phases are only 3-4 kcal mol^{-1} higher than in the corresponding isotropic phases (Table 1).

Table 1. Flow Activation Energies of Cholesteric Esters[18]

Cholesteric ester	E_{FA}(kcal/mol)	
	Mesophase[a]	Isotropic
Acetate	15.5 (c)	11.9
Myristate	11-16 (s)	8.2
Palmitate	10 (c)	8.1

[a]. c = cholesteric; s = smectic

Although k_{diff}, diffusion coefficients, and activation energies for self-diffusion in liquid crystals have been measured carefully,[20-26] the values obtained seem to depend upon the method employed.[27]   However, all of the results support the initial qualitative observations that diffusion rates in mesophases are orientation dependent.[28]   For chloroform in the mesophases of p-n-butoxybenzylidene-n-octylaniline (BBOA), diffusion coefficients parallel $(D_{||})$ and perpendicular $(D_{\perp})$ to the solvent director differ by no more than one order of magnitude. With other solutes and mesophases, both larger and smaller differences have been measured. Thus, it is difficult to generalize the magnitudes of rates of diffusion or anisotropies of solute motions in liquid-crystalline phases.

For a unimolecular process, the measured influence of liquid-crystalline order depends upon the temporal response of the solvent matrix to solute shape changes or reorientations.[30-33] This can be expressed in terms of rates of solvent relaxation[34] or solvent friction.[35]   Several types of solvent relaxation phenomena (and rates) have been identified in thermotropic and lyotropic liquid crystals.[30-33] For example, nmr linewidths depend (at least in part) on relaxation times from different nuclear and molecular motions. The proton linewidths from lyotropic aqueous dimethyldodecylamine oxide vary by over three orders of magnitude in its micellar, hexagonal, cubic, lamellar, and crystalline phases.[36] Some relaxation phenomena in liquid crystals are as rapid as in isotropic media; others are *much* slower.

The temporal response of a liquid-crystalline matrix can be influenced greatly by the degree to which a solute disturbs its local environment. In the absence of strong electrostatic interactions, solutes whose size and shape match closely those of their host molecules are incorporated into a liquid-crystalline matrix with the least disturbance.[37]   Strong polar interactions, when present, can dominate shape and size considerations. Thus, some solutes that interact strongly with solvent molecules *raise* mesophase transition temperatures.[38]   Phase *depressions* are a common qualitative measure of the disturbance caused by a solute.

Another potential complication to the interpretation of results from photochemical reactions conducted in liquid crystals (as well as other ordered media) is the influence of the temporal deposition of heat in the cybotactic region as a solute suffers nonradiative loss of energy after excitation. Recent studies have shown that the induced thermal disruption to phase order persists in the region of a solute for periods that are comparable to or greater than those associated with common excited-state reactions.[39] Consequently, photochemical reactions that exhibit no discernible excitation wavelength dependence (in either their kinetics or their product distributions) in isotropic media may do so when conducted in liquid-crystalline solvents. No such example has been reported but none appears to have been sought!

3.2. Optical Properties

A broad range of spectroscopic applications of liquid crystals has been reviewed recently.[40] Spectroscopic studies can provide a great deal of information concerning the order, orientation, and motion of liquid-crystalline molecules and their solutes. Optical studies using plane or circularly polarized light are especially useful in determining the axes of polarization of electronic transitions[41-45] in aligned mesophases.[46,47] Details of the techniques to align mesophases may be found in the references cited.

In order that spectroscopic or photochemical information be obtained, incident electromagnetic radiation must be absorbed by the molecules under study. If solute spectra are being probed, the host liquid crystal should be transparent in the wavelength regions of interest. This limits the liquid crystals that can be used for such studies since most of them contain chromophores that absorb in the near-ultraviolet/visible (uv-vis) regions. A further condition is placed upon liquid crystals as hosts for solute photochemical (or luminescence) studies: the solvent molecules should not quench the excited states of their guests.

Several families of liquid-crystalline molecules meet both of the above criteria for most applications and have been exploited rather successfully. Others have not been employed extensively since they are not commercially available and their synthesis is somewhat involved. Some liquid-crystalline molecules and phases that are transparent or nearly so to uv-vis radiation include the alkylbicyclohexylnitriles from Merck which form smectic and nematic phases[48-51]; n-butyl stearate (BS) which forms a hexatic B phase[52,53]; cholesteric and compensated nematic phases from cholesteryl chloride-cholesteryl ester mixtures[54,55]; the nematic phase from a dialkyl perhydrophenathrene[56]; discotic phases of peralkanoylated α- and β-glucopyranose[57] or $scyllo$-inositol[58]; and aqueous lyotropic liquid-crystalline phases from poly(ethyleneglycol) mono-n-alkyl-ethers $(C_n EO_m)$[59,60] or from the alkali metal salts of n-alkanoic acids.[61]

Even under optimal conditions, liquid crystals contain small amounts of absorbing and photoreactive impurities that can, in principle, compromise the quantitative measure of dopant respone to uv-vis radiation. A study of the stability of several commercially available liquid crystals to a broad band of excitation wavelengths (356-790 nm) has been conducted.[62] As expected, the relative stabilities are greatest for the liquid crystals that absorb least and least stable for those that absorb most. However, the stability of even the least absorbing liquid crystals is reduced significantly by polar impurities present in the commercial preparations. The decision to purify commercial samples of liquid crystals or use them as received must be made on a case by case basis. Economic factors are frequently an important consideration: the cost of several popular liquid crystals is ca. $ 30 per gram.

Liquid-crystalline phases can also be photoprotective. The stability of the thiacyanine dye (1) to radiation from a 100-W tungsten lamp is ca. 10-fold greater in the rodlike (N_C) and disclike (N_L) lyotropic nematic phases of sodium decyl sulfate 1-decanol H_2O mixtures than in the less organized micellar phase of sodium decyl sulfate.[63] At the same time, the fluorescence quantum yields and lifetimes of 1 are about twice the micellar values in the lyotropic nematic phases. Although some of the added stability may be due to the much larger weight fraction of hydrophobic groups in the lyotropic nematic phases, the greater organization must also provide an important contribution.

1

Most mesophases exhibit birefringence whose intensity is dependent upon the index of refraction difference parallel and perpendicular to the director,[64] the angle of incidence of light on the surface,[64,65] and the wavelength of the light.[66] As a result, the polarization of light can be altered as it passes through a mesophase. In nonoriented mesophases where domains of differing alignment are present, internal scattering can make the light path much longer than the sample thickness. To facilitate phase alignment and thereby minimize scattering, thin liquid-crystalline samples frequently are employed. The advantages of performing quantitative spectroscopic measurements on aligned phases (especially homeotropically aligned ones that allow light normal to the surface to pass through as in an isotropic environment) cannot be emphasized too strongly.

Some of the consequences of the factors mentioned above are evident in the fluorescence intensity and transparency measurements made as a function of temperature (phase) on 0.2% pyrene in cholesteryl nonanoate (CN) (Figure 3).[67] The abrupt changes in intensity occur a bit lower than the cholesteric-isotropic (T = 92°C), crystal-cholesteric (T = 81°C), and cholesteric-smectic (T = 76°C; monotropic) bulk phase transition temperatures of CN since local effects by pyrene cause microscopic pretransitional disturbances; the relative intensities do not correlate with the optical density of the sample or with changes in the fluorescence quantum yield of pyrene. As a result, quantitative interpretation of optical data from unaligned liquid-crystalline samples must be done with caution.[68]

Because of their chiral nature, cholesteric phases have special optical properties.[69] They reflect a maximum amount of light at the pitch band maximum (λ_p) that depends upon the bulk index of refraction of the medium (n) and the distance between "layers" for which the twist is π radians along a helical axis (p).[70]

$$\lambda_p = n.p \qquad (1)$$

Frequently, pressure, temperature, or composition changes can be used to move the pitch band far from the wavelength regions of interest. In the region of the pitch band, a cholesteric phase is strongly dichroic (reflected and transmitted light are circularly polarized in opposite senses[69]) and exhibits optical rotatory powers that are 10^2 - 10^3 greater than those of its isotropic phases.[71] These properties add further difficulties to absolute absorption or intensity measurements in cholesteric phases. Although less well studied, the optical characteristics of chiral smectic and discotic phases are analogous in many respects to those of cholesteric phases.

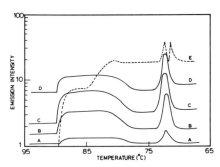

Figure 3. Fluorescence emission intensities (475 nm) as a function of temperature for 0.2% pyrene in CN.[67] Excitation wavelengths: A, 225 nm; B, 250 nm; C, 275 nm; D, 295 nm. Curve E represents transparency vs. temperature. (The ordinate of curve E is optical density multiplied by 10^2). (Reproduced with permission from ref. 67.)

3.3. Potential Problems[1]

Use and analysis of liquid-crystalline solutions for photochemical and photophysical studies are fraught with experimental difficulties. In addition to the optical problems cited above, development of methods for accurate wet analyses of irradiated mixtures can pose a formidable challenge. Also, the solubilization site of a dopant (or dopants) within a mesophase and the influence of the dopant on its cybotactic region require careful consideration: a solute may leave the bulk solvent properties virtually unperturbed while considerably altering its local environment. Alternatively, a dopant may be expelled from the solvent matrix, resulting in phase separation.[72] Unfortunately, no single recipe can be presented that allows all liquid-crystalline solutions to be characterized. The interested reader is directed to the experimental sections of the articles cited for indications of how others have proceeded.

4. Changes in Macroscopic Properties of Liquid Crystals Induced by Photochemical Processes

As early as 1969, the feasibility of forming an irreversible image upon irradiation of a cholesteric phase consisting of a mixture of cholesteryl chloride (CCl) and cholesteryl iodide was demonstrated.[73] Although the mechanism responsible for the color change has not been investigated, it probably involves cleavage of the carbon-iodine bond and, perhaps, changes in the cholesteric pitch.

In a more detailed study, the photodecarbonylation and photoracemization of optically active 2 (equation 2) as a dopant in the cholesteric (twisted nematic) phase of 4-cyano-4'-hexylbiphenyl (CNB6) was shown to lead to a nematic phase.[74]

$$(\pm)-\underline{2} \quad \xleftarrow{\;h\nu\;} \qquad \qquad \xrightarrow{\;h\nu\;} \qquad \qquad \qquad (2)$$

$$(+)-\underline{2} \qquad\qquad\qquad (\pm)-\underline{3}$$

At low dopant concentrations (usually less than 10 wt%), the induced pitch in microns can be expressed in terms of the twisting power β according to equation (3);

$$p = (\,|\,\beta\,|\,.\,c.w)^{-1} \tag{3}$$

(where c is the concentration of dopant in g/g and w is its enantiomeric purity).[71,75] As the concentration of 2(+) decreased, the pitch band maximum, λ_p, moved to longer wavelengths and eventually approached infinity. The progress of the phase transformation was monitored by the so-called "droplet" method in which aliquots of the cholesteric phase are suspended in a nondissolving isotropic solvent. The pitch can be measured from the distance between rings in the droplets that are visualized by crossed light polarization.[76] The same method and a related simpler one have been used to determine the activation barriers of a chiral dopant undergoing thermal racemization.[77]

Reversible color changes in a cholesteric phase of CCl/CN were achieved by selective trans → cis photoisomerization (313 nm) of a solute, azobenzene.[78] The trans and cis isomers affect the pitch differently, allowing changes in λ_p to serve as a quantitative measure of the extent of isomerization. In a similar approach, geometric photoisomerization of the mesomorphic trans-stilbenes (4 and 5) either neat or mixed with cholesteryl oleyl carbonate (COC), has been shown to depress the mesophase–isotropic phase transition temperatures.[79]

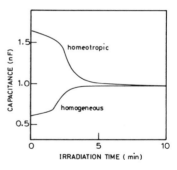

The potential of photoisomerization of 4-butyl-4'-methoxyazobenzene to induce color changes for image storage has been investigated in the nematic phases of **CNB5** [80] and poly(4-methoxyphenyl 4-acryloyloxyalkoxybenzoate).[81] In both cases, trans to cis photoisomerization at 366 nm induces a nematic–isotropic phase transition that is reversible upon photoconversion of the cis to the trans at 525 nm.

Trans → cis photoisomerization of 5 mol% 4-butyl-4'-methoxyazobenzene in the homogeneously aligned (directors parallel to sample covers) or homeotropically aligned (directors perpendicular to sample covers) nematic phase (30°C) of **CNB5** also leads to large reversible changes in the dielectric constant.[82] The capacitance changes shown in Figure 4 upon irradiation at 366 nm can be correlated with the nematic-isotropic phase transition that accompanies formation of the cis isomer.

Figure 4. Capacitance changes of homogeneously and homeotropically aligned samples of 5 mol% 4-butyl-4'-methoxyazobenzene in **CNB5** upon 366 nm irradiation. (Reproduced with permission from ref. 82.)

Figure 5. Reversible capacitance changes effected by sequential 366 nm and 525 nm irradiations of homogeneously aligned 5 mol% 4-butyl-4'-methoxyazobenzene in **CNB5**. (Reproduced with permission from ref. 82.)

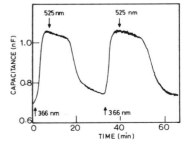

As shown in Figure 5, sequential irradiation at 366 nm and then at 525 nm (to convert cis to trans) modulates the dielectric constant as the phase is modulated. An obvious extension of work of this sort could lead to fast optical switches comprised of smectic C^* phases.

5. Photophysical Processes
5.1. Energy and Electron Hopping in Neat and Doped Mesophases

Due to their packing arrangements, oriented nematic and columnar phases are attractive candidates as "molecular wires" for the photoinitiated transport of electrons or excitation energy. While such applications are outside the scope of this chapter, their underlying principles are not.

In isotropic solutions, a mapping of electron or energy hopping looks like a random walk whose stochastic rates are determined by energetic considerations and the orientations and separations of donor–acceptor pairs. Due to the anisotropic ordering of liquid-crystalline molecules that absorb uv-vis radiation, electron and energy transfer along some directions (that can be defined with reference to the directors) may be preferred to others. Likewise, some packing arrangements may enhance complex formation (which depends upon proximity and specific orientation of the molecular partners[83]), while others may inhibit it. In both isotropic and anisotropic phases, the efficiency of hopping and the number of net forward transfers per initiated event depend upon the average rates of individual forward and reverse steps and competing deactivation rates (from emission, internal conversion, reaction, or charge annihilation).

Recently, the dependence of hopping quantum yields in mesophases on sample thickness and dopant concentration has been treated theoretically.[84] When it is assumed that the energy/electron carriers are randomly distributed dopant molecules within the matrix of inert liquid-crystalline molecules, the quantum yield becomes proportional to the inverse of sample thickness only when the dopant concentration is large. At lower concentrations, where a broad distribution of forward and backward stochastic rates must be invoked due to the random separation between hopping sites, the dependence of the quantum yield on sample thickness is more complex. This treatment is at least qualitatively consistent with electron conduction experiments in nematic[85] and columnar phases.[86,87]

One- and two-dimensional energy transfer or excimer formation in mesophases and their corresponding isotropic phases has focused mainly on two types of molecules:

1. Rodlike molecules containing an aromatic group and forming smectic A or nematic phases
2. Highly unsaturated disclike molecules forming columnar phases.

For instance, the meso- and isotropic phases of 6–8 consist of antiparallel molecular pairs which are dipole-dipole stabilized[88,89], the degree of pairing varies with temperature and phase. The microscopic changes that occur upon changing temperature (phase) are clearly observable in the fluorescence spectra of the constituent molecules. The steady-state fluorescence from neat 4-cyano-4'-octyloxybiphenyl (**6a**) in its lower temperature crystalline phase $(K)_1$, higher crystalline phase $(K)_2$, smectic A, nematic (N), and isotropic (I) phase is displayed with temperature in Figure 6.[90] The K_1 spectrum is ascribed to monomer emission, while the other spectra, which are clearly red shifted and broader, are thought to derive from excimers. When plotted versus temperature, the ratios of emission intensities at 380 and 360 nm (corresponding to the approximate wavelength maxima of the excimer and monomer, respectively) show marked changes at the phase transition temperatures and indicate that the smectic A phase can be extended into a monotropic temperature regime.[91] The fluoresence from both **7** and **8** appear to behave in a qualitatively similar way.[92,93]

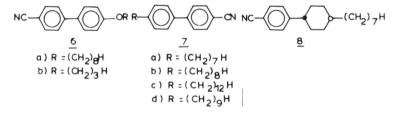

6

a) R = (CH$_2$)$_8$H
b) R = (CH$_2$)$_3$ H

7

a) R = (CH$_2$)$_7$ H
b) R = (CH$_2$)$_8$ H
c) R = (CH$_2$)$_{12}$ H
d) R = (CH$_2$)$_9$H

8

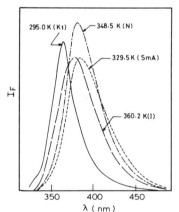

Figure 6. Fluorescence spectra of neat **6a** at various temperatures (phases). The K_2 and I phase spectra resemble each other closely. (Reproduced with permission from ref. 90.)

Consistent with the lack of molecular mobility expected from the K_1 phase of **6a**, its gated emission spectra show no discernable shape changes for 0 to 691 ps time delays and resemble the steady-state spectrum in Figure 6.[90] However, emission from the smectic A phase (329 K) evolves from a shape at 0 ps delay that resembles the K_1 spectrum to one that is like the steady-state smectic A spectrum after ca.700 ps. Since $I_{380} > I_{360}$ after 200 ps, excimer formation must be nearly complete. Qualitatively similar spectral evolutions can be observed in the even less viscous nematic and isotropic phases of **6a**.

The lifetime of the excimer formed upon irradiation of neat **6a** is somewhat phase dependent:[91] τ = 10.7 ns at 60°C (smectic); τ = 7.6 ns at 72°C (nematic). When dissolved in the smectic A phase of 4-pentylphenyl *trans*-4-pentylcyclohexanecarboxylate (**5H5**) the monomer fluorescent lifetime of **6a** was 1.1 ns.[91]

Time-correlated single-photon counting decay curves from **7** can be fit to a minimum of three phase-dependent decay constants (Table 2).[94]   The wavelength dependence on the pre-exponential factors of the decay constants and the time evolution of the emission spectra suggest the presence of three species, a monomer (τ_1) and two excimers (head-to-head, τ_2[89]; head-to-tail, τ_3).[94] The steady state emission spectra of **7a** and **7b** in their isotropic and liquid-crystalline phases are broad, structureless, and red shifted. They have been attributed to a mixture of monomer and excimer (isotropic) and pure excimer (smectic A and nematic).[92,93] Time-resolved emission spectra (120 ps windows) from crystalline and smectic A phases of **7b** (temperature not mentioned) are reproduced in Figure 7.[94]

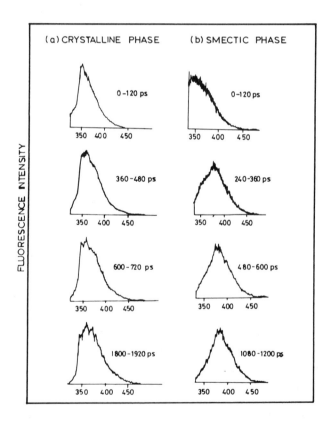

Figure 7. Time-resolved emission spectra from **7b** in its crystalline and smectic A phases. (Reproduced with permission from ref. 94.)

In the crystalline phase, the emission spectra broaden with time, but their λ_{max} remain fairly constant. In the smectic phase, the change from monomer emission ($\lambda_{max} \cong 350$ nm) to excimer emission ($\lambda_{max} \cong 380$ nm) occurs during the first nanosecond. The excimer lifetimes of **7b** are reported to be 21 ns (smectic A, 25°C), 16 ns (nematic, 33°C), and 9.6 ns (isotropic, 60°C).[92] The monomer emission from a smectic A solution of **7b** in **5H5** was found to be 1 ns, similar to the singlet lifetime of **7a** under similar experimental conditions.

Steady-state emission spectra of the longer chained homologue (**7c**), shows the presence of excimer and monomer emissions in the neat crystalline, smectic, and isotropic phases.[93] It is unclear why crystalline **7a** and **7b** do not have discernible excimer emissions[92] and **7c** does. One of the assignments may be incorrect. The proportion of excimer emission was greatest in the smectic phase of **7c** due to preordering effects of the mesophase layers. As with **7b** and **6a**, the excimer lifetime of **7c** is phase dependent: $\tau = 4$ ns for the crystal, $\tau = 18$ ns for the smectic, and $\tau = 10$ ns for the isotropic phase.[93]

Table 2. Decay Constants from Neat **7a** and **7b** Emissions at 440 nm as a Function of Phase.[94]

7	Phase	T(°C)	τ_1 (ns)	τ_2 (ns)	τ_3 (ns)
a	Nematic	35	0.1	2.8	19.7
	Isotropic	45	0.2	2.9	14.3
b	Smectic	26	0.1	2.6	21.4
	Nematic	36	0.2	3.3	19.4
	Isotropic	46	0.3	3.4	15.8

The steady-state fluorescence of **7d** has been examined in various concentrations of **5H5** as a function of temperature (phase), composition, and sample history.[95] Under the excitation conditions employed (280 nm), **5H5** neither emits nor quenches emission from **7d**. It acts as a diluting matrix to support and extend the mesomorphic range of **7d**. From differential scanning calorimetric measurements and optical micrographs, the **7d-5H5** phase diagram was determined and shown to include a monotropic smectic phase in low-wt% mixtures of **7d**. The consequences of the phase changes include striking variations in the shape and intensity of the emission spectra.

These are exemplified by data obtained from 20 wt% **7d** in **5H5**. In the first heating scan, the structured monomeric fluoresence in the crystal gives way to predominantly excimer emission from a eutectic smectic A-crystal mixture. The

618   Weiss

emission intensity and the monomer emission component increase with temperature as the excess crystalline **5H5** melts, diluting the local concentration of **7d**. Subsequent intensity increases reflect changes in mesomorphic ordering since the concentration of **7d** above 31°C is constant. The large increase in intensity and blue shift (and consequent decrease in probability) at the smectic A-nematic phase transition (40-42°C) can be ascribed to the destruction of phase layers (and consequent decrease in probability of excimer formation). However, at least some of the intensity differences may be due to differing phase orientations with respect to the incident radiation (see Section 3.2). The isoemissive point and preference for monomer emission at higher temperatures within the isotropic phase can be explained classically if the rate constant for excimer dissociation is greater than the rate constant for excimer deactivation. Rapid cooling of the isotropic phase allows formation of a monotropic smectic B phase whose emission characteristics favor the monomer. The subsequent increase in excimer emission within the smectic A phase and renewed preference for monomer fluorescence from the nematic phase follow from the explanations provided with above.

Rate constants for energy transfer from excited singlet states of **6-9** and their excimers to 1,8-diphenyl-1,3,5,7-octatetraene (**10**, a rodlike acceptor) or perylene (**11**, a platelike acceptor) have been measured as a function of phase.[91] The quenching data set using either **10** or **11** is amenable to a Stern-Volmer treatment;[96] results are summarized in Table 3. In spite of the mathematical compatability with a Stern-Volmer treatment, the derived rate constants exceed the calculated rate constants for self-diffusion. This requires that a mechanism other than simple collisional energy transfer be operative.[97] Forster energy transfer can be eliminated also since the overlap between the emission spectra of the donors and the absorption spectra of the acceptors is smaller than necessary. The favored mechanism involves excitation migration (hopping),[98] a process that is enhanced by the proximity and the alignment of **6-9** in their mesophases. Quenching occurs when an excited donor finds itself next to a molecule of **10** or **11**.

$$CH_3-\text{(O)}-\text{(O)}-(CH_2)_5H$$

9

Pulse radiolysis of neat **7b** leads to several interesting intermediates whose structures have been assigned tentatively.[99] As with electronic excitations of **7b**, the efficiency of formation and rates of disappearance of transients are acutely phase dependent. At 47° C (isotropic phase) and 27°C (smectic A phase), triplets (λ_{max} = 380 nm) with halflives of 1 and 10 ms, respectively, are produced. In the isotropic phase, an additional absorption at 480 nm with a halflife of 17 ms was ascribed to the cation radical of **7b**. Another transient at 550 nm (halflife of 2.5 ms), attributed to the anion radical, was detected in the nematic phase. No extension of this work appears to have been published to date.

Interest in the potential for exploiting electron and energy transfer across an aligned sample of a columnar discotic phase has risen steadily since the initial report in 1982.[100]

Table 3. Rate Constants for energy transfer from the excimer or monomer excited singlet state of **6–9** to **10** or **11**.[91,92]

Donor	Excited State species	Acceptor	Phase	k_{et} x 10^{-10} (1/mol/s) From static excitation	From flash excitation
6 a	Excimer	**1 0**	Smectic A	7.8	5.1
	Excimer	**1 0**	Nematic	11.0	7.2
	Excimer	**1 1**	Nematic		2.6
	Monomer	**1 0**	Smectic A[a]	110.0	
6 b	Excimer	**1 0**	Isotropic	14.0	
7 a	Excimer	**1 0**	Smectic A	2.84	0.92
	Excimer	**1 0**	Nematic	5.51	1.53
	Excimer	**1 0**	Isotropic	12.40	
	Monomer	**1 0**	Smectic A[a]	45.6	
8	Excimer	**1 0**	Nematic	1.64	0.80
9	Monomer	**1 0**	Isotropic	5.23	

[a.] 2% donor (by weight) in **5H5**

One-dimensional excitation hopping has been investigated in the ordered hexagonal columnar phase (D_{ho}; 68-97°C[101]) of hexa(dodecyloxy)triphenylene (**12**).[102] From the known columnar diameter (ca. 20 Å) and molecular core separations within a column (3.6 Å) in both the crystalline and discotic phases of **12**,[103] intracolumnar energy transfer can be calculated to be much more probable than intercolumnar transfer by 3 x 10^4 assuming a Forster mechanism[97] and by 10^7 assuming a collisional (Dexter) mechanism.[83] Only monomer emission centered at 385 nm is detected in the neat crystalline (20°C) and discotic (80°C) phases. The vibronic structure in the fluorescence from the crystalline phase is lost in the mesophase. Emission includes "prompt" fluorescence, phosphorescence, and delayed fluorescence from triplet-triplet annihilations that occur within a column [equations (4) and (5)]. Lifetimes of the various luminescence components are collected in Table 4. The dominance of delayed fluorescence over phosphorescence provides evidence for efficient energy migration within a column of **12**. The average residence time of triplet excitation energy on one molecule (i.e., the hopping rate) was not calculated.

Table 4. Excited State lifetimes from Luminescence of **12** in various Phases[102]

Phase	Temperature (°C)	$\tau_F{}^a$ (ns)	$\tau_{DF}{}^b$ (µs)	$\tau_P{}^c$ (µs)
Isotropic (2×10^{-4} M **12** in heptane)	20	5.9±0.6		
Crystal	20	7.4±0.7	0.49±0.02	1.13±0.05
Columnar discotic	80	6.5±0.7	0.45±0.02	0.85±0.05

[a] Prompt fluorescence
[b] Delayed fluorescence
[c] Phosphorescence

$$^3 12 + {}^3 12 \rightarrow {}^1 12 + 12 \qquad (4)$$

$$^1 12 \rightarrow 12 + h\nu \qquad (5)$$

The hopping rates of excitation energy and the associated activation energies were calculated in the solid and mesomorphic columnar phases (nonoriented) of six 2,4,6-tri(3,4-di-n-alkoxyphenyl)pyrylium tetrafluoroborate salts (**13**) by a different approach.[104] For several members of the series, the D_{ho} nature of the mesomorphic phases has been established by x-ray diffraction;[105] the others appear to be columnar by structural analogy and by the similarity of their spectroscopic properties to those of the D_{ho} compounds.[104] The crystalline-mesophase transition temperatures occur near room temperature for **13d** - **13f** and at ca. 100°C for **13c**. No mesophase is discernible for **13a** and **13b**.

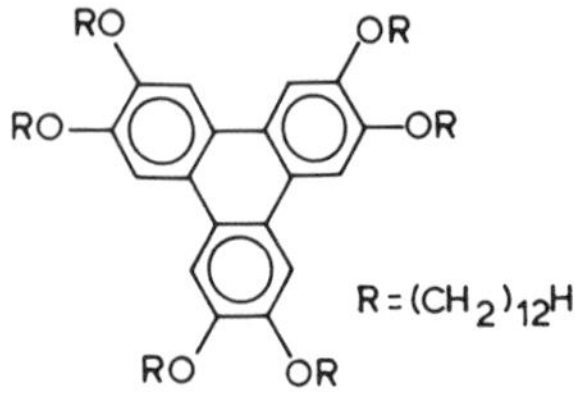

$R = (CH_2)_{12}H$

12

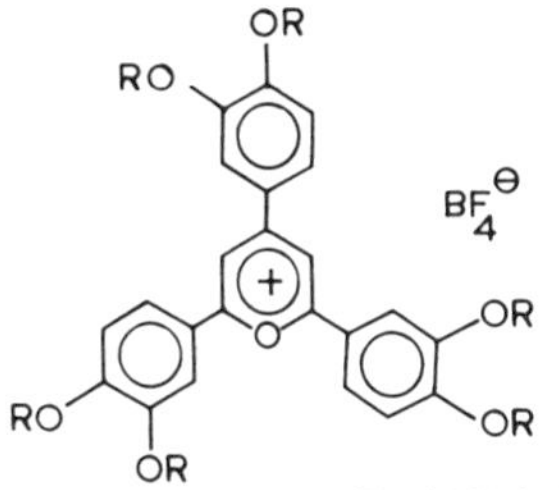

13 $R = (CH_2)_n H$

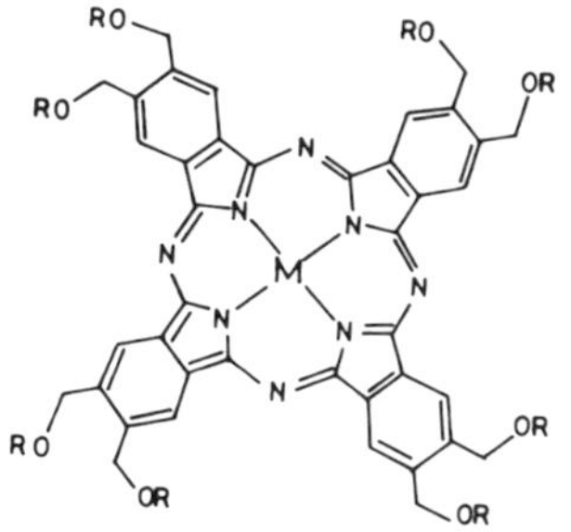

14 a) M = 2 H
    b) M = Cu(II)
    c) M = Zn(II), $R = (CH_2)_{12}H$

In dilute dichloromethane solutions of each of the **13** homologues (where ground-state aggregation is negligible), the fluorescence maximum occurs at 587 nm, $\tau_F \approx$ 1- 1.5 ns, and Φ_F increases regularly in the series from 0.7 for **13a** to 1.0 for **13f**. The shape of the emission spectra from thin films of **13** do not change with alkoxy chain length and exhibit λ_{max} at 685 nm. At 150°C, λ_{max} is blue-shifted to 673 nm. Such a shift is expected if monomer emission is preferred at elevated temperatures. Neither the shape nor the decay profiles (at one wavelength) of the room temperature film emissions varies with excitation wavelength, indicating that the emitting species emanate from a common ground state. Gated emission spectra show that the emission maximum is red shifted with time, making analysis of the early portion of decay profiles very difficult. Beyond 750 ps, the decay curves can be fit using global analysis to two decay constants (τ_1 and τ_2; Table 5). From the relative magnitude of the decays and the spectral changes with temperature, τ_1 can be assigned to monomeric emission and τ_2 to excimer emission.

Table 5. Luminescence Lifetimes of **13** Thin Films at 20° C: λ_{ex} = 580 nm.[104]

1 3	n	$\tau_1(ns)^a$	$\tau_2(ns)^a$
a	2	1.5	3.9
b	3	1.7	4.2
c	4	1.4	4.2
d	5	1.2	4.0
e	8	1.5	4.3
f	12	1.6	4.3

^{a} ± 0.2 ns

Since the columns are not aligned, energy traps may form easily at domain boundaries (column discontinuities) and appear to be responsible for most of the monomer emission. If this hypothesis is correct, the number of energy hops and, therefore, the excitation pathlength are determined by the length of a column. At low fluences the number of hops (n) can be expressed simply by equation (6);

$$n = \tau/\tau_h \qquad (6)$$

where τ is the exciton lifetime in **13** and τ_h is the hopping time.[106] Assuming that the average value of n is temperature independent, Arrhenius plots of τ_2 yield the activation energy for exciton migration, E_M. Only data from the liquid-crystalline phases of **13** follow an Arrhenius behavior. The values so obtained (E_M = 1.1 kcal mol^{-1} for **13d**, 1.4 kcal mol^{-1} for **13e**, and 2.0 kcal/mol for **13f**) are very low and similar to measured from molecular crystals.[107] They indicate that intracolumnar energy migration along oriented columns of discotic mesophases should be possible over very long distances.

Discotic phases of phthalocyanines (**14**) are attractive molecules to study intracolumnar electron and excitation energy hopping and they have been used rather extensively. For instance, in the D_{ho} phase of **14a** (78-305°C), columns are 31 Å in diameter.[108] In the solid phase at room temperature, columns are packed orthorhombically and tilted by 24°.[109] Such detailed knowledge of molecular packing facilitates interpretation of hopping results. From spectroscopic data on samples with various amounts of **14b** doped in columns of **14a**, the average hopping distance of singlet excitation energy in the solid phase at room temperature was calculated to be 100-200 Å (ca. 20-45 molecules[110]): at 16 mol% of **14b**, fluorescence from **14a** is almost completely quenched; at 0.05 mol% **14b**, fluorescence from **14a** is easily detected. However, the fluorescence of the latter solution is attenuated markedly at temperatures slightly above the crystal-D_{ho} transition (indicating more facile energy hopping).

Qualitatively similar results are found with **14c**.[111] In the solid phase, two broad absorption bands (λ_{max} = 650 and 740 nm) that change to λ_{max} = 630 and 750 nm in the D_{ho} mesophase are observed. Low-fluence pulsed excitation (532 nm) of **14c** and spectral monitoring after 0.2 μs indicate that the ratio of quantum efficiencies for triplet formation in the solid and D_{ho} phases is 3.7. The triplet lifetimes (τ_T) differ by a factor of 2 in the two phases: 21.5 μs in the crystal and 10.0 μs in the D_{ho} phase. At higher fluences, the decay of triplets becomes second order (see equation 4). Outside the phase transition region, the second-order rate constants (k_{TT}) are nearly temperature independent within a phase: k = 1.0 x 10^6 l mol^{-1}s^{-1} (crystal) and 1.3 x 10^{7}l mol^{-1}s^{-1} (D_{ho}).[111b] By comparison, the ratio of the corresponding rate constant in neat **14a** is only 1.8.[112] Thus, as expected, all intermolecular processes are faster in the more fluid mesophase. The constancy of rates within a phase indicates that they are dominated by ordering (entropic) differences rather than activated (enthalpic) processes. Using this and related data, the exciton path lengths (i.e., columnar lengths) were calculated to be >230 Å in the solid phase (25°C) and >144 Å in the mesophase (90°C).

In a related study, the luminescence properties of neat **15a** containing small amounts of **15b** were investigated over a very large temperature range (4.2 - 400 K) that includes the crystalline and D_{ho} phases (T_{K-D} = 73°C).[113] Using Q-band excitation (600 nm), the steady-state luminescence spectra of **15a** containing almost no **15b** were recorded as a function of temperature and intensity. At 15 K, two broad emission bands at λ_{max} 795 nm and 820 nm were detected. Above 80 K, only the 820 nm band is apparent. The temperature-induced relative intensity changes, followed at the emission maxima in the absence and presence of 0.5 mol% of **15b** (an efficient quencher of **15a** fluorescence), are displayed in Figure 8. As can be seen, the intensity decreases dramatically from 4.2 K (Φ_F = > 0.5) to 200 K and then again at the crystalline - D_{ho} transition. These changes have been interpreted in terms of exciton recombination at three different site types. The first, leading to emission at 795 nm, occurs along columnar stacks of **15a**. The second, responsible for the 820 nm band, occurs at longitudinal columnar boundaries or at defect sites whose energy is ca. 1.4 kcal lower than the columnar sites. The third is due to quenching traps (primarily **15b**).

R = $(CH_2)_{12}H$

<u>15</u> a) M = 2H
         b) M = Cu (II)

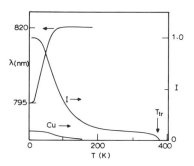

Figure 8. Emission wavelength maxima and their relative intensities from **15a** as a function of temperature.[113] T_{tr} is the solid-mesophase transition temperature. Curve Cu is for **15b** scaled to the intensity of **15a**. (Reproduced with permission from ref. 113.)

Since a major fraction (ca. 30%) of the excitons recombine radiatively along the columns of undoped **15a** at 4.2 K, exciton hopping must be somewhat restricted. However, hopping does occur even at this low temperature: addition of 0.5 mol% of **15b** results in a ca. 90% decrease in emission intensity. The activation energy for hopping in the low-temperature regime is estimated to be 0.15 kcal. The greater efficiency of exciton recombination in the mesophase can then be ascribed to a combination of faster rates at the higher temperatures[113] and greater disorganization along the columns.[109]

Relatively facile triplet exciton migration is found also in the hexagonally packed, internally disordered columnar mesophase (D_{hd}, 62-193°C) of metal-free **14c**.[112] In the solid state at 25°C, τ_T is 7.5 μs; in the mesophase at 84°C, τ_T = 4.2 μs. Using higher fluences for the laser excitation pulses (Q-band excitation at 532 nm), the rates of the triplet-triplet annihilation process of **14c** were measured as a function of temperature (Figure 9). The inflection point corresponds to the crystal-mesophase transition temperature. From this data and the average distance between stacked molecules of **14** (ca. 4.5 Å[110]), the average exciton hopping pathlength is calculated to be 640 Å in the crystal (25°C) and 670 Å in the mesophase (77°C). As expected, these lengths are longer than those found for singlet exciton hopping[109] but shorter than expected in an organic single crystal (a few microns[96]). Columnar lengths defined in this way are expected to be longer than those determined by x-ray diffraction[114] since minor dislocations detected by the diffraction technique are not registered by the triplet decay measurements unless exciton transfer is blocked.

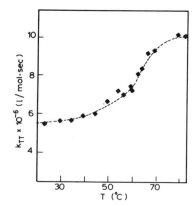

Figure 9. Temperature variation of k_{TT} of neat **14c**. (Reproduced with permission from ref. 112.)

5.2. Intermolecular Excimers and Exciplexes

Liquid crystals are excellent media for studying the properties of excimers and exciplexes.[115] Since these complexes are stable only in their excited states and are dissociative in their ground states, they have no ground-state absorption spectra and are difficult to characterize: both polarized absorption and emission spectra are desirable to assign transition polarizations. Solid media (such as doped single crystals) provide excellent orientational rigidity to solutes but are inappropriate solvents for excimers and exciplexes since diffusion-limited collisions between the molecules must occur within the excited singlet lifetime of the monomer if a complex is to form. Isotropic media allow excited complex formation, but any polarization information is lost due to the lack of orientational rigidity in the medium.

The linear polarization of fluorescence from the excimer of pyrene was examined in an electric field-oriented 1.95:1 (w/w) nematic point mixture of CCl to cholesteryl laurate (CL).[116a] The long molecular axis of pyrene (defined along carbons 2 and 7) aligns preferentially parallel to the solvent directors. While the intensity of fluorescence from monomeric pyrene was found to be more intense parallel to the solvent director than perpendicular to it ($I_{||} > I_{\perp}$), the opposite applied to the excimer emission (Figure 10). Thus, the excimer emission is polarized primarily along an axis perpendicular to the long axis. Whether that axis is in plane (x) or out of plane (y) could not be determined. Interestingly, polarization measurements on the exciplex emission band from pyrene and *N,N*- diethylaniline in the same compensated nematic mixture showed that it is polarized preferentially along the z-axis (i.e., parallel to the long molecular axis of pyrene and to the polarization of the monomer emission).[116b]

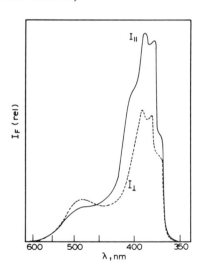

Figure 10. Polarized fluorescence spectra of 0.07 M pyrene in an oriented compensated nematic mixture of 1.95:1 (w:w) CCl/CL at 30°C. (Reproduced with permission from ref. 116a.)

To resolve the polarization ambiguity of the pyrene excimer, the monomer and excimer polarizations of 4,9-dicarboalkoxypyrene molecules whose ^1L_b and ^1L_a transitions lie along the same molecular axes as in the parent molecule were investigated.[117] When examined in the aligned nematic phase of *trans*-4-(4-

cyanophenyl)-1-n-pentylcyclohexane (**PCH5**), both diesters could be shown from the polarization of their monomer emission spectra to lie with their C_2–C_7 molecular axes nearly perpendicular to the solvent director (i.e., $I_\parallel < I_\perp$). Since $I_\parallel > I_\perp$ for the emission from the excimers of the same two diesters (Figure 11), their polarizations must be nearly parallel to the solvent director and lie principally along the x axis (in the molecular plane).

A serious complication to the calculation of order parameters from emission spectra in liquid crystals is the diffusion-induced depolarization that occurs during a lumophore's excited state lifetime: emission anisotropy, alone, is insufficient to assign correct order parameters.[118] Since viscosity and, therefore, diffusional rates are very sensitive to temperature within one phase, polarized emission intensities can also be very temperature dependent.[119] However, when the time-dependent ratios of $I_\parallel/\perp$ are analyzed carefully, detailed information concerning order parameters and the dynamics of solute motions in anisotropic environments can be obtained.[120]

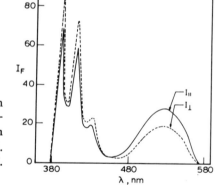

Figure 11. Fluorescence polarization spectra of the dipentyl ester of pyrene-4,9-dicarboxylic acid (0.17 mol kg^{-1}) in nematic **PCH5** at 32.5°C (λ_{ex} = 352 nm). (Reproduced with permission from ref. 117.)

Such an experiment has been performed on the fluorescence of 1,6-diphenyl-1,3,5-hexatriene in nematic **ZLI-1167**.[121] From the temperature dependence on the derived perpendicular rotational diffusion coefficients, an activation energy for this solute motion of ~6 kcal mol^{-1} was calculated. Such studies, although difficult to perform and analyze, hold promise of allowing detailed information about solute-anisotropic solvent motions, not easily accessible by other spectroscopic techniques, to be obtained. To date, most excited-state analyses have employed fluorescence rather than phosphorescence spectra since the lifetimes of excited singlet states are usually much shorter than of triplets and the degree of diffusional depolarization is dependent upon the period after excitation during which measurements are taken.

Circularly polarized fluorescence (CPF) holds complementary information concerning the polarization of solute transitions.[122] Like its linear dichroism counterparts, CPF is time dependent and, therefore, requires very detailed analyses unless the rates of solvent relaxation and of solute conformational change and reorientation after light absorption are either much faster or much slower than the rate of excited state decay.[122] The emission dissymmetry factor (g_e) has a form similar to that of the absorption factor. In equation (7), I_L and I_R are the intensities of left- and right-handed emission at wavelength λ:

$$g_e = \frac{(I_L - I_R)}{1/2(I_L + I_R)} \qquad (7)$$

The sign of g_e can be correlated to the sense of the helical pitch of a cholesteric phase and the polarization direction of emission with respect to the solvent director.[123] Circularly polarized fluorescence from cholesteryl 3-naphthylpropanoate, cholesteryl 5-naphthylpentanoate (with the point of attachment at the 1 or 2 position of naphthyl), and a 3:2 (mol/mol) mixture of cholesteryl 3-(1-pyrenyl)propanoate/cholesteryl 3-phenylpropanoate (**16/Ph-2**) has been examined.[124] In all cases, the g_e were positive, indicative of nearly perpendicular orientations for the monomer and excimer transitions with respect to the optical axis of the left-handed helical phases.[123] Although the data are not definitive, they indicate that the monomer and excimer transitions are polarized along mutually perpendicular axes. Since the $^1L_a \leftarrow {}^1A$ monomer emissions are $\pi^* \leftarrow \pi$, they must lie in the plane of the naphthyl rings. This implies that the excimer's emission polarization is along the axis normal to the molecular plane. The magnitudes of g_e in the wavelength regions of the monomer emission were smaller than in the regions of the corresponding excimers. Apparently, the aromatic groups are not oriented well within the cholesteric matrices.

$$\underline{16}$$

Circularly polarized fluorescence has also been employed to investigate the influence of electric field-induced reorientations of 20% (**16**) in a 3:5 mixture of CCl/CN.[125] The g_e variations with wavelength in the absence of an electric field were much less pronounced than in the aforementioned study. However, as the magnitude of g_e increased, its wavelength maximum changed, and the variations with wavelength became much more apparent as the electric field strength was increased from 0 to 1.6 x 10^7 V m^{-1} (Figure 12). The results suggest that the lumophores are being reoriented and their excimer structures are being changed subtly as the phase is reoriented.

The rate at which pyrene fluorescence is quenched by either pyrene[126] or 5α-cholestan-3β-yldimethylamine (17)[127] (and excited-state complexes are formed) has been measured by single photon counting techniques in the isotropic and cholesteric phases of a 59.5:15.6:24.9 (by weight) mixture of cholesteryl oleate (CO)/CN/CCl (CM).

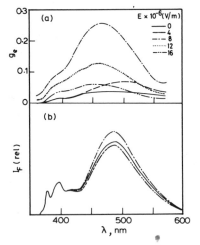

17

The data indicate that the quenching geometries for formation of the pyrene excimer and the pyrene-17 exciplex (and, presumably, the complexes themselves) are very different. Whereas the excimer has the shape of two parallel plates, the exciplex resembles a rod intersecting a plate at a near-normal angle. This geometry allows overlap between the nitrogen lone pair of electrons on 17 and the π-system of pyrene singlets.

Figure 12. (a) CPF and (b) unpolarized fluorescence spectra from 20 mol% **16** in 5:3 (mol:mol) CN/CCl at 70°C (λ_{ex} = 320 nm) under the dc electric fields indicated. (Reproduced with permission from ref. 125.)

Dynamic Stern-Volmer plots[128] of pyrene fluorescence quenching by (<0.25 M) 17 in CM were curved, indicating that 17 causes a perceptible change in cholesteric solvent organization. Similar plots for fluorescence quenching by pyrene were linear at <0.20 M pyrene. Several methods were employed to convert the experimental data into rate constants for quenching and to calculate the activation parameters. The values reported in Table 6 are derived from Eyring and Arrhenius equations.

Table 6. Activation Parameters for Quenching Pyrene Singlets by Pyrene or **17** in CM [126,127]

Quencher phase:	Pyrene		17	
	Cholesteric	Isotropic	Cholesteric	Isotropic
E_a (kcal/mol):	8.9 ± 0.4	6.6 ± 0.6	9.9 ± 0.2	5.3 ± 0.1
$\Delta S^{\ddagger}$(cal/mol/deg):	-3 ± 0.5	-4 ± 0.5	-5 ± 0.6	-10 ± 0.4

They demonstrate that, as expected from the shapes of the excited-state complexes, quenching by **17** is more difficult than by pyrene and involves a larger disruption to solvent order. The differences between each parameter in the two phases (ΔE_a and $\Delta \Delta S^{\ddagger}$) for each quencher are more diagnostic than the absolute values. Since diffusion between two molecules of pyrene or between pyrene and **17** in one phase of CM should be nearly identical, comparison between ΔE_a or $\Delta \Delta S^{\ddagger}$ for the two quenchers indicates the importance of solvent order to each process. Thus, ΔE_a (pyrene) = 2.3 kcal mol^{-1} is significantly smaller than ΔE_a (**17**) 4.6 kcal mol^{-1}: exciplex formation with **17** is energetically more disfavored than is excimer formation in the cholesteric phase. Similarly, the $\Delta \Delta S^{\ddagger}$ (pyrene) = +7 cal mol^{-1} deg^{-1} and $\Delta \Delta S^{\ddagger}$(**17**) = +15 cal mol^{-1} deg^{-1} are consistent with the exciplex being more disruptive to its local environment than the excimer.

The variation of the ratio of excimer to monomer emission intensities (I_E/I_M) from pyrenyl systems in liquid-crystalline solvents has provided interesting information also. For instance, I_E/I_M ratios for a 3:2 **16/Ph-2** mixture are summarized in Figure 13.[124,129] In absolute terms, I_E/I_M is smaller in the crystalline phase than in the cholesteric phase. However, the cholesteric ratios increase, whereas the solid-state ratios decrease with increasing temperatures. Similar plots have been made for pyrene in CM. In this case, I_E/I_M in the cholesteric phase also increases but more rapidly than in the **16/Ph-2** system. Furthermore, I_E/I_M continues to rise in the isotropic phase of CM. That I_E/I_M decreases with increasing temperature in the isotropic phase of **16/Ph-2** may be due to increased excimer dissociation at the higher temperatures (well above 100°C) where the isotropic phase exists: the origin of the differences between the two sets of isotropic data[124,126] can be traced to the photophysics of the pyrene excimer and not to differences between the arrangement of molecules in the respective isotropic phases. The same may not be true of the cholesteric phase data. The greater diffusional mobility and orientational freedom of pyrene in cholesteric **CM** allows it to attain appropriate excimer collisional geometries more easily than can **16** in its cholesteric mixture.

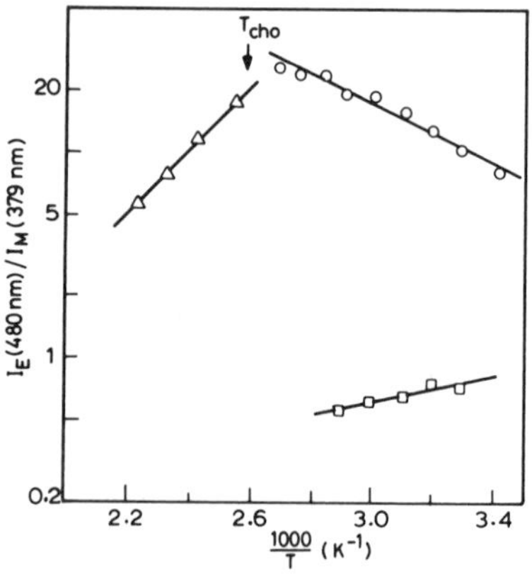

Figure 13. I_E/I_M for pyrenyl luminescence from a 3:2 (w/w) mixture of **16/Ph-4** versus temperature in the crystalline (□), cholesteric (o), and isotropic (Δ) phases. (Reproduced with permission from ref. 129.)

Fluorescence decay curves from **16** in 1/2 (mol/mol) **Ph-2**/cholesteryl 5-phenylpentanoate (**Ph-4**) have been analyzed carefully using the time-correlated single-photon counting technique.[130] In both the cholesteric and the isotropic phases and at up to 25 mol% **16**, the data could be fit to the simple mechanism in Scheme 1: the ratio of the rates for excimer rise and monomer decay remained near -1 at all temperatures and concentrations of **16**. A very fast component of decay may be associated with exciton hopping among **16** molecules at the higher concentrations.

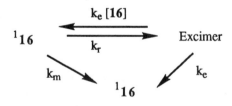

Scheme 1. Excimer formation and decay mechanism of **16** in **Ph-2–Ph.4**.[103]

Arrhenius plots of k_E (the rate constant for excimer formation) and k_r (the rate constant for excimer dissociation) at various concentrations of **16** are shown in Figures 14 and 15. For comparison purposes, k_E from pyrene in **CM** are included in Figure 14.[126] The major influence of solvent order is clearly exercised on k_r as judged by the large slope change that occurs at the cholesteric-isotropic phase transition temperature. Almost imperceptible differences in the k_E slopes for **16** are found in the cholesteric and isotropic phases of **Ph-2/Ph-4**. At all three **16**, concentrations of the activation energies associated with **16** excimer dissociation, E_r^*, are 10.4 ± 2 kcal mol^{-1} in the isotropic phase and 0 ± 2 kcal mol^{-1} in the cholesteric phase. This suggests that the excimer configuration of **16** is not incorporated well into the cholesteric matrix. The activation energies associated with excimer formation, E_a, are 8.0 ± 1.0 kcal mol^{-1} in the isotropic phase and 6.0 ± 0.8 kcal mol^{-1} in the cholesteric phase of **Ph-2/Ph-4**.

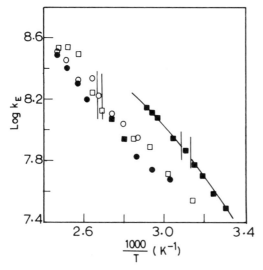

Figure 14. Arrhenius plots of rate constants for formation of excimers of **16** in **Ph-2/Ph-4**. The relative ratios of **16/Ph-2/Ph-4** are 8:31:61(O), 16:28:56 (●), and 25:25:50 (□). Included also for comparison purposes are data for pyrene excimer formation in **CM** (■). Vertical lines indicate phase transitions. (Reproduced with permission from ref. 126.)

These values are similar to the E_a for pyrene in **CM** and reported in Table 6. However, **16** excimer formation is calculated to be enthalpically easier in the cholesteric phase than in the isotropic phase but pyrene excimer formation is more difficult in the cholesteric phase. The greater ease of **16** excimer formation may be due to some pairs of **16** in their ground state being oriented somewhat near the excimer configuration in cholesteric **Ph-2/Ph-4**. If hopping to these exciton traps occurs rapidly, E_a measures local reorganizational energy, but not the energy of diffusion.

Circularly polarized fluorescence was employed to demonstrate the presence of excimer emission in cholesteric phases comprised of cholesteryl 3-(9-

carbazoyl)propanoate (**18**) or cholesteryl 3-[3-(9-ethyl carbazoyl)]propanoate (**19**) in cholesteryl 3-(2-naphthyl)propanoate (**2N-2**) or **Ph-2/Ph-4**.[131] The characteristic plane curves for g_e at long emission wavelengths serve as a diagnostic (Figure 16). From the circular dichroism and CPF spectra of **18** and **19**, the probable configuration of their excimers in the cholesteric phases was deduced (Figure 17).

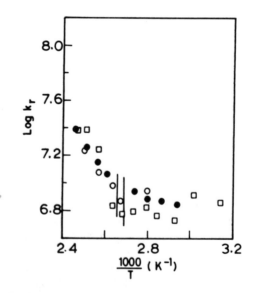

Figure 15. Arrhenius plots for rate constants associated with **16** excimer dissociation. Symbols are as in Figure 14. (Reproduced with permission from ref. 130.)

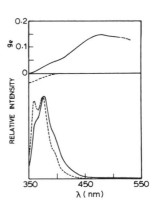

Figure 16. Fluorescence and CPF spectra: a 50:50 cholesteric mixture of **19**/Ph-2 at 60°C; fluorescence from a 1:33:66 cholesteric mixture of **19**/Ph-4/Ph-2 at 60°C and CPF from 1/99 cholesteric mixture of **19**/Ph-2 at 76°C. λ_{ex} = 340 nm. (Reproduced with permission from ref. 132.)

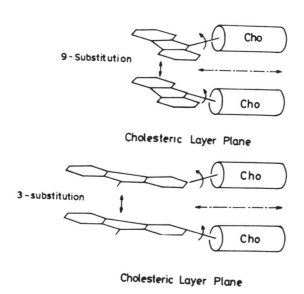

Figure 17. Probable configuration of excimers from **18** (upper) and **19** (lower) in the cholesteric phases mentioned. Head-to-tail configurations are also possible. (Reproduced with permission from ref. 131.)

Even at 50 mol% **19** its fluorescence decay in **Ph-2** or **Ph-2/Ph-4** could be described by a single exponential. At 1 mol%, only unimolecular events contribute to the fluorescence decay and an Arrhenius plot of the decay constants versus temperature showed no slope discontinuity at the cholesteric-isotropic phase transition. However, at 50 mol%, the importance of bimolecular decay processes is apparent from the slope change.[132]

Exciplexes between excited singlets of **18** or **19** and cholesteryl methyl terephthalate (**20**) have been studied in cholesteric phases, also.[133, 134]

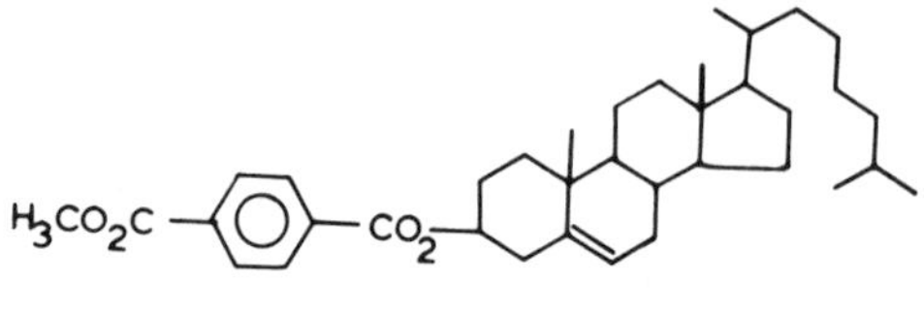

$$\underline{20}$$

From changes in the emission and CPF spectra induced by temperature or the concentration of **20**, it was concluded that the **19-20** exciplex and its components are oriented in a cholesteric phase of **2N-2** as shown in Figure 18.

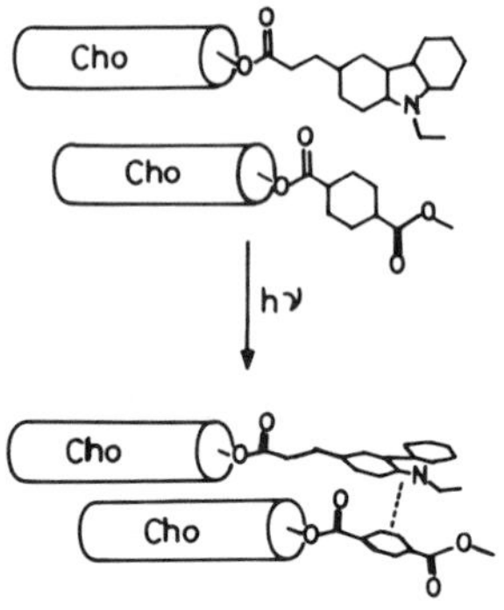

Figure 18. Probable orientations of **19** and **20** in their ground and exciplex states when dissolved in a cholesteric phase. A head-to-tail arrangement is also possible. (Reproduced with permission from ref. 133.)

From similar experiments, evidence was found for two distinct exciplexes between **18** and **20** in **2N-2**. Their probable structures are displayed in Figure 19. The geometric constraints imposed by the cholesteric matrix at lower temperatures is thought to inhibit the **18** conformations necessary for optimal carbazoyl-terephthaloyl overlap. At higher temperatures, the constraints are relaxed sufficiently to permit the more energetically favored exciplex configuration to be attained.

Evidence for the duality of exciplex geometries rests primarily in the red shift of exciplex emission at higher temperatures (Figure 20). Although the exciplex emissions in cholesteric **2N-2** at 93°C and in isotropic polyethylene glycol (**PEG**) at 26°C (that should allow the most stable configuration to form easily) are very similar, it is somewhat disturbing that the spectra appear to contain some vibronic structure. None should be present if the ground state of the exciplex (by definition) is dissociative.

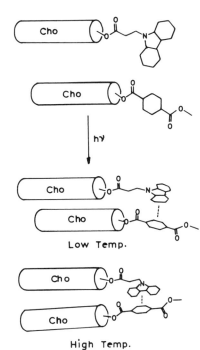

Figure 19. Probable orientation of **18** and **20** in their ground state and as exciplexes when in a cholesteric matrix at lower and higher temperatures. A head-to-tail arrangement is also possible. (Reproduced with permission from ref. 133.)

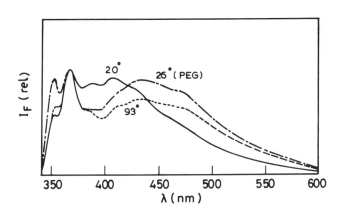

Figure 20. Fluoroscence spectra from 46.3:7.4:46.3 (mol/mol/mol) **18/20/2N-2** at the temperatures (°C) noted. The corresponding fluorescence from anisotropic solution of 46.3:7.4:46.3 (mol/mol/mol) **18/20/poly(ethylene glycol)** is included as PEG. (Reproduced with permission from ref. 133.)

Additional evidence for dual exciplexes was obtained from gated emission and time-correlated single-photon counting studies on **19** and **20** in **Ph-4**.[134] Figure 21 shows that at early times after excitation, only monomer fluorescence is present. Thus, ground-state association between **19** and **20** in this system is not important. By 13-18 ns, exciplex emission is prominent. However, it is blue shifted with respect to the static emission curves.

Figure 21. Steady-state and gated fluorescence spectra of a 18:9:73 (mol:mol:mol) mixture of **19/20/Ph-4**. Steady-state spectra in the cholesteric phase at 70°C(—) and in the isotropic phase at 130°C (-··-). Gated cholesteric phase spectra at 76°C, 0.5-1.5 ns (......) after excitation and 13-18 ns (- - -) after excitation . (Reproduced with permission from ref. 134.)

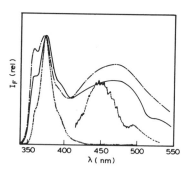

The blue shift can be due to several factors. Among these are solvent relaxation (and attendant exciplex geometry changes) and the presence of two independent exciplexes that decay over very different time scales. The waveforms for emission at >500 nm could be fit to one exponential rise function and two exponential decays. Whereas the rise time becomes shorter as the concentration of **20** increases, both of the decay constants and their preexponential factors are insensitive to **20** concentration. These results suggest that the exciplexes do not dissociate in **Ph-4** and that the mechanism of their formation may be described by Scheme 2 in which two independent exciplexes are again invoked.[134] It is also suggested that the two exciplexes represent less stable and more stable configurations that are mandated by the local order experienced by **20** and nearby **19** at the time of their excitation.[133]

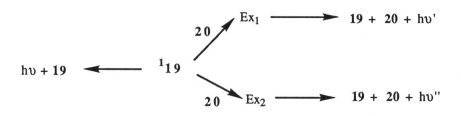

Scheme 2. Mechanism of **19/20** exciplex formation and decay.[134]

Due to the high concentrations of **19** employed and control of intermolecular orientations, exciton hopping must occur: the activation energy for formation of the exciplexes in the cholesteric phase is nearly zero while it is 3-6 kcal mol^{-1} in the corresponding isotropic phase.

Several studies have examined the photophysical properties of pyrenyl probes being quenched by a variety of species in lyotropic mesophases.[135] In essence, they show that the order and mobility afforded to solute molecules in lyotropic mesophases are intermediate between those of disorganized and fluid micelles and highly structured monolayers. They also provide evidence for pretransitional, local chain melting about probe molecules. Thus, in lyotropic phases of 1:1 (w/w) potassium (or rubidium) stearate/water or 1:1 (w/w) water and equimolar potassium stearate and 1-octadecanol, pyrenyl probes can experience extremely large *increases* in their fluorescence lifetimes as the temperature is *raised*. Apparently, the stiff polymethylene chains of the smectic B-like gel phase melt, wrapping themselves more efficiently about a guest molecule.[135b,c]

5.3. Intramolecular Excimers and Exciplexes[115]

Intramolecular studies of excimer and exciplex formation in liquid-crystalline media offer several advantages over intermolecular studies. For instance, bichromophoric dopants undergo more easily defined motions leading to the complex geometry. Also, their study in mesophases may be conducted with very low dopant concentrations (frequently $<10^{-4}$ M). Thus, the kinetic models need not consider aggregation effects, possible phase separations, and exciton-hopping or other nondiffusional energy migration processes.

The conformational bending necessary to bring into collision the head and tail groups of 1,3-bis(1-pyrenyl)propane (**P3P**)[136] and *N,N*-dimethyl-4-[3-(1-pyrenyl) propyl]aniline (**21**)[136,137] (i.e., a cyclization process as shown in equation 8) has been examined in a 59.5:15.6:24.9 (by weight) mixture of **CM**. The dynamics of the process were followed by monitoring the temporal decay of fluorescence intensity of pyrenyl singlets which do (**P3P** and **21**) and do not (**22**) have the opportunity to encounter a quenching R group.

(8)

P3P R = 1-pyrenyl
21   R = 4-(N,N-dimethylanilino)

The fluorescence decay profiles were analyzed according to Scheme 3. It was assumed, based upon several factors (including the monoexponentiality of the decay curves), that k_4' is much less than k_3' or k_5'. The sums of $k_1' + k_2'$ were obtained from the inverse of the decay constants from **22** ($1/\tau_{22}$); $k_1' + k_2' + k_3'$ were calculated in the same way from **P3P** and **21** using **22** ($1/\tau$ (**P3P** or **21**)). Thus k_3' can be determined uniquely from the difference of the inverses of the lifetimes as shown in Scheme 3.

$$\underline{22} \quad R' = C_2H_5$$
$$\underline{23} \quad R' = (CH_2)_{12}H$$

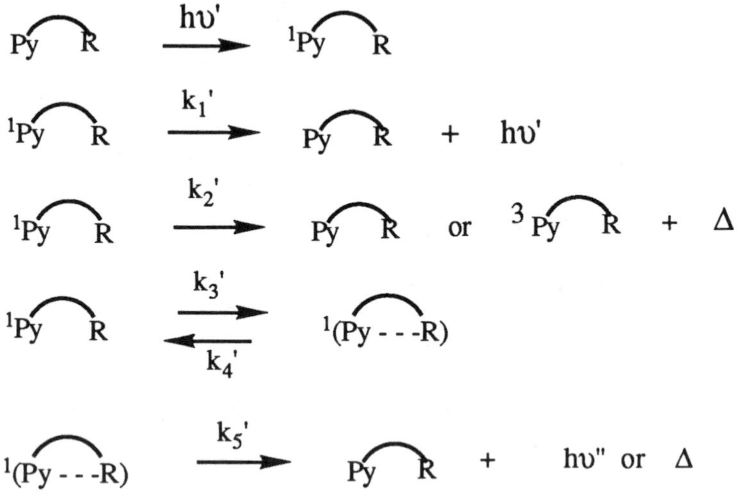

$$1/\tau_{22} = k_1' + k_2'$$

$$1/\tau_{(p3p\ or\ 21)} = k_1' + k_2' + k_3'$$

$$1/\tau_{(p3p\ or\ 21)} - 1/\tau_{22} = k_3'$$

Scheme 3. Kinetic scheme for intramolecular pyrene excimer and exciplex formation.

Activation parameters for k_3' derived from **P3P** and **21** were calculated in both the cholesteric and isotropic phases of **CM**. The results in Table 7 indicate that chain bending of neither **P3P** nor **21** is influenced significantly by cholesteric order. Furthermore, the activation energies of the corresponding intermolecular quenching processes in **CM** are only slightly smaller than the E_3' reported in Table 7.[126,127] It was suggested that the bulky pyrenyl groups may have disturbed the environment near each solute to such an extent that the macroscopic order could not be sensed by **P3P** and **21**.[136] Each solute is better viewed as a thread connectiong two large discs rather than a somewhat perturbed alkane. In spite of this characterization, the pyrenyl group bears some resemblance to a steroidal ring system (Figure 22) and, therefore, from size and shape considerations, may be expected to associate preferentially with it rather than its alkyl chains.

Table 7. Activation Parameters for Intramolecular Quenching of Pyrenyl singlets of 10^{-4} M **P3P** or **21** in **CM**[136,137]

	E_3' (kcal/mol)		$\Delta S_3^{\ddagger}$ (cal/mol/deg)	
	Cholesteric	Isotropic	Cholesteric	Isotropic
P3P	10.5 ± 0.4	10.1 ± 0.2	1 ± 1	0 ± 0.5
2 1	10.8 ± 0.2	7.9 ± 0.5	6 ± 1	-3 ± 1

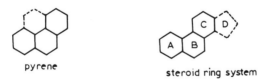

pyrene

steroid ring system

Figure 22. Flattened representations of pyrene and a steorid ring system. The common parts are shown with full lines. Parts without a common projection are given dashed lines.

To test this hypothesis and to determine definitively the influence (or lack thereof) of cholesteric mesophase order on the dynamics of polymethylene chain cyclizations, a study of intramolecular pyrenyl singlet quenching (and excimer formation) in the homologous series of α,ω-bis(1-pyrenyl)alkanes (**PnP**, where n is the number of carbons in the polymethylene chain) was undertaken.[138, 139] It can be seen immediately from Figures 23a and 23b that the enhanced ordering (and/or viscosity) experienced by the **PnP** in the cholesteric phase results in preferential excimer formation.

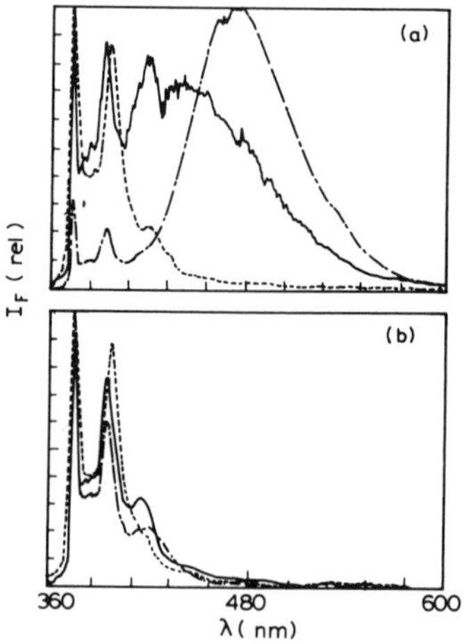

Figure 23. Comparison of emission spectra (relative intensities) from ca.10^{-4} M dodecylpyrene (- - -), **P5P** (——), and **P11P** (— - —) in (a) degassed cyclohexane and (b) **CM** at 17°C; λ_{ex} = 338 nm. (Reproduced with permission from ref. 139.)

In isotropic solvents, the activation enthalpies for chain cyclization of 'long' α,ω-diarylalkanes (measured via intramolecular electron exchange,[140,141] or excimer formation[142, 143] are nearly constant within a series and only slightly higher than the calculated potential energy barrier for hindered rotation about a single carbon-carbon bond.[144] By using the methodology employed with **P3P** and **21**, including Scheme 3 and expressions analogous to its equations, the activation parameters for **P3P** and several **PnP** (n = 5, 6, 7, 9, 10, 11, 12, 13, 22) were calculated in the cholesteric and isotropic phases of **CM**. All of the decay curves from which the constants were calculated are monoexponential. Except for **P3P**, 1-dodecylpyrene (**23**) was the model compound from which $k_1' + k_2'$ were derived.[145] The activation parameters from the **PnP** in **CM**, a 30:70 (w/w) mixture of **CCl/CN**, and **CO** are collected in Table 8.

These data and liquid crystal-induced circular dichroism (lcicd) spectra obtained in **CCl/CN** demonstrate that cholesteric mesophase order does influence strongly the dynamics of **PnP**. In addition, the degree to which solvent influence is manifested depends upon the length of the **PnP** chain and its preferred association with one component of **CM** (as evidenced by data taken in **CCl/CN** and **CO**). As with other α,ω-disubstituted alkane chain cyclization processes in isotropic media, the activation energies of the **PnP** in isotropic **CM** are nearly constant. However, in the cholesteric phase, seemingly nonsystematic, large variations occur.

Table 8. Activation Parameters for **PnP** Intramolecular Excimer Formation in Cholesteric Liquid Crystals

n	solvent	phase[a]	$\Delta H_3^\dagger$ Kcal/mole	$\Delta S_3^\ddagger$ kcal/mol/K
3	CM	chol	9.9 ± 0.4	1 ± 1
		iso	9.4 ± 0.2	0 ± 0.1
5	CM	chol	15.6 ± 1.9	16 ± 3
		iso	9.0 ± 1.8	-5 ± 3
	CN/CCl	chol	17.3 ± 0.8	20 ± 3
	CO	chol	21.9 ± 2	36 ± 3
6	CM	chol	8.0 ± 0.6	-11 ± 3
		iso	8.7 ± 2	9.4 ± 6
7	CM	chol	*b*	
		iso	*b*	
9	CM	chol	15.5 ± 1	16 ± 2
		iso	8.5 ± 2	-6 ± 4
	CO	chol	26.3 ± 0.5	48 ± 3
10	CM	chol	5.1 ± 0.2	-16 ± 1
		iso	11.1 ± 0.9	3 ± 2
11	CM	chol	27.4 ± 1	53 ± 3
		iso	11.0 ± 1.4	2.9 ± 2
12	CM	chol	24.4 ± 1.8	47 ± 3.5
		iso	11.5 ± 0.7	4.3 ± 1
	CN/CCl	chol	14.5 ± 2.5	12 ± 1
	CO	chol	26.4 ± 2.3	51 ± 4
13	CM	chol	10.9 ± 0.5	3 ± 2
		iso	11.2 ± 0.8	4 ± 2
22	CM	chol	15.5 ± 0.4	15 ± 1
		iso	12.2 ± 0.5	6 ± 2

[a] chol = cholesteric, iso = isotropic
[b] No quenching observed

A rough isokinetic relationship[146] exists for **PnP** activation data obtained in isotropic **CM**. Quite surprisingly, an excellent correlation between $\Delta H^\dagger_3$ and

$\Delta S^{\ddagger}_3$ of **PnP** (Scheme 3) exists in the cholesteric phase of **CM**, also (Figure 24). To explore the source of the activation parameter differences further, a simplified Kramers treatment[147] of the data was undertaken. In this treatment, **PnP** quenching can be divided into a part that is intrinsic to pyrene excimer formation, one that is related to solvent friction, and another that is dependent upon chain bending (E^*). It was found that E^* for all of the **PnP** in isotropic **CM** are very near the activation energy for rotation about a C—C bond.[148] This supports the contention that in isotropic media, the conformational dependency of intramolecular quenching in **PnP** has a length-independent rate-limiting step (probably rotation about one bond).

In the cholesteric phase of **CM**, the values of E^* vary widely, depending upon the length of the **PnP** chain (Table 9). These data were interpreted as further evidence that solvent order is influencing **PnP** conformational energies and the barriers which must be overcome to achieve the excimer geometry.[138,139]

Table 9. E* of PnP in **CM**[139]

	E* (kcal/mol)	
n	cholesteric	isotropic
3	1.6	3.5
5	7.8	3.2
6	-0.1	2.8
9	7.2	2.3
10	-3.2	5.1
11	19.1	5.5
12	16.1	5.7
13	2.6	5.2
22	7.2	6.3

These effects are not due to subtle changes in cholesteric pitch (which might influence the mobility of **PnP** movements).[139] Thus, **P12P** fluorescence quenching was examined at 55°C in various mixtures of **CCl** and **CN** which included very large changes in cholesteric pitch. The k_3' values, obtained as before using **56** in the same mixtures as a model for $k_1' + k_2'$, are displayed versus solvent pitch in Figure 25. All the k_3' are within experimental error of each other.

In spite of the large effort given this problem, it is obvious that several important questions remain unanswered or only partially addressed. For instance, the preference of **PnP** to interact with particular cholesteric molecules, the degree to which **PnP** molecules disturb their local environment, and the extent to which **PnP** molecules are extended before excitation all need to be explored further.

Indicative of these complications are the results obtained when **BCCN** is the solvent.[48, 49] The activation energies for intramolecular quenching in the nematic phase increase from **P3P** to **P10P** to **P22P** while the activation entropies become less negative.[138] However, only with **P10P** was the Arrhenius plot not curved when data were collected in the smectic phase. Subsequent experiments with **BCCN** indicate that its solubilization characteristics[72] may play a role in these results. Regardless, it would be unwise to generalize the cholesteric phase results to other mesophase types at this time.

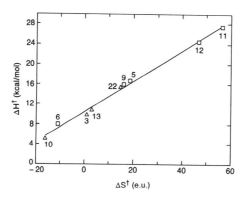

Figure 24. Isokinetic plot of activation parameters for **PnP** intramolecular excimer formation in cholesteric **CM**. Data in ref. 138 (Δ) are converted from E_3 to $\Delta H_3^{\ddagger}$ by subtraction of RT (T = 310 K) from the former. Numbers refer to n of **PnP**. (Reproduced with permission from ref. 138, 139.)

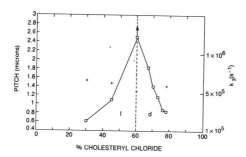

Figure 25. Cholesteric pitch maxima ($\square$) for CN/CCl mixtures and k_3 values (+) for **P12P** at 50°C. The pitch at 60% **CCl** was larger than the maximum of the instrumental range (2.5 μ). d and l refer to the dextro and levo twist direction of the helices. (Reproduced with permission from ref. 139.)

6. Unimolecular Photochemical Isomerizations and Enantiomeric Interconversion

Since attempts to attain an enantiomeric excess in products of photochemical reactions conducted in chiral isotropic solvents have been discouraging,[149] it is logical to assume that the more specific interactions between a photolabile substrate and an optically active mesophase might lead to more promising results.[150] Several efforts in this direction have been reported, but all indicate that no more than small optical inductions of little synthetic utility can be expected. Thermally induced enantiomeric interconversions in chiral mesophases have been equally discouraging. (See Chapter 6 for examples of successful asymmetric reactions in the solid state).

Some examples of photoinduced enantiomeric interconversions or isomerizations of prochiral molecules in cholesteric mesophases using unpolarized light are listed below.[151] In none has more than 1–2% enantiomeric excess been achieved.

$$\text{Ar} = 1\text{-naphthyl} \qquad (9)$$

$$(10)$$

$$(11)$$

The thermal atropisomerization of racemic 1,1'-binaphthyl (24) in several cholesteric phases led to virtually no enantiomeric excess (ee).[152] Photoresolution attempts in a 1:1 (w/w) mixture of 5α-cholestan-3β-yl acetate/5α-cholestan-3β-yl nonanoate (CHA/CHN) at 40°C were slightly more successful, yielding a 1% ee of the S atropisomer.

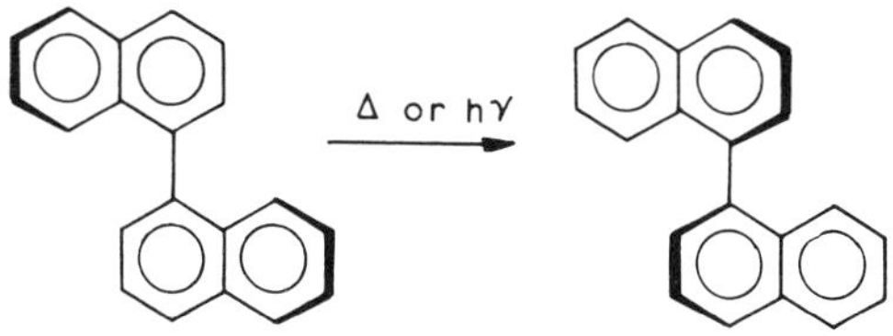

$$\xrightarrow{\Delta \text{ or } h\gamma}$$

(12)

R   24   S

Several factors may contribute to the somewhat larger ee from the photoinduced process: (1) the excited states of **24** are more polarizable than their ground states; (2) the excited states spend a longer period near the planar configuration of **24** than does the ground state as each approaches the transition-state geometry; (3) one atropisomer may be excited selectively by the partially circularly polarized light (produced by the cholesteric helix).[152] The latter factor cannot be the sole contributor since the photostationary state of **24** in dichloromethane established with circularly polarized light favors one enantiomer by no more than 0.2%.[153]

The synthesis of hexahelacene (**27**) from irradiation of **25** or **26** and subsequent oxidation has been examined in chiral mesophases. In each case, the enantiomer-determining step is excited state ring closure that forms dihydro-**27**; subsequent oxidation serves only to aromatize the new ring and those adjacent to it.

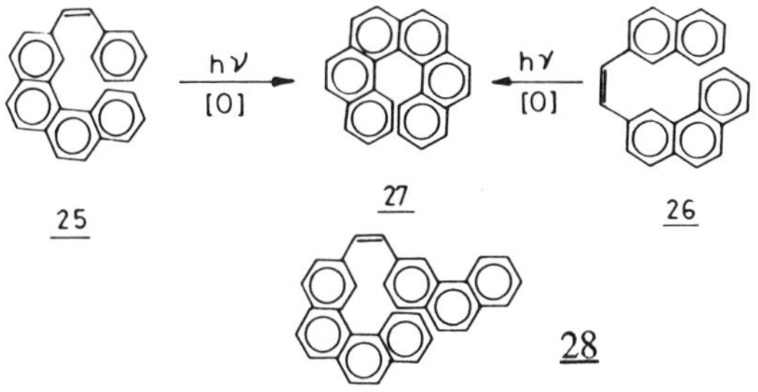

$$\xrightarrow[[O]]{h\gamma}$$ $$\xleftarrow[[O]]{h\gamma}$$

25   27   26

(13)

28

Using circularly polarized light and achiral solvents, **27** can be produced with a small (<0.2%), but real, excess of one enantiomer.[154] Unpolarized light and chiral isotropic solvents yield **27** with up to 2.1% ee.[149b] Consistent with the apparent necessity for intimate interactions between an achiral precursor of **27** and chiral solvents, irradiation of **25** or a precursor to octahelacene (**28**) in an oriented twisted nematic phase (i.e., a mechanical cholesteric phase) of 1:1 4-cyanophenyl 4-butylbenzoate-4-cyanophenyl 4-heptylbenzoate yields <0.2% ee.[155] The *local* environment experienced by **25** in this experiment is virtually achiral; the only source of chiral differentiation comes from the medium-generated polarized light. Thus, the right-hand twisted nematic phase produces an excess of (+) enantiomers

of **27** or **28**, while the left-hand twisted phase produces an excess of the (-)
enantiomers.[155]

Further evidence for the importance of direct solute–chiral solvent interactions
during the **25** → **27** photoreaction comes from experiments conducted in
cholesteric mesophases. Irradiation of **25** in the cholesteric phases of either 3:2
CN/CCl or **CNB** yields ca. 1% ee in **27**.[156] The same experiment, but
conducted in the isotropic phase of **CN/CCl**, produced no discernible optical
induction.[156] A careful study of the dependence of optical induction on cholesteric
pitch (and, therefore, the degree of circular polarization of the excitation radiation)
in **CCl**–cholesteryl myristate (**CMY**) led to no clear relationship.[157] The data,
summarized in Table 10, clearly support intermolecular interactions as being more
important than the degree to which the radiation is polarized in effecting an
enantiomeric excess. This observation should serve as a guideline for the design of
future experiments to induce solute optical activity from irradiations in cholesteric
or other chiral mesophases.

Table 10. Dependence of Enantiomeric Excess of **27** (from **25**) on Solvent
Pitch and Phase of a 1.75/1 CCl/CMY mixture.[157]

T (°C) (± 1°)	Melix handedness (or phase)	Pitch (μ)	% ee[a]
31	Left	-3.5	0.05
41	(Nematic)	∞	0.19
51	Right	-4.9	0.29
61	Right	-1.6	0.43
70	(Isotropic)		0.03
80	(Isotropic)		0.02

Potentially greater structural changes (requiring less subtle differences between
molecular interactions) may be found in photoreactions that bring about geometric
or structural isomerization of solutes. However, irradiation of either 7-
dehydrocholesterol (**29**) or lumisterol₃ (**30**) to produce, respectively, the β and α
conformational families of pre-vitamin D_3 in a 60:26:14 (by weight) mixture of
CO/CN/CCl did not result in detectably different rates for thermal conversion of
31 to vitamin D_3 (Scheme 4).[158] Since the α and β forms of **31** are
diastereomeric, differences in the rates at which they form **32** should be expected.
Either this is not the case (i.e., the rates at which they undergo [1,7]-sigmatropic

rearrangement are equal) and/or $\alpha \rightleftharpoons \beta$ equilibration is faster than $\mathbf{31} \rightarrow \mathbf{32}$. The available data do not allow the possibilities to be distinguished.

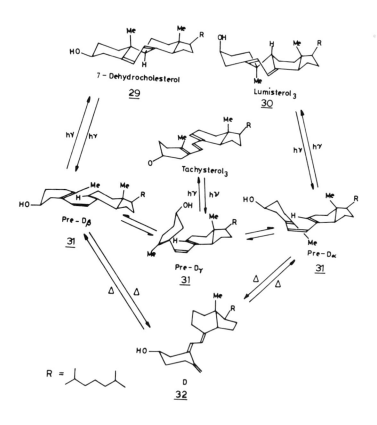

Scheme 4. Thermal and photochemical conversions between vitamin D (**32**) and related compounds.

The trans – cis photoisomerization of neat **MBBA** in its nematic phase has been investigated by uv and infrared spectroscopies and by optical birefringence changes.[159] The data indicate a possible increase in the angle between the phenyl and imino groups after irradiation.

An interesting investigation of the changes in the photostationary state (pss) ratio of all-*trans*-retinal (*trans*-**33**) induced by various solvents has been conducted.[160] Among the media selected was **BS** whose hexatic β phase[52,53] may restrict the motions required for geometrical isomerization of *trans*-**33**. Although the temperature was very near the mesophase–isotropic phase transition of **BS**, irradiations were conducted within the mesophase.[160b]

$$\underline{33}$$

Results from irradiations (>380 nm; λ_{max} = 378 nm for absorption of *trans*-33 in **BS** at 25°C) at 26°C (hexatic phase) and 40°C (isotropic phase) in **BS** and two other isotropic solvents at room temperature are collected in Table 11. Although irradiation in **BS** affords all the *cis*-33 isomers, their pss ratio is temperature (phase?) dependent. The ratio of 13-*cis* to 9-*cis* is 1.7 at 26°C and 4.35 at 40°C. Conceptually, rotation about the 9,10 double bond requires a larger sweep volume than rotation about the 13,14 double bond near the aldehyde end of the molecule. Thus, if **BS** at 25°C inhibits the torsional motions that attend isomerization of *trans*-33, processes at the 9,10 positions should be affected more than those at the 13,14 positions. The influence of the polar carboxy group of **BS** at 26 and 40° C is difficult to gauge, but the pss results in ethanol and pentane clearly indicate that specific solute-solvent interactions may be an important factor, also. Further work to separate these factors is warranted.

Table 11. Photostationary State Distributions of **33** Isomers Upon Irradiation at >380 nm at Room Temperature.[160]

Solvent	13-cis	11-cis	9-cis	7-cis	all-trans
BS (26°C)	24.8	0.95	14.1	0.65	59.1
(40°C)	32.5	1.0	7.5	0.8	57.2
hexane	44.5		4.8		50.5
ethanol	22.5	19.0	6.0	0.5	52.0

7. Unimolecular Fragmentation Reactions

The Norrish I reaction of 1-(4-methylphenyl)-3-phenylpropan-2-one (**34**) produces carbon monoxide and benzyl and (*p*-tolyl)methyl radicals that can recombine with each other (in cage) or combine indiscriminately (out of cage) (Scheme 5).[161] (For mechanistic details see Section 3. of Chapter 1). The fraction of in-cage reaction, F_c is given by Equation 14.[161] The relative product yields and, thereby, F_c from **34** were measured in the cholesteric and isotropic phases of 35:65 (w:w) **CCl/CN**, the smectic B, nematic, and isotropic phases of **BCCN**, and the solid, hexatic B, and isotropic phases of **BS**.[162]

$$F_c = \frac{[35] - [36] + [37]}{[35] + [36] + [37]} \tag{14}$$

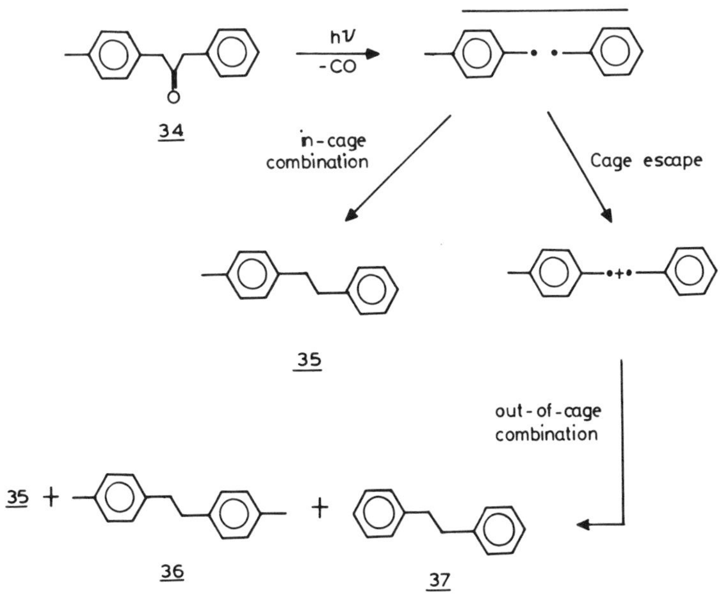

Scheme 5. In cage and out of cage combinations of radicals from **34**.

Although large changes in F_c are known to be induced by strong magnetic fields in some solvents,[163] only very small differences were observed in the isotropic and mesophases of CCl/CN or BCCN at 0 and 3 kG. Also, since the recombination of radicals depends, in part, upon the rates of intersystem crossing of in cage triplet pairs, ^{12}C and [13]C (with nuclear spins of 0 and 1/2, respectively) at the benzylic position influence product formation differently. At 30% conversion, the ^{13}C isotropic enrichment factor[164] α from irradiation of 1,3-diphenylpropan-2-one (**38**) (whose only radical recombination product is 1,2-diphenylethane) in **BS** was 1.09 ± 0.02 at 17°C (hexatic B) and 1.03 ± 0.01 at 35°C (isotropic). These values are not extraordinary for reaction in solvents of high viscosity: by comparison, $\alpha = 1.04$ in benzene and 1.37 in a cationic micelle at room temperature.[161]

The F_c from **34** in each solvent is plotted versus temperature (and phase) in Figure 26. The enhanced F_c in hexatic **BS** is easily explained if it is assumed that **34** is initially solubilized within a layer and that radicals cannot easily traverse a layer boundary. The results are suggestive but require further experimentation before more detailed explanations are advanced. This is especially true of the seeming lack of phase dependence on F_C in BCCN.[72]

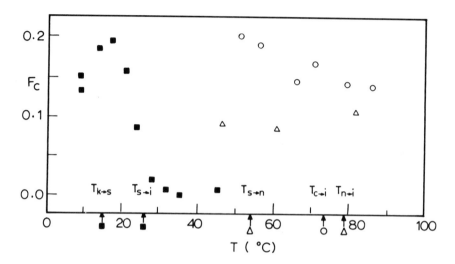

Figure 26. Fraction in-cage radical recombination (F_c) from photolyses of **34** in **BS** (■), **BCCN** (Δ), and 35:65 (w:w) **CCl/CN** (O) versus temperature. Phase transition temperatures are indicated with arrows. (Reproduced with permission from ref. 162.)

The photolysis of 2-nitroso-2-methylpropane (**39**) using red light ($\lambda > 650$ nm) and subsequent radical trapping steps have been investigated in **MBBA**[165] and **CNB5**.[166] Photolysis of **39** monomers leads to clean formation of *tert*-butyl radicals. For reasons that are not entirely clear, the formation of *tert*-butyl radicals from irradiation of **39** is more efficient in the nematic phase of **MBBA** than in its isotropic phase. The source of the phase dependence on the rates must be complex since a negative apparent activation energy for lysis is calculated from initial rates in nematic **MBBA** and no effect of phase on the rates was detected in another nematic solvent. An attempt to model phenomenologically the rates of bimolecular reactions in nematic **BCCN** has been made.[167]

Trapping rates of *tert*-butyl radicals by **39** and aromatic nitroso compounds (**40**) to form nitroxyl radicals (**41**) (Scheme 6) were measured by electron spin-resonance (esr) and converted into rate constants.[166] The rate constants, k_N, for formation of each **41**, at 25°C are included in Table 12. The data indicate that electronic differences among the nitroso compounds **40** is a dominant factor in isotropic solvents. In nematic **PCB**, orientation and viscosity probably become the controlling variables since the range of rate constants is diminished markedly.

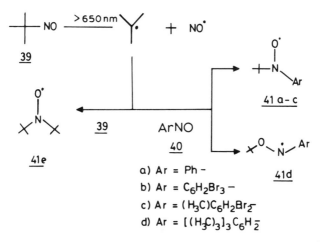

Scheme 6. Trapping of tert-butyl radicals by nitroso compounds.[166]

Table 12. Rate Constant for Additon of tert-butyl Radicals to Nitroso Compounds.[166]

4 0	k_N(rel) CNB5		k_N(1/mol/s) benzene (25°C)
	25°C (nematic)	50°C (isotropic)	
a	18.9	215	2.0×10^9
b	9.4	50.9	8.0×10^8
c	0.13	0.02	1.1×10^9
d	0.1	0.14	2.3×10^5
3 9	1	1	3.3×10^6

8. Norrish II Reactions[168]

Due to the several reactive pathways available to hydroxy-1,4-biradicals (BR), the Norrish II reactions of ketones with abstractable γ-hydrogens are attractive probes of mesophase order. A general mechanism that emphasizes the features important to reaction control by solvent anisotropy is shown in Scheme 7. The quantum yield for ketone disappearance (Φ_{II}), the ratio of elimination to cyclization products (E/C), and the ratio of diastereomeric cyclization products (t/c) can monitor separate aspects of control. Depending upon the ketone excited-state lifetimes and multiplicities, and the barriers to interconversions of BR conformers, each experimental monitor may be kinetically or thermodynamically controlled. Aspects of these concepts have been discussed at length elsewhere.[1,169,170]

The first attempt to observe restrictions to chain kinking by a liquid-crystalline matrix in the Norrish II reaction involved the α-diketones [42 and 43; equation (15)].[171] The sole products observed were the cyclobutanols (44 and 45). Since the intermediate **BR** (46) can close to yield either diastereomer, a first test of the selectivity of solute-ordered solvent interactions is the 44:45 ratio. On this basis, the influence of solvent order cannot be judged as very important: almost equal amounts of 44 and 45 were obtained from both diketones in all mesophases examined. These include a cholesteric phase and the hexatic B phase of **BS**.

Another measure of the influence of solvent order, Φ_{II}, is dependent on temperature, solvent phase, and solvent structure. These changes can be ascribed to several factors: conformational constraints on the α-diketone, which affect its ability to adopt the kinked conformation in equation (15) needed for intramolecular H abstraction; changes in the triplet lifetime of the α-diketones (from which H abstraction occurs); and variations in the fraction of **BR** that proceeds to product. Each of these must contribute to some extent to the overall results. The data are too sparse to allow definitive conclusions concerning the importance of each to be drawn but suggest that mesophase order does alter the lability of the solute chains.

$$(15)$$

$$\underline{42} \ \ R = CH_3$$
$$\underline{43} \ \ R = Ph$$
$$\underline{46}$$
$$\underline{44}$$
$$\underline{46}$$

In a more systematic attempt to determine the influence of mesophases on Norrish II reactions, a series of alkylphenones (47a–e) were irradiated in the condensed phases of **BS** and in several isotropic solvents.[172] The alkylphenones produce elimination products (48 and 49) and cyclization products 50 (equation 16) through different **BR** conformers as shown in Scheme 7. Since the quantum efficiency for intersystem crossing in alkylphenones is near unity, all reactions except final product formation must occur in the triplet manifold.

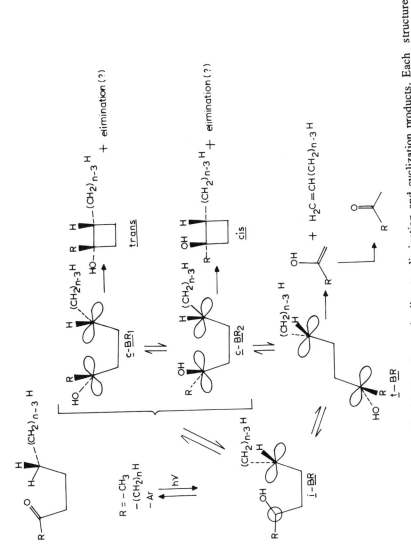

Scheme 7. Representation of the Norrish II pathways leading to elimination and cyclization products. Each structure represents a family of related conformers that undergo the same chemical processes.

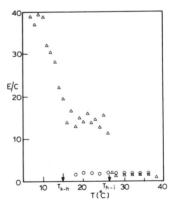

The reaction scheme (equation 16) shows structures **47** (a) n = 4, b) n = 10, c) n = 17, d) n = 19, e) n = 21), **48**, **49**, **50**, and **51**.

The ratio of E/C products (**48:50**) reflects the populations of transoid and cisoid-**BR** triplets at the moment of their intersystem crossing to singlets. The total lengths of the ketones were chosen to be much shorter than, about the same as, and longer than that of **BS**. Results of the photolyses are summarized in Table 13. It should be noted that heptadecane is a solid and **BS** is a hexatic B phase at 20°C; both solvents are isotropic liquids at 30°C. In none of the non-mesomorphic solvents is a marked temperature effect on the **48:50** ratio observed. Only for those ketones whose lengths are near that of the **BS** solvent were large changes in product ratios between 20°C (hexatic B) and 30°C (isotropic) observed. More complete profiles of the temperature dependence on the **48:50** ratios were reported for **47b** and **47d** (Figure 27). They show clearly that the influence of solvent order on the reaction of **47** in both the hexatic B phase of **BS** and its solid phases (which are organized like a smectic E phase) is large only when the solute and solvent are near in length. The **48d:50d** ratio is about constant throughout the hexatic phase of **BS** but rises rapidly in the solid phase and appears to plateau again.

Figure 27. Temperature dependence on the ratio of elimination (**48**) to cyclization (**50**) products from photolysis of **47b** (o) and **47d** (Δ) in B S. Solvent transition temperatures are shown with arrows. (Reproduced with permission from ref. 177.)

Table 13. E/C Product Ratios[a] from **47** in Several Solvents.[172]

			Solvent		
4 7		*n*-Heptane	*n*-Butyl Acetate	*n*-Heptadecane	BS
a	30°C	2.3	4.2	1.9	3.2
	20°C	2.6	4.1	2.9	3.3
	20°C/30°C	1.2	1.0	1.5	1.0
b	30°C	0.8	2.0	1.3	1.7
	20°C	0.7	1.9	1.7	2.0
	20°C/30°C	0.9	0.9	1.3	1.2
c	30°C	1.1	2.0	1.0	3.0
	20°C	1.1	2.2	1.2	21
	20°C/30°C	1.0	1.1	1.2	7.0
d	30°C	1.8	2.0	1.0	1.9
	20°C	1.5	2.0	0.7	15
	20°C/30°C	0.8	1.0	0.7	7.9
e	30°C				1.3
	20°C				9.5
	20°C/30°C				7.5

[a] Not corrected for differences in detector response (error ± 10%)

The results in Table 13 and Figure 27 were explained on the basis of **47c**–**47e** replacing effectively solvent molecules in **BS** matrices. The shorter **47**, being incapable of occupying fully the space allocated to a **BS** molecule, disrupt their environment and make it more fluid. Thus, the more ordered and selective cybotactic regions of **47c**–**47e** exercise greater control over solute motions. Since formation of **50** requires chain kinking while formation of **48** can occur from the energetically preferred transoid-**BR** (as well as the cisoid-**BR**), cyclization from **47c**–**47e** is attenuated greatly. Surprisingly, the triplet lifetime of the **BR**–**51d** was found to be the same within experimental error in the hexatic B (20°C; $\tau = 70 \pm 5$ ns) and isotropic (30°C; $\tau = 64 \pm 5$ ns) phases of **BS**: transoid and cisoid-**BR** are not expected to exhibit the same triplet lifetime.[173] Their relative populations are inferred to be very different in hexatic and isotropic **BS** on the bases of product yields and solvent order.

Complementary results were obtained with two alkanones (**52c** and **53a**).[173] Irradiation of **52c** (m = 9) allowed both its E/C ratio and the t/c ratios from **52c** and **53a** (m = 9) to be monitored (equation 17).

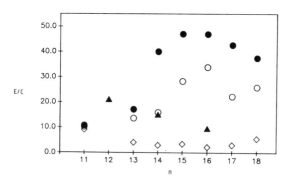

$$(17)$$

To explore the aspects of **BS** anisotropy that determine Norrish II selectivities in greater detail, the photochemistry of a series of *p-n*-pentylalkylphenones (**56** with n = 11–18) was investigated.[174] The E/C ratios from **56** are plotted versus n along with the corresponding product ratios from **47** (normalized to equal ketone lengths) in Figure 28. The effect on the E/C ratios of increasing total molecular length in the two series is very different: E/C ratios from **47** in hexatic **B S** maximize at lengths at least four C—C bonds shorter than the ratios from **56**. The large E/C increase from **56** in solid **BS** at 10°C occurs when n = 13, 14 and the total extended solute length is near that of **BS**. The t/c from **56** in hexatic **B S** vary only slightly with increasing chain length and remain near the ratios measured in the isotropic phase. This suggests that the shapes of the two cisoid-**B R** conformer families cannot be differentiated by the hexatic B phase. A mapping of E/C from **56** (n = 15) versus temperature was qualitatively similar in shape to that from **47d** (Figure 27), but with larger magnitudes in the ordered **BS** phases.

Figure 28. E/C ratios from irradiation of 20 wt% **56** as a function of phenone chain length (n) in the solid (10°C, ●), hexatic B (20°C, 0) and isotropic (30°C, ◊) phases of BS.[174] Data from **47** in the hexatic phase of BS (20°C ▲)[172] are plotted as n-5 in order to allow direct length comparisons. (Reproduced with permission from ref. 174.)

Irradiation of a series of *p*-alkyl alkylphenones (**57**) with *m* + *n* = 21 was undertaken to observe (1) the consequences of moving the biradical centers of **BR** along a **BS** layer and (2) the importance, if any, of the γ-radical site being nearer a layer boundary than the benzhydryl radical site (or vice versa). As seen in Figure 29, the ratios vary enormously: in solid **BS**, E/C > 100 with **57c** and <10 with **57j**. Based upon the data in Figure 38 and the solubilization model for **57** in layers of **BS** shown in Figure 30, product selectivity appears to depend upon the ease with which the hydroxyl of a BR can hydrogen bond to carboxyl groups of neighboring **BS** molecules. Since dipolar forces are the strongest attractions between ground-state alkyl alkylphenones and **BS** molecules, the **BR** that can hydrogen bond should be more firmly fixed along a layer than their precursors and react more selectively. Those hypotheses are supported by FTIR spectra recorded for the alcohol reduction products from **57c** and **57g** in isotropic and hexatic **BS**.

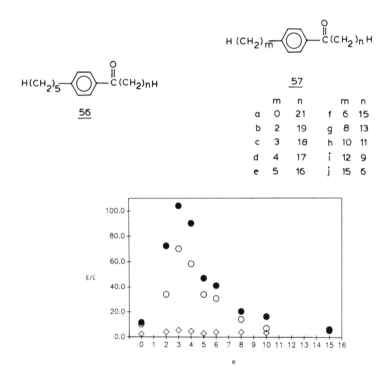

Figure 29. E/C ratios from 2 wt% **57** in the solid (10°C, ●), hexatic B (20°C, ○), and isotropic (30°C, ◊) phases of BS. n = 21 - m on the abscissa. (Reproduced with permission from ref. 172.)

The comparative proximity of the two biradical centers from a layer middle (Figure 30) also influences selectivity. The E/C ratios from **57c** and **57j** and the expected locations of the radical sites in their **BR** illustrate this point. The γ-radical site of **BR** from **57c** should be near a layer middle and its benzhydryl center anchored by hydrogen bonding. Thus, it should experience a very restrictive environment in which to react. Although the benzhydryl site of the **BR** from **57j** may be anchored by hydrogen bonding, its γ-radical center is near a layer end and subject to the disordering influences of the butoxy parts of nearby **BS** molecules. Differential scanning calorimetry histograms indicate that some **57h** is phase separated even at 2% concentrations in hexatic **BS**. This may account for its measured E/C ratio being low in spite of both its radical centers residing near a layer middle. The inability of its carboxyl (and the hydroxyl group of its **BR**) to interact with carboxyl groups when **57h** is extended as in Figure 30 probably contributes to both the lack of solubility and the low E/C ratios.

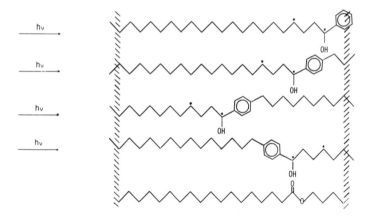

Figure 30. Representation of extended transoid hydroxy-1,4-biradicals from selected **57** and their orientations in a **BS** layer according to a "surrogate" solubilization model. Note the approximate depths of the hydroxyl groups of the biradicals and their relationship to the carboxy of **BS**. The approximate layer boundaries are shown as shaded areas.

Somewhat surprisingly, the relative Φ_{II} from **57a-57e** are almost the same, being > 80% of their isotropic values in even solid **BS**. The near zero Φ_{II} measured for **57g-57i** in solid **BS** may be related to the previously mentioned phase separation, and suggest that the ketones may be in an environment similar to that of their neat crystals.[175,176b]

After the preliminary report results from irradiation of a *sym-* and a 2-alkanone in BS,[172] a detailed investigation was conducted.[176] From a structural standpoint, alkanones should fit better into the anisotropic BS matrices than alkyl alkylphenones of similar length. However, analyses of their Norrish II ratios are complicated by the known occurrence of γ-hydrogen abstraction from both the excited singlet and triplet states.[177]

Results from the 2-alkanone (53) and [*sym-*alkanone (52)] series are collected in Tables 14 and 15, respectively. At the concentrations of 52 and 53 employed (<1.4 wt%), evidence for virtually no macroscopic disruption of solvent order was observed. The E/C and trans/cis cyclobutanol ratios respond only to mesophase and solid BS order and only for those 52 whose lengths are near that of BS. The selectivity with 53 is greatest when its extended length is near that of the stearoyl portion of BS. The source of this length dependence appears to be the same as that attributed to the alkyl alkylphenone irradiations.[174] Changes in dipolar forces (and hydrogen bonding) between the carbonyl groups of 53 (and the hydroxyl group of its BR) and the carboxyl groups of neighboring BS molecules can explain the results from the 2-alkanones: when 53 are longer than ca. 19 carbons they force either the carbonyl to be removed from the area near the carboxyl of BS or the ketone chain to kink or span a layer boundary. None of these options should be attractive to the system. The location of the carbonyl of 52, in the center of the molecules, precludes carbonyl-carboxyl interactions unless somewhat unfavorable packing arrangements or conformations of the ketones obtain. The data support strongly a solubilization environment in which the ketone chains are fully extended and lie within a solvent layer parallel to the BS molecules. They are not consistent with reaction occurring from microcrystallites of solute.[175,176b]

More complete temperature profiles for product ratios from 52d in BS versus temperature (and phase) are shown in Figure 31. That the two curves exhibit different shapes infers that BS solvent order influences differently the motions of the BR intermediates which lead to elimination or cyclization (transoid $\rightleftarrows$ cisoid conformational changes) and to trans- or cis-cyclobutanols (diastereomeric conformational changes of the cisoid-BR).

Table 14. Norrish II Product Ratios from **52** in BS.[176]

5 2	Carbon chain length	Temperature[a] (°C)	E/C	t/c
a	11	30	2.0 ± 0.4	1.8 ± 0.5
		20	1.8 ± 0.4	1.5 ± 0.5
		10	2.0 ± 0.3	1.6 ± 0.3
b	15	30	1.9 ± 0.5	2.4 ± 0.6
		20	2.3 ± 0.3	2.8 ± 0.5
		10	3.2 ± 0.6	3.5 ± 0.4
c	17	30	2.2 ± 0.6	2.2 ± 0.6
		20	3.2 ± 0.8	10.8 ± 1.6
		10	7.0 ± 1.0	11.5 ± 2.3
d	19	30	2.0 ± 0.3	1.8 ± 0.2
		20	6.2 ± 0.9	13.3 ± 0.3
		10	10.0 ± 2.3	11.3 ± 0.9
e	21	30	2.0 ± 0.2	2.5 ± 0.6
		20	5.5 ± 1.0	14.1 ± 0.1
		10	15.5 ± 5.0	12.0 ± 3.0
f	27	30	2.8 ± 0.5	2.2 ± 0.3
		20	3.2 ± 0.5	7.2 ± 0.5
		10	6.3 ± 1.3	6.1 ± 0.7
g	29	30	2.7 ± 0.7	2.2 ± 0.3
		20	3.6 ± 0.5	5.9 ± 0.1
		10	2.9 ± 1.1	7.0 ± 0.3
h	31	42	3.9 ± 0.5	2.1 ± 0.1
		20	3.0 ± 0.5	5.5 ± 0.5
		10	2.3 ± 0.5	5.0 ± 0.1

[a] 30°C is isotropic, 20°C is hexatic B.
[b] 10°C is solid

Table 15. Norrish II Product Ratios from **53** in BS.[176]

5 3	Carbon chain length	Temperature[a] (°C]	E/C	t/c
a	11	30	2.9 ± 0.3	1.3 ± 0.1
		20	3.1 ± 0.3	1.2 ± 0.1
		10	2.6 ± 0.3	1.2 ± 0.1
b	13	30	3.5 ± 0.4	1.5 ± 0.1
		20	4.0 ± 0.3	1.4 ± 0.1
		10	4.2 ± 0.4	1.3 ± 0.1
c	15	30	3.7 ± 0.2	1.3 ± 0.1
		20	12 ± 2	1.3 ± 0.3
		10	20 ± 2	1.8 ± 0.1
d	17	30	3.7 ± 0.4	1.5 ± 0.1
		20	15 ± 3	1.9 ± 0.3
		10	27 ± 3	1.9 ± 0.3
e	18	30	3.7 ± 0.4	1.1 ± 0.1
		20	15 ± 2	1.6 ± 0.2
		10	31 ± 3	1.3 ± 0.2
f	19	30	4.1 ± 0.2	1.3 ± 0.2
		20	17 ± 2	2.4 ± 0.1
		10	23 ± 4	2.5 ± 0.1
g	20	30	3.9 ± 0.5	1.0 ± 0.2
		20	10 ± 2	1.5 ± 0.1
		10	19 ± 3	1.5 ± 0.2

[a] 30°C is isotropic 20°C is hexatic B; 10°C is solid

The difference exhibited in Figure 31 between the solid smectic E-like phase of BS and its hexatic B phase[52] are intriguing since their packing arrangements are very similar. To explore the source of the differences, the Norrish II reactions of

52 and 53 (1 wt%) are being examined in the isotropic and two solid phases of heneicosane $(C_{21}H_{44})$.[178]

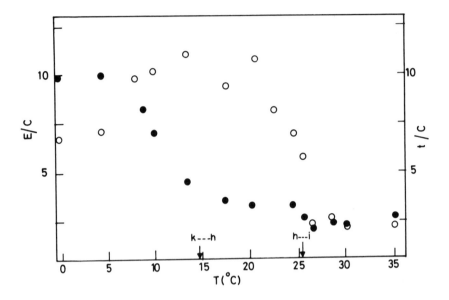

Figure 31. E/C (●) and t/c (o) ratios from irradiation of ca. 1 wt% of **52c** in BS as a function of temperature. Solvent transitions are noted with arrows. (Reproduced with permission from ref. 176.)

The organization of the solid phases of heneicosane (phase I below 32.4°C[179] and phase II up to 40.2°C[180]) are very similar to the smectic E-like and hexatic B packing, respectively, of **BS**. Heneicosane also lacks a polar carboxyl group, so that specific ketone-solvent interactions are eliminated: product ratios from **52** and **53** should be determined almost exclusively by size and shape considerations.

The phase I data in Table 16 indicate a very high selectivity for the E/C ratios from both **52** and **53** that are slightly shorter than or equal to the chain length of heneicosane. The decrease in E/C ratios from **52** longer or shorter than heneicosane is especially abrupt. The t/c ratios from **52** slightly shorter than to somewhat longer than heneicosane display large selectivities in phase II, where the E/C ratios are comparable to those in the isotropic phase. By contrast, t/c product selectivity was not noted in any **53** in either solid phase of heneicosane. The very small amounts of **54e** and the very low ratio of its diastereomers from phase I irradiations (Figure 32) may arise from **52e** located in defect sites.[167b] Furthermore, careful differential scanning calorimetry (dsc) measurements on **52** and **53** in heneicosane indicate that some solute-solvent phase separation may occur even at the low concentrations employed. Only ketones with 20 or 21 carbons show no evidence of phase separation at higher concentrations.[178b] Thus, the lack of selectivity in some of the other **52** and **53** probably is due to photoreactions occurring in a eutectic phase.[178b]

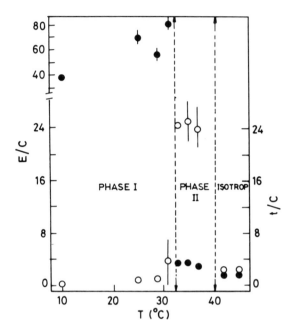

Figure 32. E/C (●) and t/c (0) photoproduct ratios from 2 wt% **52e** in heneicosane versus temperature (phase). (Reproduced with permission from ref. 178.)

The Norrish II reactions of **52** and **53** have also been examined in three lyotropic liquid-crystalline (gel) phases comprised of 50 wt% water and either potassium palmitate (**KP**), potassium stearate (**KS**), or 1:1 (mol:mol) potassium stearate/1-octadecanol (**KSO**).[181] The **KP** and **KS** gels are layered assemblies of completely interdigitated surfactant molecules, hexagonally packed with their long axes normal to the layer planes.[182] The **KSO** gel is comprised of noninterdigitated bilayers in which each of the surfactant components is alternately dispersed in a hexagonal array.[182] Layers of water containing counterions separate the surfactant layers. The packing of surfactant molecules in each of the gel phases is analogous to that in a thermotropic smectic B phase (and in the hexatic B phase of **BS**).[52,53]

Table 16. Norrish II Product Ratios from 1 Wt.% of 52 and 53 in the Isotropic (45°C), the Hexagonal Solid (35°C; Phase II), and Orthorhombic Solid (25°C; Phase I) Phases of Heneicosane.178

Chain Length of ketones	T (°C)	52		53	
		E/C	t/c	E/C	t/c
15	45	1.5 ± 0.1	2.1 ± 0.1	3.3 ± 0.1	1.8 ± 0.1
	35	1.8 ± 0.4	1.8 ± 0.1	3.7 ± 0.3	1.6 ± 0.1
	25	2.6 ± 0.1	1.5 ± 0.1	16.2 ± 0.4	1.6 ± 0.1
17	45	1.4 ± 0.1	2.4 ± 0.1	3.3 ± 0.2	1.7 ± 0.4
	35	2.4 ± 0.1	5.9 ± 0.1	4.6 ± 0.7	3.4 ± 0.2
	25	5.0 ± 0.9	1.5 ± 0.7	15 ± 4	2.4 ± 0.4
20	45			4.9 ± 0.3	1.6 ± 0.1
	35			18 ± 4	4.7 ± 0.9
	25			165 ± 15	18 ± 8
21	45	1.8 ± 0.1	2.5 ± 0.1	2.4 ± 0.6	1.0 ± 0.1
	35	3.5 ± 0.1	25 ± 3	5.7 ± 1.3	2.5 ± 0.2
	25	69 ± 7	0.9 ± 0.4	46 ± 16	3.6 ± 0.9
22	45			4.0 ± 0.2	1.7 ± 0.2
	35			8.5 ± 0.9	2.8 ± 0.1
	25			29 ± 3	4 ± 2
23	45	2.0 ± 0.1	2.0 ± 0.2	3.1 ± 0.7	1.9 ± 0.2
	35	2.9 ± 0.1	10.8 ± 0.5	1.6 ± 0.3	2.4 ± 0.2
	25	8.1 ± 0.3	9.0 ± 0.8	3.6 ± 0.1	3.0 ± 0.1
25	45	2.3 ± 0.4	2.9 ± 0.3		
	35	2.9 ± 0.5	18 ± 1		
	25	4.3 ± 0.6	15.0 ± 0.7		

On average, the polymethylene chains are all trans and rotate about the long axes. The methylenes nearest a head group are most ordered and the methyl is least ordered.[181,183] Since the solubility of long-chained alkanones in water is very low, **52** and **53** must be incorporated in the surfactant layers. To obtain a more detailed picture of the order experienced by the ketones within the layers, they were deuteriated at the carbons α to the carbonyl and their ^{2}H NMR spectra were recorded. The quadrupolar splitting for C-D bonds oriented at 90° to the external magnetic field, Δv_{90}, provides a quantitative measure of the constraints imposed by the surfactant layers on the region of the solute where reaction occurs.[184]

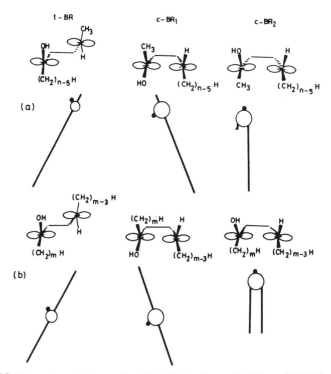

Figure 33. Preproduct hydroxy-1,4-biradicals from (a) **53** and (b) **52** in cartoon representations with extended methylene chains (—) and the relative sizes of the biradical centers (◯) and hydroxy groups (●) shown. (Reproduced with permission from ref. 169.)

The most striking observation from the E/C and t/c product ratios from 1 wt% of **53** is that they do not vary appreciably in the gel phases as a function of chain length. The t/c ratios are low and suggest that the carbonyl groups experience a polar environment as they react. The Δv_{90} values for **53b-53f** are also nearly constant (~ 1200-1300 Hz).

The results from **53** can be understood with the aid of representations of the BR in Figure 33 and as they might appear in the gels (Figure 34). The Δv_{90} values indicate that all of the carbonyl groups experience a similar environment. The very low t/c ratios further define the location of **53** and the radicaloid centers of the BR to be near a polar interface of a surfactant layer. Thus, water molecules can access the loci of reactivity in **53**, making differences among the BR shapes indistinguishable to nearby surfactant molecules.

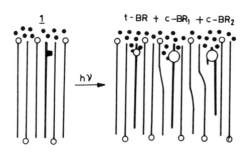

Figure 34. Shape representations of **53** and its BR in an aqueous gel layer. Water molecules are included to show how they may solvate the biradical centers. (Reproduced with permission from ref. 169.)

The t/c product ratios and Δv_{90} values follow very similar changes up to chain lengths slightly longer than the surfactant molecules (Figures 35 and 36). Thereafter, the t/c ratios are very low (symptomatic of **BR** reacting in a very polar environment), while Δv_{90} rises (to values near to those of **52c**) for **52** which approach twice the surfactant length. The maximum selectivity in the t/c ratios in each of the gels, including **KP**, is observed with the **52** whose extended length is nearest that of the polymethylene portion of a surfactant molecule.

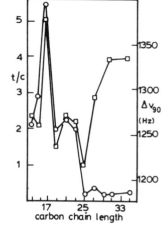

Figure 35. Deuterium quadrupolar splittings ($\square$) and t/c ratios (o) at 38°C from **52** versus ketone chain length in the K S gel phase. (Reproduced with permission from ref. 181.)

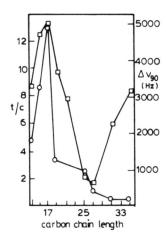

Figure 36. Deuterium quadrupolar splittings at 70°C (□) and t/c ratios at 38°C (o) from **52** versus ketone chain length in the **KSO** gel phase. (Reproduced with permission from ref. 181.)

The E/C ratios from **52** further demonstrate the dependence of product selectivity on ketone chain length. In both gel types, the E/C ratios are low and almost constant for **52** of lengths up to slightly longer than the surfactant molecules. Thereafter, the E/C ratios decrease further to <1 (indicating preferred cyclization) and then rise at still longer ketone lengths to values near those found with the short **52**.

As shown in Figure 37, the carbonyl groups of **52** (which are shorter than or comparable to the length of the surfactants) and the **BR** derived from them reside near the middle of a gel layer. Unlike in **53**, they are shielded from the aqueous portions of the phase and experience an ordered polymethylene part of a surfactant layer. However, the shape differences between *t*-**BR** and *c*-**BR** from the shorter **52** are apparently too small to be detected by even the layer middle. The expected projections of the two alkyl chains in *c*-**BR** should make it much more disruptive to the surfactant layers than either *t*-**BR** or *c*-**BR**.

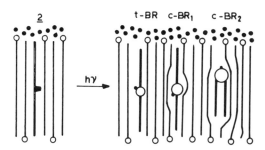

Figure 37. Shape representation of a "short" **52** and its BR in an aqueous gel layer. Water molecules are included to show that they do not have an easy access to the biradicaloid centers. (Reproduced with permission from ref. 169.)

The longer **52** (i.e., those longer than an extended surfactant molecule) must either adopt more gauche twists to stay within an interdigitated bilayer (of **KP** or **KS** gels) or traverse a gel layer boundary. For **KSO** gels, the latter option requires only that a molecule of **52** cross the bilayer midplane, without encountering a polar region hostile to its alkyl chains. For **KP** and **KS** gels, traversing a gel layer boundary places part of **52** in an unfavorable aqueous environment. Alternatively, the longer **52** may adopt a hairpin conformation that exposes the carbonyl groups to a polar layer interface but protects all but the neighboring methylenes within the hydrophobic layer interior (Figure 38).

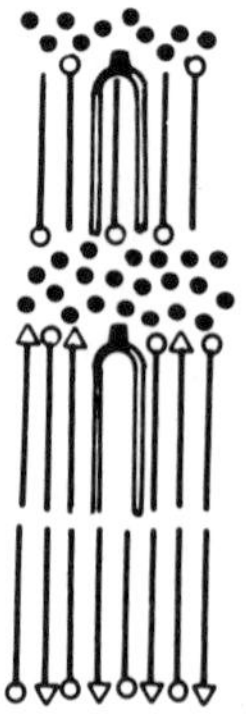

Figure 38. Representation of the solubilization sites of very long **52** in hairpin conformations in (a) **KS** and (b) **KSO** gels. Exposure of the carbonyl groups (■) to water (●) is emphasized. (Reproduced with permission from ref. 169.)

Of these, only the hairpin conformation is consistent with the product ratios and the Δv_{90} values. The very low t/c ratios from longer **52** require that their **BR** be exposed to a very polar environment. At the same time, the increase in Δv_{90} as **52** approaches twice the length of a surfactant molecule in both **KS** and **KSO** gels indicates that the methylenes alpha to the carbonyl groups become increasingly ordered (as they do when the length of shorter **52** approach that of a surfactant molecule). The **52** of intermediate length are most disordered (yielding the smallest Δv_{90} values and the lowest product selectivities) since they either contain several gauche twists or are hairpin shaped, but are too short to occupy the equivalent space of two surfactant molecules. Long **52** adopt hairpin conformations in **KSO** gels (rather than traversing the bilayer midplane) for reasons that are probably related to why phospholipids bend to form bilayers:[185] both **52** and phospholipids consist of a polar head group flanked by two alkyl chains. Also, the region near a **KSO** midplane is somewhat disordered. A molecule of **52** traversing the midplane would increase solvent order (i.e., decrease total entropy) and, thereby, increase the free energy of the system.

As mentioned previously, **BCCN** exhibits a slightly interdigitated smectic B-like mesophase[48] and a nematic phase. Several homologues form smectic and nematic phases that resemble those of **BCCN**.[48,186] For instance, the solid phases of both **BCCN** and *trans,trans*-4-ethyl[1,1'-bicyclohexyl]-4'-carbonitrile

(ECCN) are layered and interdigitated, resembling closely a very immobile smectic phase.[186,187] The smectic phase of ECCN is tilted with rhombohedral packing in the layers.[48] Several investigations of photochemical reactions have been performed in BCCN and ECCN and, prior to the discovery of solute phase separations in the mesophase temperature range of the solvent,[188] the results were interpreted on the basis of normal solutions. After some initial confusion concerning the nature of the biphasic morphology,[72a,188] it now appears fairly clear that solute-BCCN eutectics coexist with BCCN mesophases that contain their solubility limit of dopant molecules.[72c] Since the solubility limit of most of the solutes examined appears to be around 2 mol% or less, interpretation of results from experiments conducted near or above this solute concentration should be qualified. However, since the solute-rich phases are probably isotropic, the selectivities indicated from the bulk measurements must be considered lower limits to those expected from reactions within the mesophase matrices.

In this regard, intramolecular triplet quenching of 3-aryl-1-(4-alkoxyphenyl)-1-propanones (**58**), a process requiring molecular motions analogous to those leading to γ-hydrogen abstraction in the Norrish II reactions, has been studied in ECCN and the nematic and smectic phases of BCCN (equation 18).[189] The measured activation parameters varied with both solvent phase and the solute homologue.

(18)

R =	R' =
CH_3	H
C_5H_{11}	H
C_8H_{17}	H
CH_3	C_6H_{13}
CH_3	cyclohexyl
C_5H_{11}	cyclohexyl
C_8H_{17}	cyclohexyl

Large changes in the E/C ratios from irradiation of decanophenone, 4-cyclohexyl-1-phenyl-1-butanone and 5-cyclohexyl-1-phenyl-1-pentanone in BCCN also showed large variations with temperature.[190] Whether the absence of a temperature-dependent change in the E/C ratios from 2-cyclohexyl-1-(4-ethylphenyl)ethanone is due to phase separation has not been determined.[190]

The triplet lifetime of **59a** measured in acetonitrile at 30°C can be increased by 15-fold upon changing the solvent to a nematic phase mixture of 1:2 **ECCN/BCCN** and by 250-fold when the solvent is smectic **BCCN**.[191] Regardless of the **59a** solubility in the ordered media, at least some mesophase control over the motions required for remote hydrogen abstraction (the principal triplet deactivation pathway in isotropic media; Scheme 8) must be invoked to explain these results.

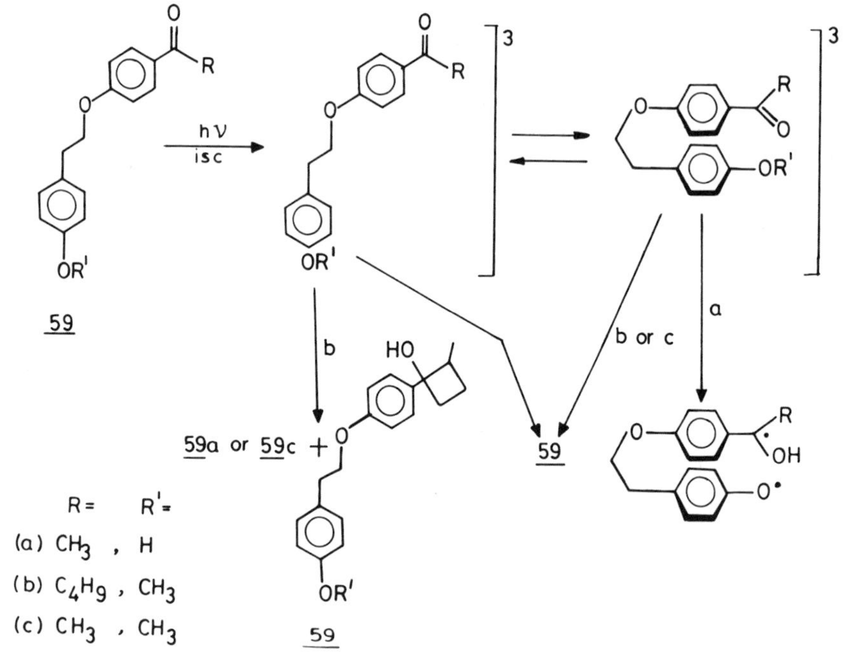

Scheme 8. Photoprocesses of **59** and related compounds.

The triplet lifetime of **59c** (that does not possess an abstractable phenolic hydrogen atom) was increased by only about five fold as solvent was changed from acetonitrile to smectic **BCCN**.[191] This result is also consistent with conformational motions analogous to those responsible for intramolecular excimer formation in **P3P** being the lifetime limiting factor. E/C ratios from **59b** in methylcyclohexane and mesophases of **BCCN** and **ECCN** showed no discernible differences: phase separation was apparent with this ketone.[191]

Thus, although a great deal of experimental work has been performed in **ECCN** and **BCCN**, few conclusions can be made with confidence at this time. Several of the reactions should be reinvestigated at dopant concentrations well below their solubility limits.

9. Photodimerizations

Irradiation of tetraphenylbutatriene (**60**) dissolved in a compensated nematic (i.e., infinite pitch cholesteric) mixture of 1.95:1 (w:w) of CCl/cholesteryl laurate (**CL**) at 30°C yielded the dimer **61** as the exclusive product (Equation 19).[192] Irradiation of **60** in the isotropic (ca. 80°C) and solid phase of CCl/CL, or in the isotropic phase (> 40°C) to which small amounts of diethylene glycol and dibutyl ether had been added (to depress the phase transition temperature), led to no dimer. Since linear dichroism spectra show that **60** does not aggregate to form microcrystals in nematic CCL/CL, the formation of **61** has been attributed to an increase in the excited state lifetime of **60** and to a limited number of collisional geometries between excited- and ground-state molecules of **60** favored by the solvent matrix.

$$Ph_2C{=}C{=}CPh_2 \xrightarrow{h\nu} \mathbf{61} \tag{19}$$

$$\underline{60}$$

$$\underline{62} \xrightarrow{h\nu} \text{syn-}\underline{63} + \text{anti-}\underline{63} \tag{20}$$

The influence of cholesteric phase order on the photodimerization of acenaphthylene [**62**, equation (20)][193] has been examined in detail.[194] The distribution of syn and anti dimers (**63**) continues, as in isotropic solvents, to be controlled by the multiplicity of the excited-state precursor.[193b] However, the efficiency of dimer formation is very dependent upon solvent phase, solute concentration, and the presence of disturbing molecules that do not interfere with the photochemistry.

As seen in Figure 39, the quantum efficiency of dimerization (Φ_D) for 0.08 M 62 in BS and toluene vary similarly with temperature and are near in magnitude. No meaningful change in Φ_D occurred between the smectic (<25°C) and isotropic phases (>25°C) of BS.

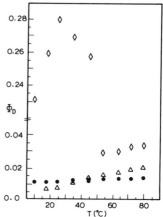

Figure 39. Quantum efficiencies for photodimerization of 0.08 M 62 in toluene (●), BS (Δ), and 1:1 (w:w) mixture of CHA/CHN (◊) as a function of temperature. (Reproduced with permission from ref. 194b.)

By contrast, irradiation of the same concentration of 62 in 1:1 (w/w) CHN/CHA led to Φ_D which are extremely sensitive to solvent phase. In the isotropic phase of CHN/CHA, Φ_D is two to three times the value in toluene. Irradiation of 0.08 M 62 in the cholesteric phase (<48°C) increases the efficiency of formation of 63 by ca. 10 fold over the isotropic phase and 20 to 25- fold over the values in toluene at the same temperature.

While it is not unreasonable that the BS smectic phase does not facilitate photodimerization since BS and 62 have very different molecular shapes, the large increase in Φ_D in cholesteric CHN/CHA is quite surprising. There is, at best, a vague similarity between the plate-like 62 and the cholesteric esters of the solvent. In spite of this, the cholesteric phase appears to control the orientations and frequency of collisions between molecules of 62: ignoring for simplicity's sake the influence of excited-state multiplicity, Φ_D can be expressed as $F_q \times F_c \times \tau$, the product of the frequency of excited 62 collisions with 62, the fraction of those collisions which lead to 63, and the measured excited lifetime of 62, respectively. The dependency of Φ_D within the cholesteric phase, as shown in Figure 39, can be understood if (1) F_q decreases and F_c and τ increase as temperature decreases and (2) below 25°C, the decrease in F_q cannot be compensated by the increases of F_c and τ.

To test this hypothesis, varying amounts of tetralin (a molecule that disrupts cholesteric order, increasing F_q and decreasing F_c, but that should not alter the and frequency of collisions between molecules of **62**: ignoring for simplicity's sake the influence of excited-state multiplicity, Φ_D can be expressed as $F_q \times F_c \times \tau$, the product of the frequency of excited **62** collisions with **62**, the fraction of those collisions which lead to **63**, and the measured excited lifetime of **62**, respectively. The dependency of Φ_D within the cholesteric phase, as shown in Figure 39, can be understood if (1) F_q decreases and F_c and τ increase as temperature decreases and (2) below 25°C, the decrease in F_q cannot be compensated by the increases of F_c and τ.

To test this hypothesis, varying amounts of tetralin (a molecule that disrupts cholesteric order, increasing F_q and decreasing F_c, but that should not alter the unimolecular decay component of τ) were added to the solutions of **62** in **CHN/CHA**. The data in Figure 40 indicate that the greater solvent fluidity induced by tetralin is accompanied by an even greater chaos in collisions between molecules of **62**: Φ_D decreases as the concentration of tetralin increases.

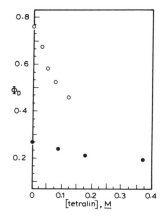

Figure 40. Quantum efficiency for photodimerization (Φ_D) of 0.02 M (O) and 0.08 M (●) **62** in 1:1 (w:w) **CHA/CHN** versus the concentration of added tetralin. (Reproduced with permission from ref. 194b.)

That these results *may* be due, at least in part, to changes in cholesteric pitch, was indicated by the observation that at 25°C and the same initial concentration of **62**, Φ_D was over eight times greater in a 70:30 mixture of CN/CCl (p ~ 2000 Å [195]) than in a 40:60 compensated nematic phase mixture of the same components (p → ∞ [195]).

Perhaps the best evidence against aggregation of **62** and for mesophase control over solute collisions being responsible for the reported changes in Φ_D was obtained from a study of the influence of initial solute concentration on dimerization efficiency in **CHN/CHA**. In an isotropic solvent, Φ_D should increase linearly with the concentration of **62** due to changes in frequency of collisions. Figure 41 shows that this is the case in isotropic **CHN/CHA** at 55°C. At 35°C, cholesteric **CHN/CHA** gives rise to Φ_D with a bizarre dependence on [**62**]: Φ_D is largest at the lowest concentration of **62** employed! Were solute aggregation within the cholesteric phase responsible for the large Φ_D, it would not have decreased at larger solute concentrations. The decrease and subsequent increase in [62] can be accommodated by a mechanistic model in which added 62, much like added tetralin (Figure 40), has a greater disruptive influence on solvent order *at low concentrations* than can be compensated by the increases in F_q. Above [62] ≈ 0.08 M, the increase in F_q becomes greater than the decrease in F_c and Φ_D increases.[196]

A related photodimerization of 1,3,5-trimethyluracil (64a) in solid, smectic, cholesteric, and isotropic phases has been reported.[197] In polar isotropic solvents, four photodimers (65-68) are produced.[198] In less polar isotropic solvents, only 65a and 66a were detected in comparable amounts (Table 17). Since high photoselectivity was observed in all the solid, cholesteric, and smectic phases employed, the origin of the product ratios is puzzling: even the smectic phase of BS whose constituent molecules differ from 64a in both size and shape, leads to a very high photodimer selectivity.

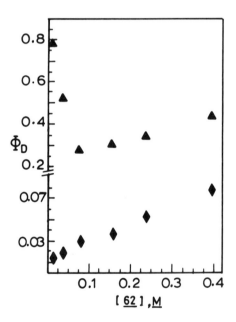

Figure 41. Quantum efficiency for photodimerization (Φ_D) of 62 in 1:1 (w:w) CHA/CHN at 35°C (▲) and 55°C (♦) as a function of the concentration of 62. (Reproduced with permission from ref. 194b.)

The photodimerization of 1,3,5,6-tetramethyluracil (64b)[199] with >300 nm radiation was also studied in liquid-crystalline solvents.[200] In isotropic media of low and high polarity, 66b and 68b are the major photodimers, being formed in a ratio of ~ 2:1.[199] At low concentrations of 64b in solid or hexatic B phases of BS, 66b and 68b were formed rapidly and in high chemical yields.[200] At the same concentrations and temperatures in isotropic solvents, irradiation led to very slow dimerization.

a) R = H
b) R = Me

$$\text{65} \qquad \text{66} \qquad \text{67} \qquad \text{68} \qquad\qquad (21)$$

Invariably, solid-state and isotropic phase irradiations of **64b** led to preferential formation of **66b** while smectic phase irradiations in **BS** or **COC** produced an excess of **68b** (Table 18).

Table 17. Photodimerization of **64a** in Various Solvents.[197]

Solvent	T(°C)	Phase	65a/66a
COC	-30	Solid	86/14
	10	Smectic	93/7
	23	Cholesteric	93/7
	43	Isotropic	71/29
Cholesteryl linoleate (CLI)	10	Solid	91/9
	30	Smectic	94/6
	43	Isotropic	77/23
BS	21	Smectic	94/6
	32	Isotropic	51/49
Dioxane	2	Solid	32/68

The solid-state results may indicate that **64b** has phase separated into solute-enriched regions or formed microcrystallites. In any case, the changes in the product ratio with phase are pronounced and dramatic. The variation of selectivity (defined as {**68b** - **66b**}/{**66b** + **68b**}) with temperature and phase in **BS** (Figure 42) is reminiscent of the changes in Φ_D observed for **62** in **CHN/CHA** (Figure 39). This may be fortuitous or mechanistically significant. It is also curious that both **64a** and **64b** yield significant amounts of *cis-anti* -**66** during

676   Weiss

isotropic phase irradiations, but form major quantities of the *cis-syn-*65a and *trans-anti-*68b dimers, respectively, during smectic phase irradiations. Why this is so remains unclear but merits of further investigation.

Table 18. Photodimerization of 64b in Various Solvents.[200]

Solvent	T(°C)	Phase	68b/66b
COC	-40	Solid	5/95
	4	Smectic	90/10
	23	Cholesteric	37/63
	40	Isotropic	29/71
B S	5	Solid	33/67
	18	Smectic	88/12
	32	Isotropic	46/54
Ethyl stearate	16	Solid	13/87
	38	Isotropic	35/65
Water	-7	Solid	3/97
	20	Isotropic	27/62[a]

[a] Plus 11% relative yield of 65b

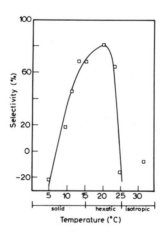

Figure 42. Photoselectivity of 64b photodimerization in B S as a function of temperature (phase).   (Reproduced with permission from ref. 200.)

The chemical and physical interactions between pairs (and larger aggregates) of cholesteryl 4-(2-anthryloxyl)butyrate (**69**) have been investigated in its solid, cholesteric, neat isotropic, solution isotropic, and gel phases.[201,202] Absorption, excitation, and induced circular dichroism spectra of 0.8 wt% of **69** in its gel phase with dodecane confirm that the gelator molecules stack with their anthracenyl groups atop one another in a nonparallel arrangement. Connectivity of the anthracenyl at its 2 position to the cholesteryl group is necessary to attain liquid crystallinity or gel formation since compounds analogous to **69** containing 9-substituted anthracenes form neither phase.[201b,c] The distribution of head-to-head/head-to-tail photodimers (hh/ht) of **70** and *syn/anti* ht-**70**, were determined in each phase (Table 19).[201b,202]

R= -O(CH$_2$)$_3$ CO$_2$- cholesteryl

(22)

69   hh -**70**   ht-**70**

Table 19. Photodimer Distributions from **69** in Various Phases.[201b, 202]

69 (wt%)	Solvent	Phase	T(°C)	**70** hh/ht	ht-**70** *syn/anti*
1.12	Dodecane	Gel	26.0	53/47	19/29
1.47	Dodecane	Gel	25.5	53/47	19/28
1.47	Toluene	Isotropic	25.5	46/54	17/37
Neat	—	Cholesteric	184.4	55/45	12/33
Neat	—	Cholesteric	183.5	55/45	13/33
Neat	—	Cholesteric	178.0	54/46	12/34
Neat	—	Isotropic	204.4	50/50	20/31
Neat	—	Isotropic	206.7	⁄ 47/53	17/36

There is little selectivity in and little difference between the distribution from the liquid-crystalline, gel, and neat isotropic phases (no dimer was detected from irradiation of solid **69**). Either the molecules of **69** pack without specific pairwise interactions in these phases or, less likely, they are able to reorient themselves according to their photophysical preferences during the lifetime of the dimer precursors: the excited singlet state of **69** has ($\tau < 20$ ns).[201b] The distribution obtained from irradiation of the isotropic toluene solution of **69** reflects the orientational preferences of **69-69** collision partners when no solvent ordering constraints are imposed. Since it differs slightly from the other ratios, they must arise from systems in which the anthracenyl groups experience slight orientational restrictions.

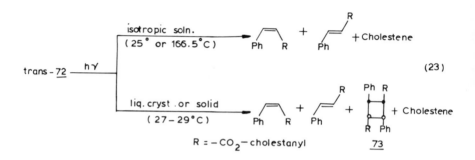

The photodimerization of cinnamoyl groups in liquid-crystalline environments has also been investigated. When neat cholesteryl *trans*-cinnamate (*trans*-**71**) was irradiated at >300 nm from 25 to 225°C (spanning its solid, cholesteric, and isotropic liquid phases) in a KBr matrix,[203] it was converted to dimers whose structures were not determined. Qualitatively, the rate of the dimerization was slowest in the solid. Competing processes included formation of the cis isomer and cinnamic acid. Subsequently evidence was obtained that irradiation of *trans*-**71** involved, at least partially, reaction of the C_5—C_6 double bond of the cholesteryl moiety.[204] To avoid this complication, the photochemistry of 5α-cholestan-3β-yl-*trans*-cinnamate (*trans*-**72**) in its solid and cholesteric phases and in isotropic solutions (*n*-hexane and *n*-hexadecane was investigated.[205] In isotropic solutions, the major photochemical processes of *trans*-**72** are photoelimination to yield *trans*-cinnamic acid and cholestene and trans $\rightleftarrows$ cis isomerization. In the neat ordered phases of *trans*-**72**, photoelimination and isomerization are much less important than phtodimerization of the cinnamoyl group [equation 23]. Only the α-truxillate diester (**73**) was detected by hplc and other analyses. It is the dimer expected if molecules *trans*-**72** are aligned in an antiparallel, pairwise fashion. It is also preferred on the basis of steric considerations.

Further evidence for this alignment and for partial interdigitation of *n*-alkyl-*trans*-cinnamates (*trans*-**74**) was obtained from their irradiation in the crystalline, hexatic B, and isotropic phases of BS.[205] The alkyl chains of *trans*-**74** were chosen to make the molecules slightly shorter than, the same length as, and slightly longer than the extended length of BS. Up to 40 wt% of *trans*-**74a** could be added to BS without significant depression of the smectic-isotropic phase transition temperature or induction of a biphasic (solid plus isotropic liquid) region. Both *trans*-**74b** and *trans*-**74c** depressed the hexatic-isotropic phase transition temperature to a much greater extent; they could not be dissolved in **BS** at greater than 10% loading without inducing a biphasic region. Thus *trans*-**74a**, although longer than **BS**, is incorporated into the hexatic matrix better than *trans*-**74b**, which is the same length as **BS**.

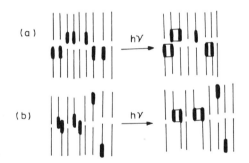

Scheme 9. Representation of the orientations of **74** in hexatic **BS**: (a) assumes alkyl cinnamates are constraine to one layer and yield only head-to-head photodimers; (b) assumes alkyl cinnamates are partially overlapping and yield only head-to-tail photodimers.[205]

Given the results obtained with *trans* -**72**, a logical interpretation is that the cinnamoyl groups of neighboring **74** are partially overlapping and in a head-to-tail arrangement: two partially overalpping *trans*-**74a** molecules can have an extended length equal to twice a **BS** layer thickness (Scheme 9). If this hypothesis is correct, irradiation of **74** in hexatic or solid **BS** should produce very large ratios of head-to-tail//head-to-head cyclobutane dimers. An alternative arrangement in which *trans*-74 molecules remain within the boarders of one **BS** layer leads to the prediction of very low head-to-tail/head-to-head ratios. In fact, irradiation of neat, solid *trans*-74a is known to produce exclusively head-to-tail, head-to-head, or no dimer depending upon the morphology of the crystal.[206] In all irradiations, only one head-to-tail dimer (**75**) and one head-to-head dimer (**76**) were observed at low percent conversions. Elimination and isomerization were competing processes whose importance increased with irradiation time. The results in Table 20 demonstrate that the **75** /**76** ratios *are* very large in the hexatic and solid phases of **BS**; they are much smaller in the isotropic phase. Thus, even at very large concentrations of a solute that is similar in structure to the solvent, the ordered solvent matrices exert significant control over solute-solute interactions. It is interesting that in spite of *trans* -74a being the more easily incorporated solute, *trans*-74b and *trans*-74c react more selectively in the ordered phases.

The sensitivity of the **BS** ordered phases to solute shape and length is reminiscent of the results obtained with *n*-alkanones and *n*-alkylphenones as solutes (see Section 8). It indicates that both the compatability of the solute in an ordered matrix and the mobility of the solute in the ordered matrix affect its overall reactivity and selectivity.

Table 20. Photodimer ratio from t-74 in Various Phases of BS.[205]

t-74	wt%	Phase (T, °C)	% Conversion	75/76
a	20	Isotropic (32)	41	2.5 ± 0.9
		Hexatic (18)	12	8.0 ± 0.6
		Solid (8)	9	8.0 ± 0.6
	40	Isotropic (32)	17	2.3 ± 0.1
		Hexatic (18)	11	9.4 ± 0.4
b	20	Isotropic (32)	40	3.3 ± 0.7
		Hexatic (18)	12	>20
		Solid (8)	12	>20
	40	Isotropic (32)	19	2.7 ± 0.1
		Hexatic (18)	14	>20
c	10	Isotropic (32)	53	No dimer detected
		Hexatic (18)	37	9.4 ± 0.2

No photocycloaddition reaction between different molecular partners has been reported in liquid-crystalline media, although they should be as feasible as photodimerization reactions.

10. Personal Postscript

In spite of the enormous effort put forth primarily during the last 20 years, *understanding how liquid-crystalline media influence photochemical processes and are influenced by them is still not well understood. Few generalizations can be made and most studies are "set pieces," with little predictive value.* At the same time, the number and types of photorelated applications in which mesophases play a central role have also increased. Photoinduced processes in (and of) polymeric liquid crystals and lyotropic phases seem like very attractive areas for future work. Applications of discotic molecules have focused on electron and energy transfers; several other uses of columnar phases, especially, should be explored.

In part, the problems associated with exploiting liquid crystals in photorelated processes and in being able to interpret results from such experiments are related to the multiplicity of disciplines and techniques that must be applied concertedly to define all of the system parameters. Photochemistry and photophysics in mesophases require study by scientists with expertise ranging from spectroscopy and dynamics to rheology and thermodynamics. Finally, there is a need for those who recognize the salient features of a liquid crystal tailored for a specific use and are capable of synthesizing it.

This chapter has attempted to present a synopsis of the photochemistry and photophyics in liquid-crystalline media. It should serve as a point of departure since only the surface of the utility of liquid crystals has been scratched, and it is beautifully birefringent!

Acknowledgments

This chapter was written during the author's sabbatical at the Sophisticated Instruments Facility of the Indian Institute of Science in Bangalore. Without the cooperation and help of the Facility Convener, Professor C.L. Khetrapal, and the other members of the SIF staff, it would not have been possible to complete the manuscript. I thank them and acknowledge their support and friendship. Mr. C.L.Viswanath is also thanked for typing (and retyping) the manuscript without complaining about my miserable handwriting. Finally the Indo-U.S. Subcommission on Education and Culture provided a fellowship for my stay in India and the National Science Foundation continues to support my research on reactions in liquid crystals.

References

1.   R. G. Weiss, *Tetrahedron*, **1988**, *44*, 3413.
2.   Seminar by G. J. T. Tiddy (Unilever Research, Port Sunlight Laboratory), Indian Institute of Science, January, 1990.

3. *Liquid Crystals : The Fourth State of Matter*, F. D. Saeva, ed., Marcel Dekker, New York, 1979.
4. H. Kelker, *Mol. Cryst. Liq. Cryst.*, **1988**, *165*, 1.
5. T. Svedberg, *Kolloid. Z.*, **1916**, *18*, 54; *Chem. Abstr.*, **1916**, *10*, 2429.
6. S. Chandrasekhar, *Liquid Crystals*, Cambridge University Press, Cambridge, 1977.
7. G.W. Gray, *Molecular Structure and Properties of Liquid Crystals*, Academic Press, New York, 1962.
8. P. G. de Gennes, *The Physics of Liquid Crystals*, Clarendon Press, Oxford, **1974**.
9. H. Kelker and R. Hatz, *The Handbook of Liquid Crystals*, Verlag Chemie,Weinheim, 1980.
10. Mesomorphic: relating to, existing in or being an intermediate state. From *Webster's New Inter-national Dictionary*, P. B. Grove, ed., Merriam-Webster, Springfield, MA, 1981.
11. D. Demus, S. Diele, S. Grande and H. Sackmann in *Advances in Liquid Crystals*, Vol. 6, G. H. Brown, ed., Academic Press, New York, 1983, p. 1.
12. S. Chandrasekhar in *Advances in Liquid Crystals*, Vol. 5, G. R. Luckhurst and G. W. Gray, eds., Academic Press, New York, 1982, p. 47.
13. V. Ramamurthy and K. Venkatesan, Chem. Rev., **1987**, *87*, 433.
14. (a) R. S. Porter and J. F. Johnson in *Rheology*, Vol. 4, F. Eirich ed., Wiley, New York, 1968, p. 317.
    (b) J. M. Pochan in *Liquid Crystals: The Fourth State of Matter*, F. D. Saeva, ed., Marcel Dekker, New York, 1979, Chap. 7.
15. P. J. Debye, *Trans. Electrochem. Soc.*, **1942**, *82* , 265.
16. R. S. Porter, E. M. Barrall II and J. F. Johnson, *J. Chem. Phys.*, **1966**, *45*, 1452.
17. R. S. Porter, A. C. Griffin, and J. F. Johnson, *Mol. Cryst. Liq. Cryst.*, **1974**, *25*, 131.
18. K. Sakamoto, R. S. Porter and J. F. Johnson, *Mol. Cryst. Liq. Cryst.*, **1969**, *8*, 443.
19. For an exception, see: E. M. Friedman and R. S. Porter, *Mol. Cryst. Liq. Cryst.*, **1975**, *31*, 47.
20. C. K. Yun and A. G. Fedrickson, *Mol. Cryst. Liq. Cryst.*, **1970**, *12*, 73.
21. R. Bline and V. Dimic, *Phys. Lett.*, **1970**, *31A*, 531.
22. J. A. Murphy, J. W. Doane, Y. Y. Hsu and D. L. Fishel, *Mol. Cryst. Liq. Cryst.*, **1973**, *22*, 133.
23. H. Hervet, W. Urbach and F. Rondolez, *J. Chem. Phys.*, **1978**, *68*, 2725.
24. W. Urbach, H. Hervet and F. Rondolez, *Mol. Cryst. Liq. Cryst.*, **1978**, *46*, 209.
25. H. Kneppe, F. Schneider and N. K. Sharma, *Ber. Bunsenges. Phys. Chem.*, **1981**, *85*, 784.
26. H. Kneppe and F. Schneider, *Mol. Cryst. Liq. Cryst.*, **1981**, *65*, 23.
27. H. Hakemi and M. M. Labes, *J. Chem. Phys.*, **1975**, *63*, 3708.
28. M. Miesowicz, *Nature*, **1946**, *158*, 27.
29. M. E. Moseley and A. Loewenstein, *Mol. Cryst. Liq. Cryst.*, **1982**, *90*, 117.
30. A. Arcioni, F. Bertinelli, R. Tarroni, and C. Zannoni, *Mol. Phys.*, **1987**, *61*, 1161.
31. D. Gutter, *Mol. Cryst. Liq. Cryst.*, **1969**, *8*, 85.
32. Ref. 8, p. 197.
33. E. D. Cehelnik, R. B. Cundall, J. R. Lockwood, and T. F. Palmer, *J. Chem. Soc., Faraday Trans. II*, **1974**, *70*, 244.
34. H. Kresse in *Advances in Liquid Crystals*, Vol. 6, G. H. Brown, ed., Academic Press, New York, 1983, p. 109.

35. G. Moro and P. L. Nordio, *Mol. Cryst. Liq. Cryst.*, **1984**, *104*, 361.
36. K. D. Lawson and T. J. Flautt, *J. Phys. Chem.*, **1968**, *72*, 2066.
37. (a) D. E. Martire in *Molecular Physics of Liquid Crystals*, G. R. Luckhurst and G.W. Gray, eds., Academic Press, London, 1979, Chap. 10.
  (b) S. Godbane and D. E. Martire, *J. Phys. Chem.*, **1987**, *91*, 6410.
  (c) M. A. Gidley and D. Stubley, *J. Chem. Thermodynamics*, **1982**, *14*, 785.
38. (a) J. W. Park, C. S. Bak, and M. M. Labes, *J. Am. Chem. Soc.*, **1975**, *97*, 4398.
  (b) G. Sigaud, M. F. Achard, F. Hardouin, and H. Gasparoux, *Chem. Phys. Lett.*, **1977**, *48*, 122.
  (c) G. Sigaud, Docteur es Sciences Thesis, University of Bordeaux 1, Talence, France 1979.
39. R. S. Becker, S. Chakravorti, and S. Das, *J. Chem. Phys.*, **1989**, *90*, 2802 and refs. cited therein.
40. C. L. Khetrapal, R. G. Weiss, and A. C. Kunwar in *Liquid Crystals - Applications and Uses,* Vol. 2, B. Bahadur, ed., World Scientific, Teaneck, N. J., 1990.
41. A. Saupe, *Mol. Cryst. Liq. Cryst.*, **1972**, *16*, 87.
42. E. Sackmann in *Applications of Liquid Crystals*, G. Meier, E. Sackmann, and J. G. Grabmaier, eds., Springer Verlag, Berlin, 1975, p. 21.
43. V. D. Neff in *Plastic Crystals and Liquid Crystals*, Vol. 2, G. W. Gray and P. A. Winsor, eds., Ellis Horwood, Chichester, U.K., 1974, Chap. 9.
44. A. V. Ivoshchenko and V. G. Rumyantsev, *Mol. Cryst. Liq. Cryst.*, **1987**, *150A*, 1.
45. *Polarized Spectroscopy of Ordered Systems*, B. Samori and E. W. Thulstrup eds., Kluwer, Dordrecht, Netherlands, 1988.
46. J. Michl and E. W. Thulstrup, *Spectroscopy with Polarized Light*, VCH Publishers, New York, 1986, Chap. 3.
47. (a) J. Cognard, *Mol. Cryst. (Suppl. Series)*, Suppl. 1, 1982, 1.
  (b) P. I. Ktorides and D. L. Uhrich, *Mol. Cryst. Liq. Cryst.*, **1982**, 87, 69.
48. (a) E. Rahimzadeh, T. Tsang, and L. Yin, *Mol. Cryst. Liq. Cryst.*, **1986**, *139*, 291.
  (b) Y. C. Chu, T. Tsang, E. Rahimzadeh, and L. Yin, *Phys. Stat. Sol. (a).*, **1988**, *105*, K1.
49. R. Eidenschinck, D. Erdmann, J.Krause, and L. Pohl, *Angew. Chem.*, **1977**, *89*, 103.
50. H. Wedel and W. Hasse, *Chem. Phys. Lett.*, **1978**, *55*, 96.
51. G. J. Brownsey and A. J. Leadbetter, *J. Phys. (Paris)*, **1981**, *42*, L135.
52. K. S. Krishnamurthy, *Mol. Cryst. Liq. Cryst.*, **1986**, *132*, 255.
53. (a) D. Krishnamurthy, K. S. Krishnamurthy, and R. Shashidar, *Mol. Cryst. Liq. Cryst.*, **1969**, *8*, 339.
  (b) J. S. Dryden, *Chem. Phys.*, **1957**, *26*, 604.
  (c) K. Sullivan, *J. Res. NBS*, **1974**, *78A*, 129.
54. E. Sackmann, *J. Am. Chem. Soc.*, **1968**, *90*, 3569.
55. E. Sackmann and H. Mohwald, *J. Chem. Phys.*, **1973**, *58*, 5407.
56. G. Gottarelli, G.P. Spada, D. Varech, and J. Jacques, *Liq. Cryst.*, **1986**, *1*, 29.
57. N. L. Morris, R. G. Zimmermann, A. W. Dalziel, G. B. Jameson, P. M. Reuss, and R. G. Weiss, *J. Am. Chem. Soc.*, **1988**, *110*, 2177.
58. B. Kohne and K. Praefcke, *Angew. Chem. Intern. Ed. Engl.*, **1984**, *23*, 82.
59. D. J. Mitchell, G. J. T. Tiddy, L. Waring, T. Bostock, and M. P. McDonald, *J. Chem. Soc., Faraday Trans, I*, **1983**, *79*, 975.
60. C. D. Adam, J. A. Durrant, M. R. Lowry, and G. J. T. Tiddy, *J. Chem. Soc., Faraday Trans. I*, **1984**, *80*, 789.
61. J. M. Vincent and A. Skoulios, *Acta Crystallorg.*, **1966**, *20*, 432, 441, 447.

684   Weiss

62. A. M. Lackner, J. D. Margerum, and C. van Ast, *Mol. Cryst. Liq. Cryst.*, **1986**, *141*, 289.
63. V. Ramesh, S. Mathew, and M. M. Labes, *Mol. Cryst. Liq. Cryst. Letters*, **1988**, *6*, 35.
64. G. Pelzl and H. Sackmann, *Sym. Faraday Soc.*, **1971**, *5*, 68.
65. R. Schaetzing and J. D. Litster in *Advances in Liquid Crystals*, G. H. Brown, ed., Vol. 4, Academic Press, New York, 1979, p. 147.
66. See Ref. 9, Chaps. 6 and 7.
67. T. J. Novak, R. A. Mackay, and E. J. Poziomek, *Mol. Cryst. Liq. Cryst.*, **1973**, *20*, 213.
68. See for instance: (a) H. Levanon, *Chem. Phys. Lett.*, **1982**, *90*, 465.
     (b) S. Das, C. Lenoble, and R. S. Becker, *J. Am. Chem. Soc.*, **1987**, *109*, 4349.
69. J. L. Fergason in *Liquid Crystals*, G. H. Brown, G. J. Dienes and M. M. Labes, eds., Gordon and Breach, New York, 1966, p. 89.
70. (a) H. L. de Vries, *Acta Crystallogr.*, **1951**, *4*, 219.
     (b) H. Baessler and M.M. Labes, *Mol. Cryst. Liq. Cryst.*, **1970**, *6*, 419.
71. R. Cano and P. Chatelain, *C.R. Acad. Sci. (Paris)*, **1964**, *259*, 352.
72. For an interesting example, see: (a) R. L. Treanor and R. G. Weiss, *J. Phys. Chem.*, **1987**, *91*, 5552.
     (b) B. Samori, P. De Maria, P. Mariani, F. Rustichelli, and P. Zani, *Tetrahedron*, **1987**, *43*, 1409.
     (c) B. J. Fahie, D. S. Mitchell, M. S. Workentin and W. J. Leigh, *J. Am. Chem. Soc.*, **1989**, *111*, 2916.
73. W. Haas, J. Adams and J. Wysocki, *Mol. Cryst. Liq. Cryst.*, **1969**, *7*, 371.
74. C. Mioskowski, J. Bourguignon, S. Candau, and G. Solladie, *Chem. Phys. Lett.*, **1976**, *38*, 456.
75. (a) G. Solladie and G. Gottarelli, *Tetrahedron*, **1987**, *43*, 1425,
     (b) R. Cano and P. Chatelain, *C. R. Acad. Sci. (Paris)*, **1961**, *253*, 1815.
76. S. Candau, P. le Roy and F. Beauvais, *Mol. Cryst. Liq. Cryst.*, **1973**, *23*, 283.
77. J. Naciri, G.P. Spada, G. Gottarelli, and R. G. Weiss, *J. Am. Chem. Soc.*, **1987**, *109*, 4352.
78. E. Sackmann, *J. Am. Chem. Soc.*, **1971**, *93*, 7088.
79. W. E. Haas, K. E. Nelson, J. E. Adams and G. A. Dir, *J. Electrochem. Soc.*, **1974**, *21*, 1667.
80. S. Tazuke, S. Kurihara, and T. Ikeda, *Chem. Lett.*, **1987**, 911.
81. T. Ikeda, S. Horiuchi, D. B. Karanjit, S. Kurihara, and S. Tazuke, *Chem. Lett.*, **1988**, 1679.
82. S. Kurihara, T. Ikeda, and S. Tazuke, *Jpn. J. Appl. Phys. (Part 2)*, **1988**, *27*, L1791.
83. (a) D. L. Dexter, *J. Chem. Phys.*, **1953**, *21*, 836.
     (b) *The Exciplex*, M. Gordon and W. R. Ware, eds., Academic Press, New York, 1975.
84. J. Arrecis, K. Kundu, and P. Phillips, *J. Phys. Chem.*, **1989**, *93*, 5981.
85. (a) L. Chapoy, D. K. Munck, K. H. Rasmussen, E. J. Diekmann, R. K. Sethi, and D. Biddle, *Mol. Cryst. Liq. Cryst.*, **1984**, *105*, 353.
     (b) L. Chapoy and D. K. Munck, *J. Phys. (Paris), C3*, **1983**, *44*, 697.
     (c) L. Chapoy, *J. Chem. Phys.*, **1985**, *84*, 1530.
86. (a) J. H. Sluyters, A. Baars, J. F. van der Pol, and W. Drenth, *J. Electroanal. Chem.*, **1989**, *271*, 41.
     (b) J. F. van der Pol, E. Neeleman, J. W. Zuikker, R. J. M. Nolte, W. Drenth, J. Aerts, R. Visser, and S. J. Picken, *Liq. Cryst.*, **1989**, *6*, 577.
87. N. Boden, R. J. Bushby, J. Clements, M. V. Jesudason, P. F. Knowles, and G. Williams, *Chem. Phys. Lett.*, **1988**, *152*, 94.
88. (a) B. R. Ratna, R. Shashidhar, *Pramana*, **1976**, *6*, 278.

(b) P. E. Cladis, D. Guillon, F.R. Bouchet and P. L. Finn, *Phys. Rev.*, **1981**, *A23*, 2594.

(c) A. J. Leadbetter, R. M. Richardson and C. N. Colling, *J. Phys. (Paris)*, **1975**, *36*, 37.

89. Recent studies suggest that the smectic A phases of alkylcyanobiphenyls may consist of a mixture of head-to-tail and head-to-head molecular pairs.
    (a) D. Guillon and A. Skoulios, *J. Phys. (Paris)*, **1984**, *45*, 607.
    (b) M. Jaffrain, G. Lacrampe and G. Martin, *J. Phys. (Paris)*, **1984**, *45*, L-1103.

90. N. Tamai, I. Yamazaki, H. Masuhara and N. Mataga, *Chem. Phys. Lett.*, **1984**, *104*, 485.

91. C. David and D. Baeyens-Volant, *Mol. Cryst. Liq. Cryst.*, **1984**, *106*, 45,

92. D. Baeyens-Volant and C. David, *Mol. Cryst. Liq. Cryst.*, **1985**, *116*, 217.

93. R. Subramanian, L. K. Patterson and H. Levanon, *Chem. Phys. Lett.*, **1982**, *93*, 578.

94. D. Markovitsi and J. P. Ide, *J. Chim. Phys.*, **1986**, *83*, 97.

95. D. Baeyens-Volant and C. David, *Mol. Cryst. Liq. Cryst.*, **1989**, *165B*, 37.

96. J. B. Birks, *Photophysics of Aromatic Molecules*, Wiley, New York, 1970.

97. Th. Forster, *Disc. Faraday. Soc.*, **1959**, *27*, 7.

98. a) R. Voltz, J. Klein, H. Lami, F. Heisel and G. Laustriat, *J. Chim. Phys.*, **1966**, *63*, 1253.
    b) R. Voltz, G. Laustriat and A. Coche, *J. Chim. Phys.*, **1966**, *63*, 1259.

99. N. Kato, Y. Kawai M. Matsushima, T. Mizazaki and K. Fueki, *Radiat. Phys. Chem.*, **1986**, *27*, 13.

100. F. D. Saeva, G. A. Reynolds and L. Kaszczuk, *J. Am. Chem. Soc.*, **1982**, *104*, 3524.

101. C. Destrade, M. C. Mondon and J. Malthete, *J. Phys. (Paris)*, **1979**, *40*, C-3.

102. D. Markovitsi, F. Rigaut, M. Mouallem and J. Malthete, *Chem. Phys. Lett.*, **1987**, *135*, 236.

103. (a) A. M. Levelut, *J. Phys. (Paris)*, **1979**, *40*, L-81.
     (b) A. M. Levelut, *J. Chim. Phys.*, **1983**, *80*, 149.
     (c) M. Cotrait, P. Marsau, M. Pesquer and V. Volpihac, *J. Phys. (Paris)*, **1982**, *43*, 355.

104. D. Markovitsi, I. Lecuyer, B. Clergeot, C. Jallabert, H. Strzelecka, and M. Veber, *Liq. Cryst.*, **1989**, *6*, 83.

105. (a) H. Strzelecka, C. Jallabert, and M. Veber, *Mol. Cryst. Liq. Cryst.*, **1988**, 156, 355.
     (b) P. Davidson, C. Jallabert, A. M. Levelut, H. Strzelecka, and M. Veber, *Liq. Cryst.*, **1988**, *3*, 133.
     (c) P. Davidson, C. Jallabert, A. M. Levelut, H. Strzelecka, and M. Veber, *Mol. Cryst. Liq. Cryst.*, **1988**, *161*, 395.

106. V. Kloppfer, H. Bauser, F. Dolezalek, and G. Naundorf, *Mol. Cryst. Liq. Cryst.*, **1972**, *16*, 229.

107. (a) V. Yakhot, M. D. Cohen, and Z. Ludmer, *Adv. Photochem.*, **1979**, *11*, 489.
     (b) B. Blanzat, C. Barthou, N. Tercier, J. J. Andre and J. Simon, *J. Am. Chem. Soc.*, **1987**, *109*, 6193.

108. C. Piechocki, J. Simon, A. Skoulios, D. Guillon and P. Weber, *Mol. Cryst., Liq. Cryst.*, **1985**, *130*, 223 and refs cited therein.

109. B. Blanzat, C. Barthou, N. Tercier, J. J. Andre, and J. Simon, *J. Am. Chem. Soc.*, **1987**, *109*, 6193.

110. D. G. Guillon, A. Skoulios, C. Piechocki, J. Simon and P. Weber, *Mol. Cryst. Liq. Cryst.*, **1983**, *100*, 275.

111. (a) D. Markovitsi and I. Lecuyer, *Chem. Phys. Lett.*, **1988**, *149*, 330.

(b) The authors use s⁻¹ as the units for their second order rate constants. It is
assumed that 1 mol⁻¹ s⁻¹ were intended.

112. D. Markovitsi, T. H. Tran-Thi, V. Briois, J. Simon and K. Ohta, *J. Am. Chem. Soc.*, **1988**, *110*, 2001.

113. G. Blasse, G. J. Dirksen, A. Meijerink, J. F. van der Pol, E. Neelman, and W. Drenth, *Chem. Phys. Lett.*, **1989**, *154*, 420.

114. A. M. Levelut, *J. Chim. Phys.*, **1983**, *80*, 149.

115. M. F. Sonnenschein and R. G. Weiss in *Photochemistry on Solid Surfaces*, M. Anpo and T. Matsuura. eds., Elsevier, Amsterdam, 1989, p. 526.

116. (a) E. Sackmann and D. Rehm, *Chem. Phys. Lett.*, **1970**, *4*, 537.

(b) H. Beens, H. Mohwald, D. Rehm, E. Sackmann, and A. Weller, *Chem. Phys. Lett.*, **1971**, *8*, 341.

117. H. Stegemeyer, J. Hasse and W. Laarhoven, *Chem. Phys. Lett.*, **1987**, *137*, 516.

118. L. B. A. Johansson *Chem. Phys. Lett.*, **1985**, *118*, 516.

119. G. Baur, A. Stieb and G. Meier, *Mol. Cryst. Liq. Cryst.*, **1973**, *22*, 261.

120. (a) A. Arcioni, R. Tarroni, and C. Zannoni, in Ref. 45, p. 421.

(b) L. B.-A. Johansson and L. D. Bergelson, *J. Am. Chem. Soc.*, **1987**, *109*, 7374.

(c) L. L. Chapoy, and D. B. DuPre, *J. Chem. Phys.*, **1979**, *70*, 2550.

121. A. Arcioni, F. Bertinelli, R. Tarroni, and C. Zannoni, *Mol. Phys.*, **1987**, *61*, 1161.

122. J. P. Riehl, and F. S. Richardson, *Chem. Rev.*, **1986**, *86*, 1.

123. K. J. Mainusch, and H. Stegemeyer, *Ber. Bunsenges. Phys. Chem.*, **1974**, *78*, 927.

124. M. Sisido, K. Takeuchi, and Y. Imanishi, *J. Phys. Chem.*, **1984**, *88*, 2893.

125. M. Sisido, K. Kawaguchi, K. Takeuchi, and Y. Imanishi, *Mol. Cryst. Liq. Cryst.*, **1988**, *162B*, 263.

126. V. C. Anderson, B. B. Craig, and R. G. Weiss, *J. Am. Chem. Soc.*, **1981**, *103*, 7169.

127. V. C. Anderson, B. B. Craig, and R. G. Weiss, *J. Am. Chem. Soc.*, **1982**, *104*, 2972.

128. I. B. Berlman, *Handbook of Fluorescence Spectra of Aromatic Molecules*, 2nd ed., Academic Press, New York, 1971.

129. M. Sisido, K. Tacheuchi, and Y. Imanishi, *Chem. Lett.*, **1983**, 961.

130. M. Sisido, X. F. Wang, K. Kawaguchi, and Y.Imanishi, *J. Phys. Chem.*, **1988**, *92*, 4797.

131. K. Kawaguchi, M. Sisido, and Y. Imanishi, *J. Phys. Chem.*, **1988**, *92*, 4806.

132. The authors did not calculate activation energies for the bimolecular process.[131] The data in Figure 27 are too imprecise to allow more than a rough estimate of E_a. However, E_a (cholesteric) is significantly larger than E_a (isotropic).

133. M. Sisido, F. Wang, K. Kawaguchi, and Y. Imanishi, *J. Phys. Chem.*, **1988**, *92*, 4801.

134. K. Kawaguchi, M. Sisido, Y. Imanishi, M. Kiguchi and Y. Taniguchi, *Bull. Chem. Soc. Jpn.*, **1989**, *62*, 2146.

135. For instance, see: (a) P. Liang and J. K. Thomas, *J. Colloid Interface Sci.*, **1988**, *124*, 358.

(b) M. F. Sonnenschein and R. G. Weiss, *Photochem. Photobiol.*, **1990**, *51*, 539.

(c) R. M. Jenkins and R. G. Weiss, *Langmuir*, **1990**, *6*, 1408.

136. V. C. Anderson, B. B. Craig and R.G. Weiss, *Mol. Cryst. Liq. Cryst.*, **1983**, *97*, 351.

137. V. C. Anderson, B. B. Craig and R.G. Weiss, *J. Phys. Chem.*, **1982**, *86*, 4642.

138. V. C. Anderson and R. G. Weiss, *J. Am. Chem. Soc.*, **1984**, *106*, 6628.
139. (a) M. F. Sonnenschein, Ph.D. Thesis, Georgetown University, 1987.
    (b) M. F. Sonnenschein and R.G. Weiss, *J. Phys. Chem.*, **1988**, *92*, 6828.
140. K. Shimada and M. Szwarc, *J. Am. Chem. Soc.*, **1975**, *97*, 3313.
141. K. Shimada, Y. Schimozato and M. Szwarc, *J. Am. Chem. Soc.*, **1975**, *97*, 5834.
142. T. Kanaya, K. Goshiki, M. Yamamoto, and Y. Nishijima, *J. Am. Chem. Soc.*, **1982**, *104*, 3580.
143. M. Yamamoto, K. Goshiki, T. Kanaya and Y. Nishijima, *Chem. Phys. Lett.*, **1978**, *56*, 333.
144. J. E. Piercy and M.G. Rao, *J. Chem. Phys.*, **1967**, *46*, 3951.
145. In Table IV of Ref. 138, the activation parameters listed for P3P in isotropic CM are those calculated using 1-dodecylpyrene as the model. The value listed in Ref. 136 (1-ethylpyrene as the model) should be used.
146. J. E. Leffler, *Rates and Equilibria of Organic Reactions*, Wiley, New York, 1963, Chaps. 6 – 9.
147. (a) H. A. Kramers, *Physica*, **1940**, *7*, 284.
    (b) S. Chandrasekhar, *Rev. Mod. Phys.*, **1943**, *5*, 1.
148. (a) P. J. Flory, *Statistical Mechanics of Chain Molecules*, Wiley-Interscience, New York, 1969.
    (b) D. M. Golden, S. Furuyama and S. W. Benson, *Int. J. Chem. Kinet.*, **1969**, *1*, 57.
    (c) K. S. Pitzer, *Disc. Faraday Soc.*, **1951**, *10*, 66.
    (d) D. R. Lide, *J. Chem. Phys.*, **1958**, 29, 1426.
149. (a) A. Faljoni, K. Zinner, and R.G. Weiss, *Tetrahedron Lett.*, **1974**, 1127.
    (b) W. H. Laarhoven and T. J. H. M. Coppen, *J. Chem. Soc., Perkin II*, **1978**, 315.
150. In the extreme, the solvent may be in its crystalline phase. See for instance: a) H.C. Chang, R. Popovitz-Biro, M. Lahav and L. Leiserowitz, *J. Am. Chem. Soc.*, **1986**, *108*, 5648.
    (b) J. van Mil, L. Addadi, M. Lahav, W. J. Boyle and S. Sifniades, *Tetrahedron.*, **1987**, *43*, 1281.
    (c) Ref. 13.
    (d) M. Gracia-Garibay, J. R. Scheffer, J. Trotter and F. Wireko, *Tetrahedron Lett.*, **1987**, *28*, 4789.
    (e) B. S. Green, M. Lahav and D. Rabinovich, *Acc. Chem. Res.*, **1979**, *12*, 191.
    (f) M. Vaida, L. J. W. Shimon, J. van Mil, K. Ernst-Cabrera, L. Addadi, L. Leiserowitz, and M. Lahav, *J. Am. Chem. Soc.*, **1989**, *111*, 1029.
    (g) S. V. Evans, M. Garcia-Garibay, N. Omkaram, J. R. Scheffer, and J. Trotter, *J. Am. Chem. Soc.*, **1986**, *108*, 5648.
151. C. Eskenazi, J. F. Nicoud, and H. B. Kagan, *J. Org. Chem.*, **1979**, *44*, 995.
152. S. Ganapathy and R. G. Weiss in *Organic Phototransformations in Nonhomogeneous Media*, M. A. Fox, ed., American Chemical Society, Washington, D.C., 1985, Chap. 10.
153. K. Hayashi and M. Irie, *Jpn. Kokai,* 7818549, 20.2.78; *Chem. Abstr.*, **1978**, *89*, 6151j.
154. A. Moradpour, J. F. Nicoud, G. Balavoine, H. Kagan and G. Tsourcaris, *J. Am. Chem. Soc.*, **1971**, *93*, 2353.
155. M. Nakazaki, K. Yamamoto, K. Fujiwara and M. Maeda, *J. Chem. Soc., Chem. Commun.*, **1979**, 1086.
156. M. Nakazaki, K. Yamamoto and K. Fujiwara, *Chem. Lett.*, **1978**, 863.
157. M. Hibert and G. Solladie, *J. Org. Chem.*, **1980**, *45*, 5393.

158. E. G. Cassis, Jr., and R. G. Weiss, *Photochem. Photobiol.*, **1982**, *35*, 439.
159. G. A. Puchkovskaya, Yu. A. Reznikov, and O. V. Yaroshchuk, *Ukr. Fiz. Zh.* (Russ.), **1989**, *34*, 1036; *Chem. Abstr.* **1989**, *111*, 163983x.
160. (a) P. Deval and A. K. Singh, *J. Photochem. Photobiol., A, Chem.*, **1988**, *42*, 329.
     (b) A. K. Singh, personal communication.
161. See for instance: N. J. Turro, D. R. Anderson, M. F. Chow, C. J. Chang and B. J. Kraeutler, *J. Am. Chem. Soc.*, **1981**, *103*, 3892.
162. D. A. Hrovat, J. H. Liu, N. J. Turro and R. G. Weiss, *J. Am. Chem. Soc.*, **1984**, *106*, 5291.
163. N. J. Turro and J. Matay, *J. Am. Chem. Soc.*, **1981**, *103*, 4200.
164. (a) R. B. Bernstein, *J. Phys. Chem.*, **1952**, *56*, 893.
     (b) R. B. Bernstein, *Science*, **1957**, *126*, 119.
165. G. B. Sergeev, V. A. Batyluk, M. B. Stepanov and T. I. Shabatina, *Dokl. Akad. Nauk, SSSR, Chem. Sect. (Engl. Transl.)*, **1979**, *246*, 552.
166. (a) V. A. Batyuk, T. I. Shabatina, Yu. N. Morosov, and G. B. Sergeev, *Mol. Cryst. Liq. Cryst.*, **1988**, *161*, 109.
     (b) A. Batyuk, T. I. Shabatina, Yu. N. Morosov and G. B. Sergeev, *Dokl. Phys. Chem.*, **1988**, *300*, 411.
167. V. A. Batyuk, T. I. Shabatina, and G. B. Sergeev, *Mol. Cryst. Liq. Cryst.*, **1989**, *166*, 105.
168. (a) P. J. Wagner, *Acc. Chem. Res.*, **1971**, *4*, 168.
     (b) N. J. Turro, J. C. Dalton, K. Dawes, G. Farrington, R. Hautala, D. Morton, M. Niemczyk, and N. Shore, *Acc. Chem. Res.*, **1972**, *5*, 92.
     (c) P. J. Wagner, *Acc. Chem. Res.*, **1983**, *16*, 461.
     (d) N. C. Yang and S. P. Elliot, *J. Am. Chem. Soc.*, **1969**, *91*, 7550.
     (e) L. M. Stephenson, P. R. Cavigli and J. L. Parlett, *J. Am. Chem. Soc.*, **1971**, *93*, 1984.
169. R. G. Weiss, R. L. Treanor and A. Nunez, *Pure Appl. Chem.*, **1988**, *60*, 999.
170. W. J. Leigh in *Photochemistry on Solid Surfaces*, M. Anpo and T. Matsuura, eds., Elsevier, Amsterdam, 1989, p. 481.
171. J. M. Nerbonne and R. G. Weiss, *Isr. J. Chem.*, **1979**, *48*, 266.
172. D. M. Hrovat, J. H. Liu, N. J. Turro and R. G. Weiss, *J. Am. Chem. Soc.*, **1984**, *106*, 7033.
173. L. Salem and C. Rowland, *Angew. Chem. Intern. Ed. Engl.*, **1972**, *11*, 92.
174. Z. He and R. G. Weiss, *J. Am. Chem. Soc.*, **1990**, *112*, 5535.
175. J. A. Slivinskas and J. E. Guillet, *J. Polym. Sci. Polym. Chem. Ed.*, **1973**, *11*, 3043.
176. (a) R. L. Treanor and R. G. Weiss, *J. Am. Chem. Soc.*, **1986**, *108*, 3137.
     (b) R. L. Treanor and R. G. Weiss, *Tetrahedron*, **1987**, *43*, 1371.
177. (a) F. J. Golemba and J. E. Guillet, *Macromolecules*, **1972**, *5*, 63.
     (b) M. V. Encina and E. A. Lissi, *J. Photochem.*, **1976/1977**, *6*, 173.
     (c) M. V. Encina, A. Nogales and E. A. Lissi, *J. Photochem.*, **1975**, *4*, 75.
178. (a) A. Nunez and R. G. Weiss, *J. Am. Chem. Soc.*, **1987**, *109*, 6215.
     (b) A. Nunez, and R. G. Weiss, *Bol. Socieda Chilena de Quimica*, **1990**, *35*, 3.
179. (a) M. Maroncelli, H. L. Strauss and R. G. Snyder, *J. Chem. Phys.*, **1985**, *82*, 2811.
     (b) M. Maroncelli, S. P. Qi, H. L. Strauss and R. G. Snyder, *J. Am. Chem. Soc.*, **1982**, *104*, 6237.
180. (a) A. A. Schaerer, C. J. Busso, A. E. Smith and L. B. Skinner, *J. Am. Chem. Soc.*, **1955**, *77*, 2017.
     (b) A. Muller, *Proc. Roy. Soc. (London)*, **1932**, *138 A*, 514.
181. (a) R. L. Treanor and R. G. Weiss, *J. Am. Chem. Soc.*, **1988**, *110*, 2170.

(b) R. L. Treanor and R. G. Weiss, *J. Am. Chem. Soc.*, **1986**, *108*, 3137.
182. J. M. Vincent and A. Skoulios, *Acta Crystallogr.*, **1966**, *20*, 432, 441 and 447.
183. M. Bloom, J. H. Davis, and M. I. Valic, *Can. J. Phys.*, **1980**, *58*, 1510.
184. (a) R. G. Griffin, *Meth. Enzymol., Lipids*, **1981**, *72 D*, 108.
(b) K. R. Jeffery, T. C. Wong, and A. P. Tulloch, *Mol. Phys.*, **1984**, *52*, 289.
(c) I. C. P. Smith in *NMR of Newly Accessible Nuclei*, Vol. 2, P. Lazlo, ed., Academic Press, New York, 1983, Chap. 1.
(d) J. Charvolin and Y. Hendrikx in *Nuclear Magnetic Resonance of Liquid Crystals*, J.W. Emsley, ed., D. Reidel, Boston, 1985, Chap. 20.
185. D. J. Mitchell and B. W. Ninham, *J. Chem. Soc., Faraday Trans. 2*, **1981**, *77*, 601.
186. See for instance: (a) G. J. Brownsey and A. Leadbetter, *J. Phys. (Paris)*, **1981**, *42*, L-135.
(b) W. Haase and H. Paulus, *Mol. Cryst. Liq. Cryst.*, **1983**, *100*, 111.
(c) L. Pohl, R. Eidenschink, J. Krause, and G. Weber, *Phys. Lett.*, **1978**, *65 A*, 169.
187. P. E. Cladis, P. L. Finn, and J. W. Goodby in *Liquid Crystals and Ordered Fluids*, A. C. Griffin and J. F. Johnson, eds., Vol. 4, Plenum, New York, 1984, p. 203.
188. (a) B. M. Fung and M. Gangoda, *J. Am. Chem. Soc.*, **1985**, *107*, 3395.
(b) M. Gangoda and B. M. Fung, *Chem. Phys. Lett.*, **1985**, *120*, 527.
189. (a) W. J. Leigh, *J. Am. Chem. Soc.*, **1985**, *107*, 6114.
(b) W. J. Leigh, *Can. J. Chem.* **1986**, *64*, 1130.
190. R. G. Zimmermann, J. H. Liu and R. G. Weiss, *J. Am. Chem. Soc.*, **1986**, *108*, 5264.
191. W. J. Leigh and S. Jakobs, *Tetrahedron*, **1987**, *43*, 1393.
192. G. Aviv, J. Sagiv and A. Yogev, *Mol. Cryst. Liq. Cryst.*, **1976**, *36*, 349.
193. (a) K. S. Wei and R. Livingston, *J. Phys. Chem.*, **1967**, *71*, 541.
(b) D. O. Cowan and J. C. Koziar, *J. Am. Chem. Soc.*, **1975**, *97*, 249.
194. (a) J. M. Nerbonne and R. G. Weiss, *J. Am. Chem. Soc.*, **1978**, *100*, 2571.
(b) J. M. Nerbonne and R. G. Weiss, *J. Am. Chem. Soc.*, **1979**, *101*, 402.
195. J. E. Adams, W. Haas and J. J. Wysocki in *Liquid Crystals and Ordered Fluids*, J. F. Johnson and R. S. Porter, eds., Plenum Press, New York, 1970, p. 463.
196. The dearth of other studies like the photodimerization of **62** in **CHN–CHA** is not accidental. Experiments of this type are very difficult to perform since they require measurement of absolute intensitites of absorbed light at constant temperature by solutions that are prone to reflection.
197. T. Kunieda, T. Takahashi and M. Hirobe, *Tetrahedron Lett.*, **1983**, *24*, 5107.
198. R. Kleopfer and H. Morrison, *J. Am. Chem. Soc.*, **1972**, *94*, 255.
199. J. G. Otten, C. S. Yeh, S. Byrn, and H. Morrison, *J. Am. Chem. Soc.*, **1977**, *99*, 6353.
200. T. Nagamatsu, C. Kawano, Y. Orita and T. Kunieda, *Tetrahedron Lett.*, **1987**, *28*, 3263.
201. (a) Y. C. Lin and R. G. Weiss, *Macromolecules*, **1987**, *20*, 414.
(b) Y. C. Lin, Ph. D. Thesis, Georgetown University, Washington, D. C., 1987.
(c) Y. C. Lin, B. Kachar, and R. G. Weiss, *J. Am. Chem. Soc.*, **1989**, *111*, 5542.
202. Y. C. Lin and R. G. Weiss, *Liq. Cryst.*, **1989**, *4*, 367.
203. (a) Y. Tanaka and H. Tsuchiya, *J. Phys. (Les Ulis, Fr.)*, **1979**, *40*, C 3-41.
(b) Y. Tanaka, H. Tsuchiya, M. Suzuki, J. Takano and H. Kurihara, *Mol. Cryst. Liq. Cryst.*, **1981**, *68*, 113.
204. V. Ramesh and R. G. Weiss, *Mol. Cryst. Liq. Cryst.*, **1986**, *135*, 13.
205. V. Ramesh and R. G. Weiss, *J. Org. Chem.*, **1986**, *51*, 2535.
206. J. Bolt, F. H. Quina and D. G. Whitten, *Tetrahedron Lett.*, **1976**, 2595.

Appendix

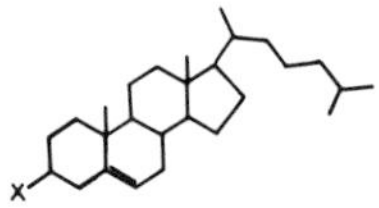

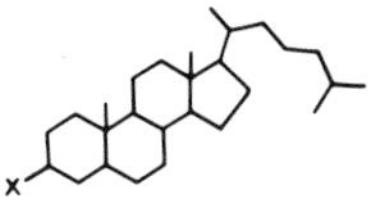

CCI X = -Cl
CL   X = -O₂C(CH₂)₁₁H
CN   X = -O₂C(CH₂)₈H
COC X = -O₂C(CH₂)₈CH=CH(CH₂)₈H
CLI X = -O₂C(CH₂)₇CH=CHCH₂CH=CH(CH₂)₅H
CO   X = -O₂C(CH₂)₇CH=CH(CH₂)₈H
CNB X = -O₂C-(C₆H₄)-NO₂
OMY X = -O₂C(CH₂)₁₃H

CHA X = -O₂CCH₃
CHN X = -O₂C(CH₂)₈H

H(CH₂)₄O₂C(CH₂)₁₇H

BS

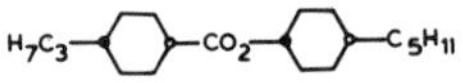

H₇C₃—◯—CO₂—◯—C₅H₁₁

OS-35

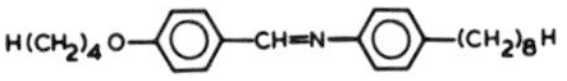

H(CH₂)₄O—◯—CH=N—◯—(CH₂)₈H

BBOA

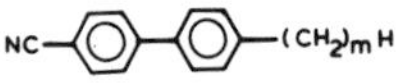

NC—◯—◯—(CH₂)ₘH

CNB 5 m = 5
CNB 6 m = 6

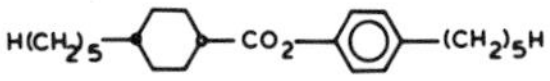

H(CH₂)₅—◯—CO₂—◯—(CH₂)₅H

5H5

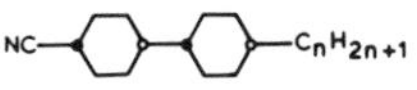

NC—◯—◯—CₙH₂ₙ₊₁

(n = 3,5,7 ; eutectic mix)
ZLI = 1167

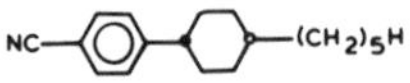

NC—◯—◯—(CH₂)₅H

PCH5

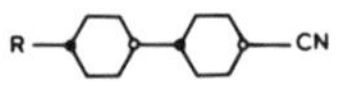

R—◯—◯—CN

ECCN R = C₂H₅-
BCCN R = C₄H₉-

H₃CO—◯—CH=N—◯—C₄H₉

MBBA

Chapter 15

Photoreactions in Monolayer Films and Langmuir-Blodgett Assemblies

Susan P. Spooner and David G. Whitten

Department of Chemistry, University of Rochester, Rochester, NY 14627

Contents

1. Introduction

The formation of films at an air–water interface has been known for centuries. The first scientific observation was recorded by Benjamin Franklin in 1774 in a report to the Royal Society of London. Franklin reported that a teaspoon of oil dropped on a pond spread out almost instantly to produce a calm over the entire pond. More than a hundred years later in 1891, Agnes Pockels reported the first use of a trough with movable barriers to control an oil film on water. Within the following two decades Rayleigh, Devaux, and Hardy made several individual observations of films at an air–water interface. They reported that the film was actually monomolecular, that it displayed properties of both a liquid and a solid, and that oils with polar groups spread differently from those without polar groups. However, it was not until 1917 that Irving Langmuir developed the theory of monolayer films that is the basis of our understanding of monolayer films today. Langmuir reported on the pressure–area relationship of molecules in compressed monolayer films and described their orientation at an air–water interface. Thus the term Langmuir films is often used to describe monolayer films at an air–water interface. Katharine Blodgett later developed the techniques for transferring monolayer films to solid substrates. Monolayer or multilayer assemblies formed using this method are known as Langmuir-Blodgett or L-B assemblies.[1,2]

2. Monolayer Films and Langmuir-Blodgett Assemblies
2.1. General Properties

Spreading of a water-insoluble amphiphile by solvent evaporation (dropwise addition of a solution of the amphiphile in a water-immiscible solvent) leads to a monolayer film that can be classified as gaseous, liquid, or solid depending upon the degree of compression and the effective area per molecule. A typical force–area isotherm measured during the compression is shown in Figure 1. Not surprisingly, the "liquid" and "solid" regions of the isotherm are strongly sensitive to molecular structure for pure materials and to composition for mixtures. Recent work by Arnett and co-workers[3-7] and Menger et al.[8,9] has demonstrated very elegantly the strong sensitivity of monolayer film behavior to stereochemistry and the direct correlation between molecular shape and specific isotherm characteristics. Clearly the liquid phase of a monolayer film and, more so, the solid represent constrained environments for individual molecules of amphiphiles. This is particularly true for mixtures where, depending upon the degree of interdigitation or interpenetration, the area per molecule can be significantly reduced from that in a monolayer film of the pure substance.

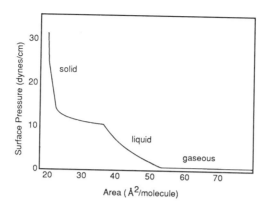

Figure 1. Typical surface pressure–area isotherm obtained upon compression of a fatty acid monolayer film at the air–water interface.

The formation of a L-B assembly by stepwise transfer of monolayer film to a rigid support can generally be accomplished only for those monolayer films which compress to a tightly packed solid phase. Although it is frequently assumed that the transfer of a monolayer film from the air–water interface to a rigid support is accomplished with minimal reorganization of the molecular rearrangement of the monolayer film, this is certainly not always the case. Transfer with little or no rearrangement is of course most likely for those monolayer films which show steep isotherms and are transferred at relatively high compression and give "transfer ratios" close to unity. The usual schemes for construction of L-B assemblies by sequential transfer of monolayer films is outlined in Figure 2.

2.2. Characterization of Langmuir-Blodgett Assemblies

A number of techniques have been used to investigate the structure and periodicity of L-B assemblies. These films are formed from compressed monolayer films of fatty acids over subphases of various metal cations at pH values where the fatty acid should be ionized. x-Ray diffraction studies, for example, indicate a near-crystalline arrangement of metal ions in the "lattice" formed by the monolayer film. Among the many techniques that have been used to develop a picture of monolayer film structure are infrared spectroscopy, ellipsometry, area per molecule calculations, electron diffraction, neutron diffraction, electron spin resonance, polarized resonance Raman spectroscopy, coherent anti-Stokes Raman spectroscopy, and optical microscopy.[10-13]

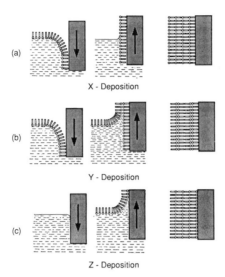

(a)

X - Deposition

(b)

Y - Deposition

(c)

Z - Deposition

Figure 2. Schematic representation of the various deposition modes for monolayer films and the resulting L-B assemblies. (Reprinted with permission from Kuhn et al.[14] Copyright 1972 John Wiley & Sons, Inc.)

Langmuir-Blodgett assemblies provide several advantages over solution's for the study of photochemical reactions. *Since L-B assemblies contain little or no solvent, their photochemical behavior is often less complicated than that observed in solution. In L-B assemblies, molecules are held in a rigid, well-defined geometry, thus aiding in the elucidation of photoreaction mechanisms.* Monolayer techniques enable the construction of assemblies in which the molecules reside in known locations and intermolecular distances may be readily varied for layers of differing composition. Since deposition occurs in steps, the interlayer distances may be controlled by varying the number of spacer layers, the surfactant chain length within spacer layers, or the mode of deposition. Thus, *L-B assemblies are ideal systems in which to study the distance dependence of energy and electron transfer.* Since L-B assemblies also resemble complex biological membranes, they are good models for studying processes that occur both within and at membrane interfaces.

3. Energy Transfer
3.1. Long-Range Singlet–Singlet Transfer with Cyanine Dyes and Dye Aggregates

It was not until the early 1960s that Hans Kuhn and co-workers[14-25] used monolayer techniques to construct highly organized L-B assemblies to study energy-transfer processes. They found that mixtures of long–chain surfactant cyanine dyes and aliphatic surfactants such as arachidic acid (C_{20}) formed stable monolayer films at the air–water interface. From surface pressure–area isotherms, it was concluded that the cyanine dye molecules were tightly packed in the monolayer film and the long axis of the dye was oriented parallel to the water surface. For the same L-B assemblies supported on glass substrates, absorption spectra indicated that the relative proportions of dye monomer, dimer, and aggregate in the L-B assemblies depended on the dye/surfactant ratio (see Figure 3).

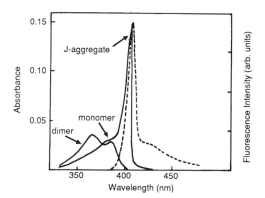

Figure 3. Absorption spectrum (———) of a pure cyanine dye (**3**) L-B assembly indicating the presence of monomers and dimers. Absorption spectrum (———) of J-aggregated **3** in a 1:1 molar ratio of **3**:hexadecane and corresponding J-aggregated fluorescence spectrum (-----). (Reprinted with permission from Möbius.[22] Copyright 1978 Verlag Chemie.)

Using other nonsurfactant additives, i.e., octadecane, J-aggregates were formed that exhibited sharp, red-shifted absorption.[26] These J-aggregates are composed of large arrays of densely packed molecules with their transition dipole moments aligned head to tail in a "brickstone work" arrangement.[27] Strong dipole–dipole interactions in the excited state lead to the formation of an exciton band.[28-30] The allowed transition to this exciton band is governed by the tilt angle, α, of the transition moments with respect to the line of centers as shown in Figure 4.[28-33] When α is less than approximately 54°, the allowed transition is to the lowest excited state level, resulting in the narrow, red-shifted absorption relative to that of the monomer which is characteristic of J aggregates.[34,35] When α is greater than approximately 54°, the allowed transition is to the highest excited state level, resulting in the blue-shifted absorption characteristic of H aggregates.[36] Recently, electron diffraction experiments were performed on monolayers of J-aggregated surfactant cyanine dye (**1**) and hexadecane in a 1:1 molar ratio.[37]

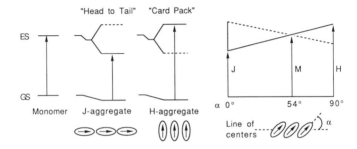

1

The crystallographic structure obtained supports the "brickstone" model for J-aggregates of **1** proposed by Bücher and Kuhn[27] more than 20 years ago, with two nonidentical molecules per unit cell.

Figure 4. Schematic representation of the allowed excited-state transitions for J-aggregated and H-aggregated molecules relative to that of the monomer. The allowed transition is dictated by α, the tilt angle of the transition moments with respect to the line of centers.

Since excited cyanine J-aggregates behave as a large array of coupled resonant oscillators, energy can migrate readily within such assemblies. Möbius and co-workers[22-25] found that a cyanine dye with a lower energy absorption band can act as a an energy trap in a J-aggregated assembly. They observed that the incorporation of a thiacyanine (low-energy) dye (**2**) into a single monolayer of J-aggregated oxacyanine (high-energy) dye (**3**) resulted in the efficient quenching of the oxacyanine fluorescence even at an oxacyanine–thiacyanine ratio of 50,000:1.

2   **3**

In a similar experiment, the J-aggregated donor **3** and acceptor **2** were located in adjacent layers with both hydrophilic chromophores in contact. The quenching of J-aggregated donor fluorescence was measured against the average distance between acceptor molecules (see Figure 5). A tenfold higher concentration of acceptor in an adjacent layer is required to duplicate the same donor quenching effect observed when the acceptor is present within the same monolayer.[21,22,25] Thus excited J-aggregates can act as antennae for isolated traps within the same monolayer film. This extremely efficient energy funnelling to lower energy trap sites observed in these mixed monolayer systems parallels the energy-harvesting cascade utilized during photosynthesis.

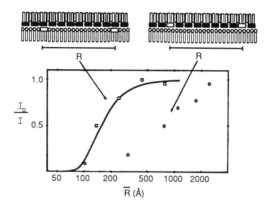

Figure 5. Plot of fluorescence intensity of J-aggregated donor cyanine dye **3** versus average distance, R, between acceptor dye molecules **2**. (O) acceptor incorporated in the same layer; (□) acceptor in an adjacent layer; solid line represents the theoretical curve obtained based on Förster theory. (Reprinted with permission from Möbius.[22] Copyright 1978 Verlag Chemie.)

In an example that elegantly demonstrated the distance dependence of energy transfer in L-B assemblies, Kuhn and co-workers[14-18] used monolayer techniques to vary the distance between donor and acceptor molecules. A monolayer of surfactant oxacyanine donor **3** was separated from a monolayer of surfactant carbooxacyanine acceptor **4** by varying numbers of arachidic acid (C_{20}) spacer monolayers, where each arachidic acid (**AA**) monolayer represents a distance of approximately 27 Å.

The donor **3** absorbs uv light and fluoresces in the blue, while the acceptor **4** absorbs light in the blue and fluoresces in the yellow. A glass slide was divided into three sections as shown in Figure 6. Section 1 consisted of the donor layer separated from the acceptor by one **AA** layer. Section 2 consisted of the donor and the acceptor separated by five layers of **AA**. Section 3 contained only a single layer of the acceptor. The slide was irradiated with uv light, which was absorbed only by the donor. In section 1, the distance between the two molecules was only 50 Å and efficient energy transfer and yellow emission from the acceptor was observed. In section 2, the donor-acceptor distance was estimated to be 150 Å and blue fluorescence was observed, thus no energy transfer occurred. No fluorescence was observed from section 3 since the acceptor did not absorb light in this region.

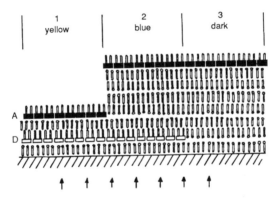

Figure 6. Schematic representation of the L-B assembly used in the study of the distance dependence of energy transfer. Donor = cyanine dye **3**; acceptor = cyanine dye **2**; spacer molecule = **AA**. (Reprinted with permission from Kuhn et al.[14] Copyright 1972 John Wiley & Sons, Inc.)

This experiment demonstrated the ease of studying the distance dependence of energy transfer in L-B assemblies that would otherwise require tremendous synthetic effort to obtain similar results from covalently linked chromophores. Several more detailed studies of a variety of donor–acceptor cyanine dye pairs in monolayers were performed. From a plot of the fluorescence intensity of the donor (I_d) vs. distance (d) between the donor and the acceptor layer, a curve fitting a fourth-power distance dependence was obtained (see Figure 7). The distance dependence was found to follow Förster theory for dipole–dipole energy transfer using a modified two-dimensional model as expressed in the following equation:

$$I_d/I = [\, 1 + (d_0/d)^4 \,]^{-1}$$

where I_d is the fluorescence intensity of the donor at distance d, I is the fluorescence intensity of the donor in the absence of acceptor and d_0 is the half quenching distance. From d_0, the quantum yield of luminescence of the donor can be calculated indirectly.[38]

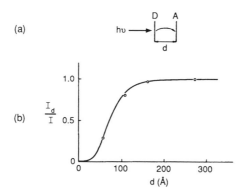

Figure 7. (a) Schematic representation of donor **3** and acceptor **2** separated by distance d. (b) Plot of fluorescence intensity of **3** versus d. (Reprinted with permission from Kuhn and Möbius[18] Copyright 1971 Verlag Chemie.)

They also demonstrated that a ternary system followed the same fourth-power distance dependence through experiments using a donor D, acceptor A_1, and acceptor A_2. It was shown that excitation energy initially transferred from D to A_1 was transferred subsequently to A_2.

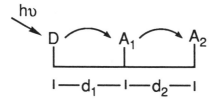

3.2. Energy Transfer with Hydrophobic Chromophores and Other Aggregates

In contrast to the above study involving J-aggregated dyes, Heesemann[39,40] examined monolayers formed from donor and acceptor para-substituted stilbenes and azobenzenes. The substituted azobenzene derivatives **5** and **6** clearly exhibit absorption that is characteristic of H-aggregates.

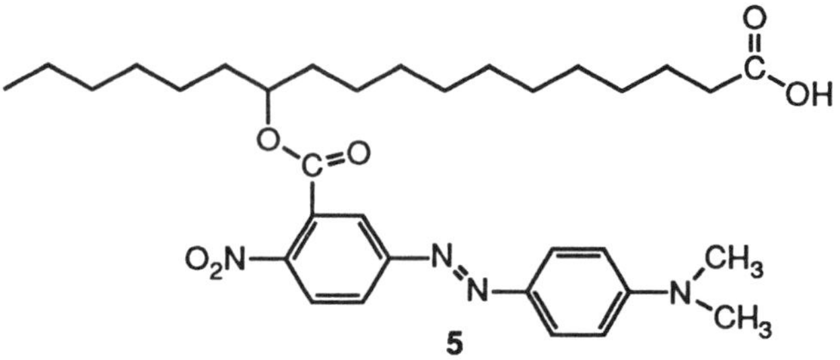

5

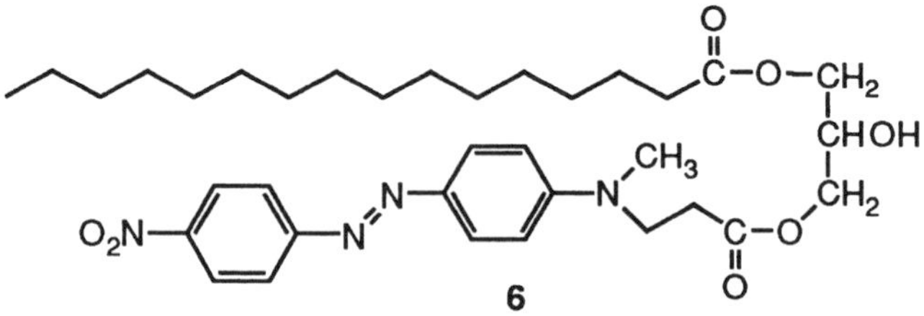

6

H-aggregates are often formed when several monomers are closely assembled in a "card pack" array (see Figure 4). Characteristics of H-aggregates are blue-shifted absorption, red-shifted fluorescence, and a long fluorescence lifetime relative to that of the monomer in solution.[36] Surface pressure–area isotherms of surfactant-linked azobenzenes (see below) suggest that the long transition moments of these molecules should be oriented nearly perpendicular to the surface of the water in the monolayer film and in a "card pack" array. Polarization behavior and the appearance of a blue-shifted absorption band indicated that these molecules form H-aggregates in L-B assemblies. It was difficult, however, for Heesemann to obtain further information about the excited states of these molecules since little or no fluorescence was observed. Nakahara and Fukada[41,42] also observed H-aggregate formation from 4,4'-disubstituted surfactant azobenzenes in L-B assemblies.

Vincett and Barlow[43] constructed monolayers of 9,10-substituted anthracene derivatives **7a– 7d** with polar head groups and alkyl chains.

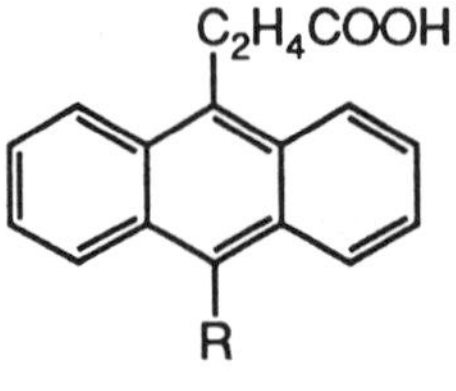

7 a	R = C_4H_9
7 b	R = C_6H_{13}
7 c	R = C_8H_{17}
7 d	R = $C_{12}H_{25}$

From absorption spectra, pressure–area isotherms and x-ray diffraction measurements of these monolayer films it is concluded that the molecules are highly organized, with their long transition moments aligned parallel to each other and slightly tilted relative to the layer plane. The excited states of these molecules are best described as "excimers" due to their broad structureless red-shifted fluorescence. Another study by Matsuki and Fukutome[44] involved a system of anthracene esters attached to surfactant molecules so that the anthracene moiety was aligned parallel to the long-chain surfactant. Surface pressure–area isotherms confirmed that the long axis of the anthracene was oriented parallel to the surfactant alkyl chain. Blue shifted-absorption suggested H-aggregate formation, while the emission indicated the presence of a mixture of monomer and aggregate.

Whitten and co-workers[45-47] studied H-aggregate formation in monolayers composed of surfactant *trans*-stilbenes **8, 9a,** and **9b.**

$CH_3(CH_2)_3$—⟨ ⟩—CH=CH—⟨ ⟩—$(CH_2)_5COOH$

8

⟨ ⟩—CH=CH—⟨ ⟩—$(CH_2)_nCOOH$

9 a n = 9
9 b n = 11

The absorption and fluorescence spectra of these stilbene derivatives are shifted in L-B assemblies when compared to monomers in solution, as shown in Figure 8. These molecules were shown to form H-aggregates differing from the ones discussed earlier in that they showed a structured aggregate fluorescence with no evidence of monomer emission. Studies of energy transfer from excited H-aggregates to surfactant cyanine dyes in L-B assemblies were performed. The surfactant thiacyanine **2** was chosen as the energy acceptor since its absorption maximum at 430 nm overlapped well with the fluorescence of the stilbenes. The distance between the stilbene donor and the acceptor was varied by up to five layers of **AA**. At short distances efficient energy transfer was observed, while at longer distances only stilbene fluorescence was seen. A plot of the distance dependence on the fluorescence of **8** or **9a** and **2** is shown in Figure 9. A half quenching distance, d_0, of 85 Å was obtained from this plot. Using methods derived by Kuhn and co-workers,[14-16] the "theoretical" half-quenching distance of monomeric stilbene was calculated to be 69 Å.

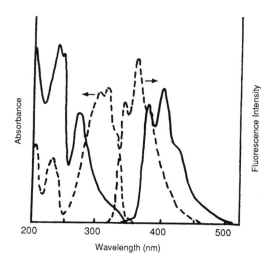

Figure 8. Comparison of the solution (----) and L-B assembly (——) absorption and fluorescence spectra for surfactant stilbene derivative **8**. (Reprinted with permission from Mooney et al.[45] Copyright 1984 American Chemical Society.)

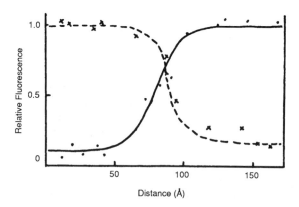

Figure 9. Plot of relative fluorescence intensity of stilbene donors **8** or **9a** (——) and cyanine dye acceptor **2** (-----) in L-B assemblies versus distance between donor and acceptor layers. (Reprinted with permission from Mooney and Whitten.[46] Copyright 1986 American Chemical Society.)

Thus Whitten and co-workers concluded that the surfactant stilbenes underwent relatively efficient energy transfer at distances of up to 100 Å. Further experiments with L-B assemblies of H-aggregated stilbene in contact with a single layer of quencher 2 indicated that little energy transfer occurred between adjacent layers of the same stilbene. This is attributed to the H aggregate stilbenes being reasonable donors of energy but poor acceptors due to a low oscillator strength of the long wavelength transition. Although the stilbenes do not function as antennae their long singlet lifetime and broad fluorescence make them efficient long-distance excitation energy donors.

Recently, several groups have reported energy transfer in L-B assemblies and monolayer films.[48-53] Fujihira et al.[51] studied the distance dependence of energy transfer between a mixture of two substituted pyrene derivatives 10a and 10b and their photooxidative products.

$$CH_3(CH_2)_{n-1} \text{—} \underset{\text{(pyrene)}}{} \text{—} (CH_2)_{m-1}COOH$$

10a   n = 6,   m = 10
10b   n = 10,   m = 6

They observed Förster-type energy-transfer quenching of the mixed pyrene layer by the pyrene photooxidation product layer. Another study by Tamai et al.[52] involved energy transfer in stearic acid (C_{18}) bilayers containing the surfactant carbazole donor 11 and surfactant anthracene acceptor 12.

$$\underset{\textbf{11}}{\text{(carbazole)} N-(CH_2)_{10}COOH} \qquad \underset{\textbf{12}}{\text{(anthracene)} COO(CH_2)_{15}COOH}$$

They investigated mixed L-B assemblies of dilute donor (1.5 mol%) with various acceptor concentrations (1.5–8.5 mol%). At high acceptor concentrations (>5 mol%), Förster-type energy transfer was observed. However, at low acceptor concentrations, electron transfer from 11 to the carboxyl group of stearic acid was competitive with energy-transfer fluorescence quenching in the mixed L-B assembly.

3.3. Distance Dependence of Singlet–Singlet Energy Transfer. Experiment and Theory

Studies of energy transfer between a J aggregated merocyanine donor and a surfactant Crystal Violet derivative acceptor in L-B assemblies were performed by Kuhn and co-workers.[54] By monitoring the fluorescence quenching of the donor as a function of the distance between the donor and the acceptor, it was observed that energy transfer occurred far more efficiently from J-aggregated dye than from a monomeric donor. The efficiency of energy transfer from the aggregated donor was found to decrease with the second power of the distance, as previously predicted for a donor in which excitation was extensively delocalized.[19]

In contrast, Penner[55] studied the distance dependence of energy transfer in a system where the donor and acceptor were both J-aggregated surfactant cyanine dyes. The efficiency of energy transfer was found to decrease with the fourth power of the distance, as expected for localized donor excitation, instead of the quadratic dependence which would be predicted for extensively delocalized excitation.[19] In this experiment, however, the acceptor was J aggregated whereas in Kuhn's experiment the acceptor was not. It was suggested that the observed fourth-power dependence was caused by strong resonance coupling between the excited J-aggregated donor and the acceptor, thus resulting in efficient energy transfer between donor and acceptor over distances comparable to that of the energy delocalization in the J aggregate.

Fromherz and Reinbold[56] also studied the distance dependence of energy transfer in L-B assemblies using surfactant cyanine dyes identical to those of Kuhn and co-workers. By varying the spacer layer composition from myristic acid (C_{14}) up to lignoceric acid (C_{24}), they were able to obtain several more points on the distance dependence curve than in Kuhn's studies. Their experimental results exhibited a slight deviation from the fourth-power dependence observed by Kuhn and coworkers. No explanation for this anomalous result was presented. (Multi-exponential fluorescence decay lifetimes could explain such deviation from fourth-power-law behavior.)

Several other experiments also have demonstrated that energy transfer does not follow Förster theory.[57-60] Leitner et al.[57] studied energy transfer in L-B assemblies of surfactant tetramethylindocarbocyanine (13) and tetramethyl-indodicarbocyanine (14) by standard continuous wave and picosecond time-resolved fluorescence measurements. They studied the concentration dependence of the donor 13 on the fluorescence decay curve in a single monolayer and the effect of a nonradiative acceptor 14 added to the same layer and to an adjacent layer. At low donor concentrations, radiationless energy transfer occurred which followed Förster theory energy transfer for two dimensions. At higher donor concentrations, the monomer fluorescence was quenched by higher order aggregates, resulting in a decay curve that did not follow Förster theory. Recently, several groups have reported similar fluorescence quenching of monomers by higher order aggregates in

L-B assemblies of surfactant pyrene,[58] carbazolyl,[59] rhodamine B,[60,61] and oxacyanine[61] derivatives in L-B assemblies.

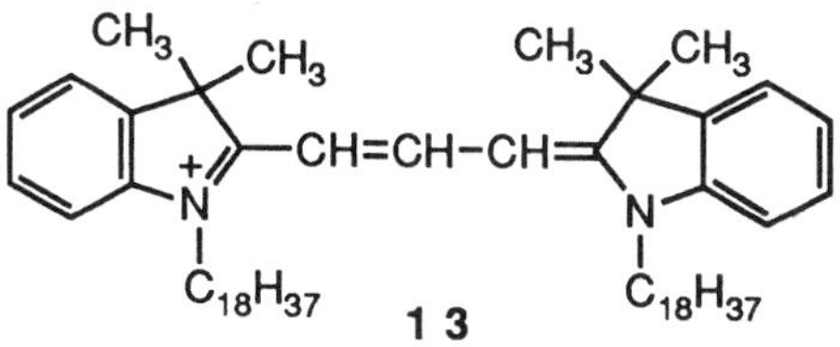

1 3

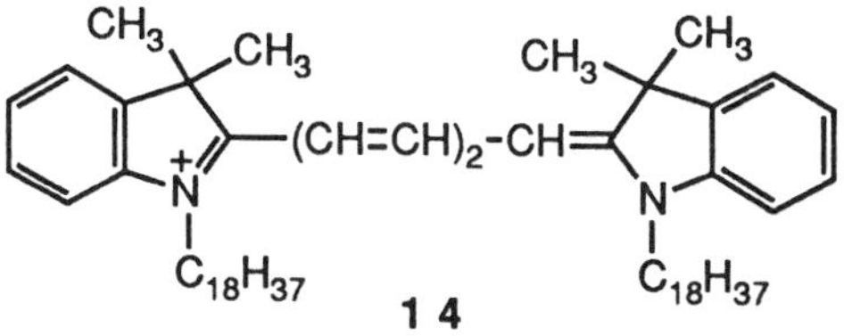

1 4

Since studies indicate that monomers and aggregates may transfer energy with different efficiencies, several experiments have focused on controlling aggregate formation in L-B assemblies. It was demonstrated by Ito et al.[62,63] that monolayers containing predominantly monomer could be formed using polymer L-B films of poly(vinyl octal). Poly(vinyl octal) was derivatized with the following chromophores: fluorene (**15a**), naphthalene (**15b**), phenanthrene (**15c**), and anthracene (**15d**), to form substituted poly(vinyl octal) polymer.

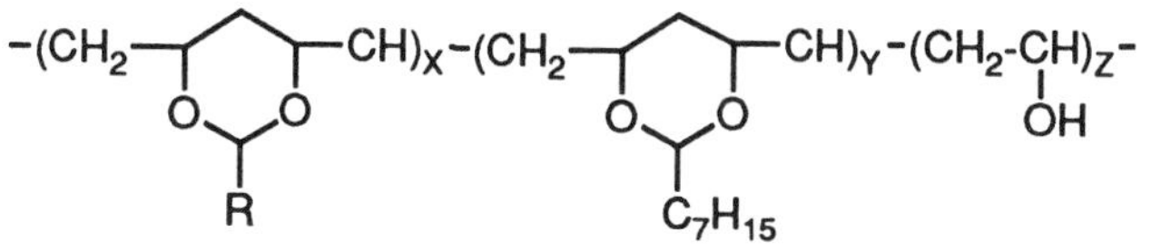

Poly(vinyl octal) polymer

where R =

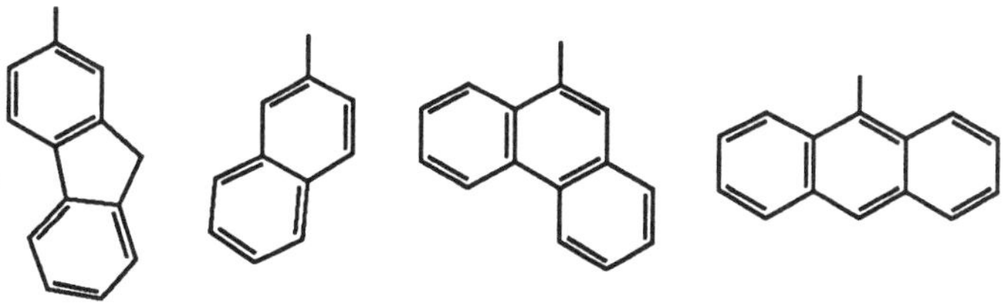

1 5 a   **1 5 b**   **1 5 c**   **1 5 d**

Energy-transfer studies were performed in L-B assemblies of these polymers where a layer containing **15b** was separated from a **15c** layer by a varying number of poly(vinyl octal) layers. The results indicated that the efficiencies of energy transfer could be controlled by varying the distance between the donor and acceptor molecules in these polymer L-B assemblies. They also demonstrated that a ternary system of poly(vinyl octal) with **15a, 15b**, and **15d** in consecutive layers exhibited stepwise energy transfer.

Kimizuka and Kunitake[64] have demonstrated that the extent of aggregation of cationic surfactant naphthalene derivatives in monolayer films can be controlled by adding polyions, such as dextran sulfate (DEX) or carboxymethylcellulose (CMC), in the subphase. The most condensed films were formed on a pure water subphase, whereas films formed on DEX- or CMC-containing subphases were more expanded, with the film formed on CMC being the most expanded. Fluorescence studies indicated that the monolayer films formed on pure water and subphases containing DEX exhibited mainly monomeric naphthalene fluorescence, while films formed on subphases containing CMC exhibited emission from naphthalene dimers and second excimers. Thus it appears that the naphthalene molecules aggregate differently depending on the subphase additives. Monolayer films were also formed from a mixture of surfactant naphthalenes and 1% cationic surfactant anthracene derivatives on subphases containing DEX or CMC. Upon irradiation of the naphthalene unit in the monolayer on the subphase containing DEX, anthracene emission was observed indicating that energy transfer occurred from the naphthalene donor to the anthracene acceptor. Irradiation on the naphthalene units of the monolayer film on the subphase containing CMC resulted in only naphthalene dimer and second excimer emission. Thus by altering the aggregation of naphthalene molecules in monolayer films using polyion complexation, energy transfer to anthracene molecules can be controlled.

On the other hand, several groups have investigated methods to enhance aggregate formation.[65-80] In particular, studies have focused on encouraging the aggregation of cyanine dyes and merocyanine dyes because of their potential as sensitizers in photographic processes.[81,82] Kawaguchi and Iwata[71,72] have found that the structure of J-aggregated merocyanine dye (**16**) could be influenced by varying the counterion used in the subphase. This is attributed to counterion chelation between the carboxylate and carbonyl group of the merocyanine dye (**17**). This chelation stabilizes the delocalized structure of the dye, forming various J-aggregates depending on the extent of stabilization. The extent of J-aggregate formation can also be controlled by varying other parameters such as the cosurfactant[65,66,73,74] and photoreactivity of the dye molecules.[77-80] Recently, cyanine dyes without long alkyl chains were found to form J-aggregates at an air–water interface when either adsorbed from the subphase onto a monolayer film of oppositely charged surfactants[83-85] or spread with oppositely charged cosurfactants.[86-89]

1 6   **1 7**

3.4. Interfacial Energy Transfer between Assembly-Bound Reagents and Reactive Supports

Fromherz and Arden[90,91] investigated energy transfer where the donor and acceptor were located in the same layer coated on a semiconductor surface (indium–tin oxide). The donor cyanine dye acts as a molecular antenna to absorb light and transfer excitation energy to the acceptor cyanine. The excited acceptor then acts as a reaction center by injecting an electron into the semiconductor. The electrons injected into the semiconductor produce a photocurrent. Fromherz and Arden later developed a "pH-modulated energy-electron transfer device", which consisted of a layer of a pH-sensitive surfactant coumarin donor molecule (18) as the antenna and a layer of surfactant cyanine acceptor (4) as reaction centers coated on a semiconductor surface.[92]

1 8 $C_{17}H_{35}$ **1 9** $C_{17}H_{35}$

When **18** that is in contact with an electrolytic solution is deprotonated, it can act as an energy donor; however, it cannot donate energy when protonated. The system can be switched off and on by changing the pH of the electrolytic solution. The excited deprotonated coumarin **19** can efficiently transfer energy to **4**, which then injects an electron into the conduction band of the semiconductor. The oxidized cyanine dye (**4**) is later regenerated by electron transfer from an electron donor in the electrolyte resulting in the generation of a photocurrent (see Figure 10). Sato et al.[93] also studied sensitized electron transfer with donor and acceptor dyes coated on a semiconductor electrode. The acceptor molecules were layered directly onto the semiconductor and the distance dependence of energy transfer was studied by varying the position of the donor layer. They observed an enhancement in the photocurrent when the donor dye was separated from the acceptor dye by more than two layers.

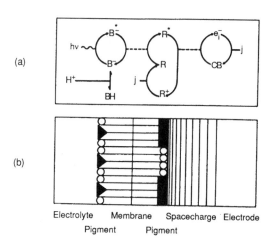

(a)

(b)

Electrolyte   Membrane   Spacecharge Electrode
     Pigment   Pigment

Figure 10. Schematic representation of the mechanism (a) and structure (b) of a "pH-modulated energy-electron transfer device." B^-/BH = deprotonated coumarin **19**/protonated coumarin **18**; R/R^+ = cyanine dye **4**/oxidized **4**. (Reprinted with permission from Fromherz and Arden[91] Copyright 1980 Verlag Chemie.)

3.5. Energy Transfer in Complex Langmuir-Blodgett Assemblies

Yamazaki et al.[94,95] have compared energy transfer that occurs in a biological photosynthetic antenna and energy transfer in L-B assemblies. They designed L-B assemblies containing carbazole (**11**) and three different cyanine dyes (**3**, **2**, and **13**), as energy donor (D) and acceptors (A_1, A_2, and A_3), respectively, such that vectorial energy transfer similar to the photosynthetic light-harvesting antennae in green plants could occur. Three different L-B assemblies were designed as shown in Figure 11. The order of dyes were chosen such that vectorial energy transfer could occur from $D \rightarrow A_1 \rightarrow A_2 \rightarrow A_3$. They observed that the vectorial energy transfer studied in L-B assemblies was similar to that of the biological system with respect to the fluorescence rise decay times and the rate constant of energy transfer. However, the transfer efficiency is higher in biological antenna than that in the L-B assembly; the efficiency in each step is 0.9 in phycobilisomes whereas it is 0.5–0.8 in L-B assemblies.[94] They attributed the lower efficiency in L-B assemblies to the loss of monomer excitation energy through energy traps composed of dimers and higher order aggregates dispersed throughout the layers.

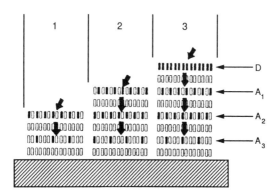

Figure 11. Schematic representation of the L-B assemblies used to study vectorial energy transfer. (Reprinted with permission from Yamazaki et al.[94] Copyright 1988 American Chemical Society.)

Recently, Spooner and Whitten[96] have constructed L-B assemblies consisting of the following alkyl-substituted surfactant *trans*-stilbene derivatives: **8, 9a, 9b, 20**, and **21** in a 1:1 molar ratio of mixed stilbenes to **AA**.

$$CH_3(CH_2)_5 \text{—} \bigcirc \text{—} \bigcirc \text{—} (CH_2)_3COOH$$

20

$$\bigcirc \text{—} \bigcirc \text{—} (CH_2)_{15}COOH$$

21

These mixed L-B assemblies exhibit similar H-aggregate absorption and emission as described earlier for L-B assemblies containing only a single type of *trans*-stilbene chromophore and **AA**. The fluorescence from these mixed stilbene L-B assemblies was strongly quenched by electron and energy acceptors. The addition of surfactant *trans*-stilbene carboxylate (**22**) into the mixed stilbene L-B assembly resulted in nearly complete fluorescence of **22**. Therefore **22** is an efficient localization site in these L-B assemblies due to its lower energy excited state relative to the stilbene derivatives lacking conjugation with the carboxylate. Electron transfer quenching in these mixed L-B assemblies, with and without **22**, has been investigated using a surfactant *trans*-stilbene carboxylate cobalt(III) complex (**23**) as the electron acceptor.

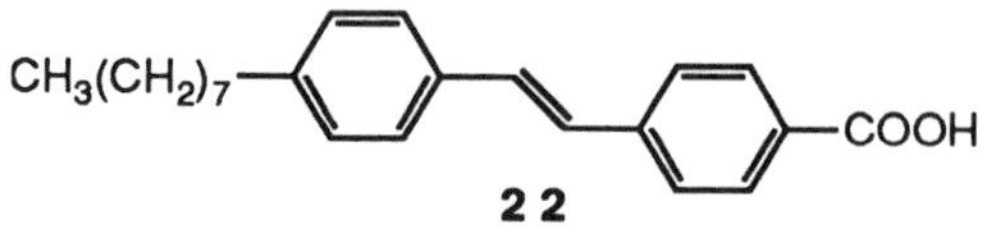

2 2

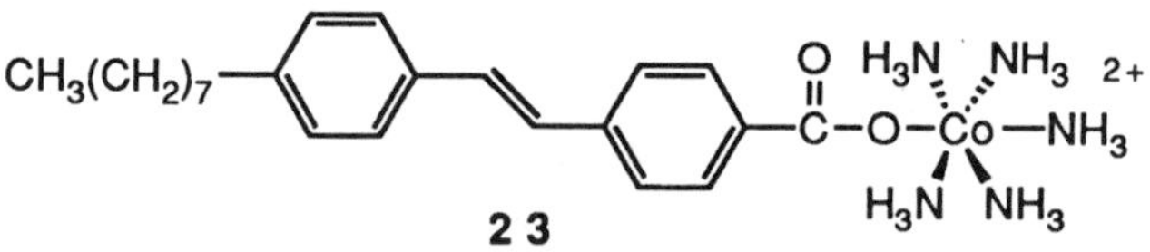

2 3

The fluorescence of a bilayer of the stilbene mixture with and without **22** was approximately 70% and 60% quenched, respectively, by the addition of an adjacent layer of **23**. Since previous studies of similar systems show a dramatic reduction in electron-transfer efficiency at donor-acceptor distances of 10–19 Å,[46,47] approximately a 30% quenching of stilbene fluorescence by electron transfer is predicted for a bilayer of mixed stilbenes. Thus the observed fluorescence quenching of 60–70% suggest that stilbene fluorescence quenching occurs through both direct electron transfer to **23** and interlayer energy transfer among the stilbene layers followed by electron transfer to **23**. Similar studies were investigated with five layers of the stilbene mixture. In these systems, the fluorescence was approximately 45% quenched by the addition of a layer of **23**. This study suggests that excitation energy in the mixed stilbene L-B assemblies is efficiently delocalized both within and between layers. Thus the mixed stilbene L-B assembly acts as an energy conductor exhibiting "wirelike" behavior. However, in the case of the mixed stilbene layers containing **22**, a 20% fluorescence quenching indicates that the "localization" of the energy in this lower energy "trap" inhibits the efficient migration from remote layers to those adjacent to the quencher.

Electron transfer quenching by **23** and the surfactant ethylenediamine cobalt(III) complex (**24**) has been investigated in similar mixed stilbene L-B assemblies.

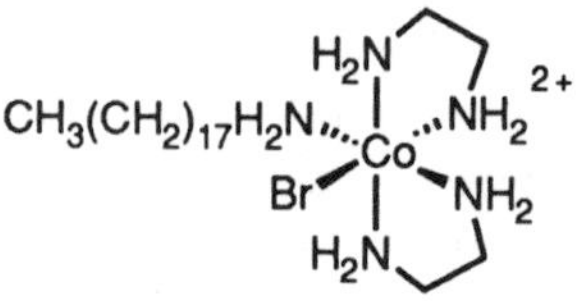

2 4

Solution studies of **23** and nonsurfactant analogue[97,98] indicate that upon irradiation with 313 nm light, intramolecular electron transfer occurs from the stilbene ligand to the Co(III) center. The reaction products are Co(II) and oxidized stilbene carboxylate ligand. In L-B assemblies, **23** does not fluoresce, indicating that excited-state electron-transfer quenching occurs.

Previous studies suggest that the aliphatic surfactant complex **24** quenches the fluorescence of surfactant stilbene derivatives via intermolecular electron transfer from stilbene to the Co(III) center. The fluorescence of a bilayer consisting of a 1:1 molar ratio of **21** to **AA** was approximately 45% quenched by the addition of an adjacent layer of **23** compared to little or no quenching observed with an adjacent layer of **24**. These results suggest that L-B assemblies of surfactant stilbene are more efficiently quenched in assemblies containing a single layer of **23** as compared to **24**. In assemblies containing **23**, energy transfer can first occur from the surfactant stilbene layer to the *trans*-stilbene carboxylate ligand of the cobalt complex followed by intramolecular electron transfer to the cobalt center. In the assemblies containing **24**, energy transfer to the quencher is not observed. Thus, **23** exhibits enhanced electron-transfer quenching via an antenna effect due to the presence of a stilbene "wire" within the electron acceptor molecule.

4. Electron Transfer
4.1. General Aspects

Photoinduced electron transfer is of great interest as it relates to understanding the steps involved in photosynthesis and facilitates the design of efficient solar devices. Extensive studies focusing on the mechanism of photoinduced electron transfer in solution indicate that this process is controlled mainly by the following factors: the donor–acceptor distance, symmetry and energetic matching of the molecular orbitals involved, the lifetime of the excited molecule, and the polarity of the medium.[99] The factors involved in electron transfer processes have been described by Marcus[100,101] in the following equation:

$$\Delta G^* = W_r + (\lambda/4)[1 + (\Delta G^\circ + W_p - W_r)/\lambda]^2$$

where ΔG^* is the free energy of activation, W_r is the coulombic energy of interaction between the reactants, W_p is the coulombic energy of interaction between the products, λ is the reorganization energy, ΔG° is the thermodynamic driving force of the reaction.

4.2. Distance Dependence Studies of Electron Transfer in Langmuir-Blodgett Assemblies

In solution, the distance dependence of electron transfer is very difficult to study because it involves a donor and an acceptor that must first diffuse together in order for electron transfer to occur. Langmuir-Blodgett assemblies have an

712   *Spooner and Whitten*

advantage over solution in that the molecules are fixed in a confined orientation at a predetermined separation distance which can be easily manipulated. Möbius and co-workers[22,99,102-104] studied electron transfer in L-B assemblies where an excited molecule can act as either an electron donor or an electron acceptor. Electron transfer was observed when an excited surfactant oxacyanine donor 3 and a surfactant bipyridinium acceptor 25 were mixed in the same layer.

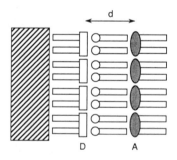

2 5

Since there was no overlap of donor fluorescence with acceptor absorption, it was suggested that electron transfer, not energy transfer, was occurring. Electron transfer was observed when the donor and acceptor were in adjacent layers with both hydrophilic chromophores in contact, however, no electron transfer was observed when the donor and acceptor layers were separated by two layers of AA. Möbius and co-workers[20,22,23,103] investigated the distance dependence of electron transfer by varying the length of the fatty acid layer, i.e., from C_{14} to C_{22}, used to separate the donor and acceptor, where the acceptor was deposited using Z deposition (see Figure 12).

Figure 12. Schematic representation of the L-B assembly of donor 2 separated from acceptor 25 by distance *d*. (Reprinted with permission from Kuhn.[103] Copyright 1979 International Union of Pure and Applied Chemistry.)

By monitoring the quenching of the donor fluorescence versus distance, a distance dependence profile was obtained for electron transfer. The rate constant k_{el} of electron transfer can be obtained from the following equation:

$$k_{el}\tau = (I_0/I) - 1$$

where I_0 is the fluorescence intensity of the donor in the absence of acceptor, I is the fluorescence intensity of the donor in the presence of acceptor, τ is the fluorescence lifetime of the donor in the absence of acceptor. The values of $(I_0/I) - 1$ were calculated and plotted logarithmically versus the thickness of the fatty acid spacer as shown in Figure 13. The rate constant k_{el} of electron transfer was found to decrease exponentially with increasing distance which is consistent with electron transfer occurring via electron tunneling across a barrier. It was noted that additional evidence for a tunneling mechanism of electron transfer was obtained from an investigation of the influence of an energy acceptor reducing the electron-transfer efficiency and from donor fluorescence decay measurements.[23] Further evidence supporting the tunneling mechanism was the lack of temperature dependence on the rate of fluorescence quenching of the donor from 293 to 77 K.[22]

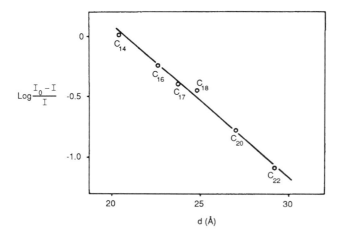

Figure 13. Plot of log $[(I_0/I) -1]$ versus the thickness of a fatty acid spacer. (Reprinted with permission from Kuhn.[103] Copyright 1979 International Union of Pure and Applied Chemistry.)

Möbius[102,104] also studied the effects of energy delocalization of the donor on the efficiency of electron transfer. Two cases were investigated in which the oxacyanine donor **3** and bipyridinium acceptor **25** were located in adjacent layers with both hydrophilic chromophores in contact and the concentration of the electron acceptor was varied. In the first case, the donor was in a monomeric state with a 1:1000 mole ratio of **3:AA**. In this case, the distance between monomers is large, so energy transfer between donor molecules is not observed. The relative intensity of the donor fluorescence I/I_0 decreased exponentially with increasing acceptor density as predicted by the hard disk model of electron transfer. In the second case, the donor was in a delocalized J-aggregated state, in which electron transfer is more efficient since fluorescence quenching of donor molecules can occur. A plot of (I_0/I) −1 against the acceptor density in the adjacent layer is shown in Figure 14. From these data it was estimated that the rate of photoinduced electron transfer was increased by a factor of more than 600 for J-aggregated molecules compared to monomers. Thus, the enhanced efficiency of electron transfer observed for J-aggregated molecules demonstrates the cooperation of donor molecules in the extended arrays. Similar experiments were performed to study electron transfer from a monomeric dye to a variety of electron acceptors with differing reduction potentials. The efficiency of electron transfer was found to increase with increasing separation between electron donor and acceptor energy levels.

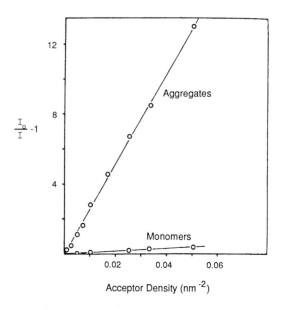

Figure 14. Plot of $[(I_0/I) - 1]$ versus acceptor **25** density comparing J-aggregated donor **3** to the corresponding monomeric state. (Reprinted with permission from Möbius.[104] Copyright 1983 Gordon and Breach, Science Publishers, Inc.)

Previous studies of electron transfer in L-B assemblies have involved chromophores located in the hydrophilic region of the surfactant. Whitten, Mooney, and co-workers[45,46] demonstrated that hydrophobic chromophores could transfer electrons to hydrophilic acceptors in adjacent layers. The distance dependence of electron transfer to either a surfactant bipyridinium (**25**) or cobalt(III) complexes (**24 and 26**) was determined using H-aggregated surfactant stilbene derivatives such as **8, 9a,** and **9b** as electron donors (see Figure 15).

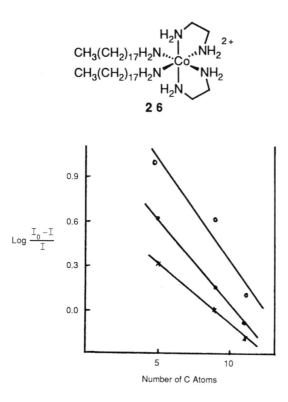

Figure 15. "Perrin" plot for the fluorescence quenching of stilbene aggregates **8**, **9a**, and **9b** by various electron acceptors. (o) quencher **25**; (•) quencher **24**; (x) quencher **26**. (Reprinted with permission from Whitten et al.[47] Copyright 1988 Plenum Publishing Corp.

In these studies the donor–acceptor distance was controlled by varying the position of the stilbene within the fatty acid chain. From the reduction potentials of the three acceptors, it was predicted that the oxidation of excited states of the H-aggregated stilbenes was energetically favorable. The addition of an adjacent layer of quencher resulted in quenching of the stilbene fluorescence. All three acceptors were found to quench the stilbene fluorescence over much shorter distances when compared to the thiacyanine acceptor **2** studied earlier. For each acceptor the degree of fluorescence quenching of the different surfactant stilbenes varied significantly. The following trend was observed for the extent of quenching: **8** > **9a** > **9b**. From the linear plot of the log $[(I_0 - I)/I]$ vs. distance, it was suggested that a static or "Perrin"-type quenching could occur. Due to the lack of overlap of the stilbene fluorescence with the viologen absorption it could be assumed that electron transfer had occurred. The same is not true of the cobalt acceptors. However, since the fluorescence quenching distances and rates are similar to those of the viologen, it appears that electron transfer also occurs with the cobalt acceptors rather than energy transfer.

4.3. Applications of Electron Transfer in Langmuir-Blodgett Assemblies. Vectorial Electron Transfer and Photovoltaic Effects.

Möbius and co-workers[20,21,23,105-107] have demonstrated that a photovoltage could be generated by vectorial photoinduced electron transfer within a L-B assembly. They designed the following system in which an aluminum electrode was coated with a layer of **AA** followed by a layer of surfactant cyanine dye (**13**), surfactant azobenzene (**5**), and surfactant bipyridinium (**25**) which is in contact with a barium electrode[23,106,107] (see Figure 16).

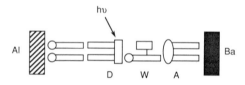

Figure 16. Schematic representation of the L-B assembly used to generate vectorial photoinduced electron transfer. (Reprinted with permission from Möbius.[23] Copyright 1981 American Chemical Society.)

When the donor (**13**) is excited, it transfers an electron to **5**, which acts as a molecular wire providing a low- lying unoccupied orbital for the transfer of electrons to the primary acceptor **5**. The primary acceptor then transfers an electron to the barium electrode, which acts as the final electron acceptor. Oxidized **13** is then regenerated by electron transfer from the aluminum electrode which allows the cycle to be repeated. When **13** is irradiated under vacuum, a photovoltage is generated which decays when the light source is turned off as shown in Figure 17. Thus it was demonstrated that unidirectional electron transfer can be facilitated through the appropriate arrangement of several molecules in a L-B assembly such that back electron transfer is minimized.

Penner and Möbius[108,109] have studied electron transfer in L-B assemblies consisting of a two-component sensitizer–acceptor system. A surfactant cyanine dye (**1**) was chosen as the electron donor (C) and a surfactant bipyridinium (**25**) as the acceptor (A). The extent of electron transfer was determined by the production of the reduced acceptor. By monitoring the absorption spectrum of the reduced acceptor, the amount of product that did not undergo back electron transfer was obtained. They were able to enhance the yield of electron transfer by the addition of an electron donor to the system.

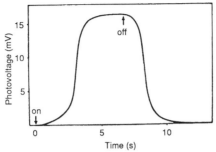

Figure 17. Plot of photovoltage versus irradiation time demonstrating the photovoltage produced upon irradiation of a L-B assembly coated between two electrodes. (Reprinted with permission from Möbius.[23] Copyright 1981 American Chemical Society.)

When an electron donor (D) such as a surfactant phenylenediamine (**27**) was introduced into the system, the yield of reduced acceptor was enhanced by a factor of 3.5 (see Figure 18).

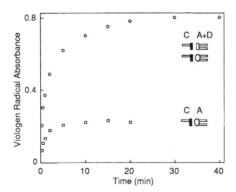

2 7

Figure 18. Plot of viologen radical absorbance versus irradiation time. (□) Two-component system of dye **1** and viologen acceptor **25**; (○) three component system of donor **27, 1**, and **25**. (Reprinted with permission from Penner and Möbius.[108] Copyright 1982 American Chemical Society.)

This enhancement or "supersensitization" is thought to occur through the following process:

$$D + C^* + A \rightarrow D^+ + C + A^-$$

It was observed that 1 must be J aggregated for supersensitization to occur. This result can be explained by enhanced electron migration in the J-aggregated dye compared to that of the monomer.

Over the last several years, Fujihira and co-workers[110-113] have published several papers dealing with the fabrication of photochemical photodiodes by the manipulation of L-B assemblies containing an electron acceptor (A), sensitizer (S), and donor (D). One of their first photochemical photodiodes was formed on an optically transparent gold electrode that was coated with a layer of a surfactant viologen acceptor, a layer of surfactant pyrene sensitizer, and a layer of surfactant ferrocene donor. Upon irradiation, a photocurrent was detected that arose from an photoinduced vectorial flow of electrons. They also fabricated a photodiode in which all three components were incorporated into the same molecule. Molecule **28** contains the viologen acceptor as the hydrophilic head group with the donor and sensitizer attached by two different surfactant chains in the order S–A–D.

2 8

A unidirectional photocurrent was also observed in these L-B assemblies. A surfactant where the three components were arranged in the following order A–S–D, having the same number of carbons separating the A–S and A–D, was later synthesized. It was observed that the A–S–D L-B assemblies showed a higher efficiency of photoinduced electron transfer than the S–A–D L-B assemblies, which indicated a better spatial alignment of A, S, and D (see Figure 19). To test the role of the donor in these systems, a compound without the D was synthesized and studied. The efficiency of the A–S–D system was much higher than the A–S system, suggesting that the donor is involved in the electron-transfer process. Other systems were devised to model photosynthesis by adding surfactant pyrene to a layer of A–S–D with perylene as the sensitizer. The pyrene acts as a light harvesting antenna and transfers its energy to the perylene of the A–S–D in which charge separation occurs. They observed a greater amount of charge separation with the added surfactant sensitizer then was obtained for the other A–S–D system.

Fujihira and co-workers[114] have also examined the photoreduction of CO_2 by L-B assemblies of nickel alkylcyclam complexes coated on electrodes.

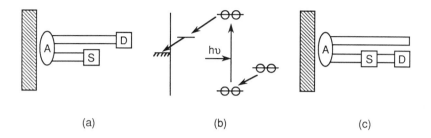

(a)   (b)   (c)

Figure 19. (a) Schematic representation of a sensitizer–acceptor–donor molecule coated on a solid support. (b) Energy diagram for the S–A–D system. (c) Schematic representation of an acceptor–sensitizer–donor molecule coated on a solid support. (Reprinted with permission from Fujihira et al.[112] Copyright 1989 Elsevier Sequoia.)

Recently, Miyashita et al.[115] have demonstrated that polymer L-B films containing carbazoles as a comonomer of alkylacrylamines could be transferred to a solid support to form stable L-B assemblies. They studied photoinduced electron-transfer quenching of the carbazole in the copolymer layer by a viologen in an adjacent layer, thus demonstrating that copolymer films can be generally used in an assortment of photofunctional devices.

5.   Photoreactions
5.1.   Photoisomerizations

The rigid, highly ordered structure of L-B assemblies provides a unique environment in which to study photoreactions requiring conformational changes of the reacting molecule and/or the resulting intermediates. Changes in product distribution are frequently observed for molecules incorporated into L-B assemblies as opposed to reactions in homogeneous solution. In solution, several photoreactions such as cis–trans isomerization of olefins are known to involve an increase in transition state volume.[116-118] Due to the limited available space between molecules in L-B assemblies, these reactions can either be slowed down or stopped completely. Thus if the favored reaction is restricted, alternative reaction pathways can be taken. Isomerization reactions have been studied in homogeneous solution and in a variety of microheterogeneous media including L-B assemblies.[116-136] Whitten and co-workers[120,121] have investigated the isomerization of surfactant thioindigo dye derivatives (**29a–29c**) in L-B assemblies.

29a *trans*-1, R = *n*-$C_6H_{13}O$; R' = H
29b *trans*-2, R = H; R' = *n*-C_4H_9
29c *trans*-3, R = H; R' = *tert*-amyl

From their isotherm data, it was apparent that the polar thioindigo portion was located at the hydrophilic interface, where the *cis*-thioindigo required a larger area per molecule than the trans isomer. In solution, the reversible isomerization of both thioindigo isomers is well known. Irradiation of *cis*-thioindigo in L-B assemblies leads to rapid, irreversible cis to trans isomerization. On the other hand, no reaction was observed when *trans*-thioindigo L-B assemblies were irradiated over prolonged periods. The fact that *cis*-thioindigo isomerizes while *trans*-thioindigo does not can be attributed to rotational restraints in the rigid L-B assembly and the larger volume occupied by the cis isomer.

In further studies Whitten and coworkers[47,127] investigated the photo-isomerizations of two thioindigo dyes, 6,6'-dihexyloxythioindigo (**30a**) and 6,6'-diethoxythioindigo (**30b**) in homogeneous solution and microheterogeneous media, in particular, monolayer films at the air–water interface.

cis-**30**   **a** R = $(CH_2)_5CH_3$ *trans*-**30**
                 b R = CH_2CH_3

In solution, both **30a** and **30b** were found to undergo reversible photoisomerization. Stable L-B films at the air–water interface were formed with a 1:1 molar ratio of AA and thioindigo. From isotherm data, it appeared that *cis*-**30a** occupied a larger area per molecule than the *trans*-**30a** isomer. Irradiation of *cis*-**30a** at the air water interface resulted in irreversible cis to trans isomerization, accompanied by a decrease in surface pressure. Isotherm data of the *cis*-**30b** and *trans*-**30b** isomers revealed that both molecules required comparable areas. Irradiation of *cis*-**30b** films over a considerable range of surface pressures resulted in *cis*-**30b** to *trans*-**30a** isomerization and an increase in surface pressure. Thus, there is not a simple relationship between the surface pressure decrease upon irradiation of a film and the extent of isomerization, since cis to trans isomerization was observed that proceeded with an increase in surface pressure. Further studies were performed where the extent of isomerization of *cis*-30a and *cis*-30b films was monitored at various surface pressures. For both *cis*-30a and *cis*-30b, it was also observed that as the surface pressure increased, the relative rates of isomerization of both compounds decreased. This result indicates that environmental restraints of the L-B assembly affect the rates and efficiency of these intramolecular isomerization reactions. Whitten and co-workers suggested that these effects could be rationalized through modifications of the triplet-state potential surface of the thioindigo (TI). Photoisomerization of TI occurs via a triplet mechanism, where upon irradiation of either *cis*- or *trans*-TI, a common triplet is achieved. Solution studies of TI lead to formation of the triplet-state potential surface shown in Figure 20.

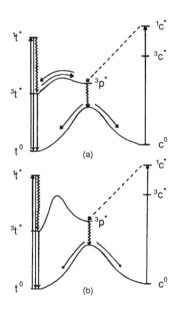

Figure 20. (a) Proposed energy profile for thioindigo in solution. (b) Proposed energy profile for thioindigo in L-B assemblies. (Reprinted with permission from Collins-Gold et al.[127] Copyright 1986 American Chemical Society.)

Irradiation of *trans*-TI results in formation of the transoid triplet, ^{3}t*, which can overcome a small activation barrier to reach the perpendicular triplet, ^{3}p*. The perpendicular triplet can deactivate via nonradiative decay to the ground state, where it partitions to both *cis*- and *trans* -TI. Thus, in solution, reversible isomerization is observed. In L-B assemblies, it was suggested that the trans isomer was excited to the same ^{3}t* state as in solution; however, an environment-contributed barrier brought about by the restricted environment of the L-B assembly acted to raise the activation energy between the ^{3}t* and ^{3}p* states. Thus, equilibration of the trans and perpendicular triplet states does not occur and excitation of *trans*-TI leads to ground- state *trans*-TI, completely "shutting off" the photoisomerization pathway to *cis*-TI. On the other hand, excitation of *cis*-TI leads to the formation of the cisoid triplet state, ^{3}c*. This cisoid triplet state can minimize its energy by rotation, resulting in population of ^{3}p*, which can then deactivate to form both ground-state *cis* - and *trans* -TI. The observation that increases in surface pressure reduce the efficiency of cis to trans isomerization suggests the possibility of a third minimum on the triplet potential surface of "near-cisoid geometry." For example, excitation of *cis*-TI L-B assemblies at high surface pressures results in no net isomerization. The above experiments therefore demonstrate that the rigid environment of L-B assemblies plays a role in modifying the excited state potential surface of the confined surfactant TI, although the underlying photophysics of the molecule remains essentially unchanged.

While the formation of monolayer films containing azobenzene derivatives has been mentioned in a previous section, Kunitake and co-workers[137] have also demonstrated the formation of stable polyion-complexed monolayer films at the air–water interface that consist of a mixture of cationic ammonium amphiphiles containing azobenzene derivative (**31**) with anionic polymers such as potassium polyvinylsulphonate (**32**).

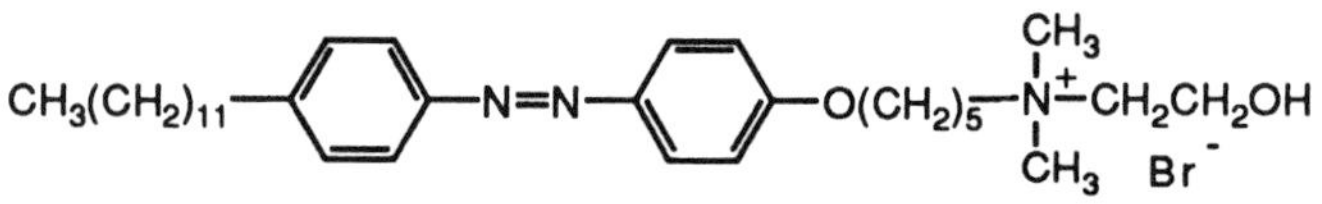

3 1

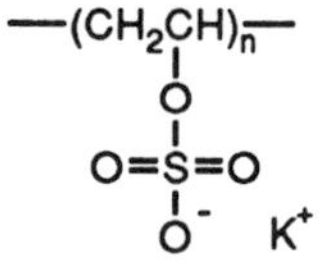

3 2

Studies of the molecular orientation of these polyion complexed films by x-ray photoelectron spectroscopy and uv spectroscopy indicate formation of a highly ordered polyion complexed film of **31** and **32**, whereas **31** alone did not form stable monolayer films due to its solubility in water. Using the polyion technique described by Kunitake, Fujihira, and their co-workers[131,132] formed stable monolayer films at the air–water interface of polyion complexes of amphiphilic azobenzene derivatives with several ionic polymers (see Figure 21). They observed reversible cis-trans photoisomerization in these films, whereas the azobenzenes in monolayer films, alone, undergo exclusively cis to trans isomerization. The reversible photoisomerization reaction in polyion complexed films is thought to occur due to the increased area per molecule provided in the film. The surface area per molecule was found to be controlled by varying the size of the monomer unit chosen in the ionic polymer.

Reversible cis–trans isomerization was also observed by Sandhu et al.[130] in monolayer films containing phospholipid molecules in which one or both acyl chains were substituted by an azobenzene moiety. Upon irradiation of a monolayer film composed of phosphatidylcholine derivatives containing *trans*-azobenzene the azobenzene portion of the molecule underwent trans to cis isomerization resulting in a photostationary state of the cis and trans isomers and an overall increase in film volume.

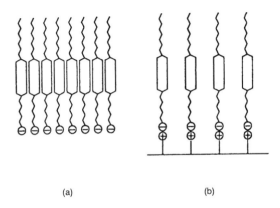

(a)   (b)

Figure 21. (a) Schematic representation of an anionic surfactant azobenzene derivative monolayer film at the air–water interface. (b) Schematic representation of the stable monolayer film formed from the polyion complex of anionic surfactant azobenzene derivatives with a cationic polymer. (Reprinted with permission from Nishiyama and Fijihira.[131] Copyright 1988 The Chemical Society of Japan.)

Fukuda and co-workers[133-136] have demonstrated reversible trans–cis isomerization of azobenzenes **33a–33c** incorporated in monolayer films of surfactant β-cyclodextrins.

3 3 a   X = H   Y = COOH
3 3 b   X = (CH₃)₂N   Y = COOH
3 3 c   X = (CH₃)₂N   Y = SO₃Na

Stable monolayer films containing amphiphilic derivatives of β-cyclodextrin were formed at the air–water interface where the base of the cyclodextrin cavity is aligned parallel to the water surface. Stable monolayer films consisting of host–guest complexes of cyclodextrin and **33a–33c** were also formed where the long axis of the azobenzene was oriented perpendicular to the water surface (see Figure 22). Irradiation of **33a–33c** of this host–guest monolayer film results in reversible cis–trans isomerization of **33a–33c**. This reversible isomerization is attributed to the large volume available for conformational changes of the molecule in the cyclodextrin cavity, whereas only cis to trans isomerization was observed in monolayer films containing only surfactant azobenzenes at the air–water interface.

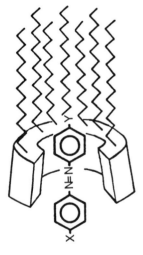

Figure 22. Schematic representation of the host–guest complex of β-cyclodextrin and azobenzene derivatives **33a–33c**. (Reprinted with permission from Tanaka et al.[135] Copyright 1987 The Chemical Society of Japan.)

5.2.   Photodimerization and Polymerization

Analogous studies[119,121, 124-126] were performed on surfactant stilbazole salts and α,ω-diphenylbutadienes. A surfactant *cis*-stilbazole derivative (**34**) displayed the same irreversible cis to trans isomerization in L-B assemblies and reversible cis–trans isomerization in solution.

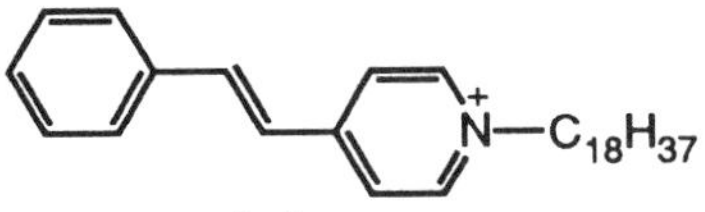

3 4

Irradiation of a *trans*-**34** L-B assembly also resulted in no isomerization. However, a 50 nm red shift was observed in the fluorescence maximum compared to that in homogeneous solution. This type of shift was not observed for *trans*-**29**, which showed solution-like fluorescence in L-B assemblies. The red-shifted stilbazole fluorescence was attributed to the formation of an excimer based on similar observations obtained in solid state. Upon prolonged irradiation, the excimer fluorescence intensity decreased due to cyclobutane dimer formation. In both L-B assemblies and in the solid state, only the head-to-head cyclobutane dimer **35** was obtained.

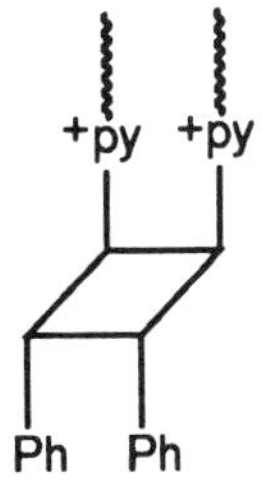

3 5

When **34** was irradiated in solution at high concentrations, only monomer fluorescence was observed. Thus, the differences in reactivity observed between L-B assembly and the solid-state studies and that in solution are mainly attributed to the rigid, highly ordered packing of molecules in the former systems. When two layers of **34** were separated by a layer of **AA**, the extent of excimer fluorescence was reduced. This result indicates that both intramolecular and intermolecular excimers are formed. Thus the close packing of the stilbazoles in L-B assemblies leads to formation of stable dimers, which results in either excimer fluorescence or photoaddition products.

    Langmuir-Blodgett assemblies of *N*-octadecyl-1-(4-pyridyl)-4-(phenyl)-1,3-butadiene (**36**) were also found to exhibit strong excimer fluorescence and photodimerization.[121, 124,126]

3 6

In solution, only monomer fluorescence was observed even at high concentrations of **36**. When two butadiene layers were separated by a layer of **AA**, no change in the extent of excimer formation was observed. Thus, the formation of excimers and photoproducts is brought about only by intralayer interactions. Several dimers were observed when **36** was irradiated in L-B assemblies or in the solid state, unlike in the stilbazole studies, where only one dimer was detected. The mixture of cyclobutane dimers formed in L-B assemblies can be explained by the initial formation of a single dimer that can undergo the following Cope rearrangement to form a mixture of interconverting isomeric products:[126]

Previous studies by Whitten and co-workers have focused only on molecules located in hydrophilic regions of L-B assemblies. More recent studies[45] involve surfactants in which the active chromophores are located in hydrophobic regions. Various surfactant stilbene derivatives have been synthesized in which the stilbene chromophores are located in the hydrophobic portion of the molecule. Surfactant *cis*-stilbenes did not form stable monolayer films at the air–water interface; as a result cis to trans isomerizations could not be investigated. Surfactant *trans*-stilbene films were found to form H-aggregates as mentioned earlier. Thus no excimer or dimer formation was observed. Surfactant styrene derivatives **37** and **38** were synthesized where the styrene unit was located at the end of a surfactant chain.[119,122,123]

3 7   **3 8**

Incorporation of these surfactants into L-B assemblies resulted in the location of styrenes in highly hydrophobic regions of the assemblies. Upon irradiation of **37**, rapid bleaching of the monomer absorption was observed due to dimer formation. The dimers were found to have a trans–syn–trans geometry due to a parallel alignment of styrene molecules in the L-B assemblies. Thus these studies suggest that molecules located in hydrophilic and hydrophobic portions of L-B assemblies exhibit similar reactivity.

Ouchi and co-workers[138] have recently studied the photodimerization of surfactant cinnamylideneacetic acid derivatives (**39a–39c**) in L-B assemblies.

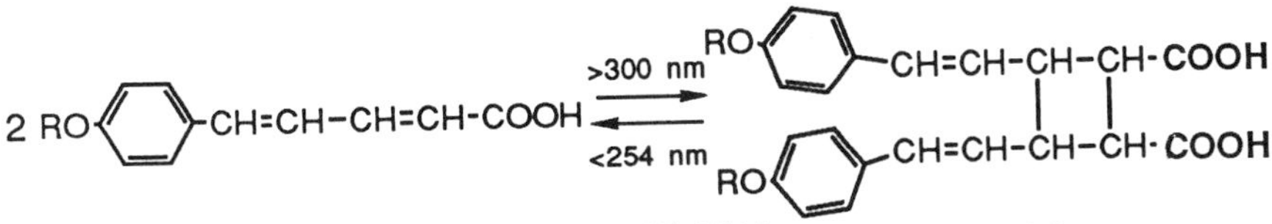

$$2 \ RO\!\!\bigcirc\!\!-CH{=}CH{-}CH{=}CH{-}COOH \underset{<254 \ nm}{\overset{>300 \ nm}{\rightleftarrows}}$$

$$RO\!\!\bigcirc\!\!-CH{=}CH{-}CH{-}CH{\cdot}COOH$$
$$RO\!\!\bigcirc\!\!-CH{=}CH{-}CH{-}CH{\cdot}COOH$$

3 9

a R = CH$_3$(CH$_2$)$_5$
b R = CH$_3$(CH$_2$)$_7$
c R = CH$_3$(CH$_2$)$_9$

4 0

Irradiation of the L-B assemblies with light of wavelengths greater than 300 nm resulted in photodimerization of cinnamylideneacetic acid. The photodimers **40a–40c** converted back to starting monomers upon irradiation with 254 nm light. Thus, they demonstrated the photoreversible [2+2] intermolecular reaction of **39a–39c** in L-B assemblies. Ouchi et al.[139] also studied the photoisomerization of anthracene derivatives in L-B assemblies, however, this reaction was not photoreversible. Several other studies have used olefins in an attempt to form polymer-linked L-B assemblies. These studies have been reviewed recently in the literature.[13, 140,141]

5.3. Other Photoreactions

Another type of photoreaction that involves conformational changes of the reactive molecule is the γ-hydrogen abstraction of ketones. Whitten and coworkers[119,123,142] studied cleavage of aliphatic and aromatic ketones containing γ-hydrogens in L-B assemblies. In solution, ketones such as butyrophenone are known to undergo Norrish type II photoelimination via the following reaction:

In benzene, this process occurs from the triplet excited state with a quantum yield of 0.36 for the disappearance of butyrophenone. When γ-hydrogens are present, this process predominates over Norrish type I fragmentation or photoreduction through hydrogen abstraction from the solvent. The surfactant ketone 16-oxo-16-*p*-tolylhexadecanoic acid (**41**) was examined in order to investigate the reactivity of molecules located in the hydrophobic portion of L-B assemblies.

4 1

In benzene, the quantum yield of type II reaction is 0.2 for **41**. This compound was found to form stable L-B assemblies, which upon irradiation, demonstrated rapid bleaching of the uv absorption band corresponding to the ketone. Infrared spectroscopic studies indicated the disappearance of the carbonyl absorption with a quantum yield of 0.06, which is 1/3 that of the solution quantum yield. However, the quantum yield of type II product formation in L-B assemblies is only 0.001, which is decreased by three orders of magnitude from that in solution. It is apparent that the type II reaction is almost "shut off" in these systems because the required transition state could not be achieved. This allows other reactions such as type I elimination and intermolecular hydrogen abstraction to compete with the type II reaction. These results demonstrate that the constraints imposed on molecules incorporated in L-B assemblies can have dramatic effects on the mechanistic course of the reaction.

Ringsdorf and co-workers[143] have studied photochemical reactions of surfactant diazo and and azido compounds in monolayer films at the air–water interface. Reactions of diazo ketones and azido ketones have been studied extensively in solution. These compounds are known to lose nitrogen upon irradiation, resulting in formation of a carbene which can further react depending on the reaction conditions. In L-B assemblies, the surfactant analogues were found to undergo the same photochemistry as in solution. Photoproducts of the reaction were detected by infrared spectroscopy and mass spectroscopy techniques. Upon irradiation of surfactant α–diazo ketone (**42**), the molecules lose nitrogen, resulting in formation of carbenes at the air–water interface.

$$\begin{array}{c} O \\ \| \\ R-C-CH=\overset{+}{N}=\overset{-}{N} \end{array} \qquad \begin{array}{l} R = \ C_{13}H_{27} \\ \ \ \ \ \ C_{17}H_{35} \\ \ \ \ \ \ C_{19}H_{39} \\ \ \ \ \ \ C_{27}H_{55} \end{array}$$

4 2

These carbenes either react with water at the interface or rearrange to form ketenes that either can be hydrolyzed or dimerize to form surfactant β-lactones. It was also observed that the yield of β-lactones decreased with increasing surfactant chain length. Since the formation of β-lactones requires a certain amount of surfactant ketene mobility, it was concluded that the longer the surfactant chain, the less mobility the molecules exhibit. Studies of surfactant diesters of 2-diazopropane-dioic acid (**43**) at the air–water interface resulted in the photochemical loss of nitrogen to form carbenes.

$$H_{37}C_{18}-\overset{\overset{\displaystyle O}{\|}}{C}-\overset{\overset{\displaystyle \|}{C}}{\underset{N_2}{C}}-\overset{\overset{\displaystyle O}{\|}}{C}-C_{18}H_{37} \qquad\qquad H_{37}C_{18}-\!\!\!\left\langle\!\!\!\bigcirc\!\!\!\right\rangle\!\!\!-\overset{\overset{\displaystyle O}{\|}}{C}-N_3$$

4 3 **4 4**

These carbenes react with water to yield the diester of 2-hydroxypropanedioic acid. The photochemistry of surfactant azido ketone (**44**) was the most complex of the reactions studied. Upon irradiation, they lose nitrogen to form the corresponding surfactant isocyanate which adds water and undergoes secondary photoreactions to produce a brown colored monolayer film. The photochemical loss of nitrogen in all of these molecules studied leads to a variety of changes in the stability of the resulting monolayer films. It was suggested that depending on factors such as "the functional group, chain length, substrate pH, and temperature it is possible to achieve changes in compressibility and collapse pressure, disappearance of expanded phases, collapse of monolayer films to give oily films, or disappearance of monolayer films by dissolution in the subphase."[143] Thus, the above experiments demonstrate that small changes in head groups for molecules incorporated into monolayer films could result in large changes in the stability of these films.

Langmuir-Blodgett assemblies of surfactant spiropyran derivatives have been investigated by several groups.[77-80, 144-151] In solution, colorless spiropyrans (SP) are known to undergo a photochromic reaction to form colored merocyanine (MC) dyes. This transformation involves "the heterolytic cleavage of the 1,2 single bond of the pyran ring followed by rotation of the two halves of the molecule to give a planar zwitterionic form which is stabilized by resonance."[151] Polymeropoulos and Möbius[147] studied mixed L-B assemblies of surfactant spiropyran derivative (**45**) and tripalmitine at the air–water interface and in L-B assemblies and reported their surface pressure isotherms, surface potential measurements, and photochromic properties.

4 5   **4 6**

Calculations of the area per molecule showed that the corresponding merocyanine (46) occupies an area three times the size of 45. Irradiation of a 45 film led to the reversible formation of 46 which resulted in an increase in surface pressure producing a photochemically generated "shock wave." When a monolayer film of 45 was transferred to a solid support, a reversible photochromic reaction was also observed. The reaction, however, was found to be considerably slower and less efficient when the molecules were incorporated deep within the L-B assemblies, suggesting that the last layer of the L-B assembly had greater mobility. Polymeropoulos and Möbius demonstrated that the fluorescence of a thiacyanine dye (TC) could be modulated by energy transfer. They coated a slide with a layer of TC followed by a layer of SP. When the SP was irradiated with 366 nm light, isomerization to MC occurred. Since the fluorescence of MC overlaps with the absorption of TC, efficient energy transfer can occur. When the system is irradiated with 545 nm light, which is absorbed by MC, the intensity of the TC fluorescence was decreased due to the photochromic reaction of MC to SP, which has no overlap of its fluorescence with the absorption of TC. Thus this experiment demonstrates that a photochromic reaction could act as a switch to turn on and off energy transfer.

Ando et al.[77-79] have investigated the possibility of controlling photochromic reactions of spiropyrans by means of L-B assemblies and monolayer films. They first investigated monolayer films of two spiropyran molecules (47a and 47b) that contain one and two alkyl chains, respectively.

$$CH_3 \quad CH_3$$

[Structure of spiropyran with N-R, O, R', and NO$_2$ substituents]

	R	R'
47a	$C_{16}H_{33}$,	H
47b	$C_{18}H_{37}$,	$CH_2OCOC_{21}H_{43}$

Spiropyran **47b** was found to form stable monolayer films at the air–water interface with a 1:1 molar ratio of spiropyran to *n*-octadecane, whereas **47a** did not form stable films. Irradiation of the double-chained spiropyran with uv light at 35°C resulted in formation of surfactant merocyanine dyes which rearrange to form J aggregates, demonstrating the utility of photochromic reactions to control J aggregate formation in monolayer films. Investigations of monolayer films containing surfactant spiropyrans and polar cosurfactants such as stearic acid were performed. The polar film matrix brought about by the the carboxylic groups of the stearic acid aided in the formation of stabilized merocyanine monolayer films that spontaneously J aggregated in the dark. Studies of the same spiropyran and nonpolar cosurfactants in monolayer films resulted in merocyanine formation upon irradiation. However, the merocyanines returned to their spiropyran form in the dark, where J aggregation was not observed. Thus these experiments demonstrate that photochromic reactions of molecules in monolayer films can be controlled by the polarity of the monolayer film matrix. Further investigations of the photochromic behavior of spiropyrans in monolayer films at the air–water interface suggest that different arrangements of molecules, which dictate whether J aggregation occurs, can be achieved during the photochromic process.

Unuma et al.[80] have recently demonstrated that J aggregation of photochromic merocyanines in L-B assemblies can be controlled by irradiating the assembly with a linearly polarized visible He–Ne laser and unpolarized uv light. Linearly polarized visible He–Ne irradiation was found to deaggregate the molecules, while irradiation with unpolarized uv light resulted in the realignment of the long axes of the molecule. Thus, J-aggregates could be deaggregated and reaggregated upon successive irradiation with linearly polarized and uv light, respectively.

Nagamura et al.[152,153] reported photochromic reactions in L–B assemblies that differed from the previously mentioned photochromic reactions, which involved changes in molecular conformation. They reported photochromic reactions brought about by electron transfer that occurred upon irradiation of an ion-pair charge-transfer complex. It was demonstrated that *N,N*'-dihexadecyl-4,4'-

bipyridinium salts with tetrakis[3,5-bis(trifluoromethyl)phenyl]borate anion (TFPB) mixed with AA formed stable monolayer films that exhibited charge-transfer characteristics. Irradiation of the charge-transfer band of the L-B assembly in vacuo or in argon above 365 nm, resulted in a color change from pale yellow to blue. This blue color was attributed to appearance of the 4,4'-bipyridinium cation radical which was monitored by uv-visible spectroscopy and esr spectroscopy. Thus, electron transfer occurred from the TFPB anion to 4,4'-bipyridinium. Photochromic reactions that occur via electron transfer can be used to design molecular switches or devices since their response time is much shorter than that of previously mentioned photochromic reactions, i.e., photoisomerizations, which must produce stable isomeric products.

Tachibana et al.[154,155] have demonstrated the use of photochromic reactions of surfactant azobenzene units as switching devices to control the conductivity of organic conductors in L-B assemblies. They designed the surfactant **48** with three sections, a switching unit, a transmission unit, and a working unit, which were an azobenzene, alkyl chain, and a TCNQ moiety (7,7,8,8-tetracyanoquinodimethane), respectively (see Figure 23).

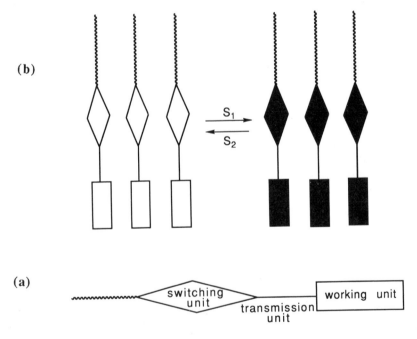

Figure 23. Schematic representation of (a) the switching device and (b) its molecular composition. (Reprinted with permission from Tachibana et al.[154] Copyright 1989 American Chemical Society.)

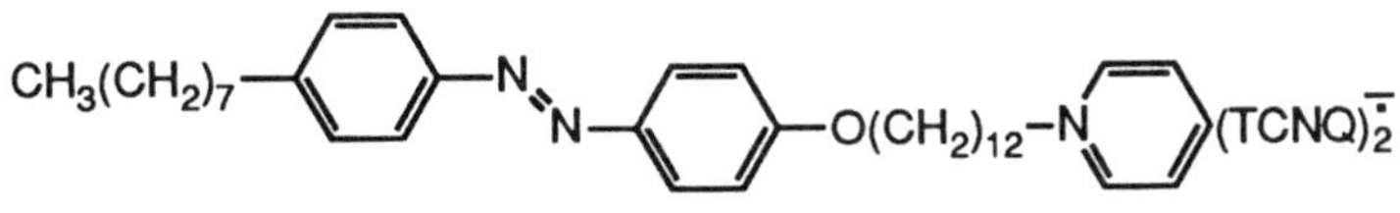

$CH_3(CH_2)_7$—⟨benzene⟩—N=N—⟨benzene⟩—$O(CH_2)_{12}$—N⟨pyridinium⟩$(TCNQ)_2^-$

4 8

Upon irradiation of the *trans*-azobenzene portion of the molecule, an increase in the lateral conductivity of the L-B assembly was observed. The lateral conductivity of the assembly decreased when the assembly was irradiated where *cis*-azobenzene absorbs. They attributed this increase in conductivity upon irradiation to conformational changes of *trans*- to *cis*-azobenzene leading to enhanced ordering of the TCNQ column, thus demonstrating that azobenzenes can act as molecular switches toward conductivity in the L-B assemblies.

Ichimura et al.[156] have shown that the photochromic reaction of surfactant azobenzene molecules in monolayer assemblies can be used to control the alignment of nematic liquid crystals (LC). Monolayer films of surfactant azobenzene **49** were transferred to quartz substrates using L-B techniques, while assemblies of surfactant azobenzene **50** were formed using adsorption techniques described by Sagiv.[157]

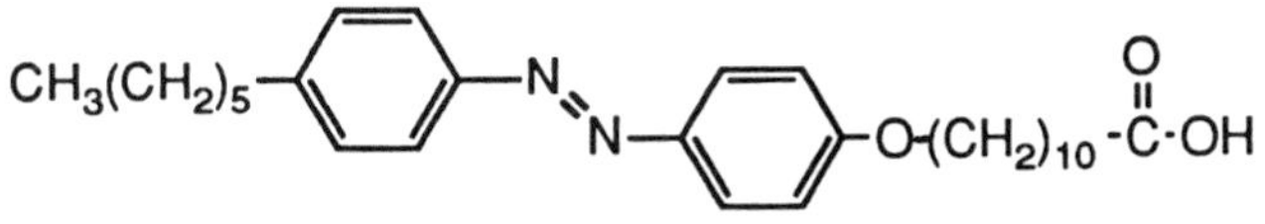

$CH_3(CH_2)_5$—⟨benzene⟩—N=N—⟨benzene⟩—$O(CH_2)_{10}$—$\overset{O}{\overset{\|}{C}}$-OH

4 9

$CH_3(CH_2)_5$—⟨benzene⟩—N=N—⟨benzene⟩—$O(CH_2)_5$-$\overset{O}{\overset{\|}{C}}$-$NH(CH_2)_3$-$Si(OC_2H_5)_3$

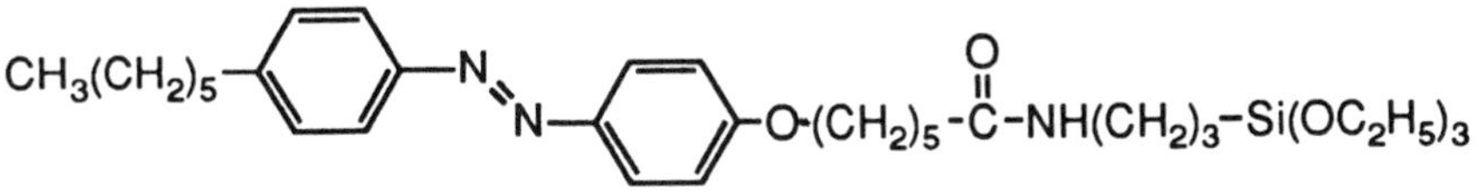

5 0

A cell was constructed consisting of nematic LC sandwiched between quartz substrates coated with either **49** or **50**. In both cells, the nematic LC were aligned parallel to the long axis of the *trans*-azobenzene molecules. Upon irradiation of the azobenzene unit in the adsorbed assembly, trans–cis isomerization occurred followed by realignment of the nematic LC such that their long axis were now parallel to the long axis of the surfactant *cis*-azobenzenes. This realignment was found to be reversible upon isomerization of the azobenzene unit, demonstrating that the reversible photoisomerization of azobenzenes can lead to reversible changes in LC alignment. It was observed, however, that assemblies of **49** formed using L-B techniques failed to isomerize from trans to cis due to the increase in surface area per molecule upon irradiation, which limited this reversible alignment to adsorbed monolayers.

6. Concluding Remarks

In summary, L-B assemblies provide a novel and frequently controllable constrained environment that can be used to study or perturb a variety of photochemical reactions. However, the small absolute amounts of material contained in most L-B assemblies and the difficulty in their construction will likely limit their employment in practical devices or in preparative scale reactions. As new and more sensitive analytical techniques are developed, together with synthetic advances, more utilization of these assemblies to obtain novel photochemical or photophysical processes from "supramolecular assemblies" will surely follow.

Acknowledgment

We are grateful to the National Science Foundation (Grant CHE-8616361) for support of this research.

References

1. G. L. Gaines, Jr., *Insoluble Monolayers At Liquid-Gas Interfaces*, Wiley-Interscience, New York, 1966.
    G. L. Gaines, Jr., *Thin Solid Films*, 1983, *ix*, 99.
2. G. Roberts, ed., Langmuir–Blodgett Films, Plenum, New York, 1990.
    H. Khun, *Thin Solid Films*, 1989, *178*, 1.
3   E. M. Arnett and J. M. Gold, *J. Am. Chem. Soc.*, 1982, *104*, 636.
4. E. M. Arnett, N. G. Harvey, and P. L. Rose, *Langmuir*, 1988, *4*, 1049.
5. N. Harvey, P. L. Rose, N. A. Porter, J. B. Huff, and E. M. Arnett, *J. Am. Chem. Soc.*, 1988, *110*, 4395.
6. N. G. Harvey, D. Mirajovsky, P. L. Rose, R. Verbiar, and E. A. Arnett, *J. Am. Chem. Soc.*, 1989, *111*, 115.
7. E. M. Arnett, N. G. Harvey, and P. L. Rose, *Acc. Chem. Res.*, 1989, *22*, 131.
8. F. M. Menger, M. G. Wood, Jr., S. Richardson, Q. Zhou, A. R. Elrington, and M. J. Sherrod, *J. Am. Chem. Soc.*, 1988, *110*, 6797.
9. F. M. Menger, S. D. Richardson, M. G. Wood, Jr., and M. J. Sherrod, *Langmuir*, 1989, *5*, 833.
10. G. G. Roberts, *Adv. Phys.*, 1985, *34*, 475.
11. R. H. Tredgold, *Rep. Prog. Phys.*, 1987, *50*, 1609.
12. J. D. Swalen, D. L. Allara, J. D. Andrade, E. A. Chandross, S. Garoff, J. Israelachvili, T. J. McCarthy, R. Murray, R. F. Pease, J. F. Rabolt, K. J. Wynne, and H. Yu, *Langmuir*, 1987, *3*, 932
13. T. L. Penner and D. G. Whitten, *J. Soc. Photogr. Sci. Technol. Jpn.*, 1990, *53*, 304.
14. H. Kuhn, D. Möbius, and H. Bücher, *Physical Methods of Chemistry*, Vol. I, Part IIIB, A. Weissberger and B. W. Rossiter, eds., John Wiley, New York, 1972, p. 577.

15. H. Bücher, K. H. Drexhage, M. Fleck, H. Kuhn, D. Möbius, F. P. Schäfer, J. Sondermann, W. Sperling, P. Tillmann, and J. Wiegand, *Mol. Cryst.*, 1967, *2*, 199.
16. H. Kuhn, H. Bücher, B. Mann, D. Möbius, L. V. Szentpaly, and P. Tillmann, *Photol Sci. Eng.*, 1967, *11*, 233.
17. K. H. Drexhage, M. M. Zwick, and H. Kuhn, *Ber. Bunsenges Phys. Chem.*, 1963, *67*, 62.
18. H. Kuhn and D. Möbius, *Angew. Chem. Int. Ed. Engl.*, 1971, *10*, 620.
19. H. Kuhn, *J. Chem. Phys.*, 1970, *53*, 101.
20. H. Kuhn, *Pure Appl. Chem.*, 1981, *53*, 2105.
21. H. Kuhn, *J. Photochem.*, 1979, *10*, 111.
22. D. Möbius, *Ber. Bunsenges. Phys. Chem.*, 1978, *82*, 848.
23. D. Möbius, *Acc. Chem. Res.*, 1981, *14*, 63.
24. D. Möbius and H. Kuhn, *Isr. J. Chem.*, 1979, *18*, 375.
25. D. Möbius and H. Kuhn, *J. Appl. Phys.*, 1988, *64*, 5138.
26. V. Czikkely, H. D. Försterling, and H. Kuhn, *Chem. Phys. Lett.*, 1970, *6*, 11.
27. H. Bücher and H. Kuhn, *Chem. Phys. Lett.*, 1970, *6*, 183.
28. E. G. McRae and M. Kasha, *J. Chem. Phys.*, 1958, *28*, 721.
29. M. Kasha, *Rad. Res.*, 1963, 20, 55.
30. R. M. Hochstrasser and M. Kasha, *Photochem. Photobiol.*, 1964, *3*, 317.
31. E. S. Emerson, M. A. Conlin, A. E. Rosenoff, K. S. Norland, H. Rodriguez, D. Chin, and G. R. Bird, *J. Phys. Chem.*, 1967, *71*, 2396.
32. A. E. Rosenoff, K. S. Norland, A. E. Ames, V. K. Walworth, and G. R. Bird, *Photo. Sci. Eng.*, 1968, *12*, 185.
33. G. R. Bird, K. S. Norland, A. E. Rosenoff, and H. B. Michaud, *Photo. Sci. Eng.*, 1968, *12*, 196.
34. E. E. Jelley, *Nature*, 1936, *138*, 1009.
35. G. Scheiber, *Angew. Chem.*, 1936, *49*, 563.
36. W. West and B. H. Carroll, *J. Chem. Phys.*, 1951, *19*, 417.
37. C. Duschl, W. Frey, and W. Knoll, *Thin Solid Films*, 1988, *160*, 251.
38. D. Möbius and G. Debuch, *Chem. Phys. Lett.*, 1974, *28*, 17.
39. J. Heesemann, *J. Am. Chem. Soc.*, 1980, *102*, 2167.
40. J. Hessemann, *J. Am. Chem. Soc.*, 1980, *102*, 2176.
41. H. Nakahara and K. Fukuda, *J. Colloid Interface Sci.*, 1981, *83*, 401.
42. K. Fukuda and H. Nakahara, *J. Colloid Interface Sci.*, 1984, *98*, 555.
43. P. S. Vincett and W. A. Barlow, *Thin Solid Films*, 1980, *71*, 305.
44. K. Matsuki and H. Fukutome, *Bull. Chem. Soc. Jpn.*, 1983, *56*, 1006.
45. W. F. Mooney, P. E. Brown, J. C. Russell, S. B. Costa, L. G. Pederson, and D. G. Whitten, *J. Am. Chem. Soc.*, 1984, *106*, 5659.
46. W. F. Mooney and D. G. Whitten, *J. Am. Chem. Soc.*, 1986, *108*, 5712.
47. D. G. Whitten, L. Collins-Gold, T. J. Dannhauser, and W. F. Mooney in *Biotechnological Applications of Lipid Microstructures*, B. P. Gaber, J. M. Schnur, and D. Chapman, eds., Plenum, New York, 1988, p. 291.
48. F. Grieser, P. Thistlethwaite, and P. Triandos, *Langmuir*, 1987, *3*, 1173.
49. N. Kimizuka and T. Kunitake, *J. Am. Chem. Soc.*, 1989, *111*, 3758.
50. M. Flörsheimer and H. Möhwald, *Thin Solid Films*, 1988, *159*, 115.
51. M. Fujihira, T. Kamei, M. Sakomura, Y. Tatsu, and Y. Kato, *Thin Solid Films*, 1989, *179*, 485.
52. N. Tamai, T. Yamazaki, and I. Yamazaki, *J. Phys. Chem.*, 1987, *91*, 841.
53. S. Draxler, M. E. Lippitsch, and F. R. Aussenegg, *Chem. Phys. Lett.*, 1989, *159*, 231.
54. H. Nakahara, K. Fukuda, D Möbius, and H. Kuhn, *J. Phys. Chem.*, 1986, *90*, 6144.
55. T. L. Penner, *Thin Solid Films*, 1988, *160*, 241.

56.   P. Fromherz and G. Reinbold, *Thin Solid Films*, **1988**, *160*, 347.
57.   A. Leitner, M. E. Lippitsch, S. Draxler, M. Riegler, and F. R. Aussenegg, *Thin Solid Films*, **1985**, *132*, 55.
58.   I. Yamazaki, N. Tamai, and T. Yamazaki, *J. Phys. Chem.*, **1987**, *91*, 3572.
59.   N. Tamai, T. Yamazaki, and I. Yamazaki, *Chem. Phys. Lett.*, **1988**, *147*, 25.
60.   N. Tamai, T. Yamazaki, and I. Yamazaki, *Thin Solid Films*, **1989**, *179*, 451.
61.   T. Nagamura, K. Toyozawa, S. Kamata, and T. Ogawa, *Thin Solid Films*, **1989**, *178*, 399.
62.   S. Ito, H. Okubo, S. Ohmori, and M. Yamamoto, *Thin Solid Films*, **1989**, *179*, 445.
63.   S. Ohmori, S. Ito, M. Yamamoto, Y. Yonezawa, and H. Hada, *J. Chem. Soc., Chem. Commun.*, **1989**, 1293.
64.   N. Kimizuka and T. Kunitake, *J. Am. Chem. Soc.*, **1989**, *111*, 3758.
65.   S. Vaidyanathan, L. K. Patterson, D. Möbius, and H. R. Gruniger, *J. Phys. Chem.*, **1985**, *89*, 491.
66.   H. Nakahara and D. Möbius, *J. Colloid Interface Sci.*, **1986**, *114*, 363.
67.   X. Xu, M. Era, T. Tsutsui, and S. Saito, *Chem. Lett.*, **1988**, 773.
68.   X. Xu, M. Era, T. Tsutsui, and S. Saito, *Thin Solid Films*, **1989**, *178*, 541.
69.   M. Sugi, M. Saito, T. Fukui, and S. Iizima, *Thin Solid Films*, **1985**, *129*, 15.
70.   S. Imazeki, M. Takeds, Y. Tomioka, A. Kakuta, A. Mukoh, and T. Narahara, *Thin Solid Films*, **1985**, *134*, 27.
71.   T. Kawaguchi and K. Iwata, *Thin Solid Films*, **1988**, *165*, 323.
72.   T. Kawaguchi and K. Iwata, *Thin Solid Films*, **1989**, *180*, 235.
73.   C. Duschl, W. Frey, C. Helm, J. Als-Nielsen, H. Möhwald, and W. Knoll, *Thin Solid Films*, **1988**, *159*, 379.
74.   H. Nakahara, H. Uchimi, K. Fukuda, N. Tamai, and I. Yamazaki, *Thin Solid Films*, **1989**, *178*, 549.
75.   M. V. Alfimov, I. K. Lednev, and D. A. Styrkas, *Thin Solid Films*, **1989**, *179*, 397.
76.   G. Biesmans, G. Verbeek, B. Verschuere, M. Van Der Auweraer, and F. C. De Schryver, *Thin Solid Films*, **1989**, *169*, 127.
77.   E. Ando, J. Miyazaki, K. Morimoto, H. Nakahara, and K. Fukuda, *Thin Solid Films*, **1985**, *133*, 21.
78.   E. Ando, J. Hibino, T. Hashida, and K. Morimoto, *Thins Solid Films*, **1988**, *160*, 279.
79.   E. Ando, M. Suzuki, K. Moriyama, and K. Morimoto, *Thin Solid Films*, **1989**, *178*, 103.
80.   Y. Unuma and A. Miyata, *Thin Solid Films*, **1989**, *179*, 497.
81.   A. H. Herz, *Photo. Sci. Eng.*, **1974**, *18*, 323.
82.   A. H. Herz, *Adv. Colloid Interface Sci.*, **1977**, *8*, 237.
83.   M. Era, S. Hayashi, T. Tsutsui, S. Saito, M. Shimomura, N. Nakashima, and T. Kunitake, *Chem. Lett.*, **1986**, 53.
84.   U. Lehmann, *Thin Solid Films*, **1988**, *160*, 257.
85.   S. Kirstein, H. Möhwald, and M. Shimomura, *Chem. Phys. Lett.*, **1989**, *154*, 303.
86.   H. Hada, R. Hanawa, A. Haraguchi, and Y. Yonezawa, *J. Phys. Chem.*, **1985**, *89*, 560.
87.   Y. Yonezawa, D. Möbius, and H. Kuhn, *Ber. Bursenges. Phys. Chem.*, **1986**, *90*, 1183.
88.   Y. Yonezawa, T. Hayashi, T. Sato, and H. Hada, *Thin Solid Films*, **1989**, *180*, 167.
89.   A. Nakano, S. Shimizu, T. Takahashi, H. Nakahara, and K. Fukuda, *Thin Solid Films*, **1988**, *160*, 303.
90.   W. Arden and P. Fromherz, *Ber. Bunsenges. Phys. Chem.*, **1978**, *82*, 868.

91. P. Fromherz and W. Arden, *Ber. Bunsenges. Phys. Chem.*, **1980**, *84*, 1045.
92. P. Fromherz and W. Arden, *J. Am. Chem. Soc.*, **1980**, *102*, 6211.
93. H. Sato, M. Kawasaki, K. Kasatani, Y. Higuchi, T. Azuma, and Y. Nishiyama, *J. Phys. Chem.*, **1988**, *92*, 754.
94. I. Yamazaki, N. Tamai, T. Yamazaki, A. Murakami, M. Mimuro, and Y. Fujita, *J. Phys. Chem.*, **1988**, *92*, 5035.
95. I. Yamazaki, N. Tamai, and T. Yamazaki, *J. Phys. Chem.*, **1990**, *94*, 516.
96. S. P. Spooner and D. G. Whitten, unpublished results.
97. A. W. Adamson, A. Vogler, and I. Lantzke, *J. Phys. Chem.*, **1969**, *73*, 4183.
98. A. Vogler and A. Kern, *Z. Naturforsch.*, **1979**, *34b*, 271.
99. R. C. Ahuja and D. Möbius, *Thin Solid Films*, **1989**, *179*, 457.
100. R. A. Marcus in *Annual Review of Physical Chemistry*, Vol. 15, H. Eyring, C. J. Christensen, and H. S. Johnston, eds., Annual Reviews, Inc., Palo Alto, CA, 1964, p. 155.
101. For a review of Marcus theory see R. D. Cannon, *Electron Transfer Reactions*, Butterworths, London, 1980.
102. D. Möbius, *Mol. Cryst. Liq. Cryst.*, **1979**, *52*, 235.
103. H. Kuhn, *Pure and Appl. Chem.*, **1979**, *51*, 341.
104. D. Möbius, *Mol. Cryst. Liq. Cryst.*, **1983**, *96*, 319.
105. M. Sugi, K. Nemback, and D. Möbius, *Thin Solid Films*, **1975**, *27*, 205.
106. E. E. Polymeropoulos, D. Möbius, and H. Kuhn, *J. Chem. Phys.*, **1978**, *68*, 3918.
107. E. E. Polymeropoulos, D. Möbius, and H. Kuhn, *Thin Solid Films*, **1980**, *68*, 173.
108. T. L. Penner and D. Möbius in *Colloids and Surfaces in Reprographic Technology*, M. L. Hair and M. C. Croucher, eds. ACS Symposium Series 200, American Chemical Society, Washington, DC, **1982**, p. 111.
109. T. L. Penner and D. Möbius, *J. Am. Chem. Soc.*, **1982**, *104*, 7407.
110. M. Fujihira, K. Nishiyama, and H. Yamada, *Thin Solid Films*, **1985**, *132*, 77.
111. M. Fujihira and H. Yamada, *Thin Solid Films*, **1988**, *160*, 125.
112. M. Fujihira and M. Sakomura, *Thin Solid Films*, **1989**, *179*, 471.
113. M. Fujihira, M. Sakomura, and T. Kamei, *Thin Solid Films*, **1989**, *180*, 43.
114. Y. Hirata, K. Suga, and M. Fujihira, *Thin Solid Films*, **1989**, *179*, 95.
115. T. Miyashita, T. Yatsue, Y. Mizuta, and M. Matsuda, *Thin Solid Films*, **1989**, *179*, 439.
116. J. Saltiel, J. D'Agostino, E. D. Megarity, L. Metts, K. R. Newberger, M. Wrighton, and O. C. Zafiriou, *Org. Photochem.*, **1973**, *3*, 1.
117. J. Saltiel and J. L. Charlton in *Rearrangements in Ground and Excited States*, Vol. 3, P. deMayo ed., Academic Press, New York, 1980. p. 25.
118. M. T. Allen and D. G. Whitten, *Chem. Rev.*, **1989**, *89*, 1691.
119. D. G. Whitten, *Angew. Chem. Int. Ed. Engl.*, **1979**, *18*, 440.
120. D. G. Whitten, *J. Am. Chem. Soc.*, **1974**, *96*, 594.
121. D. G. Whitten, F. R. Hopf, F. H. Quina, G. Sprintschnik, and H. W. Sprintschnik, *Pure and Appl. Chem.*, **1977**, *49*, 379.
122. D. G. Whitten and P. R. Worsham, *Org. Coatings and Plastics*, **1978**, *38*, 572.
123. D. G. Whitten, D. W. Eaker, B. E. Horsey, R. H. Schmehl, and P. R. Worsham, *Ber. Bunsenges. Phys. Chem.*, **1978**, *82*, 858.
124. F. H. Quina, D. Möbius, F. A. Carroll, F. R. Hopf, and D. G. Whitten, *Z. Physik. Chem.*, **1976**, *101*, 151.
125. F. H. Quina and D. G. Whitten, *J. Am. Chem. Soc.*, **1975**, *97*, 1602.
126. F. H. Quina and D. G. Whitten, *J. Am. Chem. Soc.*, **1977**, *99*, 877.
127. L. Collins-Gold, D. Möbius, and D. G. Whitten, *Langmuir*, **1986**, *2*, 191.
128. K. Kano, Y. Tanaka, T. Ogawa, M. Shimomura, Y. Okahata, and T. Kunitake, *Chem. Lett.*, **1980**, 421.

129. S.L. Regen, K. Yamaguchi, N. K. P. Samuel, and M. Singh, *J. Am. Chem. Soc.*, **1983**, *105*, 6354.
130. S. S. Sandhu, Y. P. Yianni, C. G. Morgan, D. M. Taylor, and B. Zaba, *Biochim. Biophys. Acta,* **1986**, *860*, 253.
131. K. Nishiyama and M. Fujihira, *Chem. Lett.*, **1988**, 1257.
132. K. Nishiyama, M. Kurihara, and M. Fujihira, *Thin Solid Films*, **1989**, *179*, 477.
133. A. Yabe, Y. Kawabata, H. Niino, M. Tanaka, A. Ouchi, H. Takahashi, S. Tamura, W. Tagaki, H. Nakahara, and K. Fukuda, *Chem. Lett.*, **1988**, 1.
134. A. Yabe, Y. Kawabata, H. Niino, M. Matsumoto, A. Ouchi, H. Takahashi, S. Tamura, W. Tagaki, H. Nakahara, and K. Fukuda, *Thin Solid Films*, **1988**, *160*, 33.
135. M. Tanaka, Y. Ishizuka, M. Matsumoto, T. Nakamura, A. Yabe, H. Nakanishi, Y. Kawabata, H. Takahashi, S. Tamura, W. Tagaki, H. Nakahara, and K. Fukuda, *Chem. Lett.*, **1987**, 1307.
136. Y. Kawabata, M. Matsumoto, M. Tanaka, H. Takahashi, Y. Irinatsu, S. Tamura, W. Tagaki, H. Nakahara, and K. Fukuda, *Chem. Lett.*, **1986**, 1933.
137. M. Shimomura and T. Kunitake, *Thin Solid Films*, **1985**, *132*, 243.
138. A. Yabe, Y. Kawabata, A. Ouchi, and M. Tanaka, *Langmuir*, **1987**, *3*, 405.
139. A. Ouchi, M. Tanaka, T. Nakamura, M. Matsumoto, Y. Kawabata, S. Tomimasu, and A. Yabe, *Chem. Lett.*, **1986**, 1833.
140. B. Tieke, *Adv. Polym. Sci.*, **1985**, *71*, 79.
141. C. Bubeck, *Thin Solid Films*, **1988**, *160*, 1.
142. P. R. Worsham, D. W. Eaker, and D. G. Whitten, *J. Am. Chem. Soc.*, **1978**, *100*, 7091.
143. D. A. Holden, H. Ringsdorf, and M. Haubs, *J. Am. Chem. Soc.*, **1984**, *106*, 4531.
144. H. Pommier and J. Metzger, Fr. Patent, **1972**, *2*, 181 and 208; *Chem. Abstr.* **1974**, *81*, 65255d.
145. I. Gruda and R. M. Leblanc, *Can. J. Chem.*, **1976**, *54*, 576.
146. M. Morin, R. M. Leblanc, and I. Gruda, *Can. J. Chem.*, **1980**, *58*, 2038.
147. E. E. Polymeropoulos and D. Möbius, *Ber. Bunsenges. Phys. Chem.*, **1979**, *83*, 1215.
148. H. S. Blair and I. Pogue, *Polymer*, **1979**, *20*, 99.
149. H. Gruler, R. Vilanove, and F. Rondelez, *Phys. Rev. Lett.*, **1980**, *44*, 590.
150. D. A. Holden, H. Ringsdorf, V. Deblauwe, and G. Smets, *J. Phys. Chem.*, **1984**, *88*, 716.
151. B. C. McArdle, H. Blair, A. Barraud, and A. Ruaudel-Teixier, *Thin Solid Films*, **1983**, *99*, 181.
152. T. Nagamura, K. Sakai, and T. Ogawa, *J. Chem. Soc., Chem. Commun.*, **1988**, 1035.
153. T. Nagamura, K. Sakai, and T. Ogawa, *Thin Solid Films*, **1989**, *179*, 375.
154. H. Tachibana, T. Nakamura, M. Matsumoto, H. Komizu, E. Manda, H. Niino, A. Yabe, and Y. Kawabata, *J. Am. Chem. Soc.*, **1989**, *111*, 3080.
155. H. Tachibana, A. Goto, T. Nakamura, M. Matsumoto, E. Manda, H. Niino, A. Yabe, and Y. Kawabata, *Thin Solid Films*, **1989**, *179*, 207.
156. K. Ichimura, Y. Suzuki, T. Seki, A. Hosoki, and K. Aoki, *Langumir*, **1988**, *4*, 1214.
157. J. Sagiv, *J. Am. Chem. Soc.*, **1980**, *102*, 92.

Chapter 16

Host–Guest Photochemistry in Solution

Akihiko Ueno

Department of Bioengineering,
Faculty of Bioscience and Biotechnology,
Tokyo Institute of Technology, Yokahama, Japan.

Tetsuo Osa

Pharmaceutical Institute,
Tohoku University, Sendai, Japan.

Contents

1. Introduction

Host–guest chemistry,[1] the chemistry of two or more species assembled together without a covalent bond, is the basis for a new type of photochemistry, in which photophysics and photochemistry of the guest is modified and made unique. This chapter deals with photochemistry in solutions of host–guest complexes in which cyclodextrins (CDs), crown ethers, or cryptands serve as hosts. Host–guest photochemistry in membranes is also included under some topics as an extension of host–guest photochemistry in solution (for host-guest photochemistry in the solid state see Chapter 10).

2. Cyclodextrins as Hosts

Cyclodextrins (CD) are cyclic oligosaccharides that have a central cavity capable of accommodating guest molecules in aqueous solution.[2] These molecules, containing six, seven, and eight glucose units are called α-CD (cyclohexaamylose), β-CD (cycloheptaamylose), and γ-CD (cyclooctaamylose), each having a different cavity diameter of approximately 4.5 Å, 7.0 Å, and 8.5 Å respectively.

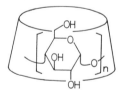

                    n = 6 α–CD
                    n = 7 β–CD
                    n = 8 γ–CD

These molecules are shaped like truncated cones, with smaller and larger opening faces at the primary and secondary hydroxyl faces, respectively. Cyclodextrins are generally soluble in water and not in organic solvents. However, several methylated derivatives have recently been made that dissolve in organic solvents. The interiors of the cavities encircled by ether oxygens provide a hydrophobic microenvironment in an aqueous solution. The guest molecules accommodated in these cavities are relatively isolated from the bulk water environment and often have enforced orientation and constrained conformation. Photochemistry of guest species included in CD cavities may be perturbed or modified. Further, CDs have been used as molecular vessels to regulate photochemical events.[3]

2.1. Photophysical Studies
2.1.1. Fluorescence

The hydrophobic microenvironment of a CD cavity is reflected in the fluorescence behavior of included guest molecules. Many aromatic compounds exhibit enhanced fluorescence in the presence of CDs. Among them, a series of compounds represented by 1-anilinonaphthalene-8-sulfonate (1, ANS) have been used extensively as fluorescence probes in biological systems.

NH

SO_3^-

1 ANS

The excited states of these molecules decay by rotation of the anilino group around the N-naphthyl bond toward an emissive "twisted intramolecular charge-transfer" state.[4] Due to the polar nature of this excited state, the wavelength of the emission maximum (λ_{max}) depends strongly on the polarity of the medium. Since the microenvironment of CD cavities is similar to that of dioxane[5] or alcohols,[3,5] the molecules included in CD cavities exhibit enhanced fluorescence with emission maxima shifted to shorter wavelength (with respect to that in the bulk water environment). In the presence of α- and β-CD (1m M) ANS shows emission maxima at 510 and 495 nm, respectively, slightly shifted to shorter wavelength from the original fluorescence (515 nm).[6] The effect of γ-CD was reported to be the same as β-CD. Another molecule, dimethylaminobenzonitrile, a molecule possessing a "twisted intramolecular charge-transfer state" also has been used as a polarity probe for CD interiors.[7] Ueno et al.[8a] observed an enhancement of fluorescence from 2-naphthyloxyacetic acid when included in γ-CD along with cyclohexanol as a coguest (Figure 1, A). Under such conditions cyclohexanol seems to act as a spacer, facilitating complexation of the naphthyl derivative by narrowing the CD cavity. Fluorescence of benzene and its derivatives is also enhanced by β-CD. Hoshino et al.[8b] attribute such an enhancement to the increase in radiative rate, the decrease in rotational freedom of the included molecule, and the elimination of quencher water molecules from the immediate surrounding.

Figure 1. 1:1:1 Complexes in which one molecule acts as a spacer (S) or a cap (C) to facilitate a guest (G) accomodation within a CD cavity

Fluorescence quenching of CD included guest molecules is also influenced by the cavity. Kano et al.[9] studied the effects of CDs on fluorescence quenching of pyrene by amines in aqueous solution. The rate of fluorescence quenching for dimethylamine, diethylamine, trimethylamine, and triethylamine was extremely high in β-CD, being 318 times higher than that without CD. In contrast, the enhancement was only threefold in γ-CD. Since pyrene cannot be included fully into the β-CD cavity due to its large size, a 1:1:1 complex in which pyrene is located at the entrance and the quencher is included in the cavity is suggested to be formed (Figure 1, B). Similar 1:1:1 complexes are suggested between 1-pyrenesulfonate, β-CD, and aniline[10] and between pyrene, β-CD, and surfactants.[11] 1-Cyanonaphthalene, pyrene, naphthalene, acenaphthene, and fluorene also form 1:1:1 complexes with β-CD in the presence of alcohols and nitriles.[12] Thus, the enhancement of fluorescence quenching observed in these systems is taken as an indication of the ability of CD to assemble both the fluorophore and the quencher within its narrow cavity.

Fluorescence behavior of molecules able to undergo adiabatic photoreaction is also affected by the CD cavity, as illustrated with an example below. Weller[13] revealed that aromatic alcohols become strong acids on photoexcitation. Dual emission (naphthol and naphthoate anion) is observed for β-naphthol in aqueous solutions of appropriate pH. The fluorescence spectrum of β-naphthol in a pH 6.2 aqueous solution shows peaks at 353 and 420 nm arising from neutral (β-naphthol) and anionic species (β-naphthaoate anion) respectively. The effect of CD on such emissions was studied with β-naphthol.[14,15] As the concentration of β-CD increases peak intensity enhancement due to neutral species occurs with a corresponding decrease in the emission from the anionic species. Such behavior indicates that the rate of excited-state proton transfer is markedly depressed for β-naphthol included in β-CD. Eaton has examined the influence of added β-CD on the excited-state lifetime of β-naphthol.[14] The excited singlet lifetime for neutral species in the presence of β-CD is 7.2 ns, which is slightly longer than that for β-naphthol in water (4.8 ns). Based on lifetime comparisons it has been concluded that the cavity microenvironment is very similar to that of alcohol solvent (methanol, 5.9 ns or ethanol, 8.9 ns), in which photodissociation of β-naphthol does not occur. Recently, emission from 1,6-naphthalenediol included in β- and γ-cyclodextrins has been analyzed on the basis of the effect of CD on proton transfer rates.[16]

2.1.2. Phosphorescence

Turro et al.[17a] investigated the phosphorescence behavior of a series of cationic detergents (**2**, **3**, and **4**) included in β-CD. No evidence for association with α-CD was detected. Phosphorescence decay has been measured using $Co(NH_3)_6^{3+}$ as an external (aqueous phase) quencher. Two types of inclusion complexes, **A** and **B**, are proposed for short- and long-lived complexes of **4** with γ-CD (Figure 2).

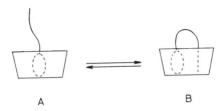

A   B

Figure 2. Equilibrium between host–guest (CD and deteregent, respectively) complexes having extended (A) and folded (B) conformations of a guest detergent molecule.

The ratio of the quenching constants for **4** in the presence and in the absence of CD are 0.57, 0.15, and 0.017 for β, γ (short-lived), and γ (long-lived) complexes, respectively. The decrease in the ratio is marked for system **B**, implying that the triplet is protected from attack by the quencher. The stable nature of the **B** complex is also reflected in the large K value (1.6×10^4 M^{-1}) as compared with the value (4.4×10^2 M^{-1}) for the **3**–γ-CD system.

$$
\begin{array}{c}
\text{Br} \\
\text{[naphthalene]} \\
| \\
\text{C}{=}\text{O} \\
| \\
(\text{CH}_2)_n \\
| \\
\text{CH}_3{-}\overset{\pm}{\text{N}}{-}\text{CH}_3 \\
| \\
\text{CH}_3 \quad \text{Br}^-
\end{array}
$$

n =	1	**2**
	5	**3**
	10	**4**

Such reductions in quenching rate by external quenchers present in aqueous media have been utilized by Turro and others to observe phosphorescence from several aromatic molecules included in CD.[17] For example, intense phosphorescence from 1-bromonaphthalene included in β-CD was seen even in the presence of oxygen. This initial observation prompted Scypinski and Cline-Love[17b] to monitor phosphorescence from a large number of organic molecules at room temperature in aqueous CD solution. Dibromoalkane included along with the organic molecule within the CD cavity provides the spin–orbit coupling needed to enhance the phosphorescence emission yield.[17c] Details of this technique are covered in Chapter 2.

2.1.3. Excimer and Exciplex

γ-Cyclodextrin has a cavity large enough to accommodate two molecules each of benzene and naphthalene. Ueno et al.[18] observed enhanced excimer formation of 1-naphthylacetic acid included in γ-CD, and this was attributed to the ability of γ-CD to include two molecules of 1-naphthylacetic acid. This finding prompted several workers to examine the effects of γ-CD on excimer formation. Particular attention has been paid to pyrene excimer in the presence of γ-CD. Yorozu et al.[19] attributed the enhanced excimer emission to the inclusion of two pyrene molecules within a single γ-CD cavity to yield a 2:1 pyrene–γ-CD complex. Similar results have been obtained by Harada et al.[20] for 1-pyrenesulfonate included in γ-CD. Kobayashi et al.[21] have pointed out that in addition to the above 2:1 complexes, pyrene excimer fluorescence is partially due to a 2:2 pyrene–γ-CD complex formed by association of 1:1 complexes. This proposal gains support from studies by Herkstroeter et al.,[22] Hamai,[23] and Ueno et al.[24,25] Based on careful fluorescence lifetime studies, Herkstroeter et al. have suggested that the excimer fluorescence in pyrene–γ-CD is due to a barrel-type 2:2 complex. Based on circular dichorism and circularly polarized fluorescence studies Kano et al.[26] showed that pyrene formed chiral excimers within the cavities of γ-CD. Excimer formation in a 2:2 complex of β-CD and naphthalene has also been reported.[27] In this case, like the above examples, a 2:2 complex is proposed to be formed by the association of 1:1 β-CD naphthalene complexes.

Turro et al.[28] reported an enhanced intramolecular excimer formation from 1,3-bichromophoric propanes (5) in aqueous γ-CD solution. Similar enhancement in intramolecular excimer formation was observed for 6 and 7 in the presence of γ-CD by Itoh and Fujiwara,[29a] for 8 and 9 by Emert et al.,[29b] for 10 by Arad-Ellin and Eaton,[30] for 11 by Ueno and co-workers,[31a] and for 12 by Hamai.[31b] In these γ-CD complexes, the bichromophoric molecules are forced to take a particular conformation with the two chromophores arranged parallel (Figure 3). Similar conformational control reflected in the intramolecular triplet quenching of *para*-methoxy-β-phenylpropiophenone included in β-CD has recently been reported.[32] These examples highlight the importance of the conformational control offered by CD cavity.

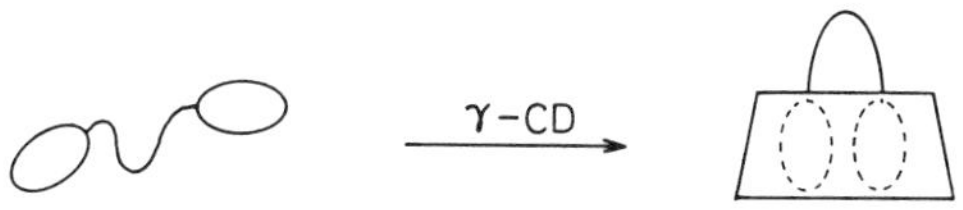

Figure 3. γ-Cyclodextrin-induced conformational change in a guest bearing two aromatic rings.

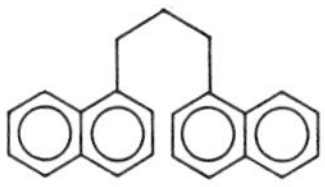

5

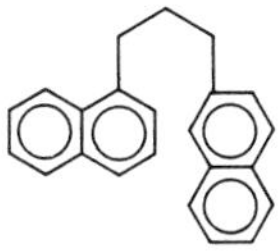

6

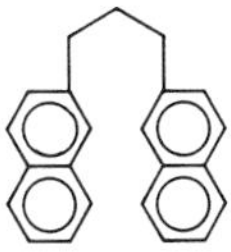

7

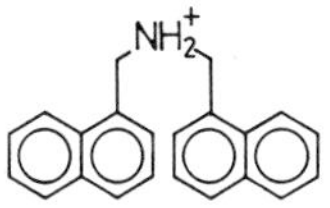

8

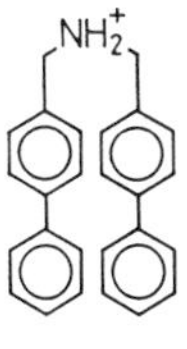

9

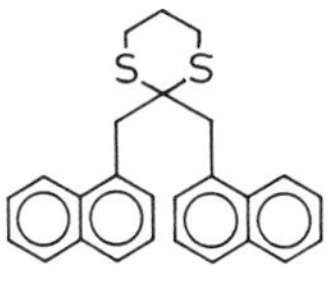

10

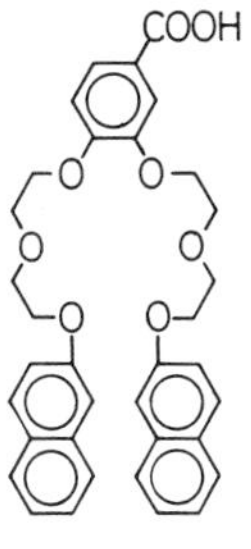

11

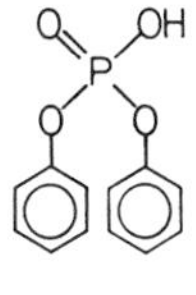

12

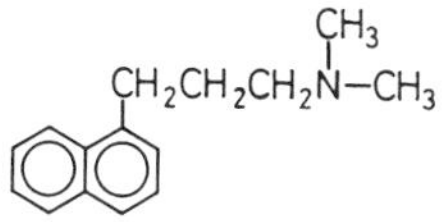

13

Exciplex emission was first observed by Hamai for a mixture of β-CD, 2-methoxynaphthalene (**MN**), and *ortho*-dicyanobenzene (**DC**).[27a] The peak wavelength of the emission is 480 nm, which is the same as that for the exciplex formed between **MN** and **DC** in dioxane. This indirectly suggested that the CD cavity might have an internal polarity similar to that of dioxane. The nature of the complex is suggested to be 1:1:2 between **MN, DC,** and β-CD. Kano et al. have reported exciplex emission (λ_{max} = 477 nm) for the β-CD–naphthalene–trimethylamine (**TMA**) system.[33a] The intensity of the emission is linearly dependent on the concentration of β-CD. The exciplex emission is suggested to arise either from aggregates of a 1:1:1 complex or from a 1:1:2 naphthalene–TMA–β-CD complex. Recently Hamai has reported similar exciplex emission from an 1:1:1 complex of pyrene, aniline, and β-CD.[27b] An intramolecular exciplex similar to the intramolecular excimer discussed above has also been reported. Turro et al.[33b] observed an intramolecular exciplex emission from 1-α-naphthyl-3-(dimethylamino)propane (**13**) included in β-CD. In the absence of β-CD, the fluorophore exhibited only naphthalene-like emission at 334 nm.

2.1.4. Energy-Transfer

Triplet energy transfer from xanthone (donor) to naphthalene (acceptor) has been observed in the cavities of γ-cyclodextrin.[34] Time-resolved esr studies showed that both xanthone and naphthalene were present in the same cavity of γ-CD and had a fixed mutual orientation. Triplet–triplet (T–T) energy transfer at 77 K is shown to be accompanied by anisotropic spin polarization. When α- or β-CD is used as the host no T–T energy transfer was observed. This was interpreted to mean that when the donor and the acceptor are present in different cavities (α and β-CD being too small to accommodate both the molecules), they do not undergo T–T energy transfer.

2.1.5. Summary

Most of the photophysical studies carried out so far have been steady state. Based on intensity and variation of λ_{max} fluorescence, the polarity of the CD cavity is proposed to be inbetween those of dioxane and ethanol. Both fluorescence and phosphorescence of the included molecules are generally enhanced as a result of tight fitting and reduced quenching by the external quenchers. Most often the cavity size can be fine tuned to yield a tight fit for the guest by having a second, inert guest molecule, such as an alcohol or a hydrocarbon, included in the cavity. Under such conditions a 1:1:1 complex results. Extensive studies related to excimers and exciplexes have clearly indicated that the structure of host–guest complexes is more complex than generally assumed (1:1). Complexes having host–guest compositions 2:1, 2:2, and 1:2 have been inferred. Studies on exciplexes have indicated the presence of 1:1:1 and 1:1:2 types of structures. The dynamic nature of these structures is revealed by a recent study[35a] on anilinonaphthalene sulfonate included in β-CD. These studies show that the fluorescence lifetimes are better represented by a distribution than by a single exponential, suggesting that the structure of host–guest complexes cannot be represented by a single structure. Not much effort has been directed toward understanding the dynamics of these complexes. In this context, a recent report by Scaiano and co-workers[35b] is important. They have shown that exit rates from the CD cavity to an aqueous exterior for triplet xanthone are higher than that for xanthone present in the ground electronic state. Photophysical studies carried out to date clearly suggest that careful attention should be paid to the nature, structure, and dynamics of cyclodextrin complexes when interpreting the photoreactions of complexes that are discussed below.

2.2. Photochemical Studies
2.2.1. Unimolecular Photoreactions

Electron-Transfer Reactions. One of the most basic photoreactions, electron transfer, is also influenced by cyclodextrin complexation. The phenothiazine triplet has been known to produce cation radicals in the presence of an appropriate electron acceptor such as viologens. However, Yonemura et al.[36a] detected no cation radicals upon photolysis of **14** in solution. This was attributed to an exceedingly short lifetime of the cation radical due to fast back electron transfer. Consistent with this rationale, **14** complexed with either α- or β-CD gave transient signals corresponding to the cation radicals of **14**. Such a difference in behavior is readily understood on the basis of structure **15** (supported by nmr studies) in which back electron transfer is expected to be slow. Structure **15** can also be expected to influence the rate of forward electron transfer. This prediction is realized on comparing the emission behavior of **14** in the presence and the absence of CD. Moderate fluorescence from **14** was observed in the presence of CD, while none was seen in its absence. Similar reduction in electron-transfer rates was reported earlier by Wilmer et al.[36b] for systems consisting of alkyl viologens, porphorins, and CDs.

14   15

Norrish Type I and II Reactions. Photolysis of α-ethyldibenzyl ketone in methanol led to type I (**AA**, 22%; **AB**, 43%; **BB**, 21%) and type II products (**DBK**, 13%) (Scheme 1).[37] Photolysis of this ketone in aqueous solution containing β-CD resulted in different product distribution (**AA**, 1%; **AB**, 40%; **BB**, 1%; **RP**, 49%). A similar effect of β-CD was also observed for several other α-alkyldibenzyl ketones. Absence of type-II products in aqueous β-CD solution is attributed to the inclusion of the substrate into the β-CD cavity in a conformation that is unfavorable for γ-hydrogen abstraction. Selective formation of AB and RP suggests that the cavity of β-CD provides a large "cage effect."

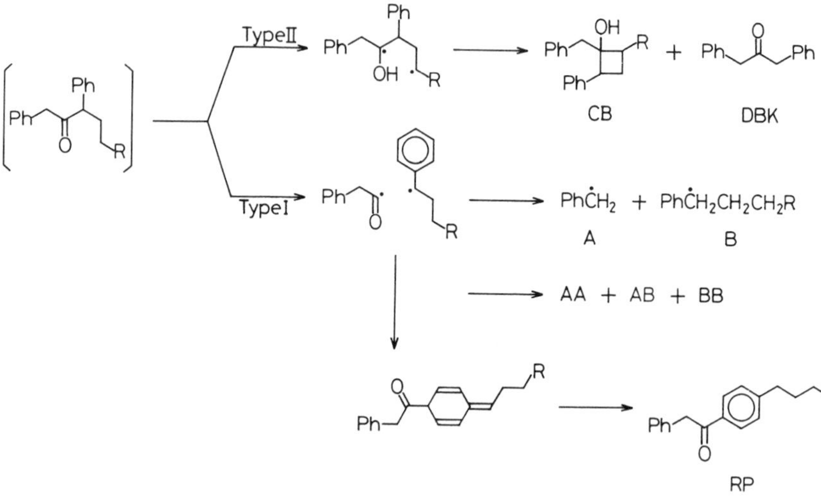

Scheme 1

Photolysis of α-ethyldeoxybenzoin in methanol (Scheme 2) resulted in type I (**A**, 12.9%; **B**, 8.0%; **C**, 46.3%) and type II products (**D**, 8.0%; **E**, 23.8%).[38] Photolysis of this compound in aqueous β-CD solution at 5°C gave the same products with a different distribution (**A**, 13.5%; **B**, 1.3%; **C**, 15.0%; **D**, 53.3%; **E**, 16.7%). The relative yield of the type-I products is decreased in the presence of β-CD. Among the type-II products, cyclobutanol, a cyclization product from the 1,4-diradical, is favored. Enhancement of cyclization process (**D** over **E**) is believed to reflect the rotational restriction brought on the 1,4-diradical intermediate by the cavity.

Product distribution upon photolysis of benzoin alkyl ethers included in β-CD depends on the length of the alkyl chain.[39] The short-chain benzoin alkyl ethers prefer to form complexes in which the alkyl chain is oriented outward from the cavity (**16a**). On the other hand, the long-chain alkyl benzoin ethers may form complexes in which the alkyl chain and the phenyl ring are within the cavity (**16b**). Since γ-hydrogen abstraction by ketone carbonyl is unable to occur in **16b**, the type-II reaction may be depressed for the long-alkyl chain benzoin ethers. Results obtained with benzoin ethers are consistent with the presence of these two distinctly different complexes.

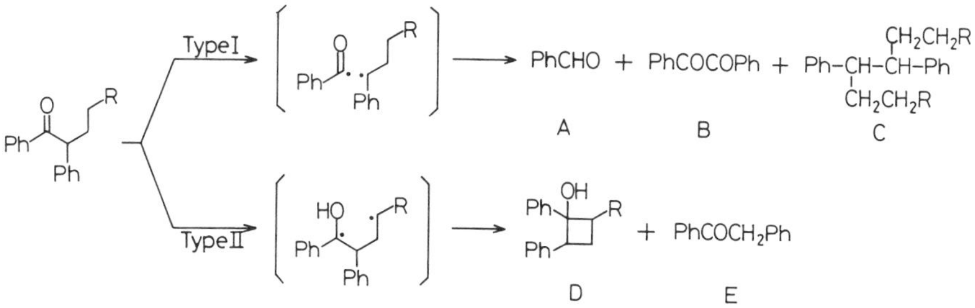

Scheme 2

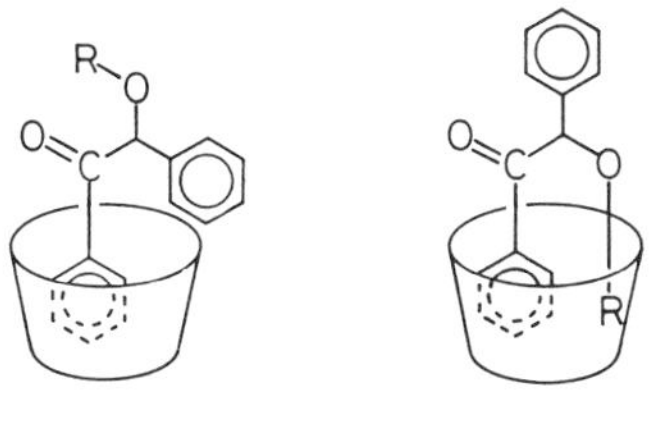

16a   16b

It is clear from the examples discussed above that the CD cavity can control the conformation of the included molecule and can restrict the intramolecular rotational motions. Details on the photobehavior of the solid CD complexes of the above molecules are provided in Chapter 7. Such a conclusion is also supported by the excimer and exciplex studies described in an earlier section.

Rearrangement Processes. Photolysis of phenyl acetate in methanol gives *ortho*-acetylphenol (28%), *para*-acetylphenol (39%), and phenol (34%) as products of photo-Fries rearrangement.[40] Product distribution is altered by the presence of β-CD, being 89% for *ortho*-acetylphenol and 11% for *para*-acetylphenol. Remarkable ortho selectivity is attributed to the formation of an inclusion complex with structure **17**. In such an arrangement β-CD protects the para position and exposes the two ortho positions of the aromatic ring for the reaction. Ortho selectivity of ~99% is obtained when the alkyl group of the ester is either phenyl, adamantyl methylene, or 2,4,6-trimethylphenyl. The same level of ortho selectivity was also observed for corresponding anilides.[41] Earlier studies had reported formation of increased para isomers for ester[42,43] and anilide.[44] This inconsistency might be due to the difference in the amount of β-CD used for the reaction.

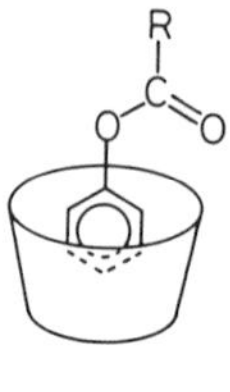

17

    Photolysis of *meta*-methoxyphenylallyl ether (**18**) in ethanol gave a mixture of three rearranged products (**19**, 22%; **20**, 28%; and **21**, 31%) along with the fragmentation product (**22**, 18%) via photo-Claisen rearrangement (Scheme 3).[45] When α-CD is added to the system, formation of **22** and the para rearrangement isomer **21** is retarded (**21**, 7%) and among the two ortho isomers **19** and **20**, an increased preference for **20** is observed (**19**, 23%; **20**, 69%). The effect of α-CD is more remarkable than that of β-CD, indicating that a "tight fit" is essential for higher selectivity.

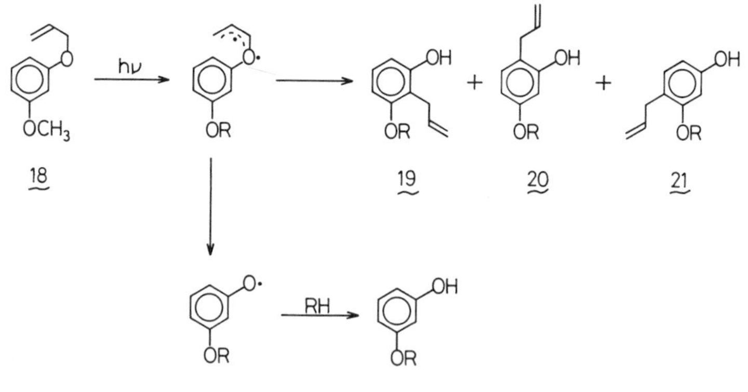

Scheme 3

    α-Cyclodextrin complexation of **23b** decreases the efficiency of base-catalyzed photo-Smiles rearrangement (Scheme 4) by 40% but enhances the efficiency of uncatalyzed one by 67%.[46]

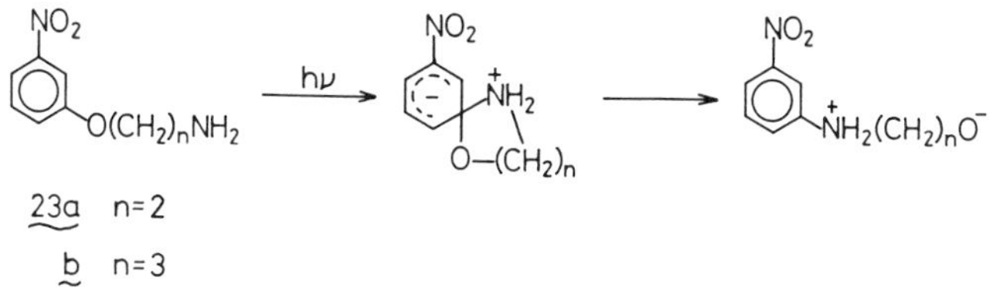

$$23a \quad n=2$$
$$b \quad n=3$$

Scheme 4

This is attributed to the hydrogen bonding by—$NH_2{}^+$—R in the intermediate to a 2- or 3-hydroxyl group in the α-CD rim of the α-CD complex. Such interaction influences the efficiency by stabilizing the nitrogen group against repulsion. The stabilizing effect of **23a** is smaller than that of **23b** due to the less stable nature of the α-CD–**23a** complex.

Examples presented in this section highlight how CD can control the accessibility of the active sites in an included guest molecule. Cyclodextrin can expose certain reactive sites selectively while protecting the others.

Geometric Isomerizaion. Geometric isomerization is one of the principal modes of deactivating excited singlet states in stilbenes and azobenzenes. The photobehavior of stilbene in a CD cavity has been examined by both steady-state and picosecond dynamic measurements.[47] While a single complex is formed between *trans*-stilbene and α-CD resulting in a single exponential decay, two distinctly different complexes which produce double exponential fluorescence decays, are formed between *trans*-stilbene and β-CD. The two decays observed are regarded as average fluorescence lifetimes of loosely bound and tightly complexed forms of *trans*-stilbene. Interconversion between the two forms is believed to occur on a time scale slower than the rate of photoisomerization. Tabushi and co-workers[48] examined the photochemistry of *trans*- and *cis*-stilbene-4,4'-disulfonyl-capped β-CDs. The trans isomer was completely converted into the cis isomer, eventually yielding a phenanthrene derivative, a product of ring closure. Such behavior is different from that of the parent *cis*- and *trans*-stilbenes included in β-CD. While *cis*-stilbene included in β-CD is converted into the trans isomer, *trans*-stilbene included in β-CD fails to isomerize upon photolysis.

Behavior similar to that of stilbene was also noted with azobenezene.[49] Both trans and cis isomers of azobenzene form inclusion complexes with α-, β- and γ-CDs.[49] Photoisomerization of the cis isomer is practically unaffected with respect to solution behavior, whereas that of the trans isomer is strongly reduced due to restricted rotational motion about the nitrogen–nitrogen double bond. Because of the difference in structure of the *trans*- and *cis*-azobenzenes (linear vs. bent), the former binds to β-CD more strongly than the latter. Such a difference in binding strength has been utilized by Ueno et al.[50] to photocontrol the hydrolysis of *para*-nitrophenyl acetate included in β-CD. In the photoregulation system, 4-carboxy azobenzene acts as a photoresponsive inhibitor, competing with the substrate for the CD cavity; this inhibitory effect weakens upon trans–cis photoisomerization because of reduced binding of the cis, resulting in the photoenhanced ester hydrolysis rate.

Several examples of azobenzene-modified CDs are known in which the binding efficiency is photoregulated. Compound **24** forms an intramolecular complex, in which the azobenzene moiety is included in the CD cavity.[51] When a guest is added, the azobenzene moiety acts as a cap and the guest is included in the CD cavity. Binding with guests is stronger when azobenzene moiety has trans geometry than when it has cis. The *trans*-azobenzene moiety of **25** acts as a spacer, narrowing the cavity for smaller guest molecules and acting as a cap for larger ones.[52] Binding ability of **25** is promoted by change from trans to cis geometry. Such a difference in binding strength allows the guest to be included or excluded from the cavity by controlling the photoisomerization through light intensity variation.

Azobenzene-capped CDs **26**, **27**, and **28** can also be used to control complex formation.[53,54] All these compounds undergo trans–cis photoisomerization upon uv irradiation, the cis content in the photostationary state being ~50%.

Since the distance between para, para' substituents of *cis*-azobenzene is shorter than between those of *trans*-azobenzene, the CD framework becomes distorted or deformed upon going from trans to cis geometry. The cis isomer of **28** may exist in two forms, the azobenzene moiety being bent outward from (**28a**) or inward to (**28b**) the cavity. With the above azobenzene-capped CDs, photocontrolled binding was achieved for many guest molecules. This is attributed to the conversion from the shallow cavity of the *trans* form to the enlarged cavity of the *cis* form. A shallow cavity does not allow deep inclusion of guest molecules, while the enlarged cavity enables the guests to be accommodated deeply.

2.2.2. Bimolecular Reactions

Results of photophysical investigations summarized in Section 2.1 confirm that two guest molecules can be accommodated within a CD cavity. Such studies also suggest the possibility of carrying out bimolecular reactions within a CD cavity. Results of such studies are summarized in the three following subsections. Studies carried out so far involve the use of either a pure CD or a CD appended with an aromatic molecule as host. Most of the studies related to CD appended with aromatic molecules originate from Ueno's and Osa's laboratories.

γ-CD

Induced-Fit Complexation. Inclusion properties of CDs appended with one and two aromatics (of the type **29** and **30**; Figures 4 and 5) have been investigated by Ueno et al.[56-59] In these modified CDs, the location of the appended aromatic molecule depends on the presence of guest within the cavity.

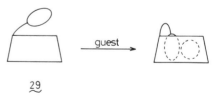

Figure 4. Induced-fit type of complex-ation in γ-CD appended with a 2-naphthyl-acetyl moiety.

29

Figure 5. Induced-fit type of complex-ation in γ-CD appended with two 2-naphthyl moieties.

30

Often the appended aromatic molecule provides a tight fit for the included guest; hence the term "induced-fit". In the absence of a guest the aromatic group in 2-naphthylacetyl appended γ-CD (**29**) resides either within or outside of the cavity. However, when guest molecules are added, it undergoes an induced-fit type of complexation by retaining the naphthyl moiety in the γ-CD cavity along with the guest. Such an arrangement provides a fairly tight fit (Figure 4).[55] In the complex, the naphthyl moiety acts as a spacer, narrowing the large γ-CD cavity to make it suitable for guest accommodation. When the guest molecules are ketones, efficient intramolecular quenching of the naphthyl fluorescence occurs.[56] The quenching rate constant depends on the size of the ketones, being in the order *l*-fenchone > diisopropyl ketone > di-*n*-propyl ketone > diethyl ketone > acetone. γ-Cyclodextrin bearing two 2-naphthylacetyl moieties (**30**) at the A and E glucose residues (for numbering see the picture below) exhibits fluorescence spectra, in which excimer emission is predominant both in the absence and the presence of guest molecules.[57,58] Circular dichorism spectra of this compound reveals a marked exciton coupling band whose intensity decreases upon guest addition. The results indicate that **30** undergoes an induced-fit type of complexation by changing the location of the napthyl moieties from inside to outside of the cavity (Figure 5). β-Cyclodextrin bearing two 2-naphthylacetyl moieties at A and D glucose residues (**31**; Figure 6) shows a fluorescence spectrum that is composed of both the monomer and the excimer emissions.[59]

Figure 6. Conformational equilibrium of β-CD bearing two 2-naphthylacetyl moieties.

31a   31b

Upon guest addition, the intensity of the excimer fluorescence increases at the expense of the monomer fluorescence. These results suggest that **31** exists in equilibrium with forms **a** and **b** (Figure 6). In **31a** one naphthyl moiety is located inside the cavity while another is located outside the cavity, whereas in **31b** two naphthyl moieties are located outside the cavity, interacting with each other. This difference in structures accounts for different emissions from **31a** (monomer fluorescence) and from **31b** (excimer fluorescence). When the **a** form complexes with a guest molecule, it is converted into the **b** form, resulting in enhanced excimer emission. Fujita et al.[60] prepared β-CD derivatives bearing 2-naphthylsulfonyl moieties at AB, AC, and AD glucose residues. Fluorescence spectra of these compounds in aqueous solution are composed of almost pure monomer emission. However, excimer emission appears as a shoulder for the AB isomer and as a predominant peak for the AC and AD isomers upon addition of 1-adamantanecarboxylate as a guest.

These results once again support the working hypothesis that two molecules of guest can be accommodated within a single cavity of CD. Since the x-ray structures of guest included CDs are not easily available we must be satisfied at this stage with conclusions based on indirect observations.

Association Dimers of Aromatic Appended CDs. Dimer in the term "association dimer" refers to the association of two host–guest pairs upon excitation and does not refer to a stable dimer obtained by dimerization of aromatics. Such association dimers, often inferred indirectly, are the precursors for dimers and/or excimers. However, association dimers may or may not yield stable dimeric products. A brief mention of such systems was made in Section 2.1.3. Examples of association dimers (from aromatic appended CDs) not yielding stable dimeric products are presented below. γ-Cyclodextrin derivatives bearing a pyrene moiety (**32, 33**) show excimer emission around 470 nm in a 10% dimethyl sulfoxide (**DMSO**)–aqueous solution.[24,25]

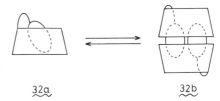

The intensity of excimer emission as well as the uv absorption spectra of **32** and **33** depend on the concentration of **32** and **33**, the temperature, and the **DMSO** content in aqueous medium. At low concentrations compound **32** gives rise to cd bands indicating that the pyrene chromophore is inserted into the γ-CD cavity and probably has structure **32a** (Figure 7). Analyses of the relationship between the 344 nm absorption and the concentration revealed that they form association dimers with association constants of 1.7 x 10^5 and 1.5 x 10^4 M^{-1} for **32** and **33**, respectively.

Figure 7. Formation of an association dimer of γ-CD bearing a pyrene moiety.

With increasing pH, the excimer intensity of **32** abruptly decreased around pH 12.5, which corresponds to the pK_a of the secondary hydroxyls of CD. This pH dependency indicates that two molecules of **32** are joined together with their secondary hydroxyl sides facing each other and come apart at high pH due to repulsion between the hydroxy anions. These results suggest that at high concentrations, the intramolecular complex (**32a**) is converted into the intermolecular association dimer (**32b**) in which two pyrene moieties interact with each other (Figure 7). The pyrene excimer formed in **32b** is almost completely protected against quenching due to isolation from the bulk aqueous environment. γ-Cyclodextrin derivatives bearing a pyrenecarbonyl moiety at C2 (**34**) and C3 (**35**) show marked differences in fluorescence spectra:[61] **34** exhibits remarkable excimer emission, whereas **35** exhibits almost pure monomer emission. The excimer emission of **34** is due to the association dimer that is formed with a large association constant of 2.7 x 10^6 M^{-1}. Compounds **34** and **35** differ only in where the pyrenecarbonyl moiety is attached (to C2 or C3). This emission behavior suggests that the pyrenecarbonyl moiety of **34** is oriented toward the interior of the cavity and so is suited to form the association dimer, whereas that of **35** is oriented in a direction that is unfavorable for dimer formation (Figure 8).

Figure 8. Proposed structures of γ-CD derivatives bearing a pyrene (Py) moiety at C2 (**34**) and C3 (**35**).

Association dimers similar to those of **32** and **33** may also be formed from β-CD derivatives with an appropriate substituent. The excimer formation found for a β-CD derivative bearing a naphthyl moiety (**36**) is one such example.[62] In this case, the excimer emission was observed at higher concentrations. Since the intensity of the excimer markedly decreased around pH 12.5 with increasing pH, the secondary hydroxyl sides of β-CD units were facing each other as observed for pyrene-modified CDs.

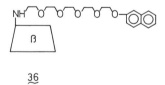

Photodimerization. Tamaki et al.[63,64] were the earliest to explore the utility of the CD cavity in effecting regio- and stereoselective dimerization of guest molecules. Photodimerization of 2-anthracenesulfonate (**2-AS**) in aqueous solution produces four photodimers: anti-head-tail (**A**), syn-head-tail (**B**), anti-head-head (**C**), and *syn-head-head* (**D**) with relative yields 1:0.8:0.4:1.1. Photolysis of a 2:2 complex of **2-AS** with β-CD gave exclusively **A**. The regioselectivity was attributed to the specific mutual orientation between the two molecules of **2-AS**

in the 2:2 complex. In contrast to the β-CD complex, photodimerization of **2-AS** included in γ-CD (2:1 complex) gave almost the same relative yields of the photodimers as those in host-free solution. However, the cd spectra of dimers **B** and **C** formed in the presence of γ-CD exhibited cd bands, indicating that the presence of the chiral cavity of γ-CD produced optically asymmetric photodimers of **B** and **C** (**B** and **C** obtained in the absence of CD did not show any cd activity). Although stereoselectivity was not high (10%) this remarkable observation has not been followed up. 1-Anthracenesulfonate (**1-AS**) forms only 1:1 complexes with β-CD and regioselective photodimerization was not observed. Photodimerization of **1-AS** was accelerated by γ-CD due to 2:1 complex formation, but the reaction proceeded nonspecifically, similarly to **2-AS** included in γ-CD.

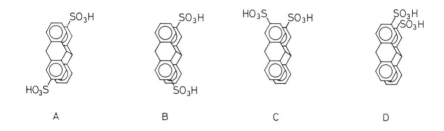

A   B   C   D

It is clear from the above studies that predictability with respect to the nature of product photodimer obtained by photolysis of aromatics included in CDs is very poor. Further, there is no guiding principle at this stage to bringing about selective geometrical orientation of guest molecules within the CD cavity. In this context the approach taken by Ueno and co-workers[65-68] is of interest. They have attempted to establish that it is possible to alter the stereochemical course of the dimerization of aromatics with the use of CD. Usually irradiation of 9-substituted anthracenes gives only the trans photodimer. Ueno et al.[65] prepared γ-CD derivatives bearing one (**37**) or two (**38 a–d**) 9-anthracenecarboxylate moieties.

37   38a   38b   38c   38d

$$X = -O-\overset{\overset{\text{O}}{\|}}{C}-$$

The host (**37**) exists as an equilibrium mixture of a monomer and an association dimer (association constant: 1.1×10^5 M^{-1}). The cd spectra of the association dimer exhibit an exciton coupling band of R helicity in the ^1B_b transition region of anthracene, suggesting an interaction between the two anthracene units. Indeed photolysis of the association dimer gives rise to anthracene dimers. The regioisomers of **38a–d** also exhibit exciton coupling patterns with a shape and magnitude that depend on the positions of the substituents.[66] Upon uv irradiation, all of them undergo intramolecular photodimerization.[67] Compounds **38c** and **38d** yield isolable trans dimers, whereas **38b** gives a trans dimer that is too unstable to isolate (Figure 9).

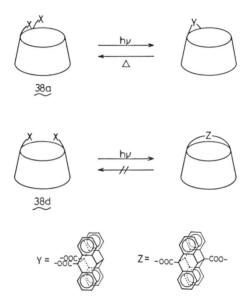

Figure 9. Photochemistry of γ-CD derivatives bearing two 9-anthracenecarboxylate units at AB (**38a**) and AE (**38d**) positions.

In the case of **38b**, formation of both the cis and the trans photodimer is stereochemically unlikely. An interesting result is obtained with **38a**. Photolysis of **38a** results in an inherently unstable cis dimer. Examination of molecular models indicates that **38a** cannot form the trans photodimer because of the geometrical arrangement between the two 9-anthracenecarboxylate moieties; that is, **38a** can give only the cis photodimer as is obtained. Results with 1-anthracene-carboxylate moieties (**39a–d**) indeed establishe the utility of CD in directing the dimerization toward a single isomer. To examine in detail the effects of orientation between two anthracene moieties on the stereochemistry of photodimers, four regioisomers of γ-CD derivatives bearing two 1-anthracenecarboxylate moieties

(39a–d) were prepared.[68]   The procedure for the photodimerization of **39a–d** includes photodimerization, hydrolysis by alkali solution and hplc analysis. The syn-head-head (**A**) photodimer is exclusively formed from **39a** and **39b**, whereas the anti-head-head (**B**) photodimer is formed from **39c** and **39d**, the results being exactly those expected from the examination of molecular models (Figure 10). In the presence of *l*-borneol as a guest, the specificity is lowered due to the locational change of the anthracene moieties from inside to outside the cavity. Tight involvement of the anthracene moiety in the cavity is needed for the high stereochemical specificity. The specificity was depressed when methanol was used as the solvent instead of 10% ethylene glycol aqueous solution, reflecting the absence of the hydrophobic driving force for CD complexation in methanol.

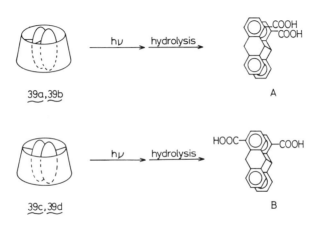

Figure 10. γ-Cyclodextrin template method for producing photodimers stereospecifically.

2.2.3. Miscellaneous Bimolecular Reactions

One of the reactions that has attracted considerable attention is photosubstitution of aromatic molecules included in CD.[69]   Quantum yields of cyanide ion substituted 4-fluoroanisole were enhanced in the presence of CD. Photooxidation of several molecules has also been investigated in the presence of CD.[70]   Miyajima et al.[70a] report that inclusion of methyl orange within CD protects it from singlet oxygen oxidation. On the other hand, Neckers and Paczkowski[70b] have utilized CD to enhance the oxidation of aromatic guest molecules. They have used Rose Bengal-appended CD as a singlet oxygen sensitizer. In this process the sensitizer and the oxidant aromatic molecules are brought closer by the CD cavity. Photolysis of benzophenone included in CD results in hydrogen abstraction.[71] Although benzophenone ketyl radical has been detected by its absorption, no stable photoproduct has been isolated; CD is presumed to serve as the hydrogen donor. Recently, remarkable selectivity in the photocycloaddition of 5-X adamantan-2-ones (X = F, Cl, Br, OH, Ph, or *t*-Bu) with fumaronitrile has been reported.[72] Irradiation in aqueous CD solution leads to a dramatic reversal in the face selectivity of the oxetane obtained in organic solvents. This is understood on the basis that complexation blocks one of the π faces and thus facilitates the fumaronitrile to approach the carbonyl from the less preferred (in the absence of CD) side.

2.3. Host–guest Sensor Systems

A methodology has been developed by Ueno et al.[73] to monitor the inclusion of guest molecules within the CD cavity. In this technique the variations in the emissions of the sensors attached to CD are monitored. This technique can be highlighted by two examples. The association dimer of pyrene-appended γ-CD (32) is converted into 1:1 host–guest complexes upon addition of external guest reagents. This change results in decreased pyrene excimer emission and enhanced monomer fluorescence (Figure 11A).

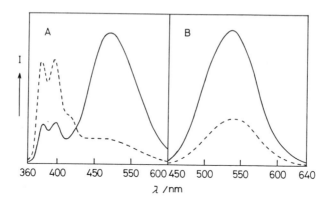

Figure 11. Fluorescence spectra of **32** (A) and **40** (B), alone (––––––) or in the presence (- - - -) of *l*-borneol as a guest (A, 2.0 mM; B, 1.74 mM).

The guest-induced decrease in the excimer emission intensity of **32** has been used to detect the inclusion of a variety of organic compounds into the CD cavity.[73] The ratio $(I_0 - I)/I_0$ where I_0 and I are the excimer emission intensities in the absence and presence of guest is used as a sensitivity factor reflecting the formation of host–guest complex. For cholic acid derivatives and several other steroidal compounds (0.2 mM), the ratio is in the following order: ursodeoxycholic acid (0.83); chenodeoxycholic acid (0.76); deoxycholic acid (0.43); progesterone (0.49); cholic acid (0.09); corticosterone (0.080); cortisone (0.038); hydrocortisone

(0.030); and prednisolone (0.024). These results indicate that remarkable molecular recognition occurs in this host–guest sensory system.

Dansyl-modified β-CD (**40**) has been used as a novel host–guest sensor.[74] The fluorescence of **40** exhibits an emission peak at 535 nm and its intensity decreases with increasing guest *l*-borneol concentration (Figure 11B). This guest-induced decrease is caused by locational change of the dansyl moiety from the interior of the hydrophobic cavity to the bulk water environment outside the cavity. This sensor has been utilized to monitor the inclusion of a large number of steroidal compounds into the CD cavity.

40

3. Crown Ethers and Other Cation Binding Hosts

Crown ethers and other ionophores bind metal and ammonium ions in nonaqueous media and in a way compliment CDs, which bind a variety of neutral organic compounds in aqueous media. Members of the crown ether family are manmade molecules, in contrast to CDs, which are naturally occurring molecules produced by enzymatic reactions from starch. Many synthetic ionophores have successfully been designed and prepared to realize particular chemical functions. In this section we discuss a few examples of interest to photochemists. However, it is not usually recognized that the compounds can become more useful in combination systems consisting of synthetic and native substances. In the final part of this section, we describe photoexcitable membranes as an example of such systems. Examples are presented to illustrate how the photochemical and photophysical behaviors of molecules that form part of an ionophoric system are altered by cation binding. Cryptands in which emission occurs from cations are not covered and recent reviews provide extensive coverage.[1]

3.1. Chromoionophores

Many ionophores bearing dyes as sensors have been prepared to detect metal ions.[75] Such systems undergo color change upon binding to metal ions; hence the name "chromoionophores." Chromoionophores described here are composed of a simple combination of a dye and an azacrown ether unit. A direct relationship exists between the sizes of the cavity and the bound cation. Cations that fit well bind strongly and provide a large shift on the absorption bands of the indicator dye. For example, largest band shifts are observed for sodium and potassium ions, when **41a** and **41b**, respectively, are used as hosts. Upon addition of a large excess of metal salts to **42** remarkable blue shifts in the absorption band occur. The extent of size-matching between the ions and the crown cavity is reflected in the shift: Li^+, 50; Na^+, 30; K^+, 17; Rb^+, 13; and Cs^+, 10 nm.[76,77] Blue shifts induced by amines on the absorption band of **42** are also remarkable.[78] The magnitude of the shift once again depends on the size of the amine. Chiral azophenolic host **43** undergoes enantiomer differential complexation.[79] Upon complexation with chiral amines, the yellow indicator **43** is converted into a purple ammonium phenolate. Significant differences up to 11 nm in λ_{max} between diastereomeric sets of the salts have been observed for several **43**-amine combinations. Chiral amines such as (S)-(-)-α-phenethylamine and (S)-(-)-1-(1-naphthyl)ethylamine induce greater blue shifts when (SSSS)-**43** is used as the host than when (RRRR)-**43** serves as the host. Chromogenic host **44** is yellow (λ_{max} = 396 nm), whereas complex of **44** with Li^+ and Na^+ is deep blue (λ_{max} = 586 nm) and violet (λ_{max} = 596 nm), respectively.[80] Thus **44** may be used as a chromogenic ion-selective indicator system capable of detecting Li^+ and Na^+.

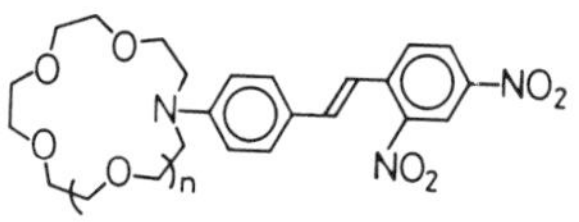

41a n=1
  b n=2

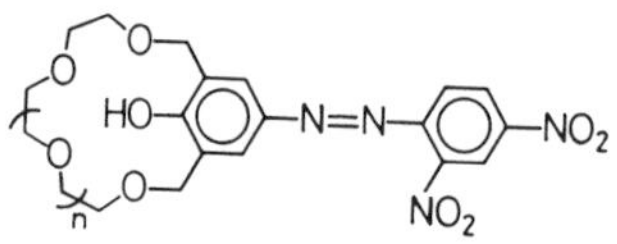

42a n=1
  b n=2
  c n=3
  d n=4

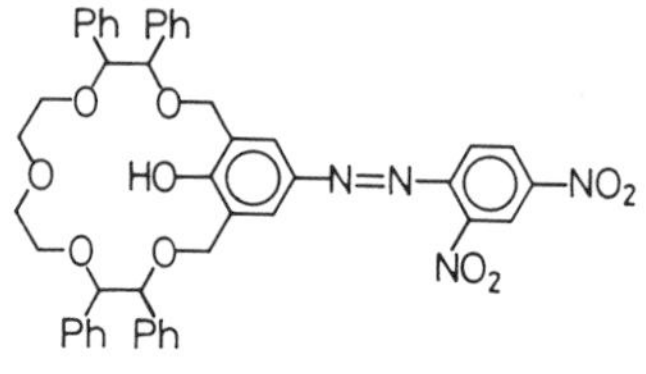

43

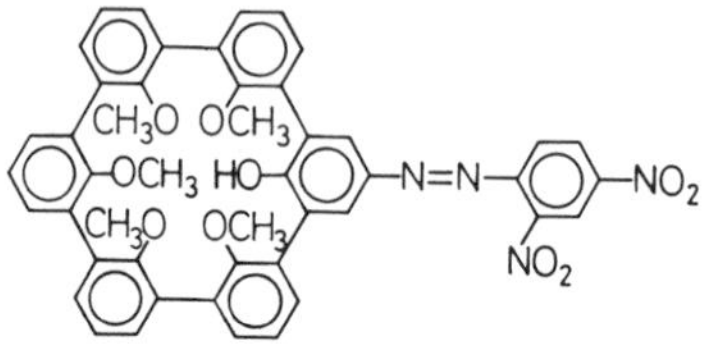

44

3.2. Fluoroionophores

Metal ions, as perturbers, can often change the emission properties of chromophoric ionophores. Such is the basis for the development of fluoroionophoric systems. In this context, results of emission studies of **45, 46,** and **47** using alkali metal ions as perturbers are interesting.[81,82] The effects of cations Li^+, Na^+, and K^+ on the photophysical properties (intersystem crossing rate and yield, fluorescence and phosphorescence yield) of **45–47** are minimal.

However, these properties are drastically altered by Rb^+, Cs^+, and Tl^+ ions. The extent of perturbation increases in the order Rb^+, Cs^+, and Tl^+, Tl^+ being the most. Furthermore, the effect of cations on **47** is much larger than on either **45** or **46**. The above effects are interpreted in terms of spin–orbit perturbation by cations on singlet–triplet crossings. Comparative study between **45**, **46**, and **47** indicates that heavy atom effects are greater when the cation is held on the π face (**47**) than at the sides of the π-system (**45** and **46**).

Fluoroionophores **48a** and **48b** show extraction ability for Li^+ and K^+ ions, respectively.[83]   Cation extraction results in shifts of the fluorescence excitation band (λ_{max} from 326 to 380 nm). Bichromophoric crown ether (**49**) is sensitive to Ca^{2+}.

Benzothiazolylphenols **50a** and **50b**, with smaller rings, exhibit fluorescence enhancement upon complexation with Li^+, while **50c** and **50d**, with larger rings, bind larger cations as well as Li^+.[84]   This is once again indicated by the change in fluorescence intensity. Absorption and emission changes caused upon binding Ca^{2+} to **51** and **52** have been examined.[85]

Binding of Ca^{2+} to **51** reduces fluorescence by a factor of 2.8. Similar binding of Ca^{2+} to **52** enhances fluorescence intensity fivefold, with a slight shift of the emission maximum (5 nm).

Fluorogenic ionophores **53** and **54** exhibit great enhancement of fluorescence intensity (410 nm) upon binding to cations.[86] Fluorophore **53** shows preference for Na^+ and **54**, for Li^+. The fluorescence band of **55** undergoes remarkable shift upon addition of cations.[87] The wavelength of emission maximum shifts from 642 nm for the free ligand to 578 nm in presence of 2×10^{-4} M calcium perchlorate. Emission intensity is also considerably enhanced upon Ca^{2+} binding (94%). On the other hand, lithium, sodium, and potassium perchlorates provide only a slight effect. Compound **56** is designed such that the anthracene fluorescence, which is efficiently quenched by electron transfer from the lone pair of nitrogen, is recovered upon incorporation of metal cations into the monoazacrown ether.[88] Fluorescence enhancement by factors up to 47 is induced by sodium and potassium ions.

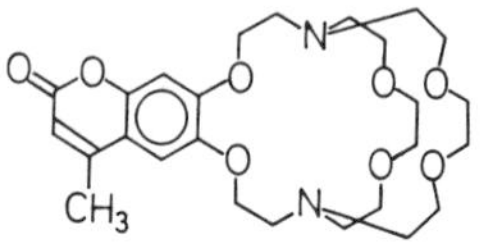

53

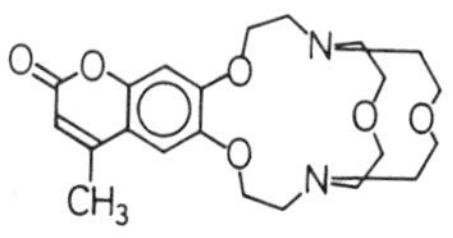

54

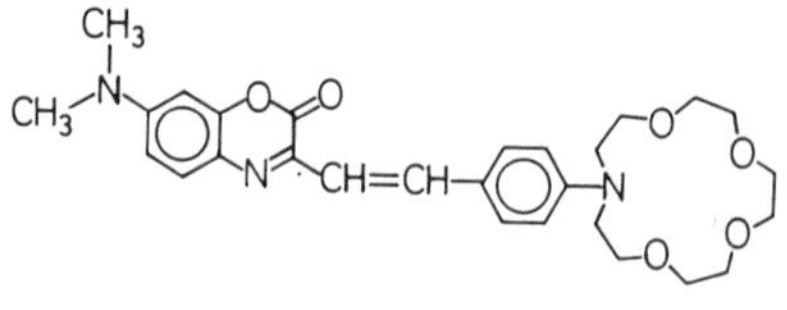

55

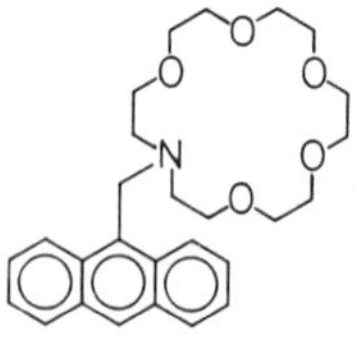

56

3.3. Exciplex

Anthraceno-cryptands **57a** and **57b** display dual fluorescence (monomer and exciplex) resulting from the intramolecular interaction between the nitrogen lone pair and the anthracene present as part of the host system.[89,90] Flexible reference compound **58** also exhibits a dual fluorescence.[90] Upon complexation with cations these molecules show strong monomer fluorescence with a corresponding decrease in the exciplex emission. Such a change is readily understood on the basis that in the presence of cations, the nitrogen lone pair is no longer available for interaction with the anthracene.

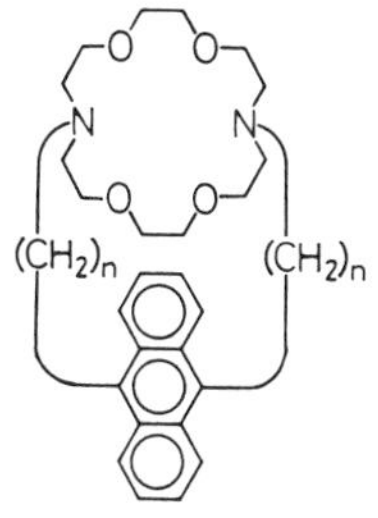

C_2H_5

$(CH_2)_n$ $(CH_2)_n$

H_5C_2— —$(CH_2)_n$

<u>57a</u>   n=2

<u>b</u>   n=3

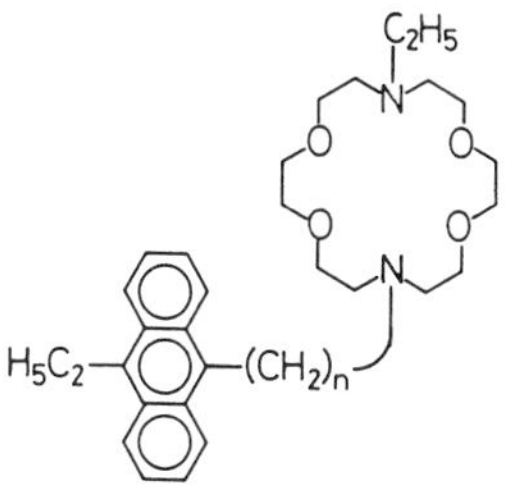

58

3.4. Photoregulated Binding and Release of Ca^{2+}

A Ca^{2+}-selective chelator **59** has a photosensitive *ortho*-nitrobenzhydryl ether moiety.[91] Photolysis converts the chelator into *ortho*-nitrosobenzophenone, whose Ca^{2+} affinity is 30-fold weaker than the original compound (Scheme 5), resulting in the release of Ca^{2+} ion. Photochemical generation of an electron-withdrawing substituent may be important in this drop in Ca^{2+} affinity. If this is the case, photochemical generation of an electron-donating group should increase the Ca^{2+} affinity and, indeed, this is the case. Photosensitive chelator **60** shows an increase in Ca^{2+} affinity upon uv irradiation; photolysis converts the diazoacetyl group into an electron-donating carboxymethyl group (Scheme 6).[92] These chelators can be used to generate controlled fast jump in intracellular concentration of Ca^{2+} ion.

3.5. Photodimerization

Bisanthracene **61**, in benzene and ether, undergoes intramolecular photodimerization with a high quantum yield. Resulting photodimer is not thermally stable and rapidly reverts to the original form with a half-life of ~3 min.[93] In contrast, irradiation of **62** only leads to the recovery of the starting material. The striking difference between **61** and **62** is related to the conformational properties of the chains interlinking the anthracene units namely polymethylene and polymethyleneoxy chains. Irradiation of **61** in the presence of lithium salt ($LiClO_4$) gives a cation locked photodimer **63** (Scheme 7). Surprisingly, the "cation locked" dimer is thermally stable up to 200–210°C!

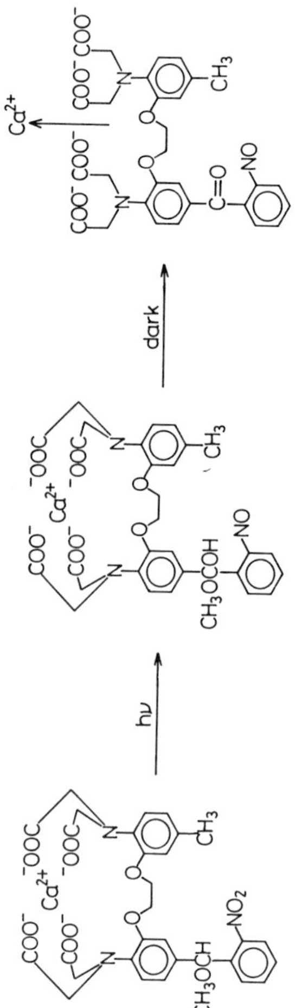

Scheme 5

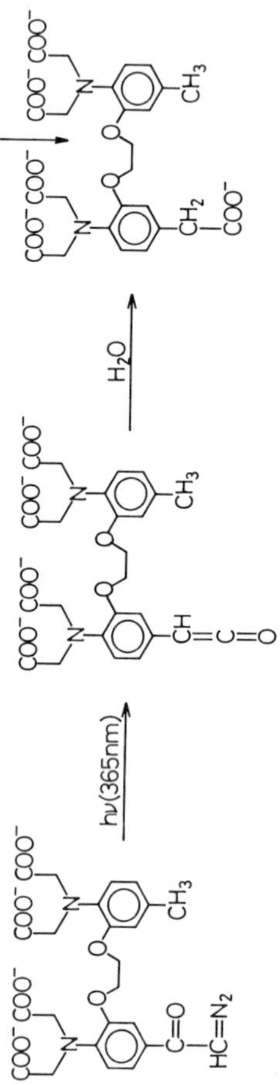

Scheme 6

Macrocyclic polyether **64a** is converted into its isomer **64b** upon irradiation (Scheme 8).[94] The half life of the dimer is prolonged in the presence of metal ions (~9.6 min for ion-free form). The effect of the ion is in the order $Na^+ > K^+ > Li^+$. Bis(anthraceno)-crown ether **65** exhibits both monomer and excimer (λ_{max} 510 nm) emissions in methanol.[95] Addition of $NaClO_4$ to the solution enhances the intensity of the excimer band and shifts it to the red (λ_{max} 570 nm). Ultraviolet photolysis of the complex **65a**·2 Na^+ produces **65b**·2 Na^+ via intramolecular dimerization. Product **65b**·2 Na^+ is thermally unstable and quantitatively reverses to **65a**.2 Na^+ (Scheme 9).

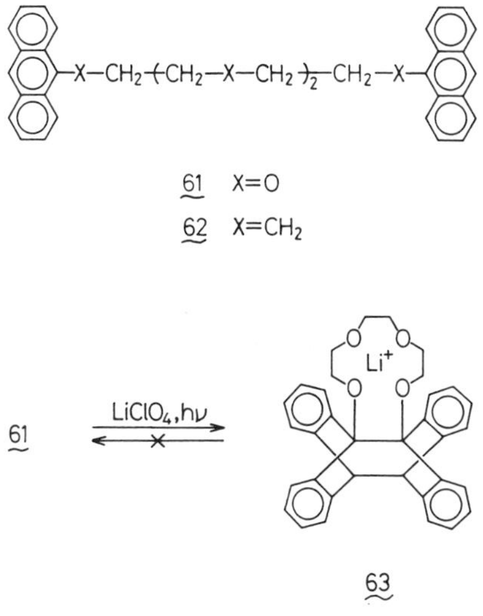
X—CH2—(CH2—X—CH2)2—CH2—X
61 X=O
62 X=CH2
Li+
LiClO4,hν
61
63

Scheme 7

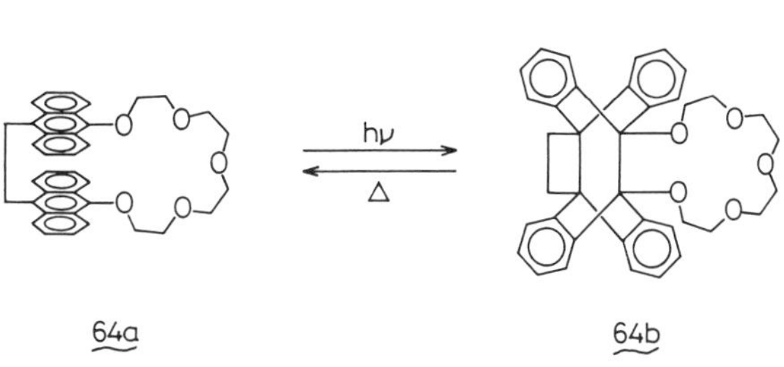
hν
Δ
64a
64b

Scheme 8

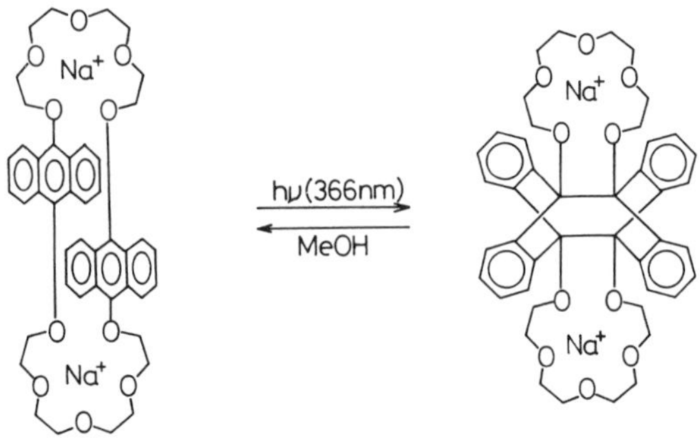

Scheme 9

Polyethylene glycol derivatives bearing cinnamoyl moieties at both ends **66a–c** are converted into crown ethers **67a–c** via intramolecular dimerization (Scheme 10).[96] The rate of intramolecular photodimerization is in the order **66b** > **66c** > **66a**. Distorted arrangements of the oxygens as well as the presence of the ester linkages contribute to the low cation extraction ability of both **66** and **67**. Upon irradiation macrocyclic compounds **68a** and **68b** are converted into the corresponding intramolecular photodimers.[97] The extraction ability of the photodimer of **68b** is higher than that of **68b** itself. Diazacrown ether bearing two cinnamoyl units (**69a** and **69b**) undergoes similar photoreaction.[98] Among **69** and its photoproducts, only the photodimer from **69a** exhibits extraction ability, the order being Li^+ > Cs^+ > K^+ > Rb^+ > Na^+.

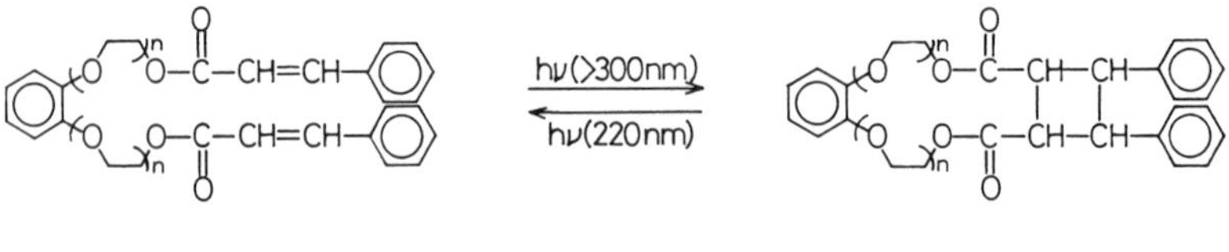

66a	n = 0		**67a**	n = 0
b	n = 1		**b**	n = 1
c	n = 2		**c**	n = 2

Scheme 10

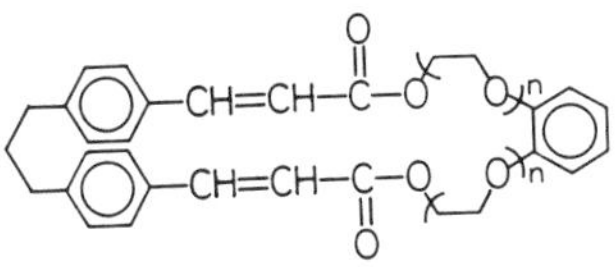

68a   n=1
b   n=2

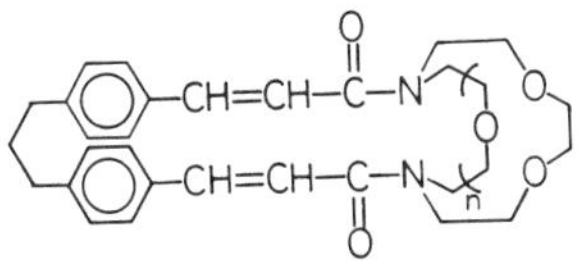

69a   n=1
b   n=2

The cinnamoyl-modified crown ethers **70** and **71** undergo geometric photoisomerization when irradiated with light above 300 nm. On the other hand, dimerized samples were obtained when photolyzed without a cutoff filter.[99] Cation transport rates through liquid membranes were examined for **70** and **71** in their trans, cis and dimeric forms. Cation transport ability of **70** is not enhanced by photoisomerization but is remarkably enhanced upon photodimerization (4.8-, 3.5-, 5.2-, and 8.1-fold increase for Na^+, K^+, Rb^+, and Ca^+, respectively), demonstrating the greater binding ability of the two crown units (by cooperative effect) in the photodimerized form. Photodimerization of **71** did not bring about a large change in cation transport rate. This is in line with the postulate that a cooperative effect is responsible for enhanced cation binding of **70** upon dimerization. In the case of **71** such cooperative binding is expected even in the trans isomer. Effects of photodimerization were also studied on some cinnamoyl group incorporated polymers.[100-102]

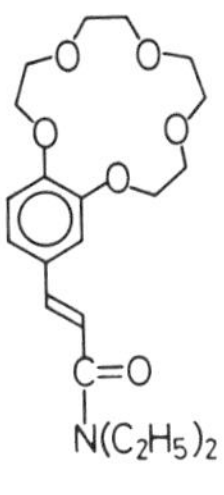

70

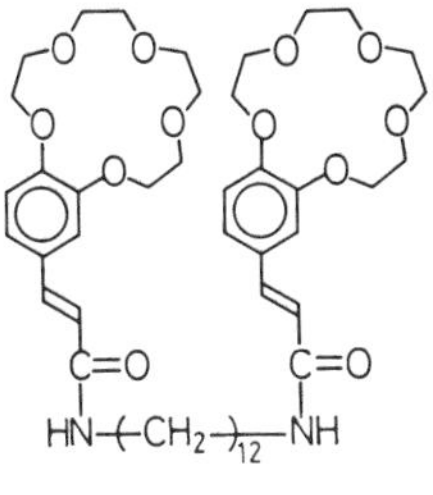

71

3.6. Geometric Photoisomerization

Geometric photoisomerization about a carbon–carbon double bond is often used to photocontrol cation extraction and transport phenomena. Basic strategy involves generating different configurations of the cation binding ligands through geometric isomerization. This is illustrated in Figure 12. A difference in cation binding ability is expected between the trans and the cis isomers if cooperative effect plays a role in cation binding by the cis isomer. This approach is illustrated with several examples below.

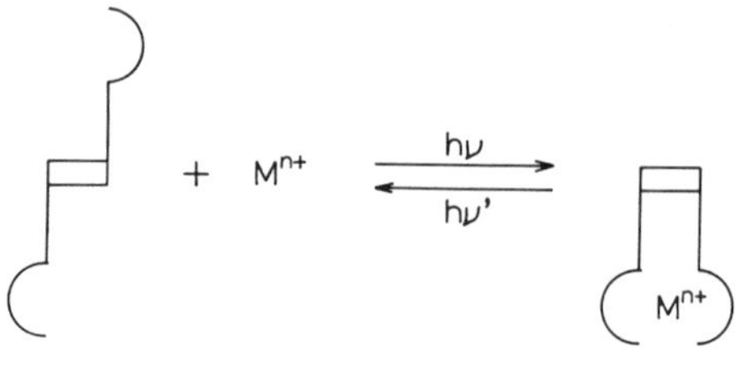

Figure 12. Promotion of metal ion binding by trans–cis photoisomerization process.

Biscrown ether **72** can be readily interconverted between cis and trans isomers upon photolysis. The two forms show a difference in cation binding ability. Bis(crown ether) derivatives of maleate (**72b**; cis isomer) shows 14-fold higher extraction ability for K^+ than that by fumarate **72a** (trans isomer).[103]

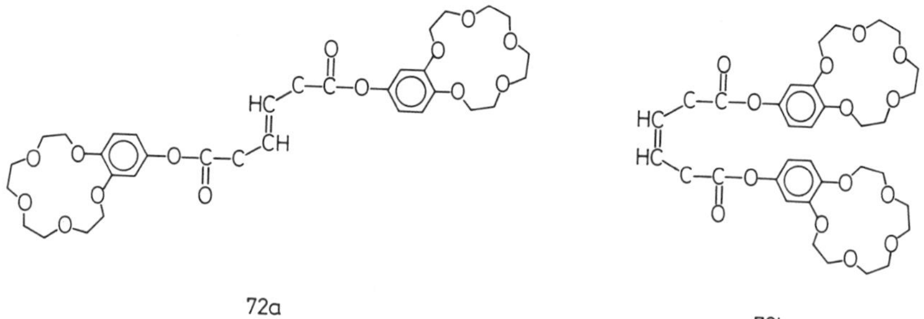

72a

72b

Thioindigo derivatives having short ligand groups such as **73a** can be photoisomerized reversibly by using 550 nm light from trans to cis and with 450 nm light from cis to trans isomer (Scheme 11).[104] Extraction experiments revealed that the trans form has no binding ability to any metal ion, whereas the cis form selectively extracts K^+, Rb^+, and Na^+. Stilbene derivative **74** exhibits similar photobehavior (Scheme 12).[105] Thirteenfold increase in extraction ability is noticed upon trans to cis conversion. However, this system is complicated by the formation of phenanthrene derivative **75** as a side product.

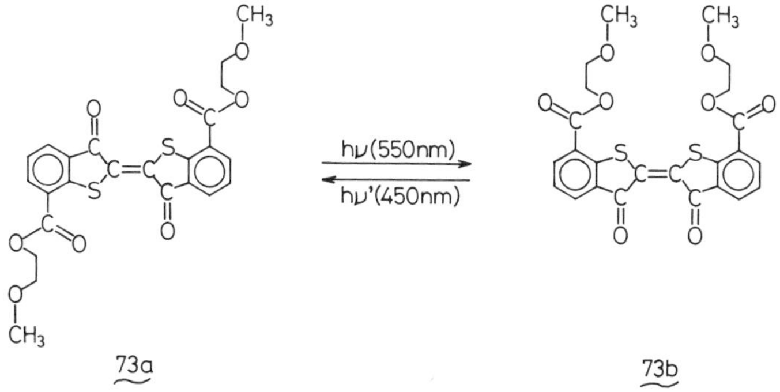

Scheme 11

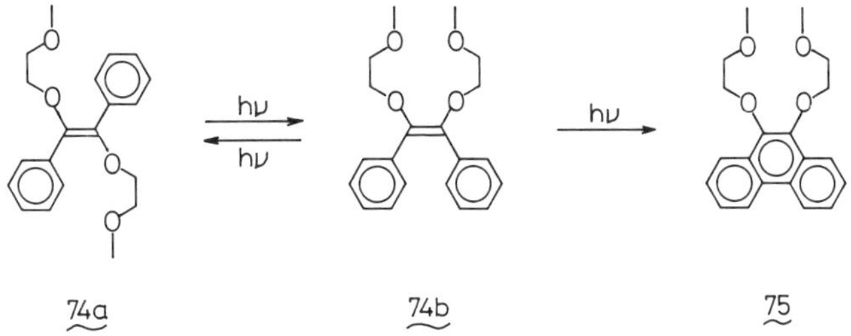

Scheme 12

Macrocyclic fumarate **76a** undergoes trans to cis photoisomerization in the presence of benzophenone as a triplet sensitizer (Scheme 13), giving maleate **76b** as the predominant component in the photostationary state (90:10).[106] The photolyzed sample exhibits enhanced extraction ability for Na$^+$.

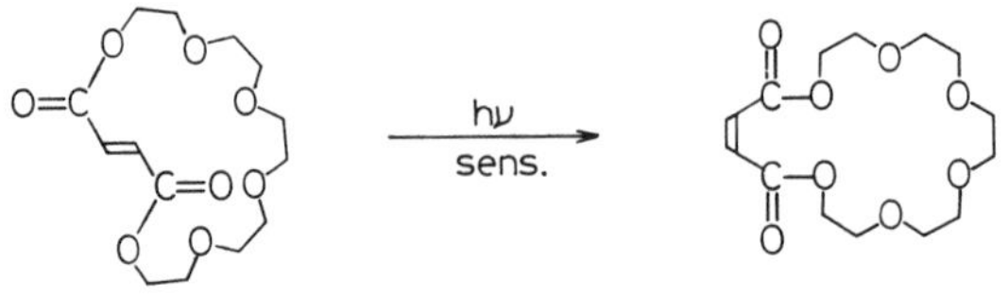

76a   76b

Scheme 13

Azobenzene undergoes reversible geometric isomerization and this structural change has been used to photocontrol cation binding. The reaction sequence shown in Figure 13 is similar to that of photoisomerization of a carbon–carbon double bond (Figure 12) except that cis to trans process occurs thermally. The trans isomer of azobenzene derivative **77** containing two iminodiacetic acid groups is converted to the cis isomer upon exposure to light of 320 nm.[107] The cis isomer binds Zn^{2+} because of cooperativity between the two iminodiacetic acid groups. Since the trans isomer does not bind Zn^{2+}, binding of Zn^{2+} can be photoregulated with **77**.

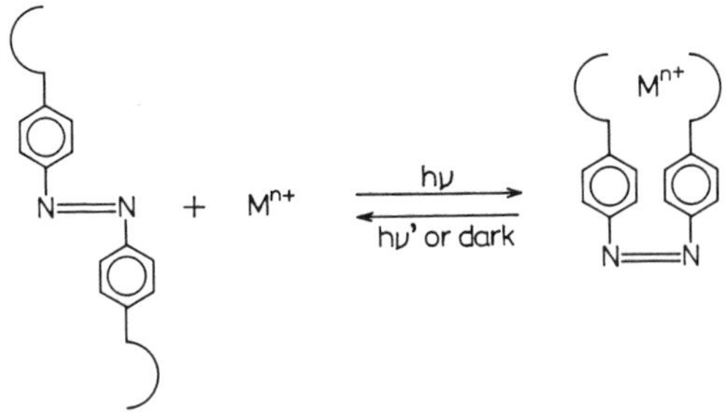

Figure 13. Photochromism of azobenzene derivatives and on–off switchable metal ion binding.

77

3.7. Photoexcitable Membranes

Several azobenzene derivatives have been investigated by Ueno and Osa[108] and by Shinkai et al.[109] in the context of using them for photodriven ion pumps across membranes. For detailed discussion on this topic readers are referred to a recent review by Shinakai and Manabe.[110] A few examples from our own investigations are presented in this section.

Extraction and transport of metal cations are facilitated by irradiation when **78** and **79** are used as hosts. The azobenzene-linked bis(crown ether) **78** shows 1.3-fold greater extraction ability and 1.5-fold greater permeation rate through a polyvinylchloride (PVC) membrane for K^+ after irradiation.[111] Since such photoinduced enhancement was not observed for Na^+, two crown ether units in *cis*-**78** must cooperate in binding of K^+ (individual crown ethers on either side can bind to Na^+ but not to K^+).

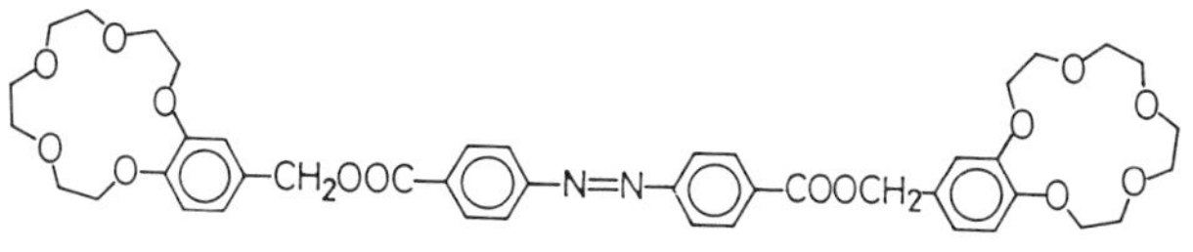

78

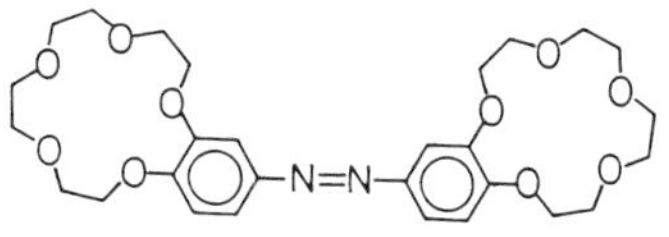

79

An experimental setup for measuring membrane potentials is shown in Figure 14. The membrane comprising of PVC, di-*n*-butylphthalate and *trans*-**78** with ca. 0.1 mm thickness was placed between two electrolyte solutions 1 and 2. Repeated reversible trans–cis isomerization was achieved by using u.v. and visible light alternately, the cis content being ca. 55% after uv irradiation. Under the conditions of $c_1 = 1$ mM and $c_2 = 500$ mM KCl, a negative shift of the membrane potential of -7 mV was induced by uv irradiation from the c_2 side and the initial potential is recovered by visible irradiation (Figure 15).[112] The photoinduced potential change depends on the concentration ratio of KCl in c_1 and c_2, being significant at higher c_2/c_1 ratios. Addition of NaCl induced no significant potential change. The photoinduced membrane potential change may arise from the fact that the irradiated surface of the PVC membrane becomes more positively charged than the opposite surface because of the enhanced binding of K^+ by *cis*-**78**.

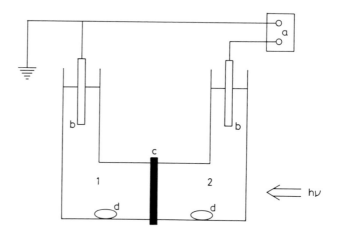

Figure 14. Schematic representation of the cell for membrane potential measurements: (a) potentiometer, (b) saturated calomel electrode, (c) membrane, and (d) stirring bar.

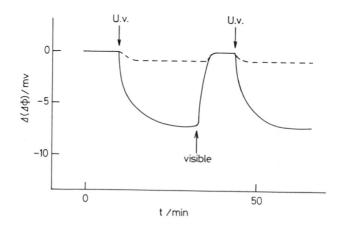

Figure 15. Photoresponse of the membrane potential: alternate uv and visible light irradiation [$c_1 = 1$ mM, $c_2 = 500$ mM NaCl (- - - -) or KCl (———)].

The asymmetric distribution of electrical charge in the membrane is supported by the observation that photoresponse of the membrane potential occurs even when $c_1 = c_2$. The imbalance of the charge increases with the thickness of the membrane.[113] Biionic potentials across the membrane were also measured using different electrolytes dissolved in sides 1 and 2.[114] The photoinduced shifts in biionic potential for 1 mM:1 mM solutions of NaCl–LiCl (-3 mV), NaCl–KCl (-12), NaCl–RbCl (-11), and NaCl–CsCl (-7.5) indicated that the selectivity of the irradiated membrane was $K^+ > Rb^+ > Cs^+ > Li^+$. Azobenzene derivatives bearing one crown unit 80a and 80b exhibit a photoinduced decrease in cation extraction ability for K^+.[115] The decreased binding may be due to the steric hindrance of the ethoxy group in the cis form. Upon uv light irradiation, the membrane potential showed a positive shift of ca. 25 mV for the system of 80b when $c_1 = c_2$ (1 mM KCl). This positive shift is consistent with the fact that the metal cation is released from the membrane surface into the solution. The photoresponse of 80a and 80b is unstable because the compounds gradually ooze from the membrane into solution due to rather low lipophilicity. This unfavorable property was eliminated with an azobenzene-modified crown ether bearing a long alkyl chain 81.[116]

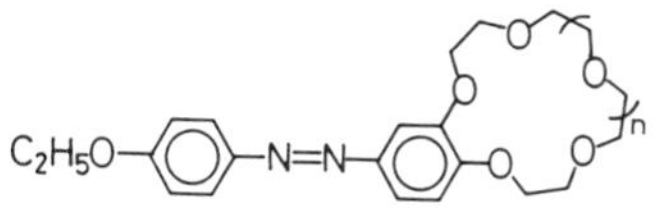

80a n=1
b n=2

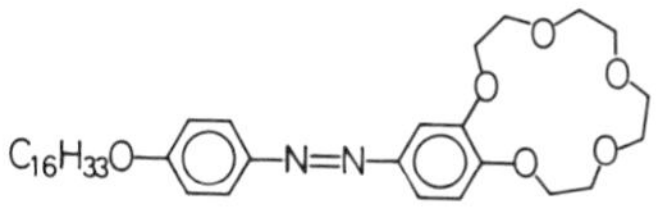

81

Ueno et al.[117] noted that spirobenzopyran-incorporated PVC membrane exhibits photoinduced membrane potential change. The spirobenzopyran derivative (82) exhibits photochromism in the membrane, existing as a closed form 82 (colorless) under visible light irradiation and an open form 83 (purple) under uv light irradiation (Scheme 14). A remarkably large negative shift in membrane potential is induced by uv light, suggesting that the irradiated membrane surface becomes positively charged. Since the magnitude of the photoinduced potential shift is pH dependent, being larger in acidic solutions than in alkaline solution, the positively charged species are likely to be protonated species 83^+.[118]

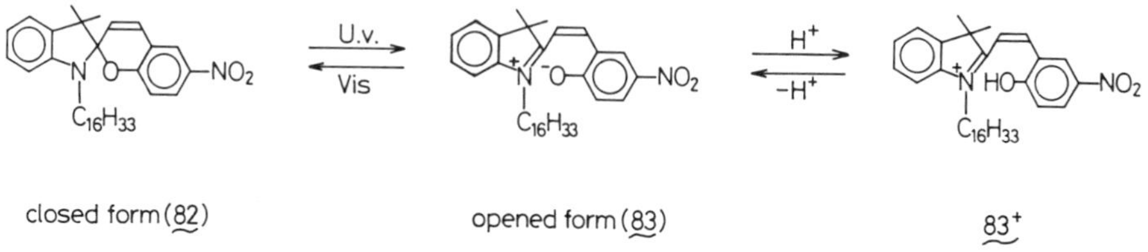

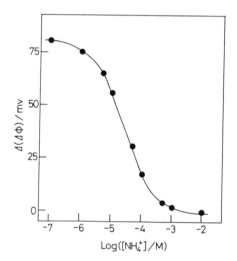

closed form (82)   opened form (83)   83⁺

Scheme 14

Nonactin is an ionophore that binds NH_4^+ ion selectively and may affect the magnitude of the photoinduced membrane potential shifts when incorporated into the membranes. In fact, the magnitude of the photoinduced membrane potential across the PVC–82–nonactin composite membrane depends remarkably on the concentration of NH_4^+ ion in the aqueous solution (Figure 16), being suppressed with increasing concentration of NH_4^+ ion.[119]

Figure 16. The effects of the NH_4Cl concentration on the photoinduced membrane potential changes.

Therefore, one may use this system as a novel ion sensor with which the concentration of NH_4^+ ion in aqueous solution can be determined. This ion sensor is unique because the output becomes greater with decreasing concentration of the ion to be measured. Ion-selectivity across PVC–82–nonactin and PVC–82–

valinomycin membranes were also examined.[120] Photoresponse is suppressed with increasing concentration of the cations. The ion selectivity in the potentiometric photoresponse of the nonactin-entrapped membrane is in the order of $NH_4^+ > K^+ > Rb^+ > Cs^+ > Na^+ > Li^+$, whereas that of the valinomycin-doped membrane is $Rb^+ > K^+ > Cs^+ > NH_4^+ > Na^+ > Li^+$.

Urease is an enzyme that decomposes urea to ammonia and carbon dioxide with the consumption of H^+ ion. Photoresponse of the PVC membrane containing **82** and crown ether (dibenzo-18-crown-6), covered with a urease layer, was studied in the presence of urea.[121] The photoresponse decreased with increasing concentration of urea in the solution in a manner similar to that shown for NH_4^+ in Figure 16. Adenosine deaminase and asparginase are enzymes each producing ammonia by decomposing adenosine or asparagine. Photoswitchable enzyme sensors were fabricated by replacing urease in the above system by deaminase or asparginase.[122]

4. Conclusion

Results reviewed in this chapter clearly demonstrate unique aspects of host–guest chemistry in solution. Many photochemical reactions of guest molecules can be modified when included in the cavities of host species such as CDs. The conformation that the guests take in the complexes is reflected in the reactions they undergo. Bimolecular reactions can also be modified. When two guest units are covalently linked to the framework of hosts, stereospecific intramolecular reactions may take place. On this basis, "topochemistry in solution" is yet to be realized. Photophysical properties of host or guest species are remarkably modified in some host–guest pairs and the absorption and the fluorescence variations induced by complexation have been used to detect particular chemical species in solution. Photochromism of host or guest molecules can be used to control chemical and physical events by light in an "on–off" fashion. Photoexcitable membranes that include photochromism units are capable not only of transducing light signals to electronic signals but also of acting as sensory systems.

Acknowledgment

We thank Dr. V. Ramamurthy for critical comments and for the help during the revision of this chapter.

References

1.   *Inclusion Compounds*, Vols. 1-3, J. L. Atwood, J . E. D. Davis and D. D. Mac Nicol, eds., Academic Press, London, 1984.
J. M. Lehn, *Angew. Chem. Int. Ed.*, **1988**, *27*, 89.
Top. Curr. Chem., E. Webber, ed., **1987** and **1988**, Vols. 140 and 149.
Top. Curr. Chem., F. Vogtle and E. Webber, eds., **1981**, **1982** and **1984**, Vols. 98, 101 and 121.

P. D. Beer, *Chem. Soc. Rev.*, **1989**, *18*, 409.
D. J. Cram, *J. Incl. Phenom.*, **1988**, *6*, 397.
C. J. Pedersen, *Science*, **1988**, *241*, 536.
F. H. Kohnke, J. P. Mathias and J. F. Stoddart, *Angew. Chem. Int. Ed. Engl.*, *Adv. Mater.*, **1989**, *28*, 1103.
I. O. Sutherland, *Chem. Soc. Rev.*, **1986**, *15*, 63.
J. Franke and F. Vogtle, *Top. Curr. Chem.*, **1986**, *132*, 136.
F. Vogtle, H. G. Lohr, J. Franke and D. Worsch, *Angew. Chem. Int. Ed.*, **1985**, *24*, 727.
F. Diederich, *Angew. Chem. Int. Ed. Engl.*, **1988**, *27*, 362.
A. Collet, *Tetrahedron*, **1987**, *43*, 5725.
I. O. Sutherland, *J. Incl. Phenon. Mol. Recog. Chem.*, **1989**, *7*, 213.
I. Koga and K. Odashima, *J. Incl. Phenom., Mol. Recog. Chem.*, **1989**, *7*, 53.
S. Shinkai, *Pure Appl. Chem.*, **1986**, *58*, 1523.
L. Mandelcorn, *Non-Stoichiometric Compounds*, Academic Press, New York, 1964.
S. M. Hagan, *Clathrate Inclusion Compounds*, Reinhold, New York, 1962.
V. Balzani, *Pure Appl. Chem.*, **1990**, *62*, 1099.
2.   W. Sanger, *Angew. Chem. Int. Ed. Engl.*, **1980**, *19*, 344.
I. Tabushi, *Acc. Chem. Res.*, **1982**, *15*, 66.
R. Breslow, *Science*, **1982**, *218*, 532.
V. T. D'souza and M. L. Bender, *Acc. Chem. Res.*, **1987**, *20*, 146.
3.   V. Ramamurthy and D. F. Eaton, *Acc. Chem. Res.*, **1988**, *21*, 300.
V. Ramamurthy, *Tetrahedron*, **1986**, *42*, 5753.
4.   W. Rettig, *Angew. Chem. Int. Ed. Engl.*, **1986**, *25*, 971.
5.   R. L. van Etten, J. F. Sebastian, G. A. Clowes and M. L. Bender, *J. Am. Chem. Soc.*, **1967**, *89*, 3242.
A. Heredia, G. Requena and F. Sanchez, *J. Chem. Soc. Chem. Commun.*, **1985**, 1814.
6.   A. Harada, M. Furue and S. Nozakura, *Macromolecules*, **1977**, *10*, 676.
7.   A. Nag and K. Bhattacharyya, *Chem. Phys. Lett.*, **1988**, *151*, 474.
A. Nag, R. Dutta, N. Chattopadhyay and K. Bhattacharyya, *Chem. Phys. Lett.*, **1989**, *157*, 83.
A. Nag and K. Bhattacharyya, *J. Chem. Soc. Faraday. Trans.*, **1990**, *86 (1)*, 53.
G. S. Cox, P. J. Hauptman and N. J. Turro, *Photochem. Photobiol.*, **1984**, *39*, 597.
8.   (a) A. Ueno, K. Takahashi, Y. Hino and T. Osa, *J. Chem. Soc., Chem. Commun.*, **1981**, 194.
(b) M. Hoshino, M. Imamura, K. Ikehara and Y. Hama, *J. Phys. Chem.*, **1981**, *85*, 1820.
9.   K. Kano, I. Takenoshita and T. Ogawa, *J. Phys. Chem.*, **1982**, *86*, 1833.
10. S. Hamai, *J. Phys. Chem.*, **1988**, *92*, 6140.
11. S. Hashimoto and J. K. Thomas, *J. Am. Chem. Soc.*, **1985**, *107*, 4655.
12. S. Hamai, *J. Am. Chem. Soc.*, **1989**, *111*, 3954.
S. Hamai, *Bull. Chem. Soc. Jpn.*, **1989**, *62*, 2763.
S. Hamai, *J. Phys. Chem.*, **1990**, *94*, 2595; **1989**, 2074.
G. Nelson, G. Patonay and I. M. Warner, *J. Incl. Phenom.*, **1988**, *6*, 277.
G. Nelson and I. M. Warner, *J. Phys. Chem.*, **1990**, *94*, 576.
13. A. Weller, *Z. Phys. Chem.*, **1958**, *17*, 224.
14. D. F. Eaton, *Tetrahedron*, **1987**, *43*, 1551.
15. T. Yorozu, M. Hoshino, M. Imamura and H. Shinozuka, *J. Phys. Chem.*, **1982**, *86*, 4422.
16. R. A. Agbaria, B. Uzan and D. Gill, *J. Phys. Chem.*, **1989**, *93*, 3855.

17. (a) N. J. Turro, T. Okubo and C. J. Chung, *J. Am. Chem. Soc.*, **1982**, *104*, 1789.
N. J. Turro, G. S. Cox and X. Li, *Photochem. Photobiol.*, **1983**, *37*, 149.
N. J. Turro, J. D. Bolt, Y. Kuroda and I. Tabushi, *Photochem. Photobiol.*, **1982**, *35*, 69.
(b) S. Scypinski and C. J. Cline Love, *Am. Lab.*, **1984**, 55.
(c) S. Hamai, *J. Am. Chem. Soc.*, **1989**, *111*, 3954.
18. A. Ueno, K. Takahashi, Y. Hino and T. Osa, *J. Chem. Soc., Chem. Commun.*, **1981**, 194.
19. T. Yorozu, M. Hoshino and M. Imamura, *J. Phys. Chem.*, **1982**, *86*, 4426.
20. A. Harada and S. Nozakura, *Polym. Bull. (Berlin)*, **1982**, *8*, 141.
21. N. Kobayashi, R. Saito, H. Hino, Y. Hino, A. Ueno and T. Osa, *J. Chem. Soc., Perkin Trans. 2*, **1983**, 1031.
22. W. G. Hersktroeter, P. A. Martic, and S. Farid, *J. Chem. Soc., Perkin Trans.*, *2*, **1984**, 1453.
W. G. Herkstroeter, P. A. Martic, T. R. Evans and S. Farid, *J. Am. Chem. Soc.*, **1986**, *108*, 3275.
23. S. Hamai, *J. Phys. Chem.*, **1989**, *93*, 6527.
24. A. Ueno, I. Suzuki and T. Osa, *J. Chem. Soc., Chem. Commun.*, **1988**, 1373.
25. A. Ueno, I. Suzuki and T. Osa, *J. Am. Chem. Soc.*, **1989**, *111*, 6391.
26. K. Kano, H. Matsumoto, Y. Yoshimura and S. Hashimoto, *J. Am. Chem. Soc.*, **1988**, *110*, 204.
27. (a) S. Hamai, *Bull. Chem. Soc. Jpn.*, **1982**, *55*, 2721.
(b) S. Hamai, *J. Phys. Chem.*, **1988**, *92*, 6140.
28. N. J. Turro, T. Okubo and G. C. Weed, *Photochem. Photobiol.*, **1982**, *35*, 325.
29. (a) M. Itoh, and Y. Fujiwara, *Bull. Chem. Soc. Jpn.*, **1984**, *57*, 2261.
(b) J. Emert, D. Kodali and R. Catena, *J. Chem. Soc., Chem. Commun.*, **1981**, 758.
30. R. Arad-Yellin and D. F. Eaton, *J. Phys. Chem.*, **1983**, *87*, 5051.
31. (a) N. Kobayashi, Y. Hino, A. Ueno and T. Osa, *Bull. Chem. Soc. Jpn.*, **1983**, *56*, 1849.
(b) S. Hamai, *Bull. Chem. Soc. Jpn.*, **1986**, *59*, 2979.
32. J. C. Netto-Ferreira and J. C. Scaiano, *J. Photochem. Photobiol. A.*, **1988**, *45*, 109.
33. (a) K. Kano, S. Hashimoto, A. Imai and T. Ogawa, *J. Inclusion Phenom.*, **1984**, *2*, 737.
(b) G. S. Cox and N. J. Turro, *J. Am. Chem. Soc.*, **1984**, *106*, 422.
34. H. Murai, Y. Mizunuma, K. Ashikawa, Y. Yamamoto and Y. J. Ihaya, *Chem. Phys. Lett.*, **1988**, *144*, 417.
H. Murai and Y. J. Ihaya, *Chem. Phys.*, **1989**, *135*, 131.
35. (a) F. V. Bright, G. C. Catena and J. Huang, *J. Am. Chem. Soc.*, **1990**, *112*, 1343.
(b) M. Barra, C. Bohne and J. C. Scaiano, *J. Am. Chem. Soc.* **1990**, *112*, 8075.
36. (a) H. Yonemura, H. Nakamura and T. Matsuo, *Chem. Phys. Lett.*, **1989**, *155*, 157.
H. Yonemura, H. Saito, S. Matsushima, H. Nakamura and T. Matsuo, *Tetrahedron Lett.*, **1989**, *30*, 3143.
(b) I. Wilmer, E. Adar, Z. Goren and B. Steinberger, *New. J. Chem.*, **1987**, *11*, 769.
37. B. N. Rao, M. S. Syamala, N. J. Turro and V. Ramamurthy, *J. Org. Chem.*, **1987**, *52*, 5517.
38. G. D. Reddy and V. Ramamurthy, *J. Org. Chem.*, **1987**, *52*, 5521.
39. G. D. Reddy, G. Usha, K. V. Ramanathan, V. Ramamurthy, *J. Org. Chem.*, **1986**, *51*, 5085.

40.   M. S. Syamala, B. N. Rao and V. Ramamurthy, *Tetrahedron*, **1988**, *44*, 7234.
       A. V. Veglia, A. M. Sanchez and R. H. de Rossi, *J. Org. Chem.*, **1990**, *55*, 4083.
41.   M. Naseeta, R. H. de Rossi and J. J. Cosa, *Can. J. Chem.*, **1988**, *66*, 2794.
42.   M. Ohara and K. Watanabe, *Angew. Chem. Int. Ed. Ingl.*, **1975**, *7*, 820.
43.   R. Chenevert and R. Plante, *Tetrahedron Lett.*, **1984**, *25*, 5007.
44.   R. Chenevert and R. Plante, *Can. J. Chem.*, **1983**, *61*, 1092.
45.   M. S. Syamala and V. Ramamurthy, *Tetrahedron*, **1988**, *44*, 7223.
46.   G. G. Wubbels and W. D. Cotter, *Tetrahedron Lett.*, **1989**, *30*, 6477.
       G. G. Wubbels, B. R. Sevetson and S. N. Kaganove, *Tetrahedron Lett.*, **1986**, *27*, 3103.
47.   M. S. Syamala, S. Devanathan and V. Ramamurthy, *J. Photochem.*, **1984**, *34*, 219.
       G. L. Duveneck, E. V. Sitzmann, K. B. Eisenthal and N. J. Turro, *J. Phys. Chem.*, **1989**, *93*, 7166.
48.   I. Tabushi and L. C. Yuan, *J. Am. Chem. Soc.*, **1981**, *103*, 3574.
49.   P. Bortolus and S. Monti, *J. Phys. Chem.*, **1987**, *91*, 5046.
50.   A. Ueno, K. Takahashi and T. Osa, *J. Chem. Soc., Chem. Commun.*, **1980**, 837.
       A. Ueno, R. Saka and T. Osa, *Chem. Lett.*, **1979**, 841.
51.   A. Ueno, M. Fukushima and T. Osa, *J. Chem. Soc., Perkin Trans. 2*, **1990**, 1067
52.   A. Ueno, Y. Tomita and T. Osa, *Tetrahedron Lett.*, **1983**, *24*, 5245.
53.   A. Ueno, H. Yoshimura, R. Saka and T. Osa, *J. Am. Chem. Soc.*, **1979**, *101*, 2779.
54.   A. Ueno, M. Fukushima and T. Osa, unpublished data.
55.   A. Ueno, Y. Tomita and T. Osa, *J. Chem. Soc., Chem. Commun.*, **1983**, 976.
56.   A. Ueno, F. Moriwaki, T. Tomita and T. Osa, *Chem. Lett.*, **1985**, 493.
57.   A. Ueno, F. Moriwaki, T. Osa, F. Hamada and K. Murai, *Tetrahedron Lett.*, **1985**, *26*, 3339.
58.   A. Ueno, F. Moriwaki, T. Osa, F. Hamada and K. Murai, *Bull. Chem. Soc. Jpn.*, **1985**, *59*, 465.
59.   F. Moriwaki, H. Kaneko, A. Ueno, T. Osa, F. Hamada and K. Murai, *Bull. Chem. Soc. Jpn.*, **1987**, *60*, 3619.
60.   K. Fujita, T. Tahara, T. Koga and T. Imoto, *Chem. Soc. Jpn.*, *(Nippon Kagakukaishi)*, **1987**, 300.
61.   I. Suzuki, A. Ueno and T. Osa, *Chem. Lett.*, **1989**, 2013.
62.   A. Ueno, F. Moriwaki, T. Osa, F. Hamada and K. Murai, *Tetrahedron*, **1987**, *43*, 1571.
63.   T. Tamaki, *Chem. Lett.*, **1984**, 53.
       T. Tamaki and T. Kokubu, *J. Incln. Phenom.*, **1984**, *2*, 815.
64.   T. Tamaki, T. Kokubu, and K. Ichimura, *Tetrahedron Lett.*, **1987**, 1484.
65.   A. Ueno, F. Moriwaki, T. Osa, F. Hamada and K. Murai, *J. Am. Chem. Soc.*, **1988**, *110*, 4323.
66.   A. Ueno, F. Moriwaki, A. Azuma and T. Osa, *Carbohydr. Res.*, **1989**, *192*, 173.
67.   F. Moriwaki, A. Ueno, T. Osa, F. Hamada and K. Murai, *Chem. Lett.*, **1986**, 1865.
       A. Ueno, F. Moriwaki, A. Azuma and T. Osa, *J. Chem. Soc., Chem. Commun.*, **1988**, 1042.
       A. Ueno, F. Moriwaki, A. Azuma and T. Osa, *J. Org. Chem.*, **1989**, *54*, 295.
68.   A. Ueno, F. Moriwaki, A. Azuma and T. Osa, *Proceedings of the International Symposium on Inclusion Phenonomena and Molecular Recognition*, Plenum, New York, in press.
69.   J. Liu and R. G. Weiss, *Isr. J. Chem.*, **1985**, *25*, 228 ; *J. Photochem.*, **1985**, *30*, 303.

70. (a) K. Miyajima, H. Komatsu, K. Inone, T. Hamada and M. Nakagaki, *Bull. Chem. Soc. Jpn.*, **1990**, *63*, 6.
     (b) D. C. Neckers and J. Paczkowski, *J. Am. Chem. Soc.*, **1986**, *108*, 291; *Tetrahedron*, *42*, 4683.
71. S. Monti, L. Flamigni, A. Martelli and P. Bortolus, *J. Phys. Chem.*, **1988**, *92*, 4447.
72. W. S. Chung, N. J. Turro, J. Silver and W. J. le Noble, *J. Am. Chem. Soc.*, **1990**, *112*, 1202.
73. A. Ueno, I. Suzuki, and T. Osa, *Chem. Lett.*, **1989**, 1059.
     A. Ueno, I. Suzuki and T. Osa, *Anal. Chem.*, **1990**, *62*, 2461.
74. A. Ueno, S. Minato, I. Suzuki, M. Fukushima, M. Ohkubo, T. Osa, F. Hamada and K. Murai, *Chem. Lett.*, **1990**, 605.
75. H. G. Lohr and F. Vogtle, *Acc. Chem. Res.*, **1985**, *18*, 65.
     M. Takagi and K. Ueno, *Top. Curr. Chem.*, **1984**, *121*, 39.
76. J. P. Dix and F. Vogtle, *Chem. Ber.*, **1980**, *113*, 457.
     J. P. Dix and F. Vogtle, *Chem. Ber.*, **1981**, *114*, 638.
     T. Kaneda, K. Sugihara, H. Kamiya and S. Misumi, *Tetrahedron Lett.*, **1981**, *22*, 4407.
77. T. Kaneda, S. Umeda, H. Tanigawa, S. Misumi, Y. Kai, H. Morii, K. Miki and N. Kasai, *J. Am. Chem. Soc.*, **1985**, *107*, 4802.
78. T. Kaneda, S. Umeda Y. Ishizaki, H. S. Kuo and S. Misumi, *J. Am. Chem. Soc.*, **1989**, *111*, 1881.
79. T. Kaneda, K. Hirose and S. Misumi, *J. Am. Chem. Soc.*, **1989**, *111*, 742.
80. D. J. Cram, R. A. Carmack and R. C. Helgeson, *J. Am. Chem. Soc.*, **1988**, *110*, 571.
81. L. R. Sousa and J. M. Larson, *J. Am. Chem. Soc.*, **1977**, *99*, 307.
     J. M. Larson and L. R. Sousa, *J. Am. Chem. Soc.*, **1978**, *100*, 1943.
82. S. Gosh, M. Petrin, A. H. Maki, L. R. Sousa, *J. Chem. Phys.*, **1987**, *87*, 4315; **1988**, *88*, 2913.
83. H. Nishida, Y. Katayama, H. Katsuki, H. Nakamura, M. Takagi and K. Ueno, *Chem. Lett.*, **1982**, 1853.
84. I. Tanigawa, K. Tsuemoto, T. Kaneda and S. Misumi, *Tetrahedron Lett.*, **1984**, *25*, 5327.
85. R. Y. Ysien, *Biochemistry*, **1980**, *19*, 2396.
86. D. M. Masilamani, M. Lucas and K. Morgan, News item in *Chem. Eng. News* **1987**, issue no.9, 26.
87. S. Fery-Forgues, M. T. Le Bris, J. P. Guette and B. Valeur, *J. Chem. Soc., Chem. Commun.*, **1988**, 384.
88. A. P. de Silva and S. A. de Silva, *J. Chem. Soc., Chem. Commun.*, **1986**, 1709.
89. J. P. Konopelski, F. K. Hilbert, J. M. Lehn, J. P. Desvergne, F. Fages, A. Castellan and H. B. Laurent, *J. Chem. Soc., Chem. Commun.*, **1985**, 433.
90. F. Fages, J. P. Desvergne, H. Bouas-Laurent, P. Marsau, J. M. Lehn, F. Kotzyba-Hilbert, A. M. Albrecht-Gary and M. Al-Joubbeh, *J. Am. Chem. Soc.*, **1989**, *111*, 8672.
91. S. R. Adams, J. P. Kao, G. Grynkiewicz, A. Minta and R. Y. Tsien, *J. Am. Chem. Soc.*, **1989**, *111*, 7957.
92. S. R. Adams, J. P. Y. Kao and R. Y. Tsien, *J. Am. Chem. Soc.*, **1989**, *111*, 7957.
93. J. P. Desvergne and H. Bauas-Laurent, *J. Chem. Soc., Chem. Commun.*, **1978**, 403.
94. I. Yamashita, M. Fujii, T. Kaneda and S. Misumi, *Tetrahedron Lett.*, **1980**, *21*, 541.

95. H. Bouas-Laurent, A. Castellan, M. Dancy, J. P. Desvergne, G. Guinand, P. Marsau and M. H. Riffaud, *J. Am. Chem. Soc.*, **1986**, *108*, 315.
96. S. Akabori, Y. Habata, M. Nakazawa, Y. Yamada, Y. Shindo, T. Sugimura and S. Sato, *Bull. Chem. Soc. Jpn.*, **1987**, *60*, 3453.
97. S. Akabori, T. Kumagai, Y. Habata and S. Sato, *Bull. Chem. Soc. Jpn.*, **1988**, *61*, 2459.
98. S. Akabori, T. Kumagai, Y. Habata and S. Sato, *J. Chem. Soc., Perkin Trans. I*, **1989**, 1497.
99. J. Anzai, Y. Suzuki, A. Ueno and T. Osa, *Polym. Commun.*, **1984**, *25*, 254.
100. J. Anzai, Y. Suzuki, A. Ueno and T. Osa, *Isr. J. Chem.*, **1985**, *26*, 60.
101. M. Shirai, T. Orikata and M. Tanaka, *Makromol. Chem. Rapid Commun.*, **1983**, *4*, 65.
102. M. Shirai, M. Kuwahara and M. Tanaka, *Eur. Polym. J.*, **1988**, *24*, 411.
103. K. Kimura, H. Tamura, T. Tsuchida and T. Shono, *Chem. Lett.*, **1979**, 611.
104. M. Irie and M. Kato, *J. Am. Chem. Soc.*, **1985**, *107*, 1024.
105. J. P. Soumillion, J. Weiler, X. De Man, R. Touillaux, J. P. Declercq, B.Tinant, *Tetrahedron Lett.*, **1989**, *30*, 4509.
106. H. Sasaki, A. Ueno and T. Osa, *Chem. Lett.*, **1986**, 1785.
107. M. Blank, L. M. Soo, N. H. Wassermann, B. F. Erlanger, *Science*, **1981**, *214*, 70.
108. A. Ueno and T. Osa, *Yuki Gosei Kagaku Kyokaishi*, **1980**, *38*, 207.
109. S. Shinkai, T. Nakaji, Y. Nishida, T. Ogawa and O. Manabe, *J. Am. Chem. Soc.*, **1980**, *102*, 5860.
110. S. Shinkai and O. Manabe, *Top. Curr. Chem.*, **1984**, *121*, 67.
111. J. Anzai, A. Ueno, H. Sasaki, K. Shimokawa and T. Osa, *Makromol. Chem., Rapid Commun.*, **1983**, *4*, 731.
112. J. Anzai, H. Sasaki, A. Ueno and T. Osa, *J. Chem. Soc., Chem. Commun.*, **1983**, 1045.
113. J. Anzai, H. Sasaki, A. Ueno and T. Osa, *Chem. Lett.*, **1984**, 1205.
114. J. Anzai, H. Sasaki, A. Ueno and T. Osa, *J. Polym. Sci., Polym. Chem. Ed.*, **1986**, *24*, 681.
115. J. Anzai, A. Ueno and T. Osa, *J. Chem. Soc., Perkin Trans. II*, **1987**, 67.
116. J. Anzai, Y. Hasebe, A. Ueno and T. Osa, *J. Polym. Sci., Polym. Chem., Ed.*, **1988**, *26*, 1519.
117. J. Anzai, A. Ueno and T. Osa, *J. Chem. Soc., Chem. Commun.*, **1984**, 688.
118. J. Anzai, Y. Hasebe, A. Ueno and T. Osa, *Bull. Chem. Soc. Jpn.*, **1988**, *61*, 2959.
119. J. Anzai, Y. Hasebe, A. Ueno and T. Osa, *Bull. Chem. Soc. Jpn.*, **1987**, *60*, 1515.
120. Y. Hasebe, J. Anzai, A. Ueno and T. Osa, *Chem. Pharm. Bull.*, **1989**, *37*, 1307.
121. Y. Hasebe, J. Anzai, A. Ueno, T. Osa and C.W. Chen, *J. Phys. Org. Chem.*, **1988**, *1*, 309.
122. C. W. Chen, Y. Sakai, Y. Hasebe, J. Anzai, A. Ueno and T. Osa, *Chem. Pharm. Bull.*, **1989**, *37*, 3316.

Chapter 17

Photoprocesses in Organized Biological Media

C. Vijaya Kumar

Department of Chemistry, University of Connecticut, Storrs, CT, USA.

Contents

1. Introduction

With external energy sources, the living world has learned to build organized assemblies for driving biological processes in a controlled manner. Such self-organized and -controlled ensembles of chemical processes are a primary, but not the only, characteristic of life. When this order or specificity is broken, the resulting chaos can terminate life processes or threaten the efficient functioning of the living system. Biological systems are probably the best examples of low-entropy and high-free energy organized media, open systems that are far from thermodynamic equilibrium. The specificity and efficiency of biological processes can be attributed to the efficient organization of biological media. Organization of these structures is controlled through a number of failsafe mechanisms, and several photoprocesses take place in these media with the high efficiency and specificity needed for the functioning of the living systems. Such processes found in nature are perhaps the best examples of how the efficiency and specificity of photoreactions can be controlled by organized media. Lipid bilayers, protein matrices that organize the active sites of enzymes, and biopolymers including nucleic acids and polysaccharides, are some of the most fascinating examples of organized biological media.[1] How do these organized media influence biological processes in general and the photobiological processes in particular? Two striking examples of photoreactions that are controlled exceptionally well in biological systems are photosynthesis and visual signal transduction.[2] Several less understood examples are the photoreactivating enzymes, phototropism, bioluminescence, etc. Not natural, although certainly very fascinating, is the use of the organized nulceotide bases of deoxyribonucleic acid (DNA) and ribonucleic acid (RNA) to control photophysical and photochemical reactions, and some details about this biological medium for the control of photoprocesses are also described here.[3]

This chapter reviews various light-induced processes that have been investigated in the organized media of proteins and nucleic acids. The choice of these two systems is essentially due to the recent progress made in these areas. The first part of the article describes the photosynthetic apparatus and the second part illustrates how the organized nucleotide bases of DNA control photophysical and photochemical processes in vitro. The review is intended not to be exhaustive but rather to provide a guiding light for the biologically inclined reader.

2. Proteins as Organized Media—Photosynthesis

Directly or indirectly, all living systems depend upon solar energy as the source of free energy. Green plants and photosynthetic bacteria convert solar energy, a renewable resource, into high-energy chemical products. Green plants have two distinct photosystems, called photosystem I and II, working in series.[4] Photosynthetic bacteria, on the other hand, have a simpler scheme of solar energy conversion, bearing similarities to the plant photosystem II.[5] The photosynthetic apparatus in both plants and bacteria consists of distinct units, called antennae, and the reaction center. Antennae are the light-gathering complexes consisting of chlorophyllous pigments as well as carotenoids arranged in the protein matrix. These pigment–protein complexes absorb light and efficiently transfer the excitation to the reaction center. They are not capable of the primary photochemical process of charge separation. Thus, the optical absorption cross section of the photosynthetic unit is increased by the light-gathering function of the antenna complexes. On an average only about 1% of the photosynthetic pigments actually carry out the primary photochemical event and the remaining function as accessory pigments.[6]

The reaction center consists of the pigments and cofactors necessary to conduct the primary photochemical reaction. The protein matrix surrounding the pigments plays an important role in organizing and controlling the photochemical as well as thermal reactions necessary for energy conversion and storage. Reaction centers are attached to the antennae and they capture the excited-state energy from the antennae and convert it into chemical potential by charge separation. In a series of events following this, protons are transported across the membrane, into the periplasm. The resulting buildup of electrical potential across the photosynthetic membrane exceeds 100 mV and a pH gradient of about 2.5 units is achieved.[7] These field and pH gradients across the photosynthetic membrane provide the necessary thermodynamic potential for the synthesis of high-energy products.

Plants use the energy stored in the charge-separation step to oxidize water and to reduce carbon dioxide through a complex series of enzymatic steps called the Calvin cycle.[8] Bacteria, on the other hand, use more easily oxidizable organic material or sulfides instead of water. Electrons released in the oxidation are used to phosphorylate adenosine diphosphate (ADP) to adenosine triphosphate (ATP) in green plants and bacteria and to produce the reduced form of nicotinamide adenine dinucleotide phosphate, in green plants. The ATP is in turn used for driving various energy-consuming processes for the functioning of the system, so that the organism can live in the absence of light. It is estimated that, in green plants, eight photons with a total energy of ~350 kcal are required to reduce one molecule of carbon dioxide and to oxidize one molecule of water to oxygen. In the process, 120 kcal is stored as chemical potential with a net efficiency of 0.34.[7] The quantum yield of the primary photoinduced charge separation was estimated to be about 1.[9] In summary, the efficient light gathering (photophysical processes) by the antenna and its conversion into useful work (photochemical processes) by the reaction center, with a high efficiency, are the basis of life on this planet.

2.1. Bacterial Photosynthetic Apparatus—The Structure

In all photosynthetic organisms, the photosynthetic apparatus is located in membranes. These membranes contain arrays of pigments, proteins, and cofactors required for light absorption, charge separation, and the subsequent redox reactions. The reaction center components are arranged in a specific geometry that allows them to carry out their functions efficiently in a predetermined sequence.[10] The size of the photosynthetic unit, consisting of the antenna complexes and the reaction center, is ~700,000 daltons in photosynthetic bacteria and ~2 million daltons in green plants.[7] In photosynthetic bacteria, photosynthetic units are located in the cytoplasmic membrane separating the periplasm from the cytoplasm (Figure 1). The special pair, where the primary photochemical event takes place, is near the periplasmic side. Cytochrome (Cyt) c_2, a water-soluble protein present in the periplasm, regenerates the special pair after the primary photochemical charge separation.

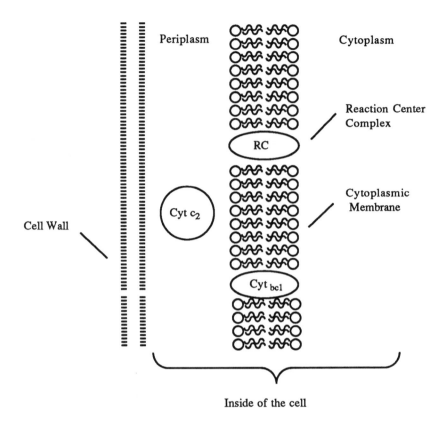

Figure 1. Schematic representation of the location of the reaction center complex in the bacterial cytoplasmic membrane.

A striking feature of the photosynthetic apparatus is that the pigments exist as distinct molecules, not covalently linked to other components, in contrast to the covalent linking of retinal in bacteriorhodopsin (see Chapter 18). For this reason, photosynthetic pigments can be extracted by organic solvents or detergents and various components of the photosynthetic apparatus can be dis-assembled in a methodical way.[11,12] It is this approach, coupled with the spectroscopic methods and x-ray diffraction studies on single crystals, that led to much of the current understanding of the structure and function of antennae and the reaction center. Since the bacterial photosystems are better understood than those of green plants, emphasis in this review is on the bacterial systems and an occasional reference to plant systems will be made when appropriate.

There are two major varieties of photosynthetic bacteria known: the green bacteria and the purple bacteria. Green bacteria are divided into two subgroups, green sulfur bacteria and green filamentous bacteria. Purple bacteria are subdivided into purple sulfur bacteria and purple nonsulfur bacteria. As the name implies, sulfur bacteria use sulfur compounds such as sulfide or thiosulfate as the source of electrons, whereas the nonsulfur bacteria use organic matter. Photosynthetic bacteria can be grown under anaerobic conditions in the presence of visible light and provide a convenient source of reaction centers and antenna complexes. A large amount of information is available on several bacterial systems, including the crystal structures of three of the bacterial reaction centers. This article is mainly concerned with the purple bacteria *Rhodopseudomonas viridis (R. viridis)* and *Rhodobactor sphaeroides (Rb. sphaeroides)* and reference to other systems will be made only when necessary.

Chlorophyllous pigments of the bacterial photosystem, the organic chromophores that are central to photosynthesis, are called bacteriochlorophylls, and five classes of these pigments have been characterized so far (Figure 2).[13] They all contain a porphyrin-type ring system capable of coordinating a Mg^{2+} ion and a hydrophobic phytyl side chain attached to the porphyrin ring through an ester link. A magnesium-less pigment identical in structure to bacteriochlorophyll called bacteriopheophytin (Figure 2) is found in all photosynthetic bacteria and is suggested to play a very important role in assisting the photosynthetic electron transfer. Bacteriochlorophylls differ from plant chlorophylls in terms of substituents on the ring and in terms of the degree of unsaturation in the ring. Ring II is partially reduced in the bacterial pigments and this ring is fully unsaturated in the corresponding plant pigments.

In bacteria, pigments a and b differ by the presence of an additional exocyclic C—C double bond in ring II. Due to the extended conjugation in the bacterial pigments, as compared to the corresponding plant pigments (Figure 2), bacterial systems absorb light at longer wavelengths than plant pigments. This ability to absorb at longer wavelengths than the plant pigments is well suited for the bacterial habitat, at the bottoms of pools and lakes. Light reaching the bottom is filtered by the plant species that live in the top layers. Since bacterial pigments absorb at longer wavelengths, bacteria can survive at the bottoms of pools.

Reaction centers in photosynthetic bacteria are located in the cytoplasmic membrane (Figure 1). The reaction center pigments are held together by the protein units, heavy (H), medium (M), and light (L), and a firmly bound cytochrome subunit with four hemes.[14] In *Rb. sphaeroides*, the reaction center consists of four BChl a molecules, two Bphe a molecules, two ubiquinone molecules, and one nonheme iron(II). In *R. viridis*, a nonsulfur purple bacterium, the reaction center contains four BChl b molecules, two Bphe b molecules, one menaquinone (Q_A), one ubiquinone (Q_B), and a nonheme iron(II).[15] In bacterial systems, the primary electron donor is a pair of bacteriochlorophylls called the "special pair" and in green plants, the primary electron donor is probably a pair of chlorophylls.

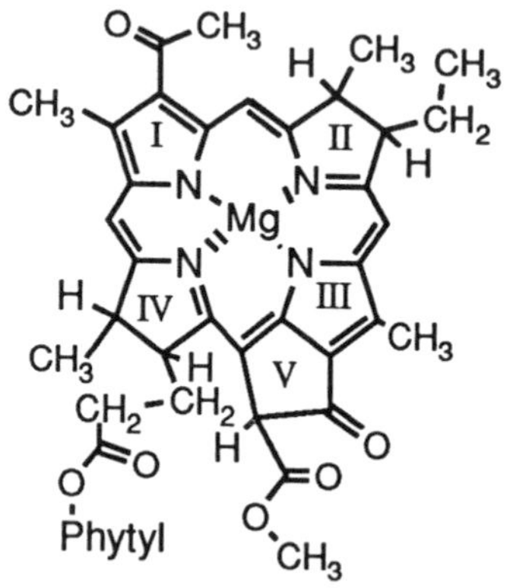

Bacteriochlorophyll a

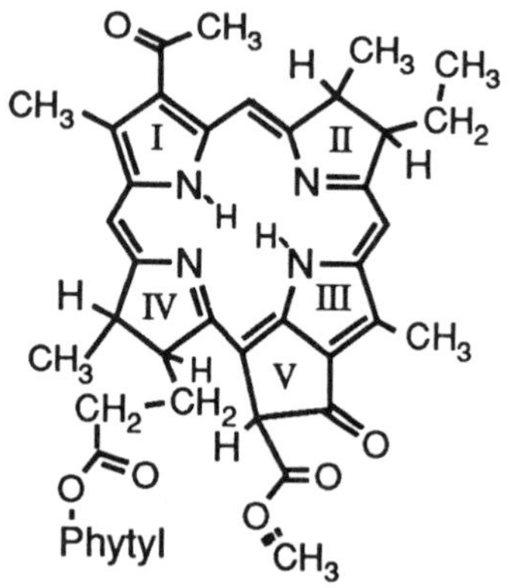

Bacteriopheophytin a

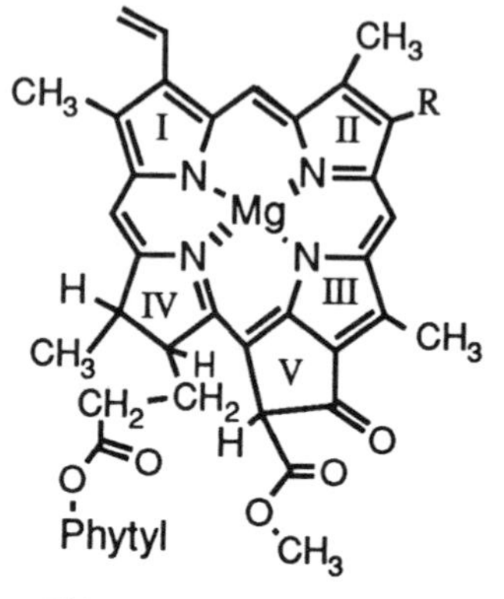

Chlorophyll a R = CH_2CH_3

Chlorophyll b R = CHO

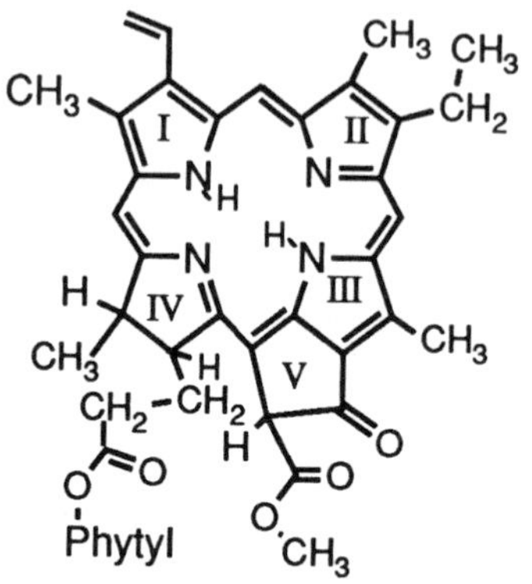

Pheophytin a

Figure 2. Structures of some of the pigments of the reaction center.

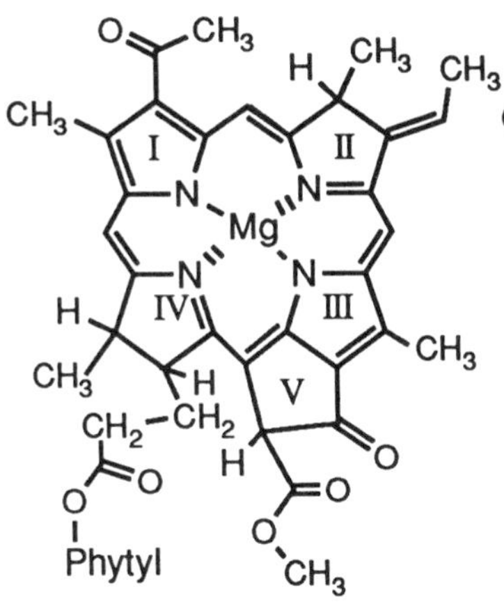

Bacteriochlorophyll b

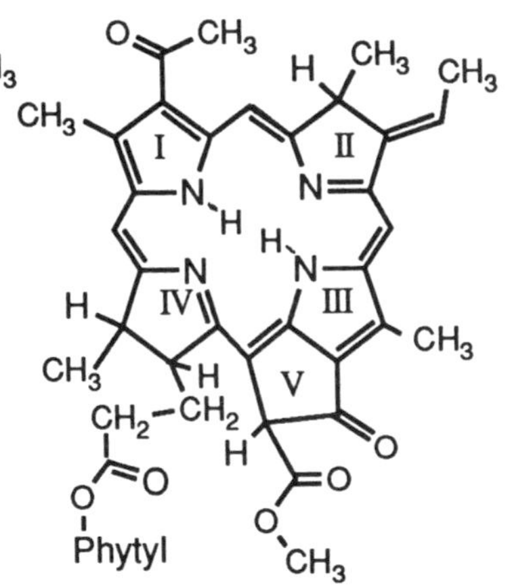

Bacteriopheophytin b

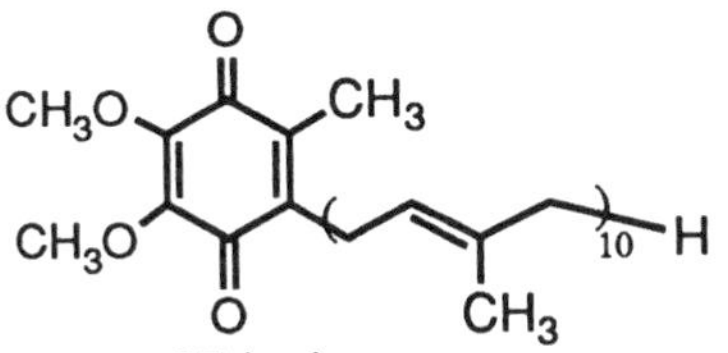

Ubiquinone

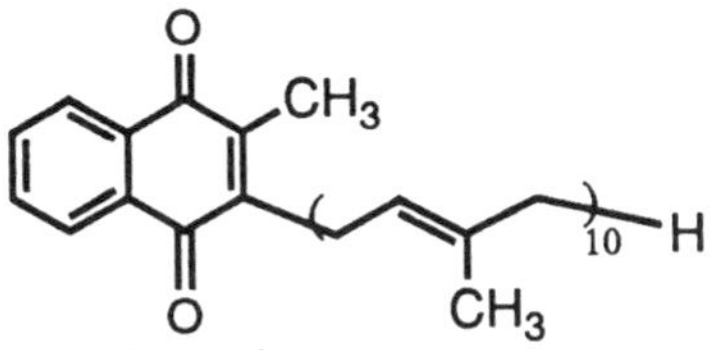

Menaquinone

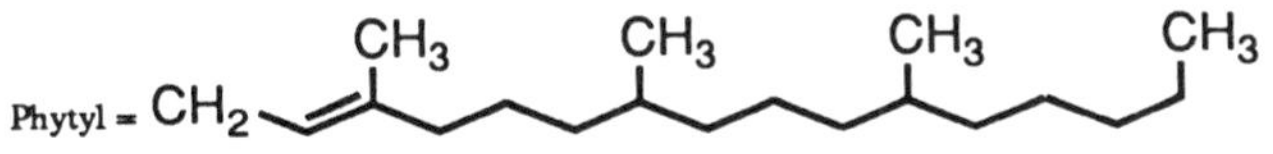

Phytyl = CH₂

Figure 2 (contd.). Structures of some of the pigments of the reaction center.

Bacteriopheophytin molecules in the reaction center have been suggested to serve as electron acceptors in the charge-separation process. The arrangement of these pigments has an approximate C_2 symmetry (if minor differences are ignored) with the symmetry axis running down through the iron atom (Figure 3). The pigments are arranged in the protein matrix of the H, L, and M subunits[16] that do not have c_2 symmetry. The special pair is located at the interface of the L and M subunits near the periplasmic side, and the iron atom connects these two subunits at the cytoplasmic side through four histidine ligands and one bidentate glutamic acid.[17] Iron (II) is approximately 2 Å closer to the quinone Q_B than to Q_A and this was suggested to favor the preferential stabilization of the electron on Q_B. In addition to this, a number of minor differences do exist between the two halves of the reaction center and are thought to be important in the functioning of the reaction center. Ordered arrangement of these various components is essential for the efficient functioning of the photosynthetic apparatus.

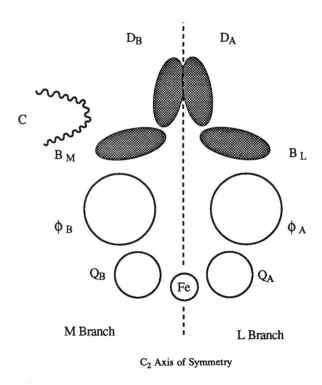

D = Primary donor, B = Accessory pigment, C = carotenoid
ϕ = Bacteriopheophytin, Q = quinone

Figure 3. A schematic representation of the components of the reaction center.

Subcellular particles containing the photosynthetic pigments, called chromatophores, can be isolated from the photosynthetic bacteria and each particle contains several photosynthetic units.[18] A single chromatophore may contain as many as 40 reaction centers, 500 light-harvesting complexes, and 1000 carotenoid and quinone molecules. These particles provide a convenient source of reaction centers and antenna complexes for experimental studies on bacterial systems.

2.2. Light Gathering in the Antenna

Even on a bright sunny day, the energy density of solar radiation is low. Since complex arrays of proteins and pigments are required to convert solar energy, photosynthetic organisms have developed antenna whose sole function is to collect the diffuse sunlight and transfer the excitation to the reaction center where the primary photochemical event can occur.[19] Thus, a few reaction centers can convert and store energy absorbed by several pigments. This is crucial for the ergonomic survival of the organism. Large antenna complexes are formed by the aggregation of small subunits and each subunit generally consists of two or three bacteriochlorophyll molecules and one or two carotenoid molecules noncovalently bound to very hydrophobic apoproteins.[19] Secondary pigments such as carotenoids, present in the antenna, have been shown to transfer excitation energy efficiently to other pigments in the antenna and thereby extend the light-gathering ability of the antenna.[20] This scheme provides an economic, yet efficient way of gathering low-density solar energy and converting it efficiently into chemical potential.

Since the efficiency of energy transfer by the Forster mechanism can be modulated by the relative orientations of transition dipoles of the participating chromophores, appropriate orientation of pigments in the protein–pigment complexes of the antenna is crucial for efficient energy transfer. The inverse sixth power distance dependence of transfer rates renders this mechanism significant only at very short distances.[21] In the antenna complex, energy is transferred rapidly between the pigments until it is trapped by the special pair or emitted as fluorescence.[22] Energy transfer in antennae is of primary importance in understanding the role of organized assemblies of the protein–pigment complexes for efficient energy transfer.

Green sulfur bacteria have large extramembrane antenna structures called chlorosomes (with BChl c and carotenoids) attached to the inside of the cytoplasmic membrane.[23] These antennae are about ten-fold larger than in purple bacteria and several times larger than in green plants.[24] The large size of these antenna is convenient in studying the ordering of the chromophores and its relation to the energy-transfer efficiency. If long-range ordering of the chromophores for efficient energy transfer between the pigments in these large antennae is important, then it is likely to be important in smaller antennae as well.

Fetisova and co-workers have shown that energy transfer between the pigments in chlorosomes of live bacteria takes place with a high conservation of polarization.[25] The BChl c pigments (similar in structure to BChl a and b) in the chlorosome were excited by a picosecond laser at 711 nm, employing live bacterial samples. Emission from BChl c in the antenna and from BChl a in the reaction center, was monitored at 730 and 820 nm, respectively. If energy transfer were to occur between randomly oriented BChl c chromophores in the antenna, then a rapid depolarization of excitation is to be expected with energy migration from pigment to pigment. On the other hand, if the participating pigments are aligned or have a long-range ordering within a cluster, then energy transfer would conserve the polarization of the initially populated state even after several transfer steps. Fluorescence emitted from the antenna BChl c was found to have a large and constant polarization value within the lifetime of the excited state, a time interval of ~230 ps. Presumably, after a large number of hops within the antenna, the excited state is still highly polarized with respect to the excitation pulse. Thus, energy transfer in the antenna among BChl c pigments occurs between ordered chromophores whose transition moments are nearly parallel. The authors conclude that the excitation migrates within a cluster of highly ordered pigments within the antenna.

On the other hand, fluorescence from the BChl a pigments in the reaction center was considerably depolarized within the excited-state lifetime and the loss of polarization was attributed to the transfer of energy from chlorosomes to uncorrelated membrane-bound BChl a pigments. Even though energy transfer from chlorosomes to the reaction center leads to depolarization, it was found to be extremely rapid[26] and competes efficiently with deactivation processes. Once trapped by the special pair, the excitation is rapidly converted into chemical potential through the primary electron-transfer step.

Energy transfer in the antennae is an excellent example of how organized media in biological systems have evolved to optimize efficiency of energy transfer over long distances. Orientation of the pigments in the antenna is controlled by the supporting protein matrix and the photosynthetic unit achieves efficient energy transfer between pigments. The next step in energy conversion after the absorption of a photon is the primary photochemical event that is initiated by the special pair.

2.3. Primary Photochemical Event
2.3.1. Regioselectivity

Photoinduced electron transfer from the special pair to the primary acceptor BChl a or b is of major interest from both theoretical as well as experimental points of view.[27] Subsequent electron transfer steps carry the electron down the electron-transfer chain, to the quinone Q_B. At every step, reverse electron transfer is inhibited such that the overall efficiency of charge separation is ~0.9. How is such a high efficiency accomplished? What are the roles of the accessory pigments in the reaction center to promote the forward electron transfer and inhibit the back electron transfer? These are some of the most interesting questions.

Excitation trapped at the special pair results in charge separation and the identity of the primary acceptor, and the participation of accessory pigments in the electron-transfer mechanism is still under intense debate. However, the formation of the anion radical of Bphe b was observed in under 4 ps[28,29] after the initial light absorption. The electron is then transferred to the quinone Q_A in about 200 ps and then on to quinone Q_B, near the nonheme iron(II), in about 100 μs[30] (Figure 4).

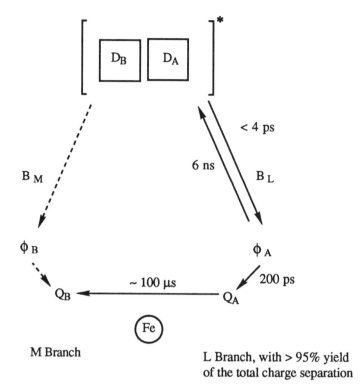

Figure 4. A schematic representation of electron transfer reaction after the initial light absorption, in the reaction center complex. Solid arrows represent the major pathway (>95%). Dotted arrows represent the minor pathway. The numbers near the arrows represent the observed inverse rate constants.

Rates of primary electron transfer in the photosynthetic reaction center have been studied very systematically by a number of physical methods. Time-resolved optical methods are very useful for studying the mechanistic details of these reactions. By the use of very short laser pulses (<100 fs), Breton et al.[31] have shown that the recovery of the special pair bleaching takes place in ~ 1.2 ps in *Rb. sphaeroides* and in 0.7 ps in *R. viridis*.[30] No evidence for the formation of BChl⁻ could be observed in these experiments or in many other related investigations. Bphe⁻ is the only intermediate that could be observed following the bleaching of the special pair, suggesting that the electron transfer takes place from the special pair to the bacteriopheophytin[32] with no kinetically resolvable intermediate involving BChl. It is difficult to explain the magnitude of the rate of electron transfer without the involvement of the intervening pigment, especially when the edge to edge distance between the donor and and acceptor is about 8 Å.[33]

A remarkable organizational feature of the reaction center is that electron transfer occurs predominantly (>95%) through the L-branch of the reaction center in comparison to the M-branch,[34] even though an approximate c_2 symmetry exists (Figure 3). The reason for this high regioselectivity is not clear and has been thought to be controlled by the supporting protein matrix as well as inter-pigment interactions that are not the same in the two halves of the reaction center.[35] Thus, despite the overall c_2 symmetry, a number of subtle differences exist between the two halves and these differences are important with reference to their possible role in the electron-transfer mechanism.

Removal of the pigment B_M from the M branch of the reaction center by sodium borohydride treatment was found to have no effect on the rate of electron transfer.[36] This is a very important observation, suggesting that the L branch of the reaction center dominates in the electron-transfer process. It is not clear what purpose this regiospecificity serves in the reaction center, but current experimental evidence suggests that the M branch of the reaction center acts to protect the pigments from light damage via triplet energy transfer. It has been suggested that one half of the reaction center is tuned for electron-transfer reactions and the other half for triplet energy transfer.[37] If this hypothesis turns out to be true, it will be a spectacular example of tuning of photoprocesses by organized biological media!

Several differences between the right and left halves of the reaction center have been pointed out by Allen et al.[38] to explain the observed regioselectivity. Electron-transfer rates are known to depend upon the electronic coupling between the donor and the acceptor as well as the Frank-Condon factors.[39] The strength of the electronic coupling depends upon the overlap of the electronic wavefunctions of the donor and the acceptor. Distances between the pigments and the extent of overlap of their van der Waals surfaces, and specific protein–pigment interactions play a major role in the regioselectivity as well as in the overall efficiency.[38] The overlap in the surface areas of the pigments D_A and B_L in the L branch was found to be 1.5 times larger than the overlap between the corresponding pigment pair on the M branch, D_B and B_M. Thus, the difference in the rates of electron transfer through the two halves of the reaction center could be partly because of the differences in the overlap between the donor and the acceptor pigments.[38]

Allen et al. pointed out that protein–pigment interactions differ among the two halves of the reaction center. In *R. sphaeroides*, hydrogen bonding between ring V of ϕ_A and glutamine L104 was suggested, whereas ϕ_B does not seem to have this specific interaction. The presence of aromatic amino acid residues near the donor and the acceptor may also influence the rate of electron transfer. Tyrosine (tyr) M210 was found to be in van der Waals contact with both D_B and ϕ_A. Recent studies by Nagarajan et al. show that replacement of Tyr M210 by phenylalanine (Phe) or isoleucine significantly alters the initial electron-transfer step emphasizing the role of Tyr M210.[40] The corresponding residue Phe L181 was suggested to be not as favorably located near D_A and ϕ_B. Furthermore, tryptophan (Trp) M252 bridges ϕ_A and Q_A, with its π plane nearly parallel to that of Q_A. Examination of the corresponding residue between ϕ_B and Q_B shows that Phe L216 has its π plane nearly perpendicular to the quinone Q_B. Positioning of these aromatic residues could be very important in the regioselective electron transfer in the reaction center. A larger number of charged residues were found near the quinone Q_B as compared to Q_A. Thus, in spite of the symmetry associated with the arrangement of the pigments, the underlying protein matrix is significantly unsymmetrical. This asymmetry was suggested to be very important in favoring electron transfer along one half and not the other.[38]

2.3.2. Temperature Dependence of Electron-Transfer Rate

Fleming et al.[41,32] monitored the rate of electron transfer from the special pair by excitation into the absorption bands of the special pair and by monitoring the stimulated emission in the near infrared from BChl a.[12] The rate of electron-transfer was monitored upon excitation at 870 nm, with a 150 fs pulse, and by monitoring the stimulated emission at 1045 nm for *R. viridis* and at 925 nm for *Rb. sphaeroides*. In both systems, emission decayed as a single exponential at temperatures ranging from 300 K to 8 K and from these experiments, rates of electron-transfer were estimated to be on the order of 4×10^{11} s^{-1} at room temperature.

A very interesting feature of this primary electron-transfer process is that the rate of the reaction increases with decrease in temperature and it increases more rapidly for *R. viridis* (4 times) than for *Rb. sphaeroides* (2.3 times). Increase in the rate of a reaction with lowering of temperature is very rare. Furthermore, the magnitudes of the rate constants, even at room temperature, are very large, given the edge to edge distance between the special pair and the acceptor, ϕ_L (~8 Å). Several interesting mechanistic models have been suggested to account for the magnitude of the rate constant as well as its temperature dependence.[42]

One of the appealing explanations for these observations centers around the participation of low-frequency polar modes of the protein in the electron-transfer step, suggesting that the charge transfer between the pigments is coupled to the surrounding protein matrix. When the reaction center was fully deuterated no change in the rate was observed, ruling out the importance of the role of intramolecular C—H modes or protein–pigment hydrogen bonds in the electron-transfer process.[40,41] The increase in the rate with decreasing temperature could be due to the changes in the electronic coupling of the pigments or changes in the coupled vibrational modes.[43]

2.4. Role of Carotenoids in the Reaction Center and in the Antenna

Carotenoids are hydrophobic polyenes and are suggested to play a dual role in photosynthesis. They function as light-harvesting molecules in the antenna by absorbing at shorter wavelengths than the chlorophyllous pigments and by transferring the energy to the pigments in the antenna. They protect the reaction center[44] from light damage by quenching the triplet states produced by the back electron transfer step (Figure 5). Carotenoids are bound noncovalently to the protein–pigment complexes near the special pair in the reaction center and they are also present in the antennae. The triplet of the special pair, if not quenched rapidly, can sensitize the formation of singlet oxygen. Singlet oxygen can in turn react with the pigments and damage the photosynthetic apparatus.[45] Carotenoid-less mutants (*Rb. sphaeroides* R26), when illuminated in the presence of oxygen, were found to sensitize their own death.[46]

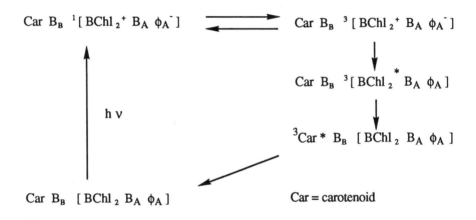

Figure 5. The proposed role of the carotenoid in the reaction center. (With permission, from ref. 61b).

Excitation into the absorption bands of the carotenoid leads to fluorescence from the chlorophylls in the antenna complexes, supporting the viability of singlet–singlet energy transfer from carotenoids to chlorophylls.[47,49] In the absence of the protein matrix (in frozen organic glasses), no energy transfer between the short-lived carotenoid singlets (~8 ps)[45] and chlorophylls can be observed. Absence of energy transfer in glassy media emphasizes the importance of the organized protein matrix in controlling the efficiency of energy transfer. In the reaction center, singlet–singlet energy-transfer efficiency between the carotenoids and chlorophyll is ~1.[48] Low-fluorescence quantum yields and short singlet lifetimes of carotenoids suggest that Forster energy transfer is not very likely in these systems.

2.4.1. Triplet Energy Traps

Carotenoids have very low triplet energies and are ideal traps for quenching even the low-energy triplets of chlorophylls and bacteriochlorophylls.[50] Model systems with carotenoids covalently linked to *meso*-tetraaryl porphyrins[51] clearly demonstrated efficient energy transfer between the triplet porphyrin and the carotene in fluid media and prevented the porphyrin triplet from sensitizing singlet oxygen formation.[52] Carotenoids have been shown to quench the triplet of the special pair in the reaction center almost as quickly as it is formed, resulting in the formation of the carotenoid triplet. The rise time of the carotenoid triplet was estimated to be between 10 and 30 ns,[53] which matches with the decay of the triplet state of the special pair (Figure 5).

Organized media can influence the spectral properties of a chromophore as well as its excited state. A good example is the carotenoid in the protein matrix of the reaction center. Absorption spectra of carotenoids in the reaction center are significantly red shifted (15–25 nm) as compared to organic solvents such as pentane.[37] The asymmetric protein environment of the reaction center induces optical activity in the carotenoid and strong induced circular dichroism (CD) bands were detected for the carotenoid.[54] The asymmetric orientation of the polyene is dictated by the protein matrix through hydrophobic binding interactions. Very little is known about the binding site and its asymmetric induction mechanism. Electron spin resonance spectroscopy of the carotenoid in the bacterial reaction centers[55-57] after illumination suggests that the carotenoid is twisted around C6—C7 or higher carbon–carbon bonds. An elaborate discussion of the carotenoid location, its geometry, and its role in the reaction centers has been published.[45]

2.4.2. Carotenoids in the Antenna

Carotenoids function as accessory pigments in the light-gathering process of the photosynthetic antennae. Absorption bands of the carotenoid in the antennae are red shifted by about 20 nm when compared to pentane solutions, similar to the carotenoid in the reaction center.[58] Carotenoids in the antennae also show strong CD bands induced by the environment, but different from the CD observed for the carotenoids in the reaction centers.[58] Thus, the environments of the carotenoid in the antennae and in the reaction center are distinct in their asymmetry. Carotenoids in the antennae complexes have been shown to be all-trans,[59,60] in contrast to the observed cis geometry in the reaction centers.

Singlet–singlet energy transfer from carotenoids to BChl in the antenna is very efficient and is independent of the structure of the carotenoid.[61] Energy transfer between the carotenoid and the bacteriochlorophyll in the antenna complexes was investigated by picosecond spectroscopy. Excitation of the carotenoid at 510 nm with a 4 ps laser pulse was followed by the ground state bleaching.[62] Formation of the bacteriochlorophyll singlet excited state by energy transfer from the carotenoid was monitored by the bleaching of the bacteriochlorophyll ground state at 860 nm. The singlet of the carotenoid was populated concomitant with the laser pulse and decayed with a time constant of about 6 ps. The rise time of the singlet bacteriochlorophyll was found to be equal to the decay rate of the carotenoid singlet excited state. Thus, within the time resolution of the experiment (4 ps) both processes are concurrent. A better time resolution may reveal intermediates that may be present in the energy-transfer mechanism.[62]

Energy-transfer efficiencies are independent of the carotenoid structure but depend upon the type of the antenna complex, and efficiencies ranging from 100 to 20% have been reported.[63] Energy transfer in the antenna is independent of temperature down to liquid helium temperatures,[63,64] in contrast to the observed temperature dependence of triplet-triplet energy transfer between the special pair and the carotenoid in the reaction center.[65] Thus, the photophysical behavior of the carotenoid in the antenna differs from its behavior in the reaction center, indicating the role of the underlying protein matrix in modulating these processes.

2.5. Summary

Current experimental and theoretical understanding of the bacterial photosynthetic unit suggests that ordering of its components is extremely important. Given the overall organization, the L branch of the reaction center is preferred for the electron transfer and the M branch is preferred for triplet energy transfer. Energy transfer in the antenna was found to depend on the long-range ordering of the pigments and nature has taken advantage of this by building ordered supramolecular structures for efficient energy pooling. Such an organizational tuning of the reaction system is extraordinary and, in principle, we can take advantage of this knowledge for building supramolecular devices for practical purposes that are highly specific and efficient. The natural polymer deoxyribonucleic acid is one such supramolecular structure readily available for the design of artificial systems that can take advantage of its ordered structure. It is with this perspective that a description of DNA structure and its interaction with small molecules and how this organized medium can influence photoprocesses, is described below.

3. Deoxyribonucleic Acid as the Organized Medium

Although nucleic acids are not used to conduct or control photoprocesses in nature, these biopolymers provide a unique organized environment for the control of photophysical and photochemical processes of DNA-bound guest molecules. A brief review of the structure of DNA and its influence on the ground- and excited-state properties of small molecules is provided. The interesting subject of DNA photodamage and photochemical reactions of DNA bases are not included here and several recent reviews can be consulted for this purpose.[66]

The organized structure of the DNA polymer provides a quasi-crystalline environment with interesting features for controlling photophysical and photochemical processes (Figure 6). These features include the arrays of the phosphate negative charges arranged in a pseudo-one-dimensional lattice, the hydrophobic core of the helix comprising the heterocyclic π systems; the outer hydrophilic surface of the helix with hydrogen-bonding acceptor and donor sites in the major and minor grooves, and the asymmetric helical, rodlike, tertiary structure. The three-dimensional arrangement of the hydrogen-bonding sites in the grooves of the helix forms the "script" for encoding the genetic information. Information stored in the helix is decoded by a complex set of proteins that make specific contacts to the helix in the major and minor grooves through hydrogen bonding as well as hydrophobic interactions.

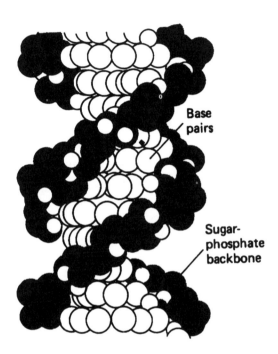

Figure 6. Structure of the DNA double helix with major and minor grooves. Reproduced with the permission from McGraw-Hill Inc, from Cell Biology by G. Karp, 1979.

Therefore, study of interactions of small molecules to DNA provides a model for protein–nucleic acid interactions[67] and progress made in this field is useful in cancer research and in biotechnology.[68] DNA-bound molecules can also be useful as probes to explore the molecular details of DNA local structure and to accelerate or to inhibit photochemical reactions. Metal complexes have been of particular interest for DNA binding studies. They provide a rigid scaffolding to orient functional groups that can make specific contacts with the double helix and with proper choice of the metal ion, such complexes can serve as artificial nucleases to cleave DNA. Thus, metal complexes and small heterocyclic cations have been extensively used as photochemical and photophysical probes in the exploration of DNA structure and function.

3.1. The Structure of DNA

The three-dimensional structure of the DNA double helix (B form) was first proposed by Watson and Crick in 1953.[69] According to them, two strands of the DNA polymer are twisted around the helix axis in a right-handed screw symmetry, with the sugar phosphate backbone on the outside and the heterocyclic bases adenine (A), thymine (T), guanine (G), and cytosine (C), inside the helix (Figure 6). π-Planes of the bases are arranged nearly perpendicular to the helix axis and are stacked one on top of another. Bases of the two strands of the helix engage in complementary hydrogen bonding, i.e., guanines hydrogen bond with cytosines and adenines hydrogen bond with thymines (Figure 7a).

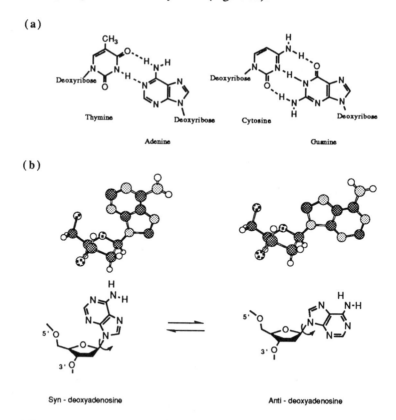

Figure 7. (a) Base pairing of thymine with adenine and cytosine with guanine. (b) Syn and anti configurations for the nucleotides of DNA.

Recognition of this H bonding complementarity between the bases was the major break-through in the proposition of the DNA structure. The long axis of the base pair is rotated in the right-handed screw direction in B-DNA, as one moves along the helix, thus leading to a right-handed helical structure. The conformation of the deoxyribose sugar in the sugar phosphate backbone varies with the conformation and the heterocyclic base can be syn or anti with respect to the ribose (Figure 7b).[70]   In the left-handed Z-DNA helix, the deoxyribose has a 2'-endo, 3'-endo conformation, different from that in the B form. Due to the variations in the orientation of heterocyclic bases and sugar puckering, the DNA double helix can adopt a variety of conformations under different experimental conditions. So far, A, B, C, D, and Z conformations have been extensively investigated and have been suggested to be biologically important. Many variants of these conformations are reported and they may be of significance in biological systems as well.[71]

The x-ray crystallography of the dodecamer $(dGdC)_{12}$ by Dickerson and co-workers[72] led to the precise dimensions of the right-handed B-form helix and these atomic coordinates are widely used for molecular modeling studies. Although these parameters may not be strictly valid with longer helices and particularly with other sequences, they provide a satisfactory working set for molecular modeling studies (Table 1).[73]   Not surprisingly, the helical structure of such a complex macromolecule is quite sensitive to the nature of the counterions, salt concentration, bound ligands, and temperature.

Table 1.   Parameters for B and Z Conformations of DNA

Conformation	Helix twist (degrees)	Groove Width Major (Å)	Minor (Å)	Rise per base pair (Å)
B conformation	36	11.4	6.0	3.38
Z conformation	-10, -50	—	—	3.7
A conformation	33	2.3	11.0	2.56

DNA conformational polymorphism is crucial for its role as an information storage molecule because different conformations may trigger different biological events. For example, information retrieval and its translation is controlled by an elaborate mechanism involving DNA binding proteins. Binding of these proteins could be sensitive to the DNA local conformation. The local conformation itself could be controlled by the binding of additional cofactors. Conformational changes may then function as biological switches. These structural differences have been suggested to serve as punctuation marks in the script of the genetic code. Thus, study of different conformations and their role in gene regulation and gene expression is central to the molecular biology of DNA. Although different conformations can be distinguished spectroscopically, local heterogeneity in the structure of the helix on the order of a few base pairs is difficult to detect. For example, spectroscopic techniques tend to measure the collective properties of an ensemble of macromolecules averaged over the time scale of the experiment. Thus, the measured property is a weighted average of the properties of various local conformations of the macromolecule. Molecular probes targeted to specific sequences of the macromolecule are very useful in the investigation of macromolecular local structure. If a few base pairs of a large DNA molecule adopt a different conformation from the rest of the molecule, then this conformational region may be recognized by a conformationally specific probe molecule. The spectroscopic and chemical properties of the conformationally sensitive probe–DNA complex may be used as a basis to determine the local structure of the biopolymer. A large number of studies targeting specific sequences of DNA by small molecules have taken this approach in the past few years.[74] Structures of a few heterocyclic DNA intercalators are shown in Figure 8. These small molecules bind to the helix in several binding modes and some important binding modes are discussed below with reference to their nature and importance.

3.2. Binding Modes

Noncovalent interaction of small molecules with DNA can be described by electrostatic binding, surface binding due to hydrophobic as well as ionic interactions, and intercalation where the probe molecule is inserted in between the base pairs (Figure 9). Alkali metal ions are electrostatically attracted to the phosphate backbone and prefer electrostatic or ionic binding, whereas transition metal ions may coordinate to the nitrogen atoms of the aromatic bases. Since the interior of the double helix is fairly hydrophobic due to the π systems of the nucleotide bases, planar aromatic heterocyclic cations prefer intercalative binding. These binding modes differ from each other, in terms of the nature of the interactions as well as the physical properties of the complex formed between the helix and the small molecule. Ground- and excited-state properties of the bound molecule or ion often differ from that of the free probe and can form a basis to identify the local conformation of the helix by spectroscopic methods.[75]

3.2.1. Electrostatic Binding

Ionized phosphate groups of the DNA backbone generate an intense charge field, which attracts metal ions to the DNA surface and results in high concentrations of counterions around the helix. The electrostatic charge density is not uniformly distributed around the helix, and it is larger in the minor groove than in the major groove.[76] Thus, stronger electrostatic interactions are to be expected in the minor groove than in the major groove. Sodium and potassium ions are attracted to the helix and the local concentrations of these counterions at the helix surface were found to be much higher than in the bulk solution.[77] This is termed as "counterion condensation" and has been covered extensively by the theoretical models of Manning[78] describing the binding of metal ions to polyelectrolytes. Electrostatic binding dominates the interaction of alkali metal ions to DNA.

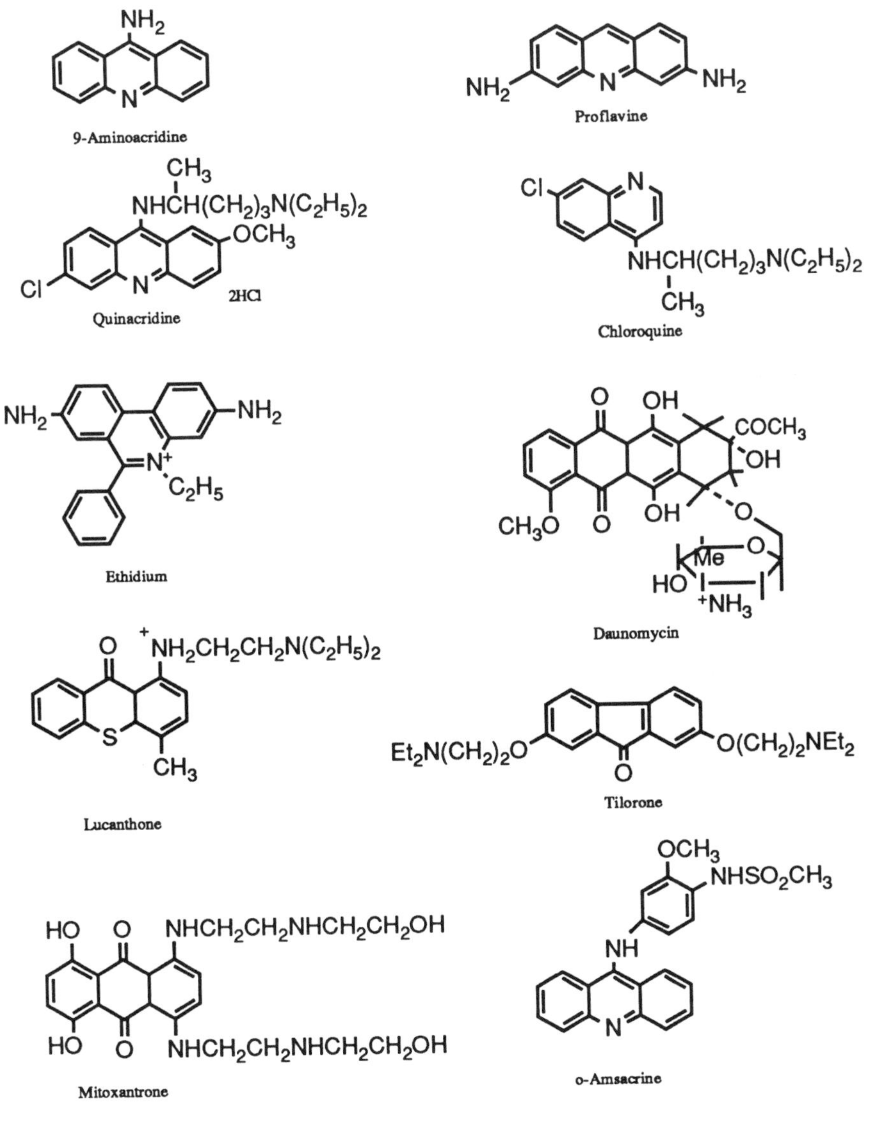

9-Aminoacridine

Proflavine

Quinacridine

Chloroquine

Ethidium

Daunomycin

Lucanthone

Tilorone

Mitoxantrone

o-Amsacrine

Figure 8. Some examples of popular DNA intercalating agents

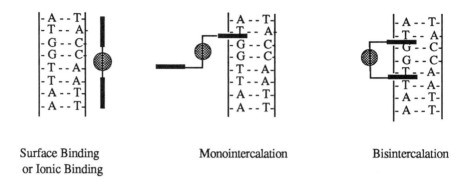

Surface Binding   Monointercalation   Bisintercalation
or Ionic Binding

Figure 9. Schematic representation of various binding modes available for small molecules with the DNA double helix.

3.2.2. Surface Binding

Hydrophobic molecules can bind on the surface of the major and minor grooves of DNA by the displacement of the resident solvent molecules from the grooves. Methyl groups on thymine and m^{-5}-cytosine point into the major groove and provide hydrophobic binding sites. On the other hand, arrays of hydrogen bonding sites, in the major and minor grooves of the helix provide hydrophilic hydrogen-bonding sites for the guest molecules (Figure 9).[79] The spine of water discovered in the minor groove of AT sequences provides additional hydrophilic sites for hydrogen bonding.[80] These various binding sites on the helix provide several possibilities for the interaction of small molecules with the helix. Binding of the natural product netropsin and the anticancer agent Hoechest 33258 are good examples of surface binding (Figure 9) to the DNA helix.[81]

3.2.2. Intercalation

The hydrophobic interior of the double helix, the space between adjacent DNA bases, can provide a nonaqueous environment for planar heterocyclic cations. The adjacent bases are pushed apart and the guest molecule is accommodated into the helix so that charged groups are exposed to the solvent whereas the hydrophobic part is intercalated into the nonpolar interior of the helix. Helix structure changes due to the intercalation of a guest molecule and these changes can be monitored in biophysical and photophysical experiments.[82] Changes in the physical properties of the guest also take place upon intercalation. For example, intercalation into the helix results in the restricted rotation of the guest, essentially due to the great length of the helix and the long residence times of the guests in the helix. In general, photophysical properties are modulated by intercalation and these modulations can be used with caution to establish binding modes.[83]

Binding of ethidium (Figure 8) to DNA is an example of intercalation. Crystal structure studies of the ethidium–DNA complex revealed that the chromophore was indeed sandwiched between the base pairs and the intercalator occupies only alternating binding sites on the helix.[84] That is, the two adjacent binding sites of an intercalated molecule are not occupied by a neighbor and this observation is generally known as the neighbor exclusion principle (Figure 9). The intercalated chromophore is in van der Waals contact with the nitrogeneous base pairs and this proximity can be expected to result in the interaction between the electronic states of the intercalator and the bases. Intercalation of ethidium increases its fluorescence quantum yield as well as its fluorescence lifetime (2.1 ns to 19 ns) and this could be due to rigidity of the environment as well as electronic interaction with the bases. Definitive information about this electronic interaction between the intercalator and the base pairs is missing. Results from optically detected magnetic resonance (ODMR) studies by Maki et al. show that intercalation of tryptophan side chains of DNA binding proteins influences the triplet sublevel splitting parameters as well as the decay rates of the triplet sublevels.[85] These studies clearly point to the possibility that electronic states of the nucleotide bases interact with the intercalator.

Very important and interesting consequences of intercalation are the stretching and unwinding of the helix. These changes in the helix structure can be readily determined in biophysical experiments to establish whether the probe molecules are intercalated into the helix or not.[86] Since the probe is transferred from an aqueous environment to a hydrophobic DNA interior, intercalation is often accompanied by changes in the energies of the probe ground and excited electronic states as well as the equilibrium nuclear configurations. These changes can be inferred from the spectroscopic investigations of probe–DNA complexes. Changes in absorption and emission maxima, emission quantum yields, oscillator strengths, and spectral bandwidths are often accompanied by intercalation and serve as identification marks for intercalative binding.

Due to the handedness of the helix, the intercalated chromophores are asymmetrically oriented along the helix, and intercalation is often accompanied by induced CD of the intercalated chromophore. Several examples of organic dyes that show induced CD upon binding to DNA are reported in the literature.[87] Several DNA binding drugs are suggested to intercalate into the DNA helix and intercalation was thought to play a role in the therapeutic action of these drugs.[88]

3.3. Control of Bimolecular Reactions with DNA

Metal ions and metal complexes bound to DNA can be used to accelerate or retard bimolecular reactions between ionic reagents. Reactions between cations can be accelerated by trapping them in the charge field of the helix and thereby promoting the bimolecular collisions. Local concentrations of these ions at the helix will be much higher, due to counterion condensation, than in the bulk and can result in accelerated bimolecular rates. On the other hand, reactions between ions of opposite charge can be retarded by trapping the cations in the negative charge field of DNA which repels the negatively charged ions, resulting in a reduction in the concentration of negatively charged ions at the helix surface. Apart from changes in the local concentrations of ions, DNA binding can also be used to increase the mobility of the bound ions to promote bimolecular reactions. Several theoretical models of ion diffusion in the restricted dimensional space surrounding the helix have predicted increased mobility of these ions at the helix surface.[89] Clear experimental evidence for this prediction is still not available in the literature.

Wensel and co-workers[90] have used energy-transfer studies between lanthanide ions in the presence of DNA to show that indeed the bimolecular collisional rate increases for cations bound to DNA. Ions used and the ratio of the energy transfer rates in the presence (k) and in the absence (k_0) of 1 mM of DNA phosphate and 2 mM sodium chloride are shown in Table 2.

Table 2. Ratio of the Collisional Energy Transfer Rate Constants (k/k_0) Measured between Donor–Acceptor Pairs in the Presence and Absence of B-DNA

Acceptor[a]	Bis(hydroxyethyl)-ethylenediamine-diacetatoTb(III), +1 donor	Hydroxyethyl-ethylenediamine-triacetatoTb(III), +0 donor	Ethylenediamine-tetraacetato-Tb(III), -1 donor
+2	29 ± 3	0.95 ± 0.1	0.24 ± 0.03
+1	6.0 ± 1.5	1.0 ± 0.1	1.0 ± 0.04
0	1.0 ± 0.1	1.0 ± 0.05	1.0 ± 0.05
-1	0.9 ± 0.1	1.0 ± 0.1	1.0 ± 0.1

[a] The acceptors used are: +2 acceptor, 2-mercaptoethylaminebis(ethylenediamine)Co(III); +1 acceptor, sym-cis-ethylenediamminediacetato(ethylenediamine)Co(III); neutral acceptor, bis(hydroxyethyl)ethylenediamine diacetatoCu(II); -1 acceptor, ethylenediaminetetraacetatoCo(III).

Analysis of these data shows that the collisional frequency between monovalent cations increases by a factor of 6 ± 1.5, and for divalent cations it increases by a factor of 29 ± 3, in the presence of DNA. On the other hand, the collisional frequency between divalent cations and monovalent anions is decreased by a factor of 0.24 ± 0.03.

The Stern-Volmer slope for the quenching of emission from DNA-bound $Ru(Phen)_3^{2+}$ by ferrocyanide was dramatically reduced when compared to the quenching in the absence of DNA.[91] This inhibition of quenching of the dication by ferrocyanide in the presence of DNA was cited as evidence for the tight association of the probe with the helix. Thus, DNA polyanion can be used to catalyze or to inhibit photochemical reactions between ionic reactants.

3.4. Energy Transfer from Nucleotides to Intercalators

Early studies on the binding properties of ethidium revealed that excitation energy can be transferred from the nucleotide bases to that of the intercalator.[92] Fluorescence intensity of DNA-bound ethidium was monitored as a function of the fraction of light absorbed by nucleotide bases. The yield of emission from ethidium increased with increased absorption by the bases. If energy transfer from the bases to ethidium is not feasible, then a decrease in fluorescence intensity with DNA concentration is expected. The authors conclude that energy is transferred from the DNA bases to ethidium and the efficiency of energy transfer depends upon the concentration of DNA. It was shown that the transfer efficiency drops with distance between the intercalator and the excited base and is nearly zero after five base pairs, a distance of 5 x 3.4 Å. This is a remarkable observation, in that the energy is transferred over a long distance in fluid media at room temperature.[93] A similar result of long-distance energy transfer was also observed with proflavin intercalated in the DNA matrix.[94] These results demonstrate that the light energy absorbed by the DNA bases can be transferred to an appropriate chromophore of low energy over a fairly long distance. The proximity of the adjacent bases (3.4 Å in the B form) as well as their systematic ordering in the helix, 36° rotation per base pair, could be important in determining the energy-transfer efficiency. Energy transfer between intercalated donors and acceptors could provide information on the energy-transfer efficiency through the DNA helix. The weakly coupled nucleotide bases may enhance energy and electron-transfer processes between intercalated chromophores. This exciting possibility is yet to be established.

3.4.1. Emission Lifetimes of Intercalators

Aminoacridines and planar heterocyclic cations bind to DNA and provide interesting examples of how the photophysical properties of the probe change upon binding to DNA. The excited-state properties of the intercalator also depends upon the local DNA sequence perhaps due to the subtle differences in the local environment and reactivity of the binding site. Summarized below are the results of Kubota et al.[94] with amino acridines as a typical example (Table 3). Fluorescence quantum yields and lifetimes increase upon binding to AT-rich DNA, whereas a decrease in these parameters was found with increasing GC content of the polymer. Quenching of probe fluorescence by GC sequences was suggested to be the reason for this trend. Guanosine monophosphate can in fact quench the fluorescence from these acridine derivatives, supporting this suggestion. Energy transfer from and to DNA bases may occur and probe fluorescence yields increase only if the DNA bases do not quench the excitation.

Table 3. Photophysical Properties of 3,6-Diaminoacridine Bound to DNA

% GC content of DNA	τ_1 (ns)	τ_2 (ns)	Quantum Yield
No DNA	4.8	—	0.440
0	6.7	—	0.477
30	6.3	—	0.211
42	6.4	—	0.066
72	6.1	0.4 (82%)	0.017
100	6.0	0.5 (99%)	0.005

3.4.2. Restricted Rotational Motions of Intercalators

Strong fluorescence from DNA-bound ethidium has been exploited to probe the rotational motion and rigidity of nucleic acid helices of varying lengths and sequences. If the intercalated probe does not come off the helix during its excited-state lifetime (~20 ns), then fluorescence depolarization of the intercalator will be primarily due to the rotational motion of the DNA helix and not the chromophore.[95] The concentration of the intercalator on the helix should be low enough to avoid intermolecular energy transfer between bound chromophores. DNA-bound ethidium samples were excited by a short laser pulse, FWHM = 3ps, and the emission from ethidium was monitored through crossed polarizers. The decay of the emission anisotropy was then fitted to a rigid rod model. The results thus obtained are shown in Table 4.

Table 4. Fluorescence Polarization of Ethidium Bound to DNA

DNA	Lifetime (ns)	Rigidity 10^{-19} erg cm
Poly-d(GC)	22.2	1.19
Poly-d(AT)	19.9	1.27
Calf thymus (CT)	23.0	1.43
Denatured CT	22.6	0.9
Poly(dGdC)–Poly(dGdC)	23.5	1.5
Poly(dAdT)–Poly(dAdT)	25.4	0.9

The rigidity of the d(GC) sequences was found to be similar to that of d(AT) sequences, but the rigidity of the alternating polymer poly(dG–dC) sequences was found to be higher than that of the alternating polymer poly(dA–dT). The rigidity of poly(dA–dT) sequences is similar to that of the denatured calf thymus DNA, indicating that these sequences are very floppy. This is consistent with the anticipated higher rigidity for the GC sequences caused by the presence of three hydrogen bonds as compared to only two hydrogen bonds between AT pairs. The longer DNA molecules were found to be more flexible than shorter helices.

3.5. Phenanthroline–Metal Complexes as Probes for DNA

Copper(II)–, Ru(II)–, Co(III)–, Cr(III)–, and Rh(III)–phenanthroline complexes have versatile spectroscopic properties that are sensitive to the environment. The ability to absorb strongly in the visible region of the electromagnetic spectrum, solubility in water, reasonable emission quantum yields (0.01 to 0.25), and the ease with which the optical isomers of the octahedral complexes can be separated, all have paved the way to study the binding of these complexes with DNA.[96] Several of these metal complexes have been successfully used to cleave the DNA polymer photochemically at the binding site and are useful as artificial nucleases. Similar results were observed with Ru(II) complexes. The lowest luminescent state in Ru(II) polypyridyl complexes is the metal to ligand charge transfer (MLCT) state. MLCT transitions are the dominant absorptions in the visible absorption spectrum of Ru(II) complexes and the non-luminescent ligand field state (LF) is close in energy to the MLCT state and acts as a thermally activated radiationless decay channel for the MLCT excited state. This low-lying deactivation channel, with a thermal barrier, is responsible for the strong temperature dependence of the excited-state properties of the Ru(II)polypyridyl complexes.

Luminescence properties of the Ru(II) complexes change upon binding to DNA, similar to the aromatic intercalators and these changes provide a spectroscopic description of their interaction with DNA.[97] The strong electrostatic field at the DNA surface attracts Ru(II) dications. Hydrophobic planar ligands on the metal complex prefer the hydrophobic interior of the double helix and tend to intercalate. Thus partial intercalation of the metal complex is possible with even the octahedral phenanthroline complexes. Electrostatic and stacking interactions, hydrogen bonding, and hydrophobic interactions influence the physicochemical properties of the bound metal complex. The predictable modulations in these properties proved to be very effective in determining the nature of binding of Ru(II) complexes to DNA[96] and in the elucidation of DNA local structure as sensed by these probe molecules.

3.5.1. Spectral Changes Due to Binding

Binding of Ru(Phen)$_3^{2+}$ to calf thymus DNA has been shown to red shift the visible absorption spectrum of the complex with an isosbestic point at 451 nm.[75]

This is a clear indication of the perturbation of the electronic states of the DNA-bound metal complex. The red shift in the absorption spectrum probably results from stabilization of the large dipole of the singlet MLCT excited state by the negatively charged DNA phosphates. Binding to DNA is also accompanied by a red shift of the emission maximum (~3 nm).[75] Since the luminescent states of these complexes are charge transfer in nature, proper location of the DNA phosphate negative charges with respect to the excited-state dipole can stabilize the MLCT state. Binding to DNA is also accompanied by increases in the emission quantum yield, emission lifetimes, and emission polarization. In principle, all these physical properties can be used to monitor binding to DNA.

Association of cationic complexes with DNA can accelerate or inhibit bimolecular reactions as described before. At low concentrations of $Co(Phen)_3^{2+}$, a very strong quenching of ruthenium emission was detected in the presence of DNA, whereas in the absence of DNA, quenching was very slow.[98] Changes in quenching slopes can be explained only if the cationic Ru(II) and Co(III) probes are intimately associated with the DNA helix and reduce the charge repulsion. Thus, electron-transfer quenching in these systems was found to be accelerated by the DNA polyanion.

3.5.2. Emission Lifetimes

Binding of the ruthenium complexes to DNA is generally accompanied by increases in their emission lifetimes due to decrease in the nonradiative rate constant of the luminescent state.[91] Since the nonradiative pathway is dominated by the LF states in these complexes, the energy gap between the triplet MLCT and LF states can dramatically influence the rate of the nonradiative rate constant. The red shift in the emission maximum suggests that the luminescent state is lowered in energy upon binding to DNA. Since the LF state does not have a dipolar character it may be unaffected by binding to DNA and thereby increase the energy gap between the MLCT and the LF states. However, there is no evidence for a greater energy gap between the MLCT and LF states for the bound complex but this can account for the observed increases in lifetimes in presence of DNA.

Luminescence lifetimes are substantially increased upon binding to DNA and the magnitude varies with the nature of the DNA, salt concentration, average molecular weight of the polymer, etc. Because of this dependence of lifetimes on various experimental factors, measured lifetimes of DNA-bound probes vary significantly, depending upon the exact conditions used for these experiments.[99] In general, the lifetime for the intercalated form was found to be substantially longer than the surface-bound complex. Due to this difference in lifetimes between bound and free complexes and due to multiple binding modes, a biexponential decay was observed for DNA-bound Ru(II) complexes.[91,97]

Binding of Ru(II) complexes to DNA invariably involves both surface as well as intercalative binding. So, one would expect at least two exponentials for the emission decay, one for the surface and another for the intercalated metal complex, if the interconversion between the two forms is slow within the excited-state lifetime. The observed emission decay curves can be fitted adequately to a sum of two exponentials for most of the complexes reported. One would expect a third exponential arising from the free complex. If surface-bound component exchanges rapidly with free probe within the excited-state lifetime, these two components cannot be time resolved. Such a rapid exchange between surface and free components is not unreasonable given the loose binding of the surface component and long lifetimes of these excited-sates. A similar exchange between the intercalated and free forms is expected to be slower due to the extra stabilization rendered for the intercalated complex. These factors may thus be responsible for the observed biexponential decays in the presence of DNA.

3.5.3. Enantiomeric Selectivity

With a right-handed double helix, a close intercalative fit is possible with the right-handed octahedral complex, whereas a less favorable fit was suggested for the corresponding left-handed complex.[100] The free energy difference between the binding of right- and left-handed complexes critically depends upon the relative dimensions of the metal complex as well as that of the DNA groove. For this reason, the observed enantiomeric selectivity is expected to increase with the size of the complex and decrease with the width of the DNA groove. Increases in the polarization upon binding to DNA were used to estimate enantiomeric selectivity. The enantiomeric selectivity was higher with larger complexes.

Higher enantiomeric selectivity can be achieved for a given octahedral complex by reducing the groove width. Groove widths of the helix can be tuned by changing the salt concentration or by changing the GC content of the polymer. Polyions such as DNA expand and widen the grooves when the ionic strength of the medium is lowered. This is because the reduction in the concentration of the counterions at the DNA helix results in increased repulsion between adjacent phosphate groups of the backbone. In order to accommodate this increased charge repulsion, the helix is stretched and the grooves widen. Conversely, increasing the salt concentration is expected to reduce the groove width. Reduced groove width was found to improve the enantiomeric selectivity for the binding of $Ru(Phen)_3^{2+}$ to the DNA helix.[97]

3.5.4. Sequence Selective Binding

A subtle way of changing the groove width and thereby influencing the enantioselectivity is to change the GC content of DNA. Because three hydrogen bonds exist between GC base pairs, as compared to two between AT base pairs, the major groove of GC sequences is narrower than that of AT sequences. The rigid, narrow major groove of GC sequences is expected to show higher enantioselectivity than the AT sequences. Luminescence polarization measurements with $Ru(Phen)_3^{2+}$ and DNAs with varying GC content demonstrated that the selectivity for intercalation increased with the GC content of the DNA polymer.

Poly(dG–dC) showed the highest enantioselectivity, whereas poly(dA–dT) showed little or no selectivity for intercalation.[97]

4.   Closing Remarks

Various photobiological processes occur in organized media. Protein matrices, nucleotide bases, and polysaccharide helices may be used to accelerate or inhibit selective photochemical reactions in a profitable way. No specific examples of artificial systems that take advantage of the organized protein matrices or nucleotide bases are known to date. Specific pendant groups in proteins have been implicated in the efficient photoinduced charge separation in the photosynthetic reaction center and energy transfer between the pigments of the photosynthetic antenna. The primary photochemical reaction that has been the basis of life on this planet serves as an excellent example of how photochemical and photophysical processes can be controlled by the organized media. The regioselectivity observed in the reaction center for the electron transfer is still unparalleled, to the best of our knowledge, in supramolecular chemistry. Energy transfer from the organized nucleotide bases of DNA to intercalated ethidium has been shown to be plausible and energy transfer for up to 17 Å along the helix is feasible. The helical structure of nucleic acids and the presence of hydrogen bonding and hydrophobic contact sites that serve to store the genetic information in the DNA molecule, located in the major and minor groove surfaces of the helix, provide very interesting environments for controlling, catalyzing, or inhibiting photochemical and photophysical processes. The developments in these areas of research can influence the design of supramolecular systems for information storage and retrieval, energy harvesting, and storage, and they are important in biomedical research as well as in biotechnology.

Acknowledgments

Thanks are due to Dr. H. A. Frank for valuable suggestions and to Merck Sharp and Dohme, American Cancer Society and the University of Connecticut Research Foundation for financial support.

References

1.   A. Trebst, *Annu. Rev. Plant Physiol.*, **1974**, *25*, 423.
2.   R. R. Rando, *Angew. Chem. Int. Ed.*, **1990**, *29*, 461, and also Chapter 18 in this monograph.
3.   J. K. Barton, *Chem. Engg. News*, **1988**, *Sep. 26*, 30.
4.   P. Mathis and A. W. Rutherford in *Photosynthesis, New Comprehensive Biochemistry*, Vol. 15, J. Amsez, ed., Elsevier, Amsterdam, 1987, p. 63.
      E. Rabinowitch and Govindjee, *Photosynthesis*, Wiley, New York, 1969.
5.   H. Michel and J. Deisenhofer in *Perspectives in Biochemsitry*, Vol. 1, H. Neurath, ed., American Chemical Society, Washington D. C. 1989, P. 30.
6.   K. Sauer, *Acc. Chem. Res.*, **1978**, *11*, 257.

7. K. Sauer in *Bioenergitics of Photosynthesis*, Govindjee, ed., Academic Press, New York, NY., 1975, p. 115.
8. A. L. Lehninger, *Biochemistry*, Worth Publishers Inc., New York, NY., 1976.
9. S. Boxer, *Ann. Rev. Biophys. Chem.*, **1990**, 19, 267.
10. J. R. Norris and M. Schiffer, *Chem. Engg. News*, **1990**, July 30, 22.
11. D. W. Reed and R. K. Clayton, *Biochem. Biophys. Res. Commun.*, **1968**, *30*, 471.
12. B. Ke and T. H. Chaney, *Biochim. Biophys. Acta*, **1971**, *226*, 341.
    R. K. Clayton, *Biochim. Biophys. Acta*, **1963**, *75* , 312.
13. A. Gloe, N. Pfennig, H. Brockmann Jr. and W. Trowitzsch, *Arch. Microbiol.*, **1975**, *102*, 103.
14. K. A. Weyer, F. Lottspeich, H. Gruenberg, F. Lang, D. Osterhelt and H. Michel, *EMBO J.*, **1987**, 6, 2197.
    J. Deisenhofer, O. Epp, K. Miki, R. Huber and H. Michel, *Nature*, **1985**, *318*, 618.
    J. P. Allen, G. Feher, T. O. Yeates, H. Komiya, and D. C. Rees, *Proc. Natl. Acad. Sci., U.S.A.*, **1987**, *84*, 6162.
15. H. Michel, *J. Mol. Biol.*, **1982**, *158*, 567.
    M. Y. Okamura, and G. Feher and N. Nelson in *Photosynthesis*, Vol. 1, Govindjee, ed., Academic Press, New York, 1982, p. 65.
16. J. Deisenhofer, O. Epp, K. Miki, R. Huber and H. Michel, *J. Mol. Biol.*, **1984**, *180*, 385.
    J. Deisenhofer, O. Epp, K. Miki, R. Huber and H. Michel, *Nature,* **1985**, *318*, 618.
17. D. J. Debus, G. Feher and M. Y. Okamura, *Biochemistry*, **1986**, *25*, 2276.
18. R. A. Niederman, and K. D. Gibson in *The Photosynthetic Bacteria*, R. K. Clayton and W. R. Sistrom, eds., Plenum Press, New York, 1978, p. 79.
    S. Kaplan and C. J. Arntzen, in *Photosynthesis*, Vol. 1, Govindjee, ed., Academic Press, New York, 1982, p. 65.
19. G. E. Gogel, P. S. Parkes, P. A. Loach, R. A. Brunisholz and G. Zuber, *Biochim. Biophys. Acta*, **1983**, 746, 32.
    R. A. Brunisholz, F. Suter and H. Zuber, *Hoppe-Seyler's Z. Physiol. Chem.*, **1984**, 365, 675.
20. P. S. Song, P. Koka, B. Prezelin and F. T. Haxo, *Biochemistry*, **1976**, *15*, 4422
    R. J. Cogdell, M. F. Hipkins, W. MacDonald and T. G. Truscott, *Biochim. Biophys. Acta*, **1981**, *634*, 191.
21. T. Forster, *Ann. Phys.* **1948**, *2*, 55.
22. R. T. Wang and R. K. Clayton, *Photochem. Photobiol.*, **1971**, *13*, 215.
    H. Kingma, R. van Grondell and L. N. M. Duysens, *Biochim. Biophys. Acta*, **1985**, *808*, 383.
23. G. E. Fox, *Science*, **1980**, *109*, 457.
    L. A. Staehelin, J. R. Golecki, R. C. Fuller and G. Drew, *Arch. Microbiol.*, **1978**, *119*, 269.
24. Z. G. Fetisova, S. G. Kharchenko and I. A. Abdourakhmanov, *FEBS Lett.*, **1986**, *199*, 234
    Z. G. Fetisova, S. G. Kharchenko and I. A. Abdourakhmanov in *Progress in Photosynthesis Research*, Vol. 1, J. Biggins, ed., Nijhoff, Dordrecht, 1987, p. 415.
25. Z. G. Fetisova, A. M. Freiberg and K. E. Timpmann, *Nature*, **1988**, 334, 633.
26. R. J. Van Dorssen and J. Amesz, *Photosynth. Res.*, **1988**, *15*, 177
    D. C. Brune, G. H. King, A. Infosino, T. Steiner, M. L. W. Thewalt and R. E. Blankenship, *Biochem.*, **1987**, 26, 8652.
27. R. A. Freisner and Y. Won, *Photochem.Photobiol.*, **1989**, *50*, 831.

814   *Kumar*

28. W. W. Parson, *Ann. Rev. Biophys. Bioengg.*, **1982**, *11*, 57.
29. C. Kirmaier and D. Holten, *Photosy. Res.*, **1987**, *13*, 225.
30. W. W. Parson, B. Ke in *Photosynthesis*, Govindjee, ed., Academic Press, New York, 1982, p. 331.
31. J. Breton, J. L. Martin, G. R. Fleming and J. C. Lambry, *Biochemistry*, **1988**, *27*, 8276.
32. J. Breton, J. L. Martin, G. R. Fleming and J. C. Lambry, *Biochemistry*, **1988**, *27*, 8276.
33. J. P. Allen, G. Feher, T. O. Yeates, H. Komiya and D. C. Rees, *Proc. Natl. Acad. Sci., U. S. A.*, **1987**, *84*, 5730.
34. M. E. Michel-Beyerle, E. M. Plato, J. Deisenhofer, H. Michel, M. Bixon and J. Jortner, *Biochim. Biophys. Acta*, **1988**, *932*, 52
    J . Deisenhofer, H. Michel and R. Huber, *Trends Biochem. Sci. Pers. Ed.*, **1985**, *10*, 243.
35. H. Michel, O. Epp and J. Diesenhofer, *EMBO J.*, **1986**, *5*, 2445.
36. S. L. Ditson, R. C. Davis and R. M. Pearlstein, *Biochim. Biophys. Acta*, **1984**, *766*, 623
    P. Maroti, C. Kirmaier, C. Wraight, D. Holten and R. M. Pearlstein, *Biochim, Biophys. Acta* , **1985**, *810*, 132.
37. H. A. Frank, B. W. Chadwick, S. Taremi, S. Kolaczkowski and M. K. Bowman, *FEBS Lett.*, **1986**, *203*, 157.
    H. A. Frank and C. A. Violette, *Biochim. Biophys. Acta* , **1989**, *976*, 222.
38. J. P. Allen, G. Feher, T. O. Yeates , H. Komiya and D. C. Rees in *The Structure of the Photosynthetic Reaction Center, NATO ASI Series*, Vol. 149, J. Breton, and A. Vermeglio, eds., Plenum Press, New York and London, 1988, p. 5.
39. R. A. Marcus and N. Sutin, *Biochim. Biophys. Acta*, **1985**, *811*, 265
    D. D. Devault in *Quantum Mechanical Tunnelling in Biological Systems*, Cambridge University Press, London, 1984.
40. V. Nagarajan, W. W. Parson, D. Gaul, and C. Schenk, *Pro. Natl. Acad. Sci., U. S. A.*, in press.
41. G. R. Fleming, L. J. Martin and J. Breton, *Nature*, **1988**, *333*, 190.
42. R. A. Marcus in *Photosynthetic Bacterial Reaction Center: Structure and Dynamics, NATO ASI Series, Series A, Life Sciences*, Vol. 149, J. Breton, and A. Vermeglio, eds., 1988, p. 389.
43. R. J. Donohoe, R. B. Dyer, B. I. Swanson, C. A. Violette, H. A. Frank and D. F. Bocian, *J. Am. Chem. Soc.*, **1990**, *112*, 6716.
44. N. I. Krinsky in *Biochemistry of Chloroplasts*, Vol. 1, T. W. Goodwin, ed., 1966, Academic Press, New York, p. 423.
45. R. J. Cogdell and H. A. Frank, *Biochim. Biophys. Acta*, **1987**, *895*, 63.
46. M. Griffith, W. R. Sistrom, G. Cohen-Bazire and R. Y. Stainer, *Nature*, **1955**, *176*, 1211.
47. P. S. Song, P. Koka, B. Prezelin and F. T. Haxo, *Biochemistry*, **1976**, *15*, 4422.
48. R. J. Cogdell, M. F. Hipkins, W. MacDonald and T. G. Truscott, *Biochim. Biophys. Acta*, **1981**, *634*, 191.
    J. C. Goedheer, *Biochim. Biophys. Acta*, **1959**, *35*, 1.
49. J. K. Trautman, A. Shreve, C. A. Violette, H. A. Frank, T. G. Owens and A. C. Albrecht, *Proc. Natl. Acad. Sci., U. S. A.*, in press.
50. C. S. Foote and R. W. Denny, *J. Am. Chem. Soc.*, **1968**, *90*, 6233.
    P. Mathis, Etudes de Formes Tranitoires des Carotenoids, Thesis, University of Paris-Orsay, 1970.
    H. Claes, *Biochem. Biophys. Res. Commun.*, *1960*, *3*, 585.
    C. V. Kumar, S. K. Chattopadhyay and P. K. Das, *J. Am. Chem. Soc.*, **1983**, *105*, 5143.

51. G. Dirks, A. L. Moore, T. A. Moore and D. Gust, *Photochem. Photobiol.*, **1980**, *32*, 277
A. L. Moore, A. Joy, R. Tom, D. Gust, T. A. Moore, R. V. Bensasson and E. J. Land, *Science*, **1982**, *216*, 982.

52. A. L. Moore, G. Dirks, D. Gust and T. A. Moore, *Photochem. Photobiol.*, **1980**, *32*, 691.

53. R. J. Cogdell, T. G. Monger and W. W. Parson, *Biochim. Biophys. Acta*, **1975**, *408*, 189.

54. R. J. Cogdell, W. W. Parson and M. A. Kerr, *Biochim. Biophys. Acta*, **1976**, *460*, 83.

55. B. W. Chadwick, and H. A. Frank, *Biochim. Biophys. Acta*, **1986**, *851*, 257.

56. W. J. McGann and H. A. Frank, *Biochim. Biophys. Acta*, **1985**, *807*, 101.

57. W. J. McGann and H. A. Frank, *Chem. Phys. Lett.*, **1985**, *121*, 253.

58. E. Davidson and R. J. Cogdell, *Biochim. Z Biophys. Acta*, **1981**, *635*, 295.

59. K. Iwata, H. Hayashi and M. Tasumi, *Biochim. Biophys. Acta*, **1985**, *810*, 269.

60. H. Hayashi, K. Iwata, T. Noguchi and M. Tasumi in *Progress in Photosynthesis Research* , Vol. 1, Biggins, J., ed., Nijhoff, Dordrecht, 1987, p. 33.

61. R. J. Cogdell, M. F. Hipkins, W. MacDonald and T. G. Truscott, *Biochim. Biophys. Acta*, **1981**, *634*, 191.

62. M. R. Wasielewski , D. M. Tiede and H. A. Frank in *Ultrafast Phenomena* Vol. 5, G. R. Fleming and A. R. Siegman, eds., Springer-Verlag, Berlin, 1986, p. 388.
J. K. Trautman, A. P. Shreve, H. A. Frank, T. G. Owens and A. C. Albrecht, *Proc. Natl. Acad. Sci. U. S. A* ., **1990**, *87*, 215.

63. A. Angerhofer, R. J. Cogdell and M. F. Hipkins, *Biochim. Biophys. Acta* , **1986**, *848*, 833
J. C. Goedheer, *Biochim. Biophys. Acta*, **1959**, *35*, 1.

64. R. Van Grondelle, H. J. J. Kramer and C. P. Rijgersberg, *Biochim. Biophys. Acta*, **1982**, *682*, 208.

65. W. W. Parson and T. G. Monger, *Brookhaven Symp. Biol.*, **1976**, *28*, 195.

66. H. Morrison, ed., *Bioorganic Photochemistry*, Vol. 1, Wiley, New York, 1990.

67. D. L. Ollis and S. W. White, *Chem. Rev.*, **1987**, *87*, 981.

68. M. L. Good, *Biotechnology and Materials Science: Chemistry for the Future*, American Chemical Society, Washington D. C., 1988.

69. J. D. Watson and F. H. C. Crick, *Nature* , **1953**, *171*, 737.

70. H. M. Berman in *Topics in Nucleic Acid Structure*, S. Neidle, ed., Wiley, New York, 1981, p. 1.

71. A. Rich, A. Nordheim A. H. J. Wang, *Ann. Rev. Biochem.*, **1984**, *53*, 791.

72. *Dickerson's crystal structure*: R. E. Dickerson and H. R. Drew, *J. Mol. Biol.*, **1981**, *149* , 761
R. E. Dickerson and H. R. Drew, *Proc. Natl. Acad. Sci., U. S. A.*, **1981**, *78*, 7318.

73. M. J. Waring, K. R. Fox, G. W. Grigg, *Biochem. J.*, **1987**, *143*, 847.
K. Kissinger, K. Krowicki, J. C. Dabrowiak and J. W. Lown, *Biochemistry*, **1987**, *26*, 5590.
H. M. Berman and P. R. Young, *Ann. Rev. Biophys. Bioengg.*, **1981**, *10*, 87.
S. K. Sim and J. W. Lown, *Biochem. Biophys. Res. Commun.*, **1978**, *81*, 99.

74. J. K. Barton, *Chem. Engg. News*, **1989**, *June 12*, 22.

75. J. K. Barton, *Science*, **1986**, *233*, 727.
J. K. Barton, *Inorg. Chem.*, **1985**, *3*, 321.

76. Proximity of the phosphate groups in the minor groove is expected to produce a stronger electrostatic field in the minor groove than in the major groove.

77. G. S. Manning, *J. Chem. Phys.*, **1969**, *51*, 924 and 3249.

78. G. S. Manning, *Acc. Chem. Res.*, **1979**, *12*, 443.

79. D. H. Ohlendorf and B. W. Matthews, *Ann. Rev. Biophys. Bioengg.*, **1983**, *12*, 259.

80. M. L. Kopka, A. V. Fratini, H. R. Drew and R. E. Dickerson, *J. Mol. Biol.*, **1983**, *163*, 129.

81. M. L. Kopka, P. E. Pjura, D. S. Goodsell, and R. E. Dickerson in *Nucleic Acids and Molecualr Biology*, Vol. 1, D. Lilley and F. Eckstein, eds., Springer-Verlag, Berlin, 1986.

82. C. Canter and P. R. Schimmel, *Biophysical Chemistry*,Vol. 3, W. H. Freeman and Co., San Fracisco, 1980.

83. C. V. Kumar, A. L. Raphael and J. K. Barton, *Biomol. Stereodynamics*, **1986**, *3*, 85.

84. H. M. Berman and P. R. Young, *Ann. Rev. Biophys. Bioengg.*, **1981**, *10*, 87.

85. T. Co and A. H. Maki, *Biochem.*, **1978**, *17*, 186.

D. H. H. Tsao, J. R. Casas-Finet, A. H. Maki and J. W. Chase, *Biophys. J.*, **1989**, *55*, 927.

86. J. C. Wang, *J. Mol. Biol.*, **1974**, *89*, 783.

T. Hsieh, and J. C. Wang, *Biochemistry*, **1975**, *14*, 527.

W. Keller, *Proc. Natl. Acad. Sci., U. S.A.*, **1975**, *72*, 4876.

87. B. Norden, F. Tjerneld and E. Palm, *Biophys. Chem.*, **1978**, *8*, 1.

88. E. F. Gale, E. Cundliffe, P. E. Reynolds, M. H. Richmond and M. J. Waring in *The Molecular Basis of Antibiotic Action*, Wiley, London, 1972.

S. Neidle and M. J. Waring, *Molecular Aspects of Anti-Cancer Drug Action*, Macmillan Press, London, 1983.

89. O. G. Berg and P. H. von Hippel, *Ann. Rev. Biophys. Biophyschem.*, **1985**, *14*, 131.

O. G. Berg, R. B. Winter and P. H. von Hippel, *Biochem.*, **1981**, *20*, 6929.

90. T. G. Wensel, C. F. Meares, V. Vlachy and J. B. Matthews, *Proc. Natl. Acad. Sci. U. S. A.*, **1986**, *83*, 3267-3271.

91. C. V. Kumar, J. K. Barton and N. J. Turro, *J. Am. Chem. Soc.*, **1985**, *107*, 5518.

92. J. B. LePecq and C. Paoletti, *J. Mol. Biol.*, **1967**, *27*, 87.

93. G. Weil and M. Calvin, *Biopolymers*, **1963**, *1*, 401.

94. Y. Kubota, Y. Motoda, Y. Kuromi and Y. Fujisaki, *Biophys. Chem.*, **1984**, *19*, 25.

95. D. P. Millar, R. J. Robbins and A. H. Zewail, *J. Chem. Phys.*, **1981**, *74*, 4200.

D. P. Millar, R. J. Robbins and A. H. Zewail, *J. Chem. Phys.*, **1982**, *76*, 2080.

96. J. K. Barton, *Chem. Engg. News*, **1989**, *July 23*, 15.

97. J. K. Barton, J. M. Goldberg, C. V. Kumar and N. J. Turro, *J. Am. Chem. Soc.*, **1986**, *108*, 2081.

98. J. K. Barton, C. V. Kumar and N. J. Turro, *J. Am. Chem. Soc.*, **1986**, *108*, 6391.

M. D. Purugganan, C. V. Kumar, N. J. Turro and J. K. Barton, *Science*, **1988**, *241*, 1645.

99. H. Gorner, A. B. Tossi, C. Stradowski and D. Shulte-Frohlinde, *J. Photochem. Photobiol., B.*, **1988**, *2*, 67.

100. J. K. Barton and A. L. Raphael, *Proc. Natl. Acad. Sci., U. S. A.*, **1985**, *82*, 6460.

H. Y. Mei and J. K. Barton, *Proc. Natl. Acad. Sci., U. S. A.*, **1988**, *85*, 1339.

J. K. Barton and A. L. Raphael, *J. Amer. Chem. Soc.*, **1984**, *106*, 2466.

Chapter 18

Protein Directed Regioselective Photoisomerization of the Retinyl Chromophore in Visual Pigment Analogs.#

Robert S. H. Liu

Department of Chemistry, University of Hawaii,
Honolulu, HI, USA.

Yoshinori Shichida

Department of Biophysics, Kyoto University,
Faculty of Science, Kyoto, Japan.

Contents

Photochemistry of polyenes 28. For a previous paper in the series, see ref. 69b.

1. Introduction

Among retinoid binding proteins,[1] only visual pigments and bacteriorhodopsin and associated pigments undergo highly regiospecific geometric isomerization upon photo-excitation. Hence the 11-cis to all-trans isomerization in visual pigments initiates visual transduction and the all-trans to 13-cis isomerization in pigments in Halobacterium halobium leads to energy storage or sensory response. The regiospecificities of these reactions differ from those of the corresponding free chromophores in solution. Hence they are clearcut examples of protein-directed selective photochemical reactions. In this chapter we wish to review the photochemistry of the visual pigment, rhodopsin, and its analogs where among all retinyl binding proteins more detailed photochemical investigations have been conducted. Rhodopsin is a general name for all visual pigments for scotopic vision that contain 11-*cis*-retinal as the chromophore. Since the visual pigment extracted from cattle retina has been studied in most detail, the rhodopsin referred to in this chapter is cattle rhodopsin unless noted otherwise.

2. Structure of Visual Pigments

An understanding of the ways that opsin, the apoprotein of rhodopsin, can affect the chemistry of the trapped retinyl chromophore requires knowledge of the protein structure in all three dimensions. Difficulties in obtaining crystals of the visual pigment have greatly hampered means of obtaining such information. The limited knowledge currently available largely derived from information accumulated on the related membrane protein bacteriorhodopsin, the structure of which is available at sufficiently high resolution to yield meaningful information on folding pattern of its protein backbone. These data are summarized below.

Bacteriorhodopsin (BR) is a polypeptide of 248 amino acids with an all-*trans*-retinyl chromophore attached to Lys-216 via a protonated Schiff base linkage.[2] Two-dimensional electron density maps[3] from electron diffraction studies of its naturally occurring trimers revealed packing of the protein into seven α-helical bundles which coincide with the seven distinct hydrophobic segments in the BR sequence. Assignment of the seven parallel helices to the seven density spots has been made.[4]

Rhodopsin is a lipoprotein containing 348 amino acids with the 11-*cis*-retinyl chromophore attached to the protein at Lys-296 also via a protonated Schiff base

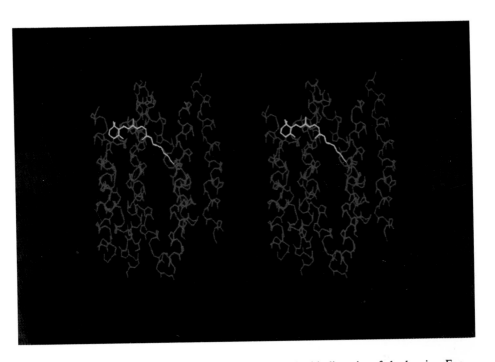

Plate 1. A stereo view of the Hargrave's model for the binding site of rhodopsin. For construction of this rhodopsin model, the seven helical axes of bacteriorhodopsin were used. The secondary structure of rhodopsin as proposed by Hargrave was used for substitution of the amino acid residues in each helix and for determining relative heights of helices. The 11-cis chromophore, anchored at two ends through a primary protonated Schiff base linkage with the butyl side chain of a lysine residue and interaction of the cyclohexenyl ring with a hydrophobic pocket, is pictured to be sandwiched between two layers of protein helices. The top is the extracellular region. [This figure was provided by Dr. T. Mirzadegan (present address: Parke Davis, 2800 Plymouth Road, Ann Arbor, MI 48105). Details related to construction of diagrams similar to this partially minimized binding site of rhodopsin are described in the Ph.D. Thesis of T. Mirzadegan, submitted to the University of Hawaii, 1989.]

linkage.[2] Seven hydrophobic segments are also known to be present. In one case (frog rhodopsin) the two-dimensional crystal structure is known but at too low a resolution to provide meaningful structural information at the molecular level.[5] The three-dimensional structures that have been postulated[6] were constructed largely by analogy with that of bacteriorhodopsin.

In Plate 1 is shown a stereo view of the Hargrave's three dimensional model of rhodopsin.[6a] In it, the 11-*cis*-retinyl chromophore is shown to lie horizontally between two layers of vertically oriented protein helices. It is primarily bonded to Lys-296 and secondarily "anchored" to the protein through interaction of the trimethylcyclohexenyl ring with a defined hydrophobic pocket within the binding site.[7]

3. Protein–Substrate Interactions

The selectivity of the binding site of opsin is reflected in the relative rates of pigment formation, favoring the native 11-cis chromophore and those isomers and analogs closest in shape, although the possible existence of many isomeric rhodopsins[8] and analogs[9] gives a good impression of its flexibility. Specific protein–substrate interactions are revealed by the strong cd (circular dichroism) activity of the retinyl chromophore when bonded to opsin in spite of the absence of a single chiral center in the chromophore.[10] The double-peaked cd band in rhodopsin is believed to be due to twisted (calculated to be 34° and 20°, respectively)[11] conformations at the 6,7- (the 487 nm α-band) and 12,13 bonds (the 335 nm β-band) as suggested by recent studies with ring-fused rhodopsin analogs.[12] Apparently interaction with the chiral cavity within the binding site forced the polyene chromophore into a chiral conformation. Chiroptical recognition of the opsin binding site toward chiral substrates is reflected in kinetic resolution of 3-substituted retinal analog (1) during its reaction with opsin[13] and different properties of pigments derived from 5,6-epoxy-3-dehydroretinal (2).[14]

11-cis-retinal

The specific hydrophobic interaction of rhodopsin with the trimethylcyclohexenyl ring has been estimated to require 4–5 kcal mol^{-1}.[15] Recognition of each of the methyl groups apparently contributes to this value because removal of any one of the methyl groups on the ring resulted in greatly reduced rates of pigment formation or yields.[16] Recognition of the 9-methyl group is reflected in another unique way. The photoproduct of rhodopsin, bathorhodopsin, produced at liquid nitrogen temperature, showed an unusual band in its vibrational spectra:[17] the hydrogen-out-of-plane (HOOP) band of H12. It derives from interaction of the protein with the 9-methyl group of the twisted, strained chromophore. Thus, for the 9-demethyl analog, the same band was absent.[18] Analog studies suggest a strong protein substrate interaction near the ring portion of the molecule because modification of the chromophore near this center drastically changes the nature of the photochemical reaction or the ensuing bleaching processes (see below) while removal of the 13-methyl group[19] has relatively little effect on these processes.

The additional 40–80 nm red shift of the absorption maxima of various visual pigments from that of the protonated Schiff base has been attributed to proximity of the counterion,[20] polarizability of media,[21] and the presence of a second point charge near the C11–C13 portion of the chromophore.[22] All these specific protein–substrate interactions are expected to have an influence on chemical processes of the chromophore, especially for photochemical reactions involving species of short lifetimes (low activation energies).

4. Photochemistry of Visual Pigments
4.1. The Primary Process of Rhodopsin

Low-temperature spectroscopy was first employed in early studies of primary photoproducts of rhodopsin. These efforts first led to the assignment of lumirhodopsin (study carried out at $-80°C$)[23] as the primary photoproduct and subsequently to bathorhodopsin at liquid nitrogen temperature.[24] Irradiation of rhodopsin at liquid helium temperature produced another photoproduct, named hypsorhodopsin,[25] but it is now believed not to be a physiological intermediate (see below). The detection of many later intermediates was also reported.

Application of fast kinetics to studies of photobleaching of rhodopsin [with the development of nanosecond (ns)[26] and picosecond (ps)[27] techniques following in rapid succession] showed that the intermediates observed at low temperature were also present at room temperature. These observations not only eliminated the worry that conclusions from low-temperature studies might be clouded by altered medium effects but also suggested that the chromophore of rhodopsin must be in a rather fluid protein pocket that does not solidify as the macroscopic medium. However, other complicating factors have also been introduced along with these new methods. In particular, the high intensity of the lasers employed in many of the picosec studies led to new excitation pathways, e.g., two-photon excitation. Only recently has this problem been fully appreciated.[28] Unfortunately, such complications and the difficulty in determining whether an observed intermediate is from an excited rhodopsin or an earlier intermediate have led to some confusion in the literature. Below we attempt to present what we believe to be the correct picture.

Based on results acquired in picosec spectroscopic studies employing lasers of low intensity, we believe there is unequivocal evidence of the presence of one additional ground-state intermediate preceding bathorhodopsin.[29] It has a red-shifted absorption maximum (570 nm) from that of bathorhodopsin (535 nm). This new intermediate, labeled photorhodopsin,[29] is believed to be conformationally, both on the part of the chromophore and the surrounding protein, a less relaxed intermediate. Another more blue-shifted intermediate (440 nm), the hypsorhodopin,[25a] reported earlier is now believed to be a secondary photoproduct from photorhodopsin.[25b,#]

It should be mentioned that preceding to the photorhodopsin study, the presence of an early intermediate had been suggested in the literature. In their low-temperature picosec studies, Peters et al.[30] observed that the formation of bathorhodopsin was accompanied by the decay of another intermediate. They suggested the latter to be the excited rhodopsin. Subsequently, these data were reinterpreted by others by assigning the early intermediate to that of a ground-state product known as batho or prebatho.[31] Kobayashi et al.[32] independently reported detection of an early intermediate, first assigned to excited rhodopsin but later reassigned to a ground-state product, "prime rhodopsin." However, all these studies were carried out under high laser intensities where two-photon excitation was likely.

A scheme incorporating photoexcitation of rhodopsin and formation and decay of subsequent intermediates and their relation to the biological properties is shown in Scheme 1.[33] We might add that the assignment of photorhodopsin as the primary photoproduct of rhodopsin is a consequence of the methods currently available for investigation. Given the complex structure of rhodopsin and the added strain as a result of photoexcitation, it should be of no surprise if in the future, a newer and more sensitive method leads to the detection of another(other) early intermediate(s).

Most of the photochemical and spectroscopic studies in the literature have been directed to the more stable (e.g., stable at liquid nitrogen temperature) and better characterized bathorhodopsin. It is the common photoproduct from rhodopsin and 9-*cis*-rhodopsin.[28,34] It displays a positive and a negative peaks in its cd spectrum instead of the two positive peaks in rhodopsin.[35] Upon irradiation, it is converted to either rhodopsin or 9-*cis*-rhodopsin in an approximate ratio of 6 to 1. For 9-*cis*-rhodopsin, the normally planar 9-*cis*-polyene sidechain[36] also exists in a twisted chiral conformation presumably forced by the chiral protein pocket as reflected in the similar cd spectra between rhodopsin and 9-*cis*-rhodopsin.[10b] An additional strain energy of 5 kcal mol^{-1} was estimated to have been introduced into 9-*cis*-rhodopsin.[37] The chirality in the polyene chromophore is retained during photoisomerization of either one of the isomeric rhodopsin analogs to the common product bathorhodopsin, as reflected in the enhanced cd spectrum.[25,35] In this regard it is interesting to note that two different pathways of decay of bathorhodopsin have been detected.[38]

This view of a singular role of hypsorhodopsin is not uniformly shared by all researchers. Kobayashi et al. suggested a dual role for hypsorhodopsin, being additionally an intermediate between "prime rhodopsin" and bathorhodopsin (ref. 32b).

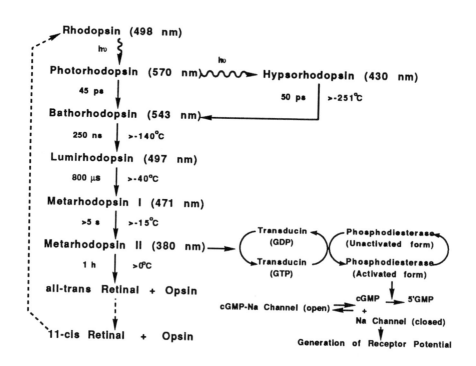

Scheme 1. Photochemical and ensuing dark processes of cattle rhodopsin. Absorption maxima of intermediates are shown in parentheses. Lifetime and transition temperature determined for each intermediate, from room-temperature fast kinetic studies or low-temperature spectroscopic measurements, are shown besides the arrows connecting the intermediates. Rhodopsin contains the 11-cis chromophore, which is converted to the all-trans geometry upon photoexcitation. The ensuing dark intermediates correspond to different stages of progressive relaxation of the initially generated, strained all-trans chromophore and the surrounding protein. These data are essentially those in the literature.[23a] The fine print in the lower right corner describes the currently accepted view of signal transduction from metarhodopsin II to generation of receptor potential. The dashed lines describe the dark isomerization process. (See, e.g., R. R. Rando, in *Chemistry and Biology of Synthetic Retinoids*, M. Dawson and W. Okamura, eds., CRC Press, Boca Raton, FL, 1990, pp. 1–26.)

Recent accumulated evidence seem to reaffirm the earlier view[24b] that the primary photochemical process of visual pigments is the geometric isomerization reaction. Specific models have been proposed to rationalize the ready occurrence of such a volume demanding process for a chromophore confined within the cavity defined by the protein helices: the stepwise or concerted bicycle pedal model of rotation of alternate bonds[39] and the CT-n (concerted twist at center *n* or Hula twist) model of concerted rotation of adjacent bonds.[40] The combined results from assignment of the 10-S-trans conformation (based on vibrational spectra) of bathorhodopsin[41] and photochemical studies of a model 9,11-ring-fused system (similar bleaching characteristics as rhodopsin)[42] appear to be incompatible with the CT-n model but in general agreement with the concerted bicycle pedal model.[39b]

4.2. Rhodopsin and Isomeric Rhodopsin Analogs

Isomerization of rhodopsin and its photoproducts follows the one-photon–one-bond isomerization process, a characteristic of excited singlet state reaction.[43] (A recent reinvestigation of 9,13-di-*cis*-rhodopsin[44] has removed the only reported case of multiple bond isomerization of a visual pigment analog.)[45] The quantum efficiency (0.67)[46] of reaction of rhodopsin is higher than that of the 9-*cis*-isomer (0.33 – 0.39 relative to that of rhodopsin).[47] At low temperatures, the values for 9-*cis*-rhodopsin (but not for rhodopsin) exhibit wavelength dependence, believed to be caused by an activated photochemical process (est. ~.2 kcal mol^{-1}).[48]

Rhodopsin $^+$NH-opsin   9-cis $^+$NH-opsin   7-cis $=$$^+$NH-opsin

The isomerization reaction from torsional relaxation of the polyene in the excited singlet state to the relaxed perpendicular structure is rapid (within a few ps)[49] rendering rhodopsin fluorescence extremely weak. Following the determination of the strain energy in bathorhodopsin[50] and the estimation of the energy of 9-*cis*-rhodopsin (see below), the relative heights of all species involved in the photochemical processes can be assigned (Figure 1).

The large strain energy (32–35 kcal mol^{-1})[50] stored in the photoproduct bathorhodopsin is believed to represent accumulated strains of conformational distortion of the polyene chain and the appended butyl group of Lys-296, charge separation, and distortion or compression strain of neighboring protein residues. Relief of these strains takes place in the form of several steps of discrete kinetic barriers accounting for the presence of the limited number of bleaching intermediates following the light excitation process. However, structural changes of the chromophore and the protein during transition of bleaching intermediates must be complex. It should be of no surprise to detect an occasional extra intermediate in the bleaching sequence of a pigment analog, such as those reported for 13-demethylrhodopsin[19] and 5,6-dihydrorhodopsin.[51]

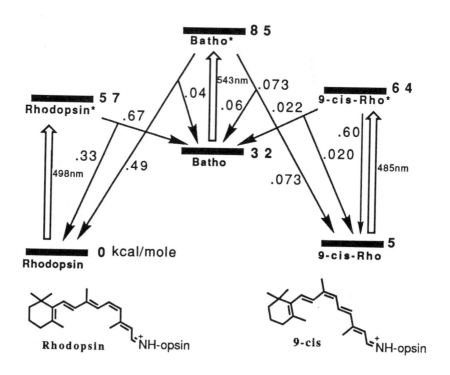

Figure 1. The state energy diagram of rhodopsin, 9-*cis*-rhodopsin, bathorhodopsin, and the corresponding excited states. Also shown are quantum yields of photoisomerization between bathorhodopsin and rhodopsin and 9-*cis*-rhodopsin. (The data are essentially those of Birge and co-workers, ref.48).

From 7-*cis*-rhodopsin,[52] the only other stable mono-*cis*-rhodopsin isomer, a batho intermediate exhibiting an absorption spectrum slightly different from that of rhodopsin, was obtained.[53] This is not surprising considering the fact that the 9-methyl group, known to interact strongly with the protein (see above), is oriented in the direction opposite to the protein for 7-*cis*-rhodopsin and other isomeric pigments. The interconversion of rhodopsin and 9-*cis*-rhodopsin does not require any significant relocation of the 9-methyl group. On the other hand, the conversion of 7-*cis*-rhodopsin to rhodopsin requires reorientation of the 9-methyl group, possibly causing local protein perturbation manifested in spectral properties of the primary photoproduct different from those of bathorhodopsin.

Among the di-cis isomeric rhodopsins, the 9,13-di-cis isomer has been shown to photo-isomerize selectively to the 13-cis isomer.[44] Hence, rhodopsin, 9-*cis*-rhodopsin, 9,13-di-cis rhodopsin and bathorhodopsin all share the common characteristics of preferred isomerization near the central portion of the polyene chromophore. The combined results probably reflect the shape of the protein cavity, more tightly surrounding the chromophore near the terminal portion of the chromophore and less so in the middle.[40]   The only exception appears to be 7-*cis*-rhodopsin. The all-trans product formed can only be due to isomerization near the terminal 7,8 bond.[54] For this more sterically crowded isomer with a highly twisted ring-chain conformation,[55] a different protein substrate interaction might be involved. It will be of interest to determine its cd spectrum. The regioselectivity of photoisomerization of the remaining known di-cis isomeric pigment analogs (9,11-di-cis and the less stable 7,13-di-cis)[56] has yet been determined.

The question of possible isomerization at the imine linkage has not been fully addressed, although bathorhodopsin appears to have retained the anti (15*E*) geometry as in rhodopsin and 9-*cis*-rhodopsin.[57]   This feature is retained even in a di-cis pigment such as 7,9-di-*cis* rhodopsin.[58]

4.3. Photochemistry of the Free Retinyl Chromophore

An appropriate model for studies of rhodopsin isomerization is the protonated retinyl Schiff base. The photochemistry of the 11-cis isomer has recently been reinvestigated by Becker and Freedman.[59]   The only product detected was the all-trans, i.e., its regioselectivity being identical to that of rhodopsin. However its quantum yield ($\sim$0.2)[59] of isomerization, found to be independent of excitation wavelength and solvent polarity, is considerably smaller than that of rhodopsin (0.67).[46]   The increased value for the latter has been attributed to reversal of the two low-lying π–$\pi*$ states with the more ionic $^1Bu^+$ state being lower in rhodopsin.[60]   Alternatively, it has been suggested that increased mixing of the upper $^1Bu^+$ state with the low-lying $^1Ag^-$ state increased the reactivity of rhodopsin.[61]   The shape of the binding cavity must also have played an important role. Photoisomerization of 9-cis protonated retinyl Schiff base also gave the all-trans isomer with a quantum efficiency ($<$ 0.05)[59] smaller than that of 9-*cis*-rhodopsin (0.2–0.3).[46,47]

There are also differences in the photochemical properties of the protonated all-trans retinyl Schiff bases[61,62] and that of bathorhodopsin (also with the all-trans geometry). The free chromophore is less selective, giving the 13-cis, 11-cis, and 9-cis isomers (1.7:6.4:1.0 in hexane and 1.8:14.3:3.0 in methanol) plus a trace amount of the 7-cis-isomer.[61]   The confined all-trans chromophore in bathorhodopsin gives only the 11-cis and 9-cis isomers in a ratio of 6:1. The quantum yield of isomerization of the all-trans chromophore in solution is 0.13 $\pm$ 0.02,[61] while that of bathorhodopsin is 0.56.[48]

Photoisomerization of other retinyl derivatives has also been examined: retinal,[63] its Schiff base,[61,62] retinonitrile,[64] and retinoates.[65] These derivatives, all containing an electron-withdrawing end group, exhibit solvent-dependent photochemical behavior: higher selectivity favoring formation of the less crowded 13-cis and 9-cis isomers from the all-trans isomer when irradiated in a nonpolar solvent and a lower selectivity leading to all four mono-cis isomers when irradiated in a polar solvent (or in silica gel[66] or with hydrogen bonding additives[67]). The quantum yield of isomerization also varied increasingly by as much as 20-fold from hexane to methanol in the case of the Schiff base.[61,62]   The results have been interpreted in terms of increased mixing of the $^1Ag^-$ and the $^1Bu^+$ states upon increasing the solvent polarity or varying degree of protonation.[61]

The retinal case is further complicated by possible involvement of a low-lying n,π^* state which in turn leads to more efficient intersystem crossing efficiency in a nonpolar solvent.[68] The quantum yield of isomerization has been shown to be concentration dependent, attributable to triplet quantum chain processes.[69] The effect is particularly large for hindered isomers. Under the condition, one-photon two-bond isomerized products were detected.[69,70] The isomerization is also less regioselective, e.g., for 11-*cis*-retinal in hexane, all-trans and 11,13-di-cis isomers were found to be the major primary products in a ratio of 2:1.[69b] A recent study reports effect of the water-soluble β-lactoglobulin protein on regioselective isomerization of the encapsulated retinal.[71]

It is interesting to note that while the protein exerts a definite effect on the conformational properties of the retinyl chromophore, the chiral cavity apparently does not greatly inhibit the ease of torsional relaxation of the excited singlet molecules as reflected in the short excited-state lifetime of the protonated 11-*cis*-retinyl chromophore in rhodopsin ($\tau_{1/2}$ = 3–6 ps)[72] as well as in solution ($\tau_{1/2}$ = < 8 ps).[73]

4.4. The Later Intermediates

Transitions from batho to lumi and subsequent meta-I intermediates are accompanied by relaxation of the polyene chromophore and the neighboring protein residues. These are reflected in the diminishing intensity of the cd bands in these intermediates[74] and changes in the resonance Raman[57] and FT-ir spectra of chromophores[75] and more remotely located amino acid residues as the pigment cascades through the bleaching intermediates.[76]

The more relaxed protein structure (probably an enlarged binding cavity)[33a] in the later intermediates is reflected in the loss of regioselectivity of photoisomerization of the retinyl chromophore. Hence, upon irradiation of the all-trans chromophore of lumirhodopsin at -78°C and -190°C,[77,78] in addition to 9-cis and 11-cis, both 13-cis and 7-cis isomers became detectable. In fact, the blue shifted absorption characteristic of the 7-cis pigment makes this isomer the major product upon extended irradiation of the lumi intermediate with light >530 nm.[74] Also, it is interesting to note that while the quantum yield of isomerization of the hindered 7-cis retinal is considerably higher than those of other retinal isomers,[69] the 7-cis rhodopsin was found to be much less photosensitive than rhodopsin or 9-cis rhodopsin.[54] These observations again suggest the importance of medium effect on direction of isomerization rather than any intrinsic properties such as steric crowding within the chromophore.

4.5. Other Naturally Occurring Visual Pigments

With the single exception of retinochrome, a cephalopod retinoid pigment that undergoes photoisomerization from the all-trans chromophore to 11-cis,[79] all visual pigments share the same primary photochemical reaction of isomerization from the 11-cis isomer to the all-trans.[33a] They include visual pigments from vertebrates and invertebrates, rod (rhodopsin) and cone (iodopsin)[80] pigments, and pigments with modified retinals (3-dehydroretinal in porphyropsin[81] and 3-hydroxyretinal in insects).[82] However, there are noticeable differences in the bleaching sequences among all the pigments.

Instead of dissociating into all-trans retinal and opsin, pigments from invertebrates are known to form a stable *meta*-rhodopsin that can revert back to rhodopsin photochemically or thermally, exchanging the all-trans chromophore with an external 11-*cis*-retinal.[83] Iodopsin exhibits a different characteristic. At room temperature it gave identical final products of all-trans retinal and opsin (known as R-photopsin) upon photoexcitation.[80] However, at low temperatures its batho intermediate exhibits a preference to revert back to iodopsin, thus significantly reducing its photosensitivity.[84] Therefore, for these pigments instead of controlling direction of photoisomerization, the protein structure plays an important role in dictating the direction of the ensuing dark processes (see discussion in Section 5).

5. Photochemistry of Visual Pigment Analogs
5.1. 10,20 Ring Fused Analogs

Several retinal analogs with fused rings at these two centers (Rho-n, 3), thus containing a "fixed" 11-cis double bond, have been prepared.

All of them gave stable visual pigment analogs (λ_{max} = 495,[12] 520,[85] 490,[86] 515, and 407 nm[9a] for, respectively, n = 1 to 5, i.e, five- to nine-membered rings). The longest wavelength absorption maximum exhibited by the six-membered ring analog could be a reflection of a smaller angle of twist in the ring-fused chromophore as well as a better fit of the analog within the binding site.

Of these, the photobleaching processes of Rho-5 (3a) and Rho-7 (3c) analogs have been examined in detail. Not too surprisingly, the five-membered ring analog was found to be stable upon irradiation at the liquid nitrogen temperature.[12] The pigment fluoresces strongly with an average lifetime (85 ps) considerably longer than that of rhodopsin. Clearly, torsional relaxation around the 11,12 bond must be the primary relaxation process in rhodopsin. Closing this channel of reaction greatly increases its excited-state lifetime. The shorter lifetime of the corresponding free protonated Schiff base (50 ps) suggests that the protein matrix inhibits other relaxation processes [possibly twisting of bond(s) other than the 11,12 bond] that are permissible for the free chromophore.[87] However, irradiation of the pigment (λ_{max} = 495 nm) at $0°C$ gave a new pigment (λ_{max} = 466 nm), which was shown by chromophore extraction experiment to have retained the original 11-cis geometry.[12] These results appear to be consistent with possible formation of the 15-syn isomer. Thus, closing of the isomerization channel of relaxation in the ring-fused pigment analog must have led to a hitherto unknown, albeit less efficient, channel of isomerization about the imine bond.

The seven-membered ring analog (λ_{max} = 490 nm) is also nonbleachable. However, recent fast kinetic studies[87,88] revealed the formation of a short-lived bathochromic intermediate[#] (λ_{max} = 570[88] or 580 nm[87]; $\tau_{1/2}$ = 20–40 ps), presumably the corresponding 11-trans isomer. It readily reverts back to the original 490 nm pigment.[87]

3c, 490nm $\longrightarrow$ NH-opsin

hv, rm temp → 580nm pigment (11-trans) + 640nm pigment (two photon)

~30ps

This analog, therefore, resembles the cone pigment iodopsin (see above). A two-photon excitation product was also detected (λ_{max} = 630–640 nm), and fluorescence was detectable only with high-intensity incident beam; hence the product is probably associated with two-photon excitation processes.[82]

It is clear that the short time needed to form bathorhodopsin does not allow any drastic compensatory protein reorganization in order to accommodate the altered chromophore, creating a situation of strained protein–substrate interactions. Whether the batho intermediate can proceed forward to the lumi intermediate or, instead, is forced by the protein back to the original pigment should be determined by the relative ease of protein reorganization. In the case of iodopsin, it appears to be more difficult for the protein to reorganize (possibly associated with a highly negative entropy of activation so as to account for the low efficiency of the forward

---

[#] In this section, no attempt will be made to distinguish between batho and photo rhodopsin.

reaction at low temperatures).[##] The highly twisted trans geometry expected for the batho intermediate of the fused seven-membered ring system apparently readily reverts back to the original cis geometry. In fact the rate of isomerization of the strained system in the protein was found to be faster than that in a free molecule,[87] suggesting that the process was assisted by the surrounding unrelaxed protein moieties. These two systems are unique, different from rhodopsin and most other visual pigment analogs.

Analogous competing dark reactions of photoproducts of simple ring-fused organic system are known. Irradiation of phenylcyclohexene gives its trans isomer, which upon warming reverts back to the cis isomer ($\tau_{1/2}$ = 4.2 m at room temperature),[89] a situation parallel to Rho-7. That similar reversal to the cis isomer was observed for iodopsin at low temperatures must mean that its protein structure is more rigid, serving the similar inhibitive role as a constrained ring. Irradiation of 1,3-cycloheptadiene at -78°C also gives a strained cis,trans isomer with a much shorter lifetime ($\tau_{1/2}$ = 7.1 m at -78°C).[90] Its ensuing thermal process, however, is not the reverse geometric isomerization process.

$$\text{(structure)} \overset{h\nu}{\underset{rm\ temp}{\rightleftarrows}} \text{(structure)}$$

$$\text{(structure)} \overset{h\nu}{\longrightarrow} \text{(structure)} \overset{-78°C}{\longrightarrow} \text{(structure)}$$

Instead, it electrocyclizes to bicyclo[3.2.0]hept-4-ene. Thus, it resembles the forward process of batho to lumi rhodopsin where continued relaxation of the all-trans chromophore is permissible by an apparently less rigid protein structure. Such similarities are shown in the reaction coordinate diagrams in Figure 2. The ready formation of the strained trans isomer of these organic systems gives credence to the assignment of the trans structure to the batho intermediate of Rho-7.

[##] A recent study of photobleaching of chloride depleted iodopsin suggests possible involvement of a specific chloride binding site on relative ease of batho- and lumi-iodopsin interconversion as well as wavelength regulation of pigment absorption (Y. Imamoto, H. Kandori, T. Okano, Y. Fukada, Y. Shichida and T. Yoshizawa, *Biochemistry*, **1989**, *28*, 9412).

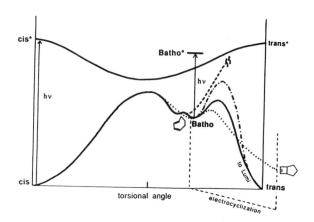

Figure 2. Postulated torsional potential curves for rhodopsin (solid lines). The protein-imposed barrier between batho (twisted trans) and lumi (relaxed trans) is sufficiently large to give batho the observed nanosecond lifetime but insufficient to force reversal of batho to rhodopsin. For the corresponding cis,trans intermediate of 1,3-cycloheptadiene, a low energy pathway involving a different reaction coordinate is also available to give the bicycloheptene (········). The protein imposed barrier for iodopsin at low temperature (–.–.–.) is higher (probably associated with the difficulty of extrusion of Cl⁻ from the protein) and that of the ring fused systems (Rho-7 or phenylcyclohexene) (----) is even more prohibitive. In both cases, the batho intermediate reverses to the original cis isomer.

The eight- and nine-membered ring analogs (Rho-8, Rho-9) are more sterically crowded and so form stable pigments at much reduced rates.[9a] No reports on their bleaching characteristics are available.

The analog with a six-membered ring fused between C9 and C11 (4) gave normal photobleaching characteristics:[42]

$$4, 485nm \xrightarrow{h\upsilon} \underset{530nm}{batho} \longrightarrow \underset{495nm}{lumi} \longrightarrow \underset{480nm}{meta}$$

5.2. Bicyclic Analogs

Two bicyclic pigment analogs (5 and 5a) are known.[91] For 5, the batho intermediate was found to be thermally more stable than that from rhodopsin. Thus, at -130°C (instead of -140°C for the batho–lumi transition in the parent system) the batho analog, presumably with the all-trans geometry, was found to rearrange to a mixture of the stable 9-cis and 11-cis pigment analogs.[91]

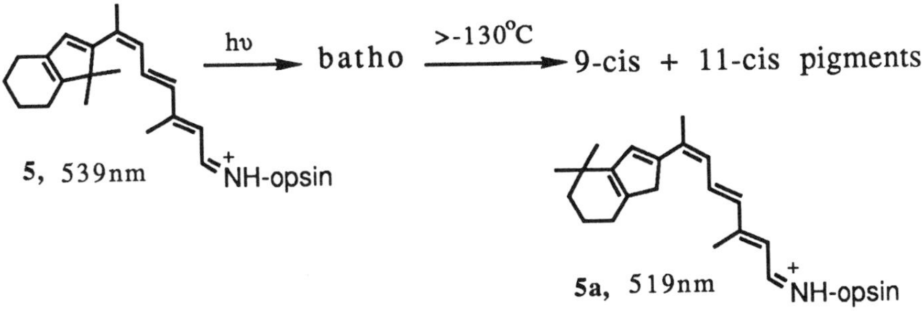

5, 539nm   NH-opsin

$>-130°C$

batho → 9-cis + 11-cis pigments

5a, 519nm   NH-opsin

These results, different from those of rhodopsin, seem to suggest that conversion to lumirhodopsin requires relaxation of the chromophore near the ring as well as the portion of the polyene close to the reactive 11-cis double bond.

5.3. The retro-γ Analogs

Exceptions to the 11-cis to all-trans (rhodopsin) and 9-cis to all-trans (9-cis rhodopsin) regioselective photoisomerization of visual pigments are found in the retro-γ analog (6). Photochemical studies conducted at room temperature showed that irradiation of its 9-cis isomer (6a) with light >420 nm gave primarily the 9,13-di-cis pigment, which upon prolonged irradiation isomerized to the 13-cis.[92]

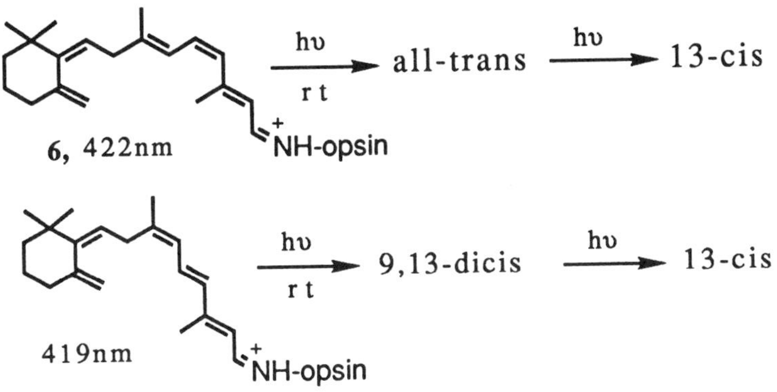

6, 422nm   NH-opsin

hυ / rt → all-trans → 13-cis

419nm   NH-opsin

hυ / rt → 9,13-dicis → 13-cis

The all-trans isomer was not detected during the course of irradiation. On the other hand, under the same conditions, the 11-cis isomer (λ_{max} = 422 nm) gave first the all-trans isomer, which subsequently gave the 13-cis isomer.[92]   The altered regioselectivity to the 9,13-di-cis isomer from the 9-cis to all-trans conversion in rhodopsin was unexpected but apparently consistent with the observed 9-cis to 9,13-di-cis isomerization of the free retro-γ-retinal analog.[93]   Thus, the truncated chromophore with a sp³ center at C8 possibly provided more flexibility to the C10 to C14 portion of the chromophore, resulting in a lack of protein-directed regioselective isomerization. The altered regioselective chemistry was further accompanied by increased stability of the isomerized products, permitting secondary photochemical reactions to the 13-cis isomer.

7, 416nm   NH-opsin   8, 468nm   NH-opsin

It is interesting to note that saturation of the ring as in 5,6-dihydrorhodopsin (7, see Section 5.5) and in the related α-rhodopsin (8)[94] does not seem to have a significant effect on the photobleaching process.

5.4. The Fluorinated Analogs

12-Fluororhodopsin was found to undergo photoisomerization in ways identical to those of rhodopsin: ready interconversion between rhodopsin, bathorhodopsin, and 9-*cis*-rhodopsin, indicating an absence of any specific protein– substrate interaction involving the fluorine substituent.[95]   The same situation is not true for the 10-fluoro analog. Low-temperature irradiation of 10-fluoror- hodopsin (9) was found to give a binary mixture of the pigment and the corresponding batho intermediate.[96]   There was no indication of involvement of the 9-cis isomer. Similarly, 9-*cis*-10-fluororhodopsin (9a) was found to be practically unreactive under the same irradiation conditions.[96]

9, 499nm   NH-opsin   batho 562nm   9a, 486nm   NH-opsin

Photosensitivity measurements of several fluorinated visual pigment analogs (bovine and gecko pigments) also showed diminished photoreactivity for the

10-fluoro analogs.[97] For this regiospecific effect of inhibition of photoisomerization, two forms of interaction between a protein residue and the fluoro substituent have been considered: attractive interaction of the electron-rich fluorine atom with an acidic proton (hydrogen bonding) or repulsive interaction of the fluorine atom with an electron-rich group such as a carboxylate or a phenoxide.[96] The former suggestion finds support in reports of weak interaction between organic acids and organo fluorides,[98] and x-ray crystal structural data.[99]

Absorption maxima of vinyl fluoro (14F, 12F, 10F, 8F = 527, 505, 495, and 485 nm respectively)[100] and trifluoro pigments (13, 9, 5 = 545, 490, and 480 nm respectively)[101] show progressive red shift as the electronegative substituent is moved closer to the polar end group. However, comparison with corresponding data for protonated Schiff bases (opsin shift) reveals that only the 10-fluoro analog exhibits a larger perturbation by the protein.[99]

5.5. Other Rhodopsin Analogs

For 9-*cis*-7,8-dihydrorhodopsin (10)[102] and 9-*cis*-5,6-dihydrorhodopsin (7),[51] low-temperature spectroscopic studies showed the presence of an additional intermediate between the batho and the lumi intermediates. As mentioned above, such a minor variance as a result of modified chromophore structure should be of no surprise considering the extensive structural changes occurred to the protein during the bleaching process.

10, 427nm $\xrightarrow[-180°C]{h\upsilon}$ batho 465nm $\xrightarrow[-170°C]{}$ BL $\xrightarrow[-140°C]{}$ lumi

7, 416nm $\xrightarrow[-180°C]{h\upsilon}$ batho 508nm $\xrightarrow{-242°C}$ BL 414nm $\xrightarrow{-160°C}$ lumi 488nm

12-Methylrhodopsin was reported to have a much reduced photosensitivity which was interpreted in terms of steric inhibition at the reaction site.[103]

13-Demethylrhodopsin exhibits photobleaching characteristics similar to that of the dihydrorhodopsins in that an additional intermediate between batho and lumi, known as BL, was detected. However, 9-demethylrhodopsin gave a primary product with absorption properties (not red shifted) similar to that of the pigment and the later lumi intermediate.[104] Also absent was the unusual HOOP band in its vibrational spectrum, normally attributed to that of a bent chromophore.[18] Hence, the absence of steric interaction between the 9-methyl group and the surrounding protein permits rapid relaxation of the primary photoproduct to one with a relatively planar chromophore.

6. Bacteriorhodopsin and Its Analogs

A different selectivity in photoisomerization is exhibited by the retinyl chromophore trapped in the binding cavity of bacteriorhodopsin[33b] and the related sensory,[105] halo-,[106] and phobobacteriorhodopsins.[107] In all cases the protonated all-trans-retinyl Schiff base undergoes regiospecific photoisomerization at the 13,14 bond. Its apparent difference from the preferred isomerization at the 11,12 bond for the free chromophore[62] and visual pigments must be due to the unique shape of the binding site, which is known to be more stereoselective. It reacts only with the all-trans and 13-cis isomers and not with the 7-cis or the centrally bent 9-cis and 11-cis isomers of retinal.[108]

The photoisomerization of *trans*-BR to 13-*cis*-BR and the ensuing result of pumping proton across the protein membrane has been examined in great detail. Scheme 2 summarizes the current understanding of the photochemical and subsequent dark processes.[33b,109]

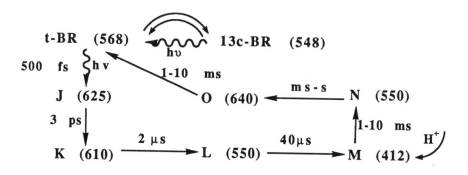

Scheme 2. Photochemical and ensuing dark processes of *trans*-BR. *trans*-BR is in equilibrium with 13-*cis*-BR in the dark. Irradiation of *trans*-BR results in formation of the J intermediate, followed by ensuing dark processes. Irradiation of 13-*cis*-BR results in its conversion to *trans*-BR. (The rates for interconversion of intermediates and their absorption maxima, in parentheses, are from ref. 109.)

The selectivity of bacterioopsin binding interactions toward retinal analogs is apparently different. Therefore, for 13-demethylretinal, the 11-cis, 9-cis, and 7-cis isomers as well as the all-trans and 13-cis isomers form pigment analogs with bacterioopsin.[110] Moreover, the longitudinal restriction of its binding site appears to be less stringent as reflected in its ability to form BR analogs with retinals of varying lengths.[111] However, direct information on regioselectivity of isomerization of the retinyl chromophore in other BR analogs are not readily available. Indirect information from extraction of final photoproducts of BR analogs indicates a general preference for isomerization at the 13,14 bonds, although relative amounts may vary among different systems.

7. Summary

In summary, the existing spectroscopic and photochemical data on visual pigments and their analogs reveal the important role of the protein structure in dictating properties of these pigments. The cd absorption spectra associated with the retinyl chromophore in the visual pigments are clear indications of the tight fitting of the chromophore within the binding site of opsin. The regioselective isomerization of such confined chromophores in different pigments is protein directed. Relaxation of such interactions could lead to a loss of regioselectivity of photoisomerization as shown in the photochemistry of the later intermediates in the photobleaching process. The direction of reaction of the photoproducts could be altered by modified protein and/or chromophore structures. A better understanding of the protein structure and the nature of specific protein substrate interactions in the future could provide a better understanding of the specific manner that the host protein controls the regioselectivity of isomerization of the substrate.[112]

References

1. *The Retionoids*, Vols. 1 and 2, M. B. Sporn, A. B. Roberts and D. S. Goodman, eds., Academic Press, New York, 1984.
2. See, e.g., H. Shichi, *Biochemistry of Vision*, Academic Press, New York, 1983.
3. (a) R. Henderson and P. N. T. Unwin, *Nature (London)*, 1975, *257*, 28.
   (b) R. Henderson, J. M. Baldwin, T. A. Ceska, F. Zemlin, E. Beckmann and K. H. Downing, *J. Mol. Biology*, 1990, *213*, 899.
4. D. M. Engelman, R. Henderson, A. D. Mclachlan and B. A. Wallace, *Proc. Natl. Acad. Sci., U. S. A.*, 1980, 77, 2023.
5. J. M. Corless, D. R. McCaslin and B. L. Scott, *Proc. Natl. Acad. Sci., U. S. A.*, 1982, *79*, 1116.
6. (a) P. A. Hargrave, J. H. McDowell, R. J. Feldman, P. H. Atkinson, R. Mohana and P. Argos, *Vision Res.*, 1984, 24, 1487.
   (b) J. B. C. Findlay, *Photobiochem. Photobiophys.*, 1986, *13*, 213.
7. H. Matsumoto and T. Yoshizawa, *Vision Res.*, 1978, *18*, 607.
8. R. S. H. Liu, H. Matsumoto, A. Kini, A. E. Asato, M. Denny, A. Kropf and W. J. deGrip, *Tetrahedron*, 1984, *40*, 473.

9. (a) V. Balogh-Nair and K. Nakanishi in *Chemistry and Biology of Synthetic Retinoids*, M. Dawson and W. Okamura, eds., CRC Press, Boca Raton, FL, 1990, p. 147.
    (b) R. S. H. Liu and A. E. Asato, in *Chemistry and Biology of Synthetic Retinoids*, M. Dawson and W. Okamura, eds., CRC Press, Boca Raton, FL, 1990, p. 51.
10. (a) F. Crescitelli, W. F. H. M. Mommarts and T. I. Shaw, *Proc. Natl. Acad. Sci., U. S. A.*, **1966**, *56*, 1729.
    (b) T. I. Shaw in *Handbook of Sensory Physiology*, Vol. 7, Part 1, H. J. A. Dartnall, ed., Springer Verlag, 1972, p. 180.
11. R. S. H. Liu and T. Mirzadegan, *J. Am. Chem. Soc.*, **1988**, *110*, 8617.
12. (a) Y. Fukada, Y. Shichida, T. Yoshizawa, M. Ito, A. Kodama and K. Tsukida, *Biochemistry*, **1984**, *23*, 5826.
    (b) M. Ito, Y. Mantani, K. Tsukida, Y. Shichida, S. Ioshida, Y. Fukada and T. Yoshizawa, *J. Nutr. Sci. Vitaminol.*, **1988**, *34*, 641.
13. R. Sen, O. Carriker, V. Balogh-Nair and K. Nakanishi, *Tetrahedron*, **1984**, *40*, 493.
14. T. Seki, T. Shingu and Y. Kito, *Eur. J. Biochem.*, **1985**, *147*, 255.
15. H. Matsumoto and T. Yoshizawa, *Nature (London)*, **1975**, *258*, 523.
16. (a) A. Kropf, P. Whittenberger, S. Goff and A. S. Waggoner, *Expt. Eye Res.*, **1973**, *17*, 591.
    (b) A. Kropf, *Nature (London)*, **1976**, *264*, 92.
17. G. Eyring, B. Curry, A. Brock, J. Lugtenburg and R. Mathies, *Biochemistry*, **1982**, *21*, 384.
18. (a) G. Eyring, B. Curry, R. Mathies, R. Fransen, I. Palings and J. Lugtenburg, *Biochemistry*, **1980**, *19*, 2410.
    (b) U. M. Ganter, R. Krautle, R. R. Rando and F. Siebert, in *Molecular Physiology of Retinal Proteins*, H. Hara, ed., Yamada Science Foundation, 1988, p. 55.
19. Y. Shichida, A. Kropf and T. Yoshizawa, *Biochemistry*, **1981**, *20*, 1962.
20. (a) P. E. Blatz, J. H. Mohler and H. V. Navangul, *Biochemistry*, **1972**, *11*, 848.
    (b) H. Suzuki, T. Komatsu and M. Kitajima, *J. Phys. Soc. Jpn.*, **1974**, *37*, 177.
21. C. S. Irving, G. W. Byers and P. A. Leermaker, *Biochemistry*, **1970**, *9*, 858.
22. (a) B. Honig, U. Dinur, K. Nakanishi, V. Balogh-Nair, M. A. Gawinowicz, M. Arnaboldi and M. G. Motto, *J. Am. Chem. Soc.*, **1979**, *101*, 7084.
    (b) J. Lugtenburg, M. Muradin-Szwaykowskev, C. Heeremens, J. A. Pardoen, G. S. Harbison, J. Herzfeld, R. G. Griffin, S. O. Smith and R. Mathies, *J. Am. Chem. Soc.*, **1986**, *108*, 3104.
23. G. Wald, *Nature*, **1968**, *219*, 800.
24. (a) T. Yoshizawa and Y. Kito, *Nature*, **1958**, *182*, 1604.
    (b) G. Wald and T. Yoshizawa, *Nature (London)*, **1963**, *197*, 1279.
25. (a) T. Yoshizawa in *Handbook of Sensory Physiology*, Vol. 7, Part 1, H. J. A. Dartnall, ed., Springer-Verlag, Berlin, 1972, p. 146.
    (b) S. Matuoka, Y. Shichida and T. Yoshizawa, *Biochim. Biophys. Acta*, **1984**, *765*, 38.
26. (a) R. Cone, *Nature, New Biol.*, **1972**, *236*, 39.
    (b) T. A. Rosenfeld, A. Alchalel and M. Ottolenghi, *Nature*, **1972**, *240*, 482.
27. G. E. Busch, M. L. Applebury, A. A. Lamola and P. M. Rentzepis, *Proc. Natl. Acad. Sci., U. S. A.*, **1972**, *69*, 2802.
28. H. Kandori, S. Matuoka, Y. Shichida and T. Yoshizawa, *Photochem. Photobiol.*, **1989**, *49*, 181.
29. (a) Y. Shichida, S. Matuoka and T. Yoshizawa, *Photobiochem. Photobiophys.*, **1984**, *7*, 221.
    (b) H. Kandori, Y. Shichida and T. Yoshizawa, *Biophys. J.*, **1989**, *56*, 453.

30. K. S. Peters, M. L. Applebury and P. M. Rentzepis, *Proc. Natl. Acad. Sci, U. S. A.*, **1977**, *74*, 3119.
31. (a) B. Honig, T. Ebrey, R. H. Callender, U. Dinur and M. Ottolenghi, *Proc. Natl. Acad. Sci., U. S. A.*, **1979**, *76*, 2503.
     (b) U. Dinur, B. Honig and M. Ottolenghi, *Photochem. Photobiol.*, **1981**, *33*, 523.
32. (a) T. Kobayashi, *Photochem. Photobiol.*, **1980**, *32*, 207.
     (b) H. Ohtani, T. Kobayashi, M. Tsuda and T. G. Ebrey, *Biophys. J.*, **1988**, *53*, 493.
33. (a) Y. Shichida, *Photobiochem. Photobiophys.*, **1986**, *4*, 287.
     (b) M. Ottolenghi and M. Sheves in *Primary Processes in Photobiology*, Springer-Verlag, Berlin, 1986, p. 144.
     (c) C. Sandorfy and D. Vocelle, *Can. J. Chem.*, **1986**, *64*, 2251.
     (d) R. Becker, *Photochem. Photobiol.*, **1988**, *48*, 369.
34. T. Ono, Y. Shichida and T. Yoshizawa, *Photochem. Photobiol.*, **1986**, *43*, 285.
35. S. Horiuchi, F. Tokunaga and T. Yoshizawa, *Biochim. Biophys. Acta*, **1980**, *591*, 445.
36. M. Takezaki and Y. Kito, *Nature (London)*, **1967**, *215*, 1197.
37. G. A. Schick, T. M. Cooper, R. A. Holloway, L. P. Murray and R. R. Birge, *Biochemistry*, **1987**, *26*, 2556.
38. (a) N. Sasaki, F. Tokunaga and T. Yoshizawa, *Photochem. Photobiol.*, **1980**, *32*, 433.
     (b) C. M. Einterz, J. W. Lewis and D. S. Kliger, *Proc. Natl. Acad. Sci., U. S. A.*, **1987**, *84*, 3699.
39. (a) A. Warshel, *Nature*, **1976**, *260*, 679.
     (b) A. Warshel and N. Barboy, *J. Am. Chem. Soc.*, **1982**, *104*, 1469.
40. R. S. H. Liu and A. E. Asato, *Proc. Natl. Acad. Sci., U. S. A.*, **1985**, *82*, 259.
41. I. Palings, J. A. Pardoen, E. van den Berg, C. Winkel, J. Lugtenburg and R. Mathies, *Biochemistry*, **1987**, *26*, 2544.
42. (a) M. Sheves, A. Albeck, M. Ottolenghi, A. H. M. Bovee-Geurts, W. J. DeGrip, C. M. Einterz, J. W. Lewis, L. E. Schaechter and D. S. Kliger, *J. Am. Chem. Soc.*, **1986**, *108*, 6440.
     (b) A. E. Asato, M. Denny and R. S. H. Liu, *J. Am. Chem. Soc.*, **1986**, *108*, 5032.
     (c) T. Yoshizawa, Y. Shichida and H. Kandori, *Proc. Yamada Conf.*, **1988**, *XXI*, 49.
43. R. S. H. Liu and A. E. Asato, *Tetrahedron*, **1984**, *40*, 1931.
44. Y. Shichida, K. Nakamura, T. Yoshizawa, A. Trehan, M. Denny and R. S. H. Liu, *Biochemistry*, **1988**, *27*, 6495.
45. R. Crouch, V. Purvin, K. Nakanishi and T. Ebrey, *Proc. Natl. Acad. Sci., U. S. A.*, **1972**, *72*, 1538.
46. H. J. A. Dartnall, *Vision Res.*, **1968**, *8*, 339.
47. (a) A. Kropf and R. Hubbard, *Ann. N. Y. Acad. Sci.*, **1958**, *74*, 266.
     (b) H. Kandori, S. Matuoka, H. Nagai, Y. Shichida and T. Yoshizawa, *Photochem. Photobiol.*, **1988**, *48*, 93.
48. R. R. Birge, C. M. Einterz, H. M. Knapp and L. P. Murray, *J. Biophys.*, **1988**, *53*, 367.
49. A. G. Doukas, M. R. Junnarkar, R. R. Alfano, R. H. Callender and V. Balogh-Nair, *J. Biophys*, **1985**, *47*, 795.
50. (a) A. Cooper, *Nature (London)*, **1979**, *282*, 531.
     (b) G. A. Schick, T. M. Cooper, R. A. Holloway, L. P. Murray and R. R. Birge, *Biophysical J.*, **1987**, *26*, 2556.
51. (a) T. Yoshizawa, Y. Shichida and S. Matuoka, *Vision Res.*, **1984**, *24*, 1455.

(b) A. Albeck, N. Friedman, M. Ottolenghi, M. Sheves, C. M. Einterz, S. J. Hug, J. W. Lewis and D. S. Kliger, *Biophys. J.*, 1989, *55*, 233.

52. W. J. DeGrip, R. S. H. Liu, A. E. Asato and V. Ramamurthy, *Nature (London)*, 1976, *262*, 416.

53. H. Kandori, S. Matuoka, Y. Shichida and T. Yoshizawa, *Proc. Yamada Conf.*, 1988, *XXI*, 383.

54. S. Kawamura, S. Miyatani, H. Matsumoto, T. Yoshizawa and R. S. H. Liu, *Biochemistry*, 1980, *19*, 1549.

55. V. Ramamurthy, T. T. Bopp and R. S. H. Liu, *Tetrahedron Lett.*, 1972, 3915.

56. A. Trehan, R. S. H. Liu, Y. Shichida, Y. Imamoto, K. Nakamura and T. Yoshizawa, *Bioorg. Chem.*, 1990, *18*, 30.

57. R. Mathies, S. O. Smith and I. Palings, *Biological Applications of Raman Spectroscopy*, Vol. 2, T. G. Spiro, ed., John Wiley and Sons, New York, 1987, p. 59.

58. G. R. Loppnow, M. E. Miley, R. A. Mathies, R. S. H. Liu, H. Kandori, Y. Shichida, Y. Fukada and T. Yoshizawa, *Biochemistry*, 1990, *29*, 1985..

59. R. S. Becker and K. Freedman, *J. Am. Chem. Soc.*, 1985, *107*, 1477.

60. R. R. Birge, L. P. Murray, B. M. Pierce, N. Akita, V. Balogh-Nair, L. A. Findsen and K. Nakanishi, *Proc. Natl. Acad. Sci., U. S. A.*, 1985, *82*, 4117.

61. K. Freedman and R. S. Becker, *J. Am. Chem. Soc.*, 1986, *108*, 1245.

62. R. F. Child and G. S. Shaw, *J. Am. Chem. Soc.*, 1988, *110*, 3013.

63. (a) A. Kropf and R. Hubbard, *Photochem. Photobiol.*, 1970, *12*, 249.

    (b) N. Jensen, R. Wilbrandt and R. Bensasson, *J. Am. Chem. Soc.*, 1989, *111*, 7877.

64. V. J. Rao, R. J. Fenstemacher and R. S. H. Liu, *Tetrahedron Lett.*, 1984, *25*, 1115.

65. B. A. Halley and E. C. Nelson, *Internat. J. Vit. Nutr. Res.*, 1979, *49*, 347.

66. M. E. Zadazki and A. B. Ellis, *J. Org. Chem.*, 1984, *48*, 3156.

67. B. W. Zhang and R. S. H. Liu, unpublished results.

68. T. Takemura, P. K. Das, G. Hug and R. Becker, *J. Am. Chem. Soc.*, 1978, *100*, 2626.

69. (a) S. Ganapathy, A. Trehan and R. S. H. Liu, *J. Chem. Soc., Chem. Comm.*, 1990, *2*, 199.

    (b) S. Ganapathy and R. S. H. Liu, Tetrahedron Lett., 1990, *31*, 6957.

70. W. H. Waddell, R. Crouch, K. Nakanishi and N. J. Turro, *J. Am. Chem. Soc.*, 1976, *98*, 4189.

71. X. Y. Li, A. E. Asato and R. S. H. Liu, *Tetrahedron Lett.*, 1990, *31*, 4841.

72. A. G. Doukas, M. R. Junnarkar, R. R. Alfano, T. Kakitani and B. Honig, *Proc. Natl. Acad. Sci., U. S. A.*, 1984, *81*, 4190.

73. R. S. Becker, K. Freedman, J. A. Hutchinson and L. Noe, *J. Am. Chem. Soc.*, 1985, *107*, 3942.

74. (a) T. G. Ebrey and T. Yoshizawa, *Exp. Eye Res.*, 1973, *17*, 545.

    (b) K. Azuma, M. Azuma and T. Suzuki, *Biochim. Biophys. Acta*, 1975, *393*, 520.

75. U. M. Ganter, W. Gartner and F. Siebert, *Biochemistry*, 1988, *27*, 7480.

76. K. J. Rothschild and W. J. DeGrip, *Photobiochem. Photobiophys.*, 1986, *13*, 245.

77. A. Maeda, T. Ogurusu, F. Tokunaga and T. Yoshizawa, *FEBS Lett.*, 1978, *92*, 77.

78. A. Maeda, Y. Shichida and T. Yoshizawa, *Biochemistry*, 1979, *18*, 1245.

79. H. Hara, R. Hara, F. Tokunaga and T. Yoshizawa, *Photochem. Photobiol.*, 1981, *33*, 883.

80. G. Wald, P. K. Brown and P. H. Smith, *J. Gen. Physiol.*, 1955, *38*, 623.

81. G. Wald, *Fed. Proc.*, 1953, *12*, 606.

82. K. Vogt and K. Kirshfeld, *Naturwissenschaften,* **1984,** *71,* 211.
83. K. Hamdorfin *Handbook of Sensory Physiology,* Vol.7, Part 6A, H. J. A. Dartnall, ed., Springer Verlag, Berlin, 1979, p. 146.
84. (a) R. Hubbard and A. Kropf, *Nature (London),* **1959,** *183,* 448.
     (b) T. Yoshizawa and G. Wald, *Nature,* **1967,** *214,* 566.
85. R. van der Steen, M. Groeskeek, L. J. P. van Amsterdam, J. Lugtenburg, J. van Oostrum and W. J. deGrip, *Recl. Trav. Chim. Pays-Bas,* **1989,** *108,* 20.
86. H. Akita, S. P. Tanis, M. Adams, V. Balogh-Nair and K. Nakanishi, *J. Am. Chem. Soc.,* **1980,** *102,* 6370.
87. H. Kandori, S. Matuoka, Y. Shichida, T. Yoshizawa, M. Ito, K. Tsukida, V. Balogh-Nair, and K. Nakanishi, *Biochemistry,* **1989,** *28,* 6460.
88. J. Buchert, V. Stanfancic, A. G. Doukas, R. R. Alfano, R. H. Callender, J. Pande, H. Akita, V. Balogh-Nair and K. Nakanishi, *Biophys. J.,* **1983,** *43,* 279.
89. R. Bonnoue, J. Joussot-Dubien, L. Salem and A. Yarwood, *J. Am. Chem. Soc.,* **1976,** *98,* 4329.
90. Y. Inoue, S. Hagiwara, Y. Daino and T. Hagushi, *J. Chem. Soc.,Chem. Comm.,* **1979,** 1307.
91. (a) M. Ito, T. Hiroshima, K. Tsukida, Y. Shichida and T. Yoshizawa, *J. Chem. Soc., Chem. Comm.,* **1985,** 1443.
     (b) S. Ioshida, Y. Shichida, T. Yoshizawa, M. Ito, T. Hiroshima and K. Tsukida, *Biophysics (Japan),* **1985,** *25,* s101.
92. S. Kawamura, T. Yoshizawa, K. Horiuchi, M. Ito, A. Kodama and K. Tsukida, *Biochim. Biophys. Acta,* **1979,** *348,* 147.
93. T. Yoshizawa, H. Matsumoto, K. Horiuchi, Y. Shichida, M. Ito, A. Kodama and K. Tsukida, in *Biophysical Studies of Retinal Proteins,* T. G. Ebrey, H. Frauenfelder, B. Honig and K. Nakanoshi., eds., Univ. Illinois Press, Champaign, USA, 1987, p. 287.
94. A. E. Asato, B. W. Zhang, M. Denny, T. Mirzadegan and R. S. H. Liu, *J. Bioorg. Chem.,* **1988,** *17,* 410.
95. R. S. H. Liu, H. Matsumoto, A. E. Asato, M. Denny, Y. Shichida, T. Yoshizawa and F. Dahlquist, *J. Am. Chem. Soc.,* **1983,** *103,* 7195.
96. Y. Shichida, T. Ono, T. Yoshizawa, H. Matsumoto, A. E. Asato, J. P. Zingoni and R. S. H. Liu, *Biochemistry,* **1987,** *26,* 4422.
97. R. S. H. Liu, F. Crescitelli, M. Denny, H. Matsumoto and A. E. Asato, *Biochemistry,* **1986,** *25,* 7026.
98. M. O. Joesten and L. J. Schaad in *Hydrogen Bonding* , Marcel Dekker, New York, 1974, p. 325.
99. (a) P. Murray-Rust, W. C. Stallings, C. T. Monti, R. K. Preston and J. P. Glusker, *J. Am. Chem. Soc.,* **1983,** *105,* 3206.
     (b) A. Karipides and C. Miller, *J. Am. Chem. Soc.,* **1984,** *106,* 1494.
100. Y. Fukada, T. Okano, Y. Shichida, T. Yoshizawa, A. Trehan, D. Mead, M. Denny, A. E. Asato and R. S. H. Liu, *Biochemistry,* **1990,** *29,* 3133.
101. R. S. H. Liu, A. E. Asato, M. Denny, D. Mead, T. Mirzadegan and B. W. Zhang, *Proc. Yamada Conf.,* **1988,** *XXI,* 43.
102. O. Muto, F. Tokunaga, T. Yoshizawa, V. Kamat, H. A. Blatchly, V. Balogh-Nair and K. Nakanishi, *Biochim. Biophys. Acta,* **1984,** *766,* 597.
103. R. S. H. Liu, A. E. Asato, M. Denny and D. Mead, *J. Am. Chem. Soc.,* **1984,** *106,* 8298.
104. U. M. Ganter, E. D. Schmid, D. Perez-Sala, R. R. Rando and F. Siebert, *Biochemistry,* **1989,** *28,* 5954.

105. (a) J. L. Spudlich and R. A. Bogomolni, *Proc. Biophys. Studies of Retinal Proteins*, T. G. Ebrey, H. Frauenfelder, B. Honig and K. Nakanoshi., eds., Univ. Illinois Press, Champaign, 1987, p. 24.
      (b) M. Ariki, Y. Shichida and T. Yoshizawa, *FEBS Letters*, **1987**, *225*, 255.
106. J. K. Lanyi, *Ann. Rev. Biophys. Chem.*, **1986**, *15*, 11.
107. Y. Shichida, Y. Imamoto, T. Yoshizawa, T. Takahashi, H. Tomioka, N. Kamo and Y. Kobatake, *FEBS Lett.*, **1988**, *236*, 333.
108. W. Stoeckenius and R. A. Bogomolni, *Annu. Rev. Biochem.*, **1982**, *51*, 587.
109. (a) S. P. A. Fodor, J. B. Ames, R. Gebhard, E. M. M. van den Berg, W. Stoeckenius, J. Lugtenburg and R. A. Mathies, *Biochemistry*, **1988**, *27*, 7097.
      (b) J. B. Ames and R. A. Mathies, *Biochemistry*, **1990**, *29*, 7181.
110. W. Gartner, P. Towner, H. Hopf and D. Oesterhelt, *Biochemistry*, **1983**, *22*, 2637.
111. (a) R. K. Crouch, S. Scott, S. Ghent, R. Govindjee, C. Chang and T. Ebrey, *Photochem. Photobiol*, **1985**, *43*, 297.
      (b) V. J. Rao, J. P. Zingoni, R. Crouch, M. Denny and R. S. H. Liu, *Photochem. Photobiol.*, **1985**, *41*, 171.
112. Supported by grants from the National Science Foundation (CHE-16500, INT-13500), U. S. Public Health Services (DK-17806), and a Special Coordination Fund of the Science and Technology Agency of the Japanese Government.

Chapter 19

Application of Organized Media: Examples

David F. Eaton

Central Research and Development Department,
The Du Pont Company, Wilmington, DE, USA.

Contents

1. Introduction: Organization as a Paradigm

Nature has developed a superb ability to organize molecular elements in order to solve problems presented during development of a physical system. One of the best known examples of a highly developed organizational approach to biological function is the photosynthetic unit. The elegant structural solution that simple bacteria have evolved for the solution of the energy collection and transport problem in photosynthesis has been elucidated by Michel, Deisenhofer, and their co-workers.[1] The beauty of this structure is readily apparent to photochemists, biologists, physical chemists, and others who study the remarkable directional electron-transfer processes initiated in this structural unit by the absorption of a photon. No system designed by scientists has the complexity or functionality present in this natural system. It would be presumptuous to suggest that science will soon be able to approach such complexity, although approaches to similar functionality are plausible and are the focus of much active research today.

In the photosynthetic reaction center of the purple bacterium *R. viridis*, it is the function of the proteins surrounding the prosthetic units to hold the electron donor, electron acceptor, and auxiliary chromophores involved in the actual electron-transfer process in position. That is, the proteins form the host structure of a complex host–guest ensemble which is, as a unit, embedded in the lipid bilayer that constitutes the membrane structure of the total reaction center. This paradigm, that the structure of a surrounding host can enforce organization on active elements so as to predispose them to a desired physical process, is the inclusion paradigm.

Another natural, light-sensitive system that also employs the inclusion paradigm is the array of rods and cones in the eye, which comprise the basic macroscopic elements of the visual system. The rod cells, responsible for vision in low light levels, consist *inter alia* of prosthetic pigments embedded in proteinaceous hosts, arranged in stacks of disk-like granules that function to increase the absorption. The chromophore is positioned within a seven-bundle protein complex that spans the membrane bilayer. Photoexcitation of the retinal–protein complex triggers a geometrical isomerization of the chromophore, which alters the physical properties of the protein–chromophore complex in such a way that membrane potentials are eventually altered, to pump ionic gradients sensed neurally as a visual impulse.[2] Both the visual system and the photosynthetic apparatus are discussed in the previous two chapters (Chapters 17 and 18) in this monograph.

It is the purpose of this chapter to report synthetic systems that employ this paradigm in order to carry out a task assigned by the scientific designer, especially with the proviso that photons be involved as the trigger to perform the task. The chapter is organized into sections that reflect the various tasks which are to be performed by the organized system. Because there are many organized systems that do not employ light in their function, a very brief introduction to ground-state chemical systems that perform some technologically important functions is first presented. Photochemical and photophysical functions are detailed in subsequent sections.

2. Inclusion Complexes and Ground-State Processes

Host–guest complexes can be used to encapsulate the guest in order to protect it physically. Complexation in this way also serves to dilute the guest. The encapsulated guest will have physical properties that are different from the pure material, and this can be used advantageously. For example, the guest may have a high vapor pressure, while the complex has a lower one, because the forces that bind host and guest lower the chemical potential of the material. The complex may have a different solubility than the pure guest and serve to transport the guest to a desired location, for example, in the case of a drug delivery system. Since the host–guest interaction is specific, the complex formation process can be used as the basis of a purification process, either "extracting" the guest from a mixture or, if the host is incorporated in a bound state on an immobile support, the process can be used chromatographically. If the host is optically active, racemic guests may be resolved during formation of the complex. In some cases (*vide infra*) spontaneous resolution is possible when both the host and the guest are initially present as racemates. As a similar, but more fundamental, consequence of the specificity of the host–guest interaction, the guest will be held in conformations that may or may not be similar to those accessible to the pure material. Reactivity can be modified in the complex. All of the utilities described above have been realized.

Szejtli has recently comprehensively reviewed the industrial applications of cyclodextrins (CDs), which constitute one of the prototype hosts for encapsulation of small molecule guests.[3] Szejtli reports that CD complexation is used to protect the guest from unwanted oxidation or other decomposition, to stabilize the material against loss by evaporation, to eliminate unwanted odors or tastes associated with the guest, to reduce hygroscopicity, and to stabilize against unwanted photochemical decomposition. The latter topic will be covered separately. The applications cited result in several technological advantages for CD-complexed formulations of the guest over the use of the pure material. For example, stable, solid, standardized formulations can be produced that are easy to handle, weigh, and dispense. The percentage composition of guest is known precisely, which makes for accurate and easy dosing of the guest.

These advantages have led to use of CD complexes in food (flavor and fragrance enhancers), cosmetics (deodorants and antiseptics as well as dental products), toiletries (fragrances, bath salts), household products (laundry bleach, defoaming agents) and medicine (drug delivery, enhanced bioavailability) or agricultural applications (stabilization of insecticides, herbicides, animal feeds, etc.). Because CDs are approved for human or animal consumption, they are the major class of hosts used in medicinal and other contact applications (although in principle other hosts could perform this function).

A variety of hosts have been used to separate mixtures by taking advantage of specific host–guest binding. Cyclodextrins have been used to separate various aromatic compounds as well as straight- and branched-chain hydrocarbons. Cyclodextrins have also been used to effect separation of racemates, although no total resolution has been reported without resort to several stages of purification. Chromatographic resolution using polymer-bound CDs can effect total separation of enantiomers. Other hosts employed in separations include ureas, thioureas, tris-ortho-thymotide (TOT), Werner complexes, and crown ethers. These applications are not reviewed here since current reviews are available.

Patents issued that relate to applications of complexation are now abstracted by the *Journal of Inclusion Phenomena and Molecular Recognition in Chemistry*. Interested readers are urged to consult this useful listing for new applications.

3. Photochemical Applications of the Inclusion Paradigm

Surprisingly, the most familiar applications of inclusion chemistry to everyday life may be those applications that involve light. Liquid crystal display devices, common in watches and computer or television screens, are anisotropic dispersions of highly birefringent dyes included in a liquid-crystalline matrix. When thin films of poly(vinyl alcohol) containing small amounts of iodine and boric acid are stretched to orient the film, efficient light polarizing films are produced. This technology was exploited by the Polaroid Corporation in the production of glare-free sunglass lenses.[4] The optical principles behind the "Polaroid lens" and the liquid crystal display are similar, since they both involve orientational anisotropy of a highly absorbing guest in a two-dimensionally organized host matrix. Another commercially important photosystem that owes its special properties to host–guest chemistry is the thin film photoconductor marketed by Kodak for the Ektaprint photocopier. In this system, an organic photoconductor is dispersed in a polymeric thin film matrix as a crystalline, phase-segregated aggregate. The aggregate is not a pure phase of the photoconductor but is in fact complexed with the polymeric backbone of the host matrix. The complexation effects the optical properties of this complex, thin film composition.

There are other noncommercial areas where host–guest complexes may play a role in modifying optical properties of materials. A wide variety of host–guest systems have been devised that function as sensors for various analytes and that also use optical phenomena as their mode of detection. Another is in nonlinear optics, especially in those areas where the symmetry properties of the optical system are critical, for example, second harmonic generation (SHG) of laser light using optical crystals. In this application, anisotropic orientation of a molecular dipole in three-dimensional space must be effected in order to manifest the appropriate properties for SHG. Each of these applications is discussed in detail below.

Many academic photochemical and photophysical studies have been undertaken in host–guest systems. Photophysical probes have been employed to examine the conformational freedom of complexed molecules both in solution and in the solid state. A large number of investigations have explored the utility of host–guest complexes as a medium in which to perform photochemical synthesis. Most studies have involved the influence of the constraining medium of the host on unimolecular chemistry, but bimolecular photochemistry and photoinduced polymerizations have also been examined. These studies, which represent potential future applications of organized media in photochemistry, are described in Chapters 7 and 16 of this volume.

3.1. The Inclusion Paradigm and Nonlinear Optics

Most of the experience of molecular photochemists resides in the linear region of optical phenomena, that is that phenomenological regime where the response of an optical system depends on the first power of the incident light intensity. It includes all the phenomena associated with photochemistry conducted at normal power intensities available in the laboratory. However, the advent of lasers with high-power densities has introduced a variety of nonlinear phenomena. Two-photon photochemistry is an expanding area of study. Two-photon spectroscopy has been an important source of new ideas concerning the application of lasers to information storage, optical switching, and optical computing.

One aspect of nonlinear optical science where inclusion complexation may play a role is in preparation of materials with specific optical properties. Second-order nonlinear optical (NLO) processes[5,6] are those which express the hyperpolarizability (β) of individual molecules in the bulk properties (χ^2) of the material. Second harmonic generation, sum-frequency mixing and various electrooptic processes are examples of such second-order processes. Materials that possess second-order NLO properties must be noncentrosymmetric, and in some cases the symmetry of efficacious crystalline materials is preferably polar. Effective molecules for second-order processes (those with large β) often exhibit large degrees of intramolecular charge-transfer and may possess a large ground-state dipole moment. Such molecules often dimerize in a head-to-tail arrangement, to minimize electrostatic repulsion, as they form crystals, and so they are rarely acentric in their crystal form. Therefore, in order to expand the range of molecules that may be used in NLO devices, it is necessary to design protocols for engineering small, dipolar organic molecules that may possess large β into polar bulk arrangements. This can be attained using the inclusion paradigm.

The first example of the use of the inclusion paradigm to design NLO materials employed cyclodextrin as host.[7,8] Cyclodextrins are chiral molecules, so crystals of the pure hosts and their complexes are noncentric. However, they need not be polar. Cyclodextrin crystal structures tend to fall into two classes (Figure 1): channel structures, which may be head-to-head or those which are head-to-tail, and cage structures containing screw-axis related guests.[9] The head-to-tail channel structure is ideal for polar alignment of guest molecules.

Tomaru et al.,[7] and we[8] found independently that hyperpolarizable molecules (nonlinear optiphores) such as p-nitroaniline (PNA), which is a prototypical small organic molecule with high β but which is itself centrosymmetric in its crystal structure, form crystalline complexes with β-cyclodextrin that are capable of SHG (two to four times that of a urea standard) when irradiated by a Nd-YAG laser (Table 1). β-Cyclodextrin is only slightly active in its uncomplexed form (0.001 x urea), in spite of its chirality. This implies that β-CD is only weakly hyperpolarizable (low β) and that the SHG of the PNA complex arises entirely from the alignment of guest molecules in the solid. Unfortunately, the crystal structure of the PNA–(β-CD) complex has not been solved to prove the point alleged above. However, other inclusion complexes have been prepared and studied that generalize the inclusion paradigm for SHG and some of their crystal structures have been solved that prove the ability of inclusion complexation to align nonlinear optiphores in a polar fashion.

Table 1. Inclusion Compounds with β-Cyclodextrin

Guest	Host–Guest	SHG relative to urea
p-Nitroaniline	1:1	2.0–4.0
p-(*N,N*-Dimethylamino)cinnamaldehyde	1:1	0.4
N-Methyl-*p*-nitroaniline	1:1	0.25
2-Amino-5-Nitropyridine	1:1	0.07
p-Dimethylaminobenzonitrile	1::1	0.015
β-Cyclodextrin (no guest)	—	0.001

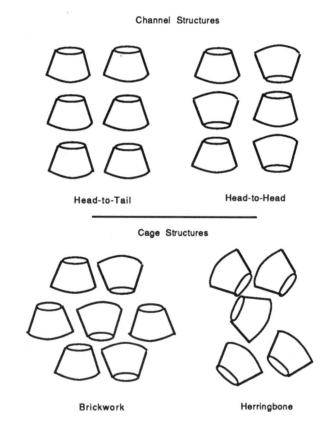

Channel Structures

Head-to-Tail   Head-to-Head

Cage Structures

Brickwork   Herringbone

Figure 1. Crystal structure motifs for cyclodextrin complexes.

Hosts such as TOT, perhydrotriphenylene (PHTP), cyclophosphazenes, deoxycholic acid, and thiourea have been shown to form polar, SHG-active solids with small organic and organometallic molecules chosen for their anticipated large β.[10] Table 2 lists selected results.

The method is judged to be a general one for the alignment of nonlinear optiphores. Neither host nor guest need be acentric or polar in its natural crystal structure, but crystals of the inclusion complex have been found to form polar, SHG-active solids with a probability of about 65%. This is an important technique for preparing new materials for nonlinear optics.

Table 2. SHG Results for Selected Host–Guest Inclusion Complexes

Complexes	Host–Guest	SHG relative to urea
With thiourea guest		
Benzenechromium tricarbonyl	3:1	2.3
(Fluorobenzene)chromium tricarbonyl	3:1	2.0
(Cyclopentadienyl)rhenium tricarbonyl	ND[a]	0.5
(1,3-Cyclohexadiene)iron tricarbonyl	3:1	0.4
(1,3-Cyclohexadienyl)manganese tricarbonyl	3:1	0.4
(Trimethylenemethane)iron tricarbonyl	3:1	0.3
(Cyclopentadienyl)manganese tricarbonyl	3:1	0.3
With tris-ortho-thymotide Guest		
p-Dimethylaminocinnamaldehyde	2:1	1.0
p-Dimethylaminobenzonitrile	2:1	0.3
(p-Cyanobenzoyl)manganese pentacarbonyl	ND[b]	0.2
(Indane)chromium tricarbonyl	1:1	0.1
(Anisole)chromium tricarbonyl	ND[b]	0.1
(Tetralin)chromium tricarbonyl	1:1	0.1
Inclusion complexes with deoxycholic acid guest		
p-Nitroaniline	ND	1.0
4-(Dicyanomethylene)-2-methyl-6-(p-dimethylaminostyryl)-4H-pyran	ND	0.4

[a] Errors in powder SHG intensity measurements can be ±50% because of particle size differences among samples, but relative rankings within a series are probably correct. ND = Not Determined.
[b] Cocrystallization of TOT with the complexes prevented accurate analysis.

For each of the host materials used in Table 2, the choice of guest is dictated by several criteria. Guests are chosen in general because of their anticipated hyperpolarizability, but within a host series suitable guests must have similar shapes. For example, TOT complexes are formed with thin, elongated guests. The structure is probably a channel one, rather than the alternative cage-type structure. Similarly, while thiourea accommodates narrow guests, Table 2 illustrates its ability to accommodate guests that are only slightly oblate. The metal carbonyls listed there are easily complexed within thiourea.

Orientation within the inclusion structure is believed to be favored because of electrostatic interactions among guests. The cartoon depiction of the phenomenon in Figure 2 illustrates the forces that may operate. In the figure, the polar guests are arranged within a channel structure such that the positive pole of one molecule is closest to the negative pole of a neighbor. This electrostatic minimum is favored as long as the channel is narrow enough so that two guests cannot occupy the channel simultaneously. If that were the case, face-to-face dimers would form that had a centrosymmetric orientation. The relation of guests channel-to-channel is also an electrostatic minimum. If the the chain of guests in one channel is translated half a molecular length down the channel, as shown, then cross-channel nearest approaches are positive to negative, and attractive, rather than repulsive, interactions. In crystal structures[10b] of SHG-active inclusion complexes which we have solved, the guests are oriented similar to that shown in Figure 2.

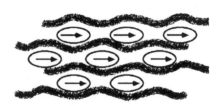

Figure 2. Schematic depiction of orientation of polar nonlinear optiphores within a guest–host ensemble.

Other inclusion hosts can also be used to effect polar alignment of organic guests. Stucky and co-workers have employed zeolite hosts to align nitroaromatics.[11] Iodoform has been found by a Swiss group to be an effective host for a variety of inorganic (sulfur) and organic (hexamethylenetetramine) compounds.[12] Layered inorganic solids such as lithium aluminates can be intercalated with nonlinear optiphores such as 4-nitrohippuric acid; the composite has been shown to be active for SHG.[13]

Several polymeric hosts have been shown to form inclusion complexes with guests. The structure of these complexes is often not known precisely. Observation of SHG not only is important from the materials perspective but also represents an important source of information about the nature of the interaction between the host and the guest. It is possible to grow oriented crystals in a polymeric matrix by doping an organic material into a matrix and then orienting the matrix, e.g., by stretching the film uniaxially.[14] The composites are SHG active. Host involvement in the orientation process is probably minimal, restricted to shear-induced alignment of the guest. Similar microcrystallite alignment was attained by heating poly(methylmethacrylate) (PMMA) thin films containing 2-methyl-4-nitroanilne (MNA) while the structure was poled in an electric field.[15] A spin-coated sample was stated to have an SHG efficacy ten times that of powdered urea. In this latter example, it is unlikely the host (PMMA) is playing a role as host. It is more likely that the crystal nucleation and growth process is initiated by softening of the matrix, and that crystallite orientation is electrostatically enforced. Liquid-crystalline polymers can be doped with nonlinear optiphores and the composites exhibit molecular alignment after electric field poling.[16]

There are several examples of polymeric optical media in which the polymer does play a genuine role as an inclusion host. Japanese workers have shown that poly(ε-caprolactone) (PCL) films containing PNA exhibit substantial SHG (115 x urea for a 500 μ film containing ~20–25% PNA by weight).[17] Spin-cast films are allowed to stand under ambient conditions for 1 d before they are active: a crystallization process occurs during this time to produce an active medium. *p*-Nitroanaline is inactive for SHG, and PCL is very weakly active (space group $P2_12_12_1$). x-Ray data confirm the presence of a new phase in the composite, but no structural data are yet available. Wang has observed that polycarbonate-based films containing the organic dye 2,6-diphenyl-4-(*p*-dimethylaminophenyl) thiopyrilium tetrafluoroborate (TPD) form exceedingly active SHG materials.[18] The SHG signal (1.9–0.95 μ) from a 5 μ film containing 1.5 wt. % of the dye was equivalent to that from a 250 μ crystal of urea. This translates roughly to a 1000-fold enhancement over urea powder. The dye phase separates from the polymer glass, forming an aggregate, fibrilar network of dye crystallites, intimately associated with the host. The inclusion complexation of the dye TPD with polycarbonate is discussed in detail in Section 6.

TPD

The examples cited above indicate that the inclusion paradigm can be used to design new ordered materials with interesting nonlinear optical properties. Time will tell whether the technique has commercial benefit.

3.2. Sensors

A variety of hosts bind guests of biological or analytical significance. It is possible to detect the binding of the analyte by optical means. Such optical sensors modify the optical properties of the host by the act of binding the guest. The earliest example of such an application was the work of Takagi, Nakamura, and Ueno,[19] who attached a crown ether to picrylamine to form a corand (1) that changed color when exposed to potassium ions. Using this sensor, it was possible to detect K^+ in the presence of Na^+ by monitoring the colorimetric changes in solutions. Tagaki and Ueno[20] and Vögtle[21] have reviewed their extensive efforts in this area.

Cram has synthesized a series of chromogenic cavitands (e.g., 2, 3) as ionophoric ligands for alkalai ions.[22,23] The selectivity and sensitivity of these sensors is such that 10^{-8} M Na^+ could be detected by a color change. In the remarkable cavitand *cum* acid–base indicator molecule 4, Cram et al.,[28] have shown that the act of binding the alkalai ion, which releases the phenolic proton associated with the color change of the indicator dye, alters the effective acidity of the medium by seven orders of magnitude. The spectral changes associated with deprotonation of 4 (which is accompanied in NaOH by complexation of the Na^+ by the crown functionality) are shown in Figure 3. This and other examples from the Cram laboratory are "extreme example(s) of the application of the principles of preorganization and complementarity to structural recognition in complexation."

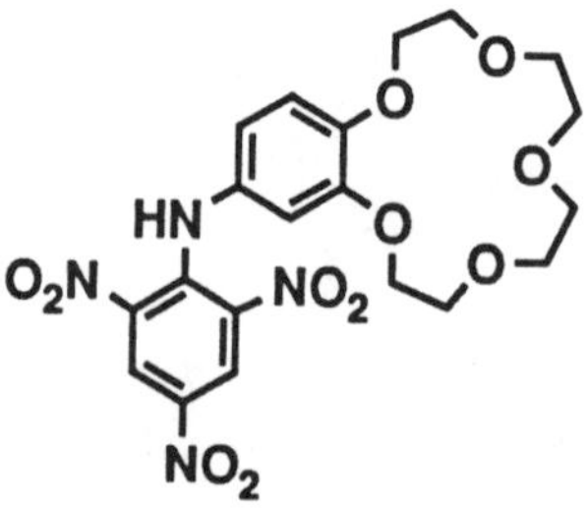

1

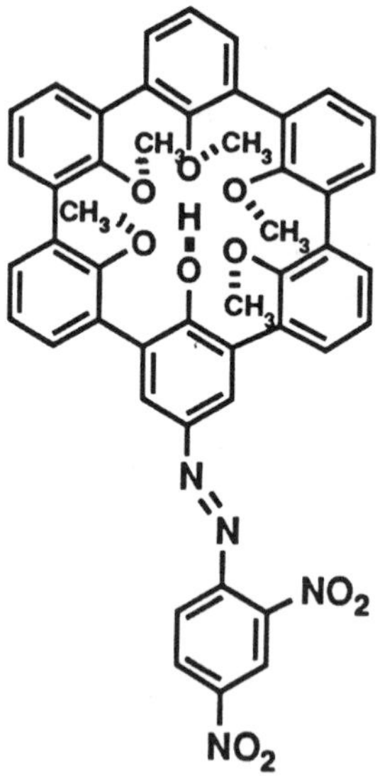

4

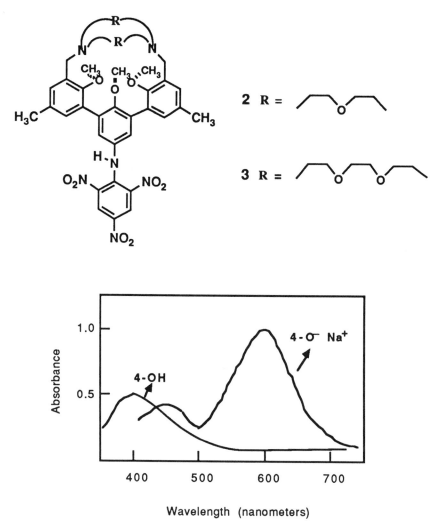

Figure 3. Spectral changes observed on deprotonation of **4** in sodium hydroxide. (After Cram et al. ref. 28)

An alternate approach involves alteration of the fluorescence of a host on complexation with the analyte as guest. A sensor for calcium at physiological concentrations has recently been reported that functions in this way.[24] Binding of calcium to the carboxylate functionality of either **5** or **6** enhances the emission intensity of the material enormously. The emission enhancement results because the normal quenching (electron transfer) of the aromatic chromophores is inhibited on binding. Emission enhancements of 16 x and 92 x are observed for **5** and **6**, respectively, on binding Ca^{2+} ions.

5

6 Ar_1 = p-NCPh
       Ar_2 = p-MeSO$_2$Ph

4. Liquid-Crystalline Media as Hosts

Liquid crystals form a class of host materials that strictly fall outside the scope of this review, because, in general, liquid-crystalline media impose only limited order on included guests. However, liquid-crystalline (LC) media are of great technological importance because they allow molecular alignment to be attained as a function of one or more of several physical parameters, e.g., temperature or applied electric field. Because of this commercial importance, some remarks will be directed toward this class of host–guest materials.

The intermolecular forces that stabilize liquid-crystalline phases are weak forces that are subject to perturbation by changes in the external environment. The ability of such systems to respond to change is the basis of their utility as display, memory, and switching media. In commercial use today, all LC media are switched by use of either electrical or thermal stimuli. Availability of other modes of inducing switching behavior would expand the uses of these novel host–guest media.

Laser-induced switching of LC media has been investigated by a number of workers as a route to optical recording media. Inclusion of an infrared absorbing dye renders the medium susceptible to rapid heating using inexpensive diode lasers emitting near 800 nm in the near infrared. Phase transition recording is a mainstay of liquid crystal memory devices.[25] Tazuke has recently reviewed the application of photochemical reactions as a trigger to induce phase transitions in LC media.[26] In this application, a guest material, usually similar in shape and size to the host mesogen, is photosensitive in a way that the absorption of light results in a large change in shape of the guest, so that irradiation will induce a structural phase change in the medium that can be detected optically. Virtually any photoreaction that results in a volume change or shape change can be used, if a suitable mesogenic sidechain can be attached synthetically. Photoisomerization is a common photoreaction used to trigger such an optical transition. The photoinduced disruption of the ordered state (Figure 4) constitutes an amplification mechanism. The sensitivity of such a system can vary depending on many factors, but the resolution is inherently high.

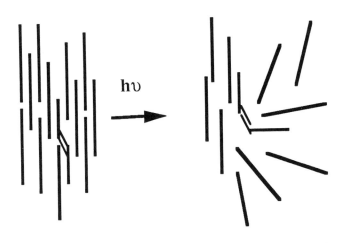

Figure 4. Schematic of a photoinduced phase-transition in liquid crystal media. (After Tazuke and Kurihara., ref. 31)

Polymeric liquid crystalline media may also be used in the same manner as monomeric LC media. In this system, sidechain liquid crystal polymers containing pendant photoresponsive units are used to alter the optical properties of thin films. The virtue of a polymeric medium is its easy use in film form. Acrylate copolymers of alkoxybenzoate mesogens and photoisomerizable mesogens such as alkoxyazobenzenes (e.g., 7) have been prepared and studied in thin films.[27] Photochemically induced isothermal phase transitions are observed in these compositions that can be read out using polarizing optics. Issues for successful use of the medium as a memory are its reversibility and stability in the on and off states, and the speed and sensitivity of the transition. In the LC polymeric media, stability can probably be solved, but long-term reversibility may be a problem. Sensitivity is an important question since a molecular change is triggered which requires, at best, one photon to initiate the change. Most important is the question of the speed of the phase transition once triggered. Generally, such transitions take seconds to minutes to occur in the viscous polymeric medium. This slowness limits the access time of devices made using such a principle, just as monomeric liquid crystal devices have relatively slow response times (milliseconds).

$$-(CH\!-\!CH_2)_x\!-\!(CH\!-\!CH_2)_y\!-$$

Structure 7: poly(vinyl) side-chain with two substituents — the left chain bears C=O, O, $(CH_2)_n$, O, a phenylene ring, C=O, O, a phenylene ring terminating in OCH_3; the right chain bears C=O, O, $(CH_2)_m$, O, a phenylene ring, an azo (N=N) linkage, a phenylene ring terminating in OCH_3.

7

5. "Polaroid" Lenses as Host–Guest Media

In any anisotropic system, inclusion of a chromophore that is ordered with respect to the host produces strong dichroism. The LC compositions discussed above either employ a mesogenic chromophore or include a dichroic dye to effect, in some cases, the optical change that is monitored. In the "Polaroid" lens popularized by Land,[4] iodine was used as the colorant. Iodine-stained films of poly(vinyl alcohol) (PVA) can be stretch oriented to order the iodine longitudinally. Commercially, boric acid was first used to stabilize the films. The oriented films act as polarizers.

The PVA molecule forms internal, hydrogen-bonded helical structures, but the helical runs are interrupted frequently by less ordered domains consisting of other conformations. The boric acid functions as a crosslinker that locally reinforces the helical conformation. Iodine (as polyiodine) occupies the interior void space of the helix. The ideal helix itself is judged to be very shallow: 12 vinylalcohol units constitute a single turn, including a single iodine atom, with one boric acid crosslink buttressing the turn to ensure its stability. The structure is illustrated in Figure 5. Early, detailed work by Zwick established that (in solution) the colors developed by treating PVA with boric acid and iodine–iodide mixtures depended on the degree of hydrolysis of the PVA, the concentration of the polymer, and the stoichiometric ratio of the boric acid and the iodine.[28] Absorption maxima varied from 580 to 700 nm, with longest absorption and highest extinction found for those systems containing the greatest amounts of bound boric acid. These data suggested to Zwick that the complex consisted of highly stereoregular segments. He also noted that many of the phenomena associated with the color changes were very slow to develop or to destroy. He therefore proposed a multiple-stage mechanism (Figure 6) for color formation that involved preequilibrium formation of helical regions (A to B in Figure 6); followed by slow inclusion formation with iodine (B to C) to form preaggregated, but included structures which are colored but not ordered; and then aggregation of helical regions (D) to form highly colored and ordered crystallites. The anisotrpy of the aggregate then arises from the crystalline orientation. In films, stretch orientation can induce crystallization. Once formed, the complex was thought to be quite stable: sensitivity to light and air was moderate, but heat caused discoloration akin to denaturation effects in proteins.[28]

The guest–host chemistry behind the "Polaroid" lens is a commercial application of a more general set of inclusion complexation phenomena exhibited by helical polymers. For example, a variety of helical polymers, including amyloses, proteins, and DNAs form inclusion complexes that are useful in medicinal therapeutic or diagnostic applications. Both general intercalation complexation and more specific inclusions are known. The blue complexes of iodine and amylose are well-known, although for many years it was promulgated that there was no relation between the amylose–I_2 complexation process and that of PVA or other film formers (e.g., cellophane or cellulosic fibers). Bishop and Dance have recently provided a review of the area of helical polymer inclusion complexation.[29]

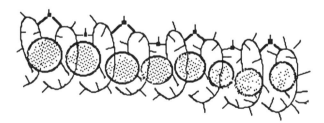

Figure 5. Helical structure of the poly(vinyl alcohol)–iodine–boric acid complex, redrawn from a depiction by Zwick.[28] Iodine is represented by the filled circles, present as a chain within the helical channel of the polymer backbone, shown with pendant OH groups drawn as sticks. Borate crosslinks are shown as trigonal units on the top of the figure; they are actually presumably randomly positioned throughout the backbone.

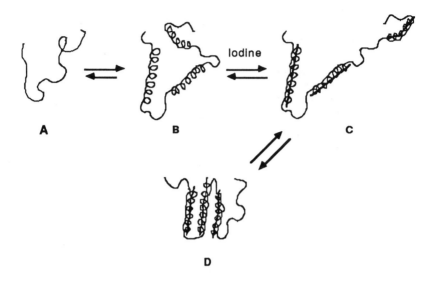

Figure 6. Proposed stages for the formation of PVA–I_2 oriented polycrystalline aggregates with polarizing properties. In the figure, uncomplexed PVA is represented by a random coil conformation; iodine complexed regions are helices in which polyiodide is depicted as linear rods. Condensation of helical regions results in formation of crystallites.

6. Kodak's Ektaprint Electrophotographic System

One of the most dramatic effects of inclusion complexation on a manmade photosystem is provided by the example of the thin film photoconductor composition originally commercialized as the photosensitive component in Kodak's Ektaprint photocopier. The thiapyrilium dye (TPD), described in an earlier section (**3.1**), is the light absorber employed in prototypical compositions.[30] Solutions (methylene chloride) of TPD, an aromatic amine such as 4,4'-diethylamino-2,2'-dimethyltriphenylmethane (LG-1) and a polymeric binder of the polycarbonate class, usually poly(bis-phenol A- diethyleneglycolcarbonate), can be cast as thin, amorphous films (several microns thick). The ratio of components is roughly 1–3:40:60 (TPD–LG-1–binder) on a weight basis. These glassy films have good optical properties and they absorb in the range 565 nm, similar to the TPD dye in methylene chloride solutions. The compositions are weakly photoconductive in the glassy state. The color of TPD originates in a strong intramolecular charge-transfer transition from the donor (dialkylamino) function to the acceptor pyrilium ring. Light absorption by the dye, in the presence of the large excess of the good electron donar amine LG-1, apparently results in electron transfer quenching of the change-transfer excited state. The ion–radicals that result (and their ionic decomposition products) are capable of charge conduction under electrophotographic conditions (voltage gradients of several hundreds of thousands of volt/cm dropped across the thin film, that is, charging levels of 100-1000 volts on a circa 10 μ film). The migration of charge in the voltage gradient within the film is the basis for image-wise modulation of surface charges which forms the basis for the electrophotographic imaging process.[31] Compared to other organic photoconductors, such as the poly(vinyl carbazole)–trinitrofluoreneone system (PVK–TNF), the TPD-based glass is not exceptional.

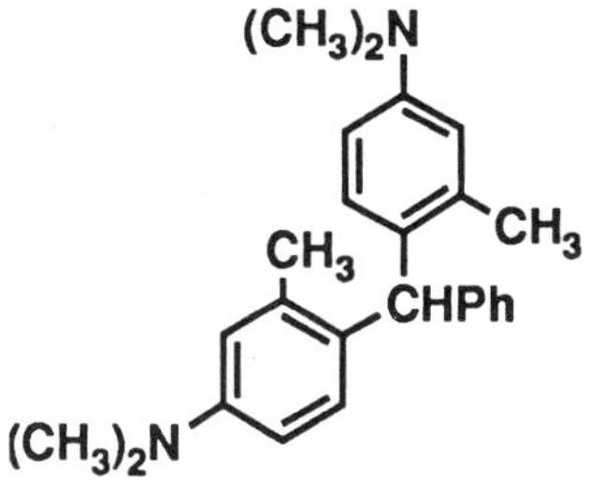

LG-1

Remarkably, however, the polycarbonate glass of TPD–LG-1 can be induced to undergo a phase transition by one of several methods. For example, exposure of glassy films to TPD–LG-1 to the vapors of methylene chloride and other solvents that can swell polycarbonate causes a softening of the glass, which allows nucleation and growth of fibrilar, dendritic clusters of microcyrstals within the film. The formation of the crystalline phase is accompanied by marked changes in the absorption spectrum of the film, and the photoconductivity of the composition improves by two orders of magnitude. The absorption spectra of aggregated and unaggregated TPD compositions are shown in Figure 7. Also shown in the figure are data on the photoconductivity of aggregated and unaggregated films. The phase structure of the film undergoes a complex change during this treatment. Studies at Du Pont[32] using dynamic loss techniques indicated the formation of three new phases in the treated films: an amorphous phase (mp 55°C) rich in the electron donor LG-1; an amorphous TPD–polycarbonate phase (T_g 125–135°C — this is higher than the T_g of the initial amorphous glass by ~ 20°C); and a crystalline, aggregate phase of 1:1 polycarbonate:TPD. No clear phase transitions were observed for this latter phase by differential scanning calorimetry. From the perspective of the inclusion paradigm, the 1:1 polycarbonate–TPD phase is of great interest.

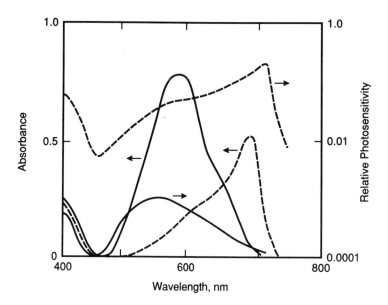

Figure 7. Spectral response of TPD–LG-1 polycarbonate (Laxan®) thin films in the glassy (———) and aggregated (- - -) states. The left axis presents absorption data for thin (1–3 μm) films containing approximately 3 wt% TPD and 40 wt% LG-1, and the right axis shows data for the electrographoc response of the composite. (Adapted from refs. 30 a and b)

Kodak workers had shown[30a] in an insightful set of experiments that the crystalline phase was most likely a complex of polycarbonate and TPD. The Kodak conclusion was based on studies using model compounds to mimic the polymeric binder. Pure crystalline complexes were obtained between TPD and 4,4'-isopropylidenedipheny-bis-phenylcarbonate, DPBC). The complex was stoichiometrically DPBC–TPD$_2$. Single crystals were grown and the structure was solved within P2$_1$/a , a centrosymmetric space group. The x-ray powder data for the model complex was compared to the powder pattern of the film-based aggregate TPD phase. The Kodak workers concluded they were not identical, but that they were structurally quite similar.

Plate 1. Presumed arrangement of TPD and polycarbonate molecules in the TPD/LG-1/polycarbonate composite. The structure is taken from the coordinate data in reference 35a for the complex DPBC:TPD$_2$. Carbon atoms are shown in dark gray, nitrogen is blue, sulfur yellow, oxygen red and chlorine yellow. Digital information for this Plate and for Plate 2 was prepared using the CHEM-X software (Chemical Design. Ltd.) under license. Color out put was generated by John Cristy of Du Pont CR&DD using Du Pont's 4CAST^{TM} System.

Plate 2. The inclusion cavity of the complex DPBC:TPD$_2$.

DPBC

The structure of the aggregate is revealing. Thiapyrilium dye molecules are arranged in chains, stacked in a "slipped deck of cards" motif, with the donor (Me_2N—) end of one molecule atop the acceptor (pyrilium ring) of a neighbor, with the intramolecular distance between dye molecules being 3.44 Å.[30a] Plate 1 illustrates the molecular arrangement within the crystalline DPBC:TPD$_2$ complex (TPD is present in this crystal as the perchlorate salt). In Plate 1, carbon atoms are shown in dark gray, nitrogen in blue, oxygen in red, sulfur in yellow, and chlorine in green. At the lower left of the plate is shown the asymmetric unit obtained crystallographically.[30a] The remainder of the plate shows the repetition of the motif in P2$_1$/a symmetry. The inclusion cavity formed by DPBC units is seen clearly in Plate 2, in which the included TPD molecules have been removed. The DPBC molecules pack in corrugated layers of chains, alternating with layers of TPD. The DPBCs have a conformation similar to that proposed for crystalline forms of polycarbonate. The inescapable conclusion is that in the thin film photoconductor, the dye is included in a crystalline phase with the binder molecules, and that it is the nature of this aggregate that alters its optical properties.

Further studies by the Kodak workers suggest that the aggregate phase is a conductor of both electrons and holes. Crystals of the DPBC–TPD$_2$ model compound have photoconductivities three orders of magnitude greater than their dark conductivities. Thus, the aggregate appears to be a Schottky barrier semiconductor, capable of producing electrons and holes on photoexcitation and separating them along the length of a crystallite within the field of an electrophotographic device. Charge injection into the matrix occurs by interaction with the electron donor LG-1, which then transports holes by dispersive hopping. Electrons are conducted electronically within the fibrilar aggregate structure to the positively biased surface of the device. Figure 8 illustrates the conduction processes schematically. The process would not be as efficient in the amorphous phase because both holes and electrons would be transported by hopping and the charge generation process would not be efficient. Within the inclusion paradigm, the complexation of TPD with the polymer backbone has organized the system to promote effective charge formation, separation, and transport. It is a remarkable effect.

hυ

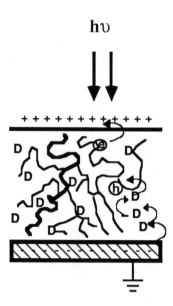

Figure 8. The charge conduction mechanism for the TPD–LG-1 photoconductor. In this side view of the thin film, a photon creates an electron (e)–hole (h) pair, which is transported in the electric field (positive charging of the device is shown) along the crystalline TPD–polycarbonate aggregate chains (one chain is shown on bold). Near the positive surface, the electron neutralizes a surface charge center. Holes within the structure are neutralized by electron hopping (arrows) via the LG-1 amine donors (D).

The structure of the model complex prepared by the Kodak workers was centrosymmetric (P2₁/a). The SHG data provided by Wang[18] (Section **3.1**) on identical film structures demand that the structure be acentric. It is interesting to speculate what the origin of this acentricity might be. The conformation of the DPCB molecules is twisted, with the carbonate function out of conjugation with the phenyl rings. Weak charge-transfer interactions with the dye layers may be responsible. In a polymeric complex, it is feasible, although not necessary, that the twist of the carbonate units could be propagated helically within small domains, so that the domain would lack a center of symmetry and be acentric by virtue of the helical pitch of the polymeric host structure. If, by a mechanism similar to that proposed for the aggregation and crystallization of the iodine–PVA–boric acid complex, helical regions of TPD–polycarbonate became associated, then the overall crystallite structure could be acentric (Figure 9). Nearby domains could form with opposite helicity, so that the overall composition would be racemic, but individual crystalline domains could be polar and acentric. The SHG properties of the films would therefore be representative of polycrystalline powder samples of traditional materials.

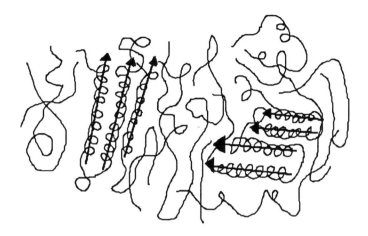

Figure 9. Proposed orientational arrangement of helical TPD–polycarbonate complex aggregates in the bulk thin film. Individual crystallites can be oriented in a polar fashion, while the overall order is random (powder-like).

It is apparent that the detailed structure of the TPD–polycarbonate is complex, but the inclusion paradigm can explain a number of physical properties not explicable if the complex were not formed. Other photoconductors behave better in their crystalline phases than as amorphous solutions in glassy polymers,[33] but to date none has required an inclusion partner to express the phase formation.

Subsequent efforts by Kodak workers have revealed further details about the nature of the excitons that are formed in the TPD aggregate.[34,35] The details are beyond the scope of this chapter.

7. Conclusion

The inclusion paradigm is a strategy employed to organize molecular partners in three dimensional-space in order to perform a physical function. In this chapter, as well as in others in this monograph, we have seen that this paradigm can effect efficient photochemistry of included guests and modify linear and nonlinear optical properties of materials. The strategy is a simple one in concept, but a very difficult one to implement at this stage of our knowledge of solid-state science. As chemists we need to understand more about the way the manifold variety of ordered phases are assembled, and the forces that influence their structural variety.[36] As our expertise increases, I predict that many more intricate and beautiful inclusion structures will be developed to perform useful functions using light as a stimulus.

References

1. (a) J. Deisenhofer, O. Epp, K. Miki, R. Huber and H. Michel, *Nature*, **1985**, *318*, 618.

   (b) H. Michel and J. Deisenhoffer, in *Progress in Photosynthesis Research*, J. Biggins, ed., Martinus Nijhoff, Boston, 1987.

2. P. S. Zurer, *Chem. Eng. News*, **1983** (Nov. 28), 24.

3. J. Szejtli in *Inclusion Compounds*, Vol. 3, J. L. Atwood, J. E. D. Davies and D. D. MacNicol, eds., Academic Press, New York, 1984, Ch. 11.

4. E. H. Land, "Light Polarizer and Process for Manufacturing Same," U. S. Patent 2,237,567 to Polaroid Corporation, Apr. 8, 1941.

5. D. J. Williams, *Angew. Chem. Int. Ed. Engl.*, **1984**, *23*, 690.

6. J. Zyss and D. S. Chemla, *Nonlinear Optical Properties of Organic Molecules and Crystals*, Academic Press, New York, 1987.

7. S. Tomaru, S. Zembutzu, M. Kawachi and M. Kobayashi, *J. Chem. Soc., Chem. Commun.*, **1984**, 1207.

8. Y. Wang and D. F. Eaton, *Chem. Phys. Lett.*, **1985**, *120*, 441.

9. (a) M. Bender, *Cyclodextrin Chemistry*, Springer-Verlag, New York, 1978.

   (b) W. Saenger, in *Inclusion Compounds*, Vol. 2, J. L. Atwood, J. E. D. Davies and D. D. MacNicol, eds., Academic Press, New York, 1984, Ch. 8.

10. (a) A. G. Anderson, D. F. Eaton, W. Tam and Y. Wang, "Optical Nonlinearity in Organic and Organometallic Molecules via Lattice Inclusion Complexation", U. S. Patent 4,818,898, Apr. 4, 1989.

    (b) W. Tam, D. F. Eaton, J. C. Calabrese, I. D. Williams, Y. Wang, and A. G. Anderson, *Chem. Mater.*, **1989**, *1*, 128.

    (c) D. F. Eaton, A. G. Anderson, W. Tam and Y. Wang, in *Polymers for High Technology Electronics and Photonics*, M. S. Bowden and R. S. Turner, eds., ACS Symposium Series No. 346, American Chemical Society, Washington, D. C., 1987, p. 381.

    (d) D. F. Eaton, A. G. Anderson, W. Tam, and Y. Wang, *J. Am. Chem. Soc.*, **1987**, *109*, 1886.

11. S. D. Cox, T. E. Gier, G. D. Stucky and J. D. Bierlein, *J. Am. Chem. Soc.*, **1988**, *110*, 2986.

12. A. Samoc, S. Samoc, J. Fänfschilling and I. Zschokke-Gränacher, *Mater. Sci.*, **1984**, *10*, 231.

13. S. Cooper and P. K. Dutta, *J. Phys. Chem.*, **1990**, *94*, 114.

14. B. D. Moyle, R. E. Ellul and P. D. Calvert, *J. Mater. Sci. Lett.*, **1987**, *6*, 167.

15. H. Daigo, N. Okamoto, H. Fujimura, *Opt. Commun.*, **1988**, *69*, 177.

16. G. R. Meredith in *Nonlinear Optical Properties of organic and Polymeric Materials*, ACS Symposium Series No. 233, D. J. Williams, ed., American Chemical Society, Washington, D.C., 1983, p. 27.

17. T. Miyazaki, T. Watanabe and S. Miyata, *Jap. J. Appl. Phys.*, **1988**, *27*, L1724.

18. (a) Y. Wang, *Chem. Phys. Letts.*, **1986**, *126*, 209.

    (b) Y. Wang, "Pyrilium Dye Nonlinear Optical Elements," U. S. Patent 4,692,636, Sept. 8, 1987.

19. M. Takagi, H. Nakamura and K. Ueno, *Anal. Lett.*, **1977**, *10*, 1115.

20. M. Takagi and K. Ueno, *Top. Curr. Chem.*, **1984**, *121*, 39.

21. H.-G. Lohr and F. Vögtle, *Acc. Chem. Res.*, **1985**, *18*, 65.

22. R. C. Helgeson, B. P. Czech, E. Chapoteau, C. R. Gebauer, A. Kumar and D. J. Cram, *J. Am. Chem. Soc.*, **1989**, *111*, 6339.

23. D. J. Cram, R. A. Carmack and R. C. Helgeson, *J. Am. Chem. Soc.*, **1988**, *110*, 571.

24. A. Prasanna de Silva and H. Q. Nimal Gunarante, *J. Chem. Soc., Chem. Commun.*, **1990**, 186.
25. R. A. Soref, *J. Appl. Phys.*, **1970**, *41*, 3022.
26. S. Tazuke and S. Kurihara in *Photochemistry on Solid Surfaces*, M. Anpo and T. Matsuura, eds., Elsevier, New York, 1989, p. 435.
27. T. Ikeda, S. Horiuchi, D. B. Karanjit, S. Kurihara and S. Tazuke, *Macromol.*, **1990**, *23*, 36.
28. M. M. Zwick, *J. Appl. Polym. Sci.*, **1965**, *9*, 2393.
29. R. Bishop and I. Dance, *Top. Curr. Chem.*, **1988**, *140*, 211.
30. (a) W. J. Dulmage, W. A. Light, S. J. Marino, C. D. Salzberger, D. L. Smith and A. Studenmeyer, *J. Appl. Phys.*, **1978**, *49*, 5543.
    (b) P.M. Borsenberger, A. Chowdry, D. C. Hoesterey and W. May, *J. Appl. Phys.*, **1978**, *49*, 5555.
    (c) W. A. Light, "Photoconductive Compositions and Elements and Method of Preparation," U.S. Patent 3,615,414, 1971.
31. For general introductions to the electrophotographic imaging process see:
    (a) M. E. Scharfe, D. M. Pai and R. J. Gruber in *Imaging Processes and Materials*, Neblette's 8 th Edition, J. Sturge, V. Walworth and A. Shepp, eds., Van Nostrand Reinhold, New York, 1989, p. 135.
    (b) R. L. Schaffert, *Electrophotography*, Focal Press, New York, 1975.
    (c) G. Pfister, *La Researche*, **1984**, *16*, 204.
32. Unpublished research by S. B. Maerov (deceased 1984); D. F. Eaton, seminar presentation at Xerox Research Center, 1985.
33. K. Y. Law, *J. Phys. Chem.*, **1988**, *92*, 4226.
34. A. P. Marchetti, M. Scozzafava and R. H. Young, *J. Chem. Phys.*, **1988**, *89*, 1827.
35. R. H. Young, A. P. Marchetti and E. I. P. Newhouse, *J. Chem. Phys.*, **1989**, *90*, 5743.
36. See G. R. Desiraju, *Crystal Engineering. The Design of Organic Solids*, Materials Science Monograph No. 54., Elsevier, Amsterdam, 1989.

Index

U, V, W

X, Y and Z

RETURN TO: CHEMISTRY LIBRARY

100 Hildebrand Hall • 510-642-3753

LOAN PERIOD	1	2	1-MONTH USE
4		5	6

ALL BOOKS MAY BE RECALLED AFTER 7 DAYS.

Renewals may be requested by phone or, ~~using GLADIS, type inv~~ ~~followed by your patron ID number.~~

DUE AS STAMPED BELOW.

FORM NO. DD 10
3M 7-08

UNIVERSITY OF CALIFORNIA, BERKELEY
Berkeley, California 94720–6000

OCT   OCT